Organohalide-Respiring Bacteria

Lorenz Adrian · Frank E. Löffler
Editors

Organohalide-Respiring Bacteria

Springer

Editors
Lorenz Adrian
Department of Isotope Biogeochemistry
Helmholtz Centre for Environmental
Research—UFZ
Leipzig
Germany

Frank E. Löffler
Department of Microbiology,
Department of Civil and Environmental
Engineering, Center for Environmental
Biotechnology
University of Tennessee
Knoxville, TN
USA

and

Biosciences Division
Oak Ridge National Laboratory
Oak Ridge
TN, USA

ISBN 978-3-662-49873-6 ISBN 978-3-662-49875-0 (eBook)
DOI 10.1007/978-3-662-49875-0

Library of Congress Control Number: 2016938398

Printed on acid-free paper

This Springer imprint is published by Springer Nature
The registered company is Springer-Verlag GmbH Berlin Heidelberg

Contents

Part VI Applications

Part VII Outlook

Part I
Introduction

Chapter 1
Organohalide-Respiring Bacteria—An Introduction

Lorenz Adrian and Frank E. Löffler

Abstract Organohalide-respiring bacteria (OHRB) "breath" halogenated compounds for energy conservation. This fascinating process has received increasing attention over the last two decades revealing the physiological, biochemical, genomic, and ecological features of this taxonomically diverse bacterial group. The discovery of OHRB enabled successful bioremediation at sites impacted with toxic chlorinated compounds, and has drawn researchers with diverse science and engineering backgrounds to study this process. Chapters discussing fundamental and applied aspects of OHRB demonstrate a vibrant research field that will continue to spur scientific discovery and innovate practice.

The realization in the 1970s of potentially harmful impacts of chlorinated organics on human and environmental health triggered extensive research on degradation mechanisms and pathways. At the time, chloroorganic compounds were considered to be predominantly of anthropogenic origin, and it was somewhat surprising that microbes capable of degrading many chlorinated chemicals were found in diverse environments. Carbon–chlorine bond breakage is mediated by dehalogenating enzyme systems (i.e., dehalogenases), and three main mechanisms were discovered: hydrolytic

L. Adrian (✉)
Department Isotope Biogeochemistry, Helmholtz Centre for Environmental Research—UFZ, Leipzig, Germany
e-mail: lorenz.adrian@ufz.de

F.E. Löffler (✉)
Biosciences Division, Oak Ridge National Laboratory, Oak Ridge, TN 37831, USA
e-mail: frank.loeffler@utk.edu

F.E. Löffler
University of Tennessee and Oak Ridge National Laboratory (UT-ORNL), Joint Institute for Biological Sciences (JIBS), Oak Ridge, TN 37831, USA

F.E. Löffler
Center for Environmental Biotechnology, Department of Microbiology, Department of Civil and Environmental Engineering, University of Tennessee, Knoxville, TN 37996, USA

L. Adrian and F.E. Löffler (eds.), *Organohalide-Respiring Bacteria*,
DOI 10.1007/978-3-662-49875-0_1

dehalogenases replace the halogen substituent with a hydroxyl group derived from water, oxygenolytic dehalogenases replace the halogen substituent with a hydroxyl group derived from oxygen, and reductive dehalogenases replace the halogen substituent with a hydrogen atom. The degree of chlorine substitution has a strong effect on hydrolytic and oxygenolytic dehalogenation, and highly chlorinated compounds such as polychlorinated dibenzodioxins, polychlorinated biphenyls, hexachlorobenzene, lindane, or tetrachloroethene appeared recalcitrant. Reductive dechlorination was an interesting alternate mechanism as it acted on polychlorinated pollutants but the process was slow, incomplete, and presumably co-metabolic. A scientific breakthrough was the discovery of the bacterium *Desulfomonile tiedjei* (DeWeerd et al. 1990), an organism that derived all its energy required for growth from the reductive dechlorination of 3-chlorobenzoate to benzoate (Suflita et al. 1982). Then, this type of metabolism was described in a mixed culture capable of reductive dechlorination of the priority groundwater contaminant tetrachloroethene (Holliger et al. 1993), which was the prelude to the discovery of a diversity of bacteria capable of using chloroorganic compounds as electron acceptors. Due to pressing environmental problems, practical interests, at least in the United States, mostly drove research on reductive dechlorination, and support from the U.S. Department of Defense was crucial for developing this field. The observation of complete reductive dechlorination of tetrachloroethene to environmentally benign ethene in a mixed culture was a seminal contribution (Freedman and Gossett 1989) that lead to the discovery of *Dehalococcoides* (Maymó-Gatell et al. 1997). A growing research community has contributed substantially to our understanding of microbial taxa capable of using halogenated compounds as terminal electron acceptors. A key feature of these organisms is their ability to couple reductive dehalogenation to energy conservation and growth. In the case of *Dehalococcoides*, no other electron acceptors support growth, which gives these organisms a selective advantage at sites impacted with chloroorganic contaminants and makes them ideal agents for bioremediation.

A few years ago, experts in the field coined the term 'Organohalide Respiration' to describe the respiratory reductive dehalogenation process. This term is analogous to terms describing other respiratory processes such as nitrate respiration, fumarate respiration, or sulfate respiration and has replaced ambiguous expressions such as "halorespiration", "dehalorespiration" and "chlororespiration". Organohalide respiration accurately describes the process under study, has been widely adopted in the peer-reviewed literature, and is used in this book.

Organohalide respiration is a mode of energy conservation under anoxic conditions. Organohalide-respiring bacteria (OHRB) "breathe" halogenated organic molecules (called organohalides or organohalogens) just like humans breathe oxygen. In biochemical terms, OHRB use organohalides as terminal electron acceptors in a respiratory chain, which is coupled to vectorial proton movement across the cell membrane and energy conservation. The required electrons stem from external electron donors such as molecular hydrogen or other oxidizable compounds. In respiratory processes, none of the participating compounds themselves provide energy but rather is energy available as an electric potential difference (voltage) between the participating redox couples. The term "substrate" is therefore imprecise when describing respiratory processes and the electron donor and the electron acceptor (i.e., the redox couple) should be indicated.

Although reductive dehalogenation is thermodynamically favorable, the actual growth yields of OHRB are low, a likely reason why obligate OHRB are minor components of natural microbial assemblies. It is now established that many organohalides occur naturally in low concentrations in diverse environments, which might explain the widespread distribution of OHRB. At contaminated sites, where chlorinated pollutants are present in elevated concentrations, a modest OHRB population size can turn over substantial amounts of the contaminants. Since many contaminated aquifers and sediments are anoxic and the priority pollutants recalcitrant to aerobic degradation, the reductive dechlorination process mediated by OHRB has had a transformative impact on remediation practice. Of note, bioaugmentation strategies with OHRB have been successfully implemented at many sites demonstrating that science-driven engineering can substantially accelerate contaminated site clean-up. The practical successes may have overshadowed the diverse contributions of the research field to basic science. Knowledge of microbial strategies for energy capture is crucial for understanding, modeling and predicting ecosystems function and responses to perturbation. For example, evidence that OHRB are members of deep subsurface environments is accumulating suggesting that natural organohalides serve as an energy source in such isolated ecosystems. OHRB are phylogenetically diverse and some taxa have deep branching points suggesting that organohalide respiration is an evolutionary old process. More than 2000 putative RDase genes are deposited in open-access databases and a dynamic classification system currently separating 46 different ortholog clusters has been proposed (Hug et al. 2013). Several of the confirmed reductive dehalogenase genes show propensity for horizontal gene transfer, an exciting field of study with great potential for discovery of novel principles of adaptation. Recently, heterologous expression of functional reductive dehalogenases (Mac Nelly et al. 2014; Parthasarathy et al. 2015) and their structural characterization (Bommer et al. 2014; Payne et al. 2015) combined with electron density modelling of dehalogenation reactions (Cooper et al. 2015) and detailed studies of the effects of the essential cobamide prosthetic group on reductive dechlorination activity (Yan et al. 2015) revealed new roles for coenzyme B_{12} in catalysis. These examples demonstrate how research driven by practical needs (e.g., contaminated site clean-up) has developed into a field of scientific endeavor with great potential for transformative discoveries.

The chapters in this book discuss environmental, physiological, biochemical, genomic, ecological, evolutionary, practical, and last but not least, historical aspects. The content of these chapters should help students to learn about OHRB and the breadth of the field, provide useful information to engineers and practitioners, and also serve as a valuable resource for experts.

References

Bommer M, Kunze C, Fesseler J, Schubert T, Diekert G, Dobbek H (2014) Structural basis for organohalide respiration. Science 346(6208):455–458. doi:10.1126/science.1258118

Cooper M, Wagner A, Wondrousch D, Sonntag F, Sonnabend A, Brehm M, Schüürmann G, Adrian L (2015) Anaerobic microbial transformation of halogenated aromatics and fate prediction using electron density modelling. Environ Sci Technol 49(10):6018–6028. doi:10.1021/acs.est.5b00303

DeWeerd KA, Mandelco L, Tanner RS, Woese CR, Suflita JM (1990) *Desulfomonile tiedjei* gen. nov. and sp. nov., a novel anaerobic dehalogenating, sulfate-reducing bacterium. Arch Microbiol 154:23–30. doi:10.1007/BF00249173

Freedman DL, Gossett JM (1989) Biological reductive dechlorination of tetrachloroethylene and trichloroethylene to ethylene under methanogenic conditions. Appl Environ Microbiol 55(9):2144–2151

Holliger C, Schraa G, Stams AJM, Zehnder AJB (1993) A highly purified enrichment culture couples the reductive dechlorination of tetrachloroethene to growth. Appl Environ Microbiol 59:2991–2997

Hug LA, Maphosa F, Leys D, Löffler FE, Smidt H, Edwards EA, Adrian L (2013) Overview of organohalide-respiring bacteria and a proposal for a classification system for reductive dehalogenases. Philos Trans R Soc Lond B Biol Sci 368(1616):20120322. doi:10.1098/rstb.2012.0322

Mac Nelly A, Kai M, Svatoš A, Diekert G, Schubert T (2014) Functional heterologous production of reductive dehalogenases from *Desulfitobacterium hafniense* strains. Appl Environ Microbiol 80(14):4313–4322. doi:10.1128/aem.00881-14

Maymó-Gatell X, Chien YT, Gossett JM, Zinder SH (1997) Isolation of a bacterium that reductively dechlorinates tetrachloroethene to ethene. Science 276:1568–1571. doi:10.1126/science.276.5318.1568

Parthasarathy A, Stich TA, Lohner ST, Lesnefsky A, Britt RD, Spormann AM (2015) Biochemical and EPR-spectroscopic investigation into heterologously expressed vinyl chloride reductive dehalogenase (VcrA) from *Dehalococcoides mccartyi* strain VS. J Am Chem Soc 137(10):3525–3532. doi:10.1021/ja511653d

Payne KAP, Quezada CP, Fisher K, Dunstan MS, Collins FA, Sjuts H, Levy C, Hay S, Rigby SEJ, Leys D (2015) Reductive dehalogenase structure suggests a mechanism for B12-dependent dehalogenation. Nature 517(7535):513–516. doi:10.1038/nature13901

Suflita JM, Horowitz A, Shelton DR, Tiedje JM (1982) Dehalogenation: a novel pathway for the anaerobic biodegradation of haloaromatic compounds. Science 218:1115–1119. doi:10.1126/science.218.4577.1115

Yan J, Simsir B, Farmer AT, Bi M, Yang Y, Campagna SR, Löffler FE (2015) The corrinoid cofactor of reductive dehalogenases affects dechlorination rates and extents in organohalide-respiring *Dehalococcoides mccartyi*. ISME J. doi:10.1038/ismej.2015.197

Chapter 2
Natural Production of Organohalide Compounds in the Environment

James A. Field

Abstract More than 5000 natural organohalogen compounds have been identified. In terrestrial environments, the bulk of the organochlorine is locked up in humic polymers, collectively accounting for a global organochlorine storage of several million Gg. Natural sources are primarily responsible for the global budget of chloromethane and chloroform. Basidiomycete fungi involved in the decomposition of forest litter produce large quantities of chlorinated phenolic methyl ethers. In marine environments naturally occurring chlorinated and brominated bipyrroles as well as methoxypolybrominated phenyl ethers biomagnify in sea mammals. There are at least five distinct halogenating enzyme systems: (1) methyl transferases; (2) heme haloperoxidases; (3) vandadium haloperoxidases; (4) flavin-dependent halogenases and (5) α-ketoglutarate/Fe(II) dependent halogenases. Natural halogenated phenolic metabolites are subject to biotransformation including *O*-demethylation and organohalide respiration. Naturally occurring phenolics are also polymerized by oxidative enzymes to dioxins and chlorohumus.

2.1 Scope

The natural production of organohalogens is an important component of the halogen biogeochemical cycles. Over 5000 natural halogenated compounds have been identified so far as of 2012 (Gribble 1996, 2010, 2012). These naturally halogenated compounds are formed by living organisms such as microalgae, sponges, fungi, bacteria, higher plants, insects, and animals. Likewise such compounds are formed by abiotic process such as volcanoes, forest fires, and abiotic oxidation of soil organic matter. Over half of the natural organohalogens

---

J.A. Field (✉)
University of Arizona, Tucson, USA
e-mail: jimfield@email.arizona.edu

L. Adrian and F.E. Löffler (eds.), *Organohalide-Respiring Bacteria*,
DOI 10.1007/978-3-662-49875-0_2

described contain chlorine, about half contain bromine, and several hundred contain both chlorine and bromine. As of 2004, there were also approximately 110 natural organoiodine and 30 organofluorine compounds described (Gribble 2004a).

The formation of organochlorine in terrestrial environments is known to be tightly linked to the decomposition of organic matter and processes of soil humus formation (Leri et al. 2007; Myneni 2002). The use of in situ X-ray spectroscopic techniques has revealed that inorganic chloride in plant leaves is initially converted to aliphatic and aromatic organochlorine structures. At advanced stages of humification, the chlorinated aromatic fraction continues to increase whereas the chlorinated aliphatic fraction is stable (Leri et al. 2007; Leri and Myneni 2010; Myneni 2002; Reina et al. 2004). The measured organohalogen or organochlorine content of soils measured around the world ranges from 12 to 340 mg organochlorine kg^{-1} soil dry weight (dwt) (Redon et al. 2013; Öberg 2003). The median organochlorine-to-organic carbon ratio in soils has been reported to be 2.3 mg organochlorine g^{-1} soil C (Öberg 2003). This value, when multiplied by the global organic carbon content of soils, indicates a global storage of organochlorine in the pedosphere of 3.35×10^6 Gg (Öberg 2003). Field measurements of organochlorine formation rates in forest soils indicate a production rate of 0.35–0.5 kg organohalogen ha^{-1} y^{-1} (Öberg and Bastviken 2012).

2.2 Evidence

There are multiple lines of evidence to support the large scope of natural organohalogen storage and production in terrestrial environments. Many of the studies evaluating soil organic matter utilize micro-coulometric measurement to assess organohalogens based on the corrosion of silver needed to replace silver ions precipitated by halides. Adsorbable organohalogens (AOX) measures the water soluble organic halogens that are adsorbable onto activated carbon and liberated by combustion (Asplund et al. 1989). Total organohalogen (TOX) measures the halogens in samples directly combusted (Hjelm et al. 1995). In both cases, the combustion is performed after rinsing away inorganic halides with a nitrate/nitric acid solution. A variation of AOX and TOX is to directly measure halides captured from the combustion with ion chromatography (Biester et al. 2004; Putschew et al. 2003). In some studies, halides are measured by neutron activation analysis in sequentially extracted samples at a research nuclear reactor facility (Redon et al. 2013). Lastly, isotopes such as radioactive $^{36}Cl^{-}$ are spiked into soil and the radioactivity is monitored over time in sequentially extracted soils (Gustavsson et al. 2012) or peat samples (Silk et al. 1997).

A critique of total organohalogen methods based on detecting halides in soil residue is that they do not account for intracellular cytoplasmic chloride of micro-organisms (Bastviken et al. 2007; Putschew et al. 2003; Rohlenova et al. 2009). Thus measures of total organohalogens in soil samples can potentially overestimate the actual value. Freezing and thawing of samples prior to extraction has been proposed as a means of lowering this type of interference by enabling the extraction of intracellular inorganic halides (Rohlenova et al. 2009). Nonetheless, there is no doubt that extensive chlorination of organic matter occurs during plant litter decay and soil humus formation. The measurement of organohalogens in water soluble humic fractions extracted from soil is not burdened by interferences caused by intracellular halides. Extensive incorporation of ^{36}Cl into humic and fulvic acids of soils has been demonstrated (Rohlenova et al. 2009). Likewise AOX measurements in natural organic matter (NOM) of pristine surface water and ancient groundwater provides additional evidence that natural occurring organohalogens are associated with humic materials (Asplund et al. 1989).

To add to the evidence, several research groups have utilized techniques to selectively fragment high molecular NOM or lignin into low molecular structures amendable to gas chromatography—mass spectrometry (GC-MS) or microwave induced plasma atomic emission detection (GC-AED). The techniques require derivatization of free OH groups. These techniques demonstrate aromatic substructures that are halogenated in decomposing wood, leaf litter, sphagnum peat, soil, or NOM recovered from surface water and groundwater. The corresponding nonderivatized substructures are shown in Fig. 2.1. The fragments reveal that 3-chloro- and 3,5-dichloro-*p*-hydroxybenzyl structures, 3,5-dichloro-*p*-anisyl, 5-chlorovanillyl as well as 2-chloro- and 2,6-dichlorosyringyl structures in addition to dichloro- and

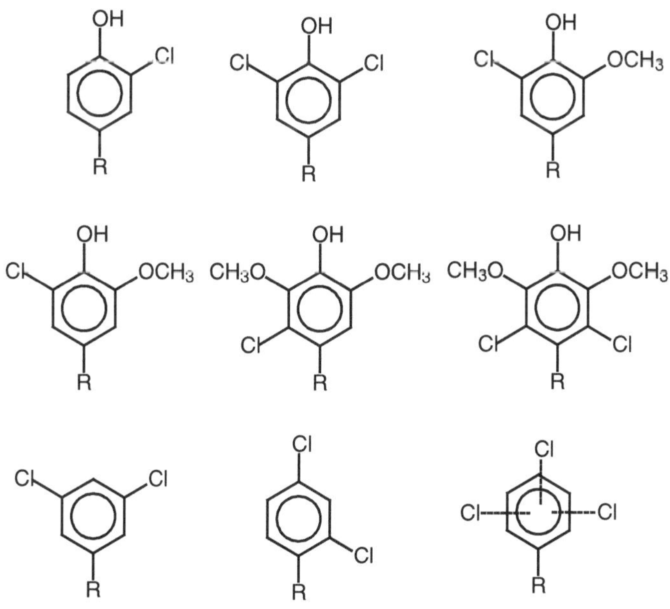

Fig. 2.1 Chlorinated aromatic moieties determined after fragmenting humus in natural organic matter or lignin subjected to microbial decomposition (Dahlman et al. 1993; Flodin et al. 1997; Ortiz-Bermudez et al. 2007). From *left* to *right*, *top row* 3-chloro- and 3,5-dichloro-*p*-hydroxybenzyl structures, 3,5-dichloro-*p*-anisyl; *middle row* 5-chlorovanillyl as well as 2-chloro- and 2,6-dichlorosyringyl. *Bottom row* to 3-5-dichloro- and 2,4-dichlorobenzyl and trichlorobenzyl

trichlorobenzyl structures are embedded in high MW NOM and decomposing lignin (Dahlman et al. 1993; Flodin et al. 1997; Ortiz-Bermudez et al. 2007).

Most recently, synchrotron enabled technique such as near edge X-ray absorption fine structure (NEXAFS) spectroscopy and extended X-ray absorption fine structure spectroscopy (EXAFS) provide spectroscopic evidence of organochlorine formation during forest litter decomposition and humus formation (Leri et al. 2007; Myneni 2002; Reina et al. 2004). These techniques can distinguish between unique spectroscopic signatures for inorganic chloride from those for aliphatic and aromatic organochlorines.

The de novo synthesis of organohalogens by living organisms provides compelling evidence for the natural formation of organohalogens. Basidiomycete fungi are the main group of organisms responsible for the decay of lignocellulosic forest litter (Frankland et al. 2009). A total of 191 Basidiomycete fungal strains were tested on defined medium for the production of AOX and half of them were shown to significantly produce AOX beyond background levels. High levels (5–67 mg AOX L^{-1}) were produced in 9 % of the strains. Fungi belonging to the genus *Hypholoma* had the highest specific production with AOX equivalent to 1.8–3.1 % of their mycelium dwt. The fungi also produced AOX on natural substrates such as forest litter and wood with levels of AOX reaching 61, 115, and 193 mg kg^{-1} dwt substrate on wood, straw, and forest litter, respectively (Verhagen et al. 1996; Öberg et al. 1997). Thus, the synthesis of organohalogen metabolites is associated with the decomposition of forest litter.

2.3 Halogenated Metabolites

Living organisms produce a multitude of halogenated metabolites (Gribble 1996, 2010). Reviews of organohalogen metabolites produced by marine algae (Ballschmiter 2003; Cabrita et al. 2010; Vetter 2006) and fungi (de Jong and Field 1997; Field and Wijnberg 2003; Rezanka and Spizek 2005) are available. The diversity of metabolites found in bacteria, sponges, lichens, higher plants, insects, and mammals can be found in reviews by Gribble (2003a, b, 2004b, 2012). Due to the large diversity of halometabolite structures, a focus will be placed on metabolites which are either identical or structurally similar to halogenated pollutants susceptible to organohalide respiration or other mechanisms of anaerobic bacterial dehalogenation. The categories of compounds to be considered will include halomethanes, chloroethenes, chloroacetic acids, chlorophenols, polychlorinated dibenzodioxins/furans, and polybrominated diphenylethers.

2.3.1 C1 and C2 Metabolites

Chloromethane Fungi, plants and marine algae are known sources of chloromethane (compound 1) (Harper 1985, 2000; Harper and Hamilton 2003).

Polypore white rot fungi are an important biological source of chloromethane, especially those of the *Hymenochaetaceae* family (Harper and Hamilton 2003). The highest producers are within the genus *Phellinus*, which convert up to 80–90 % of inorganic chloride in the growth medium to chloromethane. The rate of chloromethane formation is as high as 20 mg chloromethane kg^{-1} fresh weight (fwt) mycelium d^{-1} (Harper and Hamilton 1988). *Phellinus* spp. are also capable of methylating bromide and iodide (Harper and Hamilton 2003).

Tropical plants have also been implicated in an extensive production of chloromethane (Yokouchi et al. 2002). The flux of chloromethane from the leaves of selected tropical plants is reported to range from 1680 to 44,160 mg kg^{-1} dwt d^{-1}. Likewise microalgae and microalgae from marine environments produce low levels of chloromethane as well as bromo- and iodomethanes (Harper and Hamilton 2003; Harper 2000; Gribble 2010).

The median estimate of the global production of chloromethane and released to the atmosphere is 3000 Gg y^{-1} (Keppler et al. 2005). The anthropogenic contribution is about 5 % (industrial, coal combustion, and incineration). The contribution due to tropical plants is estimated at 910 Gg y^{-1} (or 30 % of global production). The contribution from fungi is estimated to be about 160 Gg y^{-1} (or 5 % of global production, which is equivalent to the anthropogenic contribution).

Chloroform There is strong evidence that chloroform is produced naturally. Several organisms have been shown to produce chloroform. Pure cultures of the fungus *Caldariomyces fumago*, well known for producing chloroperoxidase, were shown to produce chloroform at rates ranging from 0.07–70 μg L ^{1} culture fluid d^{-1}, depending on the culture conditions. Additionally a few basidiomycetes, *Mycena metata* and *Peniophora pseudopini*, were also shown to produce chloroform in pure cultures at rates of 0.7–40 ng L^{-1} culture fluid d^{-1} (Hoekstra et al. 1998b). Termites have been shown to produce chloroform (Khalil et al. 1990). Marine macroalgae and microalgae were also implicated in the production of chloroform (Nightingale et al. 1995; Scarratt and Moore 1999). The highest rate of chloroform production in macroalgae reaching 200 μg kg^{-1} dwt biomass d^{-1} was observed with brown seaweed, *Laminaria saccharina* (Nightingale et al. 1995). Macroalgae are however much better at producing bromoform. *L. saccharina* was also the highest producer with a bromoform production rate of 30 mg kg^{-1} dwt biomass d^{-1} (Nightingale et al. 1995). Bromoform is the most dominant volatile organohalocarbon produced by marine macroalgae accounting for 79–92 % of all volatile organohalogens this organism group produces, depending on region (Laturnus 2001).

Formation of chloroform has been noted in soil and peat in several studies (Gron et al. 2012; Haselmann et al. 2000a, b, 2002; Simmonds et al. 2010). The evidence is strengthened by de novo production when soil samples are incubated in closed bottles in the laboratory (Gron et al. 2012). In one study, the top layer of soil was spiked with $^{37}Cl^{-}$ NaCl and subsequently chloroform enriched with ^{37}Cl was detected, providing conclusive proof of de novo chloroform formation in top soil (Hoekstra et al. 1998a).

The enzyme chloroperoxidase (CPO) is responsible for the formation of chloroform when it is incubated with humic substances. Reactions of CPO with aquatic NOM and humic acids from peat produced up to 240 μg L^{-1} of chloroform (Breider and Hunkeler 2014a). The maximum rates of chloroform production with CPO incubated with humic acids were 1.4 mg L^{-1} d^{-1} (Breider and Hunkeler 2014b). Mechanistically, CPO is responsible for the formation of hypochlorous acid (HOCl) and phenolic moieties in humus become chlorinated in the chemical reaction with HOCl. Reaction schemes have been proposed for phenol (Breider and Hunkeler 2014b) and resorcinol (Hoekstra et al. 1999a) as shown in Fig. 2.2. In support of this hypothesis is the detection of CPO in forest soil and decomposing wood (Ortiz-Bermudez et al. 2007; Laturnus et al. 1995; Asplund et al. 1993).

The global production of chloroform and released to the atmosphere is estimated at 700–820 Gg y^{-1} (Laturnus et al. 2002; Gribble 2010). The known anthropogenic sources only account for 60–73 Gg y^{-1}, thus 90 % or more of the estimated annual chloroform production is natural. The most important natural sources are oceans, soil, termites, and microalgae.

Trichloroacetic acid The natural occurrence of trichloroacetic acid (TCAA) has been inferred by its presence in bog water (Niedan and Schöler 1997; Haiber et al. 1996) and in pristine forest soils (Hoekstra et al. 1999a; Frank 1988). Biological formation of TCAA occurs when CPO is incubated with humic substances or simple organic acids (Haiber et al. 1996; Niedan et al. 2000). A mechanism of TCAA formation during the chlorination of the phenolic moiety, resorcinol, is shown in Fig. 2.2.

Chloroethanes and chloroethenes Marine algae have been reported to produce trichloroethene (TCE) and tetrachloroethene (PCE) (Abrahamsson et al. 1995; Collen et al. 1994). Rates up to 81.6 and 0.2 mg kg^{-1} fresh wt d^{-1} TCE and PCE, respectively, were recorded in the highest producing red algae species. However, these results have only been reported from one research group. Another research group made an extensive attempt to confirm biogenic production from the highest producing algae but was unsuccessful (Marshall et al. 2000). TCE and PCE were also

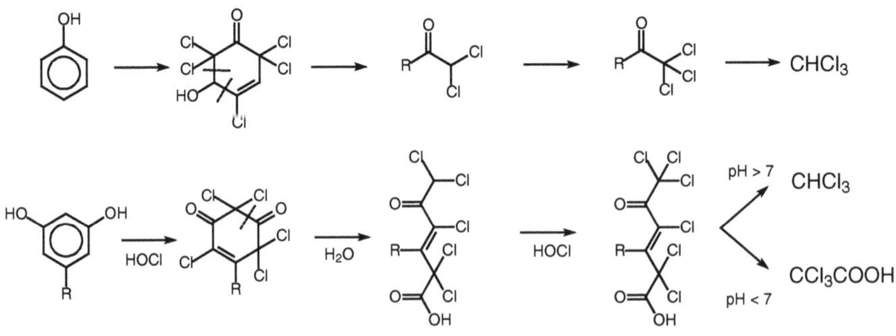

Fig. 2.2 Proposed reaction schemes for the conversion of phenolic and resorcinolic moieties in humus to chloroform and trichloroacetic acid due to their oxidation by hypochlorous acid generated by chloroperoxidase (Breider and Hunkeler 2014a; Hoekstra et al. 1999a)

reported in salt lake sediments at concentrations up to 1.65 and 8.53 μg kg^{-1} fresh wt (Weissflog et al. 2005). Incubations of salt lake sediment samples in closed bottles in the laboratory indicated de novo production was taking place, reaching levels as high as 25 and 50 μg kg^{-1} fresh wt, respectively, after 6 weeks (Weissflog et al. 2005).

2.3.2 Phenolic Compounds, Benzoates and Their Methyl Ethers

Simple phenols A selection of simple phenols that is known to be produced by living organisms are shown in Fig. 2.3. 2,6-dichlorophenol has been identified as a natural product (sex hormone) in many species of ticks (Gribble 1996). The metabolite 2,4-dichlorophenol was shown to be produced by the soil fungus *Penicillium* sp. (Ando et al. 1970). A litter-degrading fungus, *Lepista nuda*, was shown to produce low levels of 2,6-dichloroanisole (Hjelm et al. 1999). Another litter-degrading fungus from the tropics, *Mycena* sp., produced tetrachlorocatechol and tetrachloroguaiacol (Daferner et al. 1998) are as shown in Fig. 2.3. Lastly, three trichlorinated phenols were observed as metabolites in the sphagnum moss inhabiting fungus *Hypholoma elongatum* (Swarts et al. 1998). These metabolites were 2,4,6-trichloro-3-methoxyphenol, 3,5,6-trichloro-2,4-dimethoxyphenol and 3,4,6-trichloro-2,5-dimethoxyphenol (Fig. 2.3). In marine environments, red, green and brown macroalgae were shown to produce various congeners of bromophenols, with 2,4,6-tribromophenol being one of the most predominant congeners found (Paul and Pohnert 2011).

Strong evidence for the natural formation of chlorophenols was obtained by spiking forest soil with $^{37}Cl^-$ and subsequently detecting chlorophenols enriched with ^{37}Cl. Using this method, the natural formation of 4-chlorophenol, 2,4-(or 2,5-)dichlorophenol, 2,6-dichlorophenol, and 2,4,5-trichlorophenol was confirmed

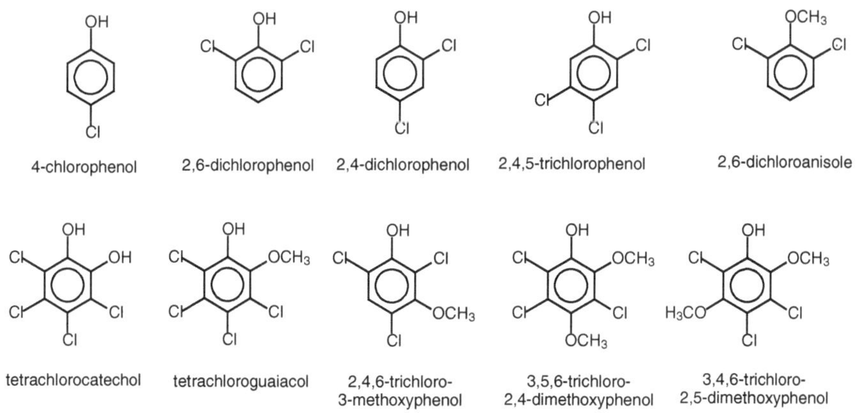

Fig. 2.3 Examples of naturally occurring chlorophenol compounds and a chloroanisole compound. References of each illustrated compound are provided in the text

(Hoekstra et al. 1999b) (Fig. 2.3). The strongest enrichment in ^{37}Cl was observed for 4-chlorophenol. The sum of all simple chlorophenols detected in forest soils ranged up to 71 μg kg^{-1} dwt.

Chlorinated anisyl metabolites A family of metabolites known as the chlorinated anisyl metabolites (CAM) are produced in large quantities by numerous basidiomycete fungi (de Jong and Field 1997; de Jong et al. 1994; Field et al. 1995; Field and Wijnberg 2003; Swarts et al. 1997). The family is composed of 3-chloro- or 3,5-dichloro- *p*-anisyl alcohols and aldehydes (Fig. 2.4). Also the benzoic acid (anisate) form of these metabolites has also been found in the litter-degrading fungus *L. nuda* (Hjelm et al. 1996) and in culture fluids of white rot fungi of to the *Bjerkandera* genus (Swarts et al. 1996) and *Hypholoma fasciculare* (Verhagen et al. 1998b). The production of CAM metabolites is significant, with levels of CAM commonly ranging between 2 and 37 mg CAM L^{-1} in the broths of pure cultures of CAM-producing basidiomycetes (de Jong et al. 1994; Verhagen et al. 1998b), with one exceptional fungus, *H. elongatum*, producing up to 108 mg CAM L^{-1} (Swarts et al. 1997). Likewise, CAM metabolites are detected at relatively high concentrations in the field. CAM concentrations of 7–180 mg kg^{-1} dwt of litter or wood colonized by CAM-producing fungi have been observed (de Jong et al. 1994; Hjelm et al. 1996). Composite forest litter has measurable concentrations of CAM up to 4.5 mg kg^{-1} dwt (de Jong et al. 1994). Estimates of CAM production indicate approximately 300 g ha^{-1} y^{-1} in Dutch forests

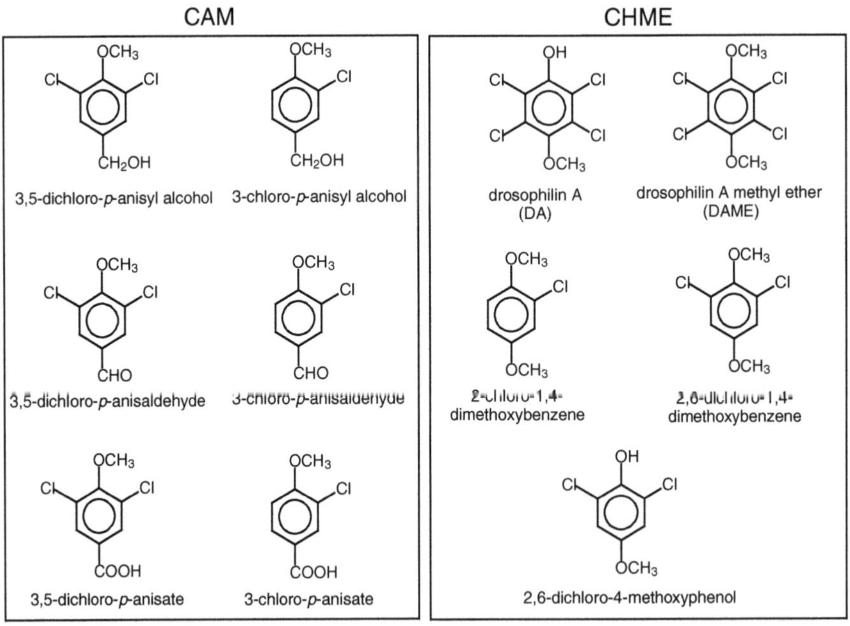

Fig. 2.4 Chlorinated anisyl metabolites (*CAM*) and chlorinated hydroquinone methyl ethers (*CHME*). References are provided in the text

(Field and Wijnberg 2003). This value corresponds to 100 g AOX ha^{-1} y^{-1} for just one family of metabolites, and provides a quantity equivalent to 20–29 % of the field measured organochlorine formation rates in forests (Öberg and Bastviken 2012).

Chlorinated hydroquinone methyl ethers Another important group of fungal chlorophenolic metabolites are the chlorinated hydroquinone methyl ethers (CHME) shown in Fig. 2.4. Tetrachlorinated drosophilin A (DA) and drosophilin A methyl ether (DAME) are found in numerous species of basidomycetes (Teunissen et al. 1997; de Jong and Field 1997), including *Agaricus bisporus*, the common store-bought white button mushroom. Concentrations of DA and DAME are typically found in the culture fluid in the range of 0.14–0.7 mg L^{-1}, with one exceptional white rot fungal species, *Phellinus fastuosus*, producing 11 mg L^{-1} (Swarts et al. 1998; Teunissen et al. 1997). Pure crystals of DAME have been found in the heartwood of mesquite trees being degraded by the basidiomycete, *Phellinus badius* (Garvie et al. 2015). In addition to DA and DAME, 2-chloro-1,4-dimethoxybenzene, 2,6-dichloro-1,4-dimethoxybenzene and 2,6-dichloro-4-methoxyphenol have been identified as fungal metabolites (Hjelm et al. 1996; Swarts et al. 1996; Spinnler et al. 1994; de Jong and Field 1997). The environmental importance of DA has recently come to light when it was reported that it potentially biomagnifies along the food chain. DA was found at concentrations of 1 mg kg^{-1} lipids in the meat of wild boars in Germany (Hiebl et al. 2011). Wild mushroom are considered an important component of the wild boar diet.

Chlorinated benzoic acids Evidence for the natural formation of chlorinated benzoic acids is limited to a few environmental measurements and several examples of chlorinated benzoic acids as metabolites. Structures of some of the naturally occurring chlorinated benzoic acids are shown in Fig. 2.5. The environmental evidence is based on the occurrence of 2,4-dichlorobenzoic acid in bog water and sediments from pristine environments (up to 0.48 μg L^{-1} and 3.4 mg kg^{-1} dwt, respectively) (Niedan and Schöler 1997). A cyanobacterium, *Fischerella ambigua*, was shown to produce 2,4-dichlorobenzoic acid with a yield of 135 mg kg^{-1} dwt cell biomass (Wright et al. 2005). Two hydroxylated chlorinated benzoic acid metabolites, 3-chloro-*p*-benzoic acid and 3,5-dichloro-*p*-benzoic acid, were observed in the culture fluids of the white rot fungus *Bjerkandera* (Swarts et al. 1996).

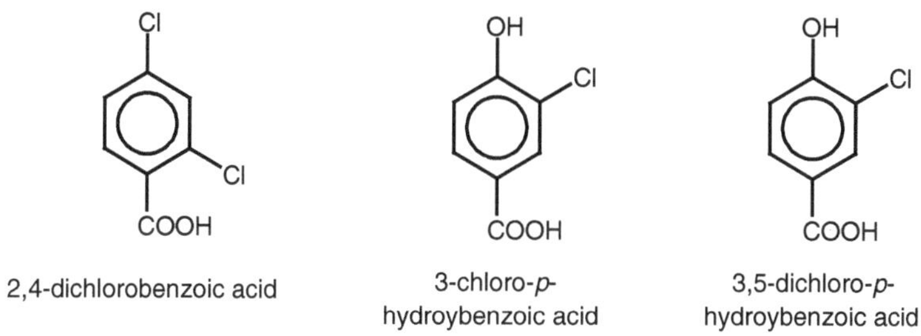

Fig. 2.5 Naturally occurring chlorinated benzoic acids. References are provided in the text

2.3.3 Multi-ring Phenolic Metabolites

Diphenyl ethers A large number of marine sponges produce polybrominated diphenyl ethers (PBDEs) (Gribble 1992, 2010). Examples of natural PBDEs from sponges such as 4,5,6,2′,4′-pentabromo-2-hydroxydiphenyl ether and 3,5,4′,5′,6′-2,2′-dimethoxydiphenyl ether are shown in Fig. 2.6. PBDEs have also been detected in macroalgae samples (Haraguchi et al. 2010). Archived samples of whale oil from 1921 contain PBDEs, providing supporting evidence of their natural occurrence since 1921 predates the industrial production of PBDEs (Gribble 2010). Methoxy-PBDEs have also been found in fish and sea mammals and strong evidence is presented that they bioaccumulate (Vetter 2006; Teuten et al. 2005). Polychlorinated diphenyl ethers are known from the freshwater fungus *Krischsteiniothelia* (Poch et al. 1992) and the cyanobacterium *F. ambigua* (Wright et al. 2005) (e.g. ambigol B in Fig. 2.6). Mushrooms of the basidomycete fungus *Russula subnigricans* contain several chlorinated polyphenyl ethers such as russuphelin-A (shown in Fig. 2.6), which was recovered at a concentration of 247 mg kg^{-1} fresh wt of mushrooms (de Jong and Field 1997; Takahashi et al. 1992).

Chlorinated xanthones and anthraquinones A large variety of chlorinated xanthones are produced by lichens (Gribble 1996, 2010). The metabolite, 4,5-dichloronorlichexanthone (Fig. 2.6) is produced by the largest diversity of lichens. Both fungi and lichens are well known for their production of chlorinated anthraquinones (Gribble 1996, 2010). The metabolite 7-chloroemodin (Fig. 2.6) is produced by the greatest diversity of fungi.

Chlorinated dibenzodioxins/furans Chlorinated dibenzodioxins/furans have been detected in archived soil samples from 1880 and in deep peat bog sediment layers, including those deposited up to 5000 years ago, indicating that dibenzodioxins/furans have a natural origin (Green et al. 2001, 2004; Silk et al. 1997). Important congeners in deep bog layers include 2468-TCDF and 1379-TCDD (Fig. 2.6) (Green et al. 2004; Silk et al. 1997). Natural chlorinated dibenzodioxin/furan formation has been attributed to the oxidation of chlorinated phenols catalyzed by peroxidases and clay. The formation of dioxins from the oxidation of simple chlorophenols has been demonstrated with chloroperoxidase (Silk et al. 1997), horseradish peroxidase (Wittsiepe et al. 1999; Öberg and Rappe 1992), lactoperoxidase (Öberg and Rappe 1992), myeloperoxidase (Wittsiepe et al. 1999) and manganese peroxidase (Munoz et al. 2014). Likewise certain clays with Fe(III) were shown to oxidize chlorophenols to chlorinated dibenzodioxins (Gu et al. 2011; Holmstrand et al. 2006; Horii et al. 2008). The best evidence for natural formation of chlorinated dibenzodioxins comes from spiking forest soil with $^{37}Cl^{-}$, and demonstrating the natural formation of ^{37}Cl enriched precursors, chlorophenols, as well as, ^{37}Cl enriched chlorinated dibenzodioxins (Hoekstra et al. 1999b). Recently, the peroxidase oxidation of bromophenols was also shown to oxidize the naturally occurring 2,4,6-tribromophenol (in marine environments) to 1,3,6,8-tertrabromodibenzodioxin by bromoperoxidase (Arnoldsson et al. 2012).

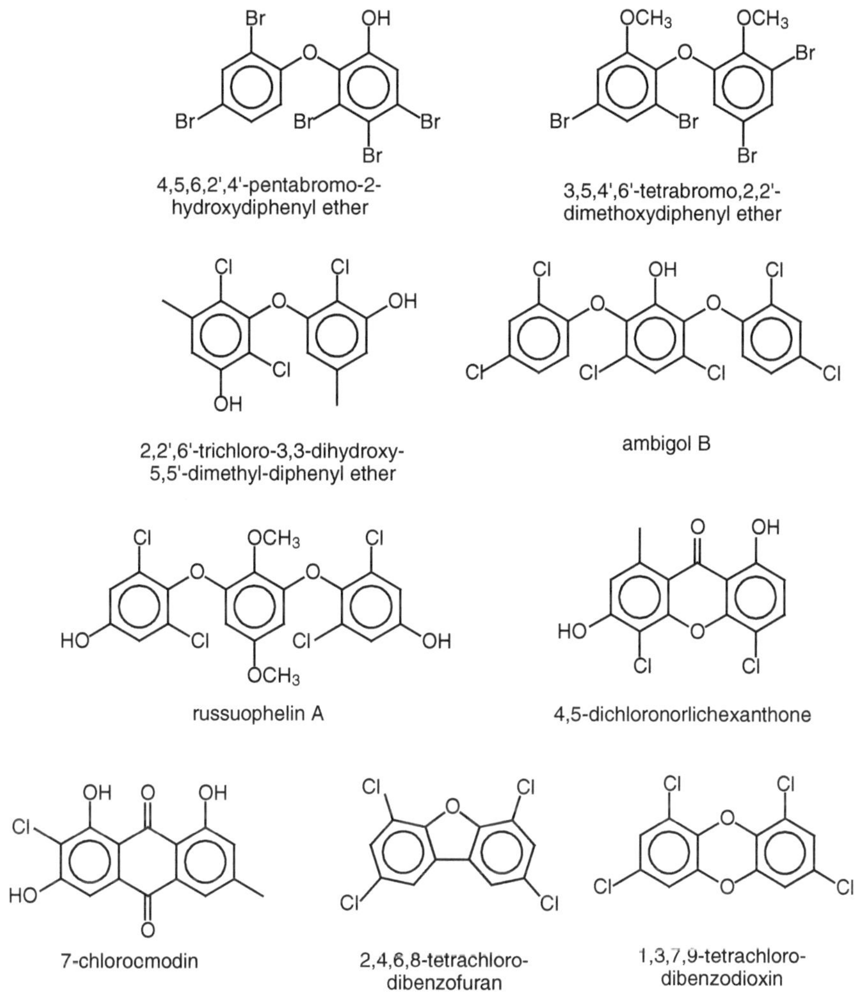

Fig. 2.6 Naturally occurring multi-ring phenolic compounds. References are provided in the text

2.3.4 Chlorinated and Brominated Bipyrroles

A heptachlorometabolite of putative natural origin has been found to accumulate in the lipids of sea mammals (Vetter et al. 2001; Vetter 2006). This compound has been designated Q1 and has been found in the blubber of dolphins at concentrations ranging from 450 to 9100 μg kg^{-1} lipids (Vetter et al. 2001). The chemical structure of Q1 has been elucidated as 2,3,3′,4,4′,5,5′-heptachloro-1′-methyl-1,2′-bipyrrole (Fig. 2.7) (Wu et al. 2002). Passive samplers placed around the Great Barrier Reef indicate that Q1 is produced naturally in that ecologically rich system (Vetter et al. 2009).

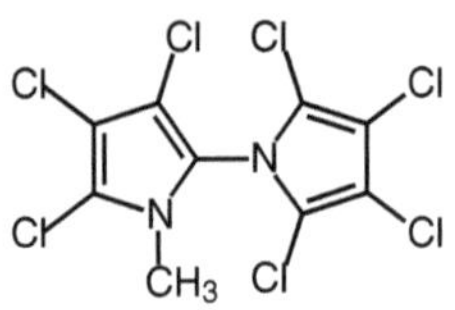

2,3,3',4,4',5,5'-heptachloro-1'-methyl-1,2'-bipyrrole (Q1)

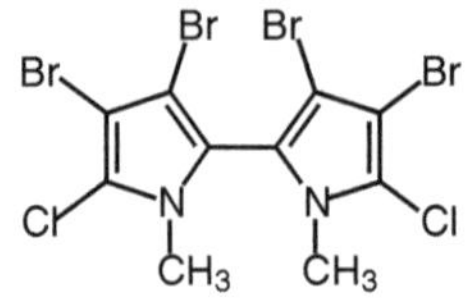

5,5'-dichloro-1,1'-dimethyl-3,3',4,4'-tetrabromo-2,2'-bipyrrole

Fig. 2.7 Halogenated bipyrrole compounds. References are provided in the text

A tetrabrominated and dichlorinated metabolite similar in structure to Q1 has also been identified in as a putative natural marine metabolite biomagnified in seabird eggs (Tittlemier et al. 1999). Its structure has been elucidated as 5,5′-dichloro-1,1′-dimethyl-3,3′,4,4′-tetrabromo-2,2′-bipyrrole (Fig. 2.7). This compound together with related halogenated dimethyl-bipyrrole compounds have been detected in a variety of sea mammals and there is evidence of their biomagnification with trophic level in the food chain (Vetter 2006).

2.4 Abiotic Formation Chlorinated Compounds in Soil

During the diagenesis of soil, there are important mechanisms leading to the abiotic formation of chlorinated compounds. A series of studies has demonstrated formation of simple chlorinated C1 and C2 compounds due to the oxidation of humus in soil by Fe(III) in the presence and absence of H_2O_2. The first of these studies demonstrated that chloromethane, bromo- and iodomethane were produced in response to the oxidation of soil organic matter with Fe(III) (Keppler et al. 2000). A similar pattern was also observed using guaiacol (2-methoxyphenol) as a model for soil organic matter together with ferrihydrite as a model for Fe(III) minerals. Chloromethane, bromomethane, or iodomethane were each produced depending on whether 10 mM KCl, KBr or KI was provided as the halide.

The natural formation of vinyl chloride (chloroethene) was indicated in a study where top soil air was observed to be highly enriched in vinyl chloride concentration compared to ambient air (60–850 times greater) (Keppler et al. 2002). In contrast, no enrichment of trichloroethene was observed in soil air. The abiotic oxidation of soil organic matter with Fe(III) was shown to form vinyl chloride (Keppler et al. 2002). Vinyl chloride, chloromethane, chloroethane and chloropropane production was also observed in a model system with Fe(III), KCl and catechol as a model for humus. The vinyl chloride production in soil or in the catechol model system increased if the reaction mixture also contained H_2O_2.

Oxidation of catechol with Fe(III) also resulted in the formation of chloroacetylene (chloroethyne) (Keppler et al. 2006) and the addition of H_2O_2 to the reaction mixture increased the amount of chloroacetylene formed. Like in the case of vinyl

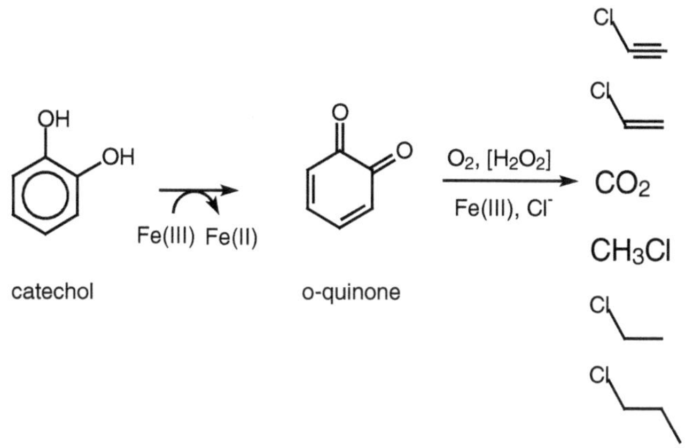

Fig. 2.8 Proposed mechanism for the abiotic formation of chloromethane, vinyl chloride, chloroacetylene, chloroethane and chloropropane by oxidation of catechol with Fe(III) (Keppler et al. 2002, 2006)

chloride, enrichment of chloroacetylene was also observed in topsoil air (Keppler et al. 2006). An overall reaction scheme for the oxidation of catechol to chlorinated alkanes, vinyl chloride, and chloroacetylene is shown in Fig. 2.8.

The addition of Fe(III) and H_2O_2 together with NaCl, humic acid or phenolic model substrates led to the formation of dichloroacetic acid (DCAA) and to a lesser extent TCAA (Fahimi et al. 2003). After the addition of Fe(III), NaCl, and H_2O_2 to soil, the collective formation of chloroacetic acids reached 1.5 mg kg^{-1} soil. In similar experiments, the formation of chloroform was also reported when oxidizing soil or phenolic model compounds with Fe(III), KCl, and H_2O_2 (Huber et al. 2009). Bromoform and trihalomethanes substituted with both chlorine and bromine were observed when KBr was added to the model systems or when soils containing bromide were used.

Plants with pectin and pectin itself serve as a methyl donor to form chloromethane when heated in the presence of chloride (Hamilton et al. 2003). In short-term temperature ramping experiments, significant formation of chloromethane starts after surpassing 170 °C. By the time the temperature was ramped up to 270 °C, 35 g CH_3Cl kg^{-1} pectin was produced while all the chloride was consumed. Long term experiments with pectin containing horse chestnut leaves allowed to slowly dry, rates of 30–50 μg CH_3Cl kg^{-1} d^{-1} dwt leaf mass were observed at 40 °C.

2.5 Biosynthesis of Natural Organohalogens

There are multiple enzyme systems responsible for the formation of organohalogens. Simple halomethanes are formed by methyl transferases, which transfer a methyl group from S-adenosylmethionine (SAM) to a chloride, bromide or, iodide ion

(Murphy 2003). Heme-containing haloperoxidases such as chloroperoxidase (CPO) uses peroxide (H_2O_2) to convert halides to hypohalous acids, such as hypochlorous acid (HOCl) which in turn chemically halogenate organic molecules (Murphy 2003; Butler and Sandy 2009). There are also haloperoxidases without heme groups, such as vanadium peroxidase, which uses H_2O_2 to oxidize bromide to form hypobromous acid (HOBr) that in turn chemically brominates organic molecules (Wever and van der Horst 2013).

In contrast to haloperoxidases, there are halogenating enzymes that catalyze more specific halogenation reactions during metabolite biosynthesis. One such group of halogenating enzymes are the flavin-dependent halogenases (Murphy 2003; Butler and Sandy 2009; Chen and van Pee 2008). Flavin-dependent halogenases are implicated in the synthesis of halogen-containing intermediates during microbial antiobiotic biosynthesis such as tryptophan 7-halogenase in the biosynthesis of pyrrolnitrin by *Pseudomonas fluorescens* (Chen and van Pee 2008). During the enzyme catalytic cycle, a flavin peroxide is formed that oxidizes chloride to HOCl. The chlorine of HOCl is enzymatically inserted into tryptophan (as opposed to the random chemical reaction induced by CPO).

Another type of specific halogenase is the α-ketoglutarate-dependent halogenase. This type of halogenase requires α-ketoglutarate and Fe(II) and is responsible for the halogenation of peptides of nonribosomal synthetic origin (Chen and van Pee 2008; Butler and Sandy 2009). The α-ketoglutarate-dependent halogenases are the only halogenases that attach chloride (or other halides) onto carbons without double bonds.

Some of the most abundant halogenated secondary metabolites in terrestrial environments belong to the CAM family produced by basiomycete fungi. Through the use of isotopically labeled precursors (Mester et al. 1997; Silk et al. 2001) or fluorinated precursors (Silk and Macaulay 2003; Lauritsen and Lunding 1998), the biosynthetic precursors of CAM have been identified. L-phenylalanine is the common precursor, which is metabolized to benzoic acid and subsequent hydroxylation in the *para* position yields 4-hydroxybenzoic acid. This intermediate becomes chlorinated in the *ortho* position(s) of the hydroxyl group and is subsequently methylated to form chlorinated *p*-anisic acids which are reduced by aryl alcohol dehydrogenases to CAM (Field and Wijnberg 2003).

2.6 Biodegradation and Biotransformation of Natural Organohalogens

Natural organohalogens are subject to biodegradation and biotransformation. The anaerobic and aerobic biodegradation of the significant naturally occurring chlorinated methanes, chloromethane, and chloroform have been reviewed elsewhere (Field and Sierra-Alvarez 2004; Harper 2000). A review of aerobic and anaerobic biodegradation of the ubiquitous and naturally occurring TCAA and DCAA has also been provided earlier (Field and Sierra-Alvarez 2004). Interestingly, TCAA is an electron acceptor for the organohalide-respiring bacterium, *Geobacter* (formerly *Trichlorobacter*) *thiogenes* (De Wever et al. 2000).

The anaerobic biotransformation of CAM to chlorophenols by anaerobic mixed cultures has been documented. The CAM 3,5-dichloro-*p*-anisyl alcohol was transformed progressively by demethylation and oxidation of the benzylic carbon leading to accumulation of 3,5-dichloro-*p*-hydroxybenzoic acid, which in turn became decarboxylated to yield 2,6-dichlorophenol (Verhagen et al. 1998a). During the biotransformation, there is a chemical coupling reaction of 3,5-dichloro-*p*-hydroxybenzyl alcohol to yield a tetrachlorinated dimer, bis(3,5-dichloro-4-hydroxyphenyl)methane (Verhagen et al. 1998a).

The anaerobic biotransformation of CHME has also been documented. An anaerobic enrichment culture was responsible for demethylating DAME and DA yielding tetrachloro-1,4-hydrquinone, this intermediate was in turn progressively dechlorinated yielding 1,4-hydroquinone (Milliken et al. 2004b). Organohalide respiring bacterial isolates of *Desulfitobacterium*, initially dechlorinated DA to 3′,5-dichloro-4-methoxyphenol. Subsequently that intermediate was demethylated to 2′,6-dichloro-1′,4-hydroquinone and the desulfitobacterium strains subsequently continued to dechlorinate the molecule, ultimately yielding 1′,4-hydroquinone (Milliken et al. 2004b). *Desulfitobacterium* sp. strain PCE1 and a microbial consortium from Baltimore Harbor sediments were also discovered that demethylate, reductively dehydroxylate, and dechlorinate DA to chlorophenols (Milliken et al. 2004a). The main chlorophenol congener that accumulated was 2,3,5,6-tetrachlorphenol and to a lesser extent 2,3,5-trichlorophenol accumulated as well.

Under aerobic conditions, the CAM, was readily mineralized to inorganic chloride and CO_2 by *Burkholderia* spp. bacterial soil isolates from an oak forest soil. The organic chlorine and carbon in 3,5-dichloro-*p*-anisyl alcohol supplied at 350 mg L^{-1} were converted to Cl^- (100 % recovery) and CO_2 (60 % recovery), respectively, within 20 days (Field and Wijnberg 2003).

Chlorinated xanthones, such as 2,7-dichloroxanthone and 5,7-dichloro-1,3-dihydroxylxanthone, were dechlorinated by soil microbial cultures in anaerobic microcosms (Krzmarzick et al. 2014). The chlorinated xanthones were dechlorinated by two of the soil samples tested with the concomitant formation of the nonchlorinated daughter products, xanthone and 1,3-dihydroxylxanthone, respectively. Bacteria belonging to the *Firmicutes* "Gopher group" were enriched in the dechlorinating cultures compared to control cultures. The "Gopher group" is a group of *Firmicutes* related to the known organohalide-respiring genera *Dehalobacter* and *Desulfitobacterium*.

Another study by the same research group found widespread occurrence of *Dehaloccocoides*-like *Chloroflexi* 16S rRNA genes in pristine soil samples (Krzmarzick et al. 2012). The numbers of *Dehaloccocoides*-like *Chloroflexi* 16S rRNA genes were positively correlated with the organochlorine/total organic carbon ratio of soil samples, whereas no such correlation could be found for universal bacterial 16S rRNA genes. Moreover, CPO-synthesized chlorohumus was shown to be reductively dechlorinated by soil microorganisms as evidenced by chloride release and enrichment of the number of *Dehaloccocoides*-like *Chloroflexi* 16S rRNA genes when the microcosms were incubated anaerobically.

During the decomposition of soil litter, humus forming reactions occur due to the polymerization of phenolic compounds originating from the decay of plants or as secondary metabolites of microorganisms. Chlorinated secondary aromatic metabolites such as CAM and CHME will become copolymerized into the chlorohumus by extracellular phenol oxidizing enzymes (peroxidases or laccases) or mineral oxidants (e.g. MnO_2) (Ruttimann-Johnson and Lamar 1996; Pizzigallo et al. 1995; Lassen et al. 1994). The microbial demethylation of methoxy-groups yielding, chlorinated phenols from CAM and CHME will activate such molecules for the oxidative polymerization reactions catalyzed by enzymes and/or MnO_2. An alternative synthesis of chlorohumus is via the activity of extracellular CPO. The chlorination of aromatic moieties in fulvic acid with CPO has been demonstrated (Niedan et al. 2000). In nature, probably both mechanisms play an important role in the formation of chlorohumus.

2.7 Conclusions

The evidence presented in this chapter indicates an extensive scope of natural organohalogen production in the earth's biosphere with over 5000 natural organohalogen compounds described and an estimated terrestrial storage of several million Gg of organochlorine. The vast majority of the identified organohalogen compounds were shown to be produced as natural metabolites by organisms such as fungi, algae, bacteria, and plants. Likewise abiotic mechanisms are also known. The types of natural organohalogens produced are halogenated methanes, halogenated acetic acids, methyl ethers of chlorophenolic compounds, polychlorinated dibenzodioxins/furans, and polybrominated diphenylethers. In the terrestrial environments, wood and litter-degrading fungi produce large quantities of chlorinated anisyl and chlorinated hydroquinone methyl ether metabolites. In marine environments, polybrominated diphenylethers and chlorinated/brominated bipyrroles of presumed natural origin are consistently detected in the lipids of sea mammals. Diverse enzymatic and abiotic mechanisms are available in nature to account for the halogenation of the natural organohalogen metabolites. The enzymatic mechanisms include: (1) methyl transferases; (2) heme haloperoxidases (3) vandadium haloperoxidases; (4) flavin-dependent halogenases and (5) α-ketoglutarate/Fe(II) dependent halogenases. The natural organohalogen metabolites are subject biotransformation and biodegradation including organisms responsible for halorespiration. Chlorinated anisyl and chlorinated hydroquinone methyl ether metabolites are subject to *O*-demethylation generating chlorinated phenols as biotransformation products that can be oxidatively polymerized into humus forming so called "chlorohumus". Chlorohumus can also be formed by direct reaction of chloride with humus by chloroperoxidases. Chlorohumus may be representative of the species responsible for the large quantities of terrestrial organochlorine in Earth's soil.

References

Abrahamsson K, Ekdahl A, Collen J, Pedersen M (1995) Marine algae—a source of trichloroethylene and perchloroethylene. Limnol Oceanogr 40(7):1321–1326

Ando K, Kato A, Suzuki S (1970) Isolation 2,4-dichorophenol from a soil fungus and its biological significance. Biochem Biophys Res Commun 39(6):1104. doi:10.1016/0006-291x(70)90672-8

Arnoldsson K, Andersson PL, Haglund P (2012) Formation of environmentally relevant brominated dioxins from 2,4,6,-tribromophenol via bromoperoxidase-catalyzed dimerization. Environ Sci Technol 46(13):7239–7244. doi:10.1021/es301255e

Asplund G, Grimvall A, Pettersson C (1989) Naturally produced adsorbable organic halogens (AOX) in humic substances from soil and water. Sci Total Environ 81–2:239–248. doi:10.1016/0048-9697(89)90130-7

Asplund G, Christiansen JV, Grimvall A (1993) A chloroperoxidase-like catalyst in soil—detection and characterization of some properties. Soil Biol Biochem 25(1):41–46. doi:10.1016/0038-0717(93)90239-8

Ballschmiter K (2003) Pattern and sources of naturally produced organohalogens in the marine environment: biogenic formation of organohalogens. Chemosphere 52(2):313–324. doi:10.1016/s0045-6535(03)00211-x

Bastviken D, Thomsen F, Svensson T, Karlsson S, Sanden P, Shaw G, Matucha M, Oberg G (2007) Chloride retention in forest soil by microbial uptake and by natural chlorination of organic matter. Geochim Cosmochim Acta 71(13):3182–3192. doi:10.1016/j.gca.2007.04.028

Biester H, Keppler F, Putschew A, Martinez-Cortizas A, Petri M (2004) Halogen retention, organohalogens, and the role of organic matter decomposition on halogen enrichment in two Chilean peat bogs. Environ Sci Technol 38(7):1984–1991. doi:10.1021/es0348492

Breider F, Hunkeler D (2014a) Investigating chloroperoxidase-catalyzed formation of chloroform from humic substances using stable chlorine isotope analysis. Environ Sci Technol 48(3):1592–1600. doi:10.1021/es403879e

Breider F, Hunkeler D (2014b) Mechanistic insights into the formation of chloroform from natural organic matter using stable carbon isotope analysis. Geochim Cosmochim Acta 125:85–95. doi:10.1016/j.gca.2013.09.028

Butler A, Sandy M (2009) Mechanistic considerations of halogenating enzymes. Nature 460(7257):848–854. doi:10.1038/nature08303

Cabrita MT, Vale C, Rauter AP (2010) Halogenated compounds from marine algae. Mar Drug 8(8):2301–2317. doi:10.3390/md8082301

Chen XP, van Pee KH (2008) Catalytic mechanisms, basic roles, and biotechnological and environmental significance of halogenating enzymes. Acta Biochim Biophys Sin 40(3):183–193. doi:10.1111/j.1745-7270.2008.00390.x

Collen J, Ekdahl A, Abrahamsson K, Pedersen M (1994) The involvement of hydrogen-peroxide in the production of volatile halogenated compounds by *Meristiella gelidium*. Phytochemistry 36(5):1197–1202. doi:10.1016/s0031-9422(00)89637-5

Daferner M, Anke T, Hellwig V, Steglich W, Sterner O (1998) Strobilurin M, tetrachloropyrocatechol and tetrachloropyrocatechol methyl ether: new antibiotics from a Mycena species. J Antibiot 51(9):816–822

Dahlman O, Morck R, Ljungquist P, Reimann A, Johansson C, Boren H, Grimvall A (1993) Chlorinated structural elements in high-molecular-weight organic-matter from unpolluted waters and bleached-kraft mill effluents. Environ Sci Technol 27(8):1616–1620. doi:10.1021/es00045a018

De Wever H, Cole JR, Fettig MR, Hogan DA, Tiedje JM (2000) Reductive dehalogenation of trichloroacetic acid by Trichlorobacter thiogenes gen. nov., sp nov. Appl Environ Microbiol 66(6):2297–2301. doi:10.1128/aem.66.6.2297-2301.2000

de Jong E, Field JA (1997) Sulfur tuft and turkey tail: biosynthesis and biodegradation of organohalogens by basidiomycetes. Annu Rev Microbiol 51:375–414. doi:10.1146/annurev.micro.51.1.375

de Jong E, Field JA, Spinnler HE, Wijnberg J, Debont JAM (1994) Significant biogenesis of chlorinated aromatics by fungi in natural environments. Appl Environ Microbiol 60(1):264–270

Fahimi IJ, Keppler F, Schöler HF (2003) Formation of chloroacetic acids from soil, humic acid and phenolic moieties. Chemosphere 52(2):513–520. doi:10.1016/s0045-6535(03)00212-1

Field JA, Sierra-Alvarez R (2004) Biodegradability of chlorinated solvents and related chlorinated aliphatic compounds. Rev Environ Sci Bio/Technol 3(3):185–254. doi:10.1007/s11157-004-4733-8

Field JA, Wijnberg JBPA (2003) An update on organohalogen metabolites produced by basidiomycetes. In: Gribble G (ed) Natural production of organohalogen compounds, vol 3. Springer, Berlin, pp 103–119

Field JA, Verhagen FJM, de Jong E (1995) Natural organohalogen production by basidiomycetes. Trends Biotechnol 13(11):451–456. doi:10.1016/s0167-7799(00)89001-0

Flodin C, Johansson E, Boren H, Grimvall A, Dahlman O, Morck R (1997) Chlorinated structures in high molecular weight organic matter isolated from fresh and decaying plant material and soil. Environ Sci Technol 31(9):2464–2468. doi:10.1021/es9603741

Frank H (1988) Trichloressigsaure im Boden: eine Ursache neuartiger Waldschaden. Nachr Chem Tech Lab 36:889

Frankland J, Hedger JN, Swift MJ (2009) Decomposer basidiomycetes: their biology and ecology. Cambridge University Press, Cambridge

Garvie LAJ, Wilkens B, Groy TL, Glaeser JA (2015) Substantial production of drosophilin A methyl ether (tetrachloro-1,4-dimethoxybenzene) by the lignicolous basidiomycete *Phellinus badius* in the heartwood of mesquite (*Prosopis juliflora*) trees. Naturwissenschaften 102(3–4):18

Green NJL, Jones JL, Johnston AE, Jones KC (2001) Further evidence for the existence of PCDD/Fs in the environment prior as 1900. Environ Sci Technol 35(10):1974–1981. doi:10.1021/es0002161

Green NJL, Hassanin A, Johnston AE, Jones KC (2004) Observations on historical, contemporary, and natural PCDD/Fs. Environ Sci Technol 38(3):715–723. doi:10.1021/es034599p

Gribble GW (1992) Naturally occurring organohalogen compounds–a survey. J Nat Prod 55(10):1353–1395. doi:10.1021/np50088a001

Gribble GW (1996) Naturally occurring organohalogen compounds–a comprehensive survey. Prog Chem Org Nat Prod 68:1–423

Gribble G (2003a) The diversity of naturally produced organohalogens. In: Gribble G (ed) Natural production of organohalogen compounds, vol 3. Springer, Berlin, pp 1–15

Gribble GW (2003b) The diversity of naturally produced organohalogens. Chemosphere 52(2):289–297. doi:10.1016/s0045-6535(03)00207-8

Gribble GW (2004a) Amazing organohalogens. Am Sci 92:342–349

Gribble GW (2004b) Natural organohalogens. Science Dossier 6. Euro Chlor, Brussels

Gribble GW (2010) Naturally occurring organohalogen compounds—a comprehensive update. Progress in the chemistry of organic natural products, vol 91, pp 1–613. doi:10.1007/978-3-211-99323-1_1

Gribble GW (2012) Recently discovered naturally occurring heterocyclic organohalogen compounds. Heterocycles 84(1):157–207. doi:10.3987/rev-11-sr(p)5

Gron C, Laturnus F, Jacobsen OS (2012) Reliable test methods for the determination of a natural production of chloroform in soils. Environ Monit Assess 184(3):1231–1241. doi:10.1007/s10661-011-2035-5

Gu C, Liu C, Ding YJ, Li H, Teppen BJ, Johnston CT, Boyd SA (2011) Clay mediated route to natural formation of polychlorodibenzo-p-dioxins. Environ Sci Technol 45(8):3445–3451. doi:10.1021/es104225d

Gustavsson M, Karlsson S, Oberg G, Sanden P, Svensson T, Valinia S, Thiry Y, Bastviken D (2012) Organic matter chlorination rates in different boreal soils: the role of soil organic matter content. Environ Sci Technol 46(3):1504–1510. doi:10.1021/es203191r

Haiber G, Jacob G, Niedan V, Nkusi G, Schöler HF (1996) The occurrence of trichloroacetic acid (TCAA)—indications of a natural production? Chemosphere 33(5):839–849. doi:10.1016/0045-6535(96)00239-1

Hamilton JTG, McRoberts WC, Keppler F, Kalin RM, Harper DB (2003) Chloride methylation by plant pectin: An efficient environmentally significant process. Science 301(5630):206–209. doi:10.1126/science.1085036

Haraguchi K, Kotaki Y, Relox JR, Romero MLJ, Terada R (2010) Monitoring of naturally produced brominated phenoxyphenols and phenoxyanisoles in aquatic plants from the philippines. J Agric Food Chem 58(23):12385–12391. doi:10.1021/jf103001n

Harper DB (1985) Halomethane from halide ion—a highly efficient fungal conversion of environmental significance. Nature 315(6014):55–57. doi:10.1038/315055a0

Harper DB (2000) The global chloromethane cycle: biosynthesis, biodegradation and metabolic role. Nat Prod Rep 17(4):337–348. doi:10.1039/a809400d

Harper DB, Hamilton JTG (1988) Biosynthesis of chloromethane in *Phellinus pomaceus*. J Gen Microbiol 134:2831–2839

Harper DB, Hamilton JTG (2003) The global cycles of the naturally-occurring monohalomethanes. In: Gribble G (ed) Natural production of organohalogen compounds, vol 3. Springer, Berlin, pp 17–41

Haselmann KF, Ketola RA, Laturnus F, Lauritsen FR, Gron C (2000a) Occurrence and formation of chloroform at Danish forest sites. Atmos Environ 34(2):187–193. doi:10.1016/s1352-2310(99)00279-4

Haselmann KF, Laturnus F, Svensmark B, Gron C (2000b) Formation of chloroform in spruce forest soil—results from laboratory incubation studies. Chemosphere 41(11):1769–1774. doi:10.1016/s0045-6535(00)00044-8

Haselmann KF, Laturnus F, Gron C (2002) Formation of chloroform in soil. A year-round study at a Danish spruce forest site. Water Air Soil Pollut 139(1–4):35–41. doi:10.1023/a:1015896719508

Hiebl J, Lehnert K, Vetter W (2011) Identification of a fungi-derived terrestrial halogenated natural product in wild boar (*Sus scrofa*). J Agric Food Chem 59(11):6188–6192. doi:10.1021/jf201128r

Hjelm O, Johansson MB, Oberg-Asplund G (1995) Organically bound halogens in coniferous forest soil—distribution pattern and evidence of in-situ production. Chemosphere 30(12):2353–2364. doi:10.1016/0045-6535(95)00107-j

Hjelm O, Boren H, Oberg G (1996) Analysis of halogenated organic compounds in coniferous forest soil from a *Lepista nuda* (wood blewitt) fairy ring. Chemosphere 32(9):1719–1728. doi:10.1016/0045-6535(96)00089-6

Hjelm O, Johansson E, Oberg G (1999) Production of organically bound halogens by the litter-degrading fungus *Lepista nuda*. Soil Biol Biochem 31(11):1509–1515. doi:10.1016/s0038-0717(99)00069-3

Hoekstra EJ, De Leer EWB, Brinkman UAT (1998a) Natural formation of chloroform and brominated trihalomethanes in soil. Environ Sci Technol 32(23):3724–3729. doi:10.1021/es980127c

Hoekstra EJ, Verhagen FJM, Field JA, De Leer EWB, Brinkman UAT (1998b) Natural production of chloroform by fungi. Phytochemistry 49(1):91–97. doi:10.1016/s0031-9422(97)00984-9

Hoekstra EJ, de Leer EWB, Brinkman UAT (1999a) Findings supporting the natural formation of trichloroacetic acid in soil. Chemosphere 38(12):2875–2883. doi:10.1016/s0045-6535(98)00487-1

Hoekstra EJ, De Weerd H, De Leer EWB, Brinkman UAT (1999b) Natural formation of chlorinated phenols, dibenzo-p-dioxins, and dibenzofurans in soil of a Douglas fir forest. Environ Sci Technol 33(15):2543–2549. doi:10.1021/es9900104

Holmstrand H, Gadomski D, Mandalakis M, Tysklind M, Irvine R, Andersson P, Gustafsson O (2006) Origin of PCDDs in ball clay assessed with compound-specific chlorine isotope analysis and radiocarbon dating. Environ Sci Technol 40(12):3730–3735. doi:10.1021/es0602142

Horii Y, van Bavel B, Kannan K, Petrick G, Nachtigall K, Yamashita N (2008) Novel evidence for natural formation of dioxins in ball clay. Chemosphere 70(7):1280–1289. doi:10.1016/j.chemosphere.2007.07.066

Huber SG, Kotte K, Schöler HF, Williams J (2009) Natural abiotic formation of trihalomethanes in soil: results from laboratory studies and field samples. Environ Sci Technol 43(13):4934–4939. doi:10.1021/es8032605

Keppler F, Eiden R, Niedan V, Pracht J, Schöler HF (2000) Halocarbons produced by natural oxidation processes during degradation of organic matter. Nature 403(6767):298–301. doi:10.1038/35002055

Keppler F, Borchers R, Pracht J, Rheinberger S, Schöler HF (2002) Natural formation of vinyl chloride in the terrestrial environment. Environ Sci Technol 36(11):2479–2483. doi:10.1021/es015611j

Keppler F, Harper DB, Rockmann T, Moore RM, Hamilton JTG (2005) New insight into the atmospheric chloromethane budget gained using stable carbon isotope ratios. Atmos Chem Phys 5:2403–2411

Keppler F, Borchers R, Hamilton JTG, Kilian G, Pracht J, Schöler HF (2006) De novo formation of chloroethyne in soil. Environ Sci Technol 40(1):130–134. doi:10.1021/es0513279

Khalil MAK, Rasmussen RA, French JRJ, Holt JA (1990) The influence of termites on atmospheric trace gases—CH_4, CO_2, $CHCl_3$, N_2O, CO, H_2, and light-hydrocarbons. J Geophys Res Atmos 95(D4):3619–3634. doi:10.1029/JD095iD04p03619

Krzmarzick MJ, Crary BB, Harding JJ, Oyerinde OO, Leri AC, Myneni SCB, Novak PJ (2012) Natural niche for organohalide-respiring chloroflexi. Appl Environ Microbiol 78(2):393–401. doi:10.1128/aem.06510-11

Krzmarzick MJ, Miller HR, Yan T, Novak PJ (2014) Novel firmicutes group implicated in the dechlorination of two chlorinated xanthones, analogues of natural organochlorines. Appl Environ Microbiol 80(3):1210–1218. doi:10.1128/aem.03472-13

Lassen P, Randall A, Jorgensen O, Warwick P, Carlsen L (1994) Enzymatically mediated incorporation of 2-chlorophenol and 4-chlorophenol into humic acids. Chemosphere 28(4):703–710. doi:10.1016/0045-6535(94)90221-6

Laturnus F (2001) Marine macroalgae in polar regions as natural sources for volatile organohalogens. Environ Sci Pollut Res 8(2):103–108. doi:10.1007/bf02987302

Laturnus F, Mehrtens G, Gron C (1995) Haloperoxidase-like activity in spruce forest soil a source of volatile halogenated organic-compounds. Chemosphere 31(7):3709–3719. doi:10.1016/0045-6535(95)00220-3

Laturnus F, Haselmann KF, Borch T, Gron C (2002) Terrestrial natural sources of trichloromethane (chloroform, CHCl(3))—an overview. Biogeochemistry 60(2):121–139. doi:10.1023/a:1019887505651

Lauritsen FR, Lunding A (1998) A study of the bioconversion potential of the fungus *Bjerkandera adusta* with respect to a production of chlorinated aromatic compounds. Enzyme Microbial Technol 22(6):459–465. doi:10.1016/s0141-0229(97)00237-8

Leri AC, Myneni SCB (2010) Organochlorine turnover in forest ecosystems: the missing link in the terrestrial chlorine cycle. Global Biogeochem Cycles 24. doi:10.1029/2010gb003882

Leri AC, Marcus MA, Myneni SCB (2007) X-ray spectromicroscopic investigation of natural organochlorine distribution in weathering plant material. Geochim Cosmochim Acta 71(23):5834–5846. doi:10.1016/j.gca.2007.09.001

Marshall RA, Hamilton JTG, Dring MJ, Harper DB (2000) The red alga *Asparagopsis taxiformis/Falkenbergia hillebrandii*—a possible source of trichloroethylene and perchloroethylene? Limnol Oceanogr 45(2):516–519

Mester T, Swarts HJ, Sole SRI, DeBont JAM, Field JA (1997) Stimulation of aryl metabolite production in the basidiomycete Bjerkandera sp. strain BOS55 with biosynthetic precursors and lignin degradation products. Appl Environ Microbiol 63(5):1987–1994

Milliken CE, Meier GP, Sowers KR, May HD (2004a) Chlorophenol production by anaerobic microorganisms: transformation of a biogenic chlorinated hydroquinone metabolite. Appl Environ Microbiol 70(4):2494–2496. doi:10.1128/aem.70.4.2494-2496.2004

Milliken CE, Meier GP, Watts JEM, Sowers KR, May HD (2004b) Microbial anaerobic demethylation and dechlorination of chlorinated hydroquinone metabolites synthesized by basidiomycete fungi. Appl Environ Microbiol 70(1):385–392. doi:10.1128/aem.70.1.385-392.2004

Munoz M, Gomez-Rico MF, Font R (2014) PCDD/F formation from chlorophenols by lignin and manganese peroxidases. Chemosphere 110:129–135. doi:10.1016/j.chemosphere.2014.02.029

Murphy CD (2003) New frontiers in biological halogenation. J Appl Microbiol 94(4):539–548. doi:10.1046/j.1365-2672.2003.01900.x

Myneni SCB (2002) Formation of stable chlorinated hydrocarbons in weathering plant material. Science 295(5557):1039–1041. doi:10.1126/science.1067153

Niedan V, Schöler HF (1997) Natural formation of chlorobenzoic acids (CBA) and distinction between PCB-degraded CBA. Chemosphere 35(6):1233–1241. doi:10.1016/s0045-6535(97)00205-1

Niedan V, Pavasars I, Oberg G (2000) Chloroperoxidase-mediated chlorination of aromatic groups in fulvic acid. Chemosphere 41(5):779–785. doi:10.1016/s0045-6535(99)00471-3

Nightingale PD, Malin G, Liss PS (1995) Production of chloroform and other low-molecular-weight halocarbons by some species of macroalgae. Limnol Oceanogr 40(4):680–689

Öberg GM (2003) The Biogeochemistry of Chlorine in Soil. In: Gribble G (ed) Natural production of organohalogen compounds, vol 3. Springer, Berlin, pp 43–62

Öberg G, Bastviken D (2012) Transformation of chloride to organic chlorine in terrestrial environments: variability, extent, and implications. Crit Rev Environ Sci Technol 42(23):2526–2545. doi:10.1080/10643389.2011.592753

Öberg LG, Rappe C (1992) Biochemical formation of PCDD/Fs from chlorophenols. Chemosphere 25(1–2):49–52. doi:10.1016/0045-6535(92)90477-9

Öberg G, Brunberg H, Hjelm O (1997) Production of organically-bound chlorine during degradation of birch wood by common white-rot fungi. Soil Biol Biochem 29(2):191–197. doi:10.1016/s0038-0717(96)00242-8

Ortiz-Bermudez P, Hirth KC, Srebotnik E, Hammel KE (2007) Chlorination of lignin by ubiq uitous fungi has a likely role in global organochlorine production. Proc Natl Acad Sci USA 104(10):3895–3900. doi:10.1073/pnas.0610074104

Paul C, Pohnert G (2011) Production and role of volatile halogenated compounds from marine algae. Nat Prod Rep 28(2):186–195. doi:10.1039/c0np00043d

Pizzigallo MDR, Ruggiero P, Crecchio C, Mininni R (1995) Manganese and iron-oxides as reactants for oxidation of chlorophenols. Soil Sci Soc Am J 59(2):444–452

Poch GK, Gloer JB, Shearer CA (1992) New bioactive metabolites from a fresh-water isolate of the fungus *Kirschsteiniothelia* sp. J Nat Prod 55(8):1093–1099. doi:10.1021/np50086a010

Putschew A, Keppler F, Jekel M (2003) Differentiation of the halogen content of peat samples using ion chromatography after combustion (TX/TOX-IC). Anal Bioanal Chem 375(6):781–785. doi:10.1007/s00216-003-1797-1

Redon PO, Jolivet C, Saby NPA, Abdelouas A, Thiry Y (2013) Occurrence of natural organic chlorine in soils for different land uses. Biogeochemistry 114(1–3):413–419. doi:10.1007/s10533-012-9771-7

Reina RG, Leri AC, Myneni SCB (2004) ClK-edge X-ray spectroscopic investigation of enzymatic formation of organochlorines in weathering plant material. Environ Sci Technol 38(3):783–789. doi:10.1021/es0347336

Rezanka T, Spizek J (2005) Griseofulvin and other biologically active, halogen containing compounds from fungi. In: UrRahman A (ed) Bioactive Natural Products, vol 32. Studies in Natural Products Chemistry. pp 471–547

Rohlenova J, Gryndler M, Forczek ST, Fuksova K, Handova V, Matucha M (2009) Microbial chlorination of organic matter in forest soil: investigation using Cl-36-chloride and its methodology. Environ Sci Technol 43(10):3652–3655. doi:10.1021/es803300f

Ruttimann-Johnson C, Lamar RT (1996) Polymerization of pentachlorophenol and ferulic acid by fungal extracellular lignin-degrading enzymes. Appl Environ Microbiol 62(10):3890–3893

Scarratt MG, Moore RM (1999) Production of chlorinated hydrocarbons and methyl iodide by the red microalga *Porphyridium purpureum*. Limnol Oceanogr 44(3):703–707

Silk PJ, Macaulay JB (2003) Stereoselective biosynthesis of chloroarylpropane diols by the basidiomycete Bjerkandera adusta: exploring the roles of amino acids, pyruvate, glycerol and phenyl acetyl carbinol. FEMS Microbiol Lett 228(1):11–19. doi:10.1016/s0378-1097(03)00725-0

Silk PJ, Lonergan GC, Arsenault TL, Boyle CD (1997) Evidence of natural organochlorine formation in peat bogs. Chemosphere 35(12):2865–2880. doi:10.1016/s0045-6535(97)00347-0

Silk PJ, Aubry C, Lonergan GC, Macaulay JB (2001) Chlorometabolite production by the ecologically important white rot fungus *Bjerkandera adusta*. Chemosphere 44(7):1603–1616. doi:10.1016/s0045-6535(00)00537-3

Simmonds PG, Derwent RG, Manning AJ, O'Doherty S, Spain G (2010) Natural chloroform emissions from the blanket peat bogs in the vicinity of Mace Head, Ireland over a 14-year period. Atmos Environ 44(10):1284–1291. doi:10.1016/j.atmosenv.2009.12.027

Spinnler HE, de Jong E, Mauvais G, Semon E, Lequere JL (1994) Production of halogenated compounds by *Bjerkandera adusta*. Appl Microbiol Biotechnol 42(2–3):212–221

Swarts HJ, Verhagen FJM, Field JA, Wijnberg J (1996) Novel chlorometabolites produced by Bjerkandera species. Phytochemistry 42(6):1699–1701. doi:10.1016/0031-9422(96)00191-4

Swarts HJ, Teunissen PJM, Verhagen FJM, Field JA, Wijnberg J (1997) Chlorinated anisyl metabolites produced by basidiomycetes. Mycol Res 101:372–374. doi:10.1017/s0953756296003036

Swarts HJ, Verhagen FJM, Field JA, Wijnberg J (1998) Trichlorinated phenols from Hypholoma elongatum. Phytochemistry 49(1):203–206. doi:10.1016/s0031-9422(97)01067-4

Takahashi A, Agatsuma T, Matsuda M, Ohta T, Nunozawa T, Endo T, Nozoe S (1992) Russuphelin-A, a new cytotoxic substance from the mushroom *Russula-subnigricans* hongo. Chem Pharm Bull 40(12):3185–3188

Teunissen PJM, Swarts HJ, Field JA (1997) The de novo production of drosophilin A (tetrachloro-4-methoxyphenol) and drosophilin A methyl ether (tetrachloro-1,4-dimethoxybenzene) by ligninolytic basidiomycetes. Appl Microbiol Biotechnol 47(6):695–700

Teuten EL, Xu L, Reddy CM (2005) Two abundant bioaccumulated halogenated compounds are natural products. Science 307(5711):917–920. doi:10.1026/science.1106882

Tittlemier SA, Simon M, Jarman WM, Elliott JE, Norstrom RJ (1999) Identification of a novel C10H6N2Br 4Cl2 heterocyclic compound in seabird eggs. A bioaccumulating marine natural product? Environ Sci Technol 33(1):26–33. doi:10.1021/es980646f

Verhagen FJM, Swarts HJ, Kuyper TW, Wijnberg J, Field JA (1996) The ubiquity of natural adsorbable organic halogen production among basidiomycetes. Appl Microbiol Biotechnol 45(5):710–718

Verhagen FJM, Swarts HJ, Wijnberg J, Field JA (1998a) Biotransformation of the major fungal metabolite 3,5-dichlorop-p-anisyl alcohol under anaerobic conditions and its role in formation of bis(3,5-dichloro-4-hydroxyphenyl)methane. Appl Environ Microbiol 64(9):3225–3231

Verhagen FJM, Van Assema FBJ, Boekema B, Swarts HJ, Wijnberg J, Field JA (1998b) Dynamics of organohalogen production by the ecologically important fungus *Hypholoma fasciculare*. FEMS Microbiol Lett 158(2):167–178. doi:10.1111/j.1574 6968.1998.tb12816.x

Vetter W (2006) Marine halogenated natural products of environmental relevance. In: Ware GW (ed) Reviews of environmental contamination and toxicology. Reviews of environmental contamination and toxicology. vol 188, pp 1–57. doi:10.1007/978-0-387-32964-2_1

Vetter W, Scholz E, Gaus C, Muller JF, Haynes D (2001) Anthropogenic and natural organohalogen compounds in blubber of dolphins and dugongs (Dugong dugon) from northeastern Australia. Arch Environ Contam Toxicol 41(2):221–231

Vetter W, Haase-Aschoff P, Rosenfelder N, Komarova T, Mueller JF (2009) Determination of halogenated natural products in passive samplers deployed along the great barrier reef, Queensland/Australia. Environ Sci Technol 43(16):6131–6137. doi:10.1021/es900928m

Weissflog L, Lange CA, Pfennigsdorff A, Kotte K, Elansky N, Lisitzyna L, Putz E, Krueger G (2005) Sediments of salt lakes as a new source of volatile highly chlorinated C1/C2 hydrocarbons. Geophys Res Lett 32(1). doi:10.1029/2004gl020807

Wever R, van der Horst MA (2013) The role of vanadium haloperoxidases in the formation of volatile brominated compounds and their impact on the environment. Dalton Trans 42(33):11778–11786. doi:10.1039/c3dt50525a

Wittsiepe J, Kullmann Y, Schrey P, Selenka F, Wilhelm M (1999) Peroxidase-catalyzed in vitro formation of polychlorinated dibenzo-p-dioxins and dibenzofurans from chlorophenols. Toxicol Lett 106(2–3):191–200. doi:10.1016/s0378-4274(99)00066-1

Wright AD, Papendorf O, Konig GM (2005) Ambigol C and 2,4-dichlorobenzoic acid, natural products produced by the terrestrial cyanobacterium *Fischerella ambigua*. J Nat Prod 68(3):459–461. doi:10.1021/np049640w

Wu J, Vetter W, Gribble GW, Schneekloth JS, Blank DH, Gorls H (2002) Angewandte Chemie-Int Ed 41(10):1740–1743. doi:10.1002/1521-3773(20020517)41:10<1740::aid-anie1740>3.0.co;2-7

Yokouchi Y, Ikeda M, Inuzuka Y, Yukawa T (2002) Strong emission of methyl chloride from tropical plants. Nature 416(6877):163–165. doi:10.1038/416163a

Chapter 3
Energetic Considerations in Organohalide Respiration

Jan Dolfing

Abstract Organohalide-respiring bacteria harness energy using halogenated organic compounds as electron acceptors. The objective of this chapter is to evaluate the thermodynamics and energetics of organohalide respiration, that is, (i) how much energy the organisms can obtain from dehalogenation, and how this energy compares to the energy available from other electron acceptors; (ii) how much energy the organisms actually harness from the dehalogenation reactions; and (iii) how much energy the organisms not only forfeit but actually dissipate when they convert halogenated compounds co-metabolically rather than metabolically. Alternative fates of organohalides—their anaerobic oxidation and fermentation—are also discussed.

3.1 Introduction

The propensity of anaerobic microorganisms to use halogenated compounds as electron acceptors in anaerobic environments is now well established (Maphosa et al. 2010; Hug et al. 2013). Indeed, the very title of this book on organohalide-respiring bacteria (McCarty 1997) implies as much. Organohalide-respiring bacteria not only use halogenated compounds as electron acceptors, they actually harness energy from this process (Dolfing and Tiedje 1987; Dolfing 1990; Leys et al. 2013), although some organohalide-respiring bacteria can be inconsistent in this respect. *Dehalococcoides mccartyi* 195 (formerly *Dehalococcoides ethenogenes* 195) performs a complete dechlorination of tetrachloroethene via trichloroethene, dichloroethene, and vinyl chloride to ethene, but the last dechlorination step from vinyl chloride to ethene is co-metabolic and not linked to

J. Dolfing (✉)
School of Civil Engineering and Geosciences, Newcastle University,
Newcastle-upon-Tyne NE1 7RU, England, UK
e-mail: jan.dolfing@ncl.ac.uk

L. Adrian and F.E. Löffler (eds.), *Organohalide-Respiring Bacteria*,
DOI 10.1007/978-3-662-49875-0_3

energy conservation (Maymo-Gatell et al. 2001). The objective of the present chapter is to evaluate the thermodynamics and energetics of organohalide respiration, and to outline how these affect the ecophysiology of the organisms involved: (a) the thermodynamics of the dehalogenation reactions, that is how much energy the organisms can obtain from dehalogenation, and how this energy compares to the energy available from other electron acceptors; (b) the energetics of the organisms, that is how much energy the organisms actually harness from the dehalogenation reactions; and (c) the co-metabolic conversion of halogenated compounds, that is how much energy the organisms not only forfeit but actually dissipate when they convert halogenated compounds co-metabolically rather than metabolically.

3.2 Chemistry of Organohalides

The halogens fluorine, chlorine, bromine, and iodine—astatine is not considered here as its total amount on earth is estimated to be about 50 mg (Kirk 1991)—are strongly electronegative, and form stable bonds with carbon atoms. Bond strength decreases markedly with increasing molecular weight: C–F > C–Cl > C–Br > C–I, while bond length decreases with increasing molecular weight (Fig. 3.1) (Reineke 1984). Both parameters as well as the reactivity of the carbon–halogen bond are

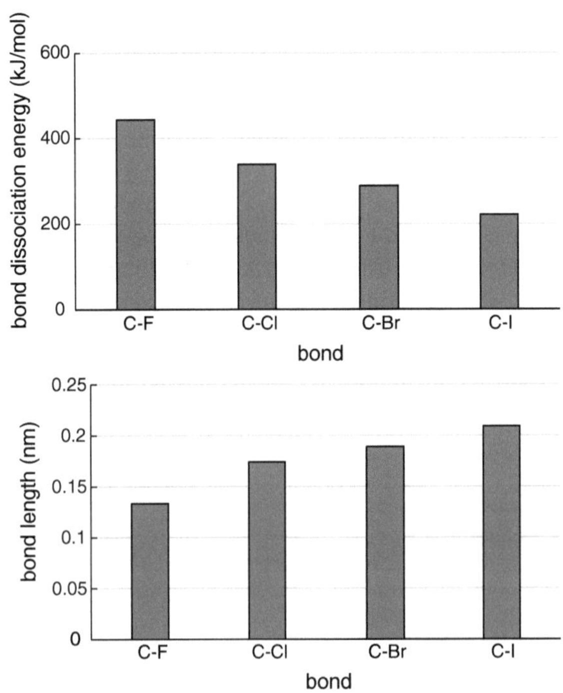

Fig. 3.1 Carbon–halogen bond dissociation energy (*upper panel*) and carbon–halogen bond length (*lower panel*) for the halogens fluorine, chlorine, bromine, and iodine. Both parameters are influenced by the type of carbon to which the halogen is attached. The example given here is for aromatic carbons

influenced by the type of carbon to which the halogen is attached. For example, nucleophilic displacement of a fluoride from a perfluoroalkane is difficult to accomplish, while aryl fluorides are more reactive than the corresponding aryl chlorides, bromides, and iodides (Wackett 1995). In this chapter, we will see to what extent organohalide respiration is affected by the type of carbon to which the halogens is attached.

3.3 Types of Organohalide Respiration

Two types of organohalide respiration can be discerned: hydrogenolysis and dihaloelimination (Fig. 3.2). In hydrogenolysis organohalide, respiration results in the formation of one mole of HX (where X = Cl, Br, F, or I) per two electrons; in dihaloelimination two moles of HX are formed per two electrons. Dihaloelimination requires the presence of two vicinal halogens and results in the formation of an unsaturated C=C bond. Due to the electron configuration of the aromatic ring, dihaloelimination does not occur with aromatic compounds.

Dehydrodehalogenation is another type of dehalogenation reaction that can occur on aliphatic compounds only. It resembles dihaloelimination in that it gives rise to the formation of an unsaturated C=C bond. Dehydrodehalogenation

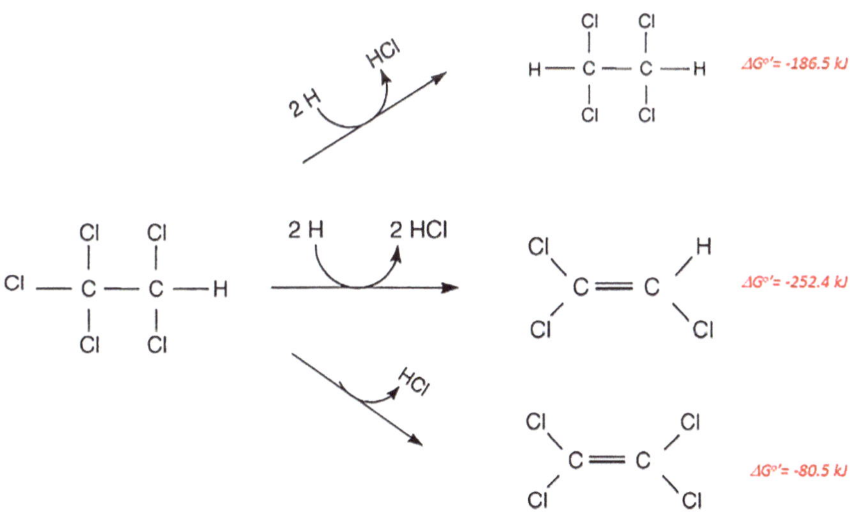

Fig. 3.2 Conceptual examples of two types of reductive dehalogenation: hydrogenolysis of pentachlorethane to tetrachloroethane (*upper pathway*), dihaloelimination of pentachloroethane to trichloroethene (*middle pathway*). In theory, both reactions can be coupled to energy conservation via organohalide respiration. The *lower pathway* shows dehydrodehalogenation of pentachloroethane to tetrachlorethene. This reaction does not consume reducing equivalents, is not a reductive dehalogenation reaction, and can therefore not be coupled to organohalide respiration

involves removal of HX and plays a role in the degradation pathway of pesticides like lindane and DDT (Mohn and Tiedje 1992). It is not a reductive reaction as no reducing equivalents are consumed. Consequently, dehydrodehalogenation is not a respiratory reaction.

3.4 Thermodynamics: The Basics

Thermodynamics can be used as a tool to gain insight into the amount of energy that is available from a reaction (Thauer et al. 1977; Zehnder and Stumm 1988; Hanselmann 1991). However, thermodynamics does not a priori give insight into how much energy an organism actually captures (Thauer et al. 1977). The conceptual framework and basic methodology to make change in Gibbs free energy calculations are textbook knowledge; Brock Biology of Microorganisms (Madigan et al. 1997) for example gives the essentials. For a more detailed description of the calculation methods the reader is referred to Dolfing (2003, 2015) and Box 1.

Box 1 Calculating Thermodynamics

Change in Gibbs free energy (ΔG) values for reactions of environmental and biotechnological interest are calculated in three steps. First the Gibbs free energy change (ΔG^{o}) is calculated for standard conditions, *id est* for a temperature of 25 °C, solutes at 1 molar concentrations, and gases at partial pressures of 1 atmosphere. Next, a temperature correction is applied if the temperature of interest is not 25 °C. Finally, a correction is applied if the actual concentrations are not 1 M and/or the actual partial pressures are not 1 atm, with special attention for speciation and pH.

1. Change in Gibbs free energy values are calculated via

$$\Delta G^{o} = \sum G_{f}{}^{o}{}_{\text{products}} - \sum G_{f}{}^{o}{}_{\text{products}} \tag{3.1}$$

   $G_f{}^o$ describe the Gibbs free energy of formation under standard conditions. $G_f{}^o$ values for halogenated compounds can be found in various sources (Table 3.1).
2. For temperatures different from 25 °C ΔG values are calculated with the Gibbs–Helmholtz equation

$$\Delta G^{o}_{T_{act}} = \Delta G^{o}_{T_{ref}}(T_{act}/T_{ref}) + \Delta H^{o}_{T_{ref}} \times (T_{ref} - T_{act})/T_{ref} \tag{3.2}$$

   where T_{ref} is 298.15 K and T_{act} is the temperature of interest, and $\Delta H^{o} = \sum G^{o}_{f\ \text{products}} - \sum G^{o}_{f\ \text{reactants}}$; for $G_f{}^o$ values see references in Table 3.1.

3. Actual concentrations are normally not 1 M or 1 atm (100 kPa). For a hypothetical reaction aA + b$B \rightarrow$ cC + dD, this is accounted for via the mass equation

$$\Delta G = \Delta G^{\mathrm{o}} + RT\ln\left([C]^{\mathrm{c}}[D]^{\mathrm{d}}/[A]^{\mathrm{a}}[B]^{\mathrm{b}}\right) \quad (3.3)$$

where R is the universal gas constant 8.314 J mol^{-1} K^{-1} and T is the temperature in Kelvin.
4. Gibbs free energy values for weak acids and their conjugated bases are calculated by using the G_{f} value of the acid, with the formula

$$G_{\mathrm{f}} = G_{\mathrm{f}}^{\mathrm{o}} + RT\ln\alpha \quad (3.4)$$

or the G_{f} value of the conjugated base with the formula

$$G_{\mathrm{f}} = G_{\mathrm{f}}^{\mathrm{o}} + RT\ln(1-\alpha) \quad (3.5)$$

where $\alpha = 10^{-\mathrm{pH}}/(10^{-\mathrm{pH}} + 10^{-\mathrm{pKa}})$.
5. The conversion between change in Gibbs free energy values (ΔG) and redox potential (ΔE) is calculated via the relationship

$$\Delta G = -nF\Delta E \quad (3.6)$$

where n is the number of electrons involved, and F is the Faraday constant (96.48 kJ/V). Note that ΔG should be for the redox couple of interest, *id est* Org-X + H^+ + e^- $\rightarrow$ Org-H + X^-, where Org-X is the organohalogen and X^- is the halide.

For detailed examples see Dolfing (2003) (overview), Dolfing et al. (2008) (temperature), Dolfing and Novak (2015) (redox potentials, speciation), and Dolfing (2015) (overview).

3.5 Gibbs Free Energy of Formation Values

Change in Gibbs free energy (ΔG) calculations hinge on Gibbs free energy of formation ($\Delta_{\mathrm{f}}G^{\mathrm{o}}$) values. Experimental methods to measure those values accurately are laborious. Consequently reliable experimental values are scarce, especially for organohalogens (Speight 2005). Until about a decade ago $\Delta_{\mathrm{f}}G^{\mathrm{o}}$ values for a wide variety of halogenated organic compounds were therefore estimated by group contribution methods. In their most basic form (Mavrovouniotis 1990, 1991; Dolfing and Janssen 1994; Jankowski et al. 2008), these methods simply attribute a value for the presence of a specific substituent; in the more sophisticated approaches, e.g., in Benson's method (Dolfing and Harrison 1992; Holmes et al. 1993; Huang et al. 1996) various group interactions are also taken into account. With the advent of reliable quantum chemical calculation methods and readily affordable computer power the landscape is changing (Dolfing et al. 2012; Dolfing and Novak 2015). Nowadays, Gibbs free energy of formation data are available for a wide variety of both aromatic and aliphatic halogenated compounds. Unfortunately, in some of these data sets $\Delta_f G^o$ values are for the gas phase while the microorganisms live in the aqueous phase. Conversion of $\Delta_f G^o{}_{gas}$ into $\Delta_f G^o{}_{aq}$ can be achieved with the Henry constant (Dolfing and Harrison 1992; Dolfing and Janssen 1994), if available, but the issues here are (i) that this approach neglects $\Delta H^o{}_{aq}$ which is needed to calculate ΔG values at temperatures other than 25 °C (e.g. Dolfing 2015; Dolfing et al. 2008), and (ii) that this approach is only justified if the compound of interest actually occurs in the gas phase under standard conditions (Dick et al. 2013). Thus, ideally the conversion of $\Delta_f G^o{}_{gas}$ into $\Delta_f G^o{}_{aq}$ should be made with a quantum chemical solvation model (Marenich et al. 2009; Dolfing and Novak 2015). Of the various models currently available, the universal solvent SMD model developed by Marenich et al. (2009) gives the most accurate results for neutral compounds (Guerard and Arey 2013).

Quantum chemical methods allow a level of precision not achievable with group contribution methods. A comparison of $\Delta_f G^o{}_{aq}$ data for chlorophenols and chlorobenzoates obtained with these respective methods showed considerable scatter; in contrast, a similar comparison of the Gibbs free energy data for chlorinated benzenes obtained with the two methods showed an excellent linear correlation between the data in the two sets (Dolfing and Novak 2015). Thus, it appears that the group contribution methods did especially poor for interactions between the carboxyl and the hydroxyl group on the one hand and the chloro-substituent on the other hand. The absolute values for the two datasets for chlorobenzenes were very different, mainly because the consensus $\Delta_f G^o{}_{aq}$ value for benzene in the 1990s (133.9 kJ/mol, Shock and Helgeson 1990) was considerably higher than the value recently proposed by Sadowsky et al. (7.1 kJ/mol).

Table 3.1 gives an overview of sources of Gibbs free energy of formation data for halogenated compounds in the aqueous phase for a wide variety of compounds.

Table 3.1 Sources of Gibbs free energy of formation data for halogenated compounds in the aqueous phase

Compound class	Methodology	Reference
Haloalkanes	Lange's handbook and Henry's constant	Dolfing and Janssen (1994)
Haloalkenes	Lange's handbook and Henry's constant	Dolfing and Janssen (1994)
Halobenzenes	Benson's method	Dolfing and Harrison (1992)
Halobenzoates	Benson's method	Dolfing and Harrison (1992)
Halophenols	Benson's method	Dolfing and Harrison (1992)
Halobenzenes	Quantum chemical	Sadowsky et al. (2013)
Halobenzoates	Quantum chemical	Dolfing and Novak (2015)
Halophenols	Quantum chemical	Dolfing and Novak (2015)
Polychlorinated biphenyls	Benson's method	Holmes et al. (1993)
Polychlorinated-dibenzo-*p*-dioxins	Benson's method	Huang et al. (1996)
DDT and its degradation products	Quantum chemical	Bylaska et al. (2004)

For various classes of compounds not delineated in Table 3.1. Gibbs free energy of formation values are only available for the gas phase; this includes polychlorinated anthracenes (Zeng et al. 2009), polychlorinated dibenzothiophenes (Chen et al. 2007), polychlorinated pyrenes (Zeng and Yu 2013), polybrominated dibenzo-*p*-dioxins (Li et al. 2003), and polybrominated naphthalenes (Yuan et al. 2006).

3.6 Hydrogenolysis Under Standard Conditions

The amount of energy available from reductive dehalogenation reactions depends on the compound, the halogen, and the reaction type. Under standard conditions (25 °C, pH 7, 1 M concentrations of substrates and products) change in Gibbs free energy ($\Delta G^{o\prime}$) values for hydrogenolysis are generally in the range of -120 to -170 kJ/Cl removed (Dolfing 2003). The differences between the various compound classes are relatively minor, and there is no a priori difference between aromatic and aliphatic compounds (Dolfing 2003). Generally, the amount of energy available from dehalogenation decreases with decreasing degree of halogenation (Table 3.2). However, this decrease is relatively minor and does not explain why dehalogenation of compounds like chlorobenzene and vinyl chloride seems more difficult (that is, it is more difficult to enrich organisms that reductively dechlorinate these compounds) than dehalogenation of their fully halogenated analogs (hexachlorobenzene and tetrachloroethene). The explanation for this phenomenon must be (bio)chemically rather than thermodynamically. The effect of the type of halogen on the change in Gibbs free energy for reductive dehalogenation has never been evaluated comprehensively, but Fig. 3.3 suggests that $\Delta G^{o\prime}$ for reductive dehalogenation has the tendency to increase in the order F < Cl < Br < I.

3.7 Dihaloelimination

The energetics of dihaloelimination are distinctly different from those of hydrogenolysis. In dichloroelimination, two moles of HCl are formed per mole of H_2 consumed, while in hydrogenolysis only one mole of HCl is formed per mole of H_2.

Table 3.2 Comparison of change in Gibbs free energy values for reductive dechlorination (hydrogenolysis) of perchlorinated compounds versus values for dechlorination of their monochlorinated analogs[a]

Reactants		Products	$\Delta G^{o\prime}$ (kJ/mol)
Hexachlorobenzene + H_2	→	Pentachlorobenzene + H^+ + Cl^-	−166.2
Chlorobenzene + H_2	→	Benzene + H^+ + Cl^-	−126.1
Tetrachloroethene + H_2	→	Trichloroethene + H^+ + Cl^-	−173.4
Vinylchloride + H_2	→	Ethene + H^+ + Cl^-	−149.4

[a]Data from Dolfing and Janssen (1994) and Dolfing and Novak (2015)

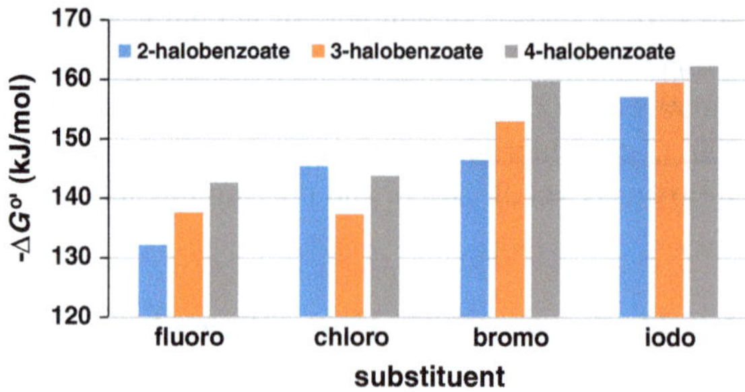

Fig. 3.3 The change in Gibbs free energy ($-\Delta G^{o\prime}$) for hydrogenolysis of halogenated benzoates tends to increase in the order F < Cl < Br < I (*Data* Dolfing and Harrison 1992)

The change in Gibbs free energy ($\Delta G^{o\prime}$) value for the formation HCl at pH 7 is −171.2 kJ/mol. Given that, as stated above $\Delta G^{o\prime}$ for hydrogenolysis is generally −120 to −170 kJ/mol, it is clear that hydrogenolysis is essentially driven by the energy released in the formation of HCl. Because in dihaloelimination two moles of HCl are formed per reaction, one would as first approximation expect that $\Delta G^{o\prime}$ for dihaloelimination is about twice that of hydrogenolysis. In reality the amount is less, as dihaloelimination results in the desaturation of a C–C bond which costs about 100 kJ/bond. Nevertheless, per mole of H_2 consumed the $\Delta G^{o\prime}$ for dihaloelimination is considerably more favorable than $\Delta G^{o\prime}$ for hydrogenolysis; however, per mole of HCl produced hydrogenolysis is more favorable (Table 3.3). Thus under conditions where organohalogens are limiting an organism that can gain energy from two sequential hydrogenolysis steps can gain more energy than an organism that can convert the same organohalogen via one dihaloelimination step.

3.8 Change in Gibbs Free Energy Under Environmentally Realistic Conditions

The change in Gibbs free energy values mentioned above are all $\Delta G^{o\prime}$ values, that is they apply to standard conditions: a temperature of 25 °C, gases (H_2 and CO_2) at partial pressures of 1 atm, and organohalogens and other solutes at a concentration of 1 M, except for H^+ which is at a concentration of 10^{-7} M (pH 7).

Table 3.3 Contrast between dihaloelimination and hydrogenolysis: change in Gibbs free energy values for reductive dechlorination of tetra- and trichloroethane[a]

Reactants		Products	$\Delta G^{o\prime}$ (kJ)
1,1,2,2-tetrachloroethane + H_2	→	1,2-dichloroethene + $2H^+$ + $2Cl^-$	−225.7
1,1,2,2-tetrachloroethane + H_2	→	1,1,2-trichloroethane + H^+ + Cl^-	−159.9
1,1,2-trichloroethane + H_2	→	chloroethene + $2H^+$ + $2Cl^-$	−205.1
1,1,2-trichloroethane + H_2	→	1,1-dichloroethane + H^+ + Cl^-	−162.3
1,1,2-trichloroethane + H_2	→	1,2-dichloroethane + H^+ + Cl^-	−166.5

[a]Data from Dolfing and Janssen (1994)

Under environmentally more realistic conditions, ΔG values are slightly lower (see Box 1 for calculation details). At H_2 partial pressures of 10 Pa (10^{-4} atm) and Cl^- concentrations of 1 mM, ΔG for hydrogenolysis will be −5.7 kJ/mol less favorable than under standard conditions. Relative to the aforementioned values of 120–170 kJ/reaction this effect is negligible. Taking physiological concentrations of organohalogens of 1–100 μM also into consideration would make hydrogenolysis −28.5 to −39.9 kJ/mol less favorable; however, the concentrations of the organic hydrogenolysis products are likely in the same range (1–100 μM), which would make hydrogenolysis −22.8 to −34.2 kJ/mol more favorable. At equimolar concentrations these effects cancel each other out.

3.9 Comparison with Other Electron Acceptors

Change in Gibbs free energy ($\Delta G^{o\prime}$) values for hydrogenolysis of −120 to −170 kJ/mol translate into redox potentials of 210–470 mV, i.e., they are comparable to the redox couple NO_3^-/NO_2^- ($E_o{}' = 433$ mV). This is substantially lower than the redox potential for O_2 ($E_o{}' = 818$ mV), but much higher than the redox potentials for sulfate reducing (SO_4^{2-}/HS^-; $E_o{}' = -217$ mV) and methanogenic (CO_2/CH_4; $E_o{}' = -244$ mV) environments. Hydrogenolysis is thus rare under aerobic conditions, where O_2 is an energetically much more favorable electron acceptor. Under sulfate reducing and methanogenic conditions, on the other hand, hydrogenolysis is the more energetically favorable process. This implies that growth of organohalide-respiring organisms can be sustained at low hydrogen concentrations where H_2-based sulfate reduction and methanogenesis would be exergonic. Observations by Löffler et al. (1999) are in line with this paradigm (Fig. 3.4). These authors showed that dehalogenating organisms can remove hydrogen to very low levels (<0.04 Pa, which at equilibrium for hydrogen between

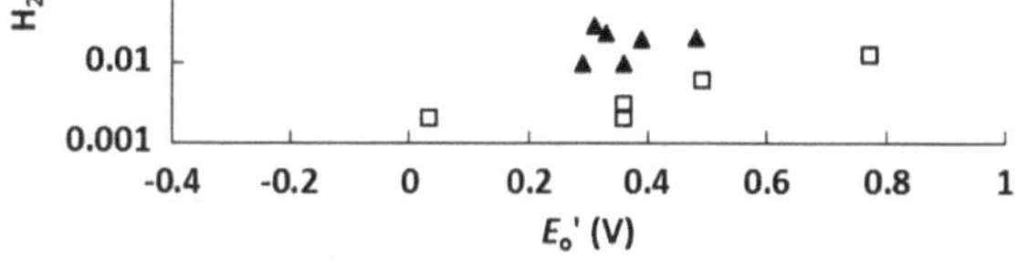

Fig. 3.4 Measured hydrogen thresholds under various electron acceptor conditions. *Filled circles*, sulfate reduction and methanogenesis; *open squares*, fumarate reduction, nitrate reduction, ammonification and iron reduction; *filled triangles*, organohalide respiration (*Data* Löffler et al. 1999)

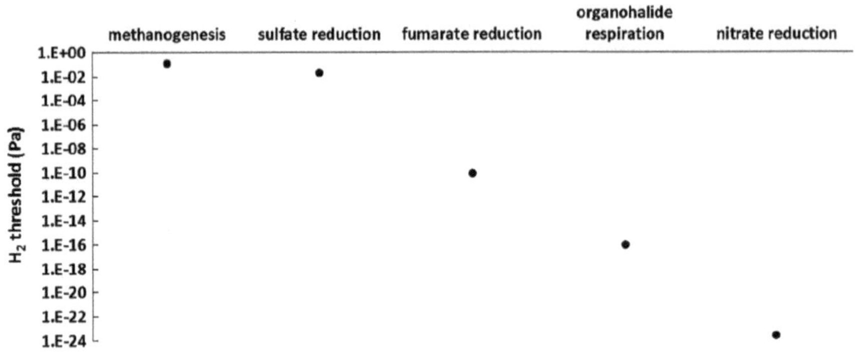

Fig. 3.5 Thermodynamic prediction of the H_2 thresholds below which methanogenesis, sulfate reduction and organohalide respiration become endergonic under otherwise standard conditions

gas phase and aqueous phase translates into a concentrations <0.3 nM) where it is out of reach for sulfate reducers and methanogens: measured H_2 thresholds for sulfate reducing and methanogenic organisms are in the range of 0.1–2 Pa (1–15 nM), and 0.6–12 Pa (5–95 nM), respectively. Interestingly, the H_2 levels where organohalide respiration would become endergonic are many orders of magnitude lower than the levels hydrogen thresholds for organohalide respiration measured experimentally (Fig. 3.5). This indicates that organohalide respiration is under kinetic rather than thermodynamic control.

3.10 Co-metabolic Reductive Dehalogenation

Given that reductive dehalogenation is a highly exergonic reaction, there is considerable energy available to the organisms that catalyze this reaction. However, not all those organisms are able to harness this energy. Methanogens are a case in point. Their metabolism involves highly reducing corrinoids and other cofactors that fortuitously dechlorinate a wide variety of halogenated compounds (Schrauzer and Katz 1978; Smith and Woods 1994; Dolfing 1995; van Eekert 1999; van Eekert et al. 1999). Not harnessing the energy liberated by the dehalogenation reaction is for the organisms involved not only a missed opportunity, it is actually a cost, as the reducing equivalents used in the process could have been used more productively elsewhere in their metabolism, *in case* to produce methane. Typical growth yields of hydrogenotrophic methanogens are in the range of 1 to 8 g dry weight per mole of methane produced (Vogels et al. 1988). Per mole of hydrogen used for reductive dechlorination methanogens thus forfeit 0.25–2 g new biomass. From an environmental engineering point of view, co-metabolic conversion of halogenated compounds can potentially lead to new technologies, especially for xenobiotic compounds. Chlordecone, a highly chlorinated pesticide, is a case in point. Currently no organisms are known that can grow with chlordecone as electron acceptor, even though its dechlorination is highly exergonic (Dolfing et al. 2012). Methanogens, however can dechlorinate this compound (Schrauzer and Katz 1978; Jablonski et al. 1996), which suggests that methanogenic biomass can be used to remove this persistent organic compound from waste streams and the environment.

Löffler et al. (1999) have pointed out an energetic/thermodynamic approach that allows distinction between metabolic and co-metabolic degradation of halogenated organics based on the premise that conservation of the considerable amount of energy available from organohalide respiration results in distinct biomass formation in and by organohalide-respiring organisms. The energy gain per electron for organohalogens as electron acceptor is higher than for sulfate or bicarbonate (methanogenesis) as electron acceptor, but lower than for nitrate or oxygen as electron acceptor. This is reflected in the fraction of electrons (f_e) that is used for biomass formation (0.30–0.37), which is higher than the f_e for methanogenesis (0.05) and sulfate reduction (0.06–0.16) but lower than the f_e for nitrate and oxygen respiration (0.43 and 0.46, respectively). The point here, as made by Löffler et al. (1999) though is that the fractions of electrons flowing to biomass formation and to reduction of the catabolic electron acceptor can be fairly easily determined, which makes for an elegant tool to distinguish between metabolic and co-metabolic organohalide respiration in pure cultures. In mixed cultures, however a substantial portion of the electrons can flow to other electron acceptors, making it difficult to accurately assess the extent of electron flow to the presumed organohalide respirers.

3.11 Anaerobic Oxidation and Fermentation of Organohalogens

The presence of a halogen substituent renders a compound more oxidized than its nonhalogenated analog, and the thermochemical characteristics of halogens are such that their presence decreases the amount of energy available when organohalogens are mineralized under aerobic conditions, i.e., with oxygen as electron acceptor (Dolfing 2003). Whether or not the presence of a halogen substituent decreases the amount of energy available when an organohalogen is mineralized under anaerobic conditions depends on the electron acceptor used. Under denitrifying conditions ($E^{0\prime}$ $NO_3^-/NO_2 = 746$ mV), this amount decreases with increasing degree of chlorination (Fig. 3.6). Coupled to sulfate reduction or methanogenesis, on the other hand, the presence of chloro-substituents increases the amount of energy available from mineralization. For most practical applications, though the main message of this evaluation is that mineralization of organohalogens is in principle always an exergonic process, irrespective of the electron acceptor used: The poor biodegradability of many of these compounds is caused by ecophysiological and evolutionary intricacies, not by a lack of potential energy.

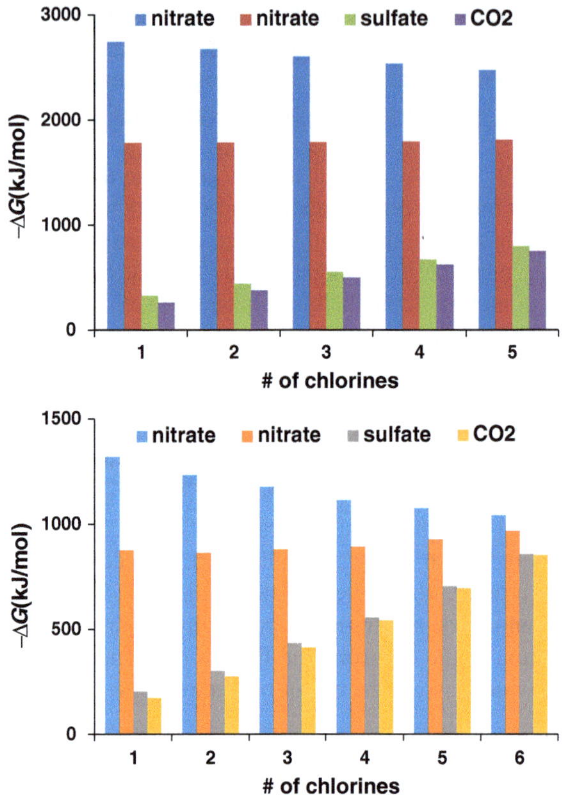

Fig. 3.6 Change in Gibbs free energy for the complete mineralization of pentachlorophenol (*upper panel*) and hexachloroethane (*lower panel*) with nitrate, sulfate or carbon dioxide as electron acceptor. Mineralisation of organics results in the generation of reducing equivalents. The thermodynamic value of these reducing equivalents depends on the electron acceptor used. Per mole of reducing equivalents (H_2) this is 224 kJ for denitrification, 150 kJ for ammonification, 38 kJ for sulfate reduction, 33 kJ for methanogenesis respectively. Each chloro-substituent reduces the potential amount of H_2 produced (cf. Table 3.4), while yielding upon mineralization about 150 kJ. Hence the above pattern: under denitrifying conditions the presence of chloro-substituents reduces the amount of energy liberated upon mineralization (224 > 150), under ammonification conditions the effect is neutral (150 = 150) and so on

One of the consequences of being good electron acceptors is that many organohalogens are fermentable, at least in theory (Dolfing 2000). Fermentations are disproportionation reactions in which a portion of the substrate is oxidized while the remainder is reduced (Madsen 2008); no external electron acceptor is involved. An example of an organohalogen that in theory can be fermented is trichloroethane. At least two pathways can be envisaged: fermentation of trichloroethane to acetate: $C_2H_3Cl_3 + 2H_2O \rightarrow CH_3COO^- + 4\ H^+ + 3Cl^-$ ($\Delta G^{o\prime} = -371$ kJ/mol), and fermentation of trichloroethane to ethane: $7C_2H_3Cl_3 + 12H_2O \rightarrow 4\ C_2H_6 + 6CO_2 + 21H^+ + 21Cl^-$ ($\Delta G^{o\prime} = -377$ kJ/mol). Ethane itself is not

Table 3.4 Change in Gibbs free energy ($\Delta G^{o\prime}$) values for the mineralization and fermentation of chlorinated phenols[a]

Reactants		Products	$\Delta G^{o\prime}$ (kJ)
Mineralization			
$C_6HOCl_5 + 11H_2O$	→	$6CO_2 + 9H_2 + 5H^+ + 5Cl^-$	−456.5
$C_6H_2OCl_4 + 11H_2O$	→	$6CO_2 + 10H_2 + 4H^+ + 4Cl^-$	−293.2
$C_6H_3OCl_3 + 11H_2O$	→	$6CO_2 + 11H_2 + 3H^+ + 3Cl^-$	−135.4
$C_6H_4OCl_2 + 11H_2O$	→	$6CO_2 + 12H_2 + 2H^+ + 2Cl^-$	17.9
$C_6H_5OCl + 11H_2O$	→	$6CO_2 + 13H_2 + H^+ + Cl^-$	169.5
$C_6H_6O + 11H_2O$	→	$6CO_2 + 14H_2$	315.6
Fermentation to acetate			
$4C_6HOCl_5 + 26H_2O$	→	$9CH_3COO^- + 29H^+ + 20Cl^- + 6CO_2$	−670.0
$4C_6H_2OCl_4 + 24H_2O$	→	$10CH_3COO^- + 26H^+ + 16Cl^- + 4CO_2$	−530.5
$4C_6H_3OCl_3 + 22H_2O$	→	$11CH_3COO^- + 23H^+ + 12Cl^- + 2CO_2$	−396.4
$4C_6H_4OCl_2 + 20H_2O$	→	$12CH_3COO^- + 20H^+ + 8Cl^-$	−266.8
$4C_6H_5OCl + 18H_2O + 2CO_2$	→	$13CH_3COO^- + 17H^+ + 4Cl^-$	−138.9
$4C_6H_6O + 16H_2O + 4CO_2$	→	$14CH_3COO^- + 14H^+$	−16.5

[a]Gibbs free energy data for halophenols are taken from Dolfing and Novak (2015)

fermentable to acetate as ethane-based acetogenesis would require reduction of CO_2: $4C_2H_6 + 2H_2O + 6CO_2 \rightarrow 7\ CH_3COO^- + 7H^+$ ($\Delta G^{o\prime} = 11$ kJ/mol). Similar games can be played with halogenated aromatics. For example, in theory fermentation of dichlorophenol could encompass formation of phenol according to $7C_6H_4OCl_2 + 11H_2O \rightarrow 6C_6H_6O + 14HCl + 6CO_2$ ($\Delta G^{o\prime} = -253$ kJ/mol) or alternatively dichlorophenol can be fermented to acetate $C_6H_4OCl_2 + 5H_2O \rightarrow 3CH_3COOH + 2HCl$ ($\Delta G^{o\prime} = -116.7$ kJ/mol). Table 3.4 lists the stoichiometry and energetics of the complete mineralization of chlorinated phenols. Combining these data with the stoichiometry and energetics of acetate formation from H_2/CO_2 ($2CO_2 + 4H_2 \rightarrow CH_3COO^- + H^+ + 2H_2O$; $\Delta G^{o\prime} = -94.9$ kJ/mol) yields an overview of the energetics of chlorophenol fermentation to acetate and HCl (Table 3.4). These calculations are purely theoretical though. The fermentation reactions presented in Table 3.4 are thermodynamically possible but have not been seen in the environment.

There are actual examples of organohalide fermentation though. In 1996, *Dehalobacterium formicoaceticum* was isolated and characterized as a strictly anaerobic bacterium utilizing dichloromethane (DCM) as source of carbon and energy, metabolizing dichloromethane according to $3CH_2Cl_2 + CO_2 \rightarrow 2HCOO^- + CH_3COO^- + 6Cl^- + 9H^+$ (Mägli et al. 1996, 1998). The organism was isolated from a mixed culture in which dichloromethane was fermented according to $2CH_2Cl_2 + 2H_2O \rightarrow CH_3COO^- + 4Cl^- + 5H^+$ (Mägli et al. 1995). More recently, Justicia-Leon et al. (2012) observed that dichloromethane can support fermentative growth of *Dehalobacter* sp.: in a mixed culture the organisms thrived on DCM, DCM was not reductively dechlorinated, and acetate was produced. Interestingly, this implies that *Dehalobacter* growth is not limited to organohalide respiration and these organisms can also shift metabolism to fermentation. Whether this is a propensity of *Dehalobacter* as a genus, or that individual species can make this shift remains to be seen. Similarly, it remains to be seen whether organisms can be found that ferment dichloromethane according to $2CH_2Cl_2 + 2H_2O \rightarrow CH_3COO^- + 4Cl^- + 5H^+$ ($\Delta G^{o\prime} = -243.7$ kJ/mol DCM) rather than $3CH_2Cl_2 + CO_2 + 4H_2O \rightarrow 2HCOO^- + CH_3COO^- + 6Cl^- + 9H^+$ ($\Delta G^{o\prime} = -225.6$ kJ/mol DCM). It should be noted that these observations support dichloromethane fermentation as a process, but not necessarily as a metabolism. There is still no proof that one organism can ferment dichloromethane without an external electron acceptor: *D. formicoaceticum* uses CO_2 as addition electron acceptor, and there still are no pure cultures of dichloromethane fermenting *Dehalobacter* species.

3.12 Defluorination

All organohalogens are characterized by strong carbon–halogen bonds, but among these, the carbon–fluorine bond is the strongest. Indeed, the C–F bond is the strongest single bond formed by carbon. This is reflected in the short bond length, the high bond dissociation energy, and due to fluorine's high electronegativity, the high polarity of the C–F bond (Reineke 1984). These characteristics may well be the reasons why microbial growth based on organofluoride respiration has not yet been observed. Thermodynamics, or better, a lack of available energy, is not the explanation. The scarce Gibbs free energy values available for organofluorides—such values are less readily available than the corresponding values for organochlorides or—bromides—indicate that the amount of energy available from defluorination is similar to (or slightly lower than) the amount of energy available from dechlorination. Thus, the reason why organofluoride respiration has never been observed is kinetic rather than thermodynamic. Though unfortunate, from an environmental point of view this conclusion is encouraging. Given that energy is available from defluorination organisms may eventually come to the fore (evolve) that are able to harness and exploit this energy, even for the notoriously persistent perfluorinated compounds (Parsons et al. 2008).

Goldman (1965) has described the first hydrolytic fluoroacetate dehalogenase already back in 1965, and similar enzymes have been described since (Murphy 2010) in aerobic bacteria. Under anaerobic conditions, defluorination has long been elusive. Vargas et al. (2000), when testing for anaerobic degradation of fluorinated aromatics, observed that these compounds were recalcitrant under sulfate reducing and methanogenic conditions; these authors isolated denitrifying organisms that could mineralize 2- and 4-fluorobenzoate, but the first step in the degradation pathway was not a reductive defluorination. Recently however Davis et al. (2012), using a novel, elegant fluoroacetate assay, isolated the first anaerobe in possession of a reductive defluorinase and presented evidence in support of the hypothesis that the organism can harness energy from defluorination. Significantly,

Davis et al. obtained their isolate from an anaerobic environment, a bovine rumen in Australia, that had been under long term exposure to fluorinated organics as the animals from which the bacteria were isolated had been eating plants that are known to produce fluoroacetate. Gribble (2002) has cataloged a wide variety of naturally occurring organofluorine compounds. Taken together, these observations suggest that it is opportune to continue screening for and studying the potential degradation of (per)fluorinated compounds under anaerobic conditions.

3.13 Organohalide Respiration and ATP Generation

It has been reiterated now several times in the preceding paragraphs that reductive dechlorination is a (very) exergonic process, also under environmentally realistic conditions and that microorganisms can potentially harness considerable energy from this reaction. Just how much energy the organisms actually harness from organohalogen respiration and how much biomass is generated based on organohalogens respiration is beyond the scope of this chapter. These aspects will be discussed in detail in a later chapter of this book (Mayer-Blackwell et al. 2015). Nevertheless, it seems prudent to point out here that organohalogen respiration yields in theory, enough energy for ATP generation via substrate level phosphorylation. Biomass yields are highly variable and—depending on the organism and the substrate used—in the range of about 0.3–5.6 g dry weight per mole of chloride produced (Ding et al. 2014). As mentioned above, Vogels et al. (1988) have summarized that biomass production by hydrogenotrophic methanogens is in the same range [1–8 g dry weight per mole of product (methane)]. But as production of one mole of methane requires four moles of H_2, versus a consumption of one mole of H_2 per mole of chloride produced, organohalogen respirers produce about four times more biomass per mole of hydrogen consumed. These trends scale with the amounts of Gibbs free energy available from these two reactions. This implies that organohalogen respiring organisms have developed efficient mechanisms to harness the energy available from this process: they are good at what they do, just as good as methanogens in what they are doing.

References

Bylaska EJ, Dixon DA, Felmy AR, Aprà E, Windus TL, Zhan C-G, Tratnyek PG (2004) The energetics of the hydrogenolysis, dehydrohalogenation, and hydrolysis of 4,4′-dichloro-diphenyl-trichloroethane from a initio electronic structure theory. J Phys Chem A 108:5883–5893

Chen SD, Liu HX, Wang ZY (2007) Study of structural and thermodynamic properties for poly-chlorinated dibenzothiophenes by density functional theory. J Chem Eng Data 52:1195–1202

Davis CK, Webb RI, Sly LI, Denman SE, McSweeney CS (2012) Isolation and survey of novel fluoroacetate-degrading bacteria belonging to the phylum *Synergistetes*. FEMS Microbiol Ecol 80:671–684

Dick JM, Evans KA, Holman AI, Jaraula CMB, Grice K (2013) Estimation and application of the thermodynamic properties of aqueous phenanthrene and isomers of methylphenanthrene at high temperature. Geochim Cosmochim Acta 122:247–266

Ding C, Siyan Zhao S, He J (2014) A *Desulfitobacterium* sp. strain PR reductively dechlorinates both 1,1,1-trichloroethane and chloroform. Environ Microbiol 16:3387–3397

Dolfing J (1990) Reductive dechlorination of 3-chlorobenzoate is coupled to ATP production and growth in an anaerobic bacterium, strain DCB-1. Arch Microbiol 153:264–266

Dolfing J (1995) Regiospecificity of chlorophenol reductive dechlorination by vitamin B_{12s}. Appl Environ Microbiol 61:2450–2451

Dolfing J (2000) Energetics of anaerobic degradation pathways of chlorinated aliphatic compounds. Microb Ecol 40:2–7

Dolfing J (2003) Thermodynamic considerations for dehalogenation. In: Häggblom MM, Bossert ID (eds) Dehalogenation: microbial processes and environmental applications. Kluwer, Boston, pp 89–114

Dolfing J (2015) Protocols for calculating reaction kinetics and thermodynamics. In: McGenity TJ, Timmis KN, Nogales B (eds) Hydrocarbon and lipid microbiology protocols: statistics, data analysis, bioinformatics and modelling. Spinger, Berlin. doi:10.1007/8623_2015_109

Dolfing J, Harrison BK (1992) The Gibbs free energy of formation of halogenated aromatic compounds and their potential role as electron acceptors in anaerobic environments. Environ Sci Technol 26:2213–2218

Dolfing J, Janssen DB (1994) Estimates of Gibbs free energies of formation of chlorinated aliphatic compounds. Biodegradation 5:21–28

Dolfing J, Novak I (2015) The Gibbs free energy of formation of halogenated benzenes, benzoates and phenols and their potential role as electron acceptors in anaerobic environments. Biodegradation 26:15–27

Dolfing J, Tiedje JM (1987) Growth yield increase linked to reductive dechlorination in a defined 3-chlorobenzoate degrading methanogenic coculture. Arch Microbiol 149:102–105

Dolfing J, Larter SR, Head IM (2008) Thermodynamic constraints on methanogenic crude oil biodegradation. ISME J 2:442–452

Dolfing J, Novak I, Archelas A, Macarie H (2012) Gibbs free energy of formation of chlordecone and potential degradation products: implications for remediation strategies and environmental fate. Environ Sci Technol 46:8131–8139

Goldman P (1965) Enzymatic cleavage of carbon-fluorine bond in fluoroacetate. J Biol Chem 240:3434–3438

Gribble GW (2002) Naturally occurring organofluorines. In: Hutzinger O (ed) Handbook of environmental chemistry, vol 3. Springer, Heidelberg, pp 121–136

Guerard JJ, Arey JS (2013) Critical evaluation of implicit solvent models for predicting aqueous oxidation potentials of neutral organic compounds. J Chem Theory Comput 9:5046–5058

Hanselmann KW (1991) Microbial energetics applied to waste repositories. Experientia 47:645–687

Holmes DA, Harrison BK, Dolfing J (1993) Estimation of Gibbs free energies of formation of polychlorinated biphenyls. Environ Sci Technol 27:725–731

Huang C-L, Harrison BK, Madura J, Dolfing J (1996) Thermodynamic prediction of dehalogenation pathways for PCDDs. Environ Toxicol Chem 15:824–836

Hug LA, Maphosa F, Leys D, Löffler FE, Smidt H, Edwards EA, Adrian L (2013) Overview of organohalide-respiring bacteria and a proposal for a classification system for reductive dehalogenases. Phil Trans R Soc B Biol Sci 368:20120322

Jablonski PE, Pheasant DJ, Ferry JG (1996) Conversion of kepone by *Methanosarcina thermophila*. FEMS Microbiol Lett 139:169–173

Jankowski MD, Henry CS, Broadbelt LJ, Hatzimanikatis V (2008) Group contribution method for thermodynamic analysis of complex metabolic networks. Biophys J 95:1487–1499

Justicia-Leon SD, Ritalahti KM, Mack EE, Löffler FE (2012) Dichloromethane fermentation by a *Dehalobacter* sp. in an enrichment culture derived from pristine river sediment. Appl Environ Microbiol 78:1288–1291

Kirk KL (1991) Biochemistry of the elemental halogens and inorganic halides. Plenum Press, New York

Leys D, Adrian L, Smidt H (2013) Organohalide respiration: microbes breathing chlorinated molcecules. Phil Trans R Soc B Biol Sci 368:20120316

Li XW, Shibata E, Nakamura T (2003) Theoretical calculation of thermodynamic properties of polybrominated dibenzo-p-dioxins. J Chem Eng Data 48:725–735

Löffler FE, Tiedje JM, Sanford RA (1999) Fraction of electrons consumed in electron acceptor reduction and hydrogen thresholds as indicators in halorespiratory physiology. Appl Environ Microbiol 65:4049–4056

Madigan MT, Martinko JM, Parker J (1997) *Brock* biology of microorganisms, 8th edn. Prentice Hall, Upper Saddle River, New Jersey

Madsen EL (2008) Environmental microbiology: from genomes to biogeochemistry. Wiley, New York

Mägli A, Rainey FA, Leisinger T (1995) Acetogenesis from dichloromethane by a two-component mixed culture comprising a novel bacterium. Appl Environ Microbiol 61:2929–2943

Mägli A, Wendt M, Leisinger T (1996) Isolation and characterization of *Dehalobacterium formicoaceticum* gen. nov. sp. nov., a strictly anaerobic bacterium utilizing dichloromethane a source of carbon and energy. Arch Microbiol 166:101–108

Mägli A, Messmer M, Leisinger T (1998) Metabolism of dichloromethane by the strict anaerobe *Dehalobacterium formicoaceticum*. Appl Environ Microbiol 64:646–650

Maphosa F, de Vos WM, Smidt H (2010) Exploiting the ecogenomics toolbox for environmental diagnostics of organohalide-respiring bacteria. Trends Biotechnol 28:308–316

Marenich AV, Cramer CJ, Truhlar DG (2009) Universal solvation model based on solute electron density and a continuum model of the solvent defined by the bulk dielectric constant and atomic surface tensions. J Phys Chem B 113:6378–6396

Mavrovouniotis ML (1990) Group contributions for estimating standard Gibbs energies of formation of biochemical compounds in aqueous solution. Biotechnol Bioeng 36:1070–1082

Mavrovouniotis ML (1991) Estimation of standard Gibbs energy changes of biotransformations. J Biol Chem 266:14440–14445

Mayer-Blackwell K et al, Spormann A (2015) Integrative and comparative physiology of OHRB's. In: Löffler FE, Adrain E (eds) Organohalide respiring bacteria. Springer, Heidelberg, pp XX–YY

Maymo-Gatell X, Nijenhuis I, Zinder SH (2001) Reductive dechlorination of *cis*-1,2-dichloroethene and vinyl chloride by "*Dehalococcoides ethenogenes*" 195. Environ Sci Technol 35:516–521

McCarty PL (1997) Microbiology—breathing with chlorinated solvents. Science 276:1521–1522

Mohn WW, Tiedje JM (1992) Microbial reductive dehalogenation. Microbiol Rev 56:482–507

Murphy CD (2010) Biodegradation and biotransformation of organofluorine compounds. Biotechnol Lett 32:351–359

Parsons JR, Sáez M, Dolfing J, de Voogt P (2008) Biodegradation of perfluorinated compounds. In: Whitacre DM (ed) Reviews of environmental contamination and toxicology, vol 196. Springer, Heidelberg, pp 53–71

Reineke W (1984) Microbial degradation of halogenated aromatic compounds. In: Gibson DT (ed) Microbial degradation of organic compounds. Marcel Dekker, pp 319–360

Sadowsky D, McNeill K, Cramer CJ (2013) Thermochemical factors affecting the dehalogenation of aromatics. Environ Sci Technol 47:14194–14203

Schrauzer GN, Katz RN (1978) Reductive dechlorination and degradation of Mirex and Kepone with vitamin B_{12s}. Bioinorg Chem 9:123–143

Shock EL, Helgeson HC (1990) Calculation of the thermodynamic and transport properties of aqueous species at high pressures and temperatures: standard partial molal properties of organic species. Geochim Cosmochim Acta 54:915–945

Smith MH, Woods SL (1994) Regiospecificity of chlorophenol reductive dechlorination by vitamin B_{12s}. Appl Environ Microbiol 60:4111–4115

Speight JG (2005) Lange's handbook of chemistry, 16th edn. McGraw-Hill, New York

Thauer RK, Jungermann K, Decker K (1977) Energy conservation in chemotrophic anaerobic bacteria. Bacteriol Rev 41:100–180

van Eekert MHA (1999) Transformation of chlorinated compounds by methanogenic granular sludge. PhD thesis, Wageningen University, Wageningen

van Eekert MHA, Stams AJM, Field JA, Schraa G (1999) Gratuitous dechlorination of chloroethanes by methanogenic granular sludge. Appl Microbiol Biotechnol 51:46–52

Vargas C, Song B, Camps M, Häggblom MM (2000) Anaerobic degradation of fluorinated aromatic compounds. Appl Microbiol Biotechnol 53:342–347

Vogels GD, Keltjens JT, van der Drift (1988) Biochemistry of methane production. In: Zehnder AJB (ed) Biology of Anaerobic Microorganisms. Wiley-Interscience, New York, pp 707–770

Wackett LP (1995) Bacterial co-metabolism of halogenated organic compounds. In: Young LY, Cerniglia CE (eds) Microbial transformation and degradation of toxic organic chemicals. Wiley-Liss, New York, pp 217–241

Yuan LX, Yu J, Wang ZY et al (2006) Thermodynamic property and relative stability of 76 polybrominated naphthalenes by density functional theory. J Chem Eng Data 51:2032–2037

Zehnder AJB, Stumm W (1988) Geochemistry and biogeochemistry of anaerobic habitats. In: Zehnder AJB (ed) Biology of anaerobic microorganisms. Wiley-Interscience, New York, pp 1–38

Zeng XL, Yu YS (2013) Theoretical study on the molecular structures and thermodynamic properties of polychlorinated pyrenes. Comp Theor Chem 1013:92–96

Zeng XL, Wang Y, Zhang XL, Yu YS (2009) Density functional theory studies on the molecular structures and thermodynamic properties of polychlorinated anthracenes. J Mol Struct THEOCHEM 906:83–86

Part II
Diversity of Organohalide-Respiring Bacteria

Chapter 4
Discovery of Organohalide-Respiring Processes and the Bacteria Involved

Perry L. McCarty

Abstract Hazardous halogenated organic compounds are produced industrially for a variety of uses, are highly resistant to degradation by aerobic organisms, and are now widely distributed throughout the natural environment. Discovered in the 1960s were anaerobic organisms that can transform chlorinated pesticides such as DDT and lindane. In the 1980s, other halogenated organics, the chlorinated solvents, were found to be major contaminants of groundwaters, and were found degradable by anaerobic organisms as well. While reductive dehalogenation, the process involved, was believed initially to be a fortuitous enzymatic or cometabolic process, organisms were found in the 1980s that could use halogenated compounds as electron acceptors in an energy-yielding process. Numerous species, both facultative and anaerobic, were then found capable of obtaining energy from reductive dehalogenation, but generally were very restricted in the particular halogens and particular halogenated compounds that they could dehalogenate. Some could dehalogenate tetrachloroethene (PCE) to trichloroethene (TCE) and cis-dichloroethene, and some even to vinyl chloride. An organism found in the late 1980s could even anaerobically convert PCE and TCE all the way to ethene. This organism was isolated in the late 1990s and named *Dehalococcoides.* Several strains of *Dehalococcoides* have now been isolated and their genomes have been sequenced. Each has different dehalogenases and dechlorinating abilities, but collectively they are capable of dehalogenating a broad variety of halogenated organic compounds. These organisms are now finding wide application in the engineered remediation of natural environments contaminated with halogenated compounds.

P.L. McCarty (✉)
Department of Civil and Environmental Engineering, Stanford University, Stanford, USA
e-mail: pmccarty@stanford.edu

L. Adrian and F.E. Löffler (eds.), *Organohalide-Respiring Bacteria*,
DOI 10.1007/978-3-662-49875-0_4

4.1 Introduction

The discovery of organohalide respiration has been of great practical significance, both for addressing the remediation of the numerous organohalide-contaminated environments that exist throughout the world and in understanding how the presence of certain toxic intermediate products in the environment came about. Finding unexpected transformation processes that occur naturally has generally been an important first step in discovery, one that has often been followed by a long circuitous route of studies by many researchers, eventually leading to a fuller understanding of the complexity of the processes, and the organisms and enzymes involved. Such has been the case with the reductive dehalogenation of organohalides.

There have been extensive reviews of reductive dehalogenation by others (Mohn and Tiedje 1992; Holliger et al. 1999), as well as by the authors of the many chapters in this book. The interested individual is encouraged to read from among them, and providing another such review is not the intent of this article. I have been one of those fortunate to have been among the early observers of this fascinating phenomenon, and welcome this opportunity to discuss from a personal perspective when and under what circumstances some of the important discoveries of reductive dehalogenation came about, and how they were subsequently investigated to gain the general understanding that we have today about these significant transformation processes.

4.2 Early Studies with Chlorinated Pesticides

My doctoral studies and that of my students when at the Massachusetts Institute of Technology from 1957–1962 were concerned with anaerobic wastewater treatment, the conversion of waste organic materials into methane gas. It was then commonly believed that this process was suitable only for more concentrated wastes such as the sludge produced from wastewater settling and biological treatment. Also believed was that treatment temperatures over 30 °C were needed for the anaerobic process to be economical, temperatures that could be obtained through use of the methane produced. Our goal was to better understand the complexity of anaerobic biological processes and the role of the organisms involved, their kinetics and nutritional requirements, so that the unique energy-producing processes might be more widely applied to treat much more dilute industrial and municipal wastewaters, and at mesophilic temperatures, perhaps as low as 10 °C. That has now been achieved, but that is another story. Another fallacy about the anaerobic process, which most of us certainly believed early on, was that anaerobic treatment was much more limited in its ability to degrade many of the compounds that could readily be degraded aerobically. If they could not be degraded aerobically, then they certainly could not be degraded anaerobically. That belief has now proven to be greatly in error.

I left MIT in 1962 to join the faculty at Stanford University. Dave Hill, one of the four students in my first year there, was studying for the MS Degree and had just read Rachel Carson's book, *Silent Spring*. He indicated an interest in pursuing a Ph.D. degree to study pesticide degradation. The chlorinated pesticides, which were the ones then in most common use, were known to be difficult to degrade aerobically. Learning of my interest in anaerobic processes, Dave said he would like to see if the chlorinated pesticides might be degraded anaerobically. I of course was sure they would not be, but thought it best not to discourage a student's enthusiasm, and so supported his desire to try. Gas chromatography (GC) was then a relatively new analytical approach, and more sensitive detectors were just being developed for chlorinated organics. One interesting approach was the microcoulometric detector, just then developed by Dr. Dale M. Coulson, a researcher at the Stanford Research Institute, as SRI was then called. Dr. Coulson generously gave Dave his original GC and detector system that he had built in his garage. It occupied a whole bench and included a column where pesticides and other organics were separated, an external furnace where the emerging pesticides were burned, releasing hydrochloric acid, and an external detection cell containing silver nitrate solution. When the HCl passed through the cell's silver-containing solution, the chloride combined with silver atoms, decreasing their solution ion concentration and changed the cell potential, causing current to flow to oxidize a silver electrode and replace the complexed silver. The current used was directly proportional to the amount of chloride complexed, and was recorded. It worked very well.

Dave obtained anaerobically digested sludge from the local wastewater treatment plant, added a range of pesticides to the sludge, and waited to see what happened. The pesticides studied were DDT, lindane, heptachlor, heptachlor epoxide, endrin, and aldrin (Hill and McCarty 1967). Analyzing the sludge for pesticides required an extensive cleanup procedure. Soon after perfecting his techniques and making initial analyses, Dave brought his results to me. Lindane, DDT, heptachlor, and endrin disappeared within a few days. I was most skeptical, so he added more to the anaerobic culture, and again within just a few days, they disappeared again. Aerobically, months were generally required before disappearance. Fortunately in Dave's studies, intermediates were formed anaerobically for DDT, heptachlor, and endrin, indicating that a transformation process was indeed occurring. With time, most, but not all, of the intermediates disappeared. Addition of mercuric chloride to samples slowed the transformation processes. There was no easy way then to positively identify the intermediates except to hypothesize what they might be, obtain potential intermediates, and compare their emergence time with that of the sample intermediates to see if a match occurred. Based upon this, it appeared that DDT was being converted initially to DDD, which also degraded, but more slowly. DDD is formed through the replacement of one chlorine atom by hydrogen, a typical reductive dehalogenation process. The rapid disappearance of all of these compounds under anaerobic conditions was a great surprise, but I questioned the practical significance of continuing studies here as pesticides were widely spread on agricultural fields, and would be exposed mainly to aerobic rather than anaerobic conditions. We unfortunately performed no further research in this area.

4.3 Chlorinated Solvent Transformation

Then in about 1975, we at Stanford were approached by the Santa Clara Valley Water District to help evaluate the presence and fate of trace organic compounds in highly reclaimed wastewater that they were to inject into the groundwater for subsequent reuse in Palo Alto near the San Francisco Bay. We later received research funds from the US Environmental Protection Agency for the study. The major trace organic compounds found remaining in the treated wastewater were the chlorinated compounds tetrachloroethene (PCE), trichloroethene (TCE), chloroform (CF), carbon tetrachloride (CT), and a range of monochloro- to trichlorobenzenes (Roberts et al. 1980). After injecting the reclaimed wastewater into groundwater for some period of time, the injection pump broke. Rather than stopping the study altogether, monitoring of the chemicals already injected into the groundwater was continued. Found was that the chlorinated compounds decreased in concentration with time (Roberts et al. 1982). Even though a recent report from the National Academy of Sciences (NAS 1978) indicated these compounds were not biodegradable, we could think of no other explanation for their disappearance. We first tried aerobic degradation without success. But similar to what we found earlier with chlorinated pesticides, when anaerobic degradation was studied using batch anaerobic cultures from a wastewater treatment plant, loss of the chlorinated aliphatic compounds resulted with time (Bouwer et al. 1981). At the same time, another group (R. R. Lang, P. R. Wood, R. A. Parsons, J. Demarco, H. J. Harween, I. L. Payan, L. M. Meyer, M. D. Ruiz, and E. D. Ravelo) indicated at a meeting of the American Water Works Association that they had found almost all chlorinated methane, ethane, and ethene compounds were susceptible to reductive dehalogenation in anaerobic cultures.

Subsequently, using continuous methanogenic column cultures with a two-day detention time and acetate as the substrate, we observed over 90 % loss for <100 μg/L concentrations of all compounds studied: chloroform, carbon tetrachloride, 1,1,2,2-tetrachloroethane (TeCA), 1,1,1-trichloroethane (TCA), trichloroethene, tetrachloroethene, 1,2-dichloroethane (1,2-DCA), and 1,2-dibromoethane (1,2-DBA) (Bouwer and McCarty 1983a). Carbon-14 evaluations with CF, CT, and 1,2-DCA indicated they were nearly completely converted to CO_2. Of particular interest was that the initial step in the transformation of TCE was found to be reductive dehalogenation to *cis*-1,2-dichloroethene (cis-DCE), and TeCA was transformed to TCE. It was hypothesized then that acetoclastic methanogens may have played a role in the transformation, but further study was suggested to be in need. Using denitrification rather than methanogenic conditions, only CT and the three brominated trihalomethanes were transformed (Bouwer and McCarty 1983b). This suggested the possibility that methanogens were involved in reductive dehalogenation.

Parsons et al. (1984) then reported that traces of cis-DCE and vinyl chloride (VC) had formed after PCE disappeared in mucks to which PCE had been added. In further studies, in the laboratory with PCE (Vogel and McCarty 1985), we found that PCE could be converted nearly quantitatively to TCE, then 1,1-dichloroethene (1,1-DCE), and then VC by reductive dehalogenation. C-14 studies suggested that VC might be slowly and partially oxidized to CO_2, but these results were not definitive.

Finding the formation of VC through reductive dehalogenation of PCE or TCE, and then VC's relative stability was rather alarming as VC was then a known carcinogen and a much more hazardous chemical than any of the mother compounds. Interest had been growing for the potential use of reductive dehalogenation to remediate contaminated groundwater, but the fact that VC appeared to be a stable end product reduced interest in pursuing that approach. Thus, when the opportunity to evaluate another approach arose, aerobic cometabolism, we quickly took it.

4.4 Early Work on Reaction Pathways and Organisms Involved

An early question concerned the process or processes involved in halogenated organic transformation. It did not occur then that reductive dehalogenation was likely to be an energy-yielding reaction for microorganisms. A more likely process was cometabolism, a process that had been described earlier, and was suggested by Horvath (1972) as being a common route by which microorganisms transformed difficult to degrade organic compounds present at trace levels. Cometabolism is the fortuitous transformation of a compound by enzymes used by organisms for other beneficial purposes. In a review article, we reported from the literature studies on the various pathways by which halogenated aliphatic compounds might be transformed aerobically and anaerobically (Vogel et al. 1987). Many of these pathways were enzymatically controlled and had been elucidated by researchers who were studying the transformations of chlorinated compounds in humans and other higher organisms in order to better understand the nature of toxicity that resulted from ingestion. With PCE, the dominant anaerobic pathway that emerged was successive reductions to TCE, cis-DCE, VC, and finally ethene. Although at times 1,1-DCE and trans-DCE might be formed from TCE, the cis form of DCE was by far the most prevalent form reported and found.

By 1993, many microbial studies had been and were being conducted on reductive dehalogenation with undefined cultures, and also by then several pure cultures of bacteria had been found that carried out reductive dehalogenation with pesticides, polychlorinated biphenyls (PCBs), and other halogenated aromatics as well as from a range of halogenated aliphatic compounds (Mohn and Tiedje 1992; Tiedje et al. 1993). Generally, each individual organism was found to carry out only limited halogen removal from a selected number of organohalides, such as removal of chlorine only from PCE and/or TCE, but not DCE nor VC. Dehalogenation from compounds with several halogens tended to be favored by anaerobes over compounds with few. Aerobic transformations, on the other hand, tended to occur to compounds that contained relatively few halogens. Among the halogens, ease and speed of transformation tended to be best with brominated, followed with chlorinated, and then iodinated compounds. Fluorinated compounds have been the most difficult to degrade, indeed fluorinated compounds are among the most persistent synthetic organic chemicals produced, either under aerobic or anaerobic conditions.

4.5 Oxidative Transformation of Chlorinated Solvents

While oxidative transformation of halogenated compounds is not the subject of this book, it is worth mentioning such reactions briefly as they are a significant alternative to reductive dehalogenation. Wilson and Wilson (1985), EPA scientists, reported on the aerobic cometabolism of TCE by methanotrophic organisms. Because of Stanford's growing experience with pilot groundwater studies, EPA sought Stanford's interest in pursuing a field pilot study of this process for destroying TCE. This was the beginning of extensive laboratory and field research on aerobic cometabolism, leading to two successful full-scale evaluations of aerobic TCE cometabolism. Methanotrophs use methane monooxygenase to initiate the conversion of methane to methanol, which they then oxidize to obtain energy for growth. Methane monooxygenase had previously been reported to fortuitously oxidize halogenated methanes (Haber et al. 1983), and this report stimulated the Wilsons to find if longer chain halogens might be transformed by this reaction. They found that TCE could be oxidized by methane monooxygenase to the end product CO_2. This was an exciting finding as it avoided the formation of VC.

Through extensive laboratory and field studies, it was found that cometabolism of TCE resulted in the formation of TCE epoxide, which was lethal to the methanotrophs. However, PCE was not cometabolized by methanotrophs or by other oxygenase-producing organisms. We found that at low TCE concentrations, the growth of methanotrophs on methane exceeded their death rate from TCE epoxide formation, but at TCE concentrations above about 1 mg/L, the death rate exceeded the growth rate, and thus the process would fail (Alvarez-Cohen and McCarty 1991; Anderson and McCarty 1996). The oxidation products formed from CF (Alvarez-Cohen and McCarty 1991) and 1,1-DCE (Dolan and McCarty 1995) through cometabolism were even more lethal. Also, the rate of TCE degradation by oxygenase was low, and the removal process was slow. However, methanotrophs were found to be quite efficient in the cometabolism of cis-DCE and VC, intermediates often found from reductive dehalogenation at PCE and TCE-contaminated sites.

In later studies, toluene was found to be a much more effective substrate than methane for TCE cometabolism (Hopkins and McCarty 1995). Applications of this substrate to a field site at Edwards Air Force Base in Southern California provided successful demonstrations of the process, where TCE concentrations ranged below 1 mg/L in an aquifer that had essentially no other organic contaminants and was aerobic (McCarty et al. 1998; Gandhi et al. 2002). However, there was concern by some over the use of toluene as a substrate, even though the remaining concentrations were orders of magnitude lower than health or ascetic concentrations. Also, to be effective, toluene had to be continuously added to the aquifer, and this involves what has been considered to be high operation and maintenance costs. Unfortunately, little interest in applying this effective process has been demonstrated by practitioners, even though it is quite a practical approach that also renders the aquifer waters aerobic.

4.6 Organohalide Respiration

All early studies on reductive dehalogenation provided no suggestion that organisms were obtaining energy for growth from these reactions. As with aerobic studies in which oxygenases were implicated in the cometabolism of TCE and other chlorinated solvents, cometabolism was generally assumed to be associated with reductive dehalogenation as well. However, this belief changed in 1986 when Dolfing and Tiedje reported on a three-organism anaerobic symbiotic conversion of 3-chlorobenzoate (Dolfing and Tiedje 1986, 1987). A new organism, strain DCB-1, was found to reductively remove the chlorine atom from the benzene ring while replacing it with hydrogen from H_2 produced by another organism that anaerobically oxidized benzoate into hydrogen, acetate, and carbon dioxide. The third organism converted the extra H_2 plus carbon dioxide into methane. The important finding was that strain DCB-1 obtained energy for growth from the reductive dehalogenation of 3-chlorobenzoate. This was the first observation of energy yield from reductive dehalogenation. The DCB-1 isolate now has the generic name *Desulfomonile tiedjei.*

Several years later, the growth of *Dehalobacter restrictus* on energy obtained from reductive dehalogenation of PCE and TCE leading to the formation of cis-DCE was reported by Holliger et al. (1993, 1998). This observation changed the outlook for reductive dehalogenation of chlorinated solvents, suggesting that the anaerobic process might be carried out much more efficiently than aerobic cometabolism as large excesses of primary substrate should not be required to drive the reaction. A broader search for organisms that could carry out reductive dehalogenation of all chlorinated aliphatic compounds was then begun.

4.7 Reductive Dehalogenation of Vinyl Chloride and the Associated Genus, *Dehalococcoides*

In 1989, a major breakthrough came with the report by Freedman and Gossett (1989) of an anaerobic mixed culture that reductively dehalogenated PCE through VC into ethene. That resulted in a strong re-interest among practitioners in the anaerobic process. Studies were begun by many to exploit this new finding.

DiStefano et al. (1991) demonstrated, with an enrichment culture grown on methanol, the complete transformation of PCE to ethene without the involvement of methanogens, thus confirming that methanogens were not involved in reductive dehalogenation. They further demonstrated that the reductive process occurred only when H_2 was available as the donor substrate (DiStefano et al. 1992).

A field study at the time allowed us to demonstrate that complete anaerobic conversion of TCE to ethene could also occur naturally. This study came about through a venture into a possible application of aerobic cometabolism to treat a TCE-contaminated aquifer in St. Joseph, Michigan, that had been brought to our attention by John Wilson, an employee of the EPA Laboratory in Ada, Oklahoma. From a few sample wells, the downgradient presence of TCE and its dechlorination products cis-DCE and VC had been found, compounds that would be readily susceptible to aerobic cometabolism using methane as a substrate. A treatment strategy was developed (McCarty et al. 1991), and a detailed site characterization was carried out by EPA in preparation for a field application of the process (Semprini et al. 1995). However, the results of the detailed characterization were unexpected, and the actual concentrations of TCE and other contaminants were an order of magnitude above those observed from the earlier limited investigation. It was apparent, first, that the TCE concentrations were well above the toxic concentrations for aerobic cometabolism to be applied, and second, that complete transformation of TCE to ethene was abundant and evident. The organic substrate that was the source of the H_2 necessary for reductive dehalogenation appeared to come from a factory dump that had been leaching organics into the groundwater for many years, forming an organic plume that intersected with the downgradient spill of TCE, thereby slowly producing through organic fermentation the H_2 that was needed for in situ remediation to occur. While aerobic cometabolism was now known not to be a viable answer at this site, the site became one of the first where "natural" attenuation through reductive dehalogenation was found to be an acceptable remediation alternative. Complete anaerobic transformation of high concentrations of TCE to ethene in the field was demonstrated not only to be possible, but was actually occurring on its own. Many such sites have now been found.

While many organisms had been identified that were capable of converting PCE and TCE to cis-DCE, isolating an organism capable of converting VC to ethene was elusive. The breakthrough came from the Cornell group with the publication in 1997 of the isolation of an organism (strain 195) that could reductively dehalogenate PCE all the way through to ethene (Maymó-Gatell et al. 1997). Hydrogen was the only substrate strain 195 was capable of using, and was oxidized through reductive dehalogenation, using the chlorinated compounds as electron acceptors. The organism's 16S ribosomal RNA gene sequence did not cluster it with any of the known phylogenetic lines, thus the authors suggested naming the organism "*Dehalococcoides ethenogenes*" strain 195. While complete conversion of PCE to ethene was possible by this organism, growth occurred only while dechlorinating PCE, TCE, and cis-DCE, but VC dehalogenation appeared to be only through cometabolism, which was relatively slow.

Another *Dehalococcoides* organism (strain CBDB1) was soon after isolated by Adrian et al. (2000) that could partially reductively dehalogenate a range of tri- and tetra-chlorobenzenes, but not PCE, TCE, DCEs, nor VC. Hendrickson et al. (2002) analyzed samples from 24 chloroethene-dechlorinating field sites throughout North America and Europe to determine the presence of the *Dehalococcoides* group using PCR analysis, and found *Dehalococcoides* to be present only at the 21 sites where complete dechlorination to ethene was occurring, but not at three others were dechlorination stopped at cis-DCE. While to date, organisms from many different genera can dehalogenate PCE and TCE to cis-DCE, only *Dehalococcoides* appears capable at this time of carrying the process all the way through to ethene.

Based upon small differences in 16S rRNA gene sequences, Hendrickson et al. (2002) noted three different clades of *Dehalococcoides*, which he designated as the Cornell group (strain 195), the Pinellas group (strains FL2 and CBDB1), and the Victoria group (strain vic). We at Stanford had been working with samples from the Victoria, TX, site for several years (Rosner et al. 1997; Yang and McCarty 1998; Haston and McCarty 1999) and eventually found from VC enrichment cultures that growth of the Victoria strain was occurring on VC reductive dehalogenation (Cupples et al. 2003), which was different from the finding that strain 195 dehalogenated VC cometabolically only. About the same time, He et al. (2003) isolated a new *Dehalococcoides* strain called BAV1, reporting that it also obtained growth from VC reduction. Somewhat differently, the Victoria strain obtained growth also from the reduction of TCE and cis-DCE, while BAV1 obtained it from all DCE isomers, but not TCE. Thus, the different *Dehalococcoides* strains all use hydrogen as their primary electron donor, but use different chlorinated compounds as their electron acceptor in energy metabolism. Also found was that the VC reductase in stain Victoria termed VcrA (Müller et al. 2004) differed from that in BAV1, which was termed BvcA (Krajmalnik-Brown et al. 2004).

Subsequently, additional *Dehalococcoides* strains have been isolated, strain FL2 that obtains energy for growth using TCE and both cis-DCE and trans-DCE as electron acceptors (He et al. 2005), and strain GT that obtains energy while using TCE, cis-DCE, and VC as electron acceptors for growth (Sung et al. 2006). The genomes of these six *Dehalococcoides* strains have all been sequenced and the differences and similarities between them have been characterized (Löffler et al. 2013). Interesting is that the various *Dehalococcoides* strains have multiple and different dehalogenation genes, but the function of many of them is not yet known. They all use hydrogen as the electron donor and organohalides as electron acceptors in energy metabolism. They are widespread throughout the world. Just what they were using as electron acceptors in energy metabolism before humans evolved to supply them with synthetic halogenated compounds is not known. While many other species are capable of reductive dehalogenation, it is the *Dehalococcoides* strains that have often been the ones capable and necessary for completing the process, allowing anaerobic dehalogenation to become the mainstream biological processes for in situ remediation of many of the organohalide-contaminated sites.

4.8 Summary Statement

The many bacterial species that are now known to reductively transform organohalides, either through energy metabolism or cometabolism, have been greatly beneficial for ridding the environment of numerous toxic organic compounds that humans have synthesized. Many such compounds tend to resist aerobic biodegradation, and thus the finding of anaerobic organisms in particular that can use organohalides as electron acceptors has been most fortunate. Such anaerobic degradation requires an electron donor, that for some species may be an organic compound such as acetate, but most often, it is H_2 that is preferred or necessary. When conditions are right and the necessary electron donor is present, complete degradation may occur. But most often, sufficient suitable electron donor is not naturally present, and thus must be added. Adding H_2 itself presents difficulties, particularly since elevated levels of H_2 presents a highly attractive energy source for many other organisms, such as homoacetogens, sulfate and Fe(III) reducers, and methanogens, all of which tend to be ubiquitous in the environment. Dehalogenating organisms may dominate if H_2 is kept at very low nanomolar concentrations, which is difficult to achieve when H_2 is added directly to groundwater, the most commonly contaminated environment. Nevertheless, such low H_2 concentrations can be achieved by adding a slowly biodegradable organic such as propionate, which is oxidized anaerobically to H_2 and acetate. Carrying out an effective in situ bioremediation system for organohalide compounds is often an engineering challenge, but this challenge is now being effectively met by careful application of the extensive knowledge about dehalogenating organisms that has been gained through the research conducted by numerous investigators over the past several decades, many of whom are authors or coauthors of the chapters in this book. I am most honored to be among them.

References

Adrian L, Szewzyk U, Wecke J, Görisch H (2000) Bacterial dehalorespiration with chlorinated benzenes. Nature 408:580–583

Alvarez-Cohen L, McCarty PL (1991) Product toxicity and cometabolic competitive inhibition modeling of chloroform and trichloroethylene by methanotrophic resting cells. Appl Environ Microbiol 57(4):1031–1037

Anderson J, McCarty PL (1996) Effect of three chlorinated ethenes on growth rates for a methanotrophic mixed culture. Environ Sci Technol 30(12):3517–3524

Bouwer EJ, McCarty PL (1983a) Transformations of 1- and 2-carbon halogenated aliphatic organic compounds under methanogenic conditions. Appl Environ Microbiol 45:1286–1294

Bouwer EJ, McCarty PL (1983b) Transformations of halogenated organic compounds under denitrification conditions. Appl Environ Microbiol 45:1295–1299

Bouwer EJ, Rittmann BE, McCarty PL (1981) Anaerobic degradation of halogenated 1- and 2-carbon organic compounds. Environ Sci Technol 15(5):596–599

Cupples AM, Spormann AM, McCarty PL (2003) Growth of a dehalococcoides-like microorganism on vinyl chloride and cis-dichloroethene as electron acceptors as determined by competitive PCR. Appl Environ Microbiol 69(2):953–959

DiStefano TD, Gossett JM, Zinder SH (1991) Reductive dechlorination of high concentrations of tetrachloroethene to ethene by an anaerobic enrichment culture in the absence of methanogenesis. Appl Environ Microbiol 57(8):2287–2292

DiStefano TD, Gossett JM, Zinder SH (1992) Hydrogen as an electron-donor for dechlorination of tetrachloroethene by an anaerobic mixed culture. Appl Environ Microbiol 58(11):3622–3629

Dolan ME, McCarty PL (1995) Methanotrophic chloroethene transformation capacities and 1,1-dichloroethene transformation product toxicity. Environ Sci Technol 29(11):2741–2747

Dolfing J, Tiedje JM (1986) Hydrogen cycling in a three-tiered food web growing on the methanogenic conversion of 3-chlorobenzoate. FEMS Microbiol Ecol 38:293–298

Dolfing J, Tiedje JM (1987) Growth yield increase linked to reductive dechlorination in a defined 3-chlorobenzoate degrading methanogenic coculture. Arch Microbiol 149:102–105

Freedman DL, Gossett JM (1989) Biological reductive dechlorination of tetrachloroethylene and trichloroethylene to ethylene under methanogenic conditions. Appl Environ Microbiol 55(9):2144–2151

Gandhi RK, Hopkins GD, Goltz MN, Gorelick SM, McCarty PL (2002) Full-scale demonstration of in situ cometabolic biodegradation of trichloroethylene in groundwater - 2. Comprehensive analysis of field data using reactive transport modeling. Water Resour Res 38(4) 11.1–11.19

Haber CL, Allen LN, Zhao S, Hanson RS (1983) Methylotrophic bacteria: biochemical diversity and genetics. Science 1147–1153

Haston ZC, McCarty PL (1999) Chlorinated ethene half-velocity coefficients (K-s) for reductive dehalogenation. Environ Sci Technol 33(2):223–226

He JZ, Ritalahti KM, Aiello MR, Löffler FE (2003) Complete detoxification of vinyl chloride by an anaerobic enrichment culture and identification of the reductively dechlorinating population as a *Dehalococcoides* species. Appl Environ Microbiol 69(2):996–1003

He J, Sung Y, Krajmalnik-Brown R, Ritalahti KM, Löffler FE (2005) Isolation and characterization of *Dehalococcoides* sp strain FL2, a trichloroethene (TCE)- and 1,2-dichloroethene-respiring anaerobe. Environ Microbiol 7(9):1442–1450

Hendrickson ER, Payne JA, Young RM, Starr MG, Perry MP, Fahnestock S, Ellis DE, Ebersole RC (2002) Molecular analysis of *Dehalococcoides* 16S ribosomal DNA from chloroethene-contaminated sites throughout north America and Europe. Appl Environ Microbiol 68(2):485–495

Hill DW, McCarty PL (1967) Anaerobic degradation of selected chlorinated hydrocarbon pesticides. J Water Pollut Control Fed 39:1259–1277

Holliger C, Schraa G, Stams AJM, Zehnder AJB (1993) A highly purified enrichment culture couples the reductive dechlorination of tetrachloroethene to growth. Appl Environ Microbiol 59(9):2991–2997

Holliger C, Hahn D, Harmsen H, Ludwig W, Schumacher W, Tindall B, Vazquez F, Weiss N, Zehnder AJB (1998) *Dehalobacter restrictus* gen. nov. and sp. nov., a strictly anaerobic bacterium that reductively dechlorinates tetra- and trichloroethene in an anaerobic respiration. Arch Microbiol 169(4):313–321

Holliger C, Wohlfarth G, Diekert G (1999) Reductive dechlorination in the energy metabolism of anaerobic bacteria. FEMS Microbiol Rev 22:383–398

Hopkins GD, McCarty PL (1995) Field-evaluation of in-situ aerobic cometabolism of trichloroethylene and 3 dichloroethylene isomers using phenol and toluene as the primary substrates. Environ Sci Technol 29(6):1628–1637

Horvath RS (1972) Microbial co-metabolism and the degradation of organic compounds in nature. Bacteriological Rev 36(2):146–155

Krajmalnik-Brown R, Holscher T, Thomson IN, Saunders FM, Ritalahti KM, Loffler FE (2004) Genetic identification of a putative vinyl chloride reductase in *Dehalococcoides* sp strain BAV1. Appl Environ Microbiol 70(10):6347–6351

Löffler FE, Yan JL, Ritalahti KM, Adrian L, Edwards EA, Konstantinidis KT, Müller JA, Fullerton H, Zinder SH, Spormann AM (2013) *Dehalococcoides mccartyi* gen. nov., sp. nov., obligately organohalide-respiring anaerobic bacteria relevant to halogen cycling and bioremediation, belong to a novel baterial class, *Dehalococcoidia* classis nov., order *Dehalococcoidales* ord nov and family *Dehalococcoidaceae* fam nov., within the phylum *Chloroflexi*. Int J Syst Evol Microbiol 63:625–635

Maymò-Gatell X, Chien YT, Gossett JM, Zinder SH (1997) Isolation of a bacterium that reductively dechlorinates tetrachloroethene to ethene. Science 276:1568–1571

McCarty PL, Semprini L, Dolan ME, Harmon TC, Tiedeman C, Gorelick SM (1991) In Situ methanotrophic bioremediation for contaminated groundwater at St. Joseph, Michigan. In: Hinchee RE, Olfenbuttel RF (eds) On-site bioreclamation, processes for xenobiotic and hydrocarbon treatment. Butterworth-Heinemann, Boston, pp 16–40

McCarty PL, Goltz MN, Hopkins GD, Dolan ME, Allan JP, Kawakami BT, Carrothers TJ (1998) Full-scale evaluation of in-situ cometabolic degradation of trichloroethylene in groundwater through toluene injection. Environ Sci Technol 32(1):88–100

Mohn WW, Tiedje JM (1992) Microbial reductive dehalogenation. Microbiol Rev 56(3):482–507

Müller JA, Rosner BM, von Abendroth G, Meshulam-Simon G, McCarty PL, Spormann AM (2004) Molecular identification of the catabolic vinyl chloride reductase from *Dehalococcoides* sp. strain VS and its environmental distribution. Appl Environ Microbiol 70(8):4880–4888

NAS (1978) Nofluorinated halomethanes in the evironment, Washington D.C, pp 36–51

Parsons F, Wood PR, DeMarco J (1984) Transformations of tetrachloroethylene and trichloroethylene in microcosms and groundwater. J Am Water Works Assoc 76:56–59

Roberts PV, Reinhard M, McCarty PL (1980) Organic contaminant behavior during groundwater recharge. J Water Pollut Control Fed 52:161–172

Roberts PV, Schreiner J, Hopkins GD (1982) Field study of organic water quality changes during groundwater recharge in the Palo Alto Baylands. Water Res 16:1025–1035

Rosner BM, McCarty PL, Spormann AM (1997) In vitro studies on reductive vinyl chloride dehalogenation by an anaerobic mixed culture. Appl Environ Microbiol 63(11):4139–4144

Semprini L, Kitanidis PK, Kampbell DH, Wilson JT (1995) Anaerobic transformation of chlorinated aliphatic-hydrocarbons in a sand aquifer based on spatial chemical-distributions. Water Resour Res 31(4):1051–1062

Sung Y, Ritalahti KM, Apkarian RP, Loffler FE (2006) Quantitative PCR confirms purity of strain GT, a novel trichloroethene-to-ethene-respiring Dehalococcoides isolate. Appl Environ Microbiol 72(3):1980–1987

Tiedje JM, Quensen JF III, Chee-Sanford J, Schimel JP, Boyd SA (1993) Microbial reductive dechlorination of PCBs. Biodegradation 4(4):231–240

Vogel TM, McCarty PL (1985) Biotransformation of tetrachloroethylene to trichloroethylene, dichloroethylene, vinyl chloride, and carbon dioxide under methanogenic conditions. Appl Environ Microbiol 49:1080–1083

Vogel TM, Criddle CS, McCarty PL (1987) Transformations of halogenated aliphatic compounds. Environ Sci Technol 21:722–736

Wilson JT, Wilson BH (1985) Biotransformation of trichloroethylene. Appl Environ Microbiol 49(1):242–243

Yang YR, McCarty PL (1998) Competition for hydrogen within a chlorinated solvent dehalogenating anaerobic mixed culture. Environ Sci Technol 32:3591–3597

Chapter 5
Overview of Known Organohalide-Respiring Bacteria—Phylogenetic Diversity and Environmental Distribution

Siavash Atashgahi, Yue Lu and Hauke Smidt

Abstract To date, organohalide respiration (OHR) has been restricted to the bacterial domain of life. Known organohalide-respiring bacteria (OHRB) are spread among several phyla comprising both Gram-positive and Gram-negative bacteria. As a unique trait, OHRB benefit from reductive dehalogenase enzymes enabling them to use different organohalides as terminal electron acceptors and occupy a wide range of terrestrial and aquatic environments. This chapter comprises three sections: First, we give an overview of phylogeny of known OHRB and briefly discuss physiological and genetic characteristics of each group. Second, the environmental distribution of OHRB is presented. Owing to the application of molecular diagnostic approaches, OHRB are being increasingly detected not only from organohalide-contaminated groundwaters and sediments but also from pristine environments, including deep oceanic sediments and soils that are ample sources of naturally occurring organohalides. Finally, we highlight important factors that impact the ecology of OHRB and their interaction with other microbial guilds.

Siavash Atashgahi and Yue Lu equally contributed.

S. Atashgahi · Y. Lu · H. Smidt (✉)
Laboratory of Microbiology, Wageningen University,
Stippeneng 4, 6708 WE Wageningen, The Netherlands
e-mail: hauke.smidt@wur.nl

S. Atashgahi
e-mail: siavash.atashgahi@wur.nl

Y. Lu
e-mail: yue.lu@wur.nl

L. Adrian and F.E. Löffler (eds.), *Organohalide-Respiring Bacteria*,
DOI 10.1007/978-3-662-49875-0_5

5.1 Introduction

The vast functional diversity of microorganisms and their metabolic capabilities have made them successful mediators of electron liberation from the oxidation of inorganic and organic matter coupled to the reduction of a wide array of inorganic and organic electron acceptors including organohalides (Leys et al. 2013). Although most well-known representatives of organohalides are considered to be man-made products of industrial origin, thousands of naturally occurring organohalides have been reported from geogenic and biogenic sources (Gribble 2010). Numerous reports exist on natural production of organohalide compounds from biogenic sources such as a broad range of seaweeds, sponges, terrestrial plants, fungi as well as through geogenic processes such as volcanic activity, forest fires and other geothermal processes, some of which predate the industrialization era (see Chap. 2). Such ancient natural production of organohalides might have contributed to the development of biochemical strategies capable of unlocking the chemically stable carbon–halogen bond in organohalides. This is particularly important for non-oxygenolytic dehalogenation processes that probably have developed in the originally oxygen-free atmosphere on Earth. Taking advantage of organohalides as thermodynamically favourable electron acceptors under anoxic conditions, reductive dehalogenation is used as a terminal electron-accepting process by organohalide-respiring bacteria (OHRB). These microbes have greatly contributed to global cycling of halogens by breathing (rather) toxic organohalides and preventing their accumulation in the environment. Exploiting reductive dehalogenases (RDase in case of functionally characterized enzymes and RdhA for yet uncharacterized reductive dehalogenases predicted from genomes and molecular surveys, see also Chap. 16) dedicated to organohalide respiration (OHR), OHRB occupy a wide range of niches/environments. Hence, understanding factors governing their evolution, distribution and ecology will help to unravel their role in the fate of organohalides. It should be noted that organohalide degradation processes have been described for a wide range of redox conditions (from highly oxidizing to strictly reducing), and mediated by a wide variety of (micro)organisms in co-metabolic and/or energy-yielding modes; however, in this chapter we specifically discuss the phylogeny and environmental distribution of OHRB that are known to gain energy and grow using organohalides as electron acceptors.

5.2 Phylogeny of OHRB

Since the description of *Desulfomonile tiedjei* as the first isolated OHRB (DeWeerd et al. 1990), numerous bacterial strains capable of OHR have been obtained in axenic culture (Fig. 5.1), providing indispensable insights into their phylogenetic, physiological and biochemical traits. Members of the genus *Dehalococcoides* comprise the biggest groups of isolates to date (19 isolates)

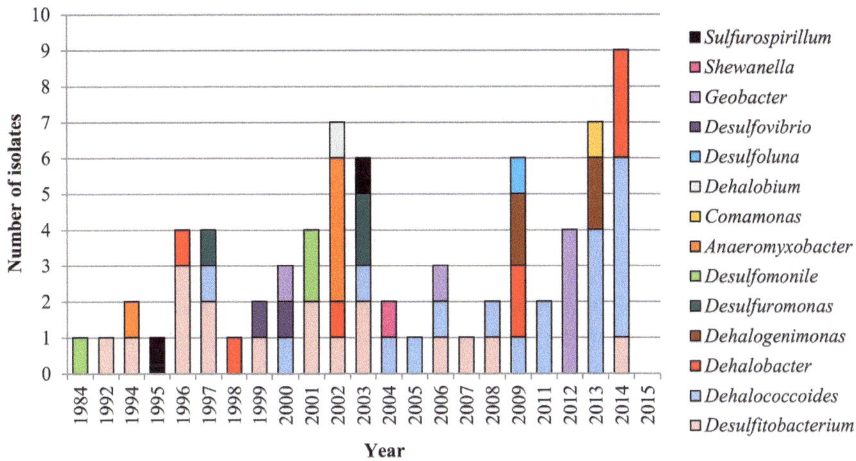

Fig. 5.1 Number of OHRB available in axenic culture. The data is based on the number of isolates as of July 2015

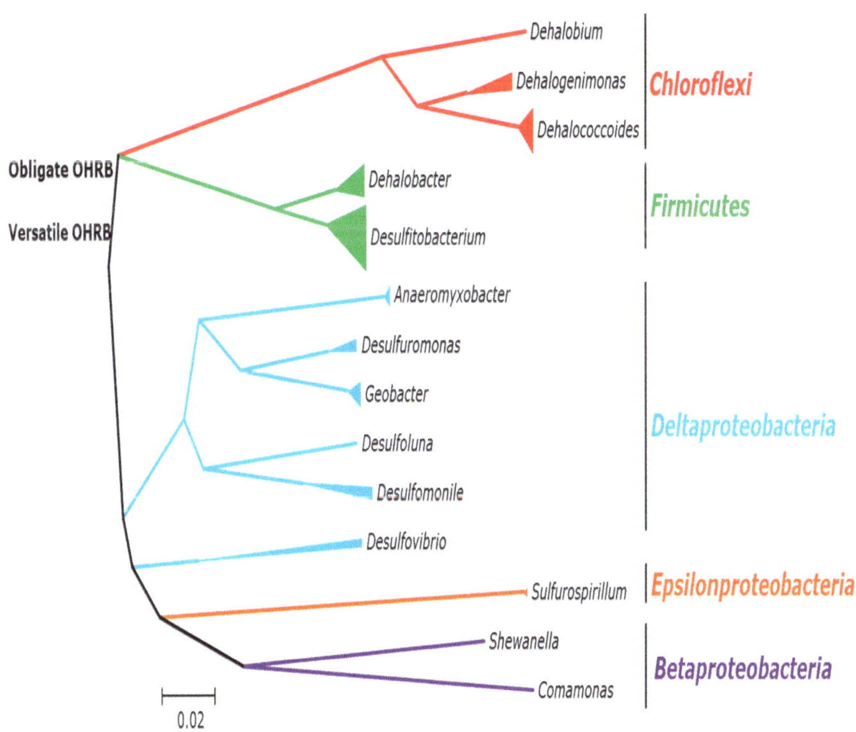

Fig. 5.2 Phylogenetic tree of known OHRB based on bacterial 16S rRNA gene sequences. Alignment and phylogenetic analysis were performed with MEGA and the tree was constructed using the neighbour-joining (NJ) method. The *reference bar at the bottom* indicates the branch length that represents 2 % sequence divergence. Colour Key: *Chloroflexi* (red), *Firmicutes* (green), *Deltaproteobacteria* (blue), *Betaproteobacteria* (violet), *Epsilonproteobacteria* (brown)

followed by strains of *Desulfitobacterium* (17 isolates) (Fig. 5.1). The known OHRB are spread among several distinct phyla comprising both Gram-positive and Gram-negative bacteria (Fig. 5.2). The known isolates can be and large be divided into facultative and obligate groups based on whether OHR is their only energy-gaining metabolism (Maphosa et al. 2010). The members of the facultative OHRB are characterized by a more versatile metabolism, in general have the ability to grow on a wide range of electron acceptors, and include proteobacterial OHRB such as *Geobacter*, *Desulfuromonas*, *Anaeromyxobacter, Desulfomonile*, *Desulfovibrio*, *Desulfoluna*, *Sulfurospirillum*, *Comamonas*, *Shewanella* as well as *Desulfitobacterium* from the phylum *Firmicutes*. The fact that some facultative OHRB such as *Comamonas*, *Geobacter* and *Shewanella* belong to phylogenetic groups that mostly comprise non-OHRB points towards horizontal acquisition of reductive dehalogenase genes. The obligate OHRB on the other hand are restricted to OHR for energy conservation and growth and include *Dehalobacter* (phylum *Firmicutes*) and the OHRB belonging to the *Dehalococcoidia* class (phylum *Chloroflexi*) including strains of *Dehalococcoides mccartyi*, *Dehalogenimonas* spp. and the single isolate '*Dehalobium chlorocoercia*' DF-1 (Maphosa et al. 2010; Löffler et al. 2013; May et al. 2008). However, recent studies showed fermentative growth of *Dehalobacter* spp. on chloromethane (Justicia-Leon et al. 2012; Lee et al. 2012), suggesting that at least some of the isolates previously considered obligate OHRB might harbour additional modes of metabolism beyond the canonical OHR. It is also interesting to note that recent single-cell genomic studies of marine *Dehalococcoidia* did not reveal any evidence for catabolic reductive dehalogenation, indicating that microorganisms closely related to known obligate OHRB do not rely on OHR for energy conservation, but rather utilize organic matter degradation pathways (Kaster et al. 2014; Wasmund et al. 2014).

With few exceptions, OHRB from different phyla, i.e. members of the *Firmicutes*, *Chloroflexi* and different classes of *Proteobacteria*, benefit from similar enzymes/pathways for OHR indicating that these genes could be acquired from common ancestors via transposon mediated dissemination (see Chap. 16 for more details). However, to date little to no correlation could be found between the type of organohalides used as electron acceptor and the phylogenetic affiliation of OHRB, and haloaliphatic and haloaromatic compounds can be dehalogenated by isolates of taxonomically different genera (Hug et al. 2013b). Furthermore, isolates with similar phylogeny and physiology have been obtained from very different environments. For example, *Desulfuromonas michiganensis* strain BB1 was isolated from pristine river sediment while the closely related strain BRS1 was obtained from chloroethene-contaminated aquifer material (Sung et al. 2003). Similarly, the closely related uncultured *Chloroflexi* Lahn and Tidal Flat Clusters were both capable of dechlorination of perchloroethene (PCE) to *trans*-DCE while being enriched from sediments from river and marine environments, respectively (Kittelmann and Friedrich 2008a, b). Another example is the case of *D. mccartyi* strains KS and RC. Both bacteria grow with 1,2-dichloropropane (1,2-D) as an electron acceptor in enrichment cultures while being derived from hydrocarbon-contaminated and pristine river sediments, respectively (Ritalahti and Löffler 2004; Löffler et al. 1997a).

In the following, we will separately discuss OHRB from different phyla.

5.2.1 *Firmicutes*

5.2.1.1 Dehalobacter

Dehalobacter has been first described as a genus in 1998, belonging to the *Firmicutes* (Phylum), *Clostridia* (Class), *Clostridiales* (Order) and *Peptococcaceae* (Family) (Holliger et al. 1998a). *Dehalobacter restrictus* PER-K23^T, the first isolate and the type species of this genus, was obtained from an anaerobic PCE-dechlorinating packed-bed column, and was found to be able to couple the reductive dehalogenation of PCE to growth and energy conversation (Holliger et al. 1993, 1998a). Subsequently, additional pure cultures were isolated and characterized from this genus (Fig. 5.1) (Wild et al. 1996; Yoshida et al. 2009; Sun et al. 2002; Wang et al. 2014b; Nelson and Zinder 2014). *Dehalobacter* spp. are known as obligate OHRB with relatively small genomes (2.60 Mb–3.09 Mb, with exception of *Dehalobacter* sp. FTH1 with a genome size of 6.33 Mb), and are restricted to OHR with H_2 as sole electron donor except for two reports on fermentative growth of *Dehalobacter* spp. with chloromethane (Justicia-Leon et al. 2012; Lee et al. 2012). The presence of up to 39 copies of putative RDH catalytic subunit encoding *rdhA* genes in currently available genomes indicates their potentially broader substrate spectrum. Cultivation and environmental studies also revealed an increasing number of halogenated compounds that can be used by *Dehalobacter* spp. as electron acceptors including PCE and trichloroethene (TCE) (Holliger et al. 1998a; Wang et al. 2014b; Wild et al. 1996), chlorinated ethanes (Grostern and Edwards 2006; Sun et al. 2002), 2,4,6-trichlorophenol (2,4,6-TCP) (Wang et al. 2014b), 2,4,6-tribromophenol (2,4,6-TBP) (Li et al. 2015), chloroform (Grostern et al. 2010), chlorinated benzenes (Nelson and Zinder 2014; Nelson et al. 2011), β-hexachlorocyclohexane (β-HCH, known as lindane) (van Doesburg et al. 2005) and phthalide (4,5,6,7-tetrachlorophthalide) (Yoshida et al. 2009). One RDase (PceA) was functionally characterized from *Dehalobacter* spp. (Maillard et al. 2003), which displayed a corrinoid as essential cofactor. Recent studies showed that most *Dehalobacter* spp. genomes have a complete anaerobic cobalamin biosynthesis pathway (Rupakula et al. 2014); however, no *Dehalobacter* spp. have been confirmed to biosynthesize cobalamin de novo (Holliger et al. 1998a; Sun et al. 2002; van Doesburg et al. 2005; Yoshida et al. 2009). Salvaging cobinamide from the environment, e.g. from cobalamin-producing non-dehalogenating community members such as *Sedimentibacter* spp. (van Doesburg et al. 2005; Maphosa et al. 2012) was suggested as the strategy *Dehalobacter* spp. employ to ensure corrinoids supply for growth and reductive dechlorination (Rupakula et al. 2013, 2014). More information of *Dehalobacter* spp. can be found in Chap. 8.

5.2.1.2 Desulfitobacterium

Desulfitobacterium is a genus belonging to the *Firmicutes* (Phylum), *Clostridia* (Class), *Clostridiales* (Order) and *Peptococcaceae* (Family). *Desulfitobacterium* spp. isolates are among the best characterized facultative OHRB that are known for their versatile metabolism with the capability to use a wide variety of electron donors and acceptors (Villemur et al. 2006). The first OHRB isolated in this genus was originally related to the genus *Clostridium* (Madsen and Licht 1992). Strain DCB-2, isolated from municipal sludge, is a Gram-positive, strictly anaerobic, *ortho-* and *meta*-chlorophenol-dechlorinating bacterium (Madsen and Licht 1992). Two years later, with the isolation of the monotypic strain *Desulfitobacterium dehalogenans* JW/IU-DC1^T, the *Desulfitobacterium* genus was proposed as a new taxon (Utkin et al. 1994). This genus was designated *Desulfitobacterium* because strain JW/IU-DC1^T grows with sulfite, but not with sulphate, as a terminal electron acceptor (Utkin et al. 1994). This was followed by re-classification of strain DCB-2 as the type strain of *Desulfitobacterium hafniense* (Christiansen and Ahring 1996). The species *Desulfitobacterium frappieri* was proposed with the isolation of strain PCP-1, an anaerobic pentachlorophenol-dechlorinating bacterium isolated from a methanogenic consortium (Bouchard et al. 1996). Based on DNA–DNA hybridization and comparative physiological studies *D. hafniense*, *D. frappieri* and the non-OHRB strain GBFH were later united into the species *D. hafniense* (Niggemyer et al. 2001). Other OHRB belonging to the genus *Desulfitobacterium* isolated in the following years are indicted in Table 5.1.

Completed (Nonaka et al. 2006; Kim et al. 2012; Goris et al. 2015) and ongoing (Kruse et al., in preparation) genome sequencing projects have identified 1–7 copies of RDH encoding genes in the genomes of *Desulfitobacterium* spp., far less than the closely related dehalogenating clade of *Dehalobacter* spp. that contain up to 39 copies (Chap. 15) (Richardson 2013). The loss of the OHR trait upon growth on alternative electron acceptors further confirmed the physiological opportunism of organohalide-respiring strains of *Desulfitobacterium* spp., and it was proposed that the RDase genes were acquired by horizontal gene transfer from closely related obligate organohalide-respiring strains of *Dehalobacter* spp. (Duret et al. 2012). Nevertheless, *Desulfitobacterium* spp. strains are of high interest in the OHR field due to their high growth rate and respiration with a diverse range of organohalides, though none of the isolates were able to dechlorinate PCE past *cis*-DCE. Diverse RDases were purified and characterized from *Desulfitobacterium* spp. including chlorophenol RDase (Thibodeau et al. 2004; van de Pas et al. 1999), PCE RDase (Suyama et al. 2002; Tsukagoshi et al. 2006; van de Pas et al. 2001; Miller et al. 1998) and dichloroethane (DCA) RDase (Marzorati et al. 2007). Additionally, a complete de novo corrinoid synthesis pathway was identified from the genome of *D. hafniense* Y51 (Nonaka et al. 2006). More information on *Desulfitobacterium* spp. can be found in Chap. 9.

Table 5.1 OHRB isolates

Strain	Source	Year	Reference
Anaeromyxobacter dehalogenans 2CP-1	A small stream sediment	1994	Cole et al. (1994), Sanford et al. (2002)
Anaeromyxobacter dehalogenans 2CP-2	Michigan pond sediment	2002	Sanford et al. (2002)
Anaeromyxobacter dehalogenans 2CP-3	Michigan yard compost	2002	Sanford et al. (2002)
Anaeromyxobacter dehalogenans 2CP-5	Michigan yard compost	2002	Sanford et al. (2002)
Anaeromyxobacter dehalogenans 2CP-C	Cameroon rain forest soil	2002	Sanford et al. (2002)
Comamonas sp. 7D-2	Bromoxynil octanoate-contaminated soil	2013	Chen et al. (2013)
Dehalobacter restrictus PER-K23	Anaerobic Rhine River sediment and ground anaerobic granular sludge	1998	Holliger et al. (1993, 1998a)
Dehalobacter sp. 12DCB1	Sediment from water-saturated drainage ditch	2014	Nelson and Zinder (2014), Nelson et al. (2011)
Dehalobacter sp. 13DCB1	Sediment from water-saturated drainage ditch	2014	Nelson and Zinder (2014), Nelson et al. (2011)
Dehalobacter sp. FTH1	Ploughed layer of a paddy field	2009	Yoshida et al. (2009)
Dehalobacter sp. FTH2	Ploughed layer of a paddy field	2009	Yoshida et al. (2009)
Dehalobacter sp. TCA1	Upper Hudson River sediment	2002	Sun et al. (2002)
Dehalobacter sp. TCP1	Digester sludge from an industrial wastewater treatment plant	2014	Wang et al. (2014b)
Dehalobacter sp. TEA	Industrially polluted, anaerobic groundwater in a fixed-bed reactor	1996	Wild et al. (1996)
Dehalobium chlorocoercia DF-1	Estuarine sediment	2002	Wu et al. (2000, 2002b)
Dehalococcoides mccartyi ANAS1	Contaminated soil	2011	Lee et al. (2011), Richardson et al. (2002)
Dehalococcoides mccartyi ANAS2	Contaminated soil	2011	Lee et al. (2011), Richardson et al. (2002)
Dehalococcoides mccartyi 11a	Sediments from Wuhan industrial districts	2013	Cheng et al. (2009), Lee et al. (2013)
Dehalococcoides mccartyi 11a5	Sediments from Wuhan industrial districts	2013	Cheng et al. (2009), Lee et al. (2013)
Dehalococcoides mccartyi BAV1	Aquifer material from the chloroethene-contaminated site	2003	He et al. (2003)
Dehalococcoides mccartyi CBDB1	Saale River sediment	2000	Adrian et al. (1998, 2000)

(continued)

Table 5.1 (continued)

Strain	Source	Year	Reference
Dehalococcoides mccartyi DCMB5	Creek Spittelwasser sediment	2008	Bunge et al. (2008)
Dehalococcoides mccartyi FL2	Non-contaminated Red Cedar River sediment	2005	He et al. (2005)
Dehalococcoides mccartyi GT	Aquifer material from a chloroethene-contaminated site	2006	Sung et al. (2006b)
Dehalococcoides mccartyi JNA	Housatonic River sediment with Aroclor 1260-contamination	2014	Bedard et al. (2006), LaRoe et al. (2014)
Dehalococcoides mccartyi VS	Aquifer material from a PCE-contaminated site	2004	Müller et al. (2004), Rosner et al. (1997)
Dehalococcoides mccartyi 195	Anaerobic sewage digestor sludge	1997	Maymo-Gatell et al. (1997)
Dehalococcoides mccartyi CG1	Sand and silt near Liangjiang River	2014	Wang and He (2013b), Wang et al. (2014a)
Dehalococcoides mcca-rtyi CG4	Sand and silt of ditch sedi-ment near electronic waste dump site	2014	Wang and He (2013b), Wang et al. (2014a)
Dehalococcoides mcca-rtyi CG5	Sand and silt of ditch sedi-ment near electronic waste dump site	2014	Wang and He (2013b), Wang et al. (2014a)
Dehalococcoides mccartyi SG1	Digester sludge of an indus-trial wastewater treatment plant	2014	Wang and He (2013b), Wang et al. (2014a)
Dehalococcoides mccartyi AD14-1	An anaerobic digester in a wastewater treatment plant	2013	Wang and He (2013a)
Dehalococcoides mccartyi AD14-2	An anaerobic digester in a wastewater treatment plant	2013	Wang and He (2013a)
Dehalococcoides mccartyi MB	Sediment in the San Francisco Bay area	2009	Cheng and He (2009)
Dehalogenimonas alkenigignens IP3-3	Contaminated groundwater	2013	Bowman et al. (2013)
Dehalogenimonas alkenigignens SBP-1	Contaminated groundwater	2013	Bowman et al. (2013)
Dehalogenimonas lykanthroporepellens BL-DC-8	Contaminated groundwater	2009	Moe et al. (2009), Yan et al. (2009)
Dehalogenimonas lykanthroporepellens BL-DC-9	Contaminated groundwater	2009	Moe et al. (2009), Yan et al. (2009)
Desulfitobacterium chlororespirans	2,3-CP dehalogenating compost soil	1996	Sanford et al. (1996)
Desulfitobacterium dehalogenans JW/IU-DC1	Freshwater sediments-pond	1994	Utkin et al. (1994)

(continued)

Table 5.1 (continued)

Strain	Source	Year	Reference
Desulfitobacterium dichloroeliminans LMG P-21439	Soil matrix of an anoxic Water-saturated layer (1 m in depth)	2003	De Wildeman et al. (2003)
Desulfitobacterium hafniense DCB-2	Municipal sludge	1992	Christiansen and Ahring (1996), Madsen and Licht (1992)
Desulfitobacterium hafniense G2	An active Dry Branch Kaolin Company mine	2003	Shelobolina et al. (2003)
Desulfitobacterium hafniense JH1	Ditch sludge (mixed with sewage)	2008	Chang et al. (1998), Fletcher et al. (2008)
Desulfitobacterium hafniense PCP-1	A mixture of PCP-contaminated soil and anerobic sewage sludge	1996	Bouchard et al. (1996)
Desulfitobacterium hafniense TCE-1	Chloroethene-contaminated soil	1999	Gerritse et al. (1999)
Desulfitobacterium hafniense TCP-A	River sediments	2001	Breitenstein et al. (2001)
Desulfitobacterium metallireducens	River sediments	2002	Finneran et al. (2002)
Desulfitobacterium sp. B31e3	Unsaturated subsurface soils	2007	Yoshida et al. (2007a)
Desulfitobacterium sp. KBC1	Crop field soil	2006	Tsukagoshi et al. (2006)
Desulfitobacterium sp. PCE1	Various freshwater sediments	1996	Gerritse et al. (1996)
Desulfitobacterium sp. PCE-S	PCE-contaminated soil	1997	Miller et al. (1997)
Desulfitobacterium sp. PR	Chloroethenes and chloroethanes-dechlorinating bioreactor	2014	Ding et al. (2014)
Desulfitobacterium sp. Viet-1	Parfume River sediment	1997	Löffler et al. (1997b)
Desulfitobacterium sp. Y51	PCE-polluted soil	2001	Suyama et al. (2001)
Desulfoluna spongiiphila AA1	Marine sponge *Aplysina aerophoba*	2009	Ahn et al. (2009)
Desulfomonile limimaris DCB-F	Marine sediments	2001	Sun et al. (2001)
Desulfomonile limimaris DCB-M	Marine sediments	2001	Sun et al. (2001)
Desulfomonile tiedjei	Sewage sludge	1984	DeWeerd et al. (1990), Shelton and Tiedje (1984)
Desulfovibrio dechloracetivorans	Marine sediment	2000	Sun et al. (2000)
Desulfovibrio sp. TBP-1	Estuarine sediments	1999	Boyle et al. (1999b)

(continued)

Table 5.1 (continued)

Strain	Source	Year	Reference
Desulfuromonas chloroethenica TT4B	Chlorinated ethylenes contaminated freshwater sediments	1997	Krumholz (1997)
Desulfuromonas michiganensis BB1	Pristine river sediment	2003	Sung et al. (2003)
Desulfuromonas michiganensis BRS1	Chloroethene-contaminated aquifer material	2003	Sung et al. (2003)
Geobacter lovleyi Geo7.1	Soil	2012	Wagner et al. (2012)
Geobacter lovleyi Geo7.2	Soil	2012	Wagner et al. (2012)
Geobacter lovleyi Geo7.3	Soil	2012	Wagner et al. (2012)
Geobacter lovleyi Geo7.4	Soil	2012	Wagner et al. (2012)
Geobacter lovleyi SZ	Freshwater sediment	2006	Sung et al. (2006a)
Geobacter thiogenes K1	Contaminated soil	2000	De Wever et al. (2000)
Shewanella sediminis HAW-EB3T	Halifax Harbour sediment	2004	Zhao et al. (2005, 2004)
Sulfurospirillum halorespirans PCE-M2	Anaerobic soil polluted with chlorinated aliphatic compounds	2003	Luijten et al. (2003)
Sulfurospirillum multivorans K	Activated sludge	1995	Scholz-Muramatsu et al. (1995)

5.2.2 *Chloroflexi*

5.2.2.1 Dehalococcoides

Dehalococcoides bacteria belong to *Chloroflexi* (Phylum), *Dehalococcoidia* (Class), *Dehalococcoidales* (Order) and *Dehalococcoidaceae* (Family) (Löffler et al. 2013), which are well known for their metabolic restriction to OHR. The genus was first established in 1997 with the isolation of *Dehalococcoides ethenogenes* 195^T capable of complete dechlorination of PCE to ethene (Maymo-Gatell et al. 1997). In following years, additional isolates of this genus were obtained in axenic culture including CBDB1, BAV1, VS, FL2 and GT strains. These five strains and the type strain *D. ethenogenes* 195 were re-classified as the new genus and species, *D. mccartyi* (Löffler et al. 2013). Recently, more strains from this species were isolated including strain JNA (LaRoe et al. 2014; Bedard et al. 2006), CG1 (Wang and He 2013b; Wang et al. 2014a), CG4 (Wang and He 2013b; Wang et al. 2014a), CG5 (Wang and He 2013b; Wang et al. 2014a), SG1 (Wang and He 2013b; Wang et al. 2014a), AD14-1 (Wang and He 2013a) and AD14-2 (Wang and He 2013a). Based on an earlier 16S rRNA gene-based classification, Hendrickson et al. (2002) divided isolates and clones available at the time into three phylogenetic subgroups, a classification which has been widely referred to ever since. Accordingly, *D. ethenogenes* 195^T belongs to the 'Cornell' subgroup, strain VS belongs to the 'Victoria' subgroup and subgroup 'Pinellas' contains most of the cultured strains such as CBDB1, BAV1, FL2 and GT and the majority of environmental sequences. Common traits of *D. mccartyi* strains are streamlined genomes, small and non-motile coccoid cells, resistant to penicillin-type antibiotics, obligately anaerobic and restricted to OHR as the only mode of metabolism using H_2 as the only electron donor (Löffler et al. 2013). Similar to the situation described above for *Dehalobacter* spp., genomic studies of known isolates showed the lack of de novo vitamin B12 synthesis pathways (Kube et al. 2005; Seshadri et al. 2005) although they are capable of corrinoid salvaging and remodelling from their growth medium (Men et al. 2014; Yan et al. 2012; Yi et al. 2012).

In spite of sharing more than 98 % identity with respect to their 16S rRNA gene sequences, different strains respire on different ranges of organohalides. For example, vinyl chloride (VC) respiration is limited to strains BAV1, GT and VS while none of them can respire on PCE. This might be due to large diversity of *rdhA* genes that are present in multiple non-identical copies on individual genomes. However, in spite of multiple homologous RDH genes the number of which currently ranges from 11 copies in strain BAV1 up to 36 copies in strain VS (McMurdie et al. 2009), only a limited number of RDases was purified and characterized from *D. mccartyi* strains including PCE RDase (Magnuson et al. 1998), TCE RDase (Magnuson et al. 2000), *trans*-DCE producing RDase (Chow et al. 2010), vinyl chloride RDase (Krajmalnik-Brown et al. 2004; Müller et al. 2004; Parthasarathy et al. 2015), chlorobenzene RDase (Adrian et al. 2007; Padilla-Crespo et al. 2014), dichloropropane RDase (Padilla-Crespo et al. 2014) and polychlorinated biphenyls (PCBs) RDase (Wang et al. 2014a). More information on *D. mccartyi* can be found in Chap. 6.

5.2.2.2 Dehalogenimonas

The genus *Dehalogenimonas* belongs to the *Chloroflexi* (Phylum), *Dehalococcoidia* (Class), *Dehalococcoidales* (Order) and *Dehalococcoidaceae* (Family) (Löffler et al. 2013). Up to now, two species were described that were isolated from groundwater contaminated with chlorinated alkanes and alkenes. *Dehalogenimonas lykanthroporepellens* was proposed as a new species in 2009 (Moe et al. 2009), with the isolation of two strains BL-DC-9^T and BL-DC-8, both Gram-negative, non-motile, non-spore-forming, polychlorinated aliphatic alkanes-dechlorinating bacteria (Yan et al. 2009). The genome of *D. lykanthroporepellens* BL-DC-9 has been released recently (Siddaramappa et al. 2012), and multiple *rdhA* genes were found simultaneously transcribed during OHR with different organohalides as the electron acceptors (Mukherjee et al. 2014). *D. lykanthroporepellens* strains show 10–11 % 16S rRNA gene sequence divergence compared to *Dehalococcoides* (Mukherjee et al. 2014), perform dihaloelimination and have a higher tolerance to oxygen exposure (Yan et al. 2009). The second species, *D. alkenigignens*, was proposed in 2013 (Bowman et al. 2013) with the isolation of two strictly anaerobic strains IP3-3^T and SBP-1. Both species (*D. alkenigignens* and *D. lykanthroporepellens*) have similar morphological features and substrate spectra with 96.2 % pairwise identity of their 16S rRNA gene sequence (Bowman et al. 2013). 1,2-DCA, 1,2-DCP, 1,2,3-DCP, 1,1,2-TCA and 1,1,2,2-TeCA can serve as electron acceptor with H_2 as sole electron donor (Bowman et al. 2013; Yan et al. 2009; Moe et al. 2009). More information on *Dehalogenimonas* spp. can be found in Chap. 7.

5.2.2.3 Dehalobium

Next to *Dehalogenimonas* and *Dehalococcoides*, the class *Dehalococcoidia* also contains the genus *Dehalobium* (Löffler et al. 2013) represented by a single isolate informally named '*D. chlorocoercia*' strain DF-1. This bacterium was highly enriched in a PCB-dechlorinating co-culture with a sulphate-reducing *Desulfovibrio* sp. from estuarine sediment of Charleston Harbor, South Carolina (Wu et al. 2000, 2002b). *D. chlorocoercia* DF-1 selectively dechlorinates PCB congeners with double-flanked chlorines with formate or H_2-CO_2 as the electron donor and carbon source (Wu et al. 2000, 2002b). Strain DF-1 is the only reported representative of the genus *Dehalobium* and the classification as '*D. chlorocoercia'* was first proposed in 2008 (Kittelmann and Friedrich 2008a); however, the current genus and species names have yet to be validated. *D. chlorocoercia* DF-1 shares 89 % 16S rRNA sequence similarity to the *ortho*-PCB-dechlorinating bacterium *o*-17 (Cutter et al. 2001) while having ~87.5 % similarity with *Dehalococcoides* (Löffler et al. 2013). Unlike *o*-17 and *D. mccartyi* strains, strain DF-1 uses formate as the electron donor, however, it requires *Desulfovibrio* spp. in co-culture or its cell extract for growth with hydrogen and PCB in mineral medium (May et al. 2008). Other phylotypes phylogenetically related to the *D. chlorocoercia* DF-1/*Bacterium o*-17 clade within the *Chloroflexi* were reported to be able to dechlorinate PCB congeners (Fagervold et al. 2005, 2007). In addition to PCB, *D. chlorocoercia* DF-1 dechlorinates highly chlorinated benzenes (Wu et al. 2002a) and PCE and TCE producing higher amounts of *trans*-DCE than *cis*-DCE compared with canonical PCE/TCE dechlorinators (Miller et al. 2005). Based on its unique dechlorinating activity, DF-1 has shown merits in PCBs bioaugmentation in contaminated soil (Kjellerup et al. 2012) and sediment (Payne et al. 2011) (Chap. 24).

5.2.3 Deltaproteobacteria

5.2.3.1 Anaeromyxobacter

Anaeromyxobacter dehalogenans belongs to the *Proteobacteria* (Phylum), *Deltaproteobacteria* (Class), *Myxococcales* (Order) and *Myxococcaceae* (Family). *Anaeromyxobacter* was established as a genus with the monotypic species *A. dehalogenans* 2CP-1^T to describe a myxobacterium that can grow without air (Sanford et al. 2002). This bacterium is a Gram-negative, motile by gliding, spore-forming, bright red pigments-producing slime rod that is capable of facultative anaerobic growth, and was isolated from stream sediment near Lansing, MI, USA (Cole et al. 1994). Subsequently, another four isolates were obtained from different sources including pond sediment, Cameroon rain forest soil and yard compost with acetate and 2-chlorophenol (2-CP) as electron donor and acceptor, respectively (Sanford et al. 2002). Halophenols such as 2-CP, 2,6-dichlorophenol (2,6-DCP), 2,5-dichlorophenol (2,5-DCP), 2-bromophenol (2-BP) as well as nitrate, fumarate and oxygen can serve as electron acceptors with acetate, H_2, succinate, pyruvate, formate and lactate as electron donors (Cole et al. 1994; Sanford et al. 2002). *A. dehalogenans* is able to grow microaerophilically (Sanford et al. 2002; Cole et al. 1994), and the genome analysis of *A. dehalogenans* 2CP-C showed that it harboured genotypic characteristics of obligately aerobic myxobacteria as well as anaerobic *Deltaproteobacteria* (Thomas et al. 2008). More information on *Anaeromyxobacter* spp. can be found in Chap. 11.

5.2.3.2 Desulfuromonas

Desulfuromonas, a genus the name of which refers to an elemental sulphur-reducing bacterium, belongs to the *Proteobacteria* (Phylum), *Deltaproteobacteria* (Class), *Desulfuromonadales* (Order) and *Desulfuromonadaceae* (Family). To date, two species of *Desulfuromonas* capable of chloroethenes respiration have been reported (Sung et al. 2003; Krumholz 1997). *Desulfuromonas chloroethenica* strain TT4B was isolated from chloroethene-contaminated freshwater sediments in eastern Massachusetts, USA (Krumholz et al. 1996; Krumholz 1997). Strain TT4B uses PCE, TCE as well as fumarate, polysulfide and Fe(III) nitriloacetate as electron acceptor with acetate or pyruvate as electron donors (Krumholz 1997). The second species, *D. michiganensis*, currently comprises two strains, BB1 and BRS1, which were isolated from pristine river sediment and contaminated aquifer materials, respectively, both collected from the state of Michigan, USA, as indicated in the species name (Sung et al. 2003). Compared to *D. chloroethenica*, strains BB1 and BRS1 use a wider range of electron donors and can grow by fermentation on fumarate or malate with succinate as major end product (Sung et al. 2003). Besides, they showed higher growth yield and PCE tolerance (Sung et al. 2003). More information on *Desulfuromonas* spp. can be found in Chap. 11.

5.2.3.3 Geobacter

Geobacter is a genus belonging to the *Proteobacteria* (Phylum), *Deltaproteobacteria* (Class), *Desulfuromonadales* (Order) and *Geobacteraceae* (Family), which is well-studied for its capabilities to oxidize organic compounds coupled to the reduction of insoluble metal oxides (Rotaru et al. 2011). This genus was established with isolation of the non-OHRB *Geobacter metallireducens* (Lovley et al. 1993). The first OHRB characterized in this genus was *Geobacter thiogenes* strain K1 (formerly described as *Trichlorobacter thiogenes*), an anaerobic trichloroacetate-dechlorinating, sulphur-producing, fast-growing (doubling time = 6 h) bacterium isolated from contaminated soil (De Wever et al. 2000). Acetate and acetoin were used as electron donor for reductive dechlorination of trichloroacetate, and fumarate served as alternative electron acceptor (De Wever et al. 2000). Subsequently, *Geobacter lovleyi* was described as a novel species with strain SZ as type strain, isolated from non-contaminated freshwater sediment collected from Su-Zi Creek near Seoul, South Korea (Sung et al. 2006a). This bacterium couples acetate oxidation to PCE dechlorination with *cis*-DCE as the end product for growth. Strain SZ can use pyruvate and H_2 as electron donors with a wide range of electron acceptors including PCE, TCE, Fe(III), Mn(IV), U(VI), malate, nitrate, fumarate and elemental sulphur and no fermentation metabolism was found (Sung et al. 2006a). Additionally, 16S rRNA gene amplicons derived from *G. lovleyi* strain SZ-like organisms were found in uranium- and chlorinated compounds polluted environmental samples using strain SZ-specific primers. Recent genomic analysis of strain SZ identified the presence of a 77 kbp plasmid, pSZ77, encoding 15 out of the 24 genes required for de novo cobalamin biosynthesis (Wagner et al. 2012). Another OHRB belonging to the *Geobacter* genus is *G. lovleyi* strain KB-1, a PCE-to-*cis*-DCE-dechlorinating bacterium identified in the PCE to ethene dechlorinating consortium KB-1 (Duhamel and Edwards 2006). More recently, four *G. lovleyi* isolates Geo7.1, Geo7.2, Geo7.3 and Geo7.4 were obtained from PCE to *cis*-DCE dechlorinating microcosms established with soil from Fort Lewis, WA, that share 99–100 % nucleotide identity with strain SZ based on 16S rRNA gene sequences (Wagner et al. 2012). More information on *Geobacter* spp. can be found in Chap. 8.

5.2.3.4 Desulfoluna

The genus *Desulfoluna* belongs to the *Proteobacteria* (Phylum), *Deltaproteobacteria* (Class), *Desulfobacterales* (Order) and *Desulfobacteraceae* (Family) (Suzuki et al. 2008). *Desulfoluna spongiiphila* strain $AA1^T$, an anaerobic Gram-negative bromophenol- dehalogenating bacterium, was isolated from the marine sponge *Aplysina aerophoba* collected from the Mediterranean Sea (Ahn et al. 2009), which is so far the only isolate of the genus *Desulfoluna* with confirmed ability to grow via OHR and the only pure OHRB culture isolated from sponge to date (Fig. 5.3). To enrich and isolate strain $AA1^T$, lactate was used as

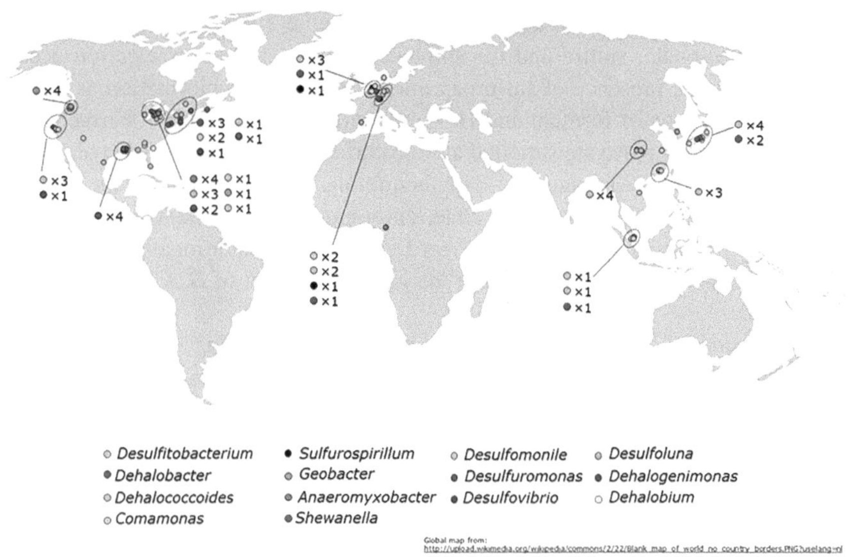

Fig. 5.3 Global distribution of OHRB isolates. The data is based on the number of isolates until July 2015. The source locations of OHRB isolates are indicated, except *Sulfurospirillum multivorans*, *Desulfitobacterium dichloroeliminans*, *Desulfitobacterium hafniense* strain DCB-2 and PCP-1, *Desulfitobacterium* sp. strain B31e3, PCE1, PCE-S and PR. For those strains without defined inoculum source, the locations of the laboratory where the strains have been isolated are shown in the global map

electron donor and carbon source and 2-BP served as sole electron acceptor (Ahn et al. 2003, 2009). Diverse halogenated compounds can be completely dehalogenated by strain AA1^T including 2-, 3- and 4-BP, 1,4- and 2,6- 2,6-DBP, 2,4,6-tribromophenol (2,4,5,TBP), 2- and 3-iodophenol, 2-bromo-4-fluorophenol and 3,5-dibromo-4-hydroxybenzoate, but not monochlorinated or fluorinated phenols. NaCl was found essential for growth and no fermentation was observed with lactate, pyruvate, fumarate or malate as carbon source (Ahn et al. 2009).

5.2.3.5 Desulfomonile

The genus *Desulfomonile* belongs to the *Proteobacteria* (Phylum), *Deltaproteobacteria* (Class), *Syntrophobacterales* (Order) and *Syntrohaceae* (Family) (DeWeerd et al. 1990). Two species of *Desulfomonile* were characterized as strictly anaerobic, 3-chlorobenzoate dechlorinating, sulfidogenic bacteria with special 'collar' girdling cells (DeWeerd et al. 1990; Sun et al. 2001). *D. tiedjei* was proposed as a novel and type species in 1990 (DeWeerd et al. 1990) with strain DCB-1^T, a Gram-negative, non-motile, non-spore-forming, anaerobic rod-shaped bacterium, which was isolated from a municipal sewage sludge collected from Adrian, MI, USA (Shelton and Tiedje 1984). Strain DCB-1 is able to reductively dehalogenate *meta*-halobenzoates (Shelton and Tiedje 1984; Linkfield and Tiedje 1990), and sulphate, sulfite and thiosulphate can also be use as electron acceptors, however, the presence of sulfuroxy anions inhibited dechlorination as well as high concentrations of formate and H_2 (Linkfield and Tiedje 1990). Fermentation was observed with pyruvate supplied as sole carbon source (DeWeerd et al. 1990). Thiamine, nicotinamide and 1,4-naphthoquinone are required in the defined media, and MoO_4^{2-},SeO_4^{2-}, tetracycline, chloramphenicol, kanamycin and streptomycin are potential inhibitors (DeWeerd et al. 1990). Additionally, cytochrome c_3 and desulfoviridin were detected in the growth medium of *D. tiedjei* (DeWeerd et al. 1990).

Desulfomonile limimaris is another organohalide-respiring species with two strains DCB-M^T and DCB-F that were isolated from marine sediments (Sun et al. 2002). These two strains share 99 % similarity based on 16S rRNA gene sequences and are both capable to dechlorinate 3-chlorobenzoate (CB) to benzoate with formate and H_2 as electron donor. A broad range of electron donors can be used such as benzoate, H_2, formate, butyrate, lactate, propionate and pyruvate. Besides 3-CB, sulfite, sulphate, thiosulphate, nitrate and fumarate can be used as electron acceptors for growth. Compared to *D. tiedjei* DCB-1 (DeWeerd et al. 1990), these two isolates of *D. limimaris* are not able to dechlorinate *meta*-halobenzoates using acetate or ferment pyruvate and NaCl (>0.32 %, w/w) is required for their growth (Sun et al. 2001). Sulfite and thiosulphate and not sulphate inhibit OHR of *D. tiedjei* DCB-1^T (DeWeerd et al. 1990, 1991) and *D. limimaris* DCB-M^T, which was explained by the competition for electron flow.

5.2.3.6 Desulfovibrio

Desulfovibrio is a genus of Gram-negative sulphate-reducing bacteria (SRB) that belongs to the *Proteobacteria* (Phylum), *Deltaproteobacteria* (Class), *Desulfovibrionales* (Order) and *Desulfovibrionaceae* (Family). Up to now, two isolates of *Desulfovibrio* spp. were published with OHR capability, and both were isolated from marine sediments with halophenols as electron acceptors. *Desulfovibrio* sp. TBP-1 was isolated from estuarine sediments of the Arthur Kill in the New York/New Jersey harbor (Fig. 5.2) with 2,4,6-TBP as electron acceptor and supplied with hydrogen and acetate (Boyle et al. 1999b). Only *ortho*-bromophenols and/or bromophenols without adjacent bromines at *para*- position were dehalogenated with lactate, pyruvate, formate or hydrogen as electron donors. Strain TBP-1 accumulated acetate when lactate and sulphate were provided, which indicated that this microorganism is an incomplete oxidizer and NaCl is required for its growth (Boyle et al. 1999b). The other isolate is strain SF3^T, the type strain of species *Desulfovibrio dechloracetivorans*, which is able to couple acetate oxidation to OHR for growth (Sun et al. 2000). Strain SF3 was isolated from San Francisco Bay sediment with acetate as electron donor and 2-CP as electron acceptor (Sun et al. 2000). Several electron donors can be used to couple to OHR for growth, such as acetate, pyruvate, fumarate, propionate, lactate, ethanol and alanine but not with sulphate, and only *ortho*-chlorophenols like 2-CP and 2,6-DCP were dechlorinated (Sun et al. 2000).

Besides, *Desulfovibrio desulfuricans*, a sulphate-reducing bacterium, pre-coated with Pd^0 was used as a biomass-supported chemical (bioinorganic) catalyst for dehalogenation of 2-CP and polychlorinated biphenyls with formate or H_2 as an electron donor (Baxter-Plant et al. 2004). Enhanced dechlorination was observed using the bioinorganic catalyst while negligible chloride release occurred from organohalides using biomass alone, chemically reduced or commercially available Pd^0 (Baxter-Plant et al. 2004). The same concept was used to dechlorinate tris(chloroisopropyl)phosphate (TCPP, a potential flame retardant) in a groundwater sample (Deplanche et al. 2009).

5.2.4 Epsilonproteobacteria

5.2.4.1 Sulfurospirillum

Sulfurospirillum, a genus of Gram-negative, elemental sulfur-reducing spirilla, belongs to the *Proteobacteria* (Phylum), *Epsilonproteobacteria* (Class), *Campylobacterales* (Order) and *Campylobacteraceae* (Family) (Schumacher et al. 1992). The first OHRB characterized in this genus was *Dehalospirillum multivorans* strain K, a strictly anaerobic PCE-dechlorinating bacterium isolated from activated sludge of a wastewater treatment plant with pyruvate and PCE (Scholz-Muramatsu et al. 1995), which was later renamed as *Sulfurospirillum multivorans* (Luijten et al. 2003). Additionally, strain PCE-M2^T, an anaerobic bacterium that is also able to reductively dechlorinate PCE to *cis*-DCE via TCE with lactate as electron donor, was isolated from a site polluted with chlorinated aliphatics in the Netherlands, and was assigned as a novel species *Sulfurospirillum halorespirans* to the genus *Sulfurospirillum* (Luijten et al. 2003). AsO_4^{3-}, SeO_4^{2-}, PCE, nitrate and fumarate can serve as electron acceptors for strains K and PCE-M2^T (Luijten et al. 2003; Holliger et al. 1998b). Strain PCE-M2^T is able to reduce nitrite while strain K is not, and the end product of nitrate reduction for strain PCE-M2^T and strain K is ammonium and nitrite, respectively (Luijten et al. 2003). Pyruvate, lactate, H_2 and formate can be used as electron donors to support PCE dechlorination, and fermentative growth was detected on pyruvate and fumarate (Luijten et al. 2003). Unlike all described RDases (Hug et al. 2013b), PCE RDase of *S. multivorans* needs a special corrinoid cofactor (norpseudo-B12) (Kräutler et al. 2003) synthesized de novo by the strain itself (Keller et al. 2013). More information on *Sulfurospirillum* spp. can be found in Chap. 7.

5.2.5 *Gammaproteobacteria*

5.2.5.1 Shewanella

Shewanella belongs to the *Proteobacteria* (Phylum), *Gammaproteobacteria* (Class), *Alteromonadales* (Order) and *Shewanellaceae* (Family). *Shewanella sediminis* strain HAW-EB3^{T} is a psychrophilic rod-shaped marine bacterium isolated from Halifax Harbour sediment, Canada (Zhao et al. 2004, 2005). This strain was originally noted for its ability to degrade hexahydro-1,3,5-trinitro-1,3,5-triazine (RDX) (Zhao et al. 2004, 2005), and its OHR capacity was later shown for PCE dechlorination to TCE (Lohner and Spormann 2013). This strain grows on several carbon sources such as N-acetyl-D-glucosamine, Tween 40, Tween 80, acetate, succinate, butyrate and serine (Zhao et al. 2005). Bacteria belonging to the *Shewanella* genus are also known for their versatile electron-accepting capabilities using a complex electron transfer network composed mainly of c-type cytochromes as well as iron–sulfur proteins (Cordova et al. 2011). To cope with low temperatures of marine environments, psychrophilic strains of *Shewanella* seem to have adapted through a decreased genome G + C content and an increased proportion of alanine, proline and arginine in proteins to increase protein structural flexibility (Zhao et al. 2010). Genomic analysis of strain HAW-EB3^{T} revealed the presence of genes encoding five putative RDHs similar to characterized RDases, of which Ssed_3769 was responsible for PCE transformation (Lohner and Spormann 2013). The OHR potential of this bacterium was proposed to play a significant role in halogen and carbon cycles as it can uniquely access organic carbon resources locked as organohalogens in sediments that may not be available to non-dehalogenating heterotrophic microbes (Lohner and Spormann 2013).

5.2.6 *Betaproteobacteria*

5.2.6.1 Comamonas

Comamonas was defined as a genus belonging to the *Proteobacteria* (Phylum), *Betaproteobacteria* (Class), *Burkholderiales* (Order) and *Comamonadaceae* (Family). *Comamonas* sp. 7D-2 is the only isolate capable of OHR. It is a strictly aerobic bacterium that can completely degrade bromoxynil (a widely used brominated aromatic herbicide). This bacterium was recently isolated from a bromoxynil octanoate-contaminated soil (Chen et al. 2013). The RDase (BhbA) of stain 7D-2, which catalyzed the reductive debromination of 3,5-dibromo-4-hydroxybenzoate (DBHB) to 4-hydroxybenzoate (HB), was characterized as the first RDase isolated from strict aerobes, with exclusive substrate specificity for haloaromatic compounds with halogen substitutions at the *ortho* position of a hydroxyl group (Chen et al. 2013). Similar to RDase genes from known OHRB, *bhbA* encodes two Fe-S cluster-binding motifs and is associated with a potential anchor protein encoding gene, however, unlike canonical RDases, it lacks a twin arginine signal peptide and B12-binding motif (Chen et al. 2013). Widespread presence of putative RDHs homologous to BhbA in marine *Proteobacteria* and the fact that the *bhbA* gene was located on a large plasmid (pBHB) and surrounded by transposase genes suggested horizontal acquisition of *bhbA* from aerobic marine *Proteobacteria* (Chen et al. 2013). Furthermore, the crystal structure of a similar aerobic RDase was recently published (Payne et al. 2014). These elegant reports point to further catabolic plasticity of OHRB and indicate that the available OHRB represent the tip of the iceberg while a huge diversity awaits discovery.

5.3 Environmental Distribution of OHRB

Our current understanding of the environmental distribution of OHRB has to some extent been obtained by domestication of microbial communities capable of OHR in the form of laboratory culturing. Furthermore, monitoring of the presence of OHRB at a particular site was mediated by application of a suit of molecular techniques targeting phylogenetic and/or functional gene markers such as 16S rRNA and *rdhA* genes that are discussed in more detail in Chap. 25. Combined with field data and compound-specific isotope analysis (CSIA) as a tool for in situ monitoring of OHR (see Chap. 26), the allied application of these tools has created a complementary toolbox for examining the dehalogenation potential mediated by OHRB. Based on the worldwide detection of OHR potential, a cosmopolitan distribution of OHRB is likely, however, an overview of the geographical distribution of isolated OHRB indicates that nearly all OHRB were isolated from a restricted number of discrete parts of the world (Fig. 5.3) though from different source environments (Fig. 5.4). Most probably, this is not due to the restricted geographical

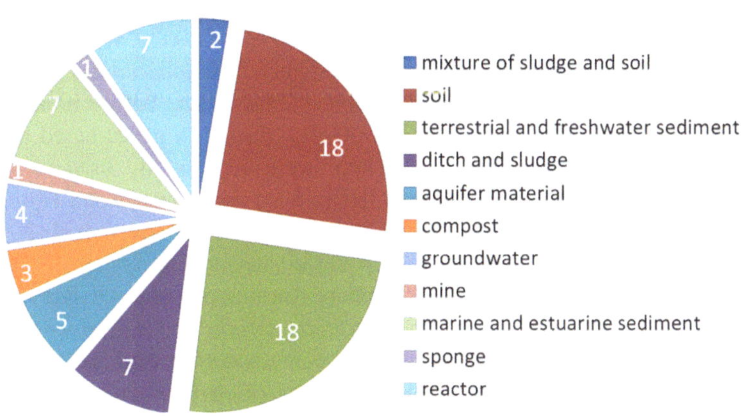

Fig. 5.4 Sources from which OHRB were isolated

distribution of OHRB, but rather due to the initiatives taken by a limited number of laboratories to isolate OHRB from these locations. Hence, a great diversity of OHRB awaits discovery that have not been studied from remote/ill-defined geographical locations. This fact is further illuminated in repeated isolation of different OHRB (Justicia-Leon et al. 2012; Sung et al. 2003; He et al. 2005) and identification of OHR potential (Futagami et al. 2009; Krzmarzick et al. 2012; Kittelmann and Friedrich 2008a; Ritalahti and Löffler 2004) from pristine environments where naturally occurring organohalides are believed to feed OHRB.

5.3.1 Groundwater and Aquifer

The infamous popularity of OHR studies centred around groundwater and aquifer is due to anthropogenic contamination of these environments with organohalides as a serious threat to aquatic as well as terrestrial ecosystems (Johnson et al. 1985; Abelson 1990; Grøn 1991; Thornton et al. 2000; Edmunds et al. 2003; Emmanuel et al. 2009).

In an early attempt to detect OHRB from aquifer samples, of the six different chloroethene-contaminated aquifer samples, only one tested positive for the presence of *Dehalococcoides*, and another tested positive for *Desulfuromonas* (Löffler et al. 2000). A similar study on a TCE-contaminated aquifer at Cape Canaveral Air Station, Florida, indicated heterogeneous distribution of dechlorinating activity and a specific dechlorinating organism, *Dehalococcoides*, at the site where the aquifer in question showed a non-uniform pattern of daughter product concentrations (Fennell et al. 2001). In a more comprehensive study, distribution of *Dehalococcoides* species was investigated in 10 groundwater samples collected from very diverse chloroethene-contaminated sites representing different geographical locations and a variety of climatic zones in North America, England and The Netherlands (Hendrickson et al. 2002). *Dehalococcoides* sequences were not detected in samples from three sites at which partial dechlorination of chloroethenes occurred (Hendrickson et al. 2002). While 16S rRNA gene-based PCR characterization failed to detect *Dehalococcoides* populations in aquifer samples from the Kelly AFB site (Texas, USA) before bioaugmentation with the KB-1 culture (Major et al. 2002), 16S rRNA genes-based qPCR of *Dehalococcoides* and *Desulfuromonas* spp. were successfully used to monitor the distribution and abundance of the OHRB from groundwater and aquifer solids in the Bachman Road Residential Wells Site, USA (Lendvay et al. 2003). In a study to track natural attenuation of a chlorinated ethene contaminated aquifer in the Bitterfeld/Wolfen area in Germany, various potential OHRB including *Dehalococcoides*, *Desulfuromonas*, *Desulfitobacterium* and *Dehalobacter* were detected providing an extra line of evidence for occurrence of natural attenuation (Nijenhuis et al. 2007). In a wider survey of 34 environmental samples from 18 different sites (including 19 aquifer samples and eight groundwater samples) in different European countries, *D. mccartyi* and *Desulfitobacterium* spp. were detected concomitantly in the majority of the samples suggesting either in parallel or stepwise catalysis of complete dehalogenation of PCE by OHR (Rouzeau-Szynalski et al. 2011).

The environmental distribution of OHRB in samples retrieved from aquifers contaminated with chlorinated ethenes was also shown by tracking the abundance and distribution of RDase genes with assigned functions, i.e. TCE-to-VC RDase gene (*tceA*) (Krajmalnik-Brown et al. 2007) and VC-to-ethene RDase gene (*vcrA*) (Müller et al. 2004). The recovered *rdhA* sequences from geographically distinct locations in the US exhibited between 95 % and 98 % identity with known *tceA* and *vcrA* genes, respectively (Krajmalnik-Brown et al. 2007; McMurdie et al. 2011; Müller et al. 2004). A similar approach was recently used for the RDase gene *dcpA* mediating 1,2-dichloropropane dichloroelimination to propene (Padilla-Crespo et al. 2014). The *dcpA* gene was detected in all aquifer samples that yielded 1,2-D-dechlorinating microcosms (Padilla-Crespo et al. 2014), and *dcpA* sequences retrieved from geographically distinct samples shared > 98 % sequence identity with *dcpA* sequences from *D. mccartyi* strains RC and KS (Ritalahti and Löffler 2004; Löffler et al. 1997a) and *D. lykanthroporepellens* BL-DC-9 (Moe et al. 2009).

Studying the abundance and environmental distribution of OHRB was propelled by application of DNA/RNA-based quantitative PCR methods (Lee et al. 2008; Ritalahti et al. 2006; Smits et al. 2004). Using this method, the 16S rRNA and *tceA*, *bvcA* and *vcrA* genes of *D. mccartyi* were detected in groundwater samples collected from a contaminated site near Montague, MI, USA and a large number of unidentified *D. mccartyi rdhA* genes were discovered (Ritalahti et al. 2006). Analysis of 45 groundwater samples taken from Rugårdsvej 234, Odense, Denmark for *D. mccartyi* 16S rRNA and *vcrA* genes indicated the presence of indigenous biomarkers at below detection limit of qPCR (Scheutz et al. 2008). Increases in ethene concentrations after lactate addition were accompanied by a 3-4 order of magnitude increase in concentrations of *vcrA* gene copies throughout the groundwater monitoring network (Scheutz et al. 2008). The oligotrophic nature of aquifers as one of the main obstacles of subsurface OHR was noted in similar studies (Lookman et al. 2007; Schneidewind et al. 2014). Quantification of transcripts of the same biomarkers in groundwater samples from the Fort Lewis military base in Tacoma, WA, USA also verified an indigenous community of *D. mccartyi* cells carrying the *tceA*, *vcrA* and *bvcA* genes (Lee et al. 2008). A less active physiological state of the *D. mccartyi* cells carrying *tceA* was noted as opposed to the cells that harbour *vcrA* and *bvcA* (Lee et al. 2008) which was similar to a DNA based survey of an alluvial plain with complex geochemical and hydraulic conditions in France (Courbet et al. 2011). In an attempt to relate molecular data to geochemical conditions, 150 monitoring wells spread over 11 chlorinated ethene polluted European locations were analyzed (van der Zaan et al. 2010). The *vcrA* gene copy numbers correlated most significantly to VC and chlorinated ethene concentrations while *bvcA* and especially *tceA* were more correlated with oxidizing conditions. The results showed geochemical conditions as the driver of geographical distribution of the RDase genes (van der Zaan et al. 2010). More recently, *Dehalobacter* was found to be more abundant than *D. mccartyi* populations in the clay till matrix collected from chlorinated ethenes and ethanes contaminated site in Vadsby, Denmark which corresponded with OHR being stalled at *cis*-DCE (Damgaard et al. 2013).

Cultivation independent approaches extended to subsurface aquifers further enhanced our understanding of metabolic plasticity of *Dehalococcoidia*-like bacteria. A recent metagenomic analysis from terrestrial aquifer sediment retrieved two *Dehalococcoidia* related pan-genomes ('RBG-2' and 'RBG-1351') (Hug et al. 2013a). Rather than OHR mediating *rdhA* genes, versatile heterotrophic life styles were found (Hug et al. 2013a). These findings indicate that OHR should not be considered as sole mode of metabolism of *Dehalococcoidia*-like bacteria.

Various OHRB capable of respiration on diverse organohalides were isolated from groundwater and aquifer material as listed in Table 5.1.

5.3.2 Marine and Estuarine Environments

Marine and estuarine environments are repositories of numerous chlorinated, brominated and iodinated organic compounds that have been produced and released both from natural and anthropogenic sources. The study of King was the first report on OHR of 2,4-dibromophenol in coastal sediments (King 1988). Since then, numerous studies reported OHR as an essential step in the marine and estuarine biodegradation of halophenols (Ahn et al. 2003), halogenated ethenes (Cheng and He 2009; Lohner and Spormann 2013), halobenzoates (Sun et al. 2001), lindane (Boyle et al. 1999a) and polychlorinated biphenyls (Berkaw et al. 1996; Øfjord et al. 1994; May et al. 2008).

Of all known OHRB, the organohalide-respiring *Chloroflexi* seem to play the most prominent role in OHR in marine, estuarine and tidal flat sediments. The *ortho*-PCB-dechlorinating bacterium *o*-17 was isolated from a PCB dechlorinating culture enriched from estuarine sediments of Baltimore Harbour (USA) (Cutter et al. 2001). Similarly, The PCB-dechlorinating *D. chlorocoercia* DF-1 is indigenous to estuarine sediment and was isolated from Charleston harbour in South Carolina (Wu et al. 2002b). An active role of these isolates and other *Chloroflexi* members that are distinct from *D. mccartyi* were shown in PCB dechlorination in sediment microcosms from Baltimore Harbour, USA (Watts et al. 2005). Similar *Chloroflexi* phylotypes were reported for PCB dechlorination in enrichment from contaminated marine sediment of the Venice lagoon, Italy (Zanaroli et al. 2012). The Tidal Flat Cluster, a deep clade in the phylum *Chloroflexi*, was characterized form North Sea tidal flat sediment of the German Wadden Sea and was found to be involved in the dechlorination of PCE to *trans*-DCE as the predominant end product (Kittelmann and Friedrich 2008b). Molecular analysis of the bacterial community in enrichment cultures obtained from estuarine sediments of the San Diego Bay, USA, indicated involvement of *Chloroflexi*-like microorganisms related to

Dehalococcoides in OHR of the model dioxin, 1,2,3,4-tetrachlorodibenzo-p-dioxin (TeCDD) (Ahn et al. 2007).

In addition to former studies where detection of organohalide-respiring *Chloroflexi* was coupled to physiological/biochemical demonstration of OHR potential, there are other reports on abundant presence of bacteria affiliated with the *Dehalococcoidia* in marine sediments (Inagaki et al. 2003; Jorgensen et al. 2012; Webster et al. 2006). Wasmund and colleagues performed a comprehensive study on distribution and diversity of *Dehalococcoidia*-affiliated bacteria along vertical profiles of various marine sediment cores obtained from Baffin Bay (Greenland), Aarhus Bay (Denmark), and tidal flat sediments of the Wadden Sea (Germany) (Wasmund et al. 2015). qPCR analysis showed that *Dehalococcoidia*-affiliated bacteria constituted only low proportions of the total bacterial in shallow sediment and generally became more prevalent in deeper sediments. Pyrosequencing of 16S rRNA genes of different clades revealed diverse and divergent co-occurring *Dehalococcoidia* populations within single biogeochemical zones. The varying local biogeochemistry was proposed as the driving force for such abrupt shifts. Hence different subgroups of *Dehalococcoidia*-affiliated bacteria with different metabolic properties likely occupy a wide range of ecological niches over sediment depth (Wasmund et al. 2015).

A number of reports exist on distribution and diversity of *rdhA* genes in marine subsurface sediments. Using an array of degenerate primer sets, 32 putative *rdhA* phylotypes were detected from marine subsurface sediments collected from the southeast Pacific off the Peruvian coast, the eastern equatorial Pacific, the Juan de Fuca Ridge flank off the coast of Oregon, and the northwest Pacific off Japan down to a depth of 358 m below the seafloor (Futagami et al. 2009). OHR was noted using 2,4,6-tribromophenol and trichloroethene in sediment slurry collected from the Nankai Trough (Futagami et al. 2009). During another study on sediments of the northwest Pacific off the Kii Peninsula of Japan, OHR was restricted to 4.7 metres below the seafloor, despite detection of *rdhA* genes in deeper sediments (Futagami et al. 2013). A recent metagenomic survey on deep sub-seafloor further broadened the diversity of currently known *rdhA* genes (Kawai et al. 2014). These results are in contrast to single-cell genomic studies of marine *Dehalococcoidia* where no evidence for OHR were found (Kaster et al. 2014; Wasmund et al. 2014). These studies indicated that microorganisms closely related to known obligate OHRB do not rely on OHR for energy conservation, but rather utilize organic matter degradation pathways which are in line with a similar study from aquifer materials (Hug et al. 2013a).

Marine and estuarine environment has been an important source for the isolation of OHRB given their likely exposure to abundant organohalides and remarkable distribution of *rdhA* genes. Various OHRB isolates with diverse OHR capabilities were obtained from marine sediment as outlined in Sect. 5.2, and a comprehensive compilation is given in Table 5.1.

5.3.3 Riverbed and Lake Sediments

Due to the high diversity of organisms, the high load of organic nutrients and strong reducing power, sediment geochemical conditions strongly differ from those observed in aquifers, and thus can support a wide diversity of OHRB. The first study on OHR potential from a lake sediment and sewage sludge was conducted by one of the pioneering labs of the OHR field, Tiedje and co-workers (Suflita et al. 1982). The observed reductive dehalogenation in the obtained enrichment culture was the first report of OHR as a novel anaerobic respiration pathway that has profoundly altered our perception of biodegradability of organohalides especially under anoxic conditions (Suflita et al. 1982). The methanogenic consortium was capable of dehalogenating and often mineralizing a variety of halobenzoates to CH_4 and CO_2 (Horowitz et al. 1983; Suflita et al. 1982). Numerous studies reported anaerobic dechlorination of chlorinated ethenes in the groundwater surface water interaction zone from various geographical locations, i.e. freshwater tidal wetlands of West Branch Canal Creek, Maryland, USA (Lorah and Olsen 1999), Lake Michigan, USA (Lendvay et al. 1998), as well as streambed sediments in Angus, Ontario, Canada (Conant et al. 2004) and in Vilvoorde, Brussels, Belgium (Hamonts et al. 2009) without identification of responsible microorganisms. With growing interest on environmental distribution of OHRB, several 16S RNA gene-based methods were applied to track their presence in organohalide impacted or pristine sediments (Abe et al. 2009; Hendrickson et al. 2002; Kittelmann and Friedrich 2008a; Löffler et al. 2000). In early attempts to detect OHRB from sediment environments, sediment samples of three pristine Michigan rivers yielded amplification products with *Desulfuromonas*- and *Dehalococcoides*-targeted primer sets independent of the sampling season (Löffler et al. 2000). RNA-based stable isotope probing was applied for detection of novel PCE to *cis*-DCE respiring OHRB from pristine river sediment of the river Lahn, Marburg, Germany (Kittelmann and Friedrich 2008a). These novel OHRB within the phylum *Chloroflexi* were only distantly related to cultivated *D. mccartyi* strains (92–94 % sequence identity) (Kittelmann and Friedrich 2008a). In a chlorinated ethene impacted streambed in Toronto, Canada, *D. mccartyi* guilds were detected where soil organic contents were above 2 % w/w and where transformation proceeded to ethene (Abe et al. 2009).

OHR in the riverbed sediments of the Zenne river, Brussels, Belgium impacted with *cis*-DCE, VC and 1,1-dichloroethane (1,1-DCA) coincided with the presence of *Dehalobacter* and *D. mccartyi* guilds (Hamonts et al. 2012). Similar presence of *Dehalobacter* (Atashgahi et al. 2014) and *D. mccartyi* (Atashgahi et al. 2013, 2014) was noticed in a number of follow up studies using microcosms prepared with material from the same site, whereas no *Desulfitobacterium* was detected. In both microcosm studies, VC RDase encoding *vcrA* and *bvcA* genes were the most abundant *rdh*A genes (Atashgahi et al. 2013, 2014). However, in spite of repeated detection of biomarkers associated with OHRB in the field, microcosm studies with Zenne riverbed sediments (Atashgahi et al. 2013, 2014; Hamonts et al. 2012) and high OHR performance in microcosm studies with material from different locations at the Zenne river site (Hamonts et al. 2009; Atashgahi et al. 2013), high spatial variability of OHR potential was observed in field measurements from the riverbed (Hamonts et al. 2012). Hence, rather than the microbial community structure of the Zenne River sediments, the CAH discharge rate was hypothesized as the limiting factor for OHR in the Zenne sediments (Hamonts et al. 2014). The high discharge rate and hence reduced residence time of discharging contaminated groundwater seems to be an important prerequisite to drive OHR within riverbed sediments (Hamonts et al. 2014; Abe et al. 2009).

The composition and abundance of active OHRB was investigated in riverbed sediments collected from the Ebro river in Spain and the Elbe river in Germany (Taş et al. 2009). *D. mccartyi* and *Desulfitobacterium* 16S rRNA were detected at the sampling locations whereas *Dehalobacter* rRNA could not be detected. *D. mccartyi* populations were more active and had a higher diversity in river sediments than in floodplain soils in the Elbe River and in general were more dominant in HCB-polluted locations within river basins than *Desulfitobacterium* spp. and *Dehalobacter* spp., albeit with significant temporal and spatial variation (Taş et al. 2009). In a recent study 18 sediment samples from three different locations in the Yangtze River, China, were tested for OHR potential (Kranzioch et al. 2013). PCE to *cis*- and *trans*-DCE dechlorination was concomitant with increasing numbers of *Dehalobacter* spp., *Desulfomonile* spp., *Desulfitobacterium* spp., or *D. mccartyi* (Kranzioch et al. 2013). Krzmarzick and co-workers detected putative organohalide-respiring *Chloroflexi* in 67 of 68 samples collected from lake sediments across a geographical sulfur gradient. Higher quantities of these bacteria were noted where lower sulphate concentration was recorded, indicating an antagonistic relationship between both functional guilds (Kranzioch et al. 2013).

5.3.4 Soil

Ubiquitous distribution and active involvement in halogen cycles was also shown for organohalide-respiring *Chloroflexi* in soil. An interesting study by Krzmarzick et al. (2012) documented widespread distribution of organohalide-respiring *Chloroflexi* in 116 pristine soil samples collected from different locations in USA where their number of 16S rRNA genes positively correlated with [organochlorine]/TOC (Krzmarzick et al. 2012). Triclosan addition to agricultural soil from Fairfax, Minnesota, USA, without previous exposure to this substrate increased abundance of *Dehalococcoides*-like *Chloroflexi* in soil microcosms (McNamara and Krzmarzick 2013). Furthermore, recently the enrichment of a TCE respiring consortium from garden soil collected from Cuzdrioara, Cluj County, Romania, was reported (Delgado et al. 2014). In spite of the presence of *D. mccartyi* and *tceA*, *vcrA* and *bvcA* genes, dechlorination stalled at *cis*-DCE even after repeated biostimulation with lactate. The authors hypothesized that this might be a consequence of the intrinsic competition for electron donor (H_2) in soils and sediments, driven by a variety of electron acceptors such as nitrate, Fe (III), sulphate and bicarbonate (Delgado et al. 2014).

Besides the abundant OHRB belonging to *Chloroflexi*, important roles played by the OHRB belonging to *Firmicutes* were reported in terrestrial soil environments. Geographic distribution of *D. hafniense* PCP-1 and other *Desulfitobacterium* spp. was studied in 44 soil samples from Quebec, Canada, using a PCR method showing ubiquitous presence of *Desulfitobacterium* spp. (Lanthier et al. 2001; Lévesque et al. 1997). A survey of 65 soil samples from chloroethene-contaminated and pristine soils in Japan showed PCE to *cis*-DCE dechlorination in 16 out of 44 chloroethene-contaminated soils and three out of 21 pristine soils (Yoshida et al. 2007a). PCE dechlorination was mostly associated with detection of *Desulfitobacterium* spp. and probably *Dehalobacter* spp. rather than *D. mccartyi* (Yoshida et al. 2007a). Complete pentachlorophenol (PCP) to phenol degradation was noted in an enrichment culture obtained from paddy field soil located at Yatomi-cho, Aichi prefecture, Japan (Yoshida et al. 2007b). The involvement of Gram-positive spore-forming bacteria belonging to the phylum *Firmicutes* was proposed, although *Desulfitobacterium* spp. were not detected by PCR using a specific primer set (Yoshida et al. 2007b). The same soil samples were later shown to be conducive to dechlorination of fthalide where *Dehalobacter* sp. strains FTH1 and FTH2 were identified as potential OHRB (Yoshida et al. 2009). Involvement of novel *Firmicutes* referred to as Gopher group was also proposed in the dechlorination of two chlorinated xanthones, analogues of natural organochlorines in pristine forest soil from New Jersey and pristine sediment samples collected from lake sediment in Minnesota, USA (Krzmarzick et al. 2014). The Gopher group was shown to be distinct from but related to *Dehalobacter* and *Desulfitobacterium* (Krzmarzick et al. 2014).

5.4 Factors Impacting Environmental Distribution of OHRB

Of particular importance for the distribution and fate of OHRB (and hence organohalides) in the environment are redox conditions, availability of micronutrients, the presence of suitable electron donors and acceptors and the interaction with chemical species.

5.4.1 Redox Conditions

With the exception of *Comamonas* sp. 7D-2 as a strict aerobic bacterium capable of OHR (Chen et al. 2013) and *A. dehalogenans* as a facultative anaerobe (Sanford et al. 2002), the currently known OHRB are generally strictly anaerobic bacteria and sensitive to oxygen exposure. This is of high importance for the more delicate OHRB such as *Dehalococcoides* and *Dehalobacter* spp. that require strict anoxic techniques and presence of reducing chemicals in laboratory cultures. Lacking the genes necessary for direct utilization of oxygen, these OHRB are highly dependent on an array of natural shields in the environment. Recent findings showed that in spite of stringent requirement for anoxic conditions in axenic cultures, members of *Dehalococcoides* might be well protected in their natural habitats such as riverbed sediments (Atashgahi et al. 2013). The organic rich matrix of riverbed sediments limits oxygen penetration to the surficial sediment layers. Moreover, sediment aggregates can host anoxic microniches inside sediment-associated biofilms. Presence of facultative aerobic bacteria colonizing the outer layers of sediment biofilms, which rapidly consume oxygen, can protect the strict anaerobes such as OHRB in core microniches (Atashgahi et al. 2013). In line with this, a former study on distribution and localization of *D. mccartyi* 195 and methanogenic populations in an enrichment culture using fluorescence in situ hybridization (FISH) showed that methanogens reside in the exterior layer of bioflocs, whereas *D. mccartyi* 195 cells were more evenly distributed within bioflocs as well as between biofloc and planktonic phases (Rowe et al. 2008). A comparative metagenomic study on three different *Dehalococcoides*-containing dechlorinating microbial consortia indicated the presence of several genes related to oxygen tolerance and scavenging from acetogens, fermenters and methanogens commonly found to co-exist with OHRB in enrichment cultures and natural environments (Hug et al. 2012). Activity of such genes in non-dehalogenating community members would enforce oxygen scavenging mechanisms, collaterally protecting the OHRB along-side.

5.4.2 Micronutrients

OHRB also depend on other non-dehalogenating community members for the supply of electron donor(s) and growth factors such as vitamins. Fermentation intermediates such as acetate, formate and H_2 are common products of anaerobic hydrocarbon degradation (Stams and Plugge 2009), and are used by other community members such as iron reducers, SRB, OHRB and methanogens. This overlapping nutritional niche caused controversial discussions in the field of OHR. Early reports considered hydrogenotrophic SRB and methanogens as competitors of OHRB for reducing equivalents (Fennell and Gossett 1998; Smatlak et al. 1996), however, the (predicted) supply of nutrients by SRB and methanogens (Hug et al. 2012) points towards interactive networks between these microbial guilds rather than pure competition for the shared resources. Indeed, the importance of symbiotic interactions between SRB and OHRB for effective OHR in their co-cultures was previously reported and was attributed to interspecies H_2 transfer (Drzyzga et al. 2001; Drzyzga and Gottschal 2002) or supply of unknown growth factors by the SRB partner (Mägli et al. 1996). Recent dedicated studies showed that supply of corrinoid cofactors by SRB and methanogens sustained OHR by *D. mccartyi*

195, although corrinoid modification and remodelling by strain 195 was essential to add 5',6'-dimethylbenzimidazole (DMB) as the lower ligand to cobamide precursors and produce cobalamin as the preferred corrinoid cofactor (Yan et al. 2012; Yi et al. 2012; Men et al. 2014). Similarly, *Geobacter lovleyi* strain SZ was shown to fulfil *Dehalococcoides'* corrinoid requirements by the exchange of DMB (Yan et al. 2012). Given the fact that all *D. mccartyi* strains for which genome sequences are available lack the capability of de novo cobalamin synthesis (McMurdie et al. 2009), and all share the same putative genes involved in corrinoid-remodelling pathways (Yi et al. 2012), it is tempting to assume that OHRB exploited the biotopes shared with non-dechlorinating community members and have lost the costly but still key metabolic trait of de novo cobalamin synthesis. In line with this, recent studies on corrinoid-auxotrophy of *Dehalobacter restrictus* showed that in spite of presence of both anaerobic cobalamin biosynthesis and cobinamide-salvaging pathways, one important protein, CbiH (precorrin-3B C17-methyltransferase), was missing from the upper cobalamin biosynthesis pathway, which was believed to be due to frame-shift mutations (Rupakula et al. 2014). Supply of the OHRB with corrinoids in their natural environment or even during long-term enrichments in the lab may enforce such a genetic trade-off that disables the unused trait. This may explain why OHRB and in particular those who rely on OHR as the sole mode of metabolism have shared/overlapping niches with non-dechlorinators that are capable of de novo corrinoid biosynthesis. Experimental evidence also suggest that OHRB are mostly enriched in methanogenic and sulphate-reducing conditions. For example, a recent batch-scale biostimulation of TCE dechlorination showed incomplete TCE dechlorination in microcosms where methanogens were absent (Schneidewind et al. 2014). In another study, inhibition of methanogenesis negatively impacted community members such as *Desulfovibrio* (Men et al. 2013) that are known to synthesize corrinoids de novo.

In natural habitats such as cell aggregates and biofilms where cells form close physical association, OHRB and methanogens may play complementary roles in alleviating thermodynamic constraints for other syntrophic microorganisms. The degradation of fatty acids such as propionate and butyrate only takes place in a thermodynamically interdependent lifestyle where degradation intermediates, usually acetate, formate and H_2, are maintained at very low concentrations (Stams and Plugge 2009). Such metabolic/physical networks may also explain why most fastidious OHRB like *D. mccartyi* strains do not submit readily to growth in the laboratory as they need metabolites, signals and structure provided by the community. Further, the structured associations would be more successful in retaining metabolites such as H_2 that would otherwise be lost by diffusion. However, in spite of the central role of microbe-microbe interactions, this subject has received surprisingly little attention in studies involved in OHR. Such information would highly assist in understanding the ecophysiology of OHRB and facilitate design of effective and sustainable bioremediation strategies.

5.4.3 Electron Acceptors

The presence of the right terminal electron acceptor is vital for the success of this process as not all OHRB are dedicated to OHR. It is not known whether facultative OHRB would prefer other electron acceptors over organohalides in their natural environment although some general assumptions can be made. At higher redox potential the facultative OHRB such as proteobacterial OHRB may prefer electron acceptors like nitrate, Mn(IV) and Fe(III). A recent study on correlations between environmental variables and bacterial community structures indicated that both VC and Fe(III) reduction were most important but antagonistic terminal electron-accepting processes (Shani et al. 2013). Once highly reducing conditions prevail, organohalides may act as electron acceptors. However, the organohalides in natural habitats may not be as readily accessible as in well-homogenized lab cultures. For example, higher degree of halogen substitution can reduce bioavailability of organohalides. This is particularly important in soil and sediment matrices that contain strong sinks of organohalides such as amorphous organic carbon and black carbon (Cornelissen et al. 2005). In such environments, highly halogenated and hydrophobic organohalides such as dioxins and PCBs adsorb strongly to organic matter, reducing their bioavailability to OHRB (Bunge and Lechner 2009). It should however be noted that such a reduced bioavailability will also lower the acute toxicity of organohalides to OHRB, hence playing a double-edged sword function. Besides, not all types of organohalides are degraded by all OHRB. The facultative OHRB known to date do not dechlorinate PCE past DCE. Even within *D. mccartyi* strains, except for genomic evidence for the strain BTF08 (Pöritz et al. 2013), no other isolate is capable of complete PCE dechlorination (Löffler et al. 2013). However, multiple OHRB may reside in close proximity and co-exist in the same habitat, which is necessary for their robust OHR activity by complementary dehalogenation activity. For example, the concomitant presence of bacteria belonging to *D. mccartyi* and *Desulfitobacterium* groups was previously reported in chloroethene-dechlorinating microbial communities (Rouzeau-Szynalski et al. 2011). Another study with an enrichment culture showed that *Sulfurospirillum*-affiliated bacteria were involved in PCE dechlorination to *cis*-DCE whereas *Dehalococcoides*-affiliated bacteria mainly dechlorinated *cis*-DCE to ethene (Maillard et al. 2011).

5.4.4 Interaction with Chemical Species

Interaction of OHRB with abiotic factors is highly important for their fate in environment. The electrophilic nature of organohalides and reductive nature of the OHR process demands efficient electron transfer mechanisms. Naturally occurring electron shuttles and redox mediators that can be reversibly oxidised and reduced have the capacity to serve as external electron carriers in redox processes such as OHR (Van der Zee and Cervantes 2009). Humic substances and their quinoid analogues are among the best studied natural redox mediators in promoting OHR (Aulenta et al. 2010; Zhang and Katayama 2012). Under oxygen-depleted conditions in the subsurface, these redox mediators can be reduced not only by combustion of organic substrates, but also by chemical reactions with reductants common to anaerobic soils and sediments such as Fe(II), sulphide and cysteine. The electron-withdrawing organohalides can in turn be reduced in subsequent OHR thereby re-oxidizing the mediators. Hence, a cocktail of biotic and abiotic reactions could indirectly stimulate OHRB. The organohalides were also shown to be abiotically degraded by naturally occurring minerals such as pyrite, magnetite and green rusts (Tobiszewski and Namieśnik 2012). Biogenic abiotic catalysts such as palladium nanoparticles can also effectively remove organohalides such as TCE (Hosseinkhani et al. 2014). The abiotic reductive dechlorination reaction could be complementary to OHR at places where the condition is less favourable for OHRB and hence compensate for their drawbacks.

5.5 Conclusions

The currently known OHRB belong to distinct phylogenetic groups of deeply branching *Chloroflexi*, Gram-negative *Proteobacteria* and Gram-positive *Firmicutes*. The ubiquitous documentation of OHR potential in aquatic as well as terrestrial environments indicates that the current collection of OHRB is only the tip of the iceberg. It is likely that a considerable portion of OHRB has to date been ignored during conventional culturing techniques. Recent advances in high-throughput culturing approaches should pave the way for isolation of novel OHRB, bypassing many obstacles of conventional cultivation. Moreover, most studies to date were performed using model chlorinated compounds of different haloaliphatic and haloaromatic classes. This limited spectrum of substrates in part is due to simplicity of handling these rather "classical" organohalides and the fact that most of complex naturally occurring organohalides are not commercially available. The known OHRB show moderate traits with regards to extremes of temperature (except for the psychrophilic marine bacterium, *Shewanella sediminis* (Zhao et al. 2005)), pH and salinity and up to now, there has been no report on OHRB isolated from (poly) extreme environments. Moreover, the OHR capacity has been restricted to the bacterial domain of life and no archaeon has been reported to be associated with OHR in axenic cultures. This is despite the fact that they have been repeatedly found in close association with OHRB in enrichment cultures and environmental samples.

Acknowledgement The authors thank the BE-Basic Foundation (project MicroControl (8.004.01)) for financial support. Yue Lu was sponsored by a CSC Fellowship.

References

Abe Y, Aravena R, Zopfi J, Parker B, Hunkeler D (2009) Evaluating the fate of chlorinated ethenes in streambed sediments by combining stable isotope, geochemical and microbial methods. J Contam Hydrol 107(1):10–21

Abelson PH (1990) Inefficient remediation of ground-water pollution. Science 250(4982)

Adrian L, Manz W, Szewzyk U, Görisch H (1998) Physiological characterization of a bacterial consortium reductively dechlorinating 1,2,3- and 1,2,4-trichlorobenzene. Appl Environ Microbiol 64(2):496–503

Adrian L, Szewzyk U, Wecke J, Görisch H (2000) Bacterial dehalorespiration with chlorinated benzenes. Nature 408:580–583

Adrian L, Rahnenführer J, Gobom J, Hölscher T (2007) Identification of a chlorobenzene reductive dehalogenase in *Dehalococcoides* sp. strain CBDB1. Appl Environ Microbiol 73(23):7717–7724

Ahn Y-B, Rhee S-K, Fennell DE, Kerkhof LJ, Hentschel U, Häggblom MM (2003) Reductive dehalogenation of brominated phenolic compounds by microorganisms associated with the marine sponge *Aplysina aerophoba*. Appl Environ Microbiol 69(7):4159–4166

Ahn Y-B, Häggblom MM, Kerkhof LJ (2007) Comparison of anaerobic microbial communities from estuarine sediments amended with halogenated compounds to enhance dechlorination of 1,2,3,4-tetrachlorodibenzo-*p*-dioxin. FEMS Microbiol Ecol 61(2):362–371

Ahn Y-B, Kerkhof LJ, Häggblom MM (2009) *Desulfoluna spongiiphila* sp. nov., a dehalogenating bacterium in the *Desulfobacteraceae* from the marine sponge *Aplysina aerophoba*. Int J Syst Evol Microbiol 59(9):2133–2139

Atashgahi S, Maphosa F, Doğan E, Smidt H, Springael D, Dejonghe W (2013) Small-scale oxygen distribution determines the vinyl chloride biodegradation pathway in surficial sediments of riverbed hyporheic zones. FEMS Microbiol Ecol 84(1):133–142

Atashgahi S, Maphosa F, De Vrieze J, Haest PJ, Boon N, Smidt H, Springael D, Dejonghe W (2014) Evaluation of solid polymeric organic materials for use in bioreactive sediment capping to stimulate the degradation of chlorinated aliphatic hydrocarbons. Appl Microbiol Biotechnol 98(5):2255–2266

Aulenta F, Di Maio V, Ferri T, Majone M (2010) The humic acid analogue antraquinone-2, 6-disulfonate (AQDS) serves as an electron shuttle in the electricity-driven microbial dechlorination of trichloroethene to *cis*-dichloroethene. Bioresour Technol 101(24):9728–9733

Baxter-Plant V, Mikheenko I, Robson M, Harrad S, Macaskie L (2004) Dehalogenation of chlorinated aromatic compounds using a hybrid bioinorganic catalyst on cells of *Desulfovibrio desulfuricans*. Biotechnol Lett 26(24):1885–1890

Bedard DL, Bailey JJ, Reiss BL, Jerzak GVS (2006) Development and characterization of stable sediment-free anaerobic bacterial enrichment cultures that dechlorinate Aroclor 1260. Appl Environ Microbiol 72(4):2460–2470

Berkaw M, Sowers KR, May HD (1996) Anaerobic ortho dechlorination of polychlorinated biphenyls by estuarine sediments from Baltimore harbor. Appl Environ Microbiol 62(7):2534–2539

Bouchard B, Beaudet R, Villemur R, McSween G, Lepine F, Bisaillon J-G (1996) Isolation and characterization of *Desulfitobacterium frappieri* sp. nov., an anaerobic bacterium which reductively dechlorinates pentachlorophenol to 3-chlorophenol. Int J Syst Bacteriol 46(4):1010–1015

Bowman KS, Nobre MF, da Costa MS, Rainey FA, Moe WM (2013) *Dehalogenimonas alkenigignens* sp. nov., a chlorinated-alkane-dehalogenating bacterium isolated from groundwater. Int J Syst Evol Microbiol 63 (Pt 4):1492–1498

Boyle AW, Häggblom MM, Young LY (1999a) Dehalogenation of lindane (γ-hexachlorocyclohexane) by anaerobic bacteria from marine sediments and by sulfate-reducing bacteria. FEMS Microbiol Ecol 29(4):379–387

Boyle AW, Phelps CD, Young L (1999b) Isolation from estuarine sediments of a *Desulfovibrio* strain which can grow on lactate coupled to the reductive dehalogenation of 2,4,6-tribromophenol. Appl Environ Microbiol 65(3):1133–1140

Breitenstein A, Saano A, Salkinoja-Salonen M, Andreesen JR, Lechner U (2001) Analysis of a 2,4,6-trichlorophenol-dehalogenating enrichment culture and isolation of the dehalogenating member *Desulfitobacterium frappieri* strain TCP-A. Arch Microbiol 175:133–142

Bunge M, Lechner U (2009) Anaerobic reductive dehalogenation of polychlorinated dioxins. Appl Microbiol Biotechnol 84(3):429–444

Bunge M, Wagner A, Fischer M, Andreesen JR, Lechner U (2008) Enrichment of a dioxin-dehalogenating *Dehalococcoides* species in two-liquid phase cultures. Environ Microbiol 10(10):2670–2683

Chang YC, Hatsu M, Jung K, Yoo YS, Takamizawa K (1998) Degradation of a variety of halogenated aliphatic compounds by an anaerobic mixed culture. J Ferment Bioeng 86(4):410–412

Chen K, Huang L, Xu C, Liu X, He J, Zinder SH, Li S, Jiang J (2013) Molecular characterization of the enzymes involved in the degradation of a brominated aromatic herbicide. Mol Microbiol 89(6):1121–1139

Cheng D, He J (2009) Isolation and characterization of "*Dehalococcoides*" sp. strain MB, which dechlorinates tetrachloroethene to *trans*-1,2-dichloroethene. Appl Environ Microbiol 75(18):5910–5918

Cheng D, Chow WL, He J (2009) A *Dehalococcoides*-containing co-culture that dechlorinates tetrachloroethene to *trans*-1,2-dichloroethene. ISME J 4(1):88–97

Chow WL, Cheng D, Wang S, He J (2010) Identification and transcriptional analysis of *trans*-DCE-producing reductive dehalogenases in *Dehalococcoides* species. ISME J 4(8):1020–1030

Christiansen N, Ahring BK (1996) *Desulfitobacterium hafniense* sp. nov., an anaerobic, reductively dechlorinating bacterium. Int J Syst Bacteriol 46(2):442–448

Cole JR, Cascarelli AL, Mohn WW, Tiedje JM (1994) Isolation and characterization of a novel bacterium growing via reductive dehalogenation of 2-chlorophenol. Appl Environ Microbiol 60(10):3536–3542

Conant B, Cherry JA, Gillham RW (2004) A PCE groundwater plume discharging to a river: influence of the streambed and near-river zone on contaminant distributions. J Contam Hydrol 73(1):249–279

Cordova CD, Schicklberger MF, Yu Y, Spormann AM (2011) Partial functional replacement of CymA by SirCD in *Shewanella oneidensis* MR-1. J Bacteriol 193(9):2312–2321

Cornelissen G, Gustafsson Ö, Bucheli TD, Jonker MT, Koelmans AA, van Noort PC (2005) Extensive sorption of organic compounds to black carbon, coal, and kerogen in sediments and soils: mechanisms and consequences for distribution, bioaccumulation, and biodegradation. Environ Sci Technol 39(18):6881–6895

Courbet C, Rivière A, Jeannottat S, Rinaldi S, Hunkeler D, Bendjoudi H, de Marsily G (2011) Complementing approaches to demonstrate chlorinated solvent biodegradation in a complex pollution plume: mass balance, PCR and compound-specific stable isotope analysis. J Contam Hydrol 126(3):315–329

Cutter LA, Watts JE, Sowers KR, May HD (2001) Identification of a microorganism that links its growth to the reductive dechlorination of 2,3,5,6-chlorobiphenyl. Environ Microbiol 3(11):699–709

Damgaard I, Bjerg PL, Bælum J, Scheutz C, Hunkeler D, Jacobsen CS, Tuxen N, Broholm MM (2013) Identification of chlorinated solvents degradation zones in clay till by high resolution chemical, microbial and compound specific isotope analysis. J Contam Hydrol 146:37–50

De Wever H, Cole JR, Fettig MR, Hogan DA, Tiedje JM (2000) Reductive dehalogenation of trichloroacetic acid by *Trichlorobacter thiogenes* gen. nov., sp. nov. Appl Environ Microbiol 66(6):2297–2301

De Wildeman S, Diekert G, Van Langenhove H, Verstraete W (2003) Stereoselective microbial dehalorespiration with vicinal dichlorinated alkanes. Appl Environ Microbiol 69(9):5643–5647

Delgado AG, Kang D-W, Nelson KG, Fajardo-Williams D, Miceli JF III, Done HY, Popat SC, Krajmalnik-Brown R (2014) Selective enrichment yields robust ethene-producing dechlorinating cultures from microcosms stalled at *cis*-dichloroethene. PLoS ONE 9(6):e100654

Deplanche K, Snape T, Hazrati S, Harrad S, Macaskie L (2009) Versatility of a new bioinorganic catalyst: palladized cells of *Desulfovibrio desulfuricans* and application to dehalogenation of flame retardant materials. Environ Technol 30(7):681–692

DeWeerd KA, Mandelco L, Tanner RS, Woese CR, Suflita JM (1990) *Desulfomonile tiedjei* gen. nov. and sp. nov., a novel anaerobic, dehalogenating, sulfate-reducing bacterium. Arch Microbiol 154(1):23–30

DeWeerd K, Concannon F, Suflita JM (1991) Relationship between hydrogen consumption, dehalogenation, and the reduction of sulfur oxyanions by *Desulfomonile tiedjei*. Appl Environ Microbiol 57(7):1929–1934

Ding C, Zhao S, He J (2014) A *Desulfitobacterium* sp. strain PR reductively dechlorinates both 1,1,1-trichloroethane and chloroform. Environ Microbiol 16(11):3387–3397

Drzyzga O, Gottschal JC (2002) Tetrachloroethene dehalorespiration and growth of *Desulfitobacterium frappieri* TCE1 in strict dependence on the activity of *Desulfovibrio fructosivorans*. Appl Environ Microbiol 68(2):642–649

Drzyzga O, Gerritse J, Dijk JA, Elissen H, Gottschal JC (2001) Coexistence of a sulphate-reducing *Desulfovibrio* species and the dehalorespiring *Desulfitobacterium frappieri* TCE1 in defined chemostat cultures grown with various combinations of sulphate and tetrachloroethene. Environ Microbiol 3(2):92–99

Duhamel M, Edwards EA (2006) Microbial composition of chlorinated ethene-degrading cultures dominated by *Dehalococcoides*. FEMS Microbiol Ecol 58(3):538–549

Duret A, Holliger C, Maillard J (2012) The physiological opportunism of *Desulfitobacterium hafniense* strain TCE1 towards organohalide respiration with tetrachloroethene. Appl Environ Microbiol 78(17):6121–6127

Edmunds W, Shand P, Hart P, Ward R (2003) The natural (baseline) quality of groundwater: a UK pilot study. Sci Total Environ 310(1):25–35

Emmanuel E, Pierre MG, Perrodin Y (2009) Groundwater contamination by microbiological and chemical substances released from hospital wastewater: Health risk assessment for drinking water consumers. Environ Int 35(4):718–726

Fagervold SK, Watts JE, May HD, Sowers KR (2005) Sequential reductive dechlorination of *meta*-chlorinated polychlorinated biphenyl congeners in sediment microcosms by two different *Chloroflexi* phylotypes. Appl Environ Microbiol 71(12):8085–8090

Fagervold SK, May HD, Sowers KR (2007) Microbial reductive dechlorination of Aroclor 1260 in Baltimore Harbor sediment microcosms is catalyzed by three phylotypes within the phylum *Chloroflexi*. Appl Environ Microbiol 73(9):3009–3018

Fennell DE, Gossett JM (1998) Modeling the production of and competition for hydrogen in a dechlorinating culture. Environ Sci Technol 32(16):2450–2460

Fennell DE, Carroll AB, Gossett JM, Zinder SH (2001) Assessment of indigenous reductive dechlorinating potential at a TCE-contaminated site using microcosms, polymerase chain reaction analysis, and site data. Environ Sci Technol 35(9):1830–1839

Finneran KT, Forbush HM, VanPraagh CVG, Lovley DR (2002) *Desulfitobacterium metallireducens* sp. nov., an anaerobic bacterium that couples growth to the reduction of metals and humic acids as well as chlorinated compounds. Int J Sys Evol Microbiol 52(6):1929–1935

Fletcher KE, Ritalahti KM, Pennell KD, Takamizawa K, Löffler FE (2008) Resolution of culture *Clostridium bifermentans* DPH-1 into two populations, a *Clostridium* sp. and tetrachloroethene-dechlorinating *Desulfitobacterium hafniense* strain JH1. Appl Environ Microbiol 74(19):6141–6143

Futagami T, Morono Y, Terada T, Kaksonen AH, Inagaki F (2009) Dehalogenation activities and distribution of reductive dehalogenase homologous genes in marine subsurface sediments. Appl Environ Microbiol 75(21):6905–6909

Futagami T, Morono Y, Terada T, Kaksonen AH, Inagaki F (2013) Distribution of dehalogenation activity in subseafloor sediments of the Nankai Trough subduction zone. Phil Trans R Soc B 368(1616):20120249

Gerritse J, Renard V, Pedro G, Lawson PA, Collins MD, Gottschal JC (1996) *Desulfitobacterium* sp. strain PCE1, an anaerobic bacterium that can grow by reductive dechlorination of tetrachloroethene or *ortho-* chlorinated phenols. Arch Microbiol 165:132–140

Gerritse J, Drzyzga O, Kloetstra G, Keijmel M, Wiersum LP, Hutson R, Collins MD, Gottschal JC (1999) Influence of different electron donors and acceptors on dehalorespiration of tetrachloroethene by *Desulfitobacterium frappieri* TCE1. Appl Environ Microbiol 65(12):5212–5221

Goris T, Hornung B, Kruse T, Reinhold A, Westermann M, Schaap PJ, Smidt H, Diekert G (2015) Draft genome sequence and characterization of *Desulfitobacterium hafniense* PCE-S. Stand Genomic Sci 10(1):15

Gribble GW (2010) Naturally occurring organohalogen compounds-a comprehensive update. Progress in the chemistry of organic natural products, vol 91. Springer/Wien, Germany

Grøn C (1991) Organic halogens in Danish groundwaters. Lect Notes in Earth Sci 33:495–506

Grostern A, Edwards EA (2006) Growth of *Dehalobacter* and *Dehalococcoides* spp. during degradation of chlorinated ethanes. Appl Environ Microbiol 72(1):428–436

Grostern A, Duhamel M, Dworatzek S, Edwards EA (2010) Chloroform respiration to dichloromethane by a *Dehalobacter* population. Environ Microbiol 12(4):1053–1060

Hamonts K, Kuhn T, Maesen M, Bronders J, Lookman R, Kalka H, Diels L, Meckenstock RU, Springael D, Dejonghe W (2009) Factors determining the attenuation of chlorinated aliphatic hydrocarbons in eutrophic river sediment impacted by discharging polluted groundwater. Environ Sci Technol 43(14):5270–5275

Hamonts K, Kuhn T, Vos J, Maesen M, Kalka H, Smidt H, Springael D, Meckenstock RU, Dejonghe W (2012) Temporal variations in natural attenuation of chlorinated aliphatic hydrocarbons in eutrophic river sediments impacted by a contaminated groundwater plume. Water Res 46(6):1873–1888

Hamonts K, Ryngaert A, Smidt H, Springael D, Dejonghe W (2014) Determinants of the microbial community structure of eutrophic, hyporheic river sediments polluted with chlorinated aliphatic hydrocarbons. FEMS Microbiol Ecol 87(3):715–732

He J, Ritalahti KM, Aiello MR, Löffler FE (2003) Complete detoxification of vinyl chloride by an anaerobic enrichment culture and identification of the reductively dechlorinating population as a *Dehalococcoides* species. Appl Environ Microbiol 69(2):996–1003

He J, Sung Y, Krajmalnik-Brown R, Ritalahti KM, Löffler FE (2005) Isolation and characterization of *Dehalococcoides* sp. strain FL2, a trichloroethene (TCE)-and 1,2-dichloroethene-respiring anaerobe. Environ Microbiol 7(9):1442–1450

Hendrickson ER, Payne JA, Young RM, Starr MG, Perry MP, Fahnestock S, Ellis DE, Ebersole RC (2002) Molecular analysis of *Dehalococcoides* 16S ribosomal DNA from chloroethene-contaminated sites throughout North America and Europe. Appl Environ Microbiol 68(2):485–495

Holliger C, Schraa G, Stams AJ, Zehnder AJB (1993) A highly purified enrichment culture couples the reductive dechlorination of tetrachloroethene to growth. Appl Environ Microbiol 59(9):2991–2997

Holliger C, Hahn D, Harmsen H, Ludwig W, Schumacher W, Tindall B, Vazquez F, Weiss N, Zehnder AJB (1998a) *Dehalobacter restrictus* gen. nov. and sp. nov., a strictly anaerobic bacterium that reductively dechlorinates tetra- and trichloroethene in an anaerobic respiration. Arch Microbiol 169(4):313–321

Holliger C, Wohlfarth G, Diekert G (1998b) Reductive dechlorination in the energy metabolism of anaerobic bacteria. FEMS Microbiol Rev 22(5):383–398

Horowitz A, Suflita JM, Tiedje JM (1983) Reductive dehalogenations of halobenzoates by anaerobic lake sediment microorganisms. Appl Environ Microbiol 45(5):1459–1465

Hosseinkhani B, Hennebel T, Van Nevel S, Verschuere S, Yakimov MM, Cappello S, Blaghen M, Boon N (2014) Biogenic nanopalladium based remediation of chlorinated hydrocarbons in marine environments. Environ Sci Technol 48(1):550–557

Hug LA, Beiko RG, Rowe AR, Richardson RE, Edwards EA (2012) Comparative metagenomics of three *Dehalococcoides*-containing enrichment cultures: the role of the non-dechlorinating community. BMC Genom 13(1):327

Hug LA, Castelle CJ, Wrighton KC, Thomas BC, Sharon I, Frischkorn KR, Williams KH, Tringe SG, Banfield JF (2013a) Community genomic analyses constrain the distribution of metabolic traits across the *Chloroflexi* phylum and indicate roles in sediment carbon cycling. Microbiome 1(1):22

Hug LA, Maphosa F, Leys D, Löffler FE, Smidt H, Edwards EA, Adrian L (2013b) Overview of organohalide-respiring bacteria and a proposal for a classification system for reductive dehalogenases. Phil Trans R Soc B 368(1616)

Inagaki F, Suzuki M, Takai K, Oida H, Sakamoto T, Aoki K, Nealson KH, Horikoshi K (2003) Microbial communities associated with geological horizons in coastal subseafloor sediments from the Sea of Okhotsk. Appl Environ Microbiol 69(12):7224–7235

Johnson RL, Brillante SM, Isabella LM, Houck JE, Pankow JF (1985) Migration of chlorophenolic compounds at the chemical waste disposal site at Alkali Lake, Oregon—2. Contaminant distributions, transport, and retardation. Groundwater 23(5):652–666

Jorgensen SL, Hannisdal B, Lanzén A, Baumberger T, Flesland K, Fonseca R, Øvreås L, Steen IH, Thorseth IH, Pedersen RB, Schleper C (2012) Correlating microbial community profiles with geochemical data in highly stratified sediments from the Arctic Mid-Ocean Ridge. Proc Natl Acad Sci USA 109(42):E2846–E2855

Justicia-Leon SD, Ritalahti KM, Mack EE, Löffler FE (2012) Dichloromethane fermentation by a *Dehalobacter* sp. in an enrichment culture derived from pristine river sediment. Appl Environ Microbiol 78(4):1288–1291

Kaster A-K, Mayer-Blackwell K, Pasarelli B, Spormann AM (2014) Single cell genomic study of *Dehalococcoidetes* species from deep-sea sediments of the Peruvian Margin. ISME J 8:1831–1842

Kawai M, Futagami T, Toyoda A, Takaki Y, Nishi S, Hori S, Arai W, Tsubouchi T, Morono Y, Uchiyama I, Ito T, Fujiyama A, Inagaki F, Takami H (2014) High frequency of phylogenetically diverse reductive dehalogenase-homologous genes in deep subseafloor sedimentary metagenomes. Front Microbiol 5:80

Keller S, Ruetz M, Kunze C, Kräutler B, Diekert G, Schubert T (2013) Exogenous 5, 6-dimethylbenzimidazole caused production of a non-functional tetrachloroethene reductive dehalogenase in *Sulfurospirillum multivorans*. Environ Microbiol 16(11):3361–3369

Kim S-H, Harzman C, Davis JK, Hutcheson R, Broderick JB, Marsh TL, Tiedje JM (2012) Genome sequence of *Desulfitobacterium hafniense* DCB-2, a Gram-positive anaerobe capable of dehalogenation and metal reduction. BMC Microbiol 12(1):21

King GM (1988) Dehalogenation in marine sediments containing natural sources of halophenols. Appl Environ Microbiol 54(12):3079–3085

Kittelmann S, Friedrich MW (2008a) Identification of novel perchloroethene-respiring microorganisms in anoxic river sediment by RNA-based stable isotope probing. Environ Microbiol 10(1):31–46

Kittelmann S, Friedrich MW (2008b) Novel uncultured *Chloroflexi* dechlorinate perchloroethene to trans-dichloroethene in tidal flat sediments. Environ Microbiol 10(6):1557–1570

Kjellerup BV, Paul P, Ghosh U, May HD, Sowers KR (2012) Spatial distribution of PCB dechlorinating bacteria and activities in contaminated soil. Appl Environ Soil Sci 2012:584970

Krajmalnik-Brown R, Hölscher T, Thomson IN, Saunders FM, Ritalahti KM, Löffler FE (2004) Genetic identification of a putative vinyl chloride reductase in *Dehalococcoides* sp. strain BAV1. Appl Environ Microbiol 70(10):6347–6351

Krajmalnik-Brown R, Sung Y, Ritalahti KM, Michael Saunders F, Löffler FE (2007) Environmental distribution of the trichloroethene reductive dehalogenase gene (*tceA*) suggests lateral gene transfer among *Dehalococcoides*. FEMS Microbiol Ecol 59(1):206–214

Kranzioch I, Stoll C, Holbach A, Chen H, Wang L, Zheng B, Norra S, Bi Y, Schramm K-W, Tiehm A (2013) Dechlorination and organohalide-respiring bacteria dynamics in sediment samples of the Yangtze Three Gorges Reservoir. Environ Sci Poll R 20(10):7046–7056

Kräutler B, Fieber W, Ostermann S, Fasching M, Ongania K-H, Gruber K, Kratky C, Mikl C, Siebert A, Diekert G (2003) The cofactor of tetrachloroethene reductive dehalogenase of *Dehalospirillum multivorans* is norpseudo-B_{12}, a new type of a natural corrinoid. Helv Chim Acta 86(11):3698–3716

Krumholz LR (1997) *Desulfuromonas chloroethenica* sp. nov. uses tetrachloroethylene and trichloroethylene as electron acceptors. Int J Syst Bacteriol 47(4):1262–1263

Krumholz LR, Sharp R, Fishbain SS (1996) A freshwater anaerobe coupling acetate oxidation to tetrachloroethylene dehalogenation. Appl Environ Microbiol 62(11):4108–4113

Krzmarzick MJ, Crary BB, Harding JJ, Oyerinde OO, Leri AC, Myneni SC, Novak PJ (2012) Natural niche for organohalide-respiring *Chloroflexi*. Appl Environ Microbiol 78(2):393–401

Krzmarzick MJ, Miller HR, Yan T, Novak PJ (2014) Novel *Firmicutes* group implicated in the dechlorination of two chlorinated xanthones, analogues of natural organochlorines. Appl Environ Microbiol 80(3):1210–1218

Kube M, Beck A, Zinder SH, Kuhl H, Reinhardt R, Adrian L (2005) Genome sequence of the chlorinated compound–respiring bacterium *Dehalococcoides* species strain CBDB1. Nat Biotechnol 23(10):1269–1273

Lanthier M, Villemur R, Lépine F, Bisaillon J-G, Beaudet R (2001) Geographic distribution of *Desulfitobacterium frappieri* PCP-1 and *Desulfitobacterium* spp. in soils from the province of Quebec, Canada. FEMS Microbiol Ecol 36(2–3):185–191

LaRoe SL, Fricker AD, Bedard DL (2014) *Dehalococcoides mccartyi* strain JNA in pure culture extensively dechlorinates aroclor 1260 according to polychlorinated biphenyl (PCB) Dechlorination Process N. Environ Sci Technol 48(16):9187–9196

Lee PK, Macbeth TW, Sorenson KS, Deeb RA, Alvarez-Cohen L (2008) Quantifying genes and transcripts to assess the in situ physiology of "*Dehalococcoides*" spp. in a trichloroethene-contaminated groundwater site. Appl Environ Microbiol 74(9):2728–2739

Lee PKH, Cheng D, Hu P, West KA, Dick GJ, Brodie EL, Andersen GL, Zinder SH, He J, Alvarez-Cohen L (2011) Comparative genomics of two newly isolated *Dehalococcoides* strains and an enrichment using a genus microarray. ISME J 5(6):1014–1024

Lee M, Low A, Zemb O, Koenig J, Michaelsen A, Manefield M (2012) Complete chloroform dechlorination by organochlorine respiration and fermentation. Environ Microbiol 14(4):883–894

Lee PKH, Cheng D, West KA, Alvarez-Cohen L, He J (2013) Isolation of two new *Dehalococcoides mccartyi* strains with dissimilar dechlorination functions and their characterization by comparative genomics via microarray analysis. Environ Microbiol 15(8):2293–2305

Lendvay JM, Dean SM, Adriaens P (1998) Temporal and spatial trends in biogeochemical conditions at a groundwater-surface water interface: Implications for natural bioattenuation. Environ Sci Technol 32(22):3472–3478

Lendvay J, Löffler FE, Dollhopf M, Aiello M, Daniels G, Fathepure B, Gebhard M, Heine R, Helton R, Shi J, Krajmalnik-Brown R, Major CL, Barcelona MJ, Petrovskis E, Hickey R, Tiedje JM, Adriaens P (2003) Bioreactive barriers: a comparison of bioaugmentation and biostimulation for chlorinated solvent remediation. Environ Sci Technol 37(7):1422–1431

Lévesque M-J, La Boissière S, Thomas J-C, Beaudet R, Villemur R (1997) Rapid method for detecting *Desulfitobacterium frappieri* strain PCP-1 in soil by the polymerase chain reaction. Appl Environ Microbiol 47(6):719–725

Leys D, Adrian L, Smidt H (2013) Organohalide respiration: microbes breathing chlorinated molecules. Phil Trans R Soc B 368(1616):20120316

Li Z, Yoshida N, Wang A, Nan J, Liang B, Zhang C, Zhang D, Suzuki D, Zhou X, Xiao Z, Katayama A (2015) Anaerobic mineralization of 2, 4, 6-tribromophenol to CO_2 by a synthetic microbial community comprising *Clostridium*, *Dehalobacter*, and *Desulfatiglans*. Bioresour Technol 176:225–232

Linkfield T, Tiedje J (1990) Characterization of the requirements and substrates for reductive dehalogenation by strain DCB-1. J Ind Microbiol 5(1):9–15

Löffler FE, Champine JE, Ritalahti KM, Sprague SJ, Tiedje JM (1997a) Complete reductive dechlorination of 1,2-dichloropropane by anaerobic bacteria. Appl Environ Microbiol 63(7):2870–2875

Löffler FE, Ritalahti KM, Tiedje JM (1997b) Dechlorination of chloroethenes is inhibited by 2-bromoethanesulfonate in the absence of methanogens. Appl Environ Microbiol 63(12):4982–4985

Löffler FE, Sun Q, Li J, Tiedje JM (2000) 16S rRNA gene-based detection of tetrachloroethene-dechlorinating *Desulfuromonas* and *Dehalococcoides* species. Appl Environ Microbiol 66(4):1369–1374

Löffler FE, Yan J, Ritalahti KM, Adrian L, Edwards EA, Konstantinidis KT, Müller JA, Fullerton H, Zinder SH, Spormann AM (2013) *Dehalococcoides mccartyi* gen. nov., sp. nov., obligately organohalide-respiring anaerobic bacteria relevant to halogen cycling and bioremediation, belong to a novel bacterial class, *Dehalococcoidia* classis nov., order *Dehalococcoidales* ord. nov. and family *Dehalococcoidaceae* fam. nov., within the phylum *Chloroflexi*. Int J Syst Evol Microbiol 63 (Pt 2):625–635

Lohner ST, Spormann AM (2013) Identification of a reductive tetrachloroethene dehalogenase in *Shewanella sediminis*. Phil Trans R Soc B 368(1616):20120326

Lookman R, Paulus D, Marnette E, Pijls C, Ryngaert A, Diels L, Volkering F (2007) Ground water transfer initiates complete reductive dechlorination in a PCE-contaminated aquifer. Ground Water Monit Remediat 27(3):65–74

Lorah MM, Olsen LD (1999) Natural attenuation of chlorinated volatile organic compounds in a freshwater tidal wetland: field evidence of anaerobic biodegradation. Water Res 35(12):3811–3827

Lovley DR, Giovannoni SJ, White DC, Champine JE, Phillips E, Gorby YA, Goodwin S (1993) *Geobacter metallireducens* gen. nov. sp. nov., a microorganism capable of coupling the complete oxidation of organic compounds to the reduction of iron and other metals. Arch Microbiol 159(4):336–344

Luijten ML, de Weert J, Smidt H, Boschker HT, de Vos WM, Schraa G, Stams AJ (2003) Description of *Sulfurospirillum halorespirans* sp. nov., an anaerobic, tetrachloroethene-respiring bacterium, and transfer of *Dehalospirillum multivorans* to the genus *Sulfurospirillum* as *Sulfurospirillum multivorans* comb. nov. Int J Syst Evol Microbiol 53 (3):787–793

Madsen T, Licht D (1992) Isolation and characterization of an anaerobic chlorophenol-transforming bacterium. Appl Environ Microbiol 58(9):2874–2878

Mägli A, Wendt M, Leisinger T (1996) Isolation and characterization of *Dehalobacterium formicoaceticum* gen. nov. sp. nov., a strictly anaerobic bacterium utilizing dichloromethane as source of carbon and energy. Arch Microbiol 166(2):101–108

Magnuson JK, Stern RV, Gossett JM, Zinder SH, Burris DR (1998) Reductive dechlorination of tetrachloroethene to ethene by a two-component enzyme pathway. Appl Environ Microbiol 64(4):1270–1275

Magnuson JK, Romine MF, Burris DR, Kingsley MT (2000) Trichloroethene reductive dehalogenase from *Dehalococcoides ethenogenes*: Sequence of *tceA* and substrate range characterization. Appl Environ Microbiol 66(12):5141–5147

Maillard J, Schumacher W, Vazquez F, Regeard C, Hagen WR, Holliger C (2003) Characterization of the corrinoid iron-sulfur protein tetrachloroethene reductive dehalogenase of *Dehalobacter restrictus*. Appl Environ Microbiol 69(8):4628–4638

Maillard J, Charnay MP, Regeard C, Rohrbach-Brandt E, Rouzeau-Szynalski K, Rossi P, Holliger C (2011) Reductive dechlorination of tetrachloroethene by a stepwise catalysis of different organohalide respiring bacteria and reductive dehalogenases. Biodegradation 22(5):949–960

Major DW, McMaster ML, Cox EE, Edwards EA, Dworatzek SM, Hendrickson ER, Starr MG, Payne JA, Buonamici LW (2002) Field demonstration of successful bioaugmentation to achieve dechlorination of tetrachloroethene to ethene. Environ Sci Technol 36(23):5106–5116

Maphosa F, de Vos WM, Smidt H (2010) Exploiting the ecogenomics toolbox for environmental diagnostics of organohalide-respiring bacteria. Trends Biotechnol 28(6):308–316

Maphosa F, van Passel MWJ, de Vos WM, Smidt H (2012) Metagenome analysis reveals yet unexplored reductive dechlorinating potential of *Dehalobacter* sp. E1 growing in co-culture with *Sedimentibacter* sp. Environ Microbiol Reports 4 (6):604–616

Marzorati M, De Ferra F, Van Raemdonck H, Borin S, Allifranchini E, Carpani G, Serbolisca L, Verstraete W, Boon N, Daffonchio D (2007) A novel reductive dehalogenase, identified in a contaminated groundwater enrichment culture and in *Desulfitobacterium dichloroeliminans* strain DCA1, is linked to dehalogenation of 1, 2-dichloroethane. Appl Environ Microbiol 73(9):2990–2999

May HD, Miller GS, Kjellerup BV, Sowers KR (2008) Dehalorespiration with polychlorinated biphenyls by an anaerobic ultramicrobacterium. Appl Environ Microbiol 74(7):2089–2094

Maymo-Gatell X, Y-t Chien, Gossett JM, Zinder SH (1997) Isolation of a bacterium that reductively dechlorinates tetrachloroethene to ethene. Science 276(5318):1568–1571

McMurdie PJ, Behrens SF, Müller JA, Göke J, Ritalahti KM, Wagner R, Goltsman E, Lapidus A, Holmes S, Löffler FE, Spormann AM (2009) Localized plasticity in the streamlined genomes of vinyl chloride respiring *Dehalococcoides*. PLoS Genet 5(11):e1000714

McMurdie PJ, Hug LA, Edwards EA, Holmes S, Spormann AM (2011) Site-specific mobilization of vinyl chloride respiration islands by a mechanism common in *Dehalococcoides*. BMC Genom 12(1):287

McNamara PJ, Krzmarzick MJ (2013) Triclosan enriches for *Dehalococcoides*-like *Chloroflexi* in anaerobic soil at environmentally relevant concentrations. FEMS Microbiol Lett 344(1):48–52

Men Y, Lee PK, Harding KC, Alvarez-Cohen L (2013) Characterization of four TCE-dechlorinating microbial enrichments grown with different cobalamin stress and methanogenic conditions. Appl Microbiol Biotechnol 97:1–12

Men Y, Seth EC, Yi S, Crofts TS, Allen RH, Taga ME, Alvarez-Cohen L (2014) Identification of specific corrinoids reveals corrinoid modification in dechlorinating microbial communities. Environ Microbiol 17(12):4873–4884

Miller E, Wohlfarth G, Diekert G (1997) Comparative studies on tetrachloroethene reductive dechlorination mediated by *Desulfitobacterium* sp. strain PCE-S. Arch Microbiol 168(6):513–519

Miller E, Wohlfarth G, Diekert G (1998) Purification and characterization of the tetrachloroethene reductive dehalogenase of strain PCE-S. Arch Microbiol 169(6):497–502

Miller GS, Milliken CE, Sowers KR, May HD (2005) Reductive dechlorination of tetrachloroethene to *trans*-dichloroethene and *cis*-dichloroethene by PCB-dechlorinating bacterium DF-1. Environ Sci Technol 39(8):2631–2635

Moe WM, Yan J, Nobre MF, da Costa MS, Rainey FA (2009) *Dehalogenimonas lykanthroporepellens* gen. nov., sp. nov., a reductively dehalogenating bacterium isolated from chlorinated solvent-contaminated groundwater. Int J Syst Evol Microbiol 59(11):2692–2697

Mukherjee K, Bowman KS, Rainey FA, Siddaramappa S, Challacombe JF, Moe WM (2014) *Dehalogenimonas lykanthroporepellens* BL-DC-9T simultaneously transcribes many rdhA genes during organohalide respiration with 1,2-DCA, 1,2-DCP and 1,2,3-TCP as electron acceptors. FEMS Microbiol Lett 351:111–118

Müller JA, Rosner BM, Von Abendroth G, Meshulam-Simon G, McCarty PL, Spormann AM (2004) Molecular identification of the catabolic vinyl chloride reductase from *Dehalococcoides* sp. strain VS and its environmental distribution. Appl Environ Microbiol 70(8):4880–4888

Nelson JJJ, Zinder S (2014) Dehalogenation of chlorobenzenes, dichlorotoluene, and tetrachloroethene by three *Dehalobacter* spp. Environ Sci Technol 48(7):3776–3782

Nelson JL, Fung JM, Cadillo-Quiroz H, Cheng X, Zinder SH (2011) A role for *Dehalobacter* spp. in the reductive dehalogenation of dichlorobenzenes and monochlorobenzene. Environ Sci Technol 45(16):6806–6813

Niggemyer A, Spring S, Stackebrandt E, Rosenzweig RF (2001) Isolation and characterization of a novel As (V)-reducing bacterium: implications for arsenic mobilization and the genus *Desulfitobacterium*. Appl Environ Microbiol 67(12):5568–5580

Nijenhuis I, Nikolausz M, Köth A, Felföldi T, Weiss H, Drangmeister J, Großmann J, Kästner M, Richnow H-H (2007) Assessment of the natural attenuation of chlorinated ethenes in an anaerobic contaminated aquifer in the Bitterfeld/Wolfen area using stable isotope techniques, microcosm studies and molecular biomarkers. Chemosphere 67(2):300–311

Nonaka H, Keresztes G, Shinoda Y, Ikenaga Y, Abe M, Naito K, Inatomi K, Furukawa K, Inui M, Yukawa H (2006) Complete genome sequence of the dehalorespiring bacterium *Desulfitobacterium hafniense* Y51 and comparison with *Dehalococcoides ethenogenes* 195. J Bacteriol 188(6):2262–2274

Øfjord G, Puhakka JA, Ferguson JF (1994) Reductive dechlorination of Aroclor 1254 by marine sediment cultures. Environ Sci Technol 28(13):2286–2294

Padilla-Crespo E, Yan J, Swift C, Wagner DD, Chourey K, Hettich RL, Ritalahti KM, Löffler FE (2014) Identification and environmental distribution of *dcpA*, which encodes the reductive dehalogenase catalyzing the dichloroelimination of 1,2-dichloropropane to propene in organohalide-respiring *Chloroflexi*. Appl Environ Microbiol 80(3):808–818

Parthasarathy A, Stich TA, Lohner ST, Lesnefsky A, Britt RD, Spormann AM (2015) Biochemical and EPR-spectroscopic investigation into heterologously expressed vinyl chloride reductive dehalogenase (VcrA) from *Dehalococcoides mccartyi* strain VS. J Am Chem Soc 137(10):3525–3532

Payne RB, May HD, Sowers KR (2011) Enhanced reductive dechlorination of polychlorinated biphenyl impacted sediment by bioaugmentation with a dehalorespiring bacterium. Environ Sci Technol 45(20):8772–8779

Payne KA, Quezada CP, Fisher K, Dunstan MS, Collins FA, Sjuts H, Levy C, Hay S, Rigby SE, Leys D (2014) Reductive dehalogenase structure suggests a mechanism for B_{12}-dependent dehalogenation. Nature 517:513–516

Pöritz M, Goris T, Wubet T, Tarkka MT, Buscot F, Nijenhuis I, Lechner U, Adrian L (2013) Genome sequences of two dehalogenation specialists–*Dehalococcoides mccartyi* strains BTF08 and DCMB5 enriched from the highly polluted Bitterfeld region. FEMS Microbiol Lett 343(2):101–104

Richardson RE (2013) Genomic insights into organohalide respiration. Curr Opin Biotechnol 24:498–505

Richardson RE, Bhupathiraju VK, Song DL, Goulet TA, Alvarez-Cohen L (2002) Phylogenetic characterization of microbial communities that reductively dechlorinate TCE based upon a combination of molecular techniques. Environ Sci Technol 36(12):2652–2662

Ritalahti KM, Löffler FE (2004) Populations implicated in anaerobic reductive dechlorination of 1,2-dichloropropane in highly enriched bacterial communities. Appl Environ Microbiol 70(7):4088–4095

Ritalahti KM, Amos BK, Sung Y, Wu Q, Koenigsberg SS, Löffler FE (2006) Quantitative PCR targeting 16S rRNA and reductive dehalogenase genes simultaneously monitors multiple *Dehalococcoides* strains. Appl Environ Microbiol 72(4):2765–2774

Rosner B, McCarty P, Spormann A (1997) In vitro studies on reductive vinyl chloride dehalogenation by an anaerobic mixed culture. Appl Environ Microbiol 63(11):4139–4144

Rotaru DEH, Franks AE, Orellana R, Risso C, Nevin KP (2011) *Geobacter*: the microbe electric's physiology, ecology, and practical applications. Adv Microb Physiol 59:1

Rouzeau-Szynalski K, Maillard J, Holliger C (2011) Frequent concomitant presence of *Desulfitobacterium* spp. and "*Dehalococcoides*" spp. in chloroethene-dechlorinating microbial communities. Appl Microbiol Biotechnol 90(1):361–368

Rowe AR, Lazar BJ, Morris RM, Richardson RE (2008) Characterization of the community structure of a dechlorinating mixed culture and comparisons of gene expression in planktonic and biofloc-associated "*Dehalococcoides*" and *Methanospirillum* species. Appl Environ Microbiol 74(21):6709–6719

Rupakula A, Kruse T, Boeren S, Holliger C, Smidt H, Maillard J (2013) The restricted metabolism of the obligate organohalide respiring bacterium *Dehalobacter restrictus*: lessons from tiered functional genomics. Phil Trans R Soc B 368 (1616)

Rupakula A, Lu Y, Kruse T, Boeren S, Holliger C, Smidt H, Maillard J (2014) Functional genomics of corrinoid starvation in the organohalide-respiring bacterium *Dehalobacter restrictus* strain PER-K23. Front Microbiol 5:751

Sanford RA, Cole JR, Löffler FE, Tiedje JM (1996) Characterization of *Desulfitobacterium chlororespirans* sp. nov., which grows by coupling the oxidation of lactate to the reductive dechlorination of 3-chloro-4-hydroxybenzoat. Appl Environ Microbiol 62(10):3800–3808

Sanford RA, Cole JR, Tiedje JM (2002) Characterization and description of *Anaeromyxobacter dehalogenans* gen. nov., sp. nov., an aryl-halorespiring facultative anaerobic myxobacterium. Appl Environ Microbiol 68(2):893–900

Scheutz C, Durant Nd, Dennis P, Hansen MH, Jørgensen T, Jakobsen R, Ee Cox, Bjerg PL (2008) Concurrent ethene generation and growth of *Dehalococcoides* containing vinyl chloride reductive dehalogenase genes during an enhanced reductive dechlorination field demonstration. Environ Sci Technol 42(24):9302–9309

Schneidewind U, Haest PJ, Atashgahi S, Maphosa F, Hamonts K, Maesen M, Calderer M, Seuntjens P, Smidt H, Springael D, Dejonghe W (2014) Kinetics of dechlorination by *Dehalococcoides mccartyi* using different carbon sources. J Contam Hydrol 157:25–36

Scholz-Muramatsu H, Neumann A, Messmer M, Moore E, Diekert G (1995) Isolation and characterization of *Dehalospirillum multivorans* gen. nov., sp. nov., a tetrachloroethene-utilizing, strictly anaerobic bacterium. Arch Microbiol 163(1):48–56

Schumacher W, Kroneck PM, Pfennig N (1992) Comparative systematic study on "Spirillum" 5175. *Campylobacter* and *Wolinella* species. Arch Microbiol 158(4):287–293

Seshadri R, Adrian L, Fouts DE, Eisen JA, Phillippy AM, Methe BA, Ward NL, Nelson WC, Deboy RT, Khouri HM, Kolonay JF, Dodson RJ, Daugherty SC, Brinkac LM, Sullivan SA, Madupu R, Nelson KE, Kang KH, Impraim M, Tran K, Robinson JM, Forberger HA, Fraser CM, Zinder SH, Heidelberg JF (2005) Genome sequence of the PCE-dechlorinating bacterium *Dehalococcoides ethenogenes*. Science 307(5706):105–108

Shani N, Rossi P, Holliger C (2013) Correlations between environmental variables and bacterial community structures suggest Fe (III) and vinyl chloride reduction as antagonistic terminal electron-accepting processes. Environ Sci Technol 47(13):6836–6845

Shelobolina ES, Vanpraagh CG, Lovley DR (2003) Use of ferric and ferrous iron containing minerals for respiration by *Desulfitobacterium frappieri*. Geomicrobiol J 20(2):143–156

Shelton DR, Tiedje JM (1984) Isolation and partial characterization of bacteria in an anaerobic consortium that mineralizes 3-chlorobenzoic acid. Appl Environ Microbiol 48(4):840–848

Siddaramappa S, Challacombe JF, Delano SF, Green LD, Daligault H, Bruce D, Detter C, Tapia R, Han S, Goodwin L, Han J, Woyke T, Pitluck S, Pennacchio L, Nolan M, Miriam L, Yun-Juan C, Kyrpides NC, Ovchinnikova G, Hauser L, Lapidus A, Yan J, Bowman KS, da Costa MS, Rainey FA, Moe WM (2012) Complete genome sequence of *Dehalogenimonas lykanthroporepellens* type strain (BL-DC-9T) and comparison to "*Dehalococcoides*" strains. Stand Genomic Sci 6(2):251

Smatlak CR, Gossett JM, Zinder SH (1996) Comparative kinetics of hydrogen utilization for reductive dechlorination of tetrachloroethene and methanogenesis in an anaerobic enrichment culture. Environ Sci Technol 30(9):2850–2858

Smits TH, Devenoges C, Szynalski K, Maillard J, Holliger C (2004) Development of a real-time PCR method for quantification of the three genera *Dehalobacter*, *Dehalococcoides*, and *Desulfitobacterium* in microbial communities. J Microbiol Methods 57(3):369–378

Stams AJ, Plugge CM (2009) Electron transfer in syntrophic communities of anaerobic bacteria and archaea. Nat Rev Microbiol 7(8):568–577

Suflita JM, Horowitz A, Shelton DR, Tiedje JM (1982) Dehalogenation: a novel pathway for the anaerobic biodegradation of haloaromatic compounds. Science 218(4577):1115–1117

Sun B, Cole JR, Sanford RA, Tiedje JM (2000) Isolation and characterization of *Desulfovibrio dechloracetivorans* sp. nov., a marine dechlorinating bacterium growing by coupling the oxidation of acetate to the reductive dechlorination of 2-chlorophenol. Appl Environ Microbiol 66(6):2408–2413

Sun B, Cole JR, Tiedje JM (2001) *Desulfomonile limimaris* sp. nov., an anaerobic dehalogenating bacterium from marine sediments. Int J Syst Evol Microbiol 51(2):365–371

Sun B, Griffin BM, Ayala-del-Rio HL, Hashsham SA, Tiedje JM (2002) Microbial dehalorespiration with 1,1,1-trichloroethane. Science 298(5595):1023–1025

Sung Y, Ritalahti KM, Sanford RA, Urbance JW, Flynn SJ, Tiedje JM, Löffler FE (2003) Characterization of two tetrachloroethene-reducing, acetate-oxidizing anaerobic bacteria and their description as *Desulfuromonas michiganensis* sp. nov. Appl Environ Microbiol 69(5):2964–2974

Sung Y, Fletcher KE, Ritalahti KM, Apkarian RP, Ramos-Hernández N, Sanford RA, Mesbah NM, Löffler FE (2006a) *Geobacter lovleyi* sp. nov. strain SZ, a novel metal-reducing and tetrachloroethene-dechlorinating bacterium. Appl Environ Microbiol 72(4):2775–2782

Sung Y, Ritalahti K, Apkarian R, Loffler F (2006b) Quantitative PCR confirms purity of strain GT, a novel trichloroethene-to-ethene-respiring *Dehalococcoides* isolate. Appl Environ Microbiol 72(3):1980–1987

Suyama A, Iwakiri R, Kai K, Tokunaga T, Sera N, Furukawa K (2001) Isolation and characterization of *Desulfitobacterium* sp strain Y51 capable of efficient dehalogenation of tetrachloroethene and polychloroethanes. Biosci Biotechnol Biochem 65(7):1474–1481

Suyama A, Yamashita M, Yoshino S, Furukawa K (2002) Molecular characterization of the PceA reductive dehalogenase of *Desulfitobacterium* sp. strain Y51. J Bacteriol 184(13):3419–3425

Suzuki D, Ueki A, Amaishi A, Ueki K (2008) *Desulfoluna butyratoxydans* gen. nov., sp. nov., a novel Gram-negative, butyrate-oxidizing, sulfate-reducing bacterium isolated from an estuarine sediment in Japan. Int J Syst Evol Microbiol 58(4):826–832

Taş N, Van Eekert MH, Schraa G, Zhou J, De Vos WM, Smidt H (2009) Tracking functional guilds:"*Dehalococcoides*" spp. in European river basins contaminated with hexachlorobenzene. Appl Environ Microbiol 75(14):4696–4704

Thibodeau J, Gauthier A, Duguay M, Villemur R, Lépine F, Juteau P, Beaudet R (2004) Purification, cloning, and sequencing of a 3, 5-dichlorophenol reductive dehalogenase from *Desulfitobacterium frappieri* PCP-1. Appl Environ Microbiol 70(8):4532–4537

Thomas SH, Wagner RD, Arakaki AK, Skolnick J, Kirby JR, Shimkets LJ, Sanford RA, Löffler FE (2008) The mosaic genome of *Anaeromyxobacter dehalogenans* strain 2CP-C suggests an aerobic common ancestor to the delta-*Proteobacteria*. PLoS ONE 3(5):e2103

Thornton SF, Bright MI, Lerner DN, Tellam JH (2000) Attenuation of landfill leachate by UK Triassic Sandstone aquifer materials: 2. Sorption and degradation of organic pollutants in laboratory columns. J Contam Hydrol 43(3):355–383

Tobiszewski M, Namieśnik J (2012) Abiotic degradation of chlorinated ethanes and ethenes in water. Environ Sci Poll R 19(6):1994–2006

Tsukagoshi N, Ezaki S, Uenaka T, Suzuki N, Kurane R (2006) Isolation and transcriptional analysis of novel tetrachloroethene reductive dehalogenase gene from *Desulfitobacterium* sp. strain KBC1. Appl Microbiol Biotechnol 69(5):543–553

Utkin I, Woese C, Wiegel J (1994) Isolation and characterization of *Desulfitobacterium dehalogenans* gen. nov., sp. nov., an anaerobic bacterium which reductively dechlorinates chlorophenolic compounds. Int J Syst Bacteriol 44(4):612–619 van de Pas BA, Smidt H, Hagen WR, van der Oost J, Schraa G, Stams AJ, de Vos WM (1999) Purification and molecular characterization of *ortho*-chlorophenol reductive dehalogenase, a key enzyme of halorespiration in *Desulfitobacterium dehalogenans*. J Biol Chem 274(29):20287–20292

van de Pas BA, Gerritse J, de Vos WM, Schraa G, Stams AJ (2001) Two distinct enzyme systems are responsible for tetrachloroethene and chlorophenol reductive dehalogenation in *Desulfitobacterium* strain PCE1. Arch Microbiol 176(3):165–169

van der Zaan B, Hannes F, Hoekstra N, Rijnaarts H, de Vos WM, Smidt H, Gerritse J (2010) Correlation of *Dehalococcoides* 16S rRNA and chloroethene-reductive dehalogenase genes with geochemical conditions in chloroethene-contaminated groundwater. Appl Environ Microbiol 76(3):843–850

Van der Zee FP, Cervantes FJ (2009) Impact and application of electron shuttles on the redox (bio) transformation of contaminants: a review. Biotechnol Adv 27(3):256–277

van Doesburg W, Eekert MH, Middeldorp PJ, Balk M, Schraa G, Stams AJ (2005) Reductive dechlorination of β-hexachlorocyclohexane (β-HCH) by a *Dehalobacter* species in coculture with a *Sedimentibacter* sp. FEMS Microbiol Ecol 54(1):87–95

Villemur R, Lanthier M, Beaudet R, Lépine F (2006) The *Desulfitobacterium* genus. FEMS Microbiol Rev 30(5):706–733

Wagner DD, Hug LA, Hatt JK, Spitzmiller MR, Padilla-Crespo E, Ritalahti KM, Edwards EA, Konstantinidis KT, Löffler FE (2012) Genomic determinants of organohalide-respiration in *Geobacter lovleyi*, an unusual member of the *Geobacteraceae*. BMC Genom 13(1):200

Wang S, He J (2013a) Dechlorination of commercial PCBs and other multiple halogenated compounds by a sediment-free culture containing *Dehalococcoides* and *Dehalobacter*. Environ Sci Technol 47(18):10526–10534

Wang S, He J (2013b) Phylogenetically distinct bacteria involve extensive dechlorination of Aroclor 1260 in sediment-free cultures. PLoS ONE 8(3):e59178

Wang S, Chng KR, Wilm A, Zhao S, Yang K-L, Nagarajan N, He J (2014a) Genomic characterization of three unique *Dehalococcoides* that respire on persistent polychlorinated biphenyls. Proc Natl Acad Sci USA 111(33):12103–12108

Wang S, Zhang W, Yang K-L, He J (2014b) Isolation and characterization of a novel *Dehalobacter* species strain TCP1 that reductively dechlorinates 2,4,6-trichlorophenol. Biodegradation 25(2):313–323

Wasmund K, Schreiber L, Lloyd KG, Petersen DG, Schramm A, Stepanauskas R, Jørgensen BB, Adrian L (2014) Genome sequencing of a single cell of the widely distributed marine subsurface *Dehalococcoidia*, phylum *Chloroflexi*. ISME J 8(2):383–397

Wasmund K, Algora C, Müller J, Krüger M, Lloyd KG, Reinhard R, Adrian L (2015) Development and application of primers for the class *Dehalococcoidia* (phylum *Chloroflexi*) reveals highly diverse and stratified sub-group distributions in the marine subsurface. Environ Microbiol 17(10):3540–3556

Watts JE, Fagervold SK, May HD, Sowers KR (2005) A PCR-based specific assay reveals a population of bacteria within the *Chloroflexi* associated with the reductive dehalogenation of polychlorinated biphenyls. Microbiol 151(6):2039–2046

Webster G, Parkes RJ, Cragg BA, Newberry CJ, Weightman AJ, Fry JC (2006) Prokaryotic community composition and biogeochemical processes in deep subseafloor sediments from the Peru Margin. FEMS Microbiol Ecol 58(1):65–85

Wild A, Hermann R, Leisinger T (1996) Isolation of an anaerobic bacterium which reductively dechlorinates tetrachloroethene and trichloroethene. Biodegradation 7(6):507–511

Wu Q, Sowers KR, May HD (2000) Establishment of a polychlorinated biphenyl-dechlorinating microbial consortium, specific for doubly flanked chlorines, in a defined, sediment-free medium. Appl Environ Microbiol 66(1):49–53

Wu Q, Milliken CE, Meier GP, Watts JE, Sowers KR, May HD (2002a) Dechlorination of chlorobenzenes by a culture containing bacterium DF-1, a PCB dechlorinating microorganism. Environ Sci Technol 36(15):3290–3294

Wu Q, Watts JE, Sowers KR, May HD (2002b) Identification of a bacterium that specifically catalyzes the reductive dechlorination of polychlorinated biphenyls with doubly flanked chlorines. Appl Environ Microbiol 68(2):807–812

Yan J, Rash B, Rainey F, Moe W (2009) Isolation of novel bacteria within the *Chloroflexi* capable of reductive dechlorination of 1,2,3-trichloropropane. Environ Microbiol 11(4):833–843

Yan J, Ritalahti KM, Wagner DD, Löffler FE (2012) Unexpected specificity of interspecies cobamide transfer from *Geobacter* spp. to organohalide-respiring *Dehalococcoides mccartyi* strains. Appl Environ Microbiol 78(18):6630–6636

Yi S, Seth EC, Men Y-J, Stabler SP, Allen RH, Alvarez-Cohen L, Taga ME (2012) Versatility in corrinoid salvaging and remodeling pathways supports corrinoid-dependent metabolism in *Dehalococcoides mccartyi*. Appl Environ Microbiol 78(21):7745–7752

Yoshida N, Asahi K, Sakakibara Y, Miyake K, Katayama A (2007a) Isolation and quantitative detection of tetrachloroethene (PCE)-dechlorinating bacteria in unsaturated subsurface soils contaminated with chloroethenes. J Biosci Bioeng 104(2):91–97

Yoshida N, Yoshida Y, Handa Y, Kim H-K, Ichihara S, Katayama A (2007b) Polyphasic characterization of a PCP-to-phenol dechlorinating microbial community enriched from paddy soil. Sci Total Environ 381(1):233–242

Yoshida N, Ye L, Baba D, Katayama A (2009) A novel *Dehalobacter* species is involved in extensive 4,5,6,7-tetrachlorophthalide dechlorination. Appl Environ Microbiol 75(8):2400–2405

Zanaroli G, Balloi A, Negroni A, Borruso L, Daffonchio D, Fava F (2012) A *Chloroflexi* bacterium dechlorinates polychlorinated biphenyls in marine sediments under in situ-like biogeochemical conditions. J Hazard Mater 209:449–457

Zhang C, Katayama A (2012) Humin as an electron mediator for microbial reductive dehalogenation. Environ Sci Technol 46(12):6575–6583

Zhao JS, Spain J, Thiboutot S, Ampleman G, Greer C, Hawari J (2004) Phylogeny of cyclic nitramine-degrading psychrophilic bacteria in marine sediment and their potential role in the natural attenuation of explosives. FEMS Microbiol Ecol 49(3):349–357

Zhao J-S, Manno D, Beaulieu C, Paquet L, Hawari J (2005) *Shewanella sediminis* sp. nov., a novel Na^+-requiring and hexahydro-1,3,5-trinitro-1,3,5-triazine-degrading bacterium from marine sediment. Int J Syst Evol Microbiol 55(4):1511–1520

Zhao J-S, Deng Y, Manno D, Hawari J (2010) *Shewanella* spp. genomic evolution for a cold marine lifestyle and in-situ explosive biodegradation. PLoS ONE 5(2):e9109

Chapter 6
The Genus *Dehalococcoides*

Stephen H. Zinder

Abstract *Dehalococcoides*, now called *Dehalococcoides mccartyi*, was first discovered in an enrichment culture from sewage sludge that reductively dechlorinated the groundwater pollutants tetrachloroethene (PCE) and trichloroethene (TCE) to vinyl chloride (VC) and ethene, in contrast to other organohalide-respiring bacteria that dechlorinated PCE and TCE only as far as dichloroethenes (DCEs). The first isolate, strain 195, was a tiny disk-shaped bacterium in the phylum *Chloroflexi* that had an S-layer protein subunit cell wall lacking peptidoglycan. It was a strict anaerobe using only H_2 as the electron donor and organohalides as respiratory electron acceptors. Other *D. mccartyi* strains are similar and use a variety of halogenated aliphatic and aromatic compounds as electron acceptors. The genomes of *D. mccartyi* are highly streamlined, varying from 1.34 to 1.5 MB, yet contain 10–36 different copies of *rdhAB* operons predicted to encode reductive dehalogenases (RDases), most with adjacent genes predicted to encode transcriptional regulators, indicating that organochloride respiration is a highly evolved and regulated process in *D. mccartyi*. The presence of *D. mccartyi* at chloroethene-contaminated groundwater sites appears necessary for dechlorination of PCE and TCE past DCEs, and molecular tests for *D. mccartyi* and its associated *rdhAB* genes have become part of contaminated site characterization. Moreover, *D. mccartyi*-containing cultures have been commercially developed for bioaugmentation of those sites to abet dechlorination to ethene, especially cultures that contain *D. mccartyi* strains that can efficiently convert VC to nontoxic ethene in a respiratory process, like strains BAV1 and VS. This tiny unusual bacterium is now considered to be an important player in the restoration of chloroethene-contaminated sites.

S.H. Zinder (✉)
Department of Microbiology, Cornell University,
Wing Hall, 123 Wing Dr., Ithaca, NY 14853, USA
e-mail: shz1@cornell.edu

L. Adrian and F.E. Löffler (eds.), *Organohalide-Respiring Bacteria*,
DOI 10.1007/978-3-662-49875-0_6

Abbreviations

1,2-DCA	1,2-dichloroethane
1,2-DCP	1,2-dichloropropane
3-Cl-4-OHPA	3-chloro-4-hydroxyphenylacetic acid
AQDS	Anthraquinone-2,6-disulfonate
BP	Bromophenol
Bromoxynil	3,5-dibromo-4-hydroxybenzonitrile
CD	Carbon dichloride
CF	Chloroform
CP	Chlorophenol
CT	Carbon tetrachloride
Cysteate	Alanine-3-sulfonate
DBP	Dibromophenol
DCA	Dichloroethane
DCHQ	Dichlorohydroquinone
DCP	Dichlorophenol
DMSO	Dimethyl sulfoxide
HCB	Hexachlorobenzene
Ioxynil	3,5-diiodo-4-hydroxybenzonitrile
Isethionate	2-hydroxyethanesulfonate
OHRB	Organohalide-respiring bacteria
PCE	Tetrachloroethene
PCP	Pentachlorophenol
RDase	Reductive dehalogenase
rdh	Reductive dehalogenase homologous genes
TCA	Trichloroethane
TCE	Trichloroethene
TCHQ	2,3,5,6-tetrachlorohydroquinone
TCMP	2,3,5,6-tetrachloro-4-methoxyphenol
TCP	Trichlorophenol
TeCP	Tetrachlorophenol
VC	Vinyl chloride

6.1 Discovery

I will begin this chapter with a more personal history of the discovery and isolation of *Dehalococcoides*, which took over 7 years and required some paradigm shifts in my and others thinking along the way, followed by more general descriptions of members of this fascinating genus.

6.1.1 In the Beginning

The discovery of *Dehalococcoides* began with my colleague, James Gossett, in the School of Civil and Environmental Engineering at Cornell. During a sabbatical in 1984–1985 at the USAF Armstrong Laboratory at Tyndall Air Force Base in Florida, USA, he became interested in chlorinated solvents, which were groundwater contaminants at numerous military bases. While there, he developed a simple protocol to determine the Henry's law constants for these solvents (Gossett 1987), information important in designing air stripping processes used to remove solvents from groundwater in pump and treat systems, the predominant remediation process used at that time.

Having done research on anaerobic digestion, Gossett also became interested in the reductive dechlorination of these solvents by anaerobes. At that time, reductive dechlorination was considered to be a slow cometabolic process carried out by reduced transition metal containing cofactors like vitamin B_{12}, coenzyme F_{430}, or hemes found in anaerobes like methanogens, acetogens, or sulfate-reducing bacteria (Fathepure and Boyd 1988; Gantzer and Wackett 1991; Wood et al. 1968). Moreover, reductive dechlorination of chloroethenes apparently stopped at vinyl chloride (VC), a known human carcinogen, so it was not considered to be a viable remediation process. Gossett's first tries at obtaining significant activity from anaerobic digestor sludge in Florida were unsuccessful.

6.1.2 Success in Ithaca

Upon his return to Cornell, Gossett had a graduate student David Freedman and started some enrichments using sludge from the Ithaca sewage treatment plant. Two types of enrichments showed activity, those with dichloromethane (Freedman and Gossett 1991), which will not be discussed here, and those with tetrachloroethene (PCE). According to the operator of the sewage plant (Gleason, personal communication), at the time of Freedman's enrichment, the anaerobic digestors were having problems with chlorinated solvents, with dry cleaners, industry, and Cornell University as potential sources, so the presence of *Dehalococcoides* in those digestors is not unreasonable. This aged sewage treatment plant was decommissioned in 1988, and its buildings now serve as a science museum.

A laboratory-scale anaerobic digestor was established by Freedman, and serum bottle microcosms containing 100 ml of the sludge were derived from it. Twenty-five PCE doses of 3–4.5 μmol/L were added to the microcosms over 55 days, and organic matter in the sludge served as the electron donor. After about 15 days, DCEs began to accumulate, followed by VC at 30 days, and after 50 days, ethene was detected. When PCE feeding was stopped, VC was slowly converted stoichiometrically to ethene. Freedman and Gossett were able to detect ethene because they were using a gas chromatograph with a flame ionization detector (FID), which is sensitive to essentially all organic compounds, while previous researchers had used electron capture detectors or electrolytic detectors, which are highly sensitive to organochlorides, but do not detect non-chlorinated compounds like ethene. Thus, if ethene was produced in previous studies, it would have been missed.

Material from these microcosms was transferred into a mineral growth medium amended with 50 mg/L yeast extract as a nutrient, and either glucose or methanol as an electron donor, and with multiple doses of ~4 μM PCE. Cultures with either electron donor converted PCE doses to VC with little buildup of TCE or DCEs, but only the methanol amended ones converted VC to ethene. Thus, began the use of methanol as an electron donor for reductive dechlorination of chloroethenes, a substrate used often in subsequent studies of chloroethene degradation (Duhamel et al. 2002; Rossetti et al. 2003). The cultures could be transferred multiple times, showing the reaction was sustainable.

Since finding PCE conversion to ethene was novel, Freedman and Gossett wanted to confirm its origin and identification. They demonstrated that label from ^{14}C-PCE could be found in VC and ethene fractions, verifying that PCE was indeed the precursor of ethene, and used the Cornell gas-chromatograph mass-spectrometry facility to indisputably identify ethene as the product. Despite these measures, the manuscript describing these findings initially had a difficult time in review because of disbelief of some of the reviewers, including the belief that an FID was unsuitable for measuring chloroethenes (Gossett, personal communication).

I served on Freedman's thesis PhD committee representing microbiology and became interested in culturing organisms potentially responsible for PCE dechlorination. Since a PCE-reducing methanogen conformed to the conventional wisdom at that time, we isolated a methanol-utilizing methanogen from a high dilution from the culture. However, it did not dechlorinate PCE to any extent, and I suspected that another type of organism was involved in dechlorination. Because the cultures were being fed micromolar amounts of PCE but were converting millimolar amounts of methanol to methane, I was concerned that the dechlorinators would be a tiny minority in the culture, and finding them would be like looking for a "needle in a haystack."

6.1.3 Ramping the Culture up

This concern became less salient when graduate student Thomas DiStefano took over the study in Gossett's lab and started increasing the PCE doses added to the cultures. Surprisingly, the more PCE that was added, the better the cultures performed (DiStefano et al. 1991) until the dose reached 0.55 mmol/L, approaching the solubility limit of PCE. These cultures produced little or no methane, presumably because of inhibition by chloroethenes and ethene, yet dechlorination continued unabated, with the PCE dose usually converted to ca. 20 % VC

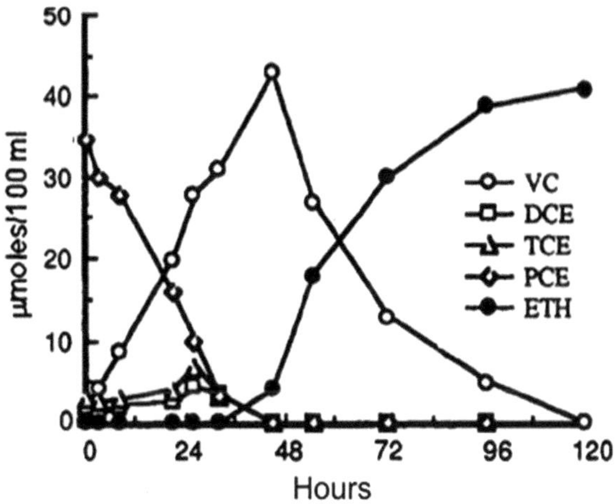

Fig. 6.1 Time course for conversion of PCE to ethene by an acetogenic methanol-PCE enrichment culture. Slow equilibration of added PCE caused its underestimation in early time points (from Tandoi et al. 1994 with permission)

and 80 % ethene within 2–3 days and nearly completely to ethene by about five days (Fig. 6.1a). Indeed, dechlorination was now consuming nearly one third of the electron equivalents added to the culture, with the remainder going to acetogenesis, instead of methanogenesis, from methanol. A methanol-utilizing *Peptostreptococcus*-like acetogen that resembled one of the main morphotypes in the culture was isolated and did not significantly dechlorinate PCE (Tandoi and Zinder, unpublished). Subsequent studies showed that H_2 rather than methanol was the direct electron donor for dechlorination (DiStefano et al. 1992).

At this point we hypothesized that rather than a cometabolic process carried out by methanogens or other anaerobes, reductive dechlorination of chloroethenes in these cultures was a respiratory process, based on the following information: (1) the high rates of and high proportion of metabolism devoted to reductive dechlorination, much higher than those described in cometabolic cultures; (2) the continuation of those rates despite the inhibition of methanogenesis by high PCE doses; (3) the low dechlorination activity of methanogenic and acetogenic isolates from the cultures; (4) the ability to transfer the activity to new cultures in which the rate of dechlorination increased over time, indicative of growth; (5) the favorable thermodynamics of reductive dehalogenation (Vogel et al. 1987); (6) the precedent of organochloride respiration by *Desulfomonile tiedjei* using chlorobenzoates (DeWeerd et al. 1990; Suflita et al. 1982).

However, this hypothesis contradicted the prevailing paradigm at the time for reductive dechlorination of chloroethenes, and an early grant Gossett and I submitted making the case for anaerobic respiration was rejected with reviewers saying that our arguments were specious and we simply did not understand the cometabolic nature of the process, and that the proposed studies would only add incrementally to the already large body of knowledge about reductive dechlorination of chloroethenes. Fortunately, the research was funded by the US Air Force Armstrong Laboratories at Tyndall AFB under the aegis of Cathy Vogel.

6.1.4 Microbiology Begins

Valter Tandoi, a microbiologist working at a water research institute in Rome, Italy came to my lab in 1991 as a visiting scientist just as DiStefano was finishing his research, and we began more concerted microbiological studies. He successfully transferred the methanol-PCE culture into a growth medium with nitrilotriacetate, the chelating agent we used in our medium for methanogens (Zinder et al. 1987), so that that the medium remained clear, instead of black with metal sulfide precipitates, facilitating microscopic observation. Tandoi examined dechlorination kinetics by the culture, demonstrating that VC and *trans*-DCE dechlorination followed first-order kinetics while the other steps occurred at similar rates and were essentially zero order (linear) (Tandoi et al. 1994). These kinetics were considerably different from those described for transition metal cofactors like corrinoids or F_{430} where each successive step was an order of magnitude slower (Gantzer and Wackett 1991).

Even in clear medium, the only organism I could see using phase-contrast microscopy that was present in numbers sufficient to account for the dechlorination activity in the culture was the *Peptostreptococcus*-like organism which we had already isolated and did not detect reductive dechlorination. However, there always seemed to be large numbers of tiny irregular particles of "junk" undergoing Brownian motion in the cultures, but they did not resemble any microbes that I had ever seen and I considered them an unusual precipitate. Tandoi collaborated with members of the lab of my colleague Bill Ghiorse, and used Bill's newly acquired scanning confocal microscope and acridine orange staining to demonstrate that the junk actually contained DNA and was in reality tiny irregular coccoid cells (Fig. 6.2).

Tandoi's last experiment before returning to Rome was to initiate tenfold serial dilutions of the methanol/PCE culture into basal medium with electron donors and

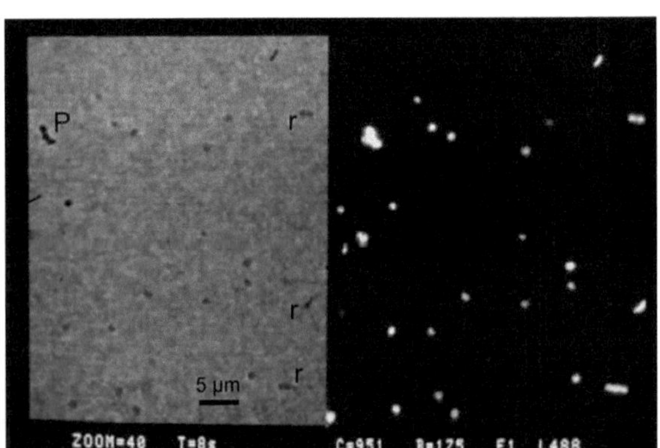

Fig. 6.2 *Left panel* phase-contrast micrograph from an acetogenic enrichment culture. We used an agar slide causing a rough background and the *D. mccartyi* cells to lie down flat making them barely phase-dark. *Right panel* fluorescence micrograph showing acridine orange fluorescence of the tiny barely visible cells. P, *Peptostreptococcus*-like organism (short chain); r, rods

acceptors designed to enumerate various physiological groups by the most probable number technique. Methanol and acetate-utilizing methanogens were not present in any dilutions, H_2–CO_2 using methanogens and acetogens were estimated at 10^4/ml, H_2-utilizing sulfate reducers at 10^6/ml, and fermentative heterotrophs and methanol-utilizing acetogens were estimated at 10^7/ml. Most importantly, in medium with H_2 as the electron donor and PCE as the electron acceptor we obtained growth and dechlorination in some of the 10^{-6} dilutions. These tubes did not appear to contain any sulfate reducers, but still contained fermentative heterotrophs that grow on the 0.2 g/L of yeast extract that we added to the culture, and methanol-utilizing acetogens. The heterotrophs remained in transfers of this culture but the methanol-utilizing acetogen could not use H_2–CO_2 and was lost on subsequent transfer.

Thus our reward from these studies was a PCE-dechlorinating culture free of other H_2 utilizing organisms such as methanogens, acetogens, or sulfate reducers. It contained those tiny organisms growing on H_2 and PCE and some rod-shaped heterotrophs, and was therefore more highly enriched. At around this time in 1992, I received by mail a PhD thesis published by Christof Holliger at Wageningen University that described his studies on reductive dechlorination and included a chapter on his pioneering studies of *Dehalobacter restrictus*, which dechlorinated PCE to *cis*-DCE (cDCE) using H_2 as the electron donor (Holliger et al. 1993, 1998). This demonstration that chloroethenes could indeed be used as electron acceptors did not surprise me. What did surprise me was that despite the long lists of electron donors and acceptors tried, *D. restrictus* could only use H_2 and PCE/TCE as its electron donor/acceptor pair. It seemed incredible to me that an organism would be so specialized in using compounds that had only been added to the environment in the past few decades.

Subsequently, several other organisms that reductively dechlorinated PCE/TCE as far as cDCE were described, including *Sulfurospirillum* (*Dehalospirillum*) *dehalogenans* (Scholz-Muramatsu et al. 1995), *Desulfitobacterium* sp. PCE and *D. hafniense* (Christianssen and Ahring 1996; Gerritse et al. 1996), *Desulfuromonas chloroethenica* (Krumholz 1997), and *Geobacter lovleyi* (Sung et al. 2006a), organisms covered in other chapters in this volume. Unlike *D. restrictus,* they could use other electron donors besides H_2 and electron acceptors besides chloroethenes. They were all members of the commonly cultivated phyla *Proteobacteria* and *Firmicutes*. Clearly the organisms in our cultures were different since they produced VC and ETH and had an unusual morphology.

6.2 Isolation and Characterization

6.2.1 Optimizing Growth

Once we had a highly enriched culture, our goal was to isolate organisms responsible for the dechlorination. Initial attempts using agar roll tubes failed, so our best chance was dilutions into liquid medium. This method involves a "numbers game," where you need somewhere near an order of magnitude greater numbers of the desired organism versus contaminants so that by the laws of probability dictated by the Poisson distribution, some of the highest dilution tubes will contain only the desired organism, hopefully a single cell. Thus, we had to optimize the medium to maximize the numbers of the dechlorinators versus other organisms.

At this time, a new graduate student, Xavier Maymó-Gatell, came to the laboratory from Barcelona, Spain and began nutritional studies on the culture (Maymó-Gatell et al. 1995) so that we could grow them in a better defined medium. He found that he could obtain PCE dechlorination to VC at rates that increased over time, indicative of growth, in medium amended with H_2 as the electron donor, 25 % v/v filter-sterilized centrifuged anaerobic digestor sludge supernatant (a source of organic nutrients we sometimes used in culturing methanogens, similar to the use of rumen fluid to culture rumen anaerobes (Hungate 1969), a vitamin solution, and 2 mM sodium acetate, instead of yeast extract. Eliminating yeast extract was a major step forward in reducing contamination since it supported growth of a "zoo" of fermentative heterotrophs, whereas organisms that could catabolically use acetate, like methanogens or sulfate reducers, were no longer present in the culture.

Once yeast extract was omitted from the medium, Maymó-Gatell found that the culture would not transfer if we omitted the vitamin solution we routinely added to the medium. By preparing vitamin stocks, each one missing one of the ten vitamins in the solution, he narrowed it down to vitamin B_{12}, which was required in amounts 50-fold higher than that by organisms that used it for biosynthetic purposes. This high requirement suggested to us that it was used instead for catabolic purposes as a prosthetic group in dechlorinating enzymes, based on vitamin B_{12}'s in vitro dechlorination activity (Gantzer and Wackett 1991). The minimum concentration for optimal dechlorination was 0.05 mg/L, identical with that subsequently determined for the pure cultures of strains 195 and BAV1 (He et al. 2007).

An interesting parallel set of experiments occurred in the Gossett lab during these studies. Graduate student Donna Fennell was switching a PCE-dechlorinating bioreactor from methanol to butyrate as the electron donor with the idea that butyrate would poise H_2 low (Schink 1997), so that dechlorination, which is more thermodynamically favorable than methanogenesis, would be more competitive. The bioreactor initially performed well, producing more ethene than VC on Day 20 (Fennell et al. 1997), but over time the VC to ethene ratio increased until VC was the only product detected, and then by Day 100 TCE and a few days later, PCE were the only chloroethenes detected—the bioreactor had gone into failure.

Fennell, stopping by our lab, learned that the enrichment required vitamins, took a bottle of our stock solution back with her, and began adding it on Day 125. Within days, the bioreactor recovered and was making VC and ethene. After we learned that large amounts of vitamin B_{12} were needed, she upped the B_{12} dose and performance improved to the point that the product was nearly all ethene. This bioreactor, dubbed Donna II, is studied in the laboratory of Ruth Richardson to this day (Rowe et al. 2012). Our interpretation of this phenomenon is that corrinoid proteins are more abundant in methanol-utilizing methanogens and acetogens, where they are integral parts of catabolic pathways, than in butyrate oxidizing syntrophs and hydrogenotrophic methanogens, and some portion of those corrinoids was excreted/leaked into the medium for use by dechlorinators. It was gratifying that our culture studies could have such a direct effect on bioreactor studies, which beautifully demonstrated that the dechlorinators were more competitive with methanogens when the H_2 concentration was poised low (Fennell et al. 1997).

6.2.2 Purification and Isolation

In the next stage of isolation, we took advantage of the surprising finding that the dechlorinators were resistant to bacterial peptidoglycan synthesis inhibitors such as 100 mg/L vancomycin (DiStefano et al. 1991) or up to 3 g/L ampicillin (Maymó-Gatell et al. 1997). Transfer of the culture into medium supplemented with acetate, a high B_{12} vitamin solution, and 25 % sewage sludge supernatant (ABSS) and containing either of these inhibitors led to growth and PCE dechlorination, and microscopically, the culture consisted of tiny coccoid organisms. However, the culture would not transfer a second time, suggesting that a sufficient amount of a limiting nutrient, perhaps provided by contaminating bacteria, was carried over with the first 2 % v/v inoculum but became too dilute on subsequent transfer. We tried various amendments known to support growth of anaerobes and culture supernatants and extracts, including branched chain fatty acids, horse serum, and extracts of *Escherichia coli* and *Clostridium pasteurianum* and obtained the best results with a filter-sterilized cell-free extract from a mixed dechlorinating culture, and for many years afterward we would obtain material from the "Donna II" butyrate-PCE culture and prepare extracts from the centrifuged pellet. As described presently, it was subsequently found that strain 195 can actually grow in defined medium.

Using this medium, we obtained growth in a 10^{-7} dilution of the culture, and this culture appeared microscopically pure (rod-shaped contaminants are easily detected when present), and contaminants were not detected in various growth media. We deemed the culture pure and later tests, including examination of its genome sequence, showed the culture was clonal. We named it "*Dehalococcoides ethenogenes*" strain 195 with the genus name signifying that it dehalogenated (like *Dehalobacter*) and was coccoid rather than a nearly spherical coccus, the species epithet signifying that it produced ethene, and the strain name signifying that the isolated culture was first obtained in January 1995. As described presently, it and all other *Dehalococcoides* strains have been named *Dehalococcoides mccartyi* (Löffler et al. 2013).

While discussing isolation it should be mentioned that while we were unable to obtain colonies of *D. mccartyi* in roll tubes, Adrian et al., in their studies on strain CBDB1 (Adrian et al. 2000), obtained colonies from diluted cultures in

Fig. 6.3 Agar shake tube showing colonies of *D. mccartyi* strain 195. The tube is 18 mm in diameter

low-melting agarose shake tubes, as we subsequently found in collaboration with Adrian, does strain 195 (Fig. 6.3). These colonies are much more likely to be clonal populations than are dilutions into liquid medium. Another approach for purifying cultures takes advantage of the small size of *D. mccartyi*. Cultures are passed through a 0.45 μm membrane filter that retains nearly all other organisms (LaRoe et al. 2014). It is possible that other small organisms and mycoplasmas also pass through these membranes, so further purification may be needed. An extensive description of techniques for culture and isolation of *D. mccartyi* and other dehalogenators is given by Löffler et al. (Löffler et al. 2005).

6.2.3 Initial Characterization: Physiology, Morphology/Ultrastructure, and Phylogeny

Our first publication on *D. mccartyi* strain 195 (Maymó-Gatell et al. 1997) described its physiology, including that it did not use any electron donor we tested other than H_2 nor any of the common electron acceptors such as oxygen, nitrate, or sulfate, a degree of specialization similar to *D. restrictus* (Holliger et al. 1998). A 100 μmol/L dose of PCE added to an active culture was stoichiometrically converted to VC within 3 h, with little accumulation of TCE or DCEs as intermediates. VC was then slowly dechlorinated to ethene in about 600 h following first-order kinetics. It was later shown that dechlorination of VC or *trans*-DCE did not support growth in strain 195 making those steps cometabolic (Maymó-Gatell et al. 1999, 2001). Doubling times near one day were found for cultures converting PCE to VC. Since then, strain 195 has been shown to use some chlorophenols

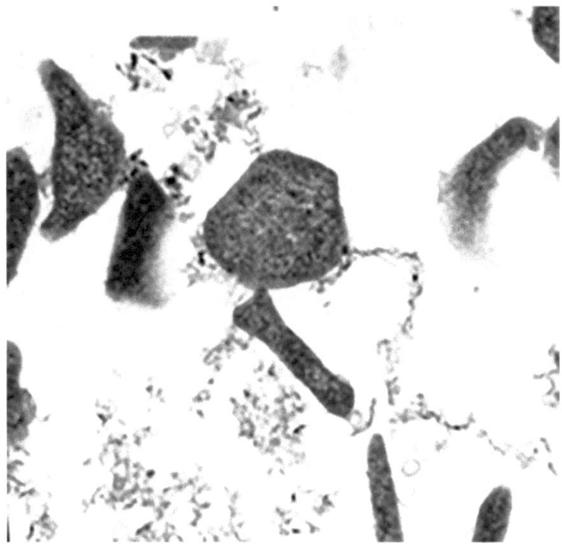

Fig. 6.4 Low power thin-section electron micrograph of *D. mccartyi* strain 195 showing disk-shaped morphology of the cells. The central cell is approximately 0.5 μm in diameter

(Adrian et al. 2007a) and highly chlorinated chlorobenzenes (Fennell et al. 2004) metabolically, and dechlorinate some chloronaphthalenes and PCBs although it is not certain whether these reactions are growth supporting (Fennell et al. 2004).

Thin-section electron micrographs of strain 195 showed an unusual cell structure. Cells appeared 0.1–0.5 μm in diameter, and it was not realized until lower power micrographs like Fig. 6.4 were examined after publication that the cells are actually curved disks 0.4–0.5 μm in diameter and 0.1–0.2 μm in height. The biovolume of these cells is estimated as 0.02 μm^3, roughly 30-fold smaller than a typical *E. coli* cell, and about twice that of *Pelagibacter ubique* in the SAR 11 cluster (Rappe et al. 2002), considered the smallest bacterium known. The small size allows a high surface area to volume ratio, useful for uptake of scarce substrates. More unusual was the cell wall structure, which resembled the S-layer protein subunit cell walls of *Archaea* (Albers and Meyer 2011), and others have also seen S-layer cell wall structures in electron micrographs of *D. mccartyi* strains, including strain CBDB1 which had a 14 nm repeating structure (Adrian et al. 2000; Löffler et al. 2013). A peptidoglycan layer was not visible in the thin sections nor was it detected using a fluorescent lectin stain for N-acetylglucosamine (Maymó-Gatell et al. 1997), and subsequent genomic studies (Seshadri et al. 2005) demonstrated the absence of peptidoglycan synthesis genes. The lack of a peptidoglycan layer readily explained strain 195's resistance to vancomycin and ampicillin. All other strains of *D. mccartyi* are also tiny disk-shaped cocci with S-layer cell walls (Löffler et al. 2013).

The cell wall structure, resistance to antibiotics, and hydrogen-based metabolism in strain 195 made us wonder whether the organism belonged to the *Archaea* rather than the *Bacteria*, and we began sequencing its 16S rRNA gene. We obtained a sequence of ~200 bp using the manual sequencing gel methods available at the time. As we got a more complete sequence it was clear that *Dehalococcoides* was in the *Bacteria*, a finding corroborated by its sensitivity to tetracycline. However, we were unable to place it in any of the known phyla (even when we included *Chloroflexus* in the analysis). Different sequence sets and algorithms caused shifts in its position on the tree. Soon after publication of the paper both Phil Hugenholtz, then at UC Boulder and Floyd Dewhirst at the Forsyth Dental Institute ran it against their more extensive databases that included uncultured sequences, and it was pulled into the *Chloroflexi* by sequences like that of SAR202. Also, a subsequent discussion with researchers at DuPont revealed that a sequence from a contaminated site in Victoria Texas, that had been discarded as a likely chimera, was over 98 % identical with that of strain 195, and eventually *D. mccartyi* strain VS was cultured from that site (Cupples et al. 2003; Müller et al. 2004). This was our first inkling that our organism from a sewage digestor might be important at contaminated groundwater sites.

6.3 Diversity and Phylogeny

6.3.1 Isolation of Other D. mccartyi Strains

Since the isolation of strain 195, several other strains of *D. mccartyi* have been isolated, and I will discuss the first three followed a briefer discussion of others as well as summarize their descriptions in Table 6.1. In 2000 Adrian et al. (Adrian et al. 2000) described *D. mccartyi* strain CBDB1 that dechlorinated chlorobenzenes with three or more chlorines (Jayachandran et al. 2003). It was shown to dechlorinate other chloroaromatics including dioxins (Bunge et al. 2003), chlorophenols including pentachlorophenol (Adrian et al. 2007a), and PCBs (Adrian et al. 2009). It also debrominates bromobenzenes to benzene (Wagner et al. 2012). Although originally reported not to use chloroethenes (Adrian et al. 2000), strain CBDB1 was subsequently shown to grow dechlorinating PCE and TCE to a 3.4/1 mixture of *trans*-DCE and *cis*-DCE (Marco-Urrea et al. 2011a), an ability it shares with strain *D. mccartyi* MB (Cheng and He 2009).

A major shortcoming of *D. mccartyi* strain 195 for use in remediation of chloroethenes is that, despite its original species name "*ethenogenes*," it produces large amounts of VC that is then only slowly cometabolized to ethene. Strains that use VC for organochloride respiration and growth are much more desirable and two such strains were described in 2003.

D. mccartyi strain BAV1 was cultured from the chloroethene-contaminated Bachman Road site in Michigan, USA (He et al. 2003a). It was enriched with pyruvate followed by H_2 with VC, and isolated using ampicillin and multiple 10^{-7} dilutions (He et al. 2003b). Besides VC, strain BAV1 could also use all three DCE isomers, 1,2-dichloroethane (DCA, dechlorinated to ethene) and vinyl bromide.

D. mccartyi strain VS was enriched on VC (Rosner et al. 1997) from a benzoate/TCE culture derived from a contaminated site in Victoria, Texas.

Table 6.1 *Dehalococcoides mccartyi* strains in pure culture

Strain	Representative substrates for OHR → products	Group	Identified RDase genes	References
195	PCE, TCE, cDCE, 1,1DCE → VC (ETH)[a] 1,2-DCA → ETH 2,3-DCPh → 3-CPh HCB → 1,2,4,5 TeCB, 1,3,5-TCB	Cornell	*pceAB*, *tceAB*	Adrian et al. (2007a), Fennell et al. (2004), Maymó-Gatell et al. (1997, 2001)
CBDB1	HCB → 1,4-DCB, 1,3-DCB, 1,3,5-DCB Dioxins PCBs[b] PeCPh → TCPhs + 2-CPh + 3-CPh Bromobenzenes → benzene PCE, TCE → tDCE	Pinellas	*cbrAB*, "*pceAB*"	Adrian et al. (2000, 2007a, 2009), Bunge et al. (2003), Jayachandran et al. (2003), Wagner et al. (2012)
BAV1	DCEs, VC → ETH VBr → ETH	Pinellas	*bvcAB*	He et al. (2003a, b)
VS	TCE, cDCE, 1,1-DCE, VC → ETH	Victoria	*vcrAB*	Cupples et al. (2003), Müller et al. (2004)
FL2	TCE, cDCE, tDCE → VC (ETH)[a]	Pinellas	*tceAB*	He et al. (2005)
GT	TCE, cDCE, 1,1-DCE → ETH	Pinellas	*vcrAB*	Sung et al. (2006b)
MB	PCE, TCE → tDCE	Cornell	*mbrA*	Cheng and He (2009)
ANAS1	TCE, cDCE, 1,1-DCE, → VC (ETH)	Cornell	*tceA*	Lee et al. (2011)
ANAS2	TCE, cDCE, 1,1-DCE, VC → ETH	Cornell	*vcrA*	Lee et al. (2011)
11a	TCE, DCEs, VC → ETH 1,2-DCA → ETH	Pinellas	*vcrA*	Lee et al. (2013)
11a5	TCE, cDCE, 1,1-DCE, tDCE → VC	Pinellas	*tceA*	Lee et al. (2013)
DCMB	Dioxins 1,2,3-TCB → 1,2-DCB	Pinellas		Pöritz et al. (2013)
JNA	PCBs (flanked meta)	Pinellas		LaRoe et al. (2014)
CG1	PCBs (meta > para) PCE → TCE	Victoria	*pcbAB1*	Wang et al. (2014)
CG4	PCBs (para > meta) PCE → tDCE	Cornell	*pcbAB4*	Wang et al. (2014)
CG5	PCBs (meta + para) PCE → cDCE + tDCE	Pinellas	*pcbAB5*	Wang et al. (2014)

Abbreviations *A* ethane; *B* benzene; *Ph* phenol; *VBr* vinyl bromide; *PCBs* polychlorinated biphenyls; *Pe* penta; *Te* tetra

[a]VC → ethene cometabolic

[b]Para and meta Cls removed from PCBs; growth using PCBs not demonstrated but likely

Table 6.2 Selected enriched cultures containing *D. mccartyi*

Culture	Description and reactions	References
Donna II	Bioreactor inoculated from an Ithaca, NY, USA sewage plant-fed butyrate that converts PCE to VC and ethene and contains a *D. mccartyi* essentially identical with strain 195	Fennell et al. (1997), Rowe et al. (2012)
RC & KS	Highly enriched cultures from the presumably pristine Red Cedar River in MI, USA and the hydrocarbon contaminated King Salmon River in AK, USA. Both use H_2 for converting 1,2-dichloropropane to propene by a dihaloelimination process	Löffler et al. (1997), Padilla-Crespo et al. (2014)
KB-1	Commercial culture from a TCE-contaminated site in Ontario, Canada used to treat chloroethenes. Contains multiple Pinellas group *D. mccartyi* strains, some using PCE and others converting VC to ethene using VcrA or BvcA	Duhamel et al. (2002, 2004), Waller et al. (2005)
ANAS	Lactate-fed reactor, inoculated from the Alameda Naval Air Station, CA, USA. Converts TCE to ethene and contains the two *D. mccartyi* strains described in Table 6.1	Johnson et al. (2005), Lee et al. (2011)
SDC-9	Commercial culture fed lactate and yeast extract and converting PCE to ethene. Contains at least two *D. mccartyi* strains	Schaefer et al. (2009)
BTF08	Highly enriched culture from Bitterfeld, Germany converting PCE to ethene. Genome sequence shows it contains homologues of PceA, TceA, and VcrA	Pöritz et al. (2013)

H_2-dependent growth of a *D. mccartyi*-like organism on VC was demonstrated using the quantitative polymerase chain reaction (qPCR) and *D. mccartyi*-specific 16S rRNA gene primers (Cupples et al. 2003). It also used cDCE and 1,1-DCE (Müller et al. 2004).

After these initial strains, several more have been isolated and are listed in Table 6.1, whereas some *D. mccartyi*-containing mixed cultures that have been intensively studied are listed in Table 6.2. Strain FL2 was isolated from dechlorinating microcosms constructed using sediments from the presumably pristine Red Cedar River in Michigan, USA. It used TCE following a pattern resembling that of strain 195 in which VC accumulated and was only slowly cometabolized to ethene, but it could not use PCE. Other *D. mccartyi* strains have been isolated that use chloroethenes, dioxins, and PCBs as electron acceptors (Table 6.1). While strain JNA was isolated directly on PCBs (LaRoe et al. 2014), a long and arduous task since PCBs are nearly insoluble and are used very slowly, strains CG1, CG4, and CG5 were isolated by shifting PCB enrichment cultures to the more readily used PCE, which was converted to TCE or DCEs, and demonstrating that the resulting isolates could still use PCBs (Wang et al. 2014). Metagenomic sequencing was used to verify that the *D. mccartyi* strains enriched on PCE were the same as those using PCBs.

All *D. mccartyi* strains described to date resemble strain 195 in that they are small disk-shaped organisms that have been found to use only H_2 as an electron donor and organohalides as electron acceptors for growth.

6.3.2 Dehalococcoides mccartyi Phylogeny

Strains 195, VS, and CBDB1/BAV1 are the founding members of three closely related clades of *D. mccartyi*, Cornell, Victoria, and Pinellas, described by Hendrickson et al. (Hendrickson et al. 2002) in their PCR studies of the distribution of *D. mccartyi* 16S rRNA genes in samples from numerous contaminated sites, naming them after where the first sequences were detected (in the latter case, Pinellas, FL, USA). These three clades have remained intact as many more sequences from cultured and uncultured *D. mccartyi* have accumulated in the databases (Fig. 6.5). Overall *D. mccartyi* strains form a tight cluster. For example, full-length 16S rRNA genes from strains 195 and CBDB1 derived from genome sequences are 98.9 % identical.

Because the *D. mccartyi* strains had such closely related 16S rRNA genes as well as considerable homology and synteny of housekeeping genes, and because the dehalogenating phenotype did not correlate with 16S rRNA sequence (see below) a group of researchers working on *D. mccartyi* decided to put all *Dehalococcoides* strains into a single species, *D. mccartyi* (Löffler et al. 2013), named after the pioneering environmental engineer Perry McCarty. Descriptions of isolated *D. mccartyi* strains and their substrates are found in Table 6.1.

Until recently, the next closest relatives to *D. mccartyi*, even including environmental sequences, were *Dehalogenimonas* (Moe et al. 2009), which uses polychlorinated aliphatic alkanes, and its relatives (Kittelmann and Friedrich 2008),

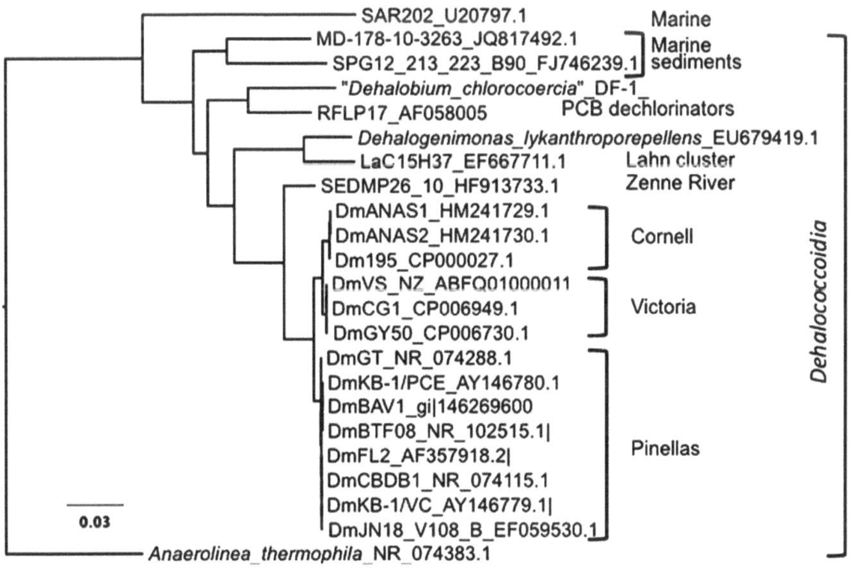

Fig. 6.5 Neighbor joining 16S rRNA phylogenetic tree for some *Dehalococcoides mccartyi* (Dm) strains and relatives along with their accession numbers. Most of the sequences used are full or near full length

at about 90 % 16S rRNA gene identity. In contrast to the tree in Fig. 6.5, most other bacterial phylogenic trees are more bush-like, with a continuum of related species. This suggests that the ancestors of *D. mccartyi* passed through a bottleneck some time in the past. However, a sequence from the chloroethene and chloroethane contaminated Zenne River in Belgium (Hamonts et al. 2014), otherwise dominated by the *D. mccartyi* Pinellas group, does have an intermediate position, with about 96 % identity with the *D. mccartyi* sequences (Fig. 6.3). Nothing is presently known about this "missing link."

More distantly related to *D. mccartyi* are the PCB dechlorinators cultured from Baltimore Harbor "*Dehalobium chlorocoercia*," which uses double-flanked chlorines, and RFLP17, which uses ortho chlorines. There is also a distinct and diverse clade of 16S rRNA sequences that are often numerous in surveys of marine sediments (Durbin and Teske 2011; Inagaki et al. 2006) that is more distantly related to *D. mccartyi*. The entire group of organisms described here forms the class *Dehalococcoidia* (Löffler et al. 2013), formerly called "*Dehalococoidetes*," a term more appropriate for a phylum than a class. While it is likely that nearly all the organisms encompassed by *D. mccartyi* and *Dehalobium* are OHRB, RDase genes were not found in two recent single-cell genomes sequences from the marine sediments cluster (Wasmund et al. 2014; Kaster et al. 2014). The genome sequences were estimated to be 61–85 % complete, so it is possible that the RDase genes were in the missed parts, but if the cluster is found not to contain OHRB, a determination should be made whether they should represent an order in the *Dehalococcoidia* distinct from the OHRB, or possibly given their own class designation.

6.4 Genomics and Physiology

6.4.1 Streamlined Genomes

Because *D. mccartyi* are fastidious anaerobes that grow slowly to low densities, they are poor candidates for study using standard biochemical and genetic techniques that require large amounts of cell material and growth as colonies on Petri dishes. Fortunately, the genomic age was dawning as the first *D. mccartyi* strains were isolated. The genome sequence of strain 195 (Seshadri et al. 2005) set the pattern that is followed by the other strains (Table 6.3). *D. mccartyi* genomes are among the smallest found in free-living organisms, 1.34–1.5 MB, less than one third the size of the *E. coli* genome. The previously mentioned *Pelagibacter ubique* has a genome size of 1.31 Mb (Giovannoni et al. 2005), and it is possible that these tiny cells cannot harbor more DNA. There were only single copies of most housekeeping genes in the *D. mccartyi* strain 195 genome, and genes for some common bacterial functions like peptidoglycan synthesis, motility, and many environmental adaptations, were absent in these stripped-down genomes, nor were there any recognizable genes for using electron acceptors other than organohalides or for most electron donors, in agreement with physiological studies.

Table 6.3 *D. mccartyi* genomes

Strain	Size (MB)	Predicted ORFs	Predicted RDase genes	Accession #	References
195	1.47	1591	17	CP000027	Seshadri et al. (2005)
CBDB1	1.39	1458	32	AJ965256	Kube et al. (2005)
BAV1	1.34	1371	11	CP000688	McMurdie et al. (2009)
VS	1.41	1447	36	CP001827	McMurdie et al. (2009)
GT	1.36	1417	20	CP001924	Unpublished
DCMB5	1.43	1477	23	CP004079	Pöritz et al. (2013)
BTF08*	1.45	1529	20	CP004080	Pöritz et al. (2013)
CG1	1.49	1557	35	CP006949	Wang et al. (2014)
CG4	1.38	1421	15	CP006950	Wang et al. (2014)
CG5	1.36	1413	28	CP006951	Wang et al. (2014)

6.4.2 Multiple RDases

Despite genome streamlining, *D. mccartyi* genomes contained 11–36 sets of *rdhAB* operons predicted to encode RDases (McMurdie et al. 2009; Seshadri et al. 2005). The large number of RDases encoded in their otherwise streamlined genomes shows that *D. mccartyi* strains are highly evolved to use organohalides as electron acceptors. Moreover, the RDase genes often have adjacent genes encoding two-component or MarR transcription regulators, indicating that they are part of a highly regulated metabolic network. The large number and phylogenetic depth of the RDase genes (Hug et al. 2013; McMurdie et al. 2009) indicate that they are ancient, probably dating back at least to the "Great Oxidation Event" over two billion years ago, since nearly all of the enzymes producing organohalides use oxygen or peroxides as reactants (Gribble 2010).

The few *D. mccartyi* RDases that have had functions assigned to them are shown in Table 6.4. PCE RDase and TCE RDase were isolated from a mixed culture containing *D. mccartyi* 195 (Magnuson et al. 1998), which grew to higher yields than the pure culture, using "brute force" biochemical purification techniques. The amino acid sequences of peptides from the proteins were used to design degenerate primers to clone and sequence the RDase genes (Magnuson et al. 2000), an approach originally used to obtain the sequence of the PCE RDase in *Sulfurospirillum multivorans* (Neumann et al. 1998). A similar approach was used to obtain the VC-reducing VcrA gene sequence from strain VS (Müller et al. 2004). Other studies have used transcriptomic (Krajmalnik-Brown et al. 2004) or proteomic techniques to infer RDase function. A useful technique to identify RDases is preparative native electrophoresis (Adrian et al. 2007b), or blue native polyacrylamide electrophoresis (BN-PAGE) which separates native proteins on the basis of size. Various fractions can be assayed for activity, and active bands can

Table 6.4 Characterized reductive dehalogenases found in *D. mccartyi* strains

Name	Reactions	Molecular wt.*	Accession #	References
TceA	TCE, cDCE, 1,1-DCE → VC 1,2-DCA → ETH + trace VC	62,025	YP_180831	Magnuson et al. (2000)
PceA	PCE → TCE 2,3-DCPh → 3-CPh	55,155	YP_181066.1	Magnuson et al. (2000)
BvcA	DCEs, VC → ETH	57,274	YP_001214307	Krajmalnik-Brown et al. (2004)
VcrA	TCE, cDCE, 1,1-DCE, VC → ETH	57,403	YP_003330719	Müller et al. (2004), Parthasarathy et al. (2015)
CbrA	1,2,3,4-TeCB → 1,2,4-TCB	54,141	CAI82345	Adrian et al. (2007b)
DcpA	1,2-DCP → propene	53,904	JX826287	Padilla-Crespo et al. (2014)

*Molecular weights are for the entire proteins before cleavage by the TAT system

be sent for proteomic identification (Tang et al. 2013). Much more detail about RDases can be obtained elsewhere in this volume.

Not only does each *D. mccartyi* strain harbor a large number of RDase genes, their distribution in those strains is disparate (Table 6.1). Each isolate described thus far has a unique complement of RDase genes with only partial overlap with other strains. This became apparent when the genome sequence of *D. mccartyi* strain CBD1 (Kube et al. 2005), which has 32 predicted RdhAB clusters, was compared with that of strain 195 (Seshadri et al. 2005). which had 17. Most of the RDase genes were found in "high-plasticity" regions near the origin of replication in these two organisms, a pattern followed by other *D. mccartyi* strains (McMurdie et al. 2009). It appears that rapid RDase gene exchange among *D. mccartyi* strains occurs in these regions (McMurdie et al. 2009, 2011). For example, strains 195 and FL2 are in the Cornell and Pinellas clades respectively, yet they contain *tceA* genes that are 99.4 % identical at the nucleotide level. In fact, the *tceA* genes from these two strains plus those amplified from several mixed cultures from various sites in North America were all greater than 96.3 % identical, and the intergenic spacer region and the *tceB* genes were 100 % identical (Krajmalnik-Brown et al. 2007). Another example is that the *vcrAB* genes can be found in all three *D. mccartyi* clades (Table 6.1). The recently described 1,2-dichloropropane (DCP) RDase in *D. mccartyi* RC (Padilla-Crespo et al. 2014) is nearly identical with that from the distantly related *Dehalogenimonas lykanthroporepellens* (Fig. 6.3), indicating that horizontal gene exchange can involve other genera. Thus, while finding *D. mccartyi* 16S rRNA genes at a contaminated site can be considered presumptive evidence for the presence of a desired dechlorinating reaction, RDase genes are considered more specific biomarkers, especially *tceA*, *bvcA*, and *vcrA* in the case of chloroethenes (Ritalahti et al. 2006; Behrens et al. 2008; Holmes et al. 2006).

6.4.3 Electron Transport

To carry out a cycle of reductive dechlorination, the Co^{3+} in the corrinoid cofactors in RDases must be reduced back to Co^{1+}. While electrons with $E^{\circ\prime}$ values near −150 mV can reduce Co^{+3} to Co^{+2}, much lower reducing potentials, −500 mV or lower, are needed to reduce Co^{+2} to Co^{+1} (Schumacher et al. 1997) (Fig. 6.5). The ultimate source of these electrons is H_2, but the path that they travel from H_2 to the RDases in *D. mccartyi* is mysterious.

The *D. mccartyi* genome is predicted to encode five distinct hydrogenase complexes, showing that it is highly adapted to use this "simple" electron donor. Of these potential hydrogenases, the one annotated as a periplasmic Ni–Fe uptake (Hup) hydrogenase is found in highest amounts in transcriptomic and proteomic studies of strain 195 (Morris et al. 2006; Rahm and Richardson 2008) (Fig. 6.6) and is considered to be responsible for catabolic H_2 uptake. Detected at lower levels was a cytoplasmic two-subunit complex annotated as Vhu and one with a membrane bound and three cytoplasmic subunits annotated as Hym, an iron hydrogenase. Closer examination of the Hym complex (Mansfeldt et al. 2014) shows that it resembles electron bifurcating hydrogenases that take electrons from H_2 and transfer them to both the low potential ferredoxin and the higher potential NAD^+ (Schut and Adams 2009). While one could imagine this complex generating both the low potential and high potential electrons needed for the RDase, it is found in much lower levels than the Hup hydrogenase, and possibly provides electrons needed for biosynthesis. The energy coupled hydrogenase (Ech) and one annotated as Hyc both contain several membrane bound subunits predicted to pump protons, and may be involved in reverse electron transport using a proton motive force to produce low potential electrons needed for biosynthesis when H_2 concentrations are very low.

A complex annotated as a formate dehydrogenase (FDH) is highly expressed, with its large subunit often visible in one dimensional polyacrylamide gels of *D. mccartyi* strain 195 extracts (Morris et al. 2006). No *D. mccartyi* has ever been found to use formate, and no FDH activity was detected in extracts from *D.*

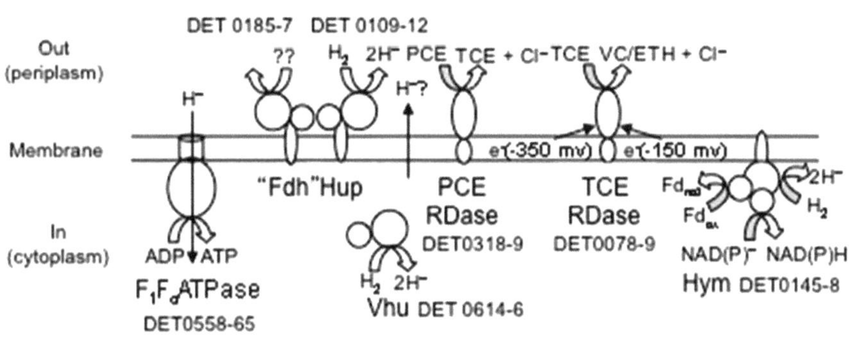

Fig. 6.6 Cartoon depicting predicted electron transport and energy conserving proteins with high coverage in proteomic studies of *D. mccartyi* strain 195 (Morris et al. 2006, 2007), their locus numbers and proposed functions

mccartyi strain 195 (Morris et al. 2006). Phylogenetically, the large subunit clusters with bona fide FDHs, but an examination of its predicted amino acid sequence showed that instead of a cysteine or selenocysteine residue considered to be essential for catalysis, the cognate amino acid in the *D. mccartyi* enzyme was predicted to be serine, which has very different chemical properties. Proteomics verified that this amino acid was indeed serine (Morris et al. 2007). Thus, the function of this highly expressed protein is unknown. Its transcription pattern under a variety of conditions was recently shown to correlate to that for the Hup hydrogenase (Mansfeldt et al. 2014), and it was noticed that both the Hup and the "FDH" complexes were missing subunits that the other possessed, suggesting that they form a hydrogenase "supercomplex," especially since FDHs can have hydrogenase activity (Soboh et al. 2011).

No physiological electron donor has been shown to supply electrons to *D. mccartyi* RDases, and viologen dyes are the only known donors that work at all (Jayachandran et al. 2004; Magnuson et al. 1998; Nijenhuis and Zinder 2005). There are some interesting differences between *D. mccartyi* and the well-characterized *S. multivorans* system (Miller et al. 1997). Similar to *D. restrictus* PCE RDase (Schumacher et al. 1997), either methyl viologen (MV, $E^{\circ\prime} = -440$ mV) or the weaker reducing agent benzyl viologen (BV, $E^{\circ\prime} = -360$ mV) can support reductive dehalogenation (Jayachandran et al. 2004; Nijenhuis and Zinder 2005), whereas only MV supports dehalogenation by the *S. multivorans* PCE RDase. Also, addition of the protonophore carbonyl cyanide *m*-chlorophenyl hydrazone (CCCP), which collapses proton gradients, inhibited dechlorination in intact *S. multivorans* cells, suggesting a need for an energized membrane to carry out reverse electron transport, CCCP did not inhibit dechlorination in *D. mccartyi* cells (Jayachandran et al. 2004; Nijenhuis and Zinder 2005) or those of *D. restrictus* (Schumacher and Holliger 1996). *D. restrictus* uses menaquinone as an electron donor, and recent genomic evidence (Goris et al. 2014) implicates quinones in *S. multivorans* electron transport. Despite an early report of the presence of ubiquinone in *D. mccartyi*, all recent genomic and biochemical evidence (Schipp et al. 2013) indicates quinones are not present in *D. mccartyi*.

Thus, our understanding is poor regarding the mechanism of electron transport from H_2 to the RDases in *D. mccartyi*, and how that electron transport is coupled to generation of a proton motive force, considerably worse than in the quinone-coupled systems in *D. restrictus* and *S. multivorans*. Energy conservation via a proton motive force in *D. mccartyi* is supported by the findings that subunits of an F_1F_0 ATPase in strain 195 are readily detected in proteomic surveys (Morris et al. 2006) and transcript levels of ATPase encoding genes correlate with respiration rates (Mansfeldt et al. 2014; Rahm and Richardson 2008). It should be noted that a gene predicted to encode a proton pumping pyrophosphatase (DET0766) resides in the *D. mccartyi* 195 genome, and peptides of this protein were detected in a proteomic survey (Morris et al. 2006), suggesting a role for pyrophosphate-coupled energy conservation. Also present in *D. mccartyi* genomes are genes predicted to encode an NADH dehydrogenase complex (Complex I, DET0923-33) that lacks an NADH receiver domain and presumably does not use a quinone electron acceptor, and a molybdopterin oxidoreductase distantly related to a tetrathionate reductase (DET0101-3) (Seshadri et al. 2005).

6.4.4 Nutrition

Since *D. mccartyi* strain 195 apparently required complex nutrients, it was hoped that its genome sequence (Seshadri et al. 2005) would provide some insight into which biosynthetic pathways were missing. In accordance with nutritional studies demonstrating a vitamin B_{12} requirement (He et al. 2007; Maymó-Gatell et al. 1995), genes involved in corrin ring synthesis were not found, whereas those encoding enzymes involved in corrinoid transport and salvage, including remodeling of the benzimidazole lower ligand, were present (Seshadri et al. 2005; Yan et al. 2013). Cells of *D. mccartyi* strain 195 provided with corrinoids with no lower ligand or the incorrect one are able to synthesize functional cofactors when provided with exogenous benzimidazoles (Yi et al. 2012), which have been detected in microbial habitats (Crofts et al. 2014). Interestingly, there is a tandem duplication of a 34 gene segment (DET0640-74 and DET0675-707) in the strain 195 genome that contains the genes for corrinoid transport, some corrinoid salvage genes, and some Wood-Ljungdahl pathway genes (Seshadri et al. 2005). This tandem duplication has not been found in other *D. mccartyi* genomes (McMurdie et al. 2009), and we speculated (Seshadri et al. 2005) that this duplication resulted from a period of corrinoid starvation in the early stages of enrichment. It is curious that *D. mccartyi* strains outsource corrin ring synthesis to other organisms, since they need corrins in large amounts to produce the RDases that are essential to their metabolism.

Only a few biosynthetic pathways seemed to be absent in the *D. mccartyi* strain 195 genome. For example there are no biotin or thiamin biosynthetic genes, and some pathways appeared incomplete, such as the methionine synthesis pathway. However, most biosynthetic pathways appeared intact. Moreover, the genome sequence of *D. mccartyi* strain CBDB1 was released soon after (Kube et al. 2005), and despite the fact that this strain could be grown in a defined medium with no organic nutrient additions except acetate and vitamins, the genome contained essentially identical biosynthetic genes. Indeed, it appears that all *D. mccartyi* strains can be grown in simple mineral defined medium in which the only organic amendments are MOPS buffer, acetate, and vitamins (Löffler et al. 2013).

An important contribution to our understanding of biosynthesis in *D. mccartyi* came from the studies of incorporation of positionally labeled ^{13}C-acetate and $^{13}CO_2$ into amino acids and peptides (Marco-Urrea et al. 2012; Tang et al. 2009) which demonstrated biosynthesis of all 17 of the readily analyzed amino acids, including methionine. In *D. mccartyi* genome sequences, there was no gene annotated as encoding citrate synthase, yet this enzyme is essential for glutamate synthesis. Positional labeling of glutamate indicated that a *re*type citrate synthase was used (Tang et al. 2009) in strain 195. That enzyme was demonstrated in strain CBDB1 (Marco-Urrea et al. 2011b) and is encoded by CbdbA1708, originally annotated as a homocitrate synthase.

A particularly interesting scenario applies to acetyl-CoA metabolism. While the genomic and biosynthetic studies showed that acetate is used as a carbon source as had been expected, it was noted early on (Seshadri et al. 2005) that *D. mccartyi* had an incomplete Wood-Ljungdahl pathway for acetyl-CoA biosynthesis/degradation, missing methylene-tetrahydrofolate reductase, and the carbon monoxide dehydrogenase (CODH) subunit of the CODH/acetyl-CoA decarbonylase-synthase (ACDS) complex, a curious finding.

It was recently demonstrated that the methyl group of methionine is derived from the methyl group of acetate (Zhuang et al. 2014) in a manner explained by splitting of acetyl-CoA by ACDS into a methyl group transferred to tetrahydrofolate, CoA, and CO. Since there is no CODH to oxidize CO to CO_2, it is released into the growth medium, where it was demonstrated to accumulate to levels as high as 0.1 % a level that can completely inhibit growth. Thus, in pure culture *D. mccartyi* eventually poisons itself. It had been known that the presence of a *Desulfovibrio* strain allowed much better growth of *D. mccartyi*, and it was found that the *Desulfovibrio* could use CO bringing its concentration down by an order of magnitude. These results may explain the common finding that *D. mccartyi* can grow to much higher densities in bioreactors than in pure cultures, and often the purer a culture becomes the worse *D. mccartyi* grows. In natural systems *D. mccartyi* can outsource CO utilization, common in many anaerobes, much as it outsources corrinoid biosynthesis, in line with it's having a stripped-down genome. A theory on the strategy of gene loss by an organism when other organisms are providing "public goods" is called the Black Queen hypothesis (Morris et al. 2012) with examples of *Prochlorococcus* and *Pelagibacter ubique*, both marine organisms with stripped-down genomes that outsource various functions, such as detoxification of radical oxygen species, to other organisms.

Strain 195, alone of all *D. mccartyi* with annotated genomes, also has a set of genes encoding the nitrogenase enzyme complex (*nif*, DET1151-8) and an associated Mo transporter (DET1159-61). The *nif* genes are in the family typically found in anaerobes, and are most closely related to those found in deltaproteobacterial sulfate reducers (Lee et al. 2009). The mol% G + C of the *nif* genes was 49–54, somewhat higher than the genomic average (49 %), suggesting a fairly recent genetic transfer. When grown in medium limited for fixed nitrogen, incorporation of $^{15}N_2$ was detected, but little growth occurred (Lee et al. 2009). Diazotrophic growth may occur under conditions more natural than in a batch culture, which may contain inhibitory levels of H_2 or CO.

6.5 Habitat and Ecology

The natural ecology of *D. mccartyi* is poorly understood. It has been cultured from ostensibly pristine sites as well as contaminated ones. As mentioned previously, there are thousands of natural organohalide compounds in terrestrial and marine habitats (Gribble 2010) so there can be a general rationale for the presence of *D. mccartyi* and other dehalogenators in anaerobic niches in these habitats. A more specific example comes from recent studies (Krzmarzick et al. 2012) in which soil humus was treated with chloroperoxidase, an enzyme commonly used by fungi for lignin degradation, which caused higher amounts of carbon-bonded chlorine in the humus.

Adding this treated humus to soils stimulated the growth of *Chloroflexi* related to *D. mccartyi*, believed to be dehalogenating, although not of *D. mccartyi* itself.

While *D. mccartyi* has been found worldwide, there is evidence that it is not present at all PCE- or TCE-contaminated sites, where its absence leads to "DCE stall" since organisms that dechlorinate as far as DCE are considered ubiquitous, whereas *D. mccartyi*, needed for full dechlorination to ethene, is not. Early results with PCR supported this hypothesis (Hendrickson et al. 2002) as have bioaugmentation studies (Ellis et al. 2000; Major et al. 2002) in which addition of *D. mccartyi*-containing cultures allowed more complete dechlorination in microcosms and at contaminated sites. PCR tests for *D. mccartyi* and other dehalogenators are now often part of contaminated site characterization, and *D. mccartyi*-containing cultures for bioaugmentation at contaminated sites have been commercially developed, a topic covered elsewhere in this volume.

The sporadic distribution of *D. mccartyi* at contaminated sites contradicts the maxim in microbial ecology attributed to Bejierinck and Baas-Becking (de Wit and Bouvier 2006) that "everything is everywhere and the environment selects." A recent finding suggests that *D. mccartyi* may be more cosmopolitan than originally thought (Delgado et al. 2014). Microcosms constructed from pristine garden soil and mangrove sediments using lactate or methanol as electron donors showed DCE stall, but when material was transferred to growth medium, *D. mccartyi* that carried out complete dechlorination to ethene grew, and carried out complete dechlorination when added back to the microcosms. The authors suggested that *D. mccartyi*, present in low numbers, was not competitive with methanogens and other hydrogenotrophs in the microcosms, but was competitive after a round of enrichment, a finding that needs to be verified in other samples.

6.6 Conclusions

The isolation of *D. mccartyi* is an example of how examining an applied problem, chloroethene contamination of groundwater, can lead to some very interesting and important fundamental science, a tradition in microbiology since Pasteur studied "diseases" of wine and beer. Had humankind avoided polluting groundwater with organohalides, the *Dehalococcoidia* would most likely be yet another small branch on the great tree of 16S rRNA sequences from uncultured organisms. It was also serendipitous that Jim Gossett, an environmental engineer with a good appreciation of microbiology, and myself, with a strong interest in chemical transformations carried out by microorganisms and a knack for culturing difficult anaerobes, were at the same institution and were friendly. However, it should be pointed out that several other groups were studying the microbiology of reductive dechlorination of chloroethenes in the early 1990s, and discovery of *Dehalococcoides* would barely have been delayed had we not begun our studies.

D. mccartyi is one of the most unusual bacteria ever described. It has a cell wall resembling those of *Archaea* and is among the smallest bacteria known with one of the smallest genomes found in a free-living bacterium. That it can use only H_2 as an electron donor and organohalides as electron acceptors makes it one of the most metabolically specialized organisms known, comparable to some methanogens that only use H_2–CO_2. Despite their streamlined genomes, *D. mccartyi* strains contain 10–36 distinct sets of *rdhAB* genes, most with adjacent transcriptional regulator genes, suggesting sophisticated regulation in this "one trick pony." We know the substrates for only a small fraction of the RDases in *D. mccartyi* and other organohalide reducing bacteria, and discerning the functions of the vast majority of RDases is a major unsolved problem.

D. mccartyi grows so poorly in pure culture that often turbidity is not visible, yet it often thrives in anaerobic groundwater habitats, driving the turnover of tons of organohalides worldwide. It has provided a fascinating glimpse into how organisms evolve in response to human activity, in this case the anthropogenic release of large amounts of organohalides. We still have much to learn about this tiny but powerful organism.

References

Adrian L, Szewzyk U, Wecke J, Görisch H (2000) Bacterial dehalorespiration with chlorinated benzenes. Nature 408(6812):580–583

Adrian L, Hansen SK, Fung JM, Görisch H, Zinder SH (2007a) Growth of *Dehalococcoides* strains with chlorophenols as electron acceptors. Environ Sci Technol 41(7):2318–2323

Adrian L, Rahnenführer J, Gobom J, Hölscher T (2007b) Identification of a chlorobenzene reductive dehalogenase in *Dehalococcoides* sp. strain CBDB1. Appl Environ Microbiol 73(23):7717–7724

Adrian L, Dudkova V, Demnerova K, Bedard DL (2009) "Dehalococcoides" sp. strain CBDB1 extensively dechlorinates the commercial polychlorinated biphenyl mixture aroclor 1260. Appl Environ Microbiol 75(13):4516–4524. doi:10.1128/AEM.00102-09

Albers SV, Meyer BH (2011) The archaeal cell envelope. Nat Rev Microbiol 9(6):414–426

Behrens S, Azizian MF, McMurdie PJ, Sabalowsky A, Dolan ME, Semprini L, Spormann AM (2008) Monitoring abundance and expression of "*Dehalococcoides*" species chloroethene-reductive dehalogenases in a tetrachloroethene-dechlorinating flow column. Appl Environ Microbiol 74(18):5695–5703. doi:10.1128/AEM.00926-08

Bunge M, Adrian L, Kraus A, Opel M, Lorenz WG, Andreesen JR, Görisch H, Lechner U (2003) Reductive dehalogenation of chlorinated dioxins by an anaerobic bacterium. Nature 421(6921):357–360

Cheng D, He J (2009) Isolation and characterization of "*Dehalococcoides*" sp. strain MB, which dechlorinates tetrachloroethene to trans-1,2-dichloroethene. Appl Environ Microbiol 75(18):5910–5918. doi:10.1128/AEM.00767-09

Christianssen N, Ahring BK (1996) *Desulfitobacterium hafniense*, sp. nov., an anaerobic reductively dechlorinating bacterium. Int J Syst Bacteriol 46:442–448

Crofts TS, Men Y, Alvarez-Cohen L, Taga ME (2014) A bioassay for the detection of benzimidazoles reveals their presence in a range of environmental samples. Front Microbiol 5:592. doi:10.3389/fmicb.2014.00592

Cupples AM, Spormann AM, McCarty PL (2003) Growth of a *Dehalococcoides*-like microorganism on vinyl chloride and *cis*-dichloroethene as electron acceptors as determined by competitive PCR. Appl Environ Microbiol 69(2):953–959

de Wit R, Bouvier T (2006) 'Everything is everywhere, but, the environment selects'; what did Baas Becking and Beijerinck really say? Environ Microbiol 8(4):755–758. doi:10.1111/j.1462-2920.2006.01017.x

Delgado AG, Kang DW, Nelson KG, Fajardo-Williams D, Miceli JF 3rd, Done HY, Popat SC, Krajmalnik-Brown R (2014) Selective enrichment yields robust ethene-producing dechlorinating cultures from microcosms stalled at cis-dichloroethene. PLoS ONE 9(6):e100654. doi:10.1371/journal.pone.0100654

DeWeerd KA, Mandelco L, Tanner RS, Woese CR, Suflita JM (1990) *Desulfomonile tiedjei* gen. nov. and sp. nov., a novel anaerobic dehalogenating, sulfate-reducing bacterium. Arch Microbiol 154:23–30

DiStefano TD, Gossett JM, Zinder SH (1991) Reductive dechlorination of high concentrations of tetrachloroethene to ethene by an anaerobic enrichment culture in the absence of methanogenesis. Appl Environ Microbiol 57(8):2287–2292

DiStefano TD, Gossett JM, Zinder SH (1992) Hydrogen as an electron donor for the dechlorination of tetrachloroethene by an anaerobic mixed culture. Appl Environ Microbiol 58:3622–3629

Duhamel M, Wehr SD, Yu L, Rizvi H, Seepersad D, Dworatzek S, Cox EE, Edwards EA (2002) Comparison of anaerobic dechlorinating enrichment cultures maintained on tetrachloroethene, trichloroethene, cis-dichloroethene and vinyl chloride. Water Res 36(17):4193–4202

Duhamel M, Mo K, Edwards EA (2004) Characterization of a highly enriched *Dehalococcoides*-containing culture that grows on vinyl chloride and trichloroethene. Appl Environ Microb 70(9):5538–5545

Durbin AM, Teske A (2011) Microbial diversity and stratification of South Pacific abyssal marine sediments. Environ Microbiol 13(12):3219–3234. doi:10.1111/j.1462-2920.2011.02544.x

Ellis DE, Lutz EJ, Odom JM, Buchanan RJ, Bartlett CL, Lee MD, Harkness MR, DeWeerd KA (2000) Bioaugmentation for accelerated in situ bioremediation. Environ Sci Technol 34:2254–2260

Fathepure BZ, Boyd SA (1988) Reductive dechlorination of perchloroethylene and the role of methanogens. FEMS Microbiol Lett 49:149–156

Fennell DE, Gossett JM, Zinder SH (1997) Comparison of butyric acid, ethanol, lactic acid, and propionic acid as hydrogen donors for the reductive dechlorination of tetrachloroethene. Environ Sci Technol 31:918–926

Fennell DE, Nijenhuis I, Wilson SF, Zinder SH, Häggblom MM (2004) *Dehalococcoides ethenogenes* strain 195 reductively dechlorinates diverse chlorinated aromatic pollutants. Envir Sci Technol 38:2075–2081

Freedman DL, Gossett JM (1991) Biodegradation of dichloromethane and its utilization as a growth substrate under methanogenic conditions. Appl Environ Microbiol 57(10):2847–2857

Gantzer CJ, Wackett LP (1991) Reductive dechlorination catalyzed by bacterial transition-metal coenzymes. Environ Sci Technol 25(4):715–722

Gerritse J, Renard V, Pedro-Gomes TM, Lawson PA, Collins MD, Gottschal JC (1996) *Desulfitobacterium* sp. strain PCE1, an anaerobic bacterium that can grow by reductive dechlorination of tetrachloroethene or *ortho*-chlorinated phenols. Arch Microbiol 165:132–140

Giovannoni SJ, Tripp HJ, Givan S, Podar M, Vergin KL, Baptista D, Bibbs L, Eads J, Richardson TH, Noordewier M, Rappe MS, Short JM, Carrington JC, Mathur EJ (2005) Genome streamlining in a cosmopolitan oceanic bacterium. Science 309(5738):1242–1245

Goris T, Schubert T, Gadkari J, Wubet T, Tarkka M, Buscot F, Adrian L, Diekert G (2014) Insights into organohalide respiration and the versatile catabolism of *Sulfurospirillum multivorans* gained from comparative genomics and physiological studies. Environ Microbiol 16(11):3562–3580. doi:10.1111/1462-2920.12589

Gossett JM (1987) Measurement of Henry's Law constants for C_1 and C_2 chlorinated hydrocarbons. Environ Sci Technol 21(12):202–208

Gribble GW (2010) Naturally occurring organohalogen compounds—a comprehensive update. Springer, New York

Hamonts K, Ryngaert A, Smidt H, Springael D, Dejonghe W (2014) Determinants of the microbial community structure of eutrophic, hyporheic river sediments polluted with chlorinated aliphatic hydrocarbons. FEMS Microbiol Ecol 87(3):715–732. doi:10.1111/1574-6941.12260

He J, Ritalahti KM, Aiello MR, Löffler FE (2003a) Complete detoxification of vinyl chloride by an anaerobic enrichment culture and identification of the reductively dechlorinating population as a *Dehalococcoides* species. Appl Environ Microbiol 69(2):996–1003

He J, Ritalahti KM, Yang KL, Koenigsberg SS, Löffler FE (2003b) Detoxification of vinyl chloride to ethene coupled to growth of an anaerobic bacterium. Nature 424(6944):62–65

He J, Sung Y, Krajmalnik-Brown R, Ritalahti KM, Löffler FE (2005) Isolation and characterization of *Dehalococcoides* sp. strain FL2, a trichloroethene (TCE)- and 1,2-dichloroethene-respiring anaerobe. Environ Microbiol 7(9):1442–1450

He J, Holmes VF, Lee PK, Alvarez-Cohen L (2007) Influence of vitamin B12 and cocultures on the growth of *Dehalococcoides* isolates in defined medium. Appl Environ Microbiol 73(9):2847–2853

Hendrickson ER, Payne JA, Young RM, Starr MG, Perry MP, Fahnestock S, Ellis DE, Ebersole RC (2002) Molecular analysis of *Dehalococcoides* 16S ribosomal DNA from chloroethene-contaminated sites throughout North America and Europe. Appl Environ Microbiol 68(2):485–495

Holliger C, Schraa G, Stams AJM, Zehnder AJB (1993) A highly purified enrichment culture couples the reductive dechlorination of tetrachloroethene to growth. Appl Environ Microbiol 59:2991–2997

Holliger C, Hahn D, Harmsen H, Ludwig W, Schumacher W, Tindall B, Vazquez F, Weiss N, Zehnder AJB (1998) *Dehalobacter restrictus* gen. nov. and sp. nov., a strictly anaerobic bacterium that reductively dechlorinates tetra- and trichloroethene in an anaerobic respiration. Arch Microbiol 169(4):313–321

Holmes VF, He J, Lee PK, Alvarez-Cohen L (2006) Discrimination of multiple *Dehalococcoides* strains in a trichloroethene enrichment by quantification of their reductive dehalogenase genes. Appl Environ Microbiol 72(9):5877–5883. doi:10.1128/AEM.00516-06

Hug LA, Maphosa F, Leys D, Löffler FE, Smidt H, Edwards EA, Adrian L (2013) Overview of organohalide-respiring bacteria and a proposal for a classification system for reductive dehalogenases. Philos Trans R Soc Lond B Biol Sci 368(1616):20120322. doi:10.1098/rstb.2012.0322 (rstb.2012.0322 [pii])

Hungate RE (1969) A roll tube method for cultivation of strict anaerobes. In: Norris JR, Ribbons DW (eds) Methods in Microbiology, vol 2B, pp 117–132

Inagaki F, Nunoura T, Nakagawa S, Teske A, Lever M, Lauer A, Suzuki M, Takai K, Delwiche M, Colwell FS, Nealson KH, Horikoshi K, D'Hondt S, Jorgensen BB (2006) Biogeographical distribution and diversity of microbes in methane hydrate-bearing deep marine sediments on the Pacific Ocean Margin. Proc Natl Acad Sci USA 103(8):2815–2820. doi:10.1073/pnas.0511033103

Jayachandran G, Görisch H, Adrian L (2003) Dehalorespiration with hexachlorobenzene and pentachlorobenzene by *Dehalococcoides* sp. strain CBDB1. Arch Microbiol 180(6):411–416

Jayachandran G, Görisch H, Adrian L (2004) Studies on hydrogenase activity and chlorobenzene respiration in *Dehalococcoides* sp. strain CBDB1. Arch Microbiol 182(6):498–504

Johnson DR, Lee PK, Holmes VF, Fortin AC, Alvarez-Cohen L (2005) Transcriptional expression of the tceA gene in a *Dehalococcoides*-containing microbial enrichment. Appl Environ Microbiol 71(11):7145–7151

Kaster AK, Mayer-Blackwell K, Pasarelli B, Spormann AM (2014) Single cell genomic study of *Dehalococcoidetes* species from deep-sea sediments of the Peruvian Margin. ISME J 8(9):1831–1842. doi:10.1038/ismej.2014.24

Kittelmann S, Friedrich MW (2008) Identification of novel perchloroethene-respiring microorganisms in anoxic river sediment by RNA-based stable isotope probing. Environ Microbiol 10(1):31–46. doi:10.1111/j.1462-2920.2007.01427.x

Krajmalnik-Brown R, Hölscher T, Thomson IN, Saunders FM, Ritalahti KM, Löffler FE (2004) Genetic identification of a putative vinyl chloride reductase in *Dehalococcoides* sp. strain BAV1. Appl Environ Microbiol 70(10):6347–6351

Krajmalnik-Brown R, Sung Y, Ritalahti KM, Michael Saunders F, Löffler FE (2007) Environmental distribution of the trichloroethene reductive dehalogenase gene (tceA) suggests lateral gene transfer among *Dehalococcoides*. FEMS Microbiol Ecol 59(1):206–214

Krumholz LR (1997) *Desulfuromonas chloroethenica* sp. nov. uses tetrachloroethylene and trichloroethylene as electron acceptors. Int J Syst Bacteriol 47:1262–1263

Krzmarzick MJ, Crary BB, Harding JJ, Oyerinde OO, Leri AC, Myneni SC, Novak PJ (2012) Natural niche for organohalide-respiring *Chloroflexi*. Appl Environ Microbiol 78(2):393–401

Kube M, Beck A, Zinder SH, Kuhl H, Reinhardt R, Adrian L (2005) Genome sequence of the chlorinated compound-respiring bacterium *Dehalococcoides* species strain CBDB1. Nat Biotechnol 23(10):1269–1273

LaRoe SL, Fricker AD, Bedard DL (2014) *Dehalococcoides mccartyi* strain JNA in pure culture extensively dechlorinates Aroclor 1260 according to polychlorinated biphenyl (PCB) dechlorination Process N. Environ Sci Technol 48(16):9187–9196. doi:10.1021/es500872t

Lee PK, He J, Zinder SH, Alvarez-Cohen L (2009) Evidence for nitrogen fixation by "*Dehalococcoides ethenogenes*" strain 195. Appl Environ Microbiol 75(23):7551–7555. doi:10.1128/AEM.01886-09

Lee PK, Cheng D, Hu P, West KA, Dick GJ, Brodie EL, Andersen GL, Zinder SH, He J, Alvarez-Cohen L (2011) Comparative genomics of two newly isolated *Dehalococcoides* strains and an enrichment using a genus microarray. ISME J 5(6):1014–1024. doi:10.1038/ismej.2010.202

Lee PK, Cheng D, West KA, Alvarez-Cohen L, He J (2013) Isolation of two new *Dehalococcoides mccartyi* strains with dissimilar dechlorination functions and their characterization by comparative genomics via microarray analysis. Environ Microbiol 15(8):2293–2305. doi:10.1111/1462-2920.12099

Löffler FE, Champine JE, Ritalahti KM, Sprague SJ, Tiedje JM (1997) Complete reductive dechlorination of 1,2-dichloropropane by anaerobic bacteria. Appl Environ Microbiol 63:2870–2875

Löffler FE, Sanford RA, Ritalahti KM (2005) Enrichment, cultivation, and detection of reductively dechlorinating bacteria. Methods Enzymol 397:77–111. doi:10.1016/S0076-6879(05)97005-5 (S0076-6879(05)97005-5 [pii])

Löffler FE, Yan J, Ritalahti KM, Adrian L, Edwards EA, Konstantinidis KT, Muller JA, Fullerton H, Zinder SH, Spormann AM (2013) *Dehalococcoides mccartyi* gen. nov., sp nov., obligately organohalide-respiring anaerobic bacteria relevant to halogen cycling and bioremediation, belong to a novel bacterial class, *Dehalococcoidia* classis nov., order *Dehalococcoidales* ord. nov and family *Dehalococcoidaceae* fam. nov., within the phylum *Chloroflexi*. Int J Syst Evol Micr 63:625–635. doi:10.1099/Ijs.0.034926-0

Magnuson JK, Stern RV, Gossett JM, Zinder SH, Burriss DR (1998) Reductive dechlorination of tetrachloroethene to ethene by a two-component enzyme pathway. Appl Environ Microbiol 64:1270–1275

Magnuson JK, Romine MF, Burris DR, Kingsley MT (2000) Trichloroethene reductive dehalogenase from *Dehalococcoides ethenogenes*: sequence of *tceA* and substrate range characterization. Appl Environ Microbiol 66(12):5141–5147

Major DW, McMaster ML, Cox EE, Edwards EA, Wworatzek SM, Hendrickson ER, Starr MG, Payne JA, Buonamici LW (2002) Field demonstration of successful bioaugmentation to achieve dechlorination of tetrachloroethene to ethene. Envir Sci Technol 36:5106–5116

Mansfeldt CB, Rowe AR, Heavner GL, Zinder SH, Richardson RE (2014) Meta-analyses of *Dehalococcoides mccartyi* strain 195 transcriptomic profiles identify a respiration rate-related gene expression transition point and interoperon recruitment of a key oxidoreductase subunit. Appl Environ Microbiol 80(19):6062–6072. doi:10.1128/AEM.02130-14

Marco-Urrea E, Nijenhuis I, Adrian L (2011a) Transformation and carbon isotope fractionation of tetra- and trichloroethene to trans-dichloroethene by *Dehalococcoides* sp. strain CBDB1. Environ Sci Technol 45(4):1555–1562. doi:10.1021/es1023459

Marco-Urrea E, Paul S, Khodaverdi V, Seifert J, von Bergen M, Kretzschmar U, Adrian L (2011b) Identification and characterization of a re-citrate synthase in *Dehalococcoides* strain CBDB1. J Bacteriol 193(19):5171–5178. doi:10.1128/JB.05120-11 (JB.05120-11 [pii])
Marco-Urrea E, Seifert J, von Bergen M, Adrian L (2012) Stable isotope peptide mass spectrometry to decipher amino acid metabolism in *Dehalococcoides* strain CBDB1. J Bacteriol 194(16):4169–4177. doi:10.1128/JB.00049-12 (JB.00049-12 [pii])
Maymó-Gatell X, Tandoi V, Gossett JM, Zinder SH (1995) Characterization of an H_2-utilizing anaerobic enrichment culture that reductively dechlorinates tetrachloroethene to vinyl chloride and ethene in the complete absence of methanogenesis and acetogenesis. Appl Environ Microbiol 61:3928–3933
Maymó-Gatell X, Chien YT, Gossett JM, Zinder SH (1997) Isolation of a bacterium that reductively dechlorinates tetrachloroethene to ethene. Science 276:1568–1571
Maymó-Gatell X, Anguish T, Zinder SH (1999) Reductive dechlorination of chlorinated ethenes and 1,2-dichloroethane by "*Dehalococcoides ethenogenes*" strain 195. Appl Environ Microbiol 65:3108–3113
Maymó-Gatell X, Nijenhuis I, Zinder SH (2001) Reductive dechlorination of *cis*-dichloroethene and vinyl chloride by "*Dehalococcoides ethenogenes*" strain 195. Environ Sci Technol 35:516–521
McMurdie PJ, Behrens SF, Muller JA, Goke J, Ritalahti KM, Wagner R, Goltsman E, Lapidus A, Holmes S, Löffler FE, Spormann AM (2009) Localized plasticity in the streamlined genomes of vinyl chloride respiring *Dehalococcoides*. PLoS Genet 5(11):e1000714
McMurdie PJ, Hug LA, Edwards EA, Holmes S, Spormann AM (2011) Site-specific mobilization of vinyl chloride respiration islands by a mechanism common in *Dehalococcoides*. BMC Genom 12:287. doi:10.1186/1471-2164-12-287 (1471-2164-12-287 [pii])
Miller E, Wohlfarth G, Diekert G (1997) Studies on tetrachloroethene respiration in *Dehalospirillum multivorans*. Arch Microbiol 166(6):379–387
Moe WM, Yan J, Nobre MF, da Costa MS, Rainey FA (2009) *Dehalogenimonas lykanthroporepellens* gen. nov., sp. nov., a reductively dehalogenating bacterium isolated from chlorinated solvent-contaminated groundwater. Int J Syst Evol Microbiol 59 (Pt 11):2692–2697. doi:10.1099/ijs.0.011502-0
Morris RM, Sowell S, Barofsky D, Zinder S, Richardson R (2006) Transcription and mass-spectroscopic proteomic studies of electron transport oxidoreductases in *Dehalococcoides ethenogenes*. Envir Microbiol 8:1499–1509
Morris RM, Fung JM, Rahm BG, Zhang S, Freedman DL, Zinder SH, Richardson RE (2007) Comparative proteomics of *Dehalococcoides* spp. reveals strain-specific peptides associated with activity. Appl Environ Microbiol 73:320–326
Morris JJ, Lenski RE, Zinser ER (2012) The Black Queen Hypothesis: evolution of dependencies through adaptive gene loss. mBio 3 (2). doi:10.1128/mBio.00036-12
Müller JA, Rosner BM, Von Abendroth G, Meshulam-Simon G, McCarty PL, Spormann AM (2004) Molecular identification of the catabolic vinyl chloride reductase from *Dehalococcoides* sp. strain VS and its environmental distribution. Appl Environ Microbiol 70(8):4880–4888
Neumann A, Wohlfarth G, Diekert G (1998) Tetrachloroethene dehalogenase from *Dehalospirillum multivorans*: cloning, sequencing of the encoding genes, and expression of the *pceA* gene in *Escherichia coli*. J Bacteriol 180:4140–4145
Nijenhuis I, Zinder SH (2005) Characterization of hydrogenase and reductive dehalogenase activities of *Dehalococcoides ethenogenes* strain 195. Appl Environ Microbiol 71(3):1664–1667
Padilla-Crespo E, Yan J, Swift C, Wagner DD, Chourey K, Hettich RL, Ritalahti KM, Löffler FE (2014) Identification and environmental distribution of dcpA, which encodes the reductive dehalogenase catalyzing the dichloroelimination of 1,2-dichloropropane to propene in organohalide-respiring chloroflexi. Appl Environ Microbiol 80(3):808–818. doi:10.1128/AEM.02927-13 (AEM.02927-13 [pii])

Parthasarathy A, Stich TA, Lohner ST, Lesnefsky A, Britt RD, Spormann AM (2015) Biochemical and EPR-spectroscopic investigation into heterologously expressed vinyl chloride reductive dehalogenase (VcrA) from *Dehalococcoides mccartyi* strain VS. J Am Chem Soc 137(10):3525–3532. doi:10.1021/ja511653d

Pöritz M, Goris T, Wubet T, Tarkka MT, Buscot F, Nijenhuis I, Lechner U, Adrian L (2013) Genome sequences of two dehalogenation specialists—*Dehalococcoides mccartyi* strains BTF08 and DCMB5 enriched from the highly polluted bitterfeld region. FEMS Microbiol Lett 343(2):101–104. doi:10.1111/1574-6968.12160

Rahm BG, Richardson RE (2008) Correlation of respiratory gene expression levels and pseudo-steady-state PCE respiration rates in *Dehalococcoides ethenogenes*. Environ Sci Technol 42(2):416–421

Rappe MS, Connon SA, Vergin KL, Giovannoni SJ (2002) Cultivation of the ubiquitous SAR11 marine bacterioplankton clade. Nature 418(6898):630–633

Ritalahti KM, Amos BK, Sung Y, Wu Q, Koenigsberg SS, Löffler FE (2006) Quantitative PCR targeting 16S rRNA and reductive dehalogenase genes simultaneously monitors multiple *Dehalococcoides* strains. Appl Environ Microbiol 72(4):2765–2774. doi:10.1128/AEM.72.4.2765-2774.2006 (72/4/2765 [pii])

Rosner BM, McCarty PL, Spormann AM (1997) In vitro studies on reductive vinyl chloride dehalogenation by an anaerobic mixed culture. Appl Environ Microbiol 63:4139–4144

Rossetti S, Blackall LL, Majone M, Hugenholtz P, Plumb JJ, Tandoi V (2003) Kinetic and phylogenetic characterization of an anaerobic dechlorinating microbial community. Microbiology 149 (Pt 2):459–469

Rowe AR, Heavner GL, Mansfeldt CB, Werner JJ, Richardson RE (2012) Relating chloroethene respiration rates in *Dehalococcoides* to protein and mRNA biomarkers. Environ Sci Technol 46(17):9388–9397. doi:10.1021/es300996c

Schaefer CE, Condee CW, Vainberg S, Steffan RJ (2009) Bioaugmentation for chlorinated ethenes using *Dehalococcoides* sp.: comparison between batch and column experiments. Chemosphere 75 (2):141–148. doi:10.1016/j.chemosphere.2008.12.041

Schink B (1997) Energetics of syntrophic cooperation in methanogenic degradation. Microbiol Rev 61:262–280

Schipp CJ, Marco-Urrea E, Kublik A, Seifert J, Adrian L (2013) Organic cofactors in the metabolism of *Dehalococcoides mccartyi* strains. Philos Trans R Soc Lond B Biol Sci 368(1616):20120321. doi:10.1098/rstb.2012.0321 (rstb.2012.0321 [pii])

Scholz-Muramatsu H, Neumann A, Messmer M, Moore E, Diekert G (1995) Isolation and characterization of *Dehalospirillum multivorans* gen nov, sp nov, a tetrachloroethene-utilizing, strictly anaerobic bacterium. Arch Microbiol 163(1):48–56

Schumacher W, Holliger C (1996) The proton/electron ratio of the menaquinone-dependent electron transport from dihydrogen to tetrachloroethylene in "*Dehalobacter restrictus*". J Bacteriol 178:2328–2333

Schumacher W, Holliger C, Zehnder AJ, Hagen WR (1997) Redox chemistry of cobalamin and iron-sulfur cofactors in the tetrachloroethene reductase of *Dehalobacter restrictus*. FEBS Lett 409(3):421–425

Schut GJ, Adams MW (2009) The iron-hydrogenase of *Thermotoga maritima* utilizes ferredoxin and NADH synergistically: a new perspective on anaerobic hydrogen production. J Bacteriol 191(13):4451–4457

Seshadri R, Adrian L, Fouts DE, Eisen JA, Phillippy AM, Methe BA, Ward NL, Nelson WC, Deboy RT, Khouri HM, Kolonay JF, Dodson RJ, Daugherty SC, Brinkac LM, Sullivan SA, Madupu R, Nelson KE, Kang KH, Impraim M, Tran K, Robinson JM, Forberger HA, Fraser CM, Zinder SH, Heidelberg JF (2005) Genome sequence of the PCE-dechlorinating bacterium *Dehalococcoides ethenogenes*. Science 307(5706):105–108

Soboh B, Pinske C, Kuhns M, Waclawek M, Ihling C, Trchounian K, Trchounian A, Sinz A, Sawers G (2011) The respiratory molybdo-selenoprotein formate dehydrogenases of *Escherichia coli* have hydrogen: benzyl viologen oxidoreductase activity. BMC Microbiol 11:173. doi:10.1186/1471-2180-11-173

Suflita JM, Horowitz A, Shelton DR, Tiedje JM (1982) Dehalogenation: a novel pathway for the anaerobic biodegradation of haloaromatic compounds. Science 218:1115–1117

Sung Y, Fletcher KE, Ritalahti KM, Apkarian RP, Ramos-Hernandez N, Sanford RA, Mesbah NM, Löffler FE (2006a) *Geobacter lovleyi* sp. nov. strain SZ, a novel metal-reducing and tetrachloroethene-dechlorinating bacterium. Appl Environ Microbiol 72(4):2775–2782

Sung Y, Ritalahti KM, Apkarian RP, Löffler FE (2006b) Quantitative PCR confirms purity of strain GT, a novel trichloroethene-to-ethene-respiring *Dehalococcoides* isolate. Appl Environ Microbiol 72(3):1980–1987. doi:10.1128/AEM.72.3.1980-1987.2006 (72/3/1980 [pii])

Tandoi V, DiStefano TD, Bowser PA, Gossett JM, Zinder SH (1994) Reductive dehalogenation of chlorinated ethenes and halogenated ethanes by a high-rate anaerobic enrichment culture. Environ Sci Technol 28:973–979

Tang YJ, Yi S, Zhuang WQ, Zinder SH, Keasling JD, Alvarez-Cohen L (2009) Investigation of carbon metabolism in "*Dehalococcoides ethenogenes*" strain 195 by use of isotopomer and transcriptomic analyses. J Bacteriol 191(16):5224–5231. doi:10.1128/JB.00085-09

Tang S, Chan WW, Fletcher KE, Seifert J, Liang X, Löffler FE, Edwards EA, Adrian L (2013) Functional characterization of reductive dehalogenases by using blue native polyacrylamide gel electrophoresis. Appl Environ Microbiol 79(3):974–981. doi:10.1128/AEM.01873-12 (AEM.01873-12 [pii])

Vogel TM, Criddle CS, McCarty PL (1987) Transformations of halogenated aliphatic compounds. Environ Sci Technol 21(8):722–736

Wagner A, Cooper M, Ferdi S, Seifert J, Adrian L (2012) Growth of *Dehalococcoides mccartyi* strain CBDB1 by reductive dehalogenation of brominated benzenes to benzene. Environ Sci Technol 46(16):8960–8968. doi:10.1021/es3003519

Waller AS, Krajmalnik-Brown R, Löffler FE, Edwards EA (2005) Multiple reductive-dehalogenase-homologous genes are simultaneously transcribed during dechlorination by *Dehalococcoides-containing* cultures. Appl Environ Microbiol 71(12):8257–8264

Wang S, Chng KR, Wilm A, Zhao S, Yang KL, Nagarajan N, He J (2014) Genomic characterization of three unique *Dehalococcoides* that respire on persistent polychlorinated biphenyls. Proc Natl Acad Sci USA 111(33):12103–12108. doi:10.1073/pnas.1404845111

Wasmund K, Schreiber L, Lloyd KG, Petersen DG, Schramm A, Stepanauskas R, Jorgensen BB, Adrian L (2014) Genome sequencing of a single cell of the widely distributed marine subsurface *Dehalococcoidia*, phylum *Chloroflexi*. ISME J 8(2):383–397. doi:10.1038/ismej.2013.143

Wood JM, Kennedy FS, Wolfe RS (1968) The reaction of multihalogenated hydrocarbons with free and bound reduced vitamin B_{12}. Biochem 7:1707–1713

Yan J, Im J, Yang Y, Löffler FE (2013) Guided cobalamin biosynthesis supports *Dehalococcoides mccartyi* reductive dechlorination activity. Philos Trans R Soc Lond B Biol Sci 368(1616):20120320. doi:10.1098/rstb.2012.0320

Yi S, Seth EC, Men YJ, Stabler SP, Allen RH, Alvarez-Cohen L, Taga ME (2012) Versatility in corrinoid salvaging and remodeling pathways supports corrinoid-dependent metabolism in *Dehalococcoides mccartyi*. Appl Environ Microbiol 78(21):7745–7752. doi:10.1128/AEM.02150-12

Zhuang WQ, Yi S, Bill M, Brisson VL, Feng X, Men Y, Conrad ME, Tang YJ, Alvarez-Cohen L (2014) Incomplete Wood-Ljungdahl pathway facilitates one-carbon metabolism in organohalide-respiring *Dehalococcoides mccartyi*. Proc Natl Acad Sci USA 111(17):6419–6424. doi:10.1073/pnas.1321542111

Zinder SH, Anguish T, Lobo T (1987) Isolation and characterization of a thermophilic acetotrophic strain of *Methanothrix*. Arch Microbiol 146:315–322

Chapter 7
The Genus *Dehalogenimonas*

William M. Moe, Fred A. Rainey and Jun Yan

Abstract The genus *Dehalogenimonas* is a relatively recently described taxonomic group whose members grow via organohalide respiration. Members of the genus are able to couple growth with the reductive dehalogenation of a variety of polychlorinated alkanes including 1,2-dichloroethane, 1,2-dichloropropane, 1,1,2-trichloroethane, 1,1,2,2-tetrachloroethane, and 1,2,3-trichloropropane. Strains of the genus are strictly anaerobic, mesophilic, Gram-negative staining, and non-spore-forming, with cells that are small, irregular cocci. This chapter reviews the current state of knowledge regarding the genus and the growing body of evidence that suggests that this genus can play an important role in dechlorinating anthropogenic contaminants in environmental systems.

7.1 Discovery

At present, the genus *Dehalogenimonas* contains two species. The first of the species to be discovered, *Dehalogenimonas lykanthroporepellens*, was isolated from groundwater contaminated by high concentrations of chlorinated alkanes and alkenes at a Superfund Site located near Baton Rouge, Louisiana (USA) (Yan et al. 2009a;

W.M. Moe (✉)
Department of Civil and Environmental Engineering,
Louisiana State University, Baton Rouge, LA 70803, USA
e-mail: moemwil@lsu.edu

F.A. Rainey
Department of Biological Sciences, University of Alaska Anchorage,
Anchorage, AK 99508, USA
e-mail: farainey@uaa.alaska.edu

J. Yan
Department of Microbiology, University of Tennessee, Knoxville, TN 37996, USA
e-mail: jyan5@utk.edu

L. Adrian and F.E. Löffler (eds.), *Organohalide-Respiring Bacteria*,
DOI 10.1007/978-3-662-49875-0_7

Moe et al. 2009). At the site where the organism was first isolated, a variety of liquid chemical manufacturing wastes were historically disposed of by direct discharge to unlined earthen lagoons, and contaminants remain in the subsurface as dense non-aqueous-phase liquids (Bowman et al. 2006). The first strains of the species to be isolated were recovered from groundwater during an investigation aimed at identifying viable bacteria able to survive the high solvent concentrations present at the site (Yan et al. 2009a).

A second species, *Dehalogenimonas alkenigignens*, was later recovered from a different portion of the same Superfund Site but in a contaminated groundwater plume containing much lower concentrations of chlorinated contaminants (tens rather than hundreds of mg/L) (Bowman et al. 2013).

7.2 Isolation

All strains of *D. lykanthroporepellens* and *D. alkenigignens* that have been isolated to purity to date were obtained using dilution-to-extinction in liquid media (Moe et al. 2009; Yan et al. 2009a; Bowman et al. 2013). After isolation in liquid media, the strains did not form visible colonies on media solidified with agar or gellan gum even after 2 months incubation (Moe et al. 2009; Yan et al. 2009a; Bowman et al. 2013). The liquid media used successfully in isolating *Dehalogenimonas* strains are similar to those reported for isolation of *Dehalococcoides* strains (Maymó-Gatell et al. 1997) and have taken advantage of the fact that representatives from the genus *Dehalogenimonas*, like the related genus *Dehalococcoides*, are resistant to the antibiotics ampicillin and vancomycin (Moe et al. 2009; Yan et al. 2009a; Bowman et al. 2013).

Because they are strict anaerobes, isolation procedures for *Dehalogenimonas* strains have employed both stringent methods to exclude oxygen and the use of reducing agents to lower oxidation-reduction potential. The first strains of *D. lykanthroporepellens* to be obtained in pure culture were isolated using medium containing a combination of sodium sulfide (1.5 mM $Na_2S{\cdot}9H_2O$) and L-cysteine hydrochloride (1.6 mM) as reducing agents (Yan et al. 2009a). Subsequently, it was found that the strains grew better in pure culture in media supplemented with titanium (III) citrate solution (Zehnder and Wuhrmann 1976) as a reducing agent at a final concentration of 1.0 mM Ti(III) and 2.0 mM citrate in place of sodium sulfide. Titanium-citrate reduced medium was later used in the successful isolation of *D. alkenigignens* strains (Bowman et al. 2013). It has been observed in our laboratory that oxidation-reduction potential lower than −450 mV is generally necessary for consistent growth of *Dehalogenimonas* strains in pure culture. Because such low redox potential can result in some abiotic dechlorination reactions, inclusion of appropriate abiotic controls is essential.

Medium formulations used for the successful isolation of *Dehalogenimonas* strains have included a suite of vitamins, trace minerals, and potential carbon sources (Yan et al. 2009a; Bowman et al. 2013). The strict need for all of these constituents has not been rigorously evaluated. Hydrogen gas has been provided as an electron donor via the gas phase and a chlorinated alkane (e.g., 1,1,2-trichloroethane) has been provided as an electron acceptor in previously successful isolation efforts. Hydrogen gas concentrations of 5–10 % (v/v) in the gas headspace have been employed previously, however, higher concentrations including pure H_2 are also effective for growth of the strains (unpublished data). Slight over-pressurization of the gas headspace is effective for preventing the unintended introduction of oxygen into serum bottles or tubes when growing the strains.

Dehalogenimonas strains have been preserved successfully in anaerobic medium amended with 5 % (v/v) filter-sterilized dimethylsulfoxide in glass serum bottles sealed with butyl rubber stoppers and aluminum crimp caps stored at −80 °C (Bowman et al. 2013, Yan et al. 2009a).

7.3 General Characteristics

Strains of *D. alkenigignens* and *D. lykanthroporepellens* share several common phenotypic features including strictly anaerobic respiration and the coupling of cell growth with reductive dehalogenation. Cells of both *D. alkenigignens* and *D. lykanthroporepellens* are relatively small (0.3–1.1 μm diameter), non-spore-forming, Gram-negative, irregular cocci (Fig. 7.1). Motility has not been observed in laboratory tests (Yan et al. 2009b; Bowman et al. 2013). All representatives of the genus *Dehalogenimonas* that have been reported thus far exhibit most rapid dechlorination in the mesophilic temperature range with optimal growth near

Fig. 7.1 Morphological features of *Dehalogenimonas lykanthroporepellens* BL-DC-9^T as observed by scanning electron microscopy. Bar: 0.5 μm (photo credit: KS Bowman and Y Xiao)

30 °C. Both species are resistance to the antibiotics ampicillin (1.0 g/L) and vancomycin (0.1 g/L) (Moe et al. 2009; Yan et al. 2009a; Bowman et al. 2013).

With respect to the functional roles that the species may play in transforming halogenated environment contaminants, the dechlorination pathways observed for *D. lykanthroporepellens* and *D. alkenigignens* strains appear to be essentially identical. All reductive dechlorination reactions determined to date for both species appear to involve an exclusively dihaloelimination reaction mechanism involving simultaneous removal of two halogens from adjacent carbon atoms and formation of a carbon–carbon double bond (Moe et al. 2009; Yan et al. 2009a; Bowman et al. 2013). 1,2-Dichloroethane (1,2-DCA) and 1,2-dichloropropane (1,2-DCP) are transformed to the nontoxic final products ethene and propene, respectively. 1,1,2-Trichloroethane is transformed to vinyl chloride. 1,1,2,2-Tetrachloroethane is transformed to dichloroethenes. While the formation of lower chlorinated ethenes, particularly vinyl chloride, in environmental settings is of potential concern from a risk-based perspective, the fact that *Dehalogenimonas* strains can grow in concert with bacteria of the dichloroethene- and vinyl chloride-respiring genus *Dehalococcoides* can decrease this risk and allow for complete 1,1,2-trichloroethane and 1,1,2,2-tetrachloroethane dehalogenation to non-halogenated, nontoxic final products.

D. lykanthroporepellens was the first species demonstrated in pure culture to reductively dehalogenate 1,2,3-trichloropropane (Yan et al. 2009a). The direct product of the dihaloelimination reaction mediated by *D. lykanthroporepellens* is allyl chloride (3-chloro-1-propene), a relatively unstable compound that can undergo relatively rapid abiotic hydrolysis [with half-life of ~4 days in water at 30 °C (Krijgsheld and van der Gen 1986)] as well as reaction with sulfides and cysteine present as reducing agents in the growth medium employed to first isolate strains of the species. Ultimately, this gave rise to the production of a variety of non-chlorinated final products including diallyl sulfide, diallyl disulfide, allyl mercaptan, and allyl methyl sulfide (Fig. 7.2) (Yan et al. 2009a). These final products exhibit a pungent, garlicky aroma, and inspired the species name, which can be translated from the Latin and Greek roots as meaning "repelling to werewolves." When grown in media lacking sulfide as a reducing agent, allyl alcohol, which forms from abiotic hydrolysis of the dihaloelimination product allyl chloride, is the primary product of 1,2,3-trichloropropane dechlorination by both *Dehalogenimonas* species (Bowman et al. 2013).

Experimental tests of the ability of *Dehalogenimonas* strains to dehalogenate compounds containing halogen substituents other than chlorine have been limited; however, it has been demonstrated that *D. lykanthroporepellens* BL-DC-9^T reductively transformed 1,2-dibromoethane to ethene and 1,2-dibromopropane to propene with concomitant release of inorganic bromide ions (unpublished data). Thus, the pathways for reductive dehalogenation of brominated compounds appears to follow the same dihaloelimination pathways observed for chlorinated analogs of the compounds tested.

Both *D. lykanthroporepellens* and *D. alkenigignens* strains have been shown to dehalogenate polychlorinated alkanes even when the contaminants are

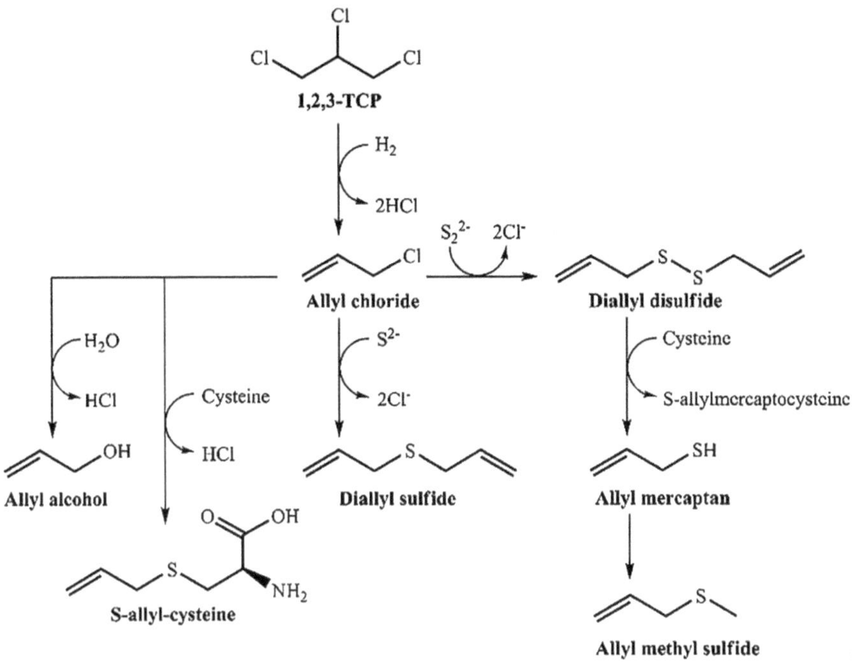

Fig. 7.2 Products and reaction pathways observed for dechlorination of 1,2,3-trichloropropane with *Dehalogenimonas lykanthroporepellens* strains grown in the presence of cysteine and sulfide as reducing agents (Yan et al. 2009a)

present at relatively high concentrations. In particular, the type strains for both *Dehalogenimonas* species were found to dechlorinate 1,2-DCA, 1,2-DCP, and 1,1,2-trichloroethane present at initial aqueous-phase concentrations at least as high as 8.7, 4.0 and 3.5 mM, respectively (Maness et al. 2012).

When provided with mixtures of multiple chlorinated alkanes, it was demonstrated that both *D. lykanthroporepellens* and *D. alkenigignens* strains exhibited preferential dechlorination of 1,1,2-trichloroethane over both 1,2-DCA and 1,2-DCP (Dillehay et al. 2014). 1,2-DCA in particular was not dechlorinated until 1,1,2-trichloroethane reached low concentrations. In contrast, both species concurrently dechlorinated 1,2-DCA and 1,2-DCP over a comparably large concentration range (Dillehay et al. 2014). Mechanistic reasons for the preferential utilization of some halogenated compounds over others have not yet been elucidated. Plausible explanations for why this may occur include the notion that bacteria may preferentially use the most thermodynamically favorable electron acceptor available (i.e., the one providing the most energy release) or that the species may respond stresses induced by the higher toxicity of 1,1,2-trichloroethane relative to 1,2-DCP and 1,2-DCA by preferentially removing the most toxic compound first (Dillehay et al. 2014).

While both *D. lykanthroporepellens* or *D. alkenigignens* strains isolated to date dechlorinate a variety of vicinally halogenated alkanes, the strains do not utilize aliphatic alkanes that contain only a single chlorine substituent (1-chloropropane, 2-chloropropane), aliphatic alkanes with multiple chlorine substituents occurring only on one carbon atom (1,1-dichloroethane, 1,1,1-trichloroethane), chlorinated methanes [dichloromethane (methylene chloride), trichloromethane (chloroform), tetrachloromethane (carbon tetrachloride)], chlorinated benzenes (1-chlorobenzene, 1,2-dichlorobenzene) or chlorinated ethenes (tetrachlorothene, trichloroethene, *cis*-1,2-dichloroethene, *trans*-1,2-dichloroethene, or vinyl chloride). H_2 serves as an electron donor for growth of both species. Growth is not supported by acetate, butyrate, citrate, ethanol, fructose, fumarate, glucose, lactate, lactose, malate, methanol, methyl ethyl ketone, propionate, pyruvate, succinate or yeast extract in the absence of H_2 (Moe et al. 2009; Yan et al. 2009a; Bowman et al. 2013).

For strains grown in pure culture, both *D. alkenigignens* and *D. lykanthroporepellens* have been shown to carry out reductive dechlorination in the pH range of 6.0–8.0, but not at pH $\leq$ 5.5 or $\geq$8.5. Given that the groundwater at the location from which *D. lykanthroporepellens* strains were first isolated had a pH of 5.1, the habitat of these species probably extends to pH levels somewhat lower than determined in the laboratory test conditions employed in the characterization of pure cultures (Moe et al. 2009). Indeed, use of qPCR has revealed the presence of 16S rRNA gene sequences sharing high identity with *Dehalogenimonas* isolates in chlorinated alkane-contaminated groundwater with pH as low as 4.2 (unpublished data).

Similarly, the oxidation-reduction potential measured on-site at the time of groundwater collection in the samples from which *D. alkenigignens* and *D. lykanthroporepellens* type strains were first isolated ranged from −46 to −124 mV. In pure culture laboratory tests, it has been observed that substantially lower oxidation-reduction potential is needed to reliably obtain growth. This may reflect the fact that lower oxidation-reduction potential niches (e.g., associated with subsurface particulates or cell aggregates) can serve as a habitat for *Dehalogenimonas* species even when "bulk" oxidation-reduction potential is considerably higher, or it may reflect differences in growth requirements for pure cultures versus mixed cultures.

Habitats of *Dehalogenimonas* appear to be chlorinated solvent contaminated soil and groundwater. Their "natural" habitat devoid of anthropogenic chlorinated solvent contamination remains unknown, as are potential metabolic pathways allowing growth in the absence of halogenated aliphatic compounds as electron acceptors.

Because of the limited range of electron donors (e.g., H_2) known to be utilized by *Dehalogenimonas*, members of the genus likely grow in syntrophic relationships with other bacteria in the environment. Several newly described fermentative and hydrogen producing species, some with an ability to grow in the presence of high concentrations of chlorinated solvents, have been isolated from the same groundwater sources as *Dehalogenimonas* strains. These include *Actinomyces naturae* (Rao et al. 2012), *Azospira restricta* (Bae et al. 2007), *Brooklawnia cerclae* (Bae et al. 2006b), *Clostridium hydrogeniformans,* and *Clostridium cavendishii* (Bowman et al. 2009, 2010), *Propionicicella superfundia* (Bae et al. 2006a), and *Pelosinus defluvii* (Moe et al. 2012). Clone library analysis suggests that some of these taxa can comprise a large fraction of the bacterial populations where *Dehalogenimonas* species are found. For example, 15.7 % of the sequences in a 16S rRNA gene library constructed using universal bacterial primers to assess bacterial community structure in the groundwater at the location where *D. lykanthroporepellens* strains were first isolated shared >99 % identity with the species later described as *A. naturae* (Bowman et al. 2006, Rao et al. 2012).

Data suggest that in at least some environments contaminated with chlorinated solvents, the genus *Dehalogenimonas* can constitute a relatively large fraction of the bacterial community. Sequences clustering within the genus *Dehalogenimonas* comprised 7.7 and 9.3 % of sequences in 16S rRNA gene libraries constructed using universal bacterial primers for groundwater from which the first two strains of *D. alkenigignens* were isolated (Bowman et al. 2013). The application of qPCR with community DNA extracted from groundwater in the vicinity of where *D. lykanthroporepellens* strains were first isolated revealed that *Dehalogenimonas* 16S rRNA gene copies ranged from 0.0014 to 18.6 % of the 16S rRNA gene copies determined using universal bacterial primers (Yan et al. 2009b).

Using an alternative qPCR method that targets both species of *Dehalogenimonas*, Chen et al. (2014) reported a substantial (more than 100-fold) increase in *Dehalogenimonas* 16S rRNA gene copies following the subsurface injection of a fermentable substrate (agricultural feed grade molasses) and alkalinity (sodium bicarbonate) in a pilot-scale study to assess the feasibility of biostimulation (i.e., supplying electron donors) as a remediation strategy. The increase in *Dehalogenimonas* 16S rRNA gene copies at the same time concentrations of 1,1,2-trichloroethane and 1,2-DCA decreased, combined with the fact that the only known electron acceptors for this bacterial genus are halogenated aliphatic compounds, suggests that *Dehalogenimonas* species may respond favorably to biostimulation remediation strategies when applied to groundwater contaminated by polychlorinated alkanes (Chen et al. 2014).

7.4 Detailed Phylogeny and Diversity

Although *D. alkenigignens* and *D. lykanthroporepellens* share many phenotypic traits, the two species can be differentiated from one another in several ways. Based on 16S rRNA gene sequence comparisons, *D. alkenigignens* IP3-3^T (JQ994266) and *D. lykanthroporepellens* BL-DC-9^T (EU679419) represent distinct lineages within the order *Dehalococcoidales* in the class *Dehalococcoidia* of the phylum *Chloroflexi* (Löffler et al. 2013). With type strains of the two species sharing only 96.2 % pairwise similarity in 16S rRNA gene sequences, sequencing can readily differentiate the two and serve as a basis for identification (Fig. 7.3) (Bowman et al. 2013).

The major cellular fatty acids determined for *D. alkenigignens* strains IP3-3^T and SBP-1 ($C_{18:1}\omega9c$, $C_{16:0}$, $C_{14:0}$, and $C_{16:1}\omega9c$) were also found to be present in *D. lykanthroporepellens* BL-DC-8 and BL-DC-9^T (Bowman et al. 2013). The

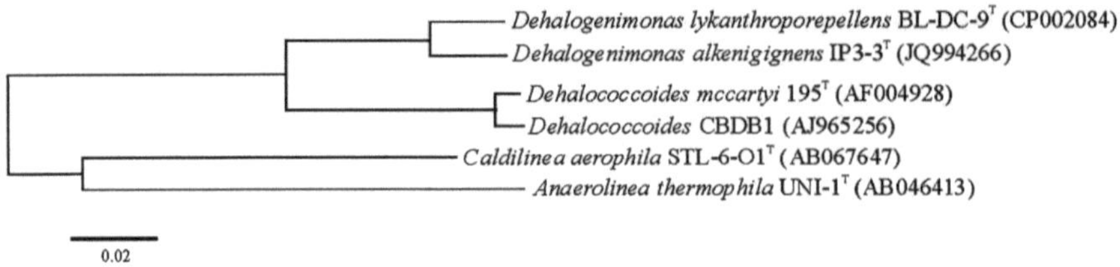

Fig. 7.3 Neighbor-joining dendrogram based on analysis of 16S rRNA gene sequences showing the phylogenetic relationship of *Dehalogenimonas* species to related genera. Scale bar: 2 substitutions per 100 nt positions

proportions of these major cellular fatty acids differ, however, with $C_{16:1}\omega9c$ higher in *D. lykanthroporepellens* strains and the proportion of C14:0 higher in *D. alkenigignens* strains. A number of minor cellular fatty acids (e.g., $C_{12:0}$, $C_{18:3}\omega6c$(6,9,12), and unidentified fatty acids with equivalent chain lengths of 11.980, 13.937, and 15.056) were also detected in only one species or the other, further serving as a basis for distinguishing *D. alkenigignens* from *D. lykanthroporepellens* (Bowman et al. 2013).

Phenotypically, *D. alkenigignens* IP3-3^T and SBP-1 also differ from *D. lykanthroporepellens* BL-DC-9^T in their inability to reductively dehalogenate in the presence of 2 % NaCl or in medium containing 3 mM sodium sulfide (Bowman et al. 2013). *D. alkenigignens* strains also contain slightly higher mol% G + C content than *D. lykanthroporepellens* as measured by HPLC (Bowman et al. 2013).

While *Dehalogenimonas* isolates obtained to date in pure culture exhibit what appears to be an exclusively dihaloelimination pathway, transforming polychlorinated alkanes but not other classes of halogenated compounds tested, a growing body of evidence suggests that there is additional metabolic and phylogenetic diversity within the genus. For example, sequences clustering most closely with the genus *Dehalogenimonas* comprised a sizeable fraction (2.16 %) of a 16S rRNA gene library prepared using high throughput sequencing (Illumina) for analysis of an enrichment culture that reductively dechlorinated a mixture of polychlorinated biphenyls (Aroclor 1260) (Wang and He 2013). Subsequent determination of a nearly complete 16S rRNA gene sequence revealed the presence of a phylotype (Dhg-CG3, GenBank accession JQ990328) that shares 99.9 % identity with the 16S rRNA gene sequence of *D. alkenigignens* IP3-3^T (differing by a single nucleotide) and 96.2 % identity with *D. lykanthroporepellens* BL-DC-9^T. Further results from qPCR analysis demonstrating a large increase in *Dehalogenimonas* 16S rRNA gene copies concomitant with Aroclor 1260 dechlorination provides convincing evidence for their involvement in PCB dechlorination (Wang and He 2013).

Similarly, Manchester et al. (2012) reported a phylotype (JQ599651) that shares 96.6 % identity with the 16S rRNA gene sequence *D. lykanthroporepellens* BL-DC-9^T and 96.5 % with *D. alkenigignens* IP3-3^T. Based on a combination of qPCR, microcosm studies, and phylogenetic analysis, they concluded that their new phylotype represented a *Dehalogenimonas* sp. able to couple growth with reductive dechlorination of *trans*-1,2-dichloroethene to vinyl chloride. Thus, it appears that the genus *Dehalogenimonas* contains as-yet uncultured representatives that can dehalogenate a broader range of chlorinated compounds that have not yet been demonstrated to be utilized by strains isolated in pure culture to date.

Numerous additional sequences of uncultured bacteria sharing >95.0 % sequence identity with *Dehalogenimonas* type strains, a level considered to represent grouping at the genus level, but only ~90 % sequence identity with *Dehalococcoides mccartyi* strains appear in the GenBank database (Chen et al. 2014). These have originated from a wide range of geographic locations including the Arctic Ocean, Baltic Sea, Canada, China, Germany, Hungary, Spain, Taiwan, and the United States, suggesting that the genus is widely distributed across the globe. A majority of these sequences originated from contaminated environments or dechlorinating enrichment cultures (Chen et al. 2014).

7.5 Genomic Features

At present, the genome of *D. lykanthroporepellens* BL-DC-9^T (CP002084, NC_014314) is the only published genome of the genus of *Dehalogenimonas*. Strain BL-DC-9^T has a 1.69 Mbp circular chromosome with 55.04 mol% G + C content and no plasmids (Sidaramappa et al. 2012). This is larger in sizc and higher in mol% G + C content than the genome sequences reported for strains of the related genus *Dehalococcoides* [e.g., *D. mccartyi* strain 195^T has a 1.47 Mbp chromosome with 48.85 mol% G + C content (Seshadri et al. 2005)]. The basic genome properties are briefly summarized in Table 7.1.

Table 1 Genome statistics for *Dehalogenimonas lykanthroporepellens* strain BL-DC-9^T (Siddaramappa et al. 2012)

Attribute	Valuc
Genome size (bp)	1,686,510
DNA coding region (bp)	1,479,636
DNA G + C content (mol%)	55.04 %
Total gcnes	1771
tRNA genes	47
rRNA operons	1
Protein-coding genes	1720
Pseudo genes	61
Genes in paralog clusters	240
Genes assigned to COGs	1257
Genes with signal peptides	417
Genes with transmembrane helices	347
CRISPR repeats	0

7.5.1 General Genomic Features

The genome of *D. lykanthroporepellens* BL-DC-9^T contains a prophage region (accounting for ~4 % of the chromosome) as well as insertion sequence elements (accounting for ~4.3 % of the chromosome) that encode 74 full-length or truncated transposases. Several of the genes unique to strain BL-DC-9^T are found in the prophage region or are associated with insertion sequence elements. Thus, it appears that horizontal gene transfer has played a major role shaping the evolution of this genus (Siddaramappa et al. 2012).

The genome of *D. lykanthroporepellens* BL-DC-9^T contains a single large subunit rRNA (23S-5S) locus and a single, distantly located, small subunit rRNA (16S) locus. *D. lykanthroporepellens* BL-DC-9^T harbors 47 tRNA genes including those that code for all 20 standard amino acids and the less common amino acid selenocysteine. A *selCDAB* operon encoding putative selenocysteine biosynthesis proteins is also present with an adjacent insertion sequence element, suggesting that this operon may have been horizontally acquired.

Comparative genomic analysis revealing that 432 core genes are shared between *D. lykanthroporepellens* BL-DC-9^T and *D. mccartyi* genomes, which is far less than the 1029 core genes conserved among *Dehalococcoides* genomes, suggests divergent evolutionary paths in these two obligate organohalide-respiring genera (McMurdie et al. 2011).

D. lykanthroporepellens BL-DC-9^T contains an *ectABC* operon for ectoine synthesis and regulation, *proVWX* and *opuABCD* operons for proline/glycine-betaine transport, and a putative bifunctional gene for mannosylglycerate synthesis. Ectoine, glycine-betaine, and proline are common compatible solutes utilized in mesophilic bacteria to overcome osmotic stress (Santos and da Costa 2002). In contrast, thermophilic prokaryotes utilize unusual compatible solutes such as mannosylglycerate (Santos and da Costa 2002). The possession of multiple osmotic stress response mechanisms provides a plausible explanation for the ability of strain BL-DC-9^T to survive moderate salinity conditions (i.e., up to 2 % NaCl, w/v) (Moe et al. 2009).

Although motility has not been observed in laboratory tests with *D. lykanthroporepellens* BL-DC-9^T, its genome contains multiple loci encoding putative type IV pili elements. Type IV pili are known to mediate twitching motility of non-flagellated Gram-negative bacteria (Mattick 2002; Wall and Kaiser 1999) and are also associated with horizontal gene transfer (Averhoff and Friedrich 2003). These loci include Dehly_1227 (*pilM*, type IV pilus assembly protein), Dehly_1228 (*pilT*, twitching motility protein), three discrete *pilT* homologs encoding putative twitching motility proteins (Dehly_0313, Dehly_0787, Dehly_0905), and a *pilD* homolog encoding a prepilin peptidase (Dehly_1130). *D. lykanthroporepellens* BL-DC-9^T also contains a locus encoding putative type II secretion system proteins (Dehly_1221, *pilB/pulE*, Dehly_1222, *pilC/pulF*). *Dehalococcoides* strains also contain homologs of *pilM*, *pilT*, *pilD*, *pilB*, and *pilC* (43–73 % identity at the predicted protein level).

7.5.2 Reductive Dehalogenases

Dihaloelimination of vicinally halogenated alkanes, putatively catalyzed by a specialized reductive dehalogenase (RDase) enzyme system, is a characteristic feature that distinguishes *Dehalogenimonas* isolates from other organohalide-respiring bacteria. A total of 17 "full-length" reductive dehalogenase homologous genes (*rdhA*) that contain conserved sequences encoding proteins with (1) a twin-arginine translocation (Tat) system signal motif in the N-terminus, (2) two iron–sulfur cluster binding motifs closer toward the C-terminus, and (3) a total length of 455–493 amino acids, in single copies, are annotated throughout the chromosome of *D. lykanthroporepellens* BL-DC-9^T (Siddaramappa et al. 2012). See Chap. 16 "Diversity, evolution and environmental distribution of RDase genes" for more information about RDase signature motifs. Eight additional *rdhA* genes either lacking the twin-arginine sequence (Dehly_0069 and 1582) or substantially truncated (Dehly_0075, 0479, 1520, 1523, 1534, and 1541) were also identified in the genome (Siddaramappa et al. 2012; Mukherjee et al. 2014).

The chromosomal location of several (13 of 25) *rdhA* genes adjacent to insertion sequence elements in *D. lykanthroporepellens* BL-DC-9^T suggests horizontal acquisition from an unknown host or hosts (Siddaramappa et al. 2012; Mukherjee et al. 2014). Putative horizontal gene acquisition has also been described for *rdhA* genes of *Dehalococcoides* strains (McMurdie et al. 2011; Krajmalnik-Brown et al. 2007). Aside from the gene with locus tag Dehly_1524 (discussed below), the translated RDases predicated from *rdhA* genes share low amino acid identities (31–78 %) to the putative or functionally confirmed RDases (e.g., TceA, VcrA, and BvcA) in *D. mccartyi* isolates (Siddaramappa et al. 2012; Löffler et al. 2013). This finding is consistent with the experimental observation that *Dehalogenimonas* isolates are unable to couple growth with reductive dechlorination of chlorinated ethenes such as trichloroethene, *cis*-1,2-dichloroethene, or vinyl chloride.

Experiments employing end-point reverse transcription-PCR (RT-PCR) in conjunction with primers targeting all 25 *rdhA* genes in *D. lykanthroporepellens* BL-DC-9^T revealed that 19 *rdhA* genes were consistently expressed in cultures actively dechlorinating three different electron acceptors, 1,2-DCA, 1,2-DCP, and 1,2,3-trichloropropane (Mukherjee et al. 2014). The identical transcription pattern, including the detection of multiple transcripts for *rdhA* genes that appear to be truncated, during growth of *D. lykanthroporepellens* BL-DC-9^T with all three electron acceptors hampers the direct implication of any one gene or group of genes with specific functionality.

Using an integrated approach that combined RNA transcripts, proteomic analyses, and enzymatic activity assays, the gene with locus tag Dehly_1524 was identified as encoding a RDase that catalyzes the dihaloelimination of 1,2-DCP-to propene and inorganic chloride (Padilla-Crespo et al. 2014). Dehly_1524 is one of only five "full-length" *rdhA* genes (Dehly_0156, 0275, 1054, 1152, and 1524) in the *D. lykanthroporepellens* BL-DC-9^T genome that is adjacent to an *rdhB* gene that encodes a putative hydrophobic RDase membrane anchor protein

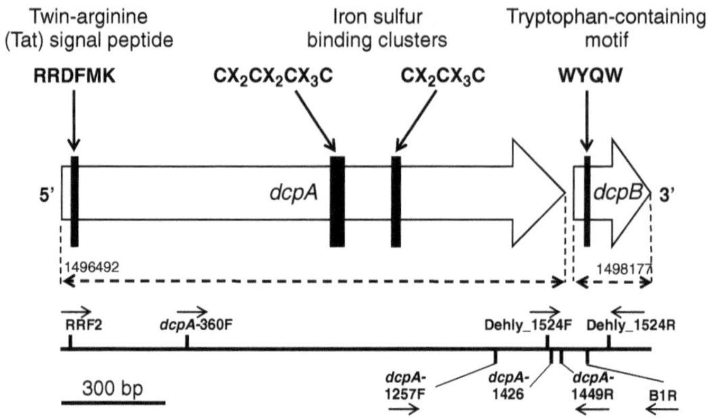

Fig. 7.4 Organization of the *dcpAB* operon and the encoded proteins in *Dehalogenimonas lykanthroporepellens* BL-DC-9^T. Characteristic dehalogenase features encoded by the *dcpA* as indicated (*top*) include a twin-arginine translocation (Tat) system signal peptide RRDFMK at the N-terminus and two iron–sulfur binding clusters closer to the C-terminus. A motif with the amino acid sequence WYQW, similar to the WYEW motif conserved in many *Dehalococcoides* strains, occurs at the edge of one of the two transmembrane elements predicated by TMMOD (Kahsay et al. 2005) for the protein encoded by *dcpB* (Padilla-Crespo et al. 2014). Numbers indicate genome coordinates spanning genes with locus tags Dehly_1524 and Dehly_1525. Annealing sites for degenerate PCR primer set RRF2/BR1 targeting RDase-like genes (Krajmalnik-Brown et al. 2004), *dcpA*-specific PCR primer set *dcpA*-360F/*dcpA*-1449R and TaqMan qPCR assay *dcpA*-1257F/*dcpA*-1426/*dcpA*-1449R (Padilla-Crespo et al. 2014), and *Dehalogenimonas dcpAB*-specific primer set Dehly_1524F/Dehly_1524R (Mukherjee et al. 2014) are labeled (*bottom*). The *dcpA*-1257F, *dcpA*-1426 (probe), *dcpA*-1449R, and B1R primers each contain one mismatch with the *dcpAB* operon of *D. lykanthroporepellens* BL-DC-9^T. Highly similar arrangements of *dcpA* and *dcpB* genes are present in *Dehalococcoides mccartyi* strains RC and KS (Padilla-Crespo et al. 2014)

(Fig. 7.4). This organization in an *rdhAB* operon is shared by other RDase genes in organohalide-respiring bacteria including *D. mccartyi* isolates, and the *dcpA* (Dehly_1524) and *dcpB* (Dehly_1525) genes were found to be co-transcribed (Padilla-Crespo et al. 2014; Mukherjee et al. 2014).

The most closely related protein sequences to the DcpA of *D. lykanthroporepellens* BL-DC-9^T (92 % amino acid identity) are two putative RDases (GenBank accession numbers JX826286 and JX826287) harbored by *D. mccartyi* strains KS and RC, respectively. These *Dehalococcoides* strains couple growth with 1,2-DCP-to-propene dechlorination in highly enriched cultures (Padilla-Crespo et al. 2014). Sequences sharing high sequence identity (91–93 %) to the *dcpA* of *D. lykanthroporepellens* BL-DC-9^T have also been detected from environmental samples and microcosms where active 1,2-DCP-to-propene dechlorination occurred, suggesting that the *dcpA* gene or DcpA RDase may serve as a suitable biomarker for detection and quantification of 1,2-DCP dechlorinators and as a diagnostic tool for monitoring in situ reductive dechlorination of 1,2-DCP (Padilla-Crespo et al. 2014). Because the catalytic activity of DcpA has not yet been tested on electron acceptors other than 1,2-DCP, it is unclear if DcpA functions as a general dihaloelimination RDase or if alternative RDases are responsible for dechlorination of different vicinally halogenated alkanes by this strain. The detection of *dcpA* gene transcripts in RT-PCR with *D. lykanthroporepellens* BL-DC-9^T cultures that were actively dechlorinating not only 1,2-DCP but also 1,2-DCA and 1,2,3-trichloropropane leaves open the possibility that the same enzyme was responsible for transformation of all three chlorinated ethanes (Mukherjee et al. 2014).

A large majority of the *rdhA* genes in *D. mccartyi* strains are associated with MarR or two-component system transcriptional regulators (Kube et al. 2005; Seshadri et al. 2005). In contrast, such transcriptional regulators are associated with only a small minority of the 25 *rdhA* genes in *D. lykanthroporepellens* BL-DC-9^T (Siddaramappa et al. 2012). The detection of amplicons from the limited number of transcription factors (four) along with transcripts from their cognate *rdhA* genes via RT-PCR in *D. lykanthroporepellens* BL-DC-9^T cultures that were actively dechlorinating 1,2-DCP, 1,2-DCA and 1,2,3-trichloropropane suggests that expression of some *rdhA* genes in this species is regulated by MarR and two-component system transcriptional regulators (Mukherjee et al. 2014). The basis of regulation for many of the *rdhA* genes, however, presently remains unknown.

References

Averhoff B, Friedrich A (2003) Type IV pili-related natural transformation systems: DNA transport in mesophilic and thermophilic bacteria. Arch Microbiol 180(6):385–393. doi:10.1007/s00203-003-0616-6

Bae HS, Moe WM, Yan J, Tiago I, da Costa MS, Rainey FA (2006a) *Propionicicella superfundia* gen. nov., sp. nov., a chlorosolvent-tolerant propionate-forming, facultative anaerobic bacterium isolated from contaminated groundwater. Syst Appl Microbiol 29(5):404–413. doi:10.1016/j.syapm.2005.11.004

Bae HS, Moe WM, Yan J, Tiago I, da Costa MS, Rainey FA (2006b) *Brooklawnia cerclae* gen. nov., sp. nov., a propionate-forming bacterium isolated from chlorosolvent-contaminated groundwater. Int J Syst Evol Microbiol 56(8):1977–1983. doi:10.1099/ijs.0.64317-0

Bae HS, Rash BA, Rainey FA, Nobre MF, Tiago I, da Costa MS, WM Moe (2007) Description of *Azospira restricta* sp. nov., a nitrogen-fixing bacterium isolated from groundwater. Int J Syst Evol Microbiol 57(7):1521–1526. doi:10.1099/ijs.0.64965-0

Bowman KS, Moe WM, Rash BA, Bae H-S, Rainey FA (2006) Bacterial diversity of an acidic Louisiana groundwater contaminated by DNAPL containing chloroethanes and other solvents. FEMS Microbiol Ecol 58(1):120–133. doi:10.1111/j.1574-6941.2006.00146.x

Bowman KS, Rainey FA, Moe WM (2009) Production of hydrogen by *Clostridium* species in the presence of chlorinated solvents. FEMS Microbiol Lett 290(2):188–194. doi:10.1111/j.1574-6968.2008.01419.x

Bowman KS, Dupré RE, Rainey FA, Moe WM (2010) *Clostridium hydrogeniformans* sp. nov. and *Clostridium cavendishii* sp. nov., hydrogen-producing bacteria from chlorinated solvent-contaminated groundwater. Int J Syst Evol Microbiol 60(2):358–363. doi:10.1099/ijs.0.013169-0

Bowman KS, Nobre MF, da Costa MS, Rainey FA, Moe WM (2013) *Dehalogenimonas alkenigignens* sp. nov., a chlorinated alkane dehalogenating bacterium isolated from groundwater. Int J Syst Evol Microbiol 63(4):1492–1498. doi:10.1099/ijs.0.045054-0

Chen J, Bowman KS, Rainey FA, Moe WM (2014) Reassessment of PCR primers targeting 16S rRNA genes of the organohalide-respiring genus *Dehalogenimonas*. Biodegradation 25:747–756. doi:10.1007/s10532-014-9696-z

Dillehay JL, Bowman KS, Rainey FA, Moe WM (2014) Substrate interactions in dehalogenation of 1,2-dichloroethane, 1,2-dichloropropane, and 1,1,2-trichloroethane mixtures by *Dehalogenimonas* spp. Biodegradation 25(2):301–312. doi:10.1007/s10532-013-9661-2

Kahsay R, Liao L, Gao G (2005) An improved hidden Markov model for transmembrane protein topology prediction and its applications to complete genomes. Bioinformatics 21(9):1853–1858. doi:10.1093/bioinformatics/bti303

Krajmalnik-Brown R, Hölscher T, Thomson IN, Saunders FM, Ritalahti KM, Löffler FE (2004) Genetic identification of a putative vinyl chloride reductase in *Dehalococcoides* sp. strain BAV1. Appl Environ Microbiol 70(10):6347–6351. doi:10.1128/AEM.70.10.6347-6351.2004

Krajmalnik-Brown R, Sung Y, Ritalahti KM, Saunders FM, Löffler FE (2007) Environmental distribution of the trichloroethene reductive dehalogenase gene (*tceA*) suggests lateral gene transfer among *Dehalococcoides*. FEMS Microbiol Ecol 59:206–214. doi:10.1111/j.1574-6941.2006.00243.x

Krijgsheld KR, van der Gen A (1986) Assessment of the impact of the emission of certain organochlorine compounds on the aquatic environment. Part II: Allylchloride, 1,3- and 2,3-dichloropropene. Chemosphere 15(7):861–880. doi:10.1016/0045-6535(86)90052-4

Kube M, Beck A, Zinder SH, Kuhl H, Reinhardt R, Adrian L (2005) Genome sequence of the chlorinated compound-respiring bacterium *Dehalococcoide*s species strain CBDB1. Nature Biotechnol 23(10):1269–1273. doi:10.1038/nbt1131

Löffler FE, Yan J, Ritalahti KM, Adrian L, Edwards EA, Konstantinidis KT, Müller JA, Fullerton H, Zinder SH, Spormann AM (2013) *Dehalococcoides mccartyi* gen. nov., sp. nov., obligate organohalide-respiring anaerobic bacteria, relevant to halogen cycling and bioremediation, belong to a novel bacterial class, *Dehalococcoidia* classis nov., order *Dehalococcoidales* ord. nov. and family *Dehalococcoidaceae* fam. nov., within the phylum *Chloroflexi*. Int J Syst Evol Microbiol 63:625–635. doi:10.1099/ijs.0.034926-0

Manchester MJ, Hug LA, Zarek M, Zila A, Edwards EA (2012) Discovery of a trans-dichloroethene-respiring *Dehalogenimonas* species in the 1,1,2,2-tetrachloroethane-dechlorinating WBC-2 consortium. Appl Environ Microbiol 78(15):5280–5287. doi:10.1128/AEM.00384-12

Maness AD, Bowman KS, Yan J, Rainey FA, Moe WM (2012) *Dehalogenimonas* spp. can reductively dehalogenate high concentrations of 1,2-dichloroethane, 1,2-dichloropropane, and 1,1,2-trichloroethane. AMB Express 2(1):54. doi:10.1186/2191-0855-2-54

Mattick JS (2002) Type IV pili and twitching motility. Annu Rev Microbiol 2002(56):289–314. doi:10.1146/annurev.micro.56.012302.160938

Maymó-Gatell X, Chien YT, Gossett JM, Zinder SH (1997) Isolation of a bacterium that reductively dechlorinates tetrachloroethene to ethene. Science 276(5318):1568–1571. doi:10.1126/science.276.5318.1568

McMurdie PJ, Hug LA, Edwards EA, Holmes S, Spormann AM (2011) Site-specific mobilization of vinyl chloride respiration islands by a mechanism common in *Dehalococcoides*. BMC Genom 12:287

Moe WM, Yan J, Nobre MF, da Costa MS, Rainey FA (2009) *Dehalogenimonas lykanthroporepellens* gen. nov., sp. nov., a reductive dehalogenating bacterium isolated from chlorinated solvent contaminated groundwater. Int J Syst Evol Microbiol 59(11):2692–2697. doi:10.1099/ijs.0.011502-0

Moe WM, Stebbing RE, Rao JU, Bowman KS, Nobre MF, da Costa MS, Rainey FA (2012) *Pelosinus defluvii* sp. nov., isolated from chlorinated solvent-contaminated groundwater, emended description of the genus *Pelosinus*, and transfer of *Sporotalea propionica* to *Pelosinus propionicus* comb. nov. Int J Syst Evol Microbiol 62(6):1369–1376. doi:10.1099/ijs.0.033753-0

Mukherjee K, Bowman KS, Rainey FA, Siddaramappa S, Challacombe J, Moe WM (2014) *Dehalogenimonas lykanthroporepellens* BL-DC-9^T simultaneously transcribes many *rdhA* genes during dehalorespiration with 1,2-DCA, 1,2-DCP and 1,2,3-TCP as electron acceptors. FEMS Microbiol Lett 345(2):111–118. doi:10.1111/1574-6968.12434

Padilla-Crespo E, Yan J, Swift C, Wagner DD, Chourey K, Hettich RL, Ritalahti KM, Löffler FE (2014) Identification and environmental distribution of *dcpA*, which encodes the reductive dehalogenase catalyzing the dichloroelimination 1,2-dichloropropane to propene in organohalide-respiring *Chloroflexi*. Appl Environ Microbiol 80(3):808–818. doi:10.1128/AEM.02927-13

Rao JU, Rash BA, Nobre MF, da Costa MS, Rainey FA, Moe WM (2012) *Actinomyces naturae* sp. nov., the first *Actinomyces* sp. isolated from a non- human or animal source. Antonie van Leeuwenhoek. J Microbiol 101(1):155–168. doi:10.1007/s10482-011-9644-4

Santos H, da Costa MS (2002) Compatible solutes of organisms that live in hot saline environments. Environ Microbiol 4:501–509. doi:10.1046/j.1462-2920.2002.00335.x

Seshadri R, Adrian L, Fouts DE, Eisen JA, Phillippy AM, Methe BA, Ward NL, Nelson WC, Deboy RT, Khouri HM, Kolonay JF, Dodson RJ, Daugherty SC, Brinkac LM, Sullivan SA, Madupu R, Nelson KE, Kang KH, Impraim M, Tran K, Robinson JM, Forberger HA, Fraser CM, Zinder SH, Heidelberg JF (2005) Genome sequence of the PCE-dechlorinating bacterium *Dehalococcoides ethenogenes*. Science 307:105–108. doi:10.1126/science.1102226

Siddaramappa S, Challacombe JF, Delano SF, Green LD, Daligault H, Bruce D, Detter C, Tapia R, Han S, Goodwin L, Han J, Woyke T, Pitluck S, Pennacchio L, Nolan M, Land M, Chang YJ, Kyrpides NC, Ovchinnikova G, Hauser L, Lapidus A, Yan J, Bowman KS, da Costa MS, Rainey FA, Moe WM (2012) Complete genome sequence of *Dehalogenimonas lykanthroporepellens* type strain (BL-DC-9^T) and comparison to "*Dehalococcoides*" strains. Stand Genomic Sci 6(2):251–264. doi:10.4056/sigs.2806097

Wall D, Kaiser D (1999) Type IV pili and cell motility. Molecular Microbiol 32(1):1–10. doi:10.1046/j.1365-2958.1999.01339

Wang S, He J (2013) Phylogenetically distinct bacteria involve extensive dechlorination of Aroclor 1260 in sediment-free cultures. PLoS ONE 8(3):e59178. doi:10.1371/journal.pone.0059178

Yan J, Rash BA, Rainey FA, Moe WM (2009a) Isolation of novel bacteria within the *Chloroflexi* capable of reductive dechlorination of 1,2,3-trichloropropane. Environ Microbiol 11(4):833–843. doi:10.1111/j.1462-2920.2008.01804.x

Yan J, Rash BA, Rainey FA, Moe WM (2009b) Detection and quantification of *Dehalogenimonas* and "*Dehalococcoides*" populations via PCR-based protocols targeting 16S rRNA genes. Appl Environ Microbiol 75(23):7560–7564. doi:10.1128/AEM.01938-09

Zehnder AJB, Wuhrmann K (1976) Titanium (III) citrate as a nontoxic oxidation-reduction buffering system for the culture of obligate anaerobes. Science 194(4270):1165–1166. doi:10.1126/science.793008

Chapter 8
The Genus *Dehalobacter*

Julien Maillard and Christof Holliger

Abstract The genus *Dehalobacter* embraces bacterial populations that seem to exclusively degrade organohalides. All isolates in pure culture and highly enriched strains are obligate organohalide-respiring bacteria that use hydrogen as energy and electron source, acetate as carbon source, and an organohalide as terminal electron acceptor. Depending on the strain, they are restricted to the use of only one or two organohalides from the same chemical group (i.e. aliphatic or aromatic organohalides), a few strains however can use several compounds and from different groups. Organohalides used by *Dehalobacter* are chlorinated methanes, ethanes, ethenes, cyclohexanes, benzenes, phenols and phthalides. However, two enrichments dominated by *Dehalobacter* spp. indicate another metabolic pathway with a specific organohalide, namely fermentation of dichloromethane. No particular habitat can be defined for this bacterial genus since the different strains have been enriched and isolated from various matrices such as sediments, aquifers and anaerobic sludge from waste treatment processes. The small motile rods (0.5 μm in diameter, 2–3 μm long) usually stain Gram-negative, contain, however, peptidoglycan features of Gram-positives, menaquinones, and cytochrome *b*, and are surrounded by proteinaceous S-layer. Phylogenetically *Dehalobacter* is affiliated to low GC Gram-positive *Firmicutes*. Recently available genome sequences revealed that *Dehalobacter* spp. harbour an unexpected large number of putative reductive dehalogenase genes (10-27 paralogs) showing a relatively high diversity, several hydrogenases of different types, an 11-subunit respiration complex I, all necessary genes for the Wood-Ljungdahl pathway and the biosynthesis pathway of corrinoids, and seemed to confirm that *Dehalobacter* spp. cannot

J. Maillard · C. Holliger (✉)
School of Architecture, Civil and Environmental Engineering,
Laboratory for Environmental Biotechnology, Ecole Polytechnique
Fédérale de Lausanne (EPFL), Station 6, 1015 Lausanne, Switzerland
e-mail: christof.holliger@epfl.ch

J. Maillard
e-mail: julien.maillard@epfl.ch

L. Adrian and F.E. Löffler (eds.), *Organohalide-Respiring Bacteria*,
DOI 10.1007/978-3-662-49875-0_8

carry out any other respiration process than organohalide respiration. Hence, the hydrogen and carbon metabolisms seem to be more complex than anticipated, and also the observed restriction to few organohalides as electron acceptor is perhaps not reflecting the real dechlorination capabilities of *Dehalobacter* strains with the numerous putative reductive dehalogenase genes in their genomes.

8.1 Enrichment and Isolation of *Dehalobacter* spp.

The first *Dehalobacter* population has been enriched from a packed-bed column that was filled with a mixture of river Rhine sediment sampled near Wageningen, The Netherlands, and with ground anaerobic granular sludge from a wastewater treatment plant of a sugar refinery (Holliger et al. 1993). The column was operated under anaerobic conditions and percolated with a mineral medium containing lactate and tetrachloroethene (PCE). PCE was completely reduced to ethane with trichloroethene (TCE), *cis*-1,2-dichloroethene (cDCE), vinyl chloride (VC) and ethene as intermediates (De Bruin et al. 1992). Initial enrichments in batch cultures containing a mineral medium low in chloride amended with lactate or H_2–CO_2 formed all PCE dechlorination products found in the column effluent, but not ethane. To provide high amounts of the electron acceptor PCE at non-toxic concentration in the water phase, a two-phase system has been used where PCE has been dissolved in hexadecane (Holliger et al. 1993). These enrichments were transferred on H_2–CO_2 only in the presence of 2-bromoethanesulfonate (BES) in order to inhibit methanogens. Adding fermented yeast extract as source of potentially needed growth factors, omitting selenium and tungsten in the trace element solution, and using dilution series to extinction led to the loss of the homoacetogenic activity. Finally, after further dilution series to extinction, a microscopically homogeneous culture of thin rods staining gram-negatively was obtained. Due to a high sensitivity of this bacterium to temperatures above 35 °C, cultivation in solidified medium was difficult but finally succeeded. Single colonies have been transferred to liquid medium which allowed isolating a pure culture and better characterizing this bacterium respiring with PCE while using hydrogen as electron donor. The organism was named *Dehalobacter restrictus* strain PER-K23 and will in the following be simply referred to as *D. restrictus* (Holliger et al. 1998).

A very close relative of *D. restrictus* has been enriched from a laboratory fixed-bed reactor that was inoculated with material from a full-scale anaerobic charcoal reactor treating groundwater contaminated with dichloromethane (DCM) and traces of TCE and PCE (Wild et al. 1996). Several dilution series on medium with TCE as electron acceptor resulted in a culture containing two morphologically distinct bacteria, a small rod and a motile vibrio. The latter could be subsequently eliminated by cultivating this enrichment in the presence of sodium molybdate, a selective inhibitor of sulfate-reducing bacteria. Three additional dilution series to extinction resulted in a pure culture of the small rod named strain TEA. The sequence of the 16S rRNA gene cloned was for 99.7 % identical with the sequence of *D. restrictus*.

Only four additional *Dehalobacter* strains have since then been obtained in pure culture. Strain TCA1 has been isolated from upper Hudson River sediment enrichments fed with lactate or H_2-acetate and with 1,1,1-trichloroethane (1,1,1-TCA) as electron acceptor, and incubated at 25 °C (Sun et al. 2002). With dilution series in deep agarose shake cultures containing low-melting agarose white faint colonies have been obtained within three weeks in tubes where dechlorination activity was observed. *Dehalobacter* sp. strain TCP1 has been enriched and isolated from cultures dechlorinating 2,4,6-trichlorophenol and inoculated with digester sludge from an industrial wastewater treatment plant in Singapore by using serial dilutions and agar shakes (Wang et al. 2014). Finally, two strains have been obtained in pure culture using dichlorobenzenes as electron acceptor, strain 12DCB1 dechlorinates 1,2-dichlorobenzene only, whereas strain 13DCB1 can use 1,2- and 1,3-dichlorobenzene (Nelson et al. 2014). Although a detailed description of the isolation of these strains has not yet been published, the enrichment procedure involved the use of BES or mevinolin to inhibit methanogenic activity, and butyrate as substrate to select for organisms with a high affinity for hydrogen (Nelson et al. 2011). In addition, the two-phase system with hexadecane was used to maximize the amount of dichlorobenzene added and the amendment of additional organics has been minimized by adding 20 mg/L casamino acids.

Two cultures containing a dechlorinating *Dehalobacter* population have been described as rather defined co-cultures harbouring another non-dechlorinating bacterium which has not been possible to eliminate. A co-culture of a *Dehalobacter* and a *Sedimentibacter* population that dechlorinated β-hexachlorocyclohexane (β-HCH) to monochlorobenzene and benzene, has been obtained from a contaminated sandy soil (van Doesburg et al. 2005). Use of BES allowed eliminating methanogenic activity after seven transfers and obtaining a culture of four morphologically distinct microorganisms. They were separated on a Percoll gradient resulting in a pure culture of a *Sedimentibacter* strain. The Percoll gradient band containing predominantly the short rod of *Dehalobacter* resulted always in a culture of two bacteria when transferred in liquid medium with H_2 and β-HCH. However, specific metagenomic approaches allowed obtaining the genome sequence of *Dehalobacter* sp. strain E1 from this co-culture (Maphosa et al. 2012).

A defined co-culture of a *Dehalobacter* and a *Acetobacterium* population has been obtained from an enrichment fed with ethanol as substrate and 1,2-dichloroethane (1,2-DCA) as electron acceptor (Grostern et al. 2009). Switching to H_2 as electron donor and acetate as carbon source, transferring regularly and carrying out dilution series resulted in the loss of methanogenic activity first followed by the loss of the earlier detected population of *Dehalococcoides*. No colony formation was observed in roll tubes amended with H_2 and 1,2-DCA despite the formation of ethene. Also the use of a medium lacking selenium and tungsten did not lead to the elimination of the homoacetogen as it was the case in the enrichment of *D. restrictus* (Grostern et al. 2009). However, it has been shown that the *Acetobacterium* population was unable to dechlorinate 1,2-DCA and that growth of *Dehalobacter* correlated with dechlorination of the chlorinated compound.

Finally, different enrichments dominated by *Dehalobacter* populations have been described but not as pure or defined co-cultures. An enrichment culture designated KFL contained two phylotypes named FTH1 and FTH2 that were affiliated with *Dehalobacter* and that increased to 30 % cell abundance upon dechlorination of 4,5,6,7-tetrachlorophthalide as determined by qPCR (Yoshida et al. 2009). The KFL culture has been enriched from paddy soil with lactate as substrate and the rather insoluble chlorinated compound as thin film on the walls of the incubation glass vials. The KFL culture dechlorinated 4,5,6,7-tetrachlorophtalide with H_2 as the electron donor in the presence of 0.01 % peptone.

An enrichment first referred to as MS culture (Grostern and Edwards 2006) and later named '*Dhb*-TCA' (Grostern et al. 2010), and containing over 60 % *Dehalobacter*, has been obtained from groundwater and solids of a site contaminated with 1,1,1-TCA and TCE as inoculum. A mixture of methanol, ethanol, acetate and lactate (MEAL) was used as electron donor. Dechlorination products were 1,1-dichloroethane (1,1-DCA) and chloroethane (CA) and methanogenesis was only observed during 1,1-DCA dechlorination. Transfers with H_2 as electron donor and acetate as carbon source sustained 1,1,1-TCA dechlorination but dechlorination of 1,1-DCA was lost after three transfers. Monitoring *Dehalobacter* 16S rRNA gene copy numbers by qPCR in cultures dechlorinating 1,1,1-TCA and 1,1-DCA showed a tenfold increase during dechlorination (Grostern and Edwards 2006). Subsequent sub-culturing under different conditions indicated that at least two different *Dehalobacter* populations were present in *Dhb*-TCA (Grostern et al. 2010). Transfers of *Dhb*-TCA with CF as electron acceptor led to a sub-culture designated *Dhb*-CF that was only able to dechlorinate CF and 1,1,1-TCA but not 1,1-DCA (Grostern et al. 2010). This is the first report of an OHRB that can use a chlorinated methane as electron acceptor. A sub-culture maintained for months on 1,1-DCA only dechlorinated 1,1-DCA but not CF and 1,1,1-TCA. A metagenomic analysis of these enrichments allowed obtaining the complete genome of both *Dehalobacter* sp. strains CF and DCA (Tang et al. 2012).

Two enrichments selected on a medium containing dichloromethane revealed a metabolic feature of *Dehalobacter* not described previously, namely the fermentation of a chlorinated compound (Justicia-Leon et al. 2012; Lee et al. 2012). In the study of Lee et al., initial cultures inoculated with samples of a core retrieved from an aquifer (state of contamination unknown) and fed with emulsified vegetable oil as electron donor transformed CF into DCM that disappeared as well (Lee et al. 2012). Subsequent transfers with H_2 and acetate dechlorinated CF to DCM but the latter was only depleted when H_2 was removed from the cultures at day 182 (Lee et al. 2012). Pyrosequencing of 16S rRNA gene amplicons showed the presence of *Dehalobacter* and *Dehalococcoides*. However, in cultures with H_2 where DCM was not degraded only *Dehalobacter* could be quantified by qPCR. Fresh cultures supplied with DCM as the sole organic carbon and energy source showed methane production with *Dehalobacter*, *Geobacter*, an unclassified *Synergistaceae* and *Methanoculleus* as the dominant microbial community members. Subsequent transfers showed the proliferation of *Dehalobacter* by qPCR quantification along with the consumption of DCM. At the same time, another enrichment was described to ferment DCM and to contain a *Dehalobacter* population that increased in abundance upon DCM consumption (Justicia-Leon et al. 2012). H_2-fed cultures showed acetate formation but lost the ability to degrade DCM and bicarbonate was needed in sediment-free cultures. This enrichment was unable to dechlorinate CF to DCM.

8.2 Morphology and Cytological Characteristics of *Dehalobacter* spp.

All reports on the morphology of *Dehalobacter* strains describe them as motile, short rods with a diameter of about 0.5 μm and a length around 2–3 μm (Wild et al. 1996; Holliger et al. 1998; Sun et al. 2002; Wang et al. 2014). The Gram staining was mostly negative except in one case where cells staining Gram-positive have been reported (Grostern and Edwards 2009). However, thin-sections of strain PER-K23 cells did not indicate the presence of an outer membrane that is typical for the cell envelope of Gram-negative bacteria. The cell envelope of PER-K23 cells was rather composed of a cytoplasmic membrane, peptidoglycan and a surface layer that was probably composed of proteins in a hexagonal arrangement and that could potentially hinder the Gram stain to reach the peptidoglycan layer (Holliger et al. 1998). The cell wall of PER-K23 contains the peptidoglycan type A3γ as found in certain groups of unequivocally Gram-positive bacteria which indicates a phylogenetic relationship between *Dehalobacter* and Gram-positive bacteria (Pickett et al. 1994). The peptidoglycan type A3γ contains LL-diaminopimelic acid that has also been detected in *Desulfitobacterium* sp. strain PCE1 that stains Gram-positive. However, the peptidoglycan type has not been identified in the latter. PER-K23 cells contain *b*-type cytochromes and menaquinones, in particular MQ-7 and MQ-8, but also some MQ-6 and MQ-9.

8.3 Habitat of *Dehalobacter* spp.

The type strain *D. restrictus* PER-K23 has been isolated from a packed-bed column that was filled with a mixture of river Rhine sediment and ground anaerobic granular sludge, two sources of environmental microorganisms that can perhaps not be considered as particularly pristine but also not as environments specifically contaminated with chlorinated compounds (Holliger et al. 1993). Also *Dehalobacter* sp. strain TCA1 has been isolated from samples of a polluted, but not specifically organohalide-contaminated river sediment (Sun et al. 2002) whereas *D. restrictus* strain TEA has been isolated from an activated carbon reactor that was colonized by bacteria originating from an aquifer contaminated with DCM (Wild et al. 1996). The most recent isolated *Dehalobacter* strain has been enriched from digester sludge of an industrial wastewater treatment plant probably purifying wastewater with a rather undefined mixture of different contaminants (Wang et al. 2014). Finally, the strains 12DCB1 and 13DCB1 have also been isolated from samples of an environment that was probably not specifically contaminated with organohalides, namely from sediment of a water-saturated drainage ditch of a chemical industry site.

Different enrichment cultures containing *Dehalobacter* and dechlorinating an organohalide have been obtained from material of sites contaminated with specific chlorinated compounds, mainly contaminated soils and aquifers (Grostern and Edwards 2006; Lacroix et al. 2014; Lee et al. 2012; Rouzeau-Szynalski et al. 2011; van Doesburg et al. 2005) whereas others enrichments have been inoculated with material that was not specifically contaminated with the organohalide used as the electron acceptor in the culture medium (Griffin et al. 2004; Yoshida et al. 2009; Zhang et al. 2012, 2013; Li et al. 2013a, b; Wang and He 2013). *Dehalobacter* has also been detected in several aquifers at sites contaminated with 1,1,1-TCA and TCE (Lowe et al. 2002; Lima et al. 2012; Damgaard et al. 2013).

The above shows that *Dehalobacter* has been enriched from many different sources of inoculum which renders difficult the definition of a typical habitat for this bacterial genus. Since *Dehalobacter* has mainly been characterized as a bacterium that respires chlorinated compounds as electron acceptor, it has been enriched from contaminated soils and aquifers, from sediments of polluted rivers and from sludge of wastewater treatment plants. Although *Dehalobacter* seems to be widespread, all the samples used as inoculum for enrichments originate from environments where *Dehalobacter* has probably already been "naturally" enriched due to exposure to organohalides. With the help of next generation sequencing it would be interesting to know how frequently *Dehalobacter* can be found in the so-called rare microbial biosphere in order to get an idea how widespread *Dehalobacter* really is.

8.4 Nutritional Requirements and Growth Conditions of *Dehalobacter* spp.

The first *Dehalobacter* strain isolated in pure culture has been enriched with fermented yeast extract as source of growth factors in addition to a mixture of vitamins. Fermented yeast extract amendment could be subsequently replaced by peptone if acetate was provided as carbon source (Holliger et al. 1998). Vitamins have also been added to the culture media in all other studies where *Dehalobacter* populations have been enriched and peptone or another oligopeptide/amino acid has often been added as well. A detailed analysis of the growth factor requirements of *D. restrictus* has shown that it depends on the amendment of the two vitamins thiamine and cyanocobalamin, and the three amino acids arginine, histidine and threonine (Holliger et al. 1998). No information is available on the growth factor requirements of other *Dehalobacter* isolates but as some strains could only be maintained in co-culture indicates that those strains depend on growth factors produced by partner organisms, e.g. *Sedimentibacter* (van Doesburg et al. 2005), *Acetobacterium* (Grostern et al. 2009) and *Desulfovibrio* (Grostern and Edwards 2006).

From the limited available information it can be concluded that *Dehalobacter* spp. are mesophilic bacteria that have an optimal growth at 25–30 °C and that are very sensitive to temperatures above 35 °C (Holliger et al. 1993; Sun et al. 2002). Whereas *D. restrictus* was not able to grow at pH < 6.5 (Holliger et al. 1993), a *Dehalobacter* population in consortium AQ-5 dechlorinated PCE to *cis*-1,2-DCE well at pH 6.5 (Lacroix et al. 2014). Since pH sensitivity is an important parameter for application of organohalide-respiring bacteria (OHRB) in bioremediation of organohalide-contaminated sites, especially when treating source zones, it would be interesting to know more about the range of pH at which different *Dehalobacter* strains are active. A characterization of different PCE-dechlorinating consortia has shown that different *Dehalococcoides* spp. can have quite distinct sensitivities to lower pH values (Lacroix et al. 2014).

8.5 Physiology and Biochemistry of *Dehalobacter* spp.

Dehalobacter spp. can only use molecular hydrogen as electron donor and source of energy, with the exception of *Dehalobacter* sp. strain TCA1 that can also use formate (Sun et al. 2002). The majority of highly enriched and pure cultures of *Dehalobacter* cannot ferment organic compounds and can only use organohalides as electron acceptors. Most *Dehalobacter* strains can use at most two different organohalides of the same chemical group, either aliphatic or aromatic organohalides. Strains PER-K23 and TEA use PCE and TCE as electron acceptor and produce *cis*-1,2-DCE (Holliger et al. 1993; Wild et al. 1996; Holliger et al. 1998) and strain TCA1 uses 1,1,1-TCA and 1,1-DCA and produces CA (Sun et al. 2002). The *Dehalobacter* population in the β-HCH dechlorinating co-culture also dechlorinated α- and γ-HCH, the former at the same rates as β-HCH whereas the latter was dechlorinated at much slower rates (van Doesburg et al. 2005). This co-culture did not dechlorinate 1,2-DCA and PCE. The *Dehalobacter* population in the 1,2-DCA dechlorinating co-culture did only dechlorinate this organohalide among all the different ones tested [PCE, TCE, *cis*-1,2-DCE, VC, 1,1,2-TCA, 1,1,1-TCA and 1,1-DCA (Grostern and Edwards 2009)]. The exceptions of this rather general *Dehalobacter* ability are the most recently isolated strains that dechlorinate aromatic as well as alkene organohalides. Strain TCP1 dechlorinates besides chlorophenols also PCE and TCE that are transformed into *cis*-1,2-DCE and *trans*-1,2,-DCE in a ratio of 5.6:1 (Wang et al. 2014), and strains 12DCB1 and 13DCB1 dechlorinate a multitude of chlorobenzenes as well as PCE and TCE to *cis*-1,2-DCE (Nelson et al. 2014). These findings are not surprising looking at the numerous reductive dehalogenase genes found in *Dehalobacter* genomes (see below), and invite the different groups working with *Dehalobacter* to reassess the dechlorination potential of their strains.

Two reports provide strong evidence that certain *Dehalobacter* strains can ferment DCM and produce acetate from this chlorinated compound, a process that requires a syntrophic relationship with a hydrogen consumer (Justicia-Leon et al. 2012; Lee et al. 2012). In a culture degrading CF completely, CF was first dechlorinated by an organohalide respiration (OHR) process to DCM which then was fermented to acetate and hydrogen (Lee et al. 2012). Although *Dehalobacter* was involved in both processes, it was not possible to determine whether two distinct *Dehalobacter* populations were responsible for the two metabolic reactions or only one. In either case, these results indicate that growth of *Dehalobacter* is not restricted to OHR only.

In contrast to hydrogenophilic homoacetogens and methanogens, *Dehalobacter* spp. are not able to grow autotrophically, i.e. they need acetate as carbon source. Succinate cannot replace acetate and one third of the carbon in newly formed biomass is coming from inorganic carbon by heterotrophic carbon fixation (Holliger et al. 1993, 1998). *Dehalobacter* has this metabolic feature, being a chemolithotroph and a heterotroph, called mixotrophy, in common with *Dehalococcoides*, the other genus of obligate organohalide-respiring bacteria (Löffler et al. 2013).

A detailed study of the creation of a proton gradient upon hydrogen oxidation and reductive dechlorination of PCE indicated a H^+/e^- ratio of 1.25 ± 0.2 which suggests that besides formation of protons due to hydrogen oxidation on the outside of the cytoplasmic membrane, vectorial translocation of protons from the inside to the outside could also occur (Schumacher and Holliger 1996). In addition, this study showed that menaquinones are involved in electron transfer from the hydrogenase to the reductive dehalogenase and that the reductive dehalogenase could be photoreversibly inactivated by 1-iodopropane, an inhibitor of corrinoid-mediated reactions.

The PCE reductive dehalogenase of *D. restrictus*, in the following referred to as PceA, indeed contains a corrinoid that is present in the protein in base-off form (Schumacher et al. 1997). In addition, PceA contains two 4Fe-4S clusters with very low redox potentials. Although initially characterized as membrane-associated protein that is cytoplasmically oriented (Schumacher and Holliger 1996), recent investigations with protoplasts and proteinase K treatment suggested that PceA is facing the periplasm (unpublished results). This topology of PceA is in agreement with the sequence information obtained some years after the first sequence of a reductive dehalogenase was published [PceA of *Sulfurospirillum multivorans*, (Neumann et al. 1998)]. The sequence of PceA of *D. restrictus* was obtained using a degenerate PCR approach that targeted a conserved amino acid stretch of PceA of *Sulfurospirillum multivorans* and CprA of *Desulfitobacterium dehalogenans* (von Wintzingerode et al. 2001) and the N-terminal sequence of PceA from *D. restrictus* (Maillard et al. 2003). The sequence of PceA of *D. restrictus* had the same features as the one of *Sulfurospirillum multivorans*, namely the absence of a corrinoid binding motif, the presence of consensus sequences for binding two 4Fe-4S clusters, and the presence of a twin-arginine motif that is usually found in proteins that contain redox cofactors and are exported across the cytoplasmic membrane in a folded conformation. This indicates that PceA should indeed rather be located at the outside of the cytoplasmic membrane.

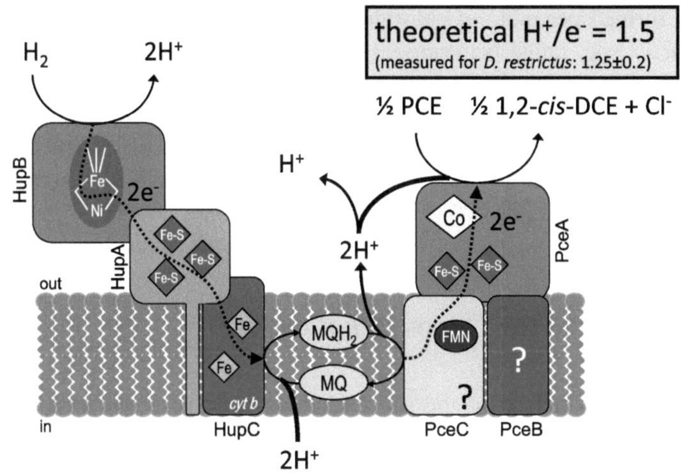

Fig. 8.1 Tentative model of the respiration chain of *Dehalobacter restrictus* involving hydrogen oxidation by a Ni-Fe hydrogenase, transfer of electrons via menaquinones from the cytochrome *b* subunit of the hydrogenase to PceC, and finally to PceA that reductively dechlorinates PCE to *cis*-1,2-DCE

The characterization of the genetic context around *pceA* resulted in the identification of the *pceABCT* gene cluster that has also been found in *Desulfitobacterium hafniense* strain TCE1 with 99 % sequence identity (Duret et al. 2012; Maillard et al. 2005). In the latter organism, this gene cluster is part of a composite transposon but not in *D. restrictus*. The product of *pceB* is predicted to be a protein with three transmembrane helices and is therefore, as it is the case for many other sequenced reductive dehalogenases such as PceA of *S. multivorans* (Neumann et al. 1998) and CprA of *D. dehalogenans* (Smidt et al. 2000), proposed to be a membrane anchor for PceA. The additional genes *pceC* and *pceT* were named according to homologous genes identified in the chlorophenol reductive dehalogenase (*cpr*) gene cluster of *D. dehalogenans* (Smidt et al. 2000). In *D. hafniense* strain TCE1, the role of PceT has been identified to be a trigger factor-like protein that seems to function as dedicated chaperone for PceA and that specifically interacts with the twin-arginine signal peptide of PceA (Maillard et al. 2011). PceC could, according to the characterization of the gene cluster of *D. dehalogenans*, be a membrane-bound regulatory protein. However, this protein contains a typical FMN binding site (Rupakula et al. 2013) and could therefore also be involved in electron transfer.

Based on physiological data and sequence information of the *pce* gene cluster, a refined version of a previously published model of the respiration chain of *D. restrictus* (Schumacher and Holliger 1996) is presented in Fig. 8.1. This model results in a theoretical H^+/e^- ratio of 1.5 which is in the range of the one experimentally determined earlier (Schumacher and Holliger 1996). Assuming the need of three protons for the formation of one molecule of ATP, about one mole of ATP would be formed per mole of chloride released. The published growth yields for *Dehalobacter* spp. range between 3.3 and 5.6 g dry weight per mole of chloride released (Holliger et al. 1998; Sun et al. 2002; Grostern et al. 2009; Wang et al. 2014). When assuming a biomass yield of 5–10 g dry weight per mole of ATP formed during catabolism, the respiration chain of *Dehalobacter* produces about half to one mole of ATP per mole of chloride released which is in agreement with the proposed model in Fig. 8.1.

8.6 Phylogeny of the Genus *Dehalobacter*

Dehalobacter spp. belong to the low GC Gram-positive *Firmicutes*. An analysis of the phylogenetic position of *Dehalobacter* spp. within OHRB is presented in Chap. 5 of this book and will not be addressed here. However, a detailed analysis of the 16S rRNA genes within the genus *Dehalobacter* reveals a relatively high heterogeneity both between and within strains. The genome sequences of *D. restrictus* (Kruse et al. 2013), *Dehalobacter* sp. strains CF, DCA (Tang et al. 2012) and TCP1 (Wang et al. 2014), all contain between 3 and 5 copies of the 16S rRNA gene. Multiple rRNA operons have already been recognized as a property of other Gram-positive OHRB, namely *Desulfitobacterium* spp. (Villemur et al. 2006). Figure 8.2 shows the phylogenetic tree of the *Dehalobacter* strains.

A closer look at the alignment of *Dehalobacter* 16S rRNA gene sequences revealed that the high diversity is due to the large degree of variability in the V1 region. Indeed this region ranges from 34 to 188 nucleotides (nt) in length and varies also in sequence. *Dehalobacter* sp. strain TCP1 harbours five 16S rRNA genes, four of them being very similar with a V1 variable region of 140 nt and one displaying a 188-nt long V1 region. In contrary, *D. restrictus* harbours four almost identical copies of the 16S rRNA gene (with a 34-nt long V1 region). The topology of the tree does not allow making any correlation between 16S rRNA sequence and the halogenated compounds that are reduced by these strains.

8.7 Metabolic Features Deduced from the Genome of *D. restrictus* and 'Omics' Studies

The almost 3-Mb long genome of *D. restrictus* PER-K23 contains 2826 protein-coding and 82 RNA genes. For a total of 76.7 % of the protein-coding genes a putative function could be identified whereas 781 genes could not be associated with any of the general COG functional categories (cluster of orthologous groups, www.ncbi.nlm.nih.gov/COG/) (Kruse et al. 2013). Numerous genes that probably play a role in ORH with H_2 as electron donor and PCE and TCE as electron acceptor have been identified. However, no functional gene for any other known

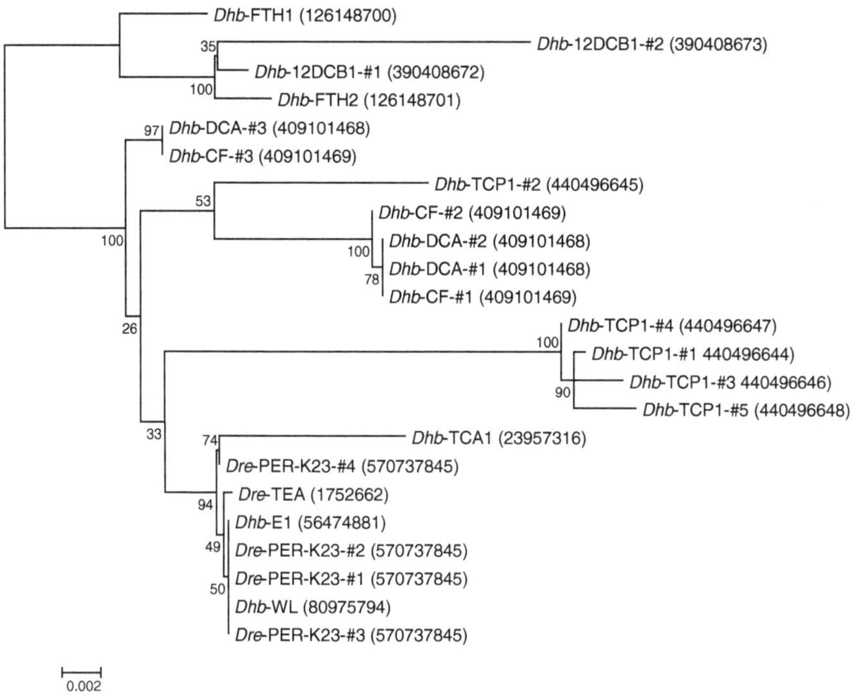

Fig. 8.2 Phylogenetic analysis of the 16S rRNA genes of *Dehalobacter* spp. For some strains, multiple 16S rRNA genes are present in the genome (noted as # followed by a number). The long branches of *Dhb*-12DCB1-#2 and *Dhb*-TCP1-#2 are due to extended V1 variable region. Legend: *Dhb*: *Dehalobacter* sp.; *Dre*: *Dehalobacter restrictus*. The abbreviation is followed by the name of the strain. The gene index (GI) reference number is given in parentheses. *Notes* the 16S rRNA gene sequence of *Dehalobacter* sp. strain 13DCB1 was excluded as it is not complete and does not contain the variable V1 region. For *Dehalobacter* sp. strain E1, only one copy of 16S rRNA gene sequence is available in databases, although 3 distinct copies have been reported (Maphosa et al. 2012)

respiration metabolism has been found, confirming the cultivation attempts that did not show growth with alternative electron donors and acceptors. Nevertheless, it cannot be excluded that the numerous unidentified genes encode for unsuspected metabolic pathways that are yet unknown and have not been tested in cultivation experiments.

Eight different hydrogenases are present on the genome of *D. restrictus* which underscores the central role of hydrogen in its metabolism. One of the three membrane-bound Ni-Fe uptake hydrogenases (Hup) has also been detected in the proteome of cells harvested during different growth phases, suggesting a major role of this hydrogenase in the core metabolism (Rupakula et al. 2013). In addition to this Hup, two of the three Fe-only hydrogenases (Hym), identified on the genome and lacking the typical membrane-associated components, have also been detected. They might be involved in generating reducing equivalents needed in anabolism in the form of NADH and FADH, or they might work with the 11-subunit respiration complex I to generate a proton motive force. Indeed, an 11-subunit respiration complex I is present in the genome and its cytoplasm-oriented subunits NuoBCD were detected in the proteome. This 11-subunit version of complex I is widely distributed, both in the archaeal and the eubacterial kingdoms, and has been proposed to be capable to function with various electron donor and acceptor proteins (Moparthi and Hägerhäll 2011). Finally, also the two large membrane-bound putatively proton-translocating hydrogenase complexes Hyc and Ech have been detected during growth of *D. restrictus* which illustrates the complex nature of hydrogen metabolism in this bacterium and possibly also the energy metabolism involving the build-up of a proton motive force.

On the electron acceptor side, a total of 25 genes predicted to encode catalytic subunits of reductive dehalogenases (*rdhA*) have been found in the genome of *D. restrictus* and a total of 86 genes potentially associated with reductive dehalogenase expression and maturation such as membrane anchors (*rdhB*), transcriptional regulators (*rdhK*) and chaperones (*rdhT*) (Kruse et al. 2013). All four proteins encoded by the *pceABCT* gene cluster have been identified in the proteome and they seemed constitutively expressed. In addition, also RdhA14 has been detected, a reductive dehalogenase with unknown substrate spectrum. A more detailed discussion of the functional diversity of the different *rdhA* genes is presented below.

The genome of *D. restrictus* encodes an intact Wood-Ljungdahl pathway that has also been reported for other OHRB such as the closely related *D. hafniense* strains Y51 (Nonaka et al. 2006), TCE1 (Prat et al. 2011) and DCB-2 (Kim et al. 2012) and the more distantly related *Dehalococcoides mccartyi* strains (Tang et al. 2009; Kruse et al. 2013). Furthermore, the genome of *D. restrictus* contains several homologues of pyruvate synthase, an enzyme that could be involved in heterotrophic CO_2 fixation (Kruse et al. 2013). For *D. mccartyi* strain 195, it has been shown that CO_2 is assimilated via two reactions, conversion of acetyl-coenzyme A to pyruvate catalyzed by pyruvate synthase and pyruvate conversion to oxaloacetate via pyruvate carboxylase and that the Wood-Ljungdahl pathway is not involved in CO_2 fixation (Tang et al. 2009). In *D. hafniense* strains, components of the Wood-Ljungdahl pathway have been shown to participate in the use of phenyl methyl esters as electron donor (Kreher et al. 2008). Although enzymes belonging to the Wood-Ljungdahl pathway and products of pyruvate synthase genes have been detected in the proteome of *D. restrictus*, it is not known at present how heterotrophic CO_2 fixation is achieved and what the role of the Wood-Ljungdahl pathway enzymes is. Cultivation attempts of *D. restrictus* with vanillate as carbon source and electron donor were so far not successful (unpublished results).

In addition to *D. restrictus*, the genome of *Dehalobacter* sp. E1 has been deduced from a metagenomic analysis of a co-culture with *Sedimentibacter* sp. (Maphosa et al. 2012). At the time it was compared to the genome of *Dehalococcoides* spp. revealing an overall richer arsenal in the metabolism of amino acids, energy and cofactor biosynthesis. Two formate dehydrogenases and one uptake hydrogenase have been also identified. Ten reductive dehalogenases have been identified in *Dehalobacter* sp. E1 (see below).

8.8 Functional Diversity of Reductive Dehalogenases in the *Dehalobacter* Genus

At the time when the sequence information of PceA was retrieved with classical molecular approaches (Maillard et al. 2003), two additional however partial *rdhA* genes have been identified from *D. restrictus* by a more extensive degenerate PCR approach (Regeard et al. 2004), suggesting that *D. restrictus* was harbouring several *rdhA* genes, although its capability of reducing organohalides is restricted to PCE and TCE. High levels of sequence conservation of the *pceABCT* gene cluster of *D. restrictus* with sequences found in several *Desulfitobacterium* strains including the 1,2-DCA reductive dehalogenase of *Desulfitobacterium dichloroeliminans* (Marzorati et al. 2007) have been found suggesting that horizontal gene transfer and adaptation to other organohalides occurred here (Duret et al. 2012). This trend was further illustrated by the identification of three *rdhA* genes from an enrichment culture dechlorinating 1,2-DCA that was dominated by *Dehalobacter* sp. strain WL. All three genes showed high sequence homology with *pceA* of *D. restrictus*, while the product of one of them (RdhA1, GI: 198404178) was proposed to be involved in 1,2-DCA dechlorination (Grostern et al. 2009).

Several recently published genomes and metagenomic analyses targeting pure and mixed cultures of *Dehalobacter* spp. revealed a much larger diversity of *rdhA* genes (Deshpande et al. 2013; Kruse et al. 2013; Maphosa et al. 2012; Tang et al. 2012), analogous to what was observed in the genus *Dehalococcoides* (see (Löffler et al. 2013) for a review). Ten *rdhA* genes were identified from a metagenomic analysis of *Dehalobacter* sp. strain E1 in a co-culture dechlorinating β-HCH (Maphosa et al. 2012). A common set of 17 *rdhA* genes was identified in the genome of the two *Dehalobacter* sp. strains CF and DCA dechlorinating chloroform and chloroethanes (Tang et al. 2012). *Dehalobacter* sp. strain UNSWDHB dechlorinating chloroform displays 17 *rdhA* genes (Deshpande et al. 2013), 14 of them identical to those found in strains CF and DCA. The genome of *D. restrictus* revealed the presence of 25 *rdhA* genes, including the well-characterized *pceABCT* gene cluster (Kruse et al. 2013). Recently, a new genome was deposited in databases (as part of the sequencing project coordinated by H. Smidt and the JGI) belonging to *Dehalobacter* sp. strain FTH1, which was isolated from a culture dechlorinating 4,5,6,7-tetrachlorophtalide (Yoshida et al. 2009). This genome harbours the highest number of *rdhA* genes in *Dehalobacter* spp. with 27 analogs.

The corresponding amino acid sequences of all the *rdhA* genes identified in *Dehalobacter* spp. were aligned and are depicted in Fig. 8.3. The RdhA sequences belonging to individual *Dehalobacter* strains appear to be relatively well distributed over the overall diversity. Most RdhA sequences are present at least in two members of the genus, while only a few sequences are exclusively found in one specific strain. Interestingly, only one RdhA sequence seems to be conserved in all strains considered here (indicated by an asterisk).

From the topology of the tree, one could consider two classes of *Dehalobacter* RdhA sequences, one that is relatively conserved and homologous to CprA of

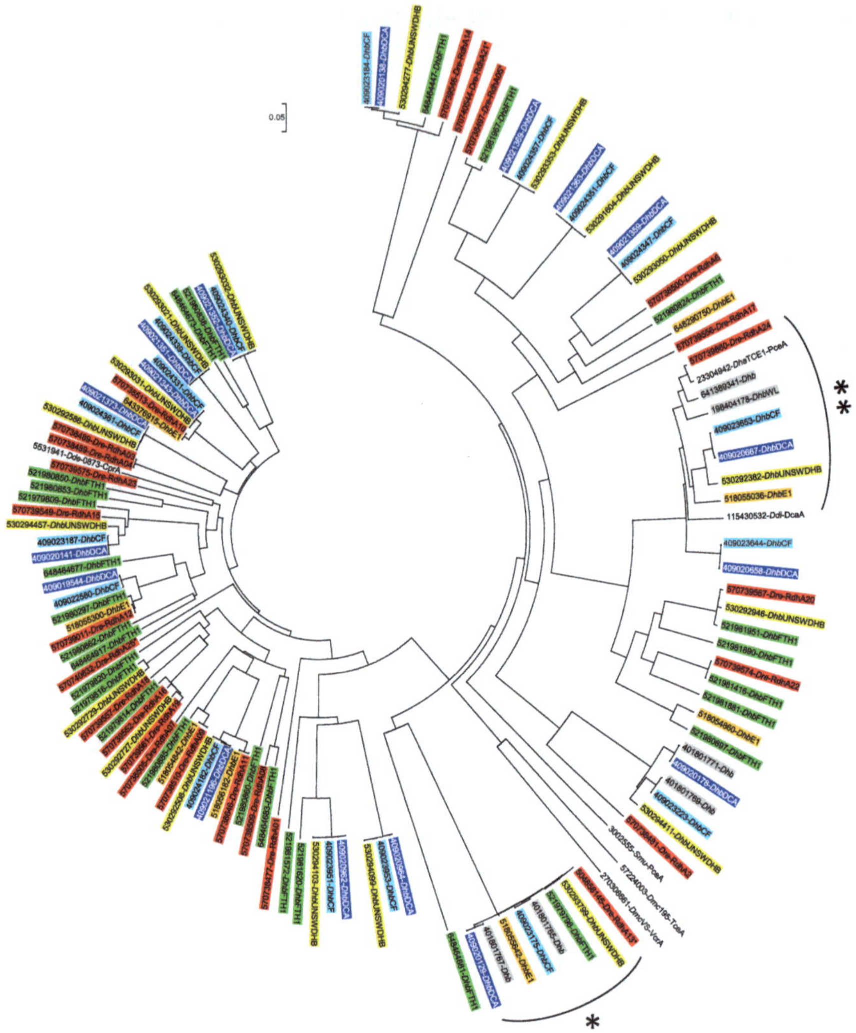

Fig. 8.3 Protein sequence likelihood tree analysis of all putative reductive dehalogenases identified in *Dehalobacter* spp. Each sequence is given by its gene index (GI) reference number and an abbreviation for the species and strain. A colour code is used to distinguish the strains: *red Dehalobacter restrictus*; *orange Dehalobacter* sp. strain E1; *yellow Dehalobacter* sp. strain UNSWDHB; *light blue Dehalobacter* sp. strain CF; *dark blue Dehalobacter* sp. strain DCA; *green Dehalobacter* sp. strain FTH1; *grey Dehalobacter* sp. in co-culture or enrichment cultures. Legend: *Dre: Dehalobacter restrictus*; *Dhb*: *Dehalobacter* sp.; *Dde*: *Desulfitobacterium dehalogenans*; *Ddi*: *Desulfitobacterium dichloroeliminans*; *Dha*: *Desulfitobacterium hafniense*; *Dmc*: *Dehalococcoides mccartyi*; *Smu*: *Sulfurospirillum multivorans*

D. dehalogenans (left side of the tree), and another class which contains many diverse RdhA sequences (right side of the tree). This distinction is further validated when considering the genetic structure of *rdh* operons in *D. restrictus*

(Rupakula et al. 2013). Indeed, a variety of gene clusters was identified there, with the *cprA*-like genes being embedded in a simple *rdhBA* operon structure, while the other ones are part of *rdhBAC*, *rdhABC* or *rdhABCT* structures. This suggests that the numerous *cprA*-like genes in *Dehalobacter* are likely the result of relatively recent gene duplications.

As for other RdhA in general (see (Hug et al. 2013) and Chap. 16), the sequence-substrate relationships in *Dehalobacter* RdhAs is impossible to establish. Too scarce knowledge is available not allowing any clear correlation between sequence features and substrate specificity. As an example, in the group of the best defined and closely related RdhA enzymes in *Dehalobacter* (indicated by a double asterisk in Fig. 8.3), two substrates were identified, PCE (and TCE) and 1,2-DCA.

8.9 Corrinoid Metabolism in *Dehalobacter* spp.

As already stated above, *D. restrictus* has been characterized as a corrinoid auxotrophic bacterium (Holliger et al. 1998). The elucidation of the genome sequence of *D. restrictus* (Kruse et al. 2013), but also other *Dehalobacter* spp. (Deshpande et al. 2013; Maphosa et al. 2012; Tang et al. 2012) gave new insights in the corrinoid metabolism and more specifically in the corrinoid biosynthesis pathway. At the very first look it was surprising to identify the complete corrinoid biosynthetic pathway in *D. restrictus*. However, a deletion mutation affecting the *cbiH* gene was then proposed to be responsible for its corrinoid auxotrophy (Kruse et al. 2013; Rupakula et al. 2013). In a proteomic study, several proteins of this pathway, including CbiH, were not detected in the proteome of *D. restrictus* cultivated in standard growth conditions (i.e. in presence of 250 μg/L of cobalamin in the medium) (Rupakula et al. 2013). A functional genomic study was then conducted to investigate the effect of corrinoid limitation on corrinoid metabolism (Rupakula et al. 2015). Five distinct operons were characterized in *D. restrictus* and two major differences were observed in the genomes of other *Dehalobacter* spp., namely the presence of an intact copy of *cbiH* and the lack of operon-2, which encodes several corrinoid transport proteins and proteins involved in corrinoid salvaging. All five operons are regulated by cobalamin riboswitches (Rupakula et al. 2015), baring similarity to *D. hafniense* (Choudhary et al. 2013). The comparison of the proteome from *D. restrictus* cells cultivated in the presence of 250, 50 and 10 μg/L of cobalamin in the medium revealed a strong up-regulation of proteins encoded in operon-2, suggesting that *D. restrictus* exploits an enhanced capacity of corrinoid transport and salvaging under corrinoid limitation. Sequence analysis further indicated that operon-2 of *D. restrictus* shows strong homology to an operon present in *Acetobacterium woodii*, suggesting that horizontal gene transfer may have occurred (Rupakula et al. 2015). So far, scarce information is available on corrinoid metabolism at the physiological level in other *Dehalobacter* strains. It was suggested that *Dehalobacter* sp. strain E1 might benefit from corrinoids provided by *Sedimentibacter* sp. present in the co-culture (Maphosa et al. 2012). When indicated, cultures of *Dehalobacter* spp. were always cultivated in the presence of cobalamin. Hence, further work is needed to test the ability of *Dehalobacter* strains that have an intact *cbiH* gene to synthesize corrinoids de novo.

8.10 Concluding Remarks

Dehalobacter in pure culture or enrichments seem to be dedicated to organohalide degradation, either by reduction in organohalide respiration or by fermentation as in the case of DCM. For the fermentative strains it is however not known whether they have additional metabolic capabilities not involving an organohalide. The recently available genome sequences and the recently isolated new strains provide new avenues of research for this bacterial genus. The increasing number of scientific publications dealing with *Dehalobacter* populations since 2009, about 70 in total, indicates that we will learn a lot more about the *Dehalobacter* genus in the near future.

References

Choudhary PK, Duret A, Rohrbach-Brandt E, Holliger C, Sigel RKO, Maillard J (2013) Diversity of cobalamin riboswitches in the corrinoid-producing organohalide respirer *Desulfitobacterium hafniense*. J Bacteriol 195(22):5186–5195. doi:10.1128/jb.00730-13

Damgaard I, Bjerg PL, Baelum J, Scheutz C, Hunkeler D, Jacobsen CS, Tuxen N, Broholm MM (2013) Identification of chlorinated solvents degradation zones in clay till by high resolution chemical, microbial and compound specific isotope analysis. J Contam Hydrol 146:37–50. doi:10.1016/j.jconhyd.2012.11.010

De Bruin WP, Kotterman MJJ, Posthumus MA, Schraa G, Zehnder AJB (1992) Complete biological reductive transformation of tetrachloroethene to ethane. Appl Environ Microbiol 58(6):1996–2000

Deshpande NP, Wong YK, Manefield M, Wilkins MR, Lee M (2013) Genome sequence of *Dehalobacter* UNSWDHB, a chloroform-dechlorinating bacterium. Genome Announc 1 (5). doi:10.1128/genomeA.00720-13

Duret A, Holliger C, Maillard J (2012) The physiological opportunism of *Desulfitobacterium hafniense* strain TCE1 towards organohalide respiration with tetrachloroethene. Appl Environ Microbiol 78(17):6121–6127. doi:10.1128/aem.01221-12

Griffin BM, Tiedje JM, Löffler FE (2004) Anaerobic microbial reductive dechlorination of tetrachloroethene to predominately *trans*-1,2-dichloroethene. Environ Sci Technol 38(16):4300–4303. doi:10.1021/es035439g

Grostern A, Duhamel M, Dworatzek S, Edwards EA (2010) Chloroform respiration to dichloromethane by a *Dehalobacter* population. Environ Microbiol 12(4):1053–1060. doi:10.1111/j.1462-2920.2009.02150.x

Grostern A, Edwards EA (2006) A 1,1,1-trichloroethane-degrading anaerobic mixed microbial culture enhances biotransformation of mixtures of chlorinated ethenes and ethanes. Appl Environ Microbiol 72(12):7849–7856. doi:10.1128/aem.01269-06

Grostern A, Edwards EA (2009) Characterization of a *Dehalobacter* coculture that dechlorinates 1,2-dichloroethane to ethene and identification of the putative reductive dehalogenase gene. Appl Environ Microbiol 75(9):2684–2693. doi:10.1128/aem.02037-08

Grostern A, Chan WW, Edwards EA (2009) 1,1,1-trichloroethane and 1,1-dichloroethane reductive dechlorination kinetics and co-contaminant effects in a *Dehalobacter*-containing mixed culture. Environ Sci Technol 43(17):6799–6807

Holliger C, Hahn D, Harmsen H, Ludwig W, Schumacher W, Tindall B, Vazquez F, Weiss N, Zehnder AJ (1998) *Dehalobacter restrictus* gen. nov. and sp. nov., a strictly anaerobic bacterium that reductively dechlorinates tetra- and trichloroethene in an anaerobic respiration. Arch Microbiol 169(4):313–321

Holliger C, Schraa G, Stams AJM, Zehnder AJB (1993) A highly purified enrichment culture couples the reductive dechlorination of tetrachloroethene to growth. Appl Environ Microbiol 59(9):2991–2997

Hug LA, Maphosa F, Leys D, Loffler FE, Smidt H, Edwards EA, Adrian L (2013) Overview of organohalide-respiring bacteria and a proposal for a classification system for reductive dehalogenases. Philos Trans R Soc Lond B Biol Sci 368(1616):20120322. doi:10.1098/rstb.2012.0322

Justicia-Leon SD, Ritalahti KM, Mack EE, Loffler FE (2012) Dichloromethane fermentation by a *Dehalobacter* sp. in an enrichment culture derived from pristine river sediment. Appl Environ Microbiol 78(4):1288–1291. doi:10.1128/aem.07325-11

Kim SH, Harzman C, Davis JK, Hutcheson R, Broderick JB, Marsh TL, Tiedje JM (2012) Genome sequence of *Desulfitobacterium hafniense* DCB-2, a gram-positive anaerobe capable of dehalogenation and metal reduction. BMC Microbiol 12(1):21 1471-2180-12-21

Kreher S, Schilhabel A, Diekert G (2008) Enzymes involved in the anoxic utilization of phenyl methyl ethers by *Desulfitobacterium hafniense* DCB2 and *Desulfitobacterium hafniense* PCE-S. Arch Microbiol 190(4):489–495. doi:10.1007/s00203-008-0400-8

Kruse T, Maillard J, Goodwin L, Woyke T, Teshima H, Bruce D, Detter C, Tapia R, Han C, Huntemann M, Wei CL, Han J, Chen A, Kyrpides N, Szeto E, Markowitz V, Ivanova N, Pagani I, Pati A, Pitluck S, Nolan M, Holliger C, Smidt H (2013) Complete genome sequence of *Dehalobacter restrictus* PER-K23(T.). Stand Genomic Sci 8 (3):375–388. doi:10.4056/sigs.3787426

Lacroix E, Brovelli A, Barry DA, Holliger C (2014) Use of silicate minerals for pH control during reductive dechlorination of chloroethenes in batch cultures of different microbial consortia. Appl Environ Microbiol 80(13):3858–3867. doi:10.1128/aem.00493-14

Lee M, Low A, Zemb O, Koenig J, Michaelsen A, Manefield M (2012) Complete chloroform dechlorination by organochlorine respiration and fermentation. Environ Microbiol 14(4):883–894. doi:10.1111/j.1462-2920.2011.02656.x

Li Z, Inoue Y, Suzuki D, Ye L, Katayama A (2013a) Long-term anaerobic mineralization of pentachlorophenol in a continuous-flow system using only lactate as an external nutrient. Environ Sci Technol 47(3):1534–1541. doi:10.1021/es303784f

Li Z, Suzuki D, Zhang C, Yoshida N, Yang S, Katayama A (2013b) Involvement of *Dehalobacter* strains in the anaerobic dechlorination of 2,4,6-trichlorophenol. J Biosci Bioeng 116(5):602–609. doi:10.1016/j.jbiosc.2013.05.009

Lima G, Parker B, Meyer J (2012) Dechlorinating microorganisms in a sedimentary rock matrix contaminated with a mixture of VOCs. Environ Sci Technol 46(11):5756–5763. doi:10.1021/es300214f

Löffler FE, Yan J, Ritalahti KM, Adrian L, Edwards EA, Konstantinidis KT, Muller JA, Fullerton H, Zinder SH, Spormann AM (2013) *Dehalococcoides mccartyi* gen. nov., sp. nov., obligately organohalide-respiring anaerobic bacteria relevant to halogen cycling and bioremediation, belong to a novel bacterial class, *Dehalococcoidia classis* nov., order *Dehalococcoidales* ord. nov. and family *Dehalococcoidaceae* fam. nov., within the phylum *Chloroflexi*. Int J Syst Evol Microbiol 63 (Pt 2):625–635. doi:10.1099/ijs.0.034926-0

Lowe M, Madsen EL, Schindler K, Smith C, Emrich S, Robb F, Halden RU (2002) Geochemistry and microbial diversity of a trichloroethene-contaminated Superfund site undergoing intrinsic in situ reductive dechlorination. FEMS Microbiol Ecol 40(2):123–134. doi:10.1111/j.1574-6941.2002.tb00944.x

Maillard J, Genevaux P, Holliger C (2011) Redundancy and specificity of multiple trigger factor chaperones in *Desulfitobacteria*. Microbiology 157(8):2410–2421. doi:10.1099/mic.0.050880-0

Maillard J, Regeard C, Holliger C (2005) Isolation and characterization of Tn-*Dha1*, a transposon containing the tetrachloroethene reductive dehalogenase of *Desulfitobacterium hafniense* strain TCE1. Environ Microbiol 7(1):107–117. doi:10.1111/j.1462-2920.2004.00671.x

Maillard J, Schumacher W, Vazquez F, Regeard C, Hagen WR, Holliger C (2003) Characterization of the corrinoid iron-sulfur protein tetrachloroethene reductive dehalogenase of *Dehalobacter restrictus*. Appl Environ Microbiol 69(8):4628–4638

Maphosa F, van Passel MW, de Vos WM, Smidt H (2012) Metagenome analysis reveals yet unexplored reductive dechlorinating potential of *Dehalobacter* sp. E1 growing in co-culture with *Sedimentibacter* sp. Environ Microbiol Rep 4(6):604–616. doi:10.1111/j.1758-2229.2012.00376.x

Marzorati M, de Ferra F, Van Raemdonck H, Borin S, Allifranchini E, Carpani G, Serbolisca L, Verstraete W, Boon N, Daffonchio D (2007) A novel reductive dehalogenase, identified in a contaminated groundwater enrichment culture and in *Desulfitobacterium dichloroeliminans* strain DCA1, is linked to dehalogenation of 1,2-dichloroethane. Appl Environ Microbiol 73(9):2990–2999. doi:10.1128/aem.02748-06

Moparthi V, Hägerhäll C (2011) The evolution of respiratory chain complex I from a smaller last common ancestor consisting of 11 protein subunits. J Mol Evol 72(5–6):484–497. doi:10.1007/s00239-011-9447-2

Nelson JL, Fung JM, Cadillo-Quiroz H, Cheng X, Zinder SH (2011) A role for *Dehalobacter* spp. in the reductive dehalogenation of dichlorobenzenes and monochlorobenzene. Environ Sci Technol 45(16):6806–6813. doi:10.1021/es200480k

Nelson JL, Jiang J, Zinder SH (2014) Dehalogenation of chlorobenzenes, dichlorotoluenes, and tetrachloroethene by three *Dehalobacter* spp. Environ Sci Technol 48(7):3776–3782. doi:10.1021/es4044769

Neumann A, Wohlfarth G, Diekert G (1998) Tetrachloroethene dehalogenase from *Dehalospirillum multivorans*: cloning, sequencing of the encoding genes, and expression of the *pceA* gene in *Escherichia coli*. J Bacteriol 180(16):4140–4145

Nonaka H, Keresztes G, Shinoda Y, Ikenaga Y, Abe M, Naito K, Inatomi K, Furukawa K, Inui M, Yukawa H (2006) Complete genome sequence of the dehalorespiring bacterium *Desulfitobacterium hafniense* Y51 and comparison with *Dehalococcoides ethenogenes* 195. J Bacteriol 188(6):2262–2274. doi:10.1128/JB.188.6.2262-2274.2006

Pickett MW, Weiss N, Kelly DJ (1994) Gram-positive cell wall structure of the A3γ type in heliobacteria. FEMS Microbiol Lett 122(1–2):7–12. doi:10.1111/j.1574-6968.1994.tb07135.x

Prat L, Maillard J, Grimaud R, Holliger C (2011) Physiological adaptation of *Desulfitobacterium hafniense* strain TCE1 to tetrachloroethene respiration. Appl Environ Microbiol 77(11):3853–3859. doi:10.1128/aem.02471-10

Regeard C, Maillard J, Holliger C (2004) Development of degenerate and specific PCR primers for the detection and isolation of known and putative chloroethene reductive dehalogenase genes. J Microbiol Methods 56(1):107–118

Rouzeau-Szynalski K, Maillard J, Holliger C (2011) Frequent concomitant presence of *Desulfitobacterium* spp. and "*Dehalococcoides*" spp. in chloroethene-dechlorinating microbial communities. Appl Microbiol Biotechnol 90(1):361–368. doi:10.1007/s00253-010-3042-0

Rupakula A, Kruse T, Boeren S, Holliger C, Smidt H, Maillard J (2013) The restricted metabolism of the obligate organohalide respiring bacterium *Dehalobacter restrictus*: lessons from tiered functional genomics. Philos Trans R Soc Lond B Biol Sci 368(1616):20120325. doi:10.1098/rstb.2012.0325

Rupakula A, Lu Y, Kruse T, Boeren S, Holliger C, Smidt H, Maillard J (2015) Functional genomics of corrinoid starvation in the organohalide-respiring bacterium *Dehalobacter restrictus* strain PER-K23. Front Microbiol 5:751. doi:10.3389/fmicb.2014.00751

Schumacher W, Holliger C (1996) The proton/electron ration of the menaquinone-dependent electron transport from dihydrogen to tetrachloroethene in "*Dehalobacter restrictus*". J Bacteriol 178(8):2328–2333

Schumacher W, Holliger C, Zehnder AJ, Hagen WR (1997) Redox chemistry of cobalamin and iron-sulfur cofactors in the tetrachloroethene reductase of *Dehalobacter restrictus*. FEBS Lett 409(3):421–425

Smidt H, van Leest M, van der Oost J, de Vos WM (2000) Transcriptional regulation of the *cpr* gene cluster in ortho-chlorophenol-respiring *Desulfitobacterium dehalogenans*. J Bacteriol 182(20):5683–5691

Sun BL, Griffin BM, Ayala-del-Rio HL, Hashsham SA, Tiedje JM (2002) Microbial dehalorespiration with 1,1,1-trichloroethane. Science 298(5595):1023–1025. doi:10.1126/science.1074675

Tang S, Gong Y, Edwards EA (2012) Semi-automatic in silico gap closure enabled de novo assembly of two *Dehalobacter* genomes from metagenomic data. PLoS ONE 7(12):e52038. doi:10.1371/journal.pone.0052038

Tang YJ, Yi S, Zhuang W-Q, Zinder SH, Keasling JD, Alvarez-Cohen L (2009) Investigation of carbon metabolism in "*Dehalococcoides ethenogenes*" strain 195 by Use of isotopomer and transcriptomic analyses. J Bacteriol 191(16):5224–5231. doi:10.1128/jb.00085-09

van Doesburg W, van Eekert MH, Middeldorp PJ, Balk M, Schraa G, Stams AJ (2005) Reductive dechlorination of beta-hexachlorocyclohexane (beta-HCH) by a *Dehalobacter* species in coculture with a *Sedimentibacter* sp. FEMS Microbiol Ecol 54(1):87–95. doi:10.1016/j.femsec.2005.03.003

Villemur R, Lanthier M, Beaudet R, Lepine F (2006) The *Desulfitobacterium* genus. FEMS Microbiol Rev 30(5):706–733

von Wintzingerode F, Schlotelburg C, Hauck R, Hegemann W, Gobel UB (2001) Development of primers for amplifying genes encoding CprA- and PceA-like reductive dehalogenases in anaerobic microbial consortia, dechlorinating trichlorobenzene and 1,2-dichloropropane. FEMS Microbiol Ecol 35(2):189–196

Wang S, He J (2013) Dechlorination of commercial PCBs and other multiple halogenated compounds by a sediment-free culture containing *Dehalococcoides* and *Dehalobacter*. Environ Sci Technol 47(18):10526–10534. doi:10.1021/es4017624

Wang S, Zhang W, Yang KL, He J (2014) Isolation and characterization of a novel *Dehalobacter* species strain TCP1 that reductively dechlorinates 2,4,6-trichlorophenol. Biodegradation 25(2):313–323. doi:10.1007/s10532-013-9662-1

Wild A, Hermann R, Leisinger T (1996) Isolation of an anaerobic bacterium which reductively dechlorinates tetrachloroethene and trichloroethene. Biodegradation 7(6):507–511. doi:10.1007/bf00115297

Yoshida N, Ye L, Baba D, Katayama A (2009) A novel *Dehalobacter* species is involved in extensive 4,5,6,7-tetrachlorophthalide dechlorination. Appl Environ Microbiol 75(8):2400–2405. doi:10.1128/aem.02112-08

Zhang C, Suzuki D, Li Z, Ye L, Katayama A (2012) Polyphasic characterization of two microbial consortia with wide dechlorination spectra for chlorophenols. J Biosci Bioeng 114(5):512–517. doi:10.1016/j.jbiosc.2012.05.025

Zhang CF, Li ZL, Suzuki D, Ye LZ, Yoshida N, Katayama A (2013) A humin-dependent *Dehalobacter* species is involved in reductive debromination of tetrabromobisphenol A. Chemosphere 92(10):1343–1348. doi:10.1016/j.chemosphere.2013.05.051

Chapter 9
The Genus *Desulfitobacterium*

Taiki Futagami and Kensuke Furukawa

Abstract To date, 22 *Desulfitobacterium* strains have been isolated. From them a total of six distinct species have been proposed: *D. hafniense*, *D. dehalogenans*, *D. chlororespirans*, *D. metallireducens*, *D. dichloroeliminans*, and *D. aromaticivorans*. The isolated strains are strictly anaerobic, mesophilic, and grow in the neutral pH range. The cells are slightly curved rods ranging from 2 to 7 μm in length and 0.3 to 1 μm in width. Most of the *Desulfitobacterium* strains have been isolated as organohalide-respiring bacteria (OHRB) and show versatile dehalogenation of both chlorinated aliphatic and aromatic compounds such as chloroethenes and chlorophenols. The *Desulfitobacterium* strains are phylogenetically classified into the phylum *Firmicutes* (Gram-positive bacteria). The closest related OHRB genus of *Desulfitobacterium* is *Dehalobacter*, the members of which are strict OHRB within the phylum *Firmicutes* (see Chap. 8). In contrast, the *Desulfitobacterium* strains isolated to date are not strict OHRB. In addition to the ability to respire with organohalides, most isolates can grow fermentatively on pyruvate and can utilize a variety of electron acceptors, including sulfite, thiosulfate, fumarate, Fe(III), and Mn(IV). Complete genome information is available for four *Desulfitobacterium* strains and draft information is available for five strains. The complete genomes range from 3.62 to 5.73 Mbp, with GC content ranging from 44.2 to 47.5 % and the number of predicted coding sequences ranging from 3340 to 5060. Consistent with their physiological diversity, the *Desulfitobacterium* genome has been shown to encode a variety of respiratory reductases, including reductive dehalogenases.

T. Futagami (✉)
Education and Research Center for Fermentation Studies, Faculty of Agriculture, Kagoshima University, 1-21-24 Korimoto, Kagoshima 890-0065, Japan
e-mail: futagami@chem.agri.kagoshima-u.ac.jp

K. Furukawa
Faculty of Food Science and Nutrition, Department of Food and Bioscience, Beppu University, Kitaishigaki 82, Beppu, Oita 874-8501, Japan
e-mail: kfurukaw@nm.beppu-u.ac.jp

L. Adrian and F.E. Löffler (eds.), *Organohalide-Respiring Bacteria*,
DOI 10.1007/978-3-662-49875-0_9

Abbreviations Used in Text

AQDS	Anthraquinone-2,6-disulfonate
3-Cl-4-OHPA	3-chloro-4-hydroxyphenylacetic acid
DCA	Dichloroethane
DCP	Dichlorophenol
DMSO	Dimethyl sulfoxide
OHRB	Organohalide-respiring bacteria
PCE	Tetrachloroethene
PCP	Pentachlorophenol
RDase	Reductive dehalogenase
TCE	Trichloroethene
TCP	Trichlorophenol

Abbreviations Used in Tables

AQDS	Anthraquinone-2,6-disulfonate
BP	Bromophenol
bromoxynil	3,5-dibromo-4-hydroxybenzonitrile
CD	Carbon dichloride
CF	Chloroform
3-Cl-4-OHPA	3-chloro-4-hydroxyphenylacetic acid
CP	Chlorophenol
CT	Carbon tetrachloride
Cysteate	Alanine-3-sulfonate
DBP	Dibromophenol
DCA	Dichloroethane
DCHQ	Dichlorohydroquinone
DCP	Dichlorophenol
HCB	Hexachlorobenzene
Ioxynil	3,5-diiodo-4-hydroxybenzonitrile
Isethionate	2-hydroxyethanesulfonate
PCE	Tetrachloroethene
PCP	Pentachlorophenol
TCA	Trichloroethane
TCE	Trichloroethene
TCHQ	2,3,5,6-tetrachlorohydroquinone
TCMP	2,3,5,6-tetrachloro-4-methoxyphenol
TCP	Trichlorophenol
TeCP	Tetrachlorophenol
VC	Vinyl chloride

9.1 Discovery

The *Desulfitobacterium* strains have been isolated primarily as organohalide-respiring bacteria (OHRB) from organohalide-contaminated environments. The first *Desulfitobacterium* isolate was strain DCB-2 (for [aryl-] dechlorinating bacterium no. 2, currently known as *Desulfitobacterium hafniense* DCB-2) (Madsen and Licht 1992). Strain DCB-2 was isolated as a chlorinated phenol-degrading bacterium from a stable trichlorophenol (TCP)-dechlorinating consortium enriched from municipal digestor sludge in Copenhagen, Denmark. At that time, strain DCB-2 was reported to dehalogenate chlorophenols at the *ortho* and *meta* positions. This activity differed from that reported for *Desulfomonile tiedjei* DCB-1, which was the first isolate of the OHRB, dechlorinating only from the *meta* position (Shelton and Tiedje 1984; DeWeerd et al. 1990). Strain DCB-2 was later designated *D. hafniense* DCB-2 (Christiansen and Ahring 1996a). The species name *hafniense* reflects that Copenhagen was the place of isolation.

The nomenclature of the genus *Desulfitobacterium* was defined upon isolation of the second strain, *D. dehalogenans* JW/IU-DC1, in Athens, Georgia, USA (Utkin et al. 1994). The strain JW/IU-DC1 can dechlorinate at the *ortho* position of both chlorophenols and chlorophenylacetates (Utkin et al. 1994, 1995) and was the first *Desulfitobacterium* strain for which energy conservation through reductive dechlorination of 3-chloro-4-hydroxyphenylacetic acid (3-Cl-4-OHPA) was shown (Mackiewicz and Wiegel 1998). The genus name *Desulfitobacterium* describes a rod-shaped bacterium that reduces sulfite. However, several *Desulfitobacterium* isolates that do not reduce sulfite have been reported, including *D. metallireducens* 853-15A and *Desulfitobacterium* sp. PR (Finneran et al. 2002; Ding et al. 2014). A total of 22 *Desulfitobacterium* strains exhibiting a wide dehalogenation spectrum have been isolated and characterized to date.

9.2 Isolation and Habitat

Desulfitobacterium strains are distributed worldwide, having been isolated from Denmark, the United States, Vietnam, The Netherlands, Germany, Canada, Japan, Finland, and Poland (Table 9.1). Most *Desulfitobacterium* strains have been isolated during the course of studies of reductive dechlorination processes. Therefore, the target organochlorines in each of these studies were generally used as the electron acceptors in enrichment cultures. A total of 15 *Desulfitobacterium* strains (*D. hafniense* strains DCB-2, TCE1, TCP-A, PCE-S, PCP-1, Y51; *D. dehalogenans* JW/IU-DC1; *D. chlororespirans* Co23; *D. dichloroeliminans* DCA1; and *Desulfitobacterium*. sp. strains Viet1, KBC1, PCE1, B31e3, JH1, PR) have been isolated from enrichment cultures containing the organochlorines tetrachloroethene (PCE), trichloroethene (TCE), 2,4,6-TCP, 2,3-dichlorophenol (2,3-DCP), pentachlorophenol (PCP), 3-Cl-4-OHPA, 1,2-dichloroethane (1,2-DCA), or 1,1,1-trichloroethane (Suyama et al. 2001; Gerritse et al. 1999; Breitenstein et al. 2001;

Table 9.1 *Desulfitobacterium* strains in order of the year they were first reported

Organism	Stock center no.	Source of isolation	Geographic origin	Reference
D. hafniense DCB-2^T	DSM 10664	Municipal digestor sludge	Copenhagen, Denmark	Madsen and Licht (1992)
D. dehalogenans JW/IU-DC1^T	DSM 9161, ATCC 51507	A freshwater sediment collected from a pond of a wooded area	The Sandy Creek Nature Park, Athens, Georgia, USA	Utkin et al. (1994)
Desulfitobacterium sp. PCE1	DSM 10344	Soil contaminated with chloroethene	The Netherlands	Gerritse et al. (1995, 1996)
D. chlororespirans Co23^T	ATCC 700175, DSM 11544	Compost soil	Lansing, Michigan, USA	Sanford et al. (1996)
D. hafniense PCP-1	ATCC 700357, DSM 12420	A mixture of anaerobic sewage sludge and soil samples that had been contaminated with PCP	Quebec, Canada	Bouchard et al. (1996)
D. hafniense PCE-S	DSM 14645	Soil contaminated with chloroethenes	Eppelheim, Germany	Miller et al. (1997), Goris et al. (2015)
Desulfitobacterium sp. Viet1	NA	Parfume River sediment	Huế, Vietnam	Löffler et al. (1997)
D. dehalogenans XZ-1	ATCC 700041	A freshwater	Athens, Georgia, USA	Wiegel et al. (1999), ATCC website
D. hafniense TCE1	DSM 12704[a]	Soil obtained from a chloroethene-polluted location	Breda, The Netherlands	Gerritse et al. (1999)
D. hafniense TCP-A	DSM 13557	Sediment of the river Saale	Germany	Breitenstein et al. (2001)
D. hafniense DP7	DSM 13498	Fresh fecal sample of a healthy 28-year-old female person	The Netherlands	van de Pas et al. (2001b)
D. hafniense Y51	NBRC 109954	Soil contaminated with PCE	Fukuoka, Japan	Suyama et al. (2001)
D. hafniense GBFH	NA	Arsenic-contaminated sediments	Coeur d'Alene River delta, Lake Coeur d'Alene, Idaho, USA	Niggemyer et al. (2001)
D. metallireducens 853-15A^T	ATCC BAA-636, DSM 15288	Uranium-contaminated anaerobic aquifer sediment	Floodplain of the San Juan River, Shiprock, NM, USA	Finneran et al. (2002)

(continued)

Table 9.1 (continued)

Organism	Stock center no.	Source of isolation	Geographic origin	Reference
D. hafniense G2	DSM-16228	Subsurface smectite bedding of the Twiggs Clay formation of late Eocene age	Georgia, USA	Shelobolina et al. (2003)
D. dichloroeliminans DCA1^{T}	BCCM/LMG P-21439	Soil matrix of an anoxic water-saturated layer (1 m in depth) that had been exclusively polluted with 50 mg/kg 1,2-DCA for 30 years	NR	De Wildeman et al. (2003)
Desulfitobacterium sp. RPf35Ei	NA	Methanogenic granular sludge and sediments	Outokumpu's Pyhäsalmi mine, Finland	Kaksonen et al. (2004)
Desulfitobacterium sp. KBC1	NA	Soil sample from crop field	Ibaraki, Japan	Tsukagoshi et al. (2006)
Desulfitobacterium sp. B31e3	NA	Unsaturated subsurface soils contaminated with chloroethene	NR	Yoshida et al. (2007)
D. hafniense JH1	NA	Ditch sludge (mixed with sewage) contaminated with PCE and halogenated aliphatic compounds	Gifu, Japan	Chang et al. (2000), Fletcher et al. (2008)
D. aromaticivorans UKTL^{T}	DSM 19510, JCM 15765	Soil of a former coal gasification site	Gliwice, Poland	Kunapuli et al. (2010)
Desulfitobacterium sp. PR	NA	An anaerobic mixed culture enriched from a bioreactor maintained to perform dechlorination of chloroethenes and chloroethanes	NR	Ding et al. (2014)

[a] Currently not available. Type strains are indicated by a superscript capital T. *NA* not available; *NR* not reported; *PCP* pentachlorophenol; *PCE* tetrachloroethene, *DCA* dichloroethane

Miller et al. 1997; Bouchard et al. 1996; Madsen and Licht 1992; Sanford et al. 1996; Löffler et al. 1997; Tsukagoshi et al. 2006; Utkin et al. 1994; Gerritse et al. 1996; De Wildeman et al. 2003; Yoshida et al. 2007; Fletcher et al. 2008; Ding et al. 2014). In contrast, six *Desulfitobacterium* strains (*D. hafniense* strains G2, DP7, and GBFH; *Desulfitobacterium* sp. RPf35Ei; *D. metallireducens* 853-15A; and *D. aromaticivorans* UKTL) were isolated in the absence of organohalides (Shelobolina et al. 2003; van de Pas et al. 2001b; Niggemyer et al. 2001; Kaksonen et al. 2004; Finneran et al. 2002; Kunapuli et al. 2010).

Desulfitobacterium strains have been isolated from a variety of anaerobic environments, including digestor sludge, sewage sludge, ditch sludge mixed with sewage, methanogenic granular sludge, freshwater sediment, river sediment, arsenic-contaminated sediments, uranium mill tailing site floodplain, active mines, compost soil, soil samples from crop fields, soil from a former coal gasification site, and human feces (Table 9.1). Studies indicate that *Desulfitobacterium* spp. are widely distributed in anaerobic terrestrial environments, but not in marine environments. Molecular ecological studies have confirmed the ubiquitous distribution of *Desulfitobacterium* spp. For example, in one study, *Desulfitobacterium* spp. were detected in 31 of 48 soil samples from Canada (mainly from Quebec), which included 24 samples from contaminated industrial sites (Lanthier et al. 2001). In addition, *Desulfitobacterium* spp. as well as other OHRB have been detected with a high frequency in samples from chloroethenes-contaminated sites (Davis et al. 2002; Yang et al. 2005; Yoshida et al. 2007; Dowideit et al. 2010; Rouzeau-Szynalski et al. 2011; Kranzioch et al. 2013). Thus, the *Desulfitobacterium* spp. are one of the most abundant OHRB in the natural environment.

Methods involving serial dilution or single colony isolation on agar plates or in agar shake cultures have been used to isolate *Desulfitobacterium* strains. In addition, flow cytometric sorting was utilized in the isolation of *Desulfitobacterium* sp. KBC1 (Tsukagoshi et al. 2006). Anaerobic media containing target organohalides in combination with an electron donor such as pyruvate works well when attempting to isolate organohalide-respiring *Desulfitobacterium* strains. Yeast extract can be used as a source of vitamins and trace elements when culturing *Desulfitobacterium* strains. Antimicrobial agents may also be useful for *Desulfitobacterium* enrichment, as evidenced by the isolation of *D. hafniense* DP7 using the antibacterial agent aztreonam (50 μg/L), which is a β-lactam compound that inhibits the growth of aerobic Gram-negative bacteria (van de Pas et al. 2001b). Sequence analyses identified a vancomycin resistance gene cluster in the genome of *D. hafniense* Y51 (Kalan et al. 2009; Kruse et al. 2014a). Thus, exploiting vancomycin resistance might also be useful for the enrichment and isolation of *Desulfitobacterium* strains. Pasteurization is also useful for the isolation of spore-forming *Desulfitobacterium* strains (see Sect. 9.3).

The *Desulfitobacterium* strains can be preserved as frozen or freeze-dried stocks (Spring and Rosenzweig 2006). For long-term storage, anaerobic 50 % glycerol stocks are kept at −80 °C or in liquid nitrogen. Glycerol stocks remain viable for at least 5 years when stored at −80 °C (authors' unpublished data). The procedures used for manipulating the stocks under anaerobic conditions are important for the long-term preservation of *Desulfitobacterium* strains. Although freeze-drying is also suitable for long-term storage, several *Desulfitobacterium* strains are sensitive to the freeze-drying process and display low survival rates after lyophilization. For these strains, preparation of cell suspensions in freshly prepared medium supplemented with 5 % dimethyl sulfoxide (DMSO) as a cryoprotectant and subsequent storage in liquid nitrogen is recommended (Spring and Rosenzweig 2006).

9.3 Morphology, Physiology, and Growth Characteristics

Morphologically, the *Desulfitobacterium* strains appear as slightly curved rods (Fig. 9.1). The cells range from 2 to 7 μm in length and 0.3 to 1 μm in width (Table 9.2). Variation in cell length (exceeding 10 μm) has been reported for *D. dichloroeliminans* DCA1 (De Wildeman et al. 2003). In addition, a coupled cell structure was reported for *Desulfitobacterium* sp. PR (Ding et al. 2014), which may form filaments composed of up to 10 cells. Most *Desulfitobacterium* isolates are motile and have from 1 to 6 flagella. Nonmotile cells without flagella have been reported for *D. hafniense* PCP-1 (Bouchard et al. 1996). The motility characteristics of *D. hafniense* JH1 and *Desulfitobacterium* sp. PR have not been described, but these cells also do not have flagella (Fletcher et al. 2008; Ding et al. 2014).

The *Desulfitobacterium* strains are mesophilic and grow at neutral pH. The optimum growth temperature and pH are in the ranges of 25–40 °C and 6.5–7.8, respectively (Table 9.2). Ten *Desulfitobacterium* strains are capable of forming terminal endospores. The capability of forming spores is likely associated with ensuring survival during heat stress, as evidenced by the observation that *D. hafniense* strains DCB-2, PCP-1, and TCP-A survived pasteurization during the isolation process (Madsen and Licht 1992; Bouchard et al. 1996; Breitenstein et al. 2001); during the isolation of *D. hafniense* DCB-2, the dechlorinating enrichment consortium was incubated at 80 °C for 60 min (Madsen and Licht 1992). Other research has confirmed that *D. chlororespirans* Co23 and *D. hafniense* GBFH can survive heat treatment (Sanford et al. 1996; Niggemyer et al. 2001). On the other hand, the non-spore forming strains *Desulfitobacterium* sp. PCE1 and *D. metallireducens* 853-15A do not survive heat treatment (Gerritse et al. 1996; Finneran et al. 2002); incubation at 85 °C for 20 min irreversibly inhibits the growth of *Desulfitobacterium* sp. PCE1 (Gerritse et al. 1996).

Although the known *Desulfitobacterium* strains are strict anaerobes, they are slightly oxygen tolerant. For example, *D. dehalogenans* JW/IU-DC1 was shown to dechlorinate 3-Cl-4-OHPA under aerobic condition with 2 % air, indicating its tolerance for oxygen (Utkin et al. 1994). It has also been reported that *D. hafniense* strains DCB-2 and GBFH, *D. dichloroeliminans* DCA1, and *Desulfitobacterium* sp. strain PR could survive aerobic conditions for 24 h (Madsen and Licht 1992; Niggemyer et al. 2001; De Wildeman et al. 2003; Ding et al. 2014).

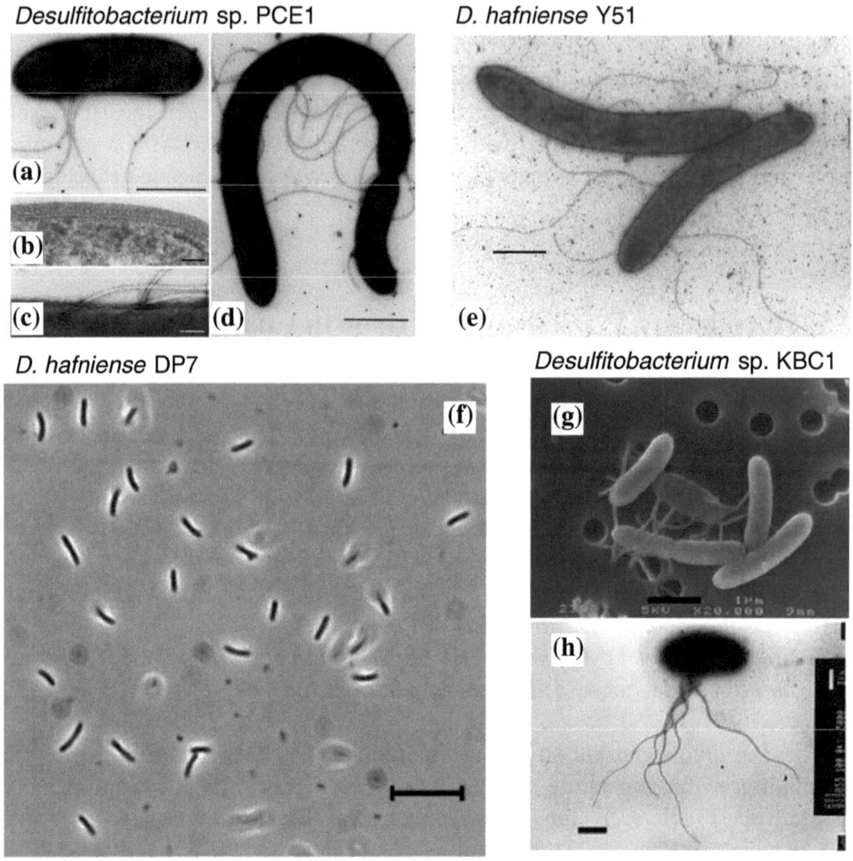

Fig. 9.1 Morphology of *Desulfitobacterium* strains. **a–d** Electron micrographs of negatively stained exponential-phase cells of *Desulfitobacterium* sp. PCE1. **a** Cell with four laterally attached flagella. **b** Ultrathin section revealing the thick Gram-positive cell wall. **c** S-layer surrounding the cell wall and **d** dividing long, curved cell of *Desulfitobacterium* sp. PCE1. Bars indicate 1 μm in (**a**) and (**d**) and 0.1 μm in (**b**) and (**c**). **e** Electron micrograph of negatively stained exponential-phase cells of *D. hafniense* Y51. Bar, 1 μm. **f** Phase contrast light micrograph of *D. hafniense* DP7. Bar, 10 μm. (**g–h**). **g** Scanning electron micrograph and **h** transmission electron micrograph of exponential-phase cells of *Desulfitobacterium* sp. KBC1. Bar, 1 μm. Photos were taken from Gerritse et al. (1996), van de Pas et al. (2001b), Furukawa et al. (2005), and Tsukagoshi et al. (2006) with permission

Characterization of the cellular fatty acid composition is frequently used in microbial taxonomy. The fatty acid composition has been reported for *D. dehalogenans* JW/IU-DC1, *D. hafniense* strains DCB-2 and PCP-1, *D. chlororespirans* Co23, *D. metallireducens* 853-15A, and *D. aromaticivorans* UKTL. Spring and Rosenzweig (2006) reported that the major fatty acids in *D. dehalogenans* JW/IU-DC1, *D. hafniense* strains DCB-2 and PCP-1, *D. chlororespirans* Co23, and *D. metallireducens* 853-15A are 14:0 (4.0–22.8 %), 16:1 *cis*9 (6.4–13.0 %),

Table 9.2 Physiological features of *Desulfitobacterium* strains

Organism	Cell size (long × wide μm)	Motility	Flagella	Sporulation	GC content (%)[a]	Optimum temperature	Optimum pH	Reference
D. hafniense DCB-2	3.3–6 × 0.6–0.7	+	+	+	47.5	37	ND	Madsen and Licht (1992), Christiansen and Ahring (1996a), Kim et al. (2012)
D. dehalogenans JW/IU-DC1	2.5–4 × 0.7	+	+	–	45	38	7.5	Utkin et al. (1994), van de Pas et al. (2001b), Kruse et al. (2014b)
Desulfitobacterium sp. PCE1	2–7 × 0.6–0.8	+	+	–	45	34–37	7.2	Gerritse et al. (1996)
D. chlororespirans Co23	3–5 × 0.5–1	+	ND	+	48.8	37	6.8–7.5	Sanford et al. (1996), van de Pas et al. (2001b)
D. hafniense PCP-1	2–4.5 × 0.7	–	–	+	47.5	38	7.5	Bouchard et al. (1996)
D. hafniense PCE-S	6 × 0.6	+	ND	+	47.3	37	ND	Miller et al. (1997), Goris et al. (2015)
Desulfitobacterium sp. Viet1	ND	ND	ND	ND	ND	ND	7.5	Löffler et al. (1997)
D. dehalogenans XZ-1	ND	ND	ND	ND	ND	ND	ND	Wiegel et al. (1999)
D. hafniense TCE1	2–4 × 0.6–0.8	+	+	–	47.5	35	7.2	Gerritse et al. (1997, 1999), van de Pas et al. (2001b)
D. hafniense TCP-A	2.5–5 × 0.6	+	ND	+	47.3	ND	ND	Breitenstein et al. (2001)
D. hafniense DP7	4–6 × 0.6	ND	+	–	47.6	34–40	7.2–7.4	van de Pas et al. (2001b)
D. hafniense Y51	5–7 × 0.8–1	+	+	–	47.4	37	6.5–7.5	Suyama et al. (2001), Nonaka et al. (2006)
D. hafniense GBFH	2–4 × 0.3–0.5	+	+	+	ND	37–38	7.5	Niggemyer et al. (2001)

(continued)

Table 9.2 (continued)

Organism	Cell size (long × wide μm)	Motility	Flagella	Sporulation	GC content (%)[a]	Optimum temperature	Optimum pH	Reference
D. metallireducens 853-15A	2–5 × 0.5	–	ND	–	41.8	30	7	Finneran et al. (2002)
D. hafniense G2	2–4 × 0.6–0.8	+	+	+	ND	ND	ND	Shelobolina et al. (2003)
D. dichloroeliminans DCA1	2–5 × 0.5–0.7	+	ND	ND	44.2	25–30	7.2–7.8	De Wildeman et al. (2003)
Desulfitobacterium sp. RPf35Ei	2.5–4 × 0.3–0.5	ND	ND	+	ND	35	ND	Kaksonen et al. (2004)
Desulfitobacterium sp. KBC1	2–3	+	+	ND	46	34	7.5	Tsukagoshi et al. (2006)
Desulfitobacterium sp. B31e3	2–3 × 0.4–0.5	+	ND	+		ND	7.0	Yoshida et al. (2007)
D. hafniense JH1	ND	ND	ND	–	ND	ND	ND	Chang et al. (2000), Fletcher et al. (2008)
D. aromaticivorans UKTL	2.3–4 × 0.5	+	ND	+	47.4	30	6.6–7.0	Kunapuli et al. (2010)
Desulfitobacterium sp. PR	2–5 × 0.3–0.35	–	ND	–	43.5	27–33	7.1-7.5	Ding et al. (2014)

[a]GC content of strains DCA1, PCP-1, PCE-S, TCP-A, DP7, 853-15A, and PCE1 were from their genome information. *ND* not determined

16:0 (4.4–25.6 %), and 18:1 *cis*11 (0.5–13.6 %). The fatty acid profiles of *D. hafniense* strains DCB-2 and PCP-1 are similar. In contrast, the fatty acid compositions of *D. dehalogenans* JW/IU-DC1 and *D. metallireducens* 853-15A differ. The characteristic fatty acids were identified as 16:0, 14:0, and 16:1 *cis*9 in *D. dehalogenans* JW/IU-DC1; 18:1 *cis*11 dimethylacetal, 18:1 *cis*1, and 16:1 *cis*9 in *D. hafniense* strains DCB-2 and PCP-1; 16:0, 16:1 *cis*9, and 14:0 in *D. chlororespirans* Co23; and 14:0, 16:1 *cis*9 dimethylacetal, and an unidentified fatty acid with an equivalent chain length of 13.52 in *D. metallireducens* 853-15A (Spring and Rosenzweig 2006). In addition, Kunapuli et al. (2010) investigated the fatty acid composition of *D. aromaticivorans* UKTL, with *D. chlororespirans* Co23 serving as a control, and found that strain UKTL contains 15:0 iso, but not 18:1ω7*c* fatty acids, in contrast to strain Co23. These data have to be used with a caution because fatty acid composition can change with factors such as medium composition and growth phase.

The cytochrome and quinone contents have been reported for several *Desulfitobacterium* strains. The *D. hafniense* strains DCB-2 and TCE1, *D. metallireducens* 853-15A, and *Desulfitobacterium* sp. PCE1 were shown to contain cytochrome c (Christiansen and Ahring 1996a; Gerritse et al. 1996, 1999; Finneran et al. 2002). Menaquinone-7 was identified as the primary menaquinone of *D. hafniense* TCP-A and *D. aromaticivorans* UKTL (Breitenstein et al. 2001; Kunapuli et al. 2010).

The *Desulfitobacterium* strains can also grow fermentatively; fermentative growth on pyruvate has been reported for most *Desulfitobacterium* isolates except for *D. metallireducens* 853-15A (Finneran et al. 2002) (Table 9.3). The fermentative growth yield of *D. dehalogenans* JW/IU-DC1 on pyruvate is approximately 14 g of dry cell weight per mole of pyruvate (van de Pas et al. 2001a). Tryptophan and serine also reportedly support the fermentative growth of *D. hafniense* DCB-2 and *D. hafniense* TCE1, respectively (Christiansen and Ahring 1996a; Gerritse et al. 1999).

Formate, lactate, and pyruvate generally serve as electron donors for *Desulfitobacterium* strains. In addition, most *Desulfitobacterium* strains exhibit *O*-demethylation activity. The *O*-demethylation was reported to be involved in the intermediary metabolism for methoxylated organochlorines such as tetrachloroguaiacol, tetrachloroveratrole, pentachloroanisole, and 3,5-dichloro-4-methoxyphenol in *D. hafniense* strains PCP-1 and DCB-2, *D. chlororespirans* Co23, and *D. dehalogenans* JW/IU-DC1 (Dennie et al. 1998; Milliken et al. 2004b). Then, the utilization of phenyl methyl ethers, vanillate and syringate, as electron donors via *O*-demethylation was reported for *D. hafniense* strains DCB-2, PCE-S, DP7, G2, PCP-1, TCP-A, and Y51, *D. chlororespirans* Co23, and *D. dehalogenans* JW/IU- 1 (Neumann et al. 2004; Mingo et al. 2014). Enzymes involved in *O*-demethylation have been biochemically characterized in *D. hafniense* strains DCB-2 and PCE-S (Kreher et al. 2008; Studenik et al. 2012). The methyl group from phenyl methyl ethers is transferred to tetrahydrofolate and considered to be further used as an electron donor via acetyl-CoA formation. Because phenyl methyl ethers are lignin decomposition products, *Desulfitobacterium* spp. are

Table 9.3 Primary metabolic featu̲res of *Desulfitobacterium* isolates

Organism		Electron donors						Electron acceptors														Reference
	Fermentatoin of pyruvate	Formate	Hydrogen	Lactate	Pyruvate	Vanillate	Syringate	Fumarate	Sulfate	Sulfur	Sulfite	Thiosulfate	AQDS	Isethionate	Cysteate	Nitrate	Fe(III)	As(V)	Se(VI)	Mn(IV)	U(VI)	
D. hafniense DCB-2	+	+	–	+	+	+	+	+	–	+	+	+	+	+	–	+	+	+	+	+	+	Madsen and Licht (1992), Christiansen and Ahring (1996a), Lie et al. (1999), Niggemyer et al. (2001), Neumann et al. (2004), Milliken and May (2007), Kim et al. (2012), Mingo et al. (2014)
D. dehalogenans JW/IU-DC1	+	+	+	+	+	+	+	+	–	+	+	+	+	+	+	+	+	–	+	+	+	Utkin et al. (1994), Lovley et al. (1998), Lie et al. (1999), Niggemyer et al. (2001), Cervantes et al. (2002), Luijten et al. (2004), Fletcher et al. (2010), Mingo et al. (2014)
Desulfitobacterium sp. PCE1	+	+	–	+	+	ND	ND	+	–	ND	+	+	+	+	+	+	–	–	+	–	+	Gerritse et al. 1996), Lie et al. (1999), Gerritse et al. (1999), Cervantes et al. (2002), Luijten et al. (2004), Fletcher et al. (2010)

(continued)

Table 9.3 (continued)

Organism	Fermentatoin of pyruvate	Electron donors						Electron acceptors														Reference
		Formate	Hydrogen	Lactate	Pyruvate	Vanillate	Syringate	Fumarate	Sulfate	Sulfur	Sulfite	Thiosulfate	AQDS	Isethionate	Cysteate	Nitrate	Fe(III)	As(V)	Se(VI)	Mn(IV)	U(VI)	
D. chlororespirans Co23	+	+	+	+	+	+	+	–	–	+	+	+	+	–	–	–	+	–	+	+	+	Sanford et al. (1996), Lie et al. (1999), Niggemyer et al. (2001), Luijten et al. (2004), Fletcher et al. (2010), Mingo et al. (2014)
D. hafniense PCP-1	+	+	–	+	+	+	+	+	–	+	+	+	ND	ND	ND	+	+	+	+	+	ND	Bouchard et al. (1996), Niggemyer et al. (2001), Mingo et al. (2014)
D. hafniense PCE-S	+	+	ND	ND	+	+	–	+	ND	ND	ND	ND	ND	ND	ND	ND	ND	ND	ND	ND	ND	Miller et al. (1997), Neumann et al. (2004), Ye et al. (2010), Mingo et al. (2014)
Desulfitobacterium sp. Viet1	+	ND	+	ND	ND	ND	ND	ND	ND	ND	ND	ND	ND	–	–	ND	ND	ND	ND	ND	+	Löffler et al. (1997), Lie et al. (1999), Fletcher et al. (2010)
D. dehalogenans XZ-1	ND	ND	ND	ND	+	ND	ND	ND	ND	ND	ND	ND	ND	ND	ND	ND	ND	ND	ND	ND	ND	Wiegel et al. (1999)
D. hafniense TCE1	+	+	+	+	+	ND	ND	+	–	ND	+	+	ND	ND	ND	+	+	+	+	+	ND	Gerritse et al. (1999), Luijten et al. (2004)
D. hafniense TCP-A	+	+	+	+	+	+	+	+	–	ND	+	+	ND	ND	ND	+	ND	ND	ND	ND	ND	Breitenstein et al. (2001), Mingo et al. (2014)
D. hafniense DP7	+	+	+	+	+	+	+	+	–	ND	+	+	+	ND	ND	+	+	–		+	ND	van de Pas et al. (2001b), Luijten et al. (2004), Mingo et al. (2014)

(continued)

Table 9.3 (continued)

Organism	Fermentatoin of pyruvate	Electron donors						Electron acceptors														Reference
		Formate	Hydrogen	Lactate	Pyruvate	Vanillate	Syringate	Fumarate	Sulfate	Sulfur	Sulfite	Thiosulfate	AQDS	Isethionate	Cysteate	Nitrate	Fe(III)	As(V)	Se(VI)	Mn(IV)	U(VI)	
D. hafniense Y51	ND	+	ND	+	+	+	+	+	+	ND	+	ND	ND	ND	ND	+	ND	ND	ND	ND	ND	Suyama et al. (2001), Peng et al. (2012), Mingo et al. (2014)
D. hafniense GBFH	+	+	–	+	+	ND	ND	+	–	+	+	+	ND	ND	ND	–	+	+	+	+	ND	Niggemyer et al. (2001)
D. metallireducens 853-15A	–	+	–	+	+	–	–	–	–	+	–	+	+	ND	ND	–	+	ND	–	+	ND	Finneran et al. (2002), Mingo et al. (2014)
D. hafniense G2	+	+	–	+	+	+	+	+	–	–	+	+	+	ND	ND	+	+	ND	ND	–	+	Shelobolina et al. (2003), Mingo et al. (2014)
D. dichloroeliminans DCA1	ND	+	–	+	ND	ND	ND	–	–	ND	+	+	ND	ND	ND	+	ND	ND	ND	ND	ND	De Wildeman et al. (2003)
Desulfitobacterium sp. RPf35Ei	ND	ND	ND	+	ND	ND	ND	ND	–	ND	+	ND	ND	ND	ND	ND	ND	ND	ND	ND	ND	Kaksonen et al. (2004)
Desulfitobacterium sp. KBC1	+	+	–	+	+	ND	ND	+	–	ND	+	+	ND	ND	ND	–	ND	ND	ND	ND	ND	Tsukagoshi et al. (2006)
Desulfitobacterium sp. B31e3	ND	ND	+	+	+	ND	ND	ND	ND	ND	ND	ND	ND	ND	ND	ND	ND	ND	ND	ND	ND	Yoshida et al. (2007)
D. hafniense JH1	+	+	+	ND	+	ND	ND	ND	–	ND	+	ND	ND	ND	ND	+	+	ND	ND	ND	+	Fletcher et al. (2008, 2010)
D. aromaticivorans UKTL	+	+	–	–	+	ND	ND	–	–	–	+	+	ND	ND	ND	–	+	ND	ND	–	ND	Kunapuli et al. (2010)
Desulfitobacterium sp. PR	+	+	+	–	+	ND	ND	ND	–	ND	–	–	ND	ND	ND	–	ND	ND	ND	ND	ND	Ding et al. (2014)

AQDS anthraquinone-2,6-disulfonate; *Cysteate* alanine-3-sulfonate; *Isethionate* 2-hydroxyethanesulfonate; *ND* not determined thought to be involved in biological lignin transformation. *D. hafniense* strains DCB-2 and PCE-S grown on fumarate and vanillate or syringate yield approximately 15 g of dry cell weight per mole of methyl moiety converted (Neumann et al. 2004).

Compounds capable of supporting the growth of *Desulfitobacterium* strains as electron acceptor include sulfite, thiosulfate, fumarate, nitrate, anthraquinone-2,6-disulfonate (AQDS, a humic acid analog), isethionate (2-hydroxyethanesulfonate), and cysteate (adenine-3-sulfonate). *Desulfitobacterium* strains generally cannot use sulfate and nitrite as electron acceptors with the exception that *D. hafniense* Y51 reduces sulfate (Suyama et al. 2001). Inorganic metals such as Fe(III), Mn(IV), Se(VI), As(V), and U(VI) can also be used as electron acceptors by *Desulfitobacterium* spp., and strains GBFH, G2, 853-15A, and UKTL have been isolated as metal reducers without demonstrable organohalide respiration. *D. hafniense* GBFH was isolated as an As(V)-reducing bacterium from arsenic-contaminated sediments (Niggemyer et al. 2001). *D. hafniense* G2 can utilize U(VI), Fe(III) and AQDS as electron acceptors (Shelobolina et al. 2003). *D. metallireducens* 853-15A can utilize AQDS, chelated Fe(III) (not crystalline Fe[III] oxide), humic acids, Mn(IV), colloidal sulfur, Se(IV), and Cr(VI) as electron acceptors (Finneran et al. 2002). *D. aromaticivorans* UKTL was isolated as an iron-reducing bacterium capable of anaerobic degradation of monoaromatic hydrocarbons, including toluene, phenol, and *p*-cresol (Kunapuli et al. 2010). The *Desulfitobacterium* strains initially isolated as OHRB also have the potential to reduce a variety of electron acceptors including metals (Table 9.3). Thus, the *Desulfitobacterium* spp. play important roles in the natural cycles of a variety of compounds other than organohalides.

9.4 Phylogeny

The genus *Desulfitobacterium* belongs to the phylum *Firmicutes*, class *Clostridia*, order *Clostridiales*, and family *Peptococcaceae* (Lupa and Wiegel 2009). Members of the family *Peptococcaceae* are anaerobes, and the *Desulfitobacterium* spp. which are strictly anaerobic. However, in contradiction to the name *Peptococcaceae*, the known *Desulfitobacterium* strains are not cocci, but curved rods throughout all growth phases, similar to some other *Peptococcaceae* genera (e.g., *Dehalobacter*; see Chap. 8) (Fig. 9.1). This apparent contradiction derives from the fact that the family *Peptococcaceae* was composed of three genera of cocci (*Peptococcus*, *Peptostreptococcus*, and *Ruminococcus*) when it was proposed (Rogosa 1971). As the phylum *Firmicutes* is composed of Gram-positive bacteria, the *Desulfitobacterium* spp. are recognized as Gram positive. However, Gram stain analyses determined that five isolates, including *D. hafniense* strains DCB-2, PCP-1, and Y51, *D. chlororespirans* Co23, and *Desulfitobacterium* sp. B31e3, are actually staining Gram negative (Bouchard et al. 1996; Christiansen and Ahring 1996a; Sanford et al. 1996; Suyama et al. 2001) (Table 9.2). In the case of *D. hafniense* PCP-1, the cells stain Gram negative, but observation of ultrathin cross section of a strain PCP-1 by electron microscopic analysis indicated that this strain is Gram positive. The thick Gram-positive cell wall structure is also evident in ultrathin sections of *Desulfitobacterium* sp. PCE1 (Fig. 9.1b). Thus, it appears that staining is insufficient for determining whether *Desulfitobacterium* strains are Gram positive or Gram negative. On the other hand, the cell envelope

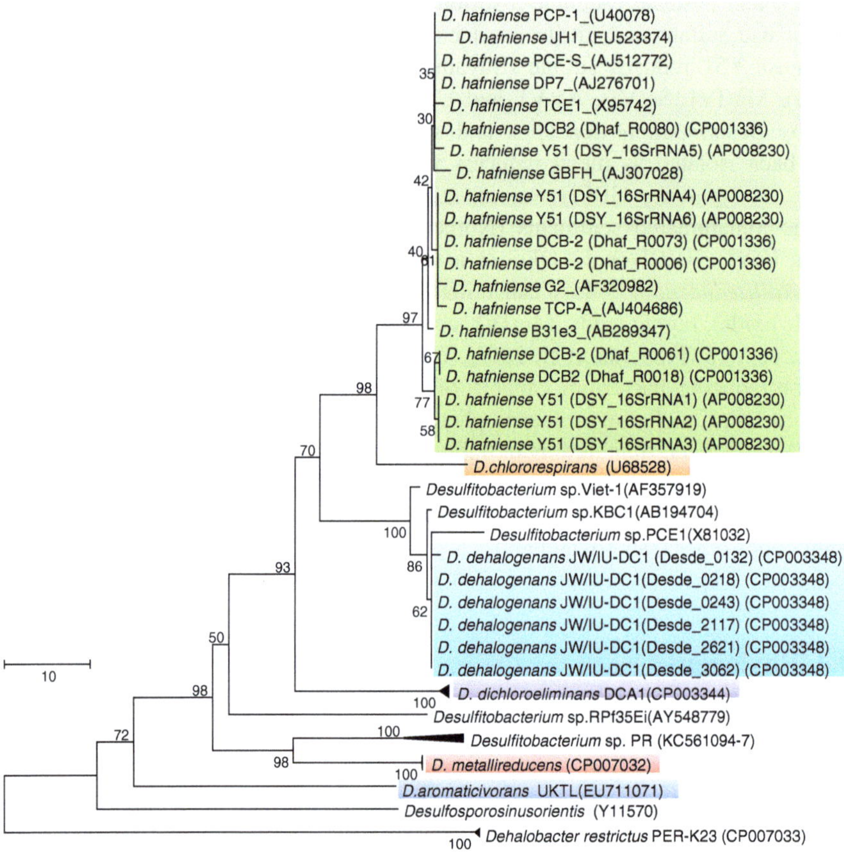

Fig. 9.2 Phylogenetic relationships of *Desulfitobacterium* isolates based on 16S rRNA gene sequences. Sequence accession numbers are indicated in parentheses. Branch supporting values (%) were evaluated with 1000 bootstrap replications. The tree was constructed using the neighbor-joining method based on an alignment of almost-complete 16S rRNA gene sequences. The 100- to 200-bp insertion sequence harboring the 5′ end (Villemur et al. 2007) was excluded using the complete gap deletion option in MEGA version 6 software (Tamura et al. 2013). *Desulfosporosinus orientis* and *Dehalobacter restrictus* PER-K23 were used as outgroups. Multiple copies of 16S rRNA gene sequences of *D. hafniense* strains Y51 and DCB-2, *D. dehalogenans* JW/IU-DC-1, *D. dichloroeliminans* DCA1, *D. metallireducens* 853-15A, and *Dehalobacter restrictus* PER-K23 were identified based on genome information. In those cases, locus tags are indicated in the parentheses. Different species are highlighted in different colors. The scale bar represents 10 % estimated sequence divergence

architecture can give a clearer definition than the staining properties (Sutcliffe 2010). From this point of view, the *Desulfitobacterium* spp. should be recognized as monoderm that have an envelope with one membrane.

The phylogenetic relationships of *Desulfitobacterium* isolates based on 16S rRNA gene sequences are shown in Fig. 9.2. The *Desulfitobacterium* isolates are classified into six species: *D. hafniense*, *D. dehalogenans*, *D. dichloroeliminans*, *D. chlororespirans*, *D. metallireducens*, and *D. aromaticivorans*. Currently, no species name has been assigned to six strains: Viet1, PCE1, RPf35Ei, KBC1, B31e3, and PR. The *D. hafniense* strains PCP-1, DP7, TCP-A, G2, and TCE1 were previously classified as *D. frappieri*, but the species name *frappieri* is no longer used based on detailed 16S rRNA gene analyses. As a result, *D. frappieri* has been reclassified as *D. hafniense*. This confusion can be attributed to the size of the 16S rRNA gene in *D. hafniense* PCP-1 (formerly *D. frappieri* PCP-1) (Bouchard et al. 1996, reviewed in Villemur et al. 2006). The 16S rRNA gene of strain PCP-1 is 1655-bp long, whereas that of *D. hafniense* DCB-2 is 1530-bp long. It is now known that the *Desulfitobacterium* genome encodes multiple 16S rRNA genes and shows intra-genomic heterogeneity (Villemur et al. 2007). Denaturing gradient gel electrophoresis (DGGE) analysis indicated that 16S rRNA gene copy numbers in *Desulfitobacterium* strains vary from 2 to 7. Heterogeneity with respect to 16S rRNA gene copies is caused by 100- to 200-bp insertions in the 5′ region, the region that differs between strains PCP-1 and DCB-2. The 16S rRNA gene of strain PCP-1 has a 128-nt insertion that is absent in the 16S rRNA gene of strain DCB-2 (Bouchard et al. 1996). The insertions are predicted to form an energetically stable loop when they are transcribed. In addition, reverse transcriptase-PCR analyses have demonstrated that most of the observed insertions in the 16S rRNA gene of *Desulfitobacterium* strains are excised from the mature 16S rRNA transcripts. Although such insertion sequences are rarely observed in bacterial 16S rRNA genes, they are also found in several other bacterial genera, such as *Desulfotomaculum* and *Anaerospora* (Patel et al. 1992; Woo et al. 2005).

9.5 Reductive Dehalogenation Characteristics

The reductive dehalogenation activities of *Desulfitobacterium* isolates are summarized in Table 9.4. The reductive dehalogenation spectrum of the strains does not correlate with their phylogenetic relationships and seems to depend heavily upon the enzyme reductive dehalogenase (RDase), which is a terminal reductase in the organohalide respiratory pathway expressed in each *Desulfitobacterium* isolate (Hug et al. 2013).

The OHRB exhibits two types of reductive dehalogenation: hydrogenolysis and dihaloelimination (Mohn and Tiedje 1992; Holliger et al. 1998; Smidt and de Vos, 2004). In hydrogenolysis, the halogen substituents of alkyl or aryl halides are replaced with hydrogen atoms. In contrast, in dihaloelimination, two halogen substituents of alkyl halides are removed from adjacent carbon atoms. Both types of reductive dehalogenation have been observed in *Desulfitobacterium* strains.

Table 9.4 Dehalogenation activity of *Desulfitobacterium* strains

Organism	Dehalogenated	Not dehalogenated	Reference
D. hafniense DCB-2	PCE (weak activity), 3-Cl-4-OHPA, PCP, 2,3,4,5-TeCP, 2,4,5- and 2,4,6-TCPs, 2,3-, 2,4-, and 3,5-DCPs, TCMP, TCHQ	TCE, 3,4,5-TCP, 3,4-DCP, CPs	Madsen and Licht (1992), Christiansen and Ahring (1996a), Gerritse et al. (1999), Milliken et al. (2004b), Mac Nelly et al. (2014)
D. dehalogenans JW/IU-DC1	PCE (weak activity), 3-Cl-4-OHPA, PCP, TeCPs, 2,3,4-, 2,3,6-, and 2,4,6-TCPs, 2,3-, 2,4-, and 2,6-DCPs, TCMP, TCHQ, 2,6-dichloro-4-R-phenol (R: -H, -Cl, -F, $-NO_2$, -COOH, $-COOCH_3$), 2-chloro-4-R-phenol (R: -Cl, -F, -Br, $-NO_2$, -COOH, $-COOCH_3$, $-CH_2COOH$), 3,3″5,5″ tetrachloro-4,4″dihydroxybiphenyl congeners, 3,4″5-trichloro-4- hydroxybiphenyl, 3,5-dichloro-4-hydroxybiphenyl, 2,6-DBP, 2-BP, 2-bromo-4-CP, 2-bromo-4-methylphenol	TCE, 2,3,5-, 2,4,5-, and 3,4,5-TCPs, 3,4-, 3,5-, and 2,5-DCPs, CPs, 3-fluoro-4-hydroxyphenylacetate	Utkin et al. (1994, 1995), Wiegel et al. (1999), Gerritse et al. (1999), Milliken et al. (2004b)
Desulfitobacterium sp. PCE1	PCE, TCE (weak activity), 3-Cl-4-OHPA, 2,3,5,6-TeCP, 2,4,6-TCP, 2,4-DCP, 2-CP,TCMP, TCHQ, 2,3,5-trichlorohydroquinone	DCE, CF, CD, PCP, 2,3,5-TCP, HCB, 2,5- and 3,4-dichlorobenzoates, 2,3,6-trichlorobenzoate, trichloroacetate, hexachloroethane, 4-chlorophenylacetate, 2,5-DCHQ, 2,6-DCHQ, 2-chloro-1,4-hydroquinone	Gerritse et al. (1996), Milliken et al. (2004a, b)
D. chlororespirans Co23	3-Cl-4-OHPA, 2,4,6-TCP, 2,3,- and 2,6-DCPs, TCMP, TCHQ, 3-chloro-4-hydroxybenzoate, 3,5-dibromo-4-hydroxybenzoate, 2,4,6-tribromophenol, bromoxynil, ioxynil	PCE, PCP, 2,3,5-TCP, 2,4- and 2,5-DCPs, CPs, 3-chlorobenzoate, 3-chloro-L-tyrosine, 3-chloroanisaldehyde, 2-BP, 2-iodophenol, 2-fluorophenol	Sanford et al. (1996), Milliken et al. (2004b), Cupples et al. (2005)

(continued)

Table 9.4 (continued)

Organism	Dehalogenated	Not dehalogenated	Reference
D. hafniense PCP-1	PCE, PCP, 2,3,4,5- and 2,3,5,6-TeCPs, 2,3,4-, 2,3,5-, 2,3,6-, 2,4,5-, 2,4,6-, and 3,4,5-TCPs, 2,4-, 2,6-, and 3,5-DCPs, TCMP, TCHQ, 3,5-dichloro-4-hydroxybiphenyl; polychlorinated catechol, guaiacol, veratrole, anisole, aniline, nitrobenzene and pyridine; 2,4,6-tribromophenol, 2,4-DBP	TCE, 3-Cl-4-OHPA, 2,3-, 2,5-, and 3,4-DCPs. CPs, 2,2″5″trichloro-4- hydroxybiphenyl, 3,5-dichloro-2-hydroxybiphenyl, 2″5″dichloro-2-hydroxybiphenyl, 2,3,5- and 2,4,6- trichlorobenzoates, 2,3,5- and 2,4,6-trifluorophenols	Bouchard et al. (1996), Dennie et al. (1998), Milliken et al. (2004b)
D. hafniense PCE-S	PCE, TCE, 2,3-dichloropropene, tribromoethene, *cis*- and *trans*- dibromoethenes, 2,3-dibromopropene	DCEs, 3-Cl-4-OHPA, PCP, 2,4,6-TCP	Miller et al. (1997), Ye et al. (2010)
Desulfitobacterium sp. Viet1	PCE, 2,4,6-TCP, 2,4-DCP	TCE, 2,3,5- and 2,4,5-TCPs, 2,5-, 2,6-, and 3,4-DCPs	Löffler et al. (1997), Tront et al. (2006)
D. dehalogenans XZ-1	3,3″5,5″tetrachloro-4,4″dihydroxybiphenyl congeners, 3,4″5-trichloro-4-hydroxybiphenyl, 3,5-dichloro-4-hydroxybiphenyl	ND	Wiegel et al. (1999)
D. hafniense TCE1	PCE, TCE, CT (weak activity), CF (weak activity)	*cis*- and *trans*-DCEs, CD, 3-Cl-4-OHPA, 2,4,6-TCP, 1,2-dichloropropane, TCMP, TCHQ	Gerritse et al. (1999), Milliken et al. (2004b)
D. hafniense TCP-A	PCE (weak activity), PCP, 2,3,4,5- and 2,3,5,6-TeCPs, 2,3,5- and 2,4,6-TCPs, 2,3-, and 2,4-, 3,5-DCPs, 2-CP (weak activity)	3-CP, 4-CP	Breitenstein et al. (2001)
D. hafniense DP7	ND	PCE, 3-Cl-4-OHPA, PCP, 2,3-, 2,4-, 2,5-, and 2,6-DCPs, CPs, TCMP, TCHQ	van de Pas et al. (2001b), Milliken et al. (2004b)
D. hafniense Y51	PCE, TCE, 1,1,1,2,2,3,3-heptachloropropane, hex- and pentachloroethanes, 1,1,1,2- and 1,1,2,2-tetrachloroethane	DCEs, hexachloro-1,3-butadiene, 1,1,1- and 1,1,2-TCAs, 1,1,2,3-tetrachloropropane, *o*-, *m*-, and *p*-chlorobenzoates, CF, CT, CD	Suyama et al. (2001), Futagami et al. (2006b)

(continued)

Table 9.4 (continued)

Organism	Dehalogenated	Not dehalogenated	Reference
D. hafniense GBFH	ND	3-Cl-4-OHPA	Niggemyer et al. (2001)
D. metallireducens 853-15A	PCE, TCE, 3-Cl-4-OHPA	ND	Finneran et al. (2002)
D. hafniense G2	PCE, TCE	ND	Shelobolina et al. (2003)
D. dichloroeliminans DCA1	1,2-DCA, 1,1,2-TCA, *meso* and DL stereoisomers of 2,3- dichlorobutane, 1,2-dichlorobutane, 1,2- dichloropropane	VC, hexa-, penta-, and tetrachloroethanes, chlorinated methanes, monochloroalkanes, nonvicinal dichloroalkanes	De Wildeman et al. (2003)
Desulfitobacterium sp. RPf35Ei	ND	ND	Kaksonen et al. (2004)
Desulfitobacterium sp. KBC1	PCE, 2,4,6-TCP, 2,4-DCP	TCE, CPs	Tsukagoshi et al. (2006)
Desulfitobacterium sp. B31e3	PCE, TCE	*cis*-DCE, PCP, 2,4,6-TCP, 2,4-DCP, CPs	Yoshida et al. (2007)
D. hafniense JH1	PCE, TCE	*cis*- and *trans*-DCEs, VC, 1,1,2-TCA 1,1- and 1,2-DCAs, CT, CF, CD, 1,2,3-trichloropropane, 1,2-dichloropropane, 2-chlorotoluene, HCB, 3-chloro-4-hydroxybenzoate	Fletcher et al. (2008)
D. aromaticivorans UKTL	ND	ND	Kunapuli et al. (2010)
Desulfitobacterium sp. PR	PCE, 1,1,1- and 1,1,2-TCA, CF	TCE, PCP, 2,4,6-TCP	Ding et al. (2014), Zhao et al. (2015)

PCE tetrachloroethene; *3-Cl-4-OHPA* 3-chloro-4-hydroxyphenylacetic acid; *PCP* pentachlorophenol; *TeCP* tetrachlorophenol; *TCP* trichlorophenol; *DCP* dichlorophenol; *TCMP* 2,3,5,6-tetrachloro-4-methoxyphenol; *TCHQ* 2,3,5,6-tetrachlorohydroquinone; *TCE* trichloroethene; *CP* chlorophenol; *DBP* dibromophenol; *BP* bromophenol; *CF* chloroform; *CD* carbon dichloride; *HCB* hexachlorobenzene; *DCHQ* dichlorohydroquinone; *bromoxynil* 3,5-dibromo-4-hydroxybenzonitrile; *ioxynil* 3,5-diiodo-4-hydroxybenzonitrile; *TCA* trichloroethane; *CT* carbon tetrachloride; *DCA* dichloroethane; *VC* vinyl chloride; *ND* not determined

Most *Desulfitobacterium* strains can dechlorinate chloroaliphatic and/or chloroaromatic compounds. In addition, several *Desulfitobacterium* strains were found to dehalogenate organobromine and organoiodine compounds. However, to date, no dehalogenation activity has been reported for four *Desulfitobacterium* strains: *D. hafniense* DP7 and GBFH, *Desulfitobacterium* sp. RPf35Ei, and *D. aromaticivorans* UKTL. Of these strains, *D. hafniense* DP7 was characterized as a non-dechlorinator. Strain DP7 was isolated from human feces (nonpolluted environment) and does not dechlorinate monochlorophenols, 2,3-, 2,4-, 2,5-, and 2,6-DCPs, PCP, PCE, 3-Cl-4-OHPA, 2,3,5,6-tetrachloro-4-methoxyphenol, or tetrachlorohydroquinone (van de Pas et al. 2001b; Milliken et al. 2004a).

9.5.1 Dehalogenation of Chloroaliphatic Compounds

As it is one of the most common environmental pollutants, bacterial dechlorination of chloroethenes has been widely studied. The *Desulfitobacterium* spp. identified to date do not dechlorinate DCE or vinyl chloride and can be classified into one of two categories: PCE to *cis*-DCE dechlorinators or PCE to TCE dechlorinators (Table 9.4). The former category includes *D. hafniense* strains TCE1, PCE-S, Y51, G2, JH1, *D. metallireducens* 853-15A, *Desulfitobacterium* sp. B31e3, *D. hafniense* DCB-2, and *D. dehalogenans* JW/IU-DC1, whereas the latter includes *D. hafniense* PCP-1 and *Desulfitobacterium* sp. strains Viet1, PCE1, and KBC1. In the case of strains DCB-2 and JW/IU-DC1, they dechlorinate PCE to TCE slowly via cometabolic processes, which do not provide energy for growth, when the cells are cultured in the presence of 3-Cl-4-OHPA as an inducer (Gerritse et al. 1999). In addition, *D. hafniense* TCP-A and *Desulfitobacterium* sp. PR also dechlorinate PCE to TCE slowly, but the reaction does not require inducers (Breitenstein et al. 2001; Ding et al. 2014).

The unique reaction utilized by *Desulfitobacterium* strains for the dechlorination of chloroaliphatic compounds is as follow. *D. hafniense* Y51 dechlorinates hexachloroethane, pentachloroethane, 1,1,1,2-tetrachloroethane, and 1,1,2,2-tetrachloroethane (Suyama et al. 2001). *D. dichloroeliminans* DCA1 can dechlorinate 1,1,2-trichloroethane, 1,2-DCA, *meso* and DL stereoisomers of 2,3-dichlorobutane, 1,2-dichlorobutane, and 1,2-dichloropropane (De Wildeman et al. 2003). Strain DCA1 dechlorinate these compounds via the dichloroelimination system. *Desulfitobacterium* sp. PR reductively dechlorinates both 1,1,1-trichloroethane and chloroform (Ding et al. 2014). Strain PR can dechlorinate 1,1,1-trichloroethane completely to monochloroethane and dechlorinates chloroform (trichloromethane) to predominantly carbon dichloride (dichloromethane) and trace amounts of monochloromethane. Efficient degradation of chloroform has been reported only in strain PR among the *Desulfitobacterium* strains described to date.

The PCE/TCE RDase PceA has thus far been purified and characterized in *D. hafniense* strains PCE-S, Y51, and TCE1 (Miller et al. 1998; Suyama et al. 2002; Maillard et al. 2003). The PceA-encoding gene is found in the conserved *pceA-pceB-pceC-pceT* gene cluster in *D. hafniense* strains TCE1 and Y51. PceB is predicted to be a membrane anchor protein for PceA. The *pceT* gene encodes a trigger factor protein that binds to the twin arginine translocation (Tat) signal sequence of PceA (Morita et al. 2009; Maillard et al. 2011). The degradation gene cluster is sometimes located on a mobile element, and in fact, the *pce* gene cluster forms a catabolic transposon in *D. hafniense* strains TCE1 and Y51 (Maillard et al. 2005; Futagami et al. 2006a, b; Duret et al. 2012).

The activity of the PCE RDase PrdA of *Desulfitobacterium* sp. KBC1 is controlled at the transcriptional level (Tsukagoshi et al. 2006). Transcription of the *prdA* gene in strain KBC1 is induced by PCE. In contrast, the cultivation conditions reportedly affect PceA activity in *D. hafniense* Y51, without transcriptional regulation. Because the *pceA* gene is located on the transposable element in strain Y51, the activity of PceA decreases when cells are grown in the absence of an organochlorine substrate due to loss of the *pce* gene cluster through transposition (Reinhold et al. 2012). Vitamin B_{12} (a corrinoid) also reportedly affects the stability of the *pce* genes (Reinhold et al. 2012). Corrinoids play a significant role in regulating RDase catalytic activity as cofactors. Exogenous vitamin B_{12} hampers the transposition of the *pce* gene cluster, although the exogenous vitamin B_{12} does not appear to be incorporated into the PceA precursor. In addition, when strain Y51 is grown in the absence of an organochlorine substrate, the intracellular corrinoid level decreases and the PceA precursor forms catalytically inactive and corrinoid-free aggregates.

9.5.2 Dehalogenation of Chloroaromatic Compounds

The dechlorination of chlorophenols and 3-Cl-4-OHPA has also been well studied. Dechlorination of these chloroaromatic compounds has been reported in *D. hafniense* strains DCB-2 and TCP-A, *D. dehalogenans* strains JW/IU-DC1 and XZ-1, *Desulfitobacterium* sp. strains Viet1, PCE1, and KBC1, *D. chlororespirans* Co23, and *D. metallireducens* 853-15A (Table 9.4). The spectrum of haloaromatic compound dechlorination has been characterized most thoroughly in *D. hafniense* PCP-1 and *D. dehalogenans* JW/IU-DC1 (reviewed in Villemur 2013), e.g., *D. hafniense* PCP-1 can dechlorinate the *ortho*, *meta*, and *para* positions of chlorophenols, in this strain, PCP is converted to 3-chlorophenol via 2,3,4,5-tetrachlorophenol, 3,4,5-TCP, and 3,5-DCP (Bouchard et al. 1996). In strain PCP-1, *ortho*-dechlorination of PCP to 3,4,5-TCP proceeds rapidly, and after sufficient accumulation, 3,4,5-DCP is *meta*- and *para*-dechlorinated to 3,5-DCP and 3-chlorophenol. Thus, two different RDases are involved in PCP dechlorination in strain PCP-1. *Ortho*-dechlorination activity is induced by PCP, 2,4,6-TCP, 2,3,4-TCP, 2,3,5-TCP, 2,6-DCP, and 2,4-DCP, whereas *meta*- and *para*-dechlorination activities are induced by 3,4,5-TCP and 3,5-DCP (reviewed in Villemur 2013).

In general, RDases preferentially target highly halogenated compounds as substrates for reductive attack over compounds exhibiting lower degrees of halogenation. For example, to date, no DCE- or vinyl chloride-respiring *Desulfitobacterium* strains have been isolated, as mentioned above (see Sect. 9.5.1). In the case of chlorophenols, nine strains have been tested for dechlorination of monochlorophenols, but only *Desulfitobacterium* sp. PCE1 was found to dechlorinate 2-chlorophenol efficiently (Gerritse et al. 1996). Dechlorination of 2,3,5-trichlorohydroquinone has been reported only in strain PCE1.

With respect to chloroaromatic compounds other than chlorophenols and 3-Cl-4-OHPA, *D. hafniense* PCP-1 dechlorinates organochlorides containing a phenyl or pyridine ring substituted by hydroxyl, nitro, methoxy, or amino groups (Bouchard et al. 1996; Dennie et al. 1998). Polychlorinated nitrobenzene is dechlorinated after reduction of the nitro group to an amino group. In addition, *D. dehalogenans* strains JW/IU-DC1 and XZ-1 and *D. hafniense* strain PCP-1 were reported to dechlorinate chlorinated hydroxybiphenyls such as 3,5-dichloro-4-hydroxybiphenyl (Dennie et al. 1998; Wiegel et al. 1999).

The enzyme chlorophenol-RDase (CprA) has been purified and characterized in *D. hafniense* strains PCP-1 and DCB-2, *D. dehalogenans* JW/IU-DC1, *Desulfitobacterium* sp. PCE1, and *D. chlororespirans* Co23 (Christiansen et al. 1998; van de Pas et al. 1999, 2001a; Krasotkina et al. 2001; Löffler et al. 1996; Boyer et al. 2003; Thibodeau et al. 2004; Bisaillon et al. 2010). These enzymes exhibit features similar to the RDases PceA, such as the presence of a Tat signal sequence, two Fe–S clusters, and a corrinoid cofactor. However, the chlorophenol-RDase CrdA purified from *D. hafniense* PCP-1 exhibits different features (Boyer et al. 2003). CrdA *ortho*-dechlorinates PCP and 2,4,6-TCP. Like other RDases, CrdA contains a corrinoid cofactor, but this enzyme also has a LysM domain that may be involved in binding to peptidoglycan (Bateman et al. 2000, reviewed in Villemur 2013). The *crdA* gene has been found in *D. hafniense* strains DCB-2, DP7, TCP-1, Y51 and TCE1, *Desulfitobacterium* sp. PCE1, *D. dehalogenans* JW/IU-DC1, and *D. chlororespirans* Co23 (Gauthier et al. 2006; Nonaka et al. 2006). Transcription of *crdA* in *D. hafniense* Y51 was shown to be upregulated 225-fold by TCE (Peng et al. 2012), although strain Y51 has not been tested for chlorophenol dechlorination activity.

The larger components of the RDase gene cluster have been identified in the upstream and downstream region of the *cprA* gene in *D. dehalogenans* JW/IU-DC1. This gene cluster consists of *cprA*, *cprB*, *cprC*, *cprD*, *cprT*, *cprK*, *cprZ*, and *cprE* (Smidt et al. 2000). In contrast to the *pce* gene cluster, the *cpr* cluster encodes the transcriptional regulator CprK, a member of the CRP-FNR (cAMP-binding protein/fumarate nitrate reduction regulatory protein) family. Transcription of the *cprA* gene is strictly regulated by CprK (Smidt et al. 2000; Gábor et al. 2008). CprK is the most thoroughly studied transcriptional regulator in the OHRB (see Chap. 15).

9.5.3 Dehalogenation of Organobromine and Organoiodine Compounds

Little information is available regarding the reductive dehalogenation of organobromine and organoiodine compounds as compared with organochlorine compounds (Table 9.4). *D. dehalogenans* JW/IU-DC1 can debrominate 2-bromo-4-chlorophenol, 2-bromo-4-methylphenol, 2,6-dibromophenol, and 2-bromophenol (Utkin et al. 1995). *D. hafniense* PCP-1 can debrominate 2,4,6-tribromophenol into 4-bromophenol in the absence of inducers (Dennie et al. 1998). *D. chlororespirans* Co23 can debrominate 2,4,6-tribromophenol, bromoxynil (3,5-dibromo-4-hydroxybenzonitrile), ioxynil (3,5-diiodo-4-hydroxybenzonitrile), and the bromoxynil metabolite 3,5-dibromo-4-hydroxybenzoate (Sanford et al. 1996; Cupples et al. 2005). 2,4,6-tribromophenol is debrominated into 4-bromophenol by strain Co23 (Sanford et al. 1996). Bromoxynil and ioxynil are herbicides used for the postemergence control of dicotyledonous weeds in cereal crops. Bromoxynil is stoichiometrically converted into 4-cyanophenol and the metabolite 3,5-dibromo-4-hydroxybenzoate, which is then converted to 4-hydroxybenzoate. On the other hand, ioxynil is stoichiometrically converted into 4-cyanophenol in the presence of 3-chloro-4-hydroxybenzoate as an inducer. Strain Co23 can utilize bromoxynil and 3,5-dibromo-4-hydroxybenzoate as growth-supporting electron acceptors (Cupples et al. 2005). *D. hafniense* PCE-S can debrominate and grow on *cis*- and *trans*-dibromoethenes (Ye et al. 2010). Strain PCE-S converts *cis*- and *trans*-dibromoethenes to vinyl bromide and ethene. The PceA of strain PCE-S was initially purified as a PCE/TCE RDase (Miller et al. 1998), but the enzyme was later found to also catalyze the debromination of tribromoethene, *cis*- and *trans*-dibromoethenes (Ye et al. 2010).

Defluorination has not been observed by *Desulfitobacterium* spp. or any other OHRB. *Desulfitobacterium dehalogenans* JW/IU-DC1 does not defluorinate 3-fluoro-4-hydroxyphenylacetate (Utkin et al. 1995). *Desulfitobacterium hafniense* PCP-1 does not defluorinate 2,4,6-trifluorophenol or 2,3,5-trifluorophenol (Dennie et al. 1998), and *D. chlororespirans* Co23 does not grow in medium containing 2-fluorophenol as an electron acceptor (Sanford et al. 1996).

9.6 Bioremediation

The isolation of and ecological studies pertaining to *Desulfitobacterium* spp. has demonstrated that these organisms play a significant role in the natural attenuation of organohalide contaminants such as chloroethenes (see Sect. 9.2). For the bioremediation of chloroethenes, however, it is noteworthy that the genus *Dehalococcoides* is known as a key player. This is due to their ability to respire with DCE and vinyl chloride, a potential that has thus far been found solely in the genus *Dehalococcoides* (see Chap. 6).

Desulfitobacterium strains have been examined in both laboratory- and field-scale bioremediation studies. For example, the applicability of bioaugmentation with *D. hafniense* strains PCP-1 and Y51, *D. dehalogenans* JW/IU-DC1, or *D. dichloroeliminans* DCA1 for the degradation of PCP, PCE, 3-Cl-4-OHPA, and 1,2-DCA, respectively, has been demonstrated (El Fantroussi et al. 1997; Beaudet et al. 1998; Lanthier et al. 2000; Lee et al. 2001; De Wildeman et al. 2004). In addition, bioaugmentation using strain DCA1 for the removal of 1,2-DCA at an industrial site in Belgium was also demonstrated (Maes et al. 2006). The success of bioremediation efforts, however, depends largely on environmental factors, such as soil composition (reviewed in Villemur et al. 2006). For example, Lanthier et al. (2000) reported the dechlorination of PCP in a PCP-amended soil rich in organic matter, but no dechlorination was observed in sandy soil amended with PCP, highlighting the impact of factors such as organic matter and toxicity associated with other pollutants on the viability of OHRB. Geochemical properties such as redox conditions and pH also significantly impact the success of in situ bioremediation efforts.

Desulfitobacterium strains have also been used in investigations aimed at increasing bioremediation efficiency. The effect of PCE dense nonaqueous phase liquid on dechlorination by *Desulfitobacterium* strains Viet1 and PCE1 was examined via mathematical modeling (Amos et al. 2007; Huang and Becker 2011). Strain Viet1 was used to investigate the environmental fate of 2,4-DCP sequestered by the aquatic plant *Lemna minor* (Tront et al. 2006). In the environment, aquatic plants are known to take up and sequester various organohalines, such as chlorophenols (Newman and Reynolds 2004).

Desulfitobacterium strains have been utilized in bioreactors for treating organohaline-contaminated wastes. For example, *D. hafniense* strains DCB-2 and PCP-1 were inoculated into an up-flow anaerobic-sludge bed reactor used for PCP degradation (Christiansen and Ahring 1996b). This PCP degrading bioreactor using strain PCP-1 is one of the most thoroughly studied systems of its type (e.g., the localization of strain PCP-1 in the reactor granules and biofilm was demonstrated by fluorescence in situ hybridization) (Tartakovsky et al. 1999; Lanthier et al. 2000, 2002, 2005). Degradation of 2,4,6-TCP in anaerobic granular sludge bioaugmented with *D. hafniense* strains PCP-1 and TCP-A and *D. chlororespirans* Co23 has also been reported (Puyol et al. 2011). The *Desulfitobacterium* spp. were also found to be enriched in a deiodination bioreactor for 5-amino-2,4,6-triiodoisophthalic acid (a core structure of X-ray contrast media) (Lecouturier et al. 2003). These studies have demonstrated that *Desulfitobacterium* spp. are key components in dechlorination bioreactors.

Desulfitobacterium spp. can be employed not only for organohalide detoxification but also for the removal of metals. The solubility of a metal significantly affects its mobility in the environment. Therefore, altering a metal contaminant's solubility through bioreduction is one approach for removing it. For example, several reports have described metal bioremediation strategies employing *Desulfitobacterium* strains to reduce soluble U(VI) to insoluble U(IV) (Fletcher et al. 2010; Boyanov et al. 2011).

Tools such as real-time PCR targeting the 16S rRNA and RDase genes of OHRB, and stable-isotope fractionation analysis are important for monitoring bioremediation efforts and evaluating their success. These tools were also developed for *Desulfitobacterium* spp. as one of important class of OHRB (see Chaps. 25 and 26).

9.7 Genomic Features

9.7.1 Basic Genome Facts

Complete genome sequences have been published for *D. hafniense* strains Y51 and DCB-2 and *D. dehalogenans* JW/IU-DC1 (Nonaka et al. 2006; Kim et al. 2012; Kruse et al. 2014b). In addition, the complete genome sequence for *D. dichloroeliminans* DCA1 and draft genome sequences for *D. hafniense* strains PCE-S (Goris et al. 2015), PCP-1, TCP-A, and DP7, *D. metallireducens* 853-15A, and *Desulfitobacterium* sp. PCE1 are available in the National Center for Biotechnology Information (NCBI) database.

The complete genomes of *Desulfitobacterium* strains vary widely in size, ranging from 3.62 to 5.73 Mbp, with similar GC content, ranging from 44.2 to 47.5 % (Table 9.5). No plasmids have been identified by genomic analyses. The number of coding sequences (CDSs) identified ranges from 3340 to 5060, which is consistent with the differences in genome size. As described in Sect. 9.4, multiple copies of rRNA operons (in the range from 2 to 7) are found in *Desulfitobacterium* genomes.

The number of RDase genes differs in each strain (ranging from 2 to 7), which explains the differences in their dehalogenation spectra. Analyses of the *D. hafniense* strains Y51 and DCB-2 genomes revealed the existence of versatile terminal reductases, such as DMSO reductase superfamily proteins, sulfite

Table 9.5 *Desulfitobacterium* complete genome information

Organism	*D. hafniense* Y51	*D. hafniense* DCB-2	*D. dehalogenans* JW/IU-DC1	*D. dichloroeliminans* DCA1
Size (Mb)	5.73	5.28	4.32	3.62
GC%	47.4	47.5	45	44.2
CDSs	5060	4883	4011	3340
rRNA operon	6	5	6	6
tRNA	59	74	75	73
Rdase genes	2	7	6	1
RefSeq accession number	NC_007907.1	NC_011830.1	NC_018017.1	NC_019903.1
INSDC accession number	AP008230.1	CP001336.1	CP003348.1	CP003344.1
Reference	Nonaka et al. (2006)	Kim et al. (2012)	Kruse et al. (2014b)	Unpublished

reductase, nitrate reductase, and fumarate reductase, in addition to RDases (Nonaka et al. 2006; Kim et al. 2012). This finding is consistent with the growth phenotype of the *Desulfitobacterium* strains, with the exception that the nitrate reductase gene cluster (Nap system) was not identified in the genome of strain DCB-2, even though it can grow using nitrate as an electron acceptor.

As the *Desulfitobacterium* strains are slightly oxygen tolerant, the genomes of *D. hafniense* strains Y51 and DCB-2 encode putative catalases, superoxide dismutases, and rubrerythrin-rubredoxin enzymes that may be involved in imparting oxygen tolerance (Nonaka et al. 2006; Kim et al. 2012). In addition, consistent with its ability to form spores, genes involved in spore formation are present in strain DCB-2 (Kim et al. 2012).

9.7.2 Functional Genes and Metabolic Pathways

Genomic information has enabled the prediction of metabolic pathways in *Desulfitobacterium* strains. Central metabolic pathways for *D. hafniense* strains Y51 and DCB-2 have been proposed (Nonaka et al. 2006; Kim et al. 2012). Both strains Y51 and DCB-2 have a functional Embden-Meyerhof-Parnas pathway. On the other hand, the tricarboxylic acid cycle is characterized by the absence of 2-oxoglutarate dehydrogenase and the anaplerotic glyoxylate bypass, indicating an incomplete cycle. The reductive acetyl-CoA pathway (Wood-Ljungdahl pathway) for carbon dioxide fixation has also been found in both strains Y51 and DCB-2. Moreover, autotrophic growth of strain DCB-2 in a carbon dioxide fixation medium was confirmed (Kim et al. 2012).

Genome analyses of *D. hafniense* DCB-2 confirmed the presence of the complete synthetic pathways for the cofactors flavin adenine dinucleotide, nicotinamide adenine dinucleotide, menaquinone, heme, and cobalamin (Kim et al. 2012). The cobalamin biosynthetic pathway has also been identified in *D. hafniense* strains Y51 and TCE1 (Nonaka et al. 2006; Choudhary et al. 2013). Genomic analyses also revealed the presence of diverse cobalamin riboswitches that may be involved in regulating corrinoid metabolism in these *D. hafniense* strains (Choudhary et al. 2013). In addition, 14 putative ABC-type corrinoid transporter-encoding genes were identified in the genome of *D. hafniense* Y51 (Nonaka et al. 2006), but exogenous corrinoid may not be incorporated into the PceA precursor in strain Y51 (Reinhold et al. 2012). Menaquinone is believed to serve as an electron carrier in organohalide respiration. Electron paramagnetic resonance spectroscopy, visible spectroscopy, and proteomic analyses of *D. dehalogenans* JW/IU-DC1 suggested that electrons are transferred from menaquinones to RDase CprA via an as yet unidentified membrane complex and potentially by an extracellular flavoprotein (Kruse et al. 2014b).

9.7.3 Protein Characterization Based on Genome Data

Genomic data have also enabled other "omics" studies. For example, high expression of the unusual rhodanese protein PhsE during respiration with PCE in *D. hafniense* TCE1 as revealed by a proteomic study led to the characterization of this protein (Prat et al. 2011, 2012). PhsE was shown to have two unusual rhodanese domains that usually function to bind sulfane sulfur and catalyze sulfur transfer. The expression of PhsE was found to be induced in the presence of sulfide in the medium and is believed to play a role in sulfur homeostasis. In addition, genomic studies have also opened the door to the discovery of new scientific knowledge from perspectives other than organohalide respiration. For example, the crystal structure of Dhaf4260 from *D. hafniense* DCB-2 was determined as the first representative of the Pfam family PF04016 (DUF364) (DUF: domains of unknown function) in the course of a study aimed at enhancing knowledge regarding the structures of proteins of unknown biological function (Miller et al. 2010). Moreover, DSY3156 (MtgB) from *D. hafniense* Y51 was characterized as a non-pyrrolysine member of the widely distributed trimethylamine methyltransferase family and was the first glycine betaine:corrinoid methyltransferase characterized (Ticak et al. 2014).

Desulfitobacterium strains are among the ideal model microorganisms for the study of organohalide respiration. Molecular biology tools will become increasingly important in future studies of *Desulfitobacterium* spp. Transposon mutagenesis using Tn*916* and gene disruption using a thermosensitive plasmid have been used in studies of *D. dehalogenans* JW/IU-DC1, and functional heterologous expression of RDase in *D. hafniense* strains Y51 and DCB-2 in the Gram-negative gamma-proteobacterium *Shimwellia blattae* ATCC 33,430 has been demonstrated using molecular tools (Smidt et al. 1999, 2001; Mac Nelly et al. 2014). Recent sequencing efforts of further *Desulfitobacterium* genomes will also attract researchers to the study of *Desulfitobacterium* species from perspectives other than organohalide respiration. Combining molecular biology tools with current genomic information will reveal novel insights into the genus *Desulfitobacterium*.

References

Amos BK, Christ JA, Abriola LM, Pennell KD, Löffler FE (2007) Experimental evaluation and mathematical modeling of microbially enhanced tetrachloroethene (PCE) dissolution. Environ Sci Technol 41:963–970. doi:10.1021/es061438n

Beaudet R, Lévesque MJ, Villemur R, Lanthier M, Chénier M, Lépine F, Bisaillon JG (1998) Anaerobic biodegradation of pentachlorophenol in a contaminated soil inoculated with a methanogenic consortium or with *Desulfitobacterium frappieri* strain PCP-1. Appl Microbiol Biotechnol 50:135–141. doi:10.1007/s002530051268

Bisaillon A, Beaudet R, Lépine F, Déziel E, Villemur R (2010) Identification and characterization of a novel CprA reductive dehalogenase specific to highly chlorinated phenols from *Desulfitobacterium hafniense* strain PCP-1. Appl Environ Microbiol 76:7536–7540. doi:10.1128/AEM.01362-10

Bouchard B, Beaudet R, Villemur R, McSween G, Lépine F, Bisaillon JG (1996) Isolation and characterization of *Desulfitobacterium frappieri* sp. nov., an anaerobic bacterium which reductively dechlorinates pentachlorophenol to 3-chlorophenol. Int J Syst Bacteriol 46:1010–1015

Boyanov MI, Fletcher KE, Kwon MJ, Rui X, O'Loughlin EJ, Löffler FE, Kemner KM (2011) Solution and microbial controls on the formation of reduced U(IV) species. Environ Sci Technol 45:8336–8344. doi:10.1021/es2014049

Boyer A, Pagé-BéLanger R, Saucier M, Villemur R, Lépine F, Juteau P, Beaudet R (2003) Purification, cloning and sequencing of an enzyme mediating the reductive dechlorination of 2,4,6-trichlorophenol from *Desulfitobacterium frappieri* PCP-1. Biochem J 373:297–303. doi:10.1042/BJ20021837

Breitenstein A, Saano A, Salkinoja-Salonen M, Andreesen JR, Lechner U (2001) Analysis of a 2,4,6-trichlorophenol-dehalogenating enrichment culture and isolation of the dehalogenating member *Desulfitobacterium frappieri* strain TCP-A. Arch Microbiol 175:133–142

Cervantes FJ, de Bok FAM, Duong-Dac T, Stams AJM, Lettinga G, Field JA (2002) Reduction of humic substances by halorespiring, sulphate-reducing and methanogenic microorganisms. Environ Microbiol 4:51–57

Chang YC, Hatsu M, Jung K, Yoo YS, Takamizawa K (2000) Isolation and characterization of a tetrachloroethylene dechlorinating bacterium, *Clostridium bifermentans* DPH-1. J Biosci Bioeng 89:489–491

Choudhary PK, Duret A, Rohrbach-Brandt E, Holliger C, Sigel RKO, Maillard J (2013) Diversity of cobalamin riboswitches in the corrinoid-producing organohalide respirer *Desulfitobacterium hafniense*. J Bacteriol 195:5186–5195. doi:10.1128/JB.00730-13

Christiansen N, Ahring BK (1996a) *Desulfitobacterium hafniense* sp. nov., an anaerobic, reductively dechlorinating bacterium. Int J Syst Bacteriol 46:442–448. doi:10.1099/00207713-46-2-442

Christiansen N, Ahring BK (1996b) Introduction of a de novo bioremediation activity into anaerobic granular sludge using the dechlorinating bacterium DCB-2. Antonie Van Leeuwenhoek 69:61–66. doi:10.1007/BF00641612

Christiansen N, Ahring BK, Wohlfarth G, Diekert G (1998) Purification and characterization of the 3-chloro-4-hydroxy-phenylacetate reductive dehalogenase of *Desulfitobacterium hafniense*. FEBS Lett 436:159–162. doi:10.1016/S0014-5793(98)01114-4

Cupples AM, Sanford RA, Sims GK (2005) Dehalogenation of the herbicides bromoxynil (3,5-dibromo-4-hydroxybenzonitrile) and ioxynil (3,5-diiodino-4-hydroxybenzonitrile) by *Desulfitobacterium chlororespirans*. Appl Environ Microbiol 71:3741–3746. doi:10.1128/AEM.71.7.3741-3746.2005

Davis JW, Odom JM, DeWeerd KA, Stahl DA, Fishbain SS, West RJ, Klecka GM, DeCarolis JG (2002) Natural attenuation of chlorinated solvents at Area 6, Dover Air Force Base: characterization of microbial community structure. J Contam Hydrol 57:41–59. doi:10.1016/S0169-7722(01)00217-0

De Wildeman S, Diekert G, Van Langenhove H, Verstraete W (2003) Stereoselective microbial dehalorespiration with vicinal dichlorinated alkanes. Appl Environ Microbiol 69:5643–5647

De Wildeman S, Linthout G, Van Langenhove H, Verstraete W (2004) Complete lab-scale detoxification of groundwater containing 1,2-dichloroethane. Appl Microbiol Biotechnol 63:609–612. doi:10.1007/s00253-003-1363-y

Dennie D, Gladu I, Lepine F, Villemur R, Bisaillon J, Beaudet R (1998) Spectrum of the reductive dehalogenation activity of *Desulfitobacterium frappieri* PCP-1. Appl Environ Microbiol 64:4603–4606

DeWeerd K, Mandelco L, Tanner R, Woese C, Suflita J (1990) *Desulfomonile tiedjei* gen. nov. and sp. nov., a novel anaerobic, dehalogenating, sulfate-reducing bacterium. Arch Microbiol. doi:10.1007/BF00249173

Ding C, Zhao S, He J (2014) A desulfitobacterium sp. strain PR reductively dechlorinates both 1,1,1-trichloroethane and chloroform. Environ Microbiol 11:3387–97. doi: 10.1111/1462-2920.12387

Dowideit K, Scholz-Muramatsu H, Miethling-Graff R, Vigelahn L, Freygang M, Dohrmann AB, Tebbe CC (2010) Spatial heterogeneity of dechlorinating bacteria and limiting factors for in situ trichloroethene dechlorination revealed by analyses of sediment cores from a polluted field site. FEMS Microbiol Ecol 71:444–459. doi:10.1111/j.1574-6941.2009.00820.x

Duret A, Holliger C, Maillard J (2012) The physiological opportunism of *Desulfitobacterium hafniense* strain TCE1 towards organohalide respiration with tetrachloroethene. Appl Environ Microbiol 78:6121–6127. doi:10.1128/AEM.01221-12

El Fantroussi S, Mahillon J, Naveau H, Agathos SN (1997) Introduction of anaerobic dechlorinating bacteria into soil slurry microcosms and nested-PCR monitoring. Appl Environ Microbiol 63:806–811

Finneran KT, Forbush HM, VanPraagh CVG, Lovley DR (2002) *Desulfitobacterium metallireducens* sp. nov., an anaerobic bacterium that couples growth to the reduction of metals and humic acids as well as chlorinated compounds. Int J Syst Evol Microbiol 52:1929–1935

Fletcher KE, Ritalahti KM, Pennell KD, Takamizawa K, Löffler FE (2008) Resolution of culture *Clostridium bifermentans* DPH-1 into two populations, a *Clostridium* sp. and tetrachloroethene-dechlorinating *Desulfitobacterium hafniense* strain JH1. Appl Environ Microbiol 74:6141–6143. doi:10.1128/AEM.00994-08

Fletcher KE, Boyanov MI, Thomas SH, Wu Q, Kemner KM, Löffler FE (2010) U(VI) reduction to mononuclear U(IV) by *Desulfitobacterium* species. Environ Sci Technol 44:4705–4709. doi:10.1021/es903636c

Furukawa K, Suyama A, Tsuboi Y, Futagami T, Goto M (2005) Biochemical and molecular characterization of a tetrachloroethene dechlorinating *Desulfitobacterium* sp. strain Y51: a review. J Ind Microbiol Biotechnol 32:534–541. doi:10.1007/s10295-005-0252-z

Futagami T, Tsuboi Y, Suyama A, Goto M, Furukawa K (2006a) Emergence of two types of nondechlorinating variants in the tetrachloroethene-halorespiring *Desulfitobacterium* sp. strain Y51. Appl Microbiol Biotechnol 70:720–728. doi:10.1007/s00253-005-0112-9

Futagami T, Yamaguchi T, Nakayama S-I, Goto M, Furukawa K (2006b) Effects of chloromethanes on growth of and deletion of the *pce* gene cluster in dehalorespiring *Desulfitobacterium hafniense* strain Y51. Appl Environ Microbiol 72:5998–6003. doi:10.1128/AEM.00979-06

Gábor K, Hailesellasse Sene K, Smidt H, de Vos WM, van der Oost J (2008) Divergent roles of CprK paralogues from *Desulfitobacterium hafniense* in activating gene expression. Microbiology 154:3686–3696. doi:10.1099/mic.0.2008/021584-0

Gauthier A, Beaudet R, Lépine F, Juteau P, Villemur R (2006) Occurrence and expression of *crdA* and *cprA5* encoding chloroaromatic reductive dehalogenases in *Desulfitobacterium* strains. Can J Microbiol 52:47–55. doi:10.1139/w05-111

Gerritse J, Renard V, Visser J, Gottschal JC (1995) Complete degradation of tetrachloroethene by combining anaerobic dechlorinating and aerobic methanotrophic enrichment cultures. Appl Microbiol Biotechnol 43:920–928

Gerritse J, Renard V, Pedro Gomes TM, Lawson PA, Collins MD, Gottschal JC (1996) *Desulfitobacterium* sp. strain PCE1, an anaerobic bacterium that can grow by reductive dechlorination of tetrachloroethene or *ortho*-chlorinated phenols. Arch Microbiol 165:132–140

Gerritse J, Kloetstra G, Borger A, Dalstra G, Alphenaar A, Gottschal JC (1997) Complete degradation of tetrachloroethene in coupled anoxic and oxic chemostats. Appl Microbiol Biotechnol 48:553–562

Gerritse J, Drzyzga O, Kloetstra G, Keijmel M, Wiersum LP, Hutson R, Collins MD, Gottschal JC (1999) Influence of different electron donors and acceptors on dehalorespiration of tetrachloroethene by *Desulfitobacterium frappieri* TCE1. Appl Environ Microbiol 65:5212–5221

Goris T, Hornung B, Kruse T, Reinhold A, Westermann M, Schaap PJ, Smidt H, Gabriele Diekert G (2015) Draft genome sequence and characterization of *Desulfitobacterium hafniense* PCE-S. Stand Genomic Sci 10:15. doi:10.1186/1944-3277-10-15

Holliger C, Wohlfarth G, Diekert G (1998) Reductive dechlorination in the energy metabolism of anaerobic bacteria. FEMS Microbiol Rev 22:383–398. doi:10.1016/S0168-6445(98)00030-8

Huang D, Becker JG (2011) Dehalorespiration model that incorporates the self-inhibition and biomass inactivation effects of high tetrachloroethene concentrations. Environ Sci Technol 45:1093–1099. doi:10.1021/es102729v

Hug LA, Maphosa F, Leys D, Löffler FE, Smidt H, Edwards EA, Adrian L (2013) Overview of organohalide-respiring bacteria and a proposal for a classification system for reductive dehalogenases. Philos Trans R Soc Lond B Biol Sci 368:20120322. doi:10.1098/rstb.2012.0322

Kaksonen AH, Plumb JJ, Robertson WJ, Franzmann PD, Gibson JAE, Puhakka JA (2004) Culturable diversity and community fatty acid profiling of sulfate-reducing fluidized-bed reactors treating acidic, metal-containing wastewater. Geomicrobiol J 21:469–480. doi:10.1080/01490450490505455

Kalan L, Ebert S, Kelly T, Wright GD (2009) Noncanonical vancomycin resistance cluster from *Desulfitobacterium hafniense* Y51. Antimicrob Agents Chemother 53:2841–2845. doi:10.1128/AAC.01408-08

Kim S-H, Harzman C, Davis JK, Hutcheson R, Broderick JB, Marsh TL, Tiedje JM (2012) Genome sequence of *Desulfitobacterium hafniense* DCB-2, a Gram-positive anaerobe capable of dehalogenation and metal reduction. BMC Microbiol 12:21. doi:10.1186/1471-2180-12-21

Kranzioch I, Stoll C, Holbach A, Chen H, Wang L, Zheng B, Norra S, Bi Y, Schramm K-W, Tiehm A (2013) Dechlorination and organohalide-respiring bacteria dynamics in sediment samples of the Yangtze Three Gorges Reservoir. Environ Sci Pollut Res Int 20:7046–7056. doi:10.1007/s11356-013-1545-9

Krasotkina J, Walters T, Maruya KA, Ragsdale SW (2001) Characterization of the B_{12}- and iron-sulfur-containing reductive dehalogenase from *Desulfitobacterium chlororespirans*. J Biol Chem 276:40991–40997. doi:10.1074/jbc.M106217200

Kreher S, Schilhabel A, Diekert G (2008) Enzymes involved in the anoxic utilization of phenyl methyl ethers by *Desulfitobacterium hafniense* DCB2 and *Desulfitobacterium hafniense* PCE-S. Arch Microbiol 190:489–495. doi:10.1007/s00203-008-0400-8

Kruse T, Levisson M, de Vos WM, Smidt H (2014a) *vanI*: a novel D-Ala-D-Lac vancomycin resistance gene cluster found in *Desulfitobacterium hafniense*. Microb Biotechnol 7:456–466. doi:10.1111/1751-7915.12139

Kruse T, van de Pas BA, Atteia A, Krab K, Hagen WR, Goodwin L, Chain P, Boeren S, Maphosa F, Schraa G, de Vos WM, van der Oost J, Smidt H, Stams AJM (2014b) Genomic, proteomic and biochemical analysis of the organohalide respiratory pathway in *Desulfitobacterium dehalogenans*. J Bacteriol JB.02370-14-. doi:10.1128/JB.02370-14

Kunapuli U, Jahn MK, Lueders T, Geyer R, Heipieper HJ, Meckenstock RU (2010) *Desulfitobacterium aromaticivorans* sp. nov. and *Geobacter toluenoxydans* sp. nov., iron-reducing bacteria capable of anaerobic degradation of monoaromatic hydrocarbons. Int J Syst Evol Microbiol 60:686–695. doi:10.1099/ijs.0.003525-0

Lanthier M, Villemur R, Lépine F, Bisaillon JG, Beaudet R (2000) Monitoring of *Desulfitobacterium frappieri* PCP-1 in pentachlorophenol-degrading anaerobic soil slurry reactors. Environ Microbiol 2:703–708. doi:10.1046/j.1462-2920.2000.00144.x

Lanthier M, Villemur R, Lépine F, Bisaillon JG, Beaudet R (2001) Geographic distribution of *Desulfitobacterium frappieri* PCP-1 and *Desulfitobacterium* spp. in soils from the province of Quebec. Canada. FEMS Microbiol Ecol 36:185–191

Lanthier M, Tartakovsky B, Villemur R, DeLuca G, Guiot SR (2002) Microstructure of anaerobic granules bioaugmented with *Desulfitobacterium frappieri* PCP-1. Appl Environ Microbiol 68:4035–4043. doi:10.1128/AEM.68.8.4035-4043.2002

Lanthier M, Juteau P, Lépine F, Beaudet R, Villemur R (2005) *Desulfitobacterium hafniense* is present in a high proportion within the biofilms of a high-performance pentachlorophenol-degrading, methanogenic fixed-film reactor. Appl Environ Microbiol 71:1058–1065. doi:10.1128/AEM.71.2.1058-1065.2005

Lecouturier D, Godon JJ, Lebeault JM (2003) Phylogenetic analysis of an anaerobic microbial consortium deiodinating 5-amino-2,4,6-triiodoisophthalic acid. Appl Microbiol Biotechnol 62:400–406. doi:10.1007/s00253-003-1278-7

Lee T, Tokunaga T, Suyama A, Furukawa K (2001) Efficient dechlorination of tetrachloroethylene in soil slurry by combined use of an anaerobic *Desulfitobacterium* sp. strain Y-51 and zero-valent iron. J Biosci Bioeng 92:453–458. doi:10.1016/S1389-1723(01)80295-4

Lie TJ, Godchaux W, Leadbetter ER (1999) Sulfonates as terminal electron acceptors for growth of sulfite-reducing bacteria (*Desulfitobacterium* spp.) and sulfate-reducing bacteria: effects of inhibitors of sulfidogenesis. Appl Environ Microbiol 65:4611–4617

Löffler F, Sanford R, Tiedje J (1996) Initial characterization of a reductive dehalogenase from *Desulfitobacterium chlororespirans* Co23. Appl Envir Microbiol 62:3809–3813

Löffler FE, Ritalahti KM, Tiedje JM (1997) Dechlorination of chloroethenes is inhibited by 2-bromoethanesulfonate in the absence of methanogens. Appl Environ Microbiol 63:4982–4985

Lovley DR, Fraga JL, Blunt-Harris EL, Hayes LA, Phillips EJP, Coates JD (1998) Humic substances as a mediator for microbially catalyzed metal reduction. Acta Hydrochim Hydrobiol 26:152–157. doi:10.1002/(SICI)1521-401X(199805)26:3<152:AID-AHEH152>3.0.CO;2-D

Luijten ML, Weelink SA, Godschalk B, Langenhoff AA, van Eekert MH, Schraa G, Stams AJ (2004) Anaerobic reduction and oxidation of quinone moieties and the reduction of oxidized metals by halorespiring and related organisms. FEMS Microbiol Ecol 49:145–150. doi:10.1016/j.femsec.2004.01.015

Lupa B, Wiegel J (2009) Genus IV: *Desulfitobacterium*. In: Vos P, Garity G, Jones D, Krieg NR, Ludwig W, Rainey FA, Schleifer K-H, Whitman W (eds) Bergey's manual of systematic bacteriology, vol 3, 2nd edn. Springer, New York, pp 975–982

Mac Nelly A, Kai M, Svatoš A, Diekert G, Schubert T (2014) Functional heterologous production of reductive dehalogenases from *Desulfitobacterium hafniense* strains. Appl Environ Microbiol 80:4313–4322. doi:10.1128/AEM.00881-14

Mackiewicz M, Wiegel J (1998) Comparison of energy and growth yields for *Desulfitobacterium dehalogenans* during utilization of chlorophenol and various traditional electron acceptors. Appl Environ Microbiol 64:352–355

Madsen T, Licht D (1992) Isolation and characterization of an anaerobic chlorophenol-transforming bacterium. Appl Environ Microbiol 58:2874–2878

Maes A, Van Raemdonck H, Smith K, Ossieur W, Lebbe L, Verstraete W (2006) Transport and activity of *Desulfitobacterium dichloroeliminans* strain DCA1 during bioaugmentation of 1,2-DCA-contaminated groundwater. Environ Sci Technol 40:5544–5552. doi:10.1021/es060953i

Maillard J, Schumacher W, Vazquez F, Regeard C, Hagen WR, Holliger C (2003) Characterization of the corrinoid iron-sulfur protein tetrachloroethene reductive dehalogenase of *Dehalobacter restrictus*. Appl Environ Microbiol 69:4628–4638

Maillard J, Regeard C, Holliger C (2005) Isolation and characterization of *Tn*-Dha1, a transposon containing the tetrachloroethene reductive dehalogenase of *Desulfitobacterium hafniense* strain TCE1. Environ Microbiol 7:107–117. doi:10.1111/j.1462-2920.2004.00671.x

Maillard J, Genevaux P, Holliger C (2011) Redundancy and specificity of multiple trigger factor chaperones in Desulfitobacteria. Microbiology 157:2410–2421. doi:10.1099/mic.0.050880-0

Miller E, Wohlfarth G, Diekert G (1997) Comparative studies on tetrachloroethene reductive dechlorination mediated by *Desulfitobacterium* sp. strain PCE-S. Arch Microbiol 168:513–519

Miller E, Wohlfarth G, Diekert G (1998) Purification and characterization of the tetrachloroethene reductive dehalogenase of strain PCE-S. Arch Microbiol 169:497–502

Miller MD, Aravind L, Bakolitsa C, Rife CL, Carlton D, Abdubek P, Astakhova T, Axelrod HL, Chiu HJ, Clayton T, Deller MC, Duan L, Feuerhelm J, Grant JC, Han GW, Jaroszewski L, Jin KK, Klock HE, Knuth MW, Kozbial P, Krishna SS, Kumar A, Marciano D, McMullan D, Morse AT, Nigoghossian E, Okach L, Reyes R, van den Bedem H, Weekes D, Xu Q, Hodgson KO, Wooley J, Elsliger MA, Deacon AM, Godzik A, Lesley SA, Wilson IA (2010) Structure of the first representative of Pfam family PF04016 (DUF364) reveals enolase and Rossmann-like folds that combine to form a unique active site with a possible role in heavy-metal chelation. Acta Crystallogr, Sect F: Struct Biol Cryst Commun 66:1167–1173. doi:10.1107/S1744309110007517

Milliken CE, May HD (2007) Sustained generation of electricity by the spore-forming, Gram-positive, *Desulfitobacterium hafniense* strain DCB2. Appl Microbiol Biotechnol 73:1180–1189. doi:10.1007/s00253-006-0564-6

Milliken CE, Meier GP, Sowers KR, May HD (2004a) Chlorophenol production by anaerobic microorganisms: transformation of a biogenic chlorinated hydroquinone metabolite. Appl Environ Microbiol 70:2494–2496

Milliken CE, Meier GP, Watts JEM, Sowers KR, May HD (2004b) Microbial anaerobic demethylation and dechlorination of chlorinated hydroquinone metabolites synthesized by basidiomycete fungi. Appl Environ Microbiol 70:385–392

Mingo FS, Studenik S, Diekert G (2014) Conversion of phenyl methyl ethers by *Desulfitobacterium* spp. and screening for the genes involved. FEMS Microbiol Ecol 90:783–790. doi:10.1111/1574-6941.12433

Mohn WW, Tiedje JM (1992) Microbial reductive dehalogenation. Microbiol Rev 56:482–507

Morita Y, Futagami T, Goto M, Furukawa K (2009) Functional characterization of the trigger factor protein PceT of tetrachloroethene-dechlorinating *Desulfitobacterium hafniense* Y51. Appl Microbiol Biotechnol 83:775–781. doi:10.1007/s00253-009-1958-z

Neumann A, Engelmann T, Schmitz R, Greiser Y, Orthaus A, Diekert G (2004) Phenyl methyl ethers: novel electron donors for respiratory growth of *Desulfitobacterium hafniense* and *Desulfitobacterium* sp. strain PCE-S. Arch Microbiol 181:245–249. doi:10.1007/s00203-004-0651-y

Newman LA, Reynolds CM (2004) Phytodegradation of organic compounds. Curr Opin Biotechnol 15:225–230. doi:10.1016/j.copbio.2004.04.006

Niggemyer A, Spring S, Stackebrandt E, Rosenzweig RF (2001) Isolation and characterization of a novel As(V)-reducing bacterium: implications for arsenic mobilization and the genus *Desulfitobacterium*. Appl Environ Microbiol 67:5568–5580. doi:10.1128/AEM.67.12.5568-5580.2001

Nonaka H, Keresztes G, Shinoda Y, Ikenaga Y, Abe M, Naito K, Inatomi K, Furukawa K, Inui M, Yukawa H (2006) Complete genome sequence of the dehalorespiring bacterium *Desulfitobacterium hafniense* Y51 and comparison with *Dehalococcoides ethenogenes* 195. J Bacteriol 188:2262–2274. doi:10.1128/JB.188.6.2262-2274.2006

Patel BK, Love CA, Stackebrandt E (1992) Helix 6 of the 16S rRNA of the bacterium *Desulfotomaculum australicum* exhibits an unusual structural idiosyncrasy. Nucleic Acids Res 20:5483

Peng X, Yamamoto S, Vertès AA, Keresztes G, Inatomi K, Inui M, Yukawa H (2012) Global transcriptome analysis of the tetrachloroethene-dechlorinating bacterium *Desulfitobacterium hafniense* Y51 in the presence of various electron donors and terminal electron acceptors. J Ind Microbiol Biotechnol 39:255–68. doi:10.1007/s10295-011-1023-7

Prat L, Maillard J, Grimaud R, Holliger C (2011) Physiological adaptation of *Desulfitobacterium hafniense* strain TCE1 to tetrachloroethene respiration. Appl Environ Microbiol 77:3853–3859. doi:10.1128/AEM.02471-10

Prat L, Maillard J, Rohrbach-Brandt E, Holliger C (2012) An unusual tandem-domain rhodanese harbouring two active sites identified in *Desulfitobacterium hafniense*. FEBS J 279:2754–2767. doi:10.1111/j.1742-4658.2012.08660.x

Puyol D, Mohedano AF, Rodriguez JJ, Sanz JL (2011) Effect of 2,4,6-trichlorophenol on the microbial activity of adapted anaerobic granular sludge bioaugmented with *Desulfitobacterium* strains. N Biotechnol 29:79–89. doi:10.1016/j.nbt.2011.06.011

Reinhold A, Westermann M, Seifert J, von Bergen M, Schubert T, Diekert G (2012) Impact of vitamin B_{12} on formation of the tetrachloroethene reductive dehalogenase in *Desulfitobacterium hafniense* strain Y51. Appl Environ Microbiol 78:8025–8032. doi:10.1128/AEM.02173-12

Rogosa M (1971) *Peptococcaceae*, a new Family to include the Gram-positive, anaerobic cocci of the Genera *Peptococcus*, *Peptostreptococcus*, and *Ruminococcus*. Int J Syst Bacteriol 21:234–237. doi:10.1099/00207713-21-3-234

Rouzeau-Szynalski K, Maillard J, Holliger C (2011) Frequent concomitant presence of *Desulfitobacterium* spp. and "*Dehalococcoides*" spp. in chloroethene-dechlorinating microbial communities. Appl Microbiol Biotechnol 90:361–368. doi:10.1007/s00253-010-3042-0

Sanford RA, Cole JR, Löffler FE, Tiedje JM (1996) Characterization of *Desulfitobacterium chlororespirans* sp. nov., which grows by coupling the oxidation of lactate to the reductive dechlorination of 3-chloro-4-hydroxybenzoate. Appl Environ Microbiol 62:3800–3808

Shelobolina ES, VanPraagh CG, Lovley DR (2003) Use of ferric and ferrous iron containing minerals for respiration by *Desulfitobacterium frappieri*. Geomicrobiol J 20:143–156. doi:10.1080/01490450390193255

Shelton DR, Tiedje JM (1984) Isolation and partial characterization of bacteria in an anaerobic consortium that mineralizes 3-chlorobenzoic Acid. Appl Environ Microbiol 48:840–848

Smidt H, de Vos WM (2004) Anaerobic microbial dehalogenation. Annu Rev Microbiol 58:43–73. doi:10.1146/annurev.micro.58.030603.123600

Smidt H, Song D, van Der Oost J, de Vos WM (1999) Random transposition by Tn*916* in *Desulfitobacterium dehalogenans* allows for isolation and characterization of halorespiration-deficient mutants. J Bacteriol 181:6882–6888

Smidt H, van Leest M, van der Oost J, de Vos WM (2000) Transcriptional regulation of the *cpr* gene cluster in *ortho*-chlorophenol-respiring *Desulfitobacterium dehalogenans*. J Bacteriol 182:5683–5691

Smidt H, van der Oost J, de Vos WM (2001) Development of a gene cloning and inactivation system for halorespiring *Desulfitobacterium dehalogenans*. Appl Environ Microbiol 67:591–597. doi:10.1128/AEM.67.2.591-597.2001

Spring S, Rosenzweig F (2006) The genera *Desulfitobacterium* and *Desulfosporosinus*: taxonomy. In: Dworkin M, Falkow S, Rosenberg E, Schleifer K, Stackebrandt E (eds) The prokaryotes, Volume 4: bacteria: firmicutes, cyanobacteria. Springer, New York, pp 771–786. doi:10.1007/0-387-30744-3_240

Studenik S, Vogel M, Diekert G (2012) Characterization of an *O*-demethylase of *Desulfitobacterium hafniense* DCB-2. J Bacteriol 194:3317–3326. doi:10.1128/JB.00146-12

Sutcliffe IC (2010) A phylum level perspective on bacterial cell envelope architecture. Trends Microbiol 18:464–470. doi:10.1016/j.tim.2010.06.005

Suyama A, Iwakiri R, Kai K, Tokunaga T, Sera N, Furukawa K (2001) Isolation and characterization of *Desulfitobacterium* sp. strain Y51 capable of efficient dehalogenation of tetrachloroethene and polychloroethanes. Biosci Biotechnol Biochem 65:1474–1481. doi:10.1271/bbb.65.1474

Suyama A, Yamashita M, Yoshino S, Furukawa K (2002) Molecular characterization of the PceA reductive dehalogenase of *Desulfitobacterium* sp. strain Y51. J Bacteriol 184:3419–3425

Tamura K, Stecher G, Peterson D, Filipski A, Kumar S (2013) MEGA6: molecular evolutionary genetics analysis version 6.0. Mol Biol Evol 30:2725–2729. doi:10.1093/molbev/mst197

Tartakovsky B, Levesque MJ, Dumortier R, Beaudet R, Guiot SR (1999) Biodegradation of pentachlorophenol in a continuous anaerobic reactor augmented with *Desulfitobacterium frappieri* PCP-1. Appl Environ Microbiol 65:4357–4362

Thibodeau J, Gauthier A, Duguay M, Villemur R, Lépine F, Juteau P, Beaudet R (2004) Purification, cloning, and sequencing of a 3,5-dichlorophenol reductive dehalogenase from *Desulfitobacterium frappieri* PCP-1. Appl Environ Microbiol 70:4532–4537. doi:10.1128/AEM.70.8.4532-4537.2004

Ticak T, Kountz DJ, Girosky KE, Krzycki JA, Ferguson DJ (2014) A nonpyrrolysine member of the widely distributed trimethylamine methyltransferase family is a glycine betaine methyltransferase. Proc Natl Acad Sci USA 111:E4668–E4676. doi:10.1073/pnas.1409642111

Tront JM, Amos BK, Löffler FE, Saunders FM (2006) Activity of *Desulfitobacterium* sp. strain Viet1 demonstrates bioavailability of 2,4-dichlorophenol previously sequestered by the aquatic plant *Lemna minor*. Environ Sci Technol 40:529–535. doi:10.1021/es0514545

Tsukagoshi N, Ezaki S, Uenaka T, Suzuki N, Kurane R (2006) Isolation and transcriptional analysis of novel tetrachloroethene reductive dehalogenase gene from *Desulfitobacterium* sp. strain KBC1. Appl Microbiol Biotechnol 69:543–553. doi:10.1007/s00253-005-0022-x

Utkin I, Woese C, Wiegel J (1994) Isolation and characterization of *Desulfitobacterium dehalogenans* gen. nov., sp. nov., an anaerobic bacterium which reductively dechlorinates chlorophenolic compounds. Int J Syst Bacteriol 44:612–619

Utkin I, Dalton DD, Wiegel J (1995) Specificity of reductive dehalogenation of substituted ortho-chlorophenols by *Desulfitobacterium dehalogenans* JW/IU-DC1. Appl Environ Microbiol 61:346–351

van de Pas BA, Smidt H, Hagen WR, van der Oost J, Schraa G, Stams AJ, de Vos WM (1999) Purification and molecular characterization of ortho-chlorophenol reductive dehalogenase, a key enzyme of halorespiration in *Desulfitobacterium dehalogenans*. J Biol Chem 274:20287–20292

van de Pas BA, Gerritse J, de Vos WM, Schraa G, Stams AJ (2001a) Two distinct enzyme systems are responsible for tetrachloroethene and chlorophenol reductive dehalogenation in *Desulfitobacterium* strain PCE1. Arch Microbiol 176:165–169

van de Pas BA, Harmsen HJ, Raangs GC, de Vos WM, Schraa G, Stams AJ (2001b) A *Desulfitobacterium* strain isolated from human feces that does not dechlorinate chloroethenes or chlorophenols. Arch Microbiol 175:389–394

Villemur R (2013) The pentachlorophenol-dehalogenating *Desulfitobacterium hafniense* strain PCP-1. Philos Trans R Soc Lond B Biol Sci 368:20120319. doi:10.1098/rstb.2012.0319

Villemur R, Lanthier M, Beaudet R, Lépine F (2006) The *Desulfitobacterium* genus. FEMS Microbiol Rev 30:706–733. doi:10.1111/j.1574-6976.2006.00029.x

Villemur R, Constant P, Gauthier A, Shareck M, Beaudet R (2007) Heterogeneity between 16S ribosomal RNA gene copies borne by one *Desulfitobacterium* strain is caused by different 100-200 bp insertions in the 5′ region. Can J Microbiol 53:116–128. doi:10.1139/w06-111

Wiegel J, Zhang X, Wu Q (1999) Anaerobic dehalogenation of hydroxylated polychlorinated biphenyls by *Desulfitobacterium dehalogenans*. Appl Environ Microbiol 65:2217–2221

Woo PCY, Teng JLL, Leung K-W, Lau SKP, Woo GKS, Wong ACY, Wong MKM, Yuen K-Y (2005) *Anaerospora hongkongensis* gen. nov. sp. nov., a novel genus and species with ribosomal DNA operon heterogeneity isolated from an intravenous drug abuser with pseudobacteremia. Microbiol Immunol 49:31–39

Yang Y, Pesaro M, Sigler W, Zeyer J (2005) Identification of microorganisms involved in reductive dehalogenation of chlorinated ethenes in an anaerobic microbial community. Water Res 39:3954–3966. doi:10.1016/j.watres.2005.07.010

Ye L, Schilhabel A, Bartram S, Boland W, Diekert G (2010) Reductive dehalogenation of brominated ethenes by *Sulfurospirillum multivorans* and *Desulfitobacterium hafniense* PCE-S. Environ Microbiol 12:501–509. doi:10.1111/j.1462-2920.2009.02093.x

Yoshida N, Asahi K, Sakakibara Y, Miyake K, Katayama A (2007) Isolation and quantitative detection of tetrachloroethene (PCE)-dechlorinating bacteria in unsaturated subsurface soils contaminated with chloroethenes. J Biosci Bioeng 104:91–97. doi:10.1263/jbb.104.91

Zhao S, Ding C, He J (2015) Detoxification of 1,1,2-trichloroethane to ethene by desulfitobacterium and identification of its functional reductase gene. PLoS ONE 10:e0119507. doi:10.1371/journal.pone.0119507

Chapter 10
The Genus *Sulfurospirillum*

Tobias Goris and Gabriele Diekert

Abstract The only organohalide-respiring Epsilonproteobacteria (ε-proteobacteria) described so far are found in the genus *Sulfurospirillum*. This genus consists of versatile, often microaerophilic bacteria, growing with many different growth substrates. Only a few of these organisms use halogenated compounds, mainly chlorinated ethenes, as electron acceptors. Organohalide respiration was extensively studied in *Sulfurospirillum multivorans*, but seems to be similar in other reductively dehalogenating Sulfurospirilla like *Sulfurospirillum halorespirans*. While most *Sulfurospirillum* species are unable to utilize organohalides as electron acceptors, many of them grow with other toxic substrates such as arsenate or selenate. Other typical electron acceptors are nitrate and sulfur compounds. Electron donors used are pyruvate, hydrogen and formate. The anaerobic respiratory chains of *Sulfurospirillum* spp. involve most likely menaquinones and cytochromes for most electron donor/acceptor combinations. The growth substrate range which includes many toxic compounds enables many *Sulfurospirillum* species to thrive in polluted habitats, which is reflected by the presence of these bacteria in many contaminated sites. The genomes of *Sulfurospirillum* spp. are small to average in size (about 2.5–3 Mbp) and the genes necessary for organohalide respiration, if present, are clustered in one area, including corrinoid biosynthesis genes responsible for production of the unique norpseudovitamin B_{12}. The gene inventory in this area differs from that of other organohalide-respiring bacterial classes in that a putative quinol dehydrogenase and other accessory proteins are encoded. This points to a respiratory chain differing from other organohalide-respiring bacteria.

T. Goris (✉) · G. Diekert
Department of Applied and Ecological Microbiology, Institute of Microbiology, Friedrich Schiller University, 07743 Jena, Germany
e-mail: tobias.goris@uni-jena.de

G. Diekert
e-mail: gabriele.diekert@uni-jena.de

L. Adrian and F.E. Löffler (eds.), *Organohalide-Respiring Bacteria*,
DOI 10.1007/978-3-662-49875-0_10

10.1 Discovery

10.1.1 A History of the Genus Sulfurospirillum and Its Isolates

The genus *Sulfurospirillum* was established in 1992 with the taxonomic description of *Sulfurospirillum deleyianum*, which was originally isolated as Spirillum 5175 from a *Desulfuromonas acetoxidans*-containing mixed culture enriched from anoxic mud of a forest pond near Heiningen, Braunschweig area, Germany (Wolfe and Penning 1977; Schumacher et al. 1992). Physiological and molecular biological investigations pointed toward a relationship of this isolate to *Wolinella succinogenes* and *Campylobacter jejuni*, but at the same time it could be clearly distinguished from both genera by a higher GC content and through DNA–DNA hybridization. Therefore, the new genus *Sulfurospirillum* with its type strain *S. deleyianum* was established (Schumacher et al. 1992). In the same study, *Campylobacter* species DSM 806 was described as similar in GC content to *S. deleyianum*, but it was not included in the *Sulfurospirillum* genus, as several physiological characteristics were different. *S. deleyianum* is microaerophilic, able to reduce and oxidize sulfur compounds and to carry out dissimilatory nitrate reduction (Schumacher and Kroneck 1992). Dechlorination of organohalides was not tested in these studies; later experiments with tetrachloroethene (PCE) or trichloroethene (TCE) as potential electron acceptors showed that *S. deleyianum* was unable to dechlorinate these compounds. *Campylobacter* species DSM 806 was shown to be a member of the *Sulfurospirillum* genus (Gevertz et al. 2000), but its taxonomic affiliation has not been validly published by now. In the mid to late nineties, three additional *Sulfurospirillum* species were isolated. One of these, *Sulfurospirillum arcachonense,* was a marine isolate from oxidized surface sediment of the French Atlantic Coast (Finster et al. 1997). The other two were isolated from a selenate-contaminated freshwater marsh in western Nevada (*Sulfurospirillum barnesii* (Oremland et al. 1994)) and from arsenic contaminated sediments in eastern Massachusetts (*S. arsenophilum* (Ahmann et al. 1994)). Both were capable of selenate and arsenate reduction (Stolz et al. 1999). *S. barnesii* was initially described as *Geospirillum barnesii* (Stolz et al. 1997). Another bacterium, preliminary named *Geospirillum* species SM-5, was isolated in the early 1990s (Stolz et al. 1999), but is not further discussed here due to scarce literature and missing deposition in any of the major culture collections. None of these isolates were described as reductively dechlorinating.

Sulfurospirillum multivorans (Fig. 10.1) was first described as *Dehalospirillum multivorans* because of its ability to grow with chlorinated ethenes as electron acceptors. It was isolated from activated sludge of a wastewater treatment plant in Stuttgart-Büsnau, Germany, which was not known to have been exposed to chlorinated ethenes. The sample was collected around 1990 and a methanogenic enrichment culture was established in the following years, which was able of dechlorinating PCE completely to ethene. Using pyruvate as electron donor and PCE as electron

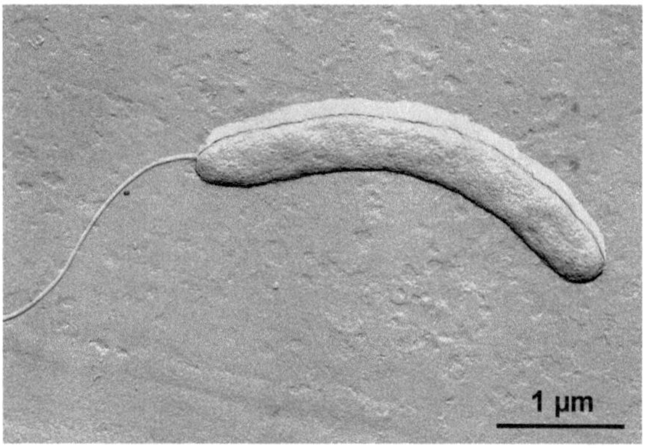

Fig. 10.1 Electron micrograph of *S. multivorans* using unidirectional surface shadowing with platinum. Picture courtesy of Martin Westermann (FSU Jena, Electron Microscopy Center of the University Hospital Jena)

acceptor, a highly enriched culture dechlorinating PCE to *cis*-1,2-dichloroethene was obtained. In 1993, the isolation of *D. multivorans* from agar roll tubes was successful (Scholz-Muramatsu et al. 1995). Another strain of *D. multivorans*, the non-dechlorinating strain N, was isolated from the same enrichment culture (Siebert et al. 2002). Later, *D. multivorans* was reclassified as a member of the genus *Sulfurospirillum*, together with the isolation of the physiologically similar, also PCE-dechlorinating, *Sulfurospirillum halorespirans* (Luijten et al. 2003). *S. halorespirans* was isolated from flow through soil columns which were inoculated with a sample taken from a site heavily polluted with chlorinated aliphatics in Maassluis near the Rotterdam harbour, the Netherlands (Middeldorp et al. 1998). The reclassification of *D. multivorans* was justified by a phylogenetic analysis of the 16S rRNA genes of all *Sulfurospirillum* species known to that date. The isolation of another dehalogenating *Sulfurospirillum* species using an inoculate from groundwater in the USA (*Sulfurospirillum* sp. JPD-1, also referred to as *Sulfurospirillum tacomaensis*) was described in a PhD thesis from 2002 (Pietari 2002).

Two further marine strains (*Sulfurospirillum* sp. 18.1 and AM-N) were isolated from deep sea habitats. *Sulfurospirillum* species AM-N was found to be living symbiotically with the tubeworm *Alvinella pompejana* (Campbell et al. 2001). Another marine strain, *Sulfurospirillum carboxydovorans*, was isolated from marine sediment of the North Sea and described to be capable of PCE dechlorination (Jensen and Finster 2005). A non-dechlorinating *Sulfurospirillum* species, *Sulfurospirillum cavolei* was found in a crude oil storage cave in Japan and described in 2007 (Kodama et al. 2007). Another strain (*Sulfurospirillum* spp. strain NP4), characterized through its growth in the presence of high arsenate concentrations and with arsenate as electron acceptor, was isolated from groundwater with a high arsenic concentration (MacRae et al. 2007).

Recently, the first alkaliphilic *Sulfurospirillum* species was isolated as *Sulfurospirillum alkalitolerans* (Sorokin et al. 2013) and another *Sulfurospirillum* strain was isolated from the feces of a malaysian giant turtle in the Netherlands (Gilbert et al. 2014). The latter was the only *Sulfurospirillum* species found in a vertebrate until now.

10.1.2 Isolation

Most *Sulfurospirillum* species are physiologically versatile, microaerophilic organisms that are growing relatively fast with a variety of electron donors and acceptors. Thus, anaerobic enrichment with specific electron acceptors and, e.g., pyruvate as electron donor, dilution and picking of single colonies from soft agar/rolltubes is in general a promising isolation approach. Organohalide-respiring species are enriched with the addition of higher concentrations of PCE (~200 μM) to the medium and the use of lactate or pyruvate as electron donor. *S. multivorans* was isolated as follows. From PCE-dechlorinating, methanogenic enrichment cultures, the fastest PCE-dechlorinating cultures were separated via several (15) frequent (every 24 h) transfers into freshly prepared, PCE-containing minimal medium (Scholz-Muramatsu et al. 1995) with hydrogen, pyruvate, or lactate as electron donor. When hydrogen was used, acetate served as carbon source. The resulting enrichment culture did no longer produce methane and dechlorinated PCE to *cis*-DCE. For isolation of the dechlorinating bacterium, dilution series in rolltubes with pyruvate or lactate (10 mM) containing 100 μM PCE were used and several single colonies were picked and characterized.

10.2 General Characteristics

10.2.1 Morphology and Growth Characteristics

Most *Sulfurospirillum* isolated so far are mesophilic (growth optimum between 26 and 33 °C), neutrophilic (pH 7–7.5), slightly curved, thin, medium-sized rods (0.3–0.5 μm in diameter and 2–5 μm in length), which are motile through a single polar flagellum (see Fig. 10.1). They usually grow to an OD of about 0.5 with a doubling time of about 2–10 h in a minimal medium containing pyruvate and fumarate (20 mM each) or 4–14 h with, e.g., formate and nitrate as energy substrates (unpublished results; see also (Eisenmann et al. 1995; Scholz-Muramatsu et al. 1995)). When formate or hydrogen is used as electron donors in combination with an inorganic electron acceptor, acetate has to be added as a carbon source. Marine species usually grow slower and to a lower cell density than terrestrial species (Finster et al. 1997; Campbell et al. 2001; Jensen and Finster 2005). Most species, except the halophilic marine species (see 10.3.2), are moderately halotolerant (up to about 1 % NaCl), but grow best with low salt concentrations (<0.2 %). Up to date, no *Sulfurospirillum* species was found to be pathogenic or living commensally with a mammalian host (including humans). The fatty acid composition of the membrane lipids consists mostly of 16:0, 16:1 × 7*c*, and 18:1 × 7*c* fatty acids (Finster et al. 1997; Luijten et al. 2003; Sorokin et al. 2013).

10.2.2 Habitat

Sulfurospirillum spp. seem to be globally distributed bacteria mostly found in sediments, groundwater, or soil, often in environments polluted with organohalides, arsenate, or selenate (see Table 10.1) or in habitats rich in sulfur compounds. The few studies referring to marine habitats in which Sulfurospirilla were found, led to the conclusion that *Sulfurospirillum* spp. are also found in either marine surface sediments, deep sea habitats, or in symbiotic relationship with tubeworms (Finster et al. 1997; Campbell et al. 2001; Takai et al. 2004; Jensen and Finster 2005), Areas, in which *Sulfurospirillum* is often the dominating species, are oil-contaminated sites or oil fields (Hubert and Voordouw 2007; Kodama et al. 2007; Grigoryan et al. 2008). Here, the high abundance of sulfur compounds, several of which can be used as electron donor or acceptor by *Sulfurospirillum* spp., might have led to an enrichment. Wastewater, often rich in eventually toxic, industrial residues, and/or organic matter are also a habitat, where *Sulfurospirillum* spp. is frequently abundant (Table 10.1).

Many environmental studies on organohalide-respiring organisms focus on screening for reductive dehalogenase genes and of those, a major part deals with *Dehalococcoides rdhA* genes. Primer design for *Sulfurospirillum rdhA* genes is difficult, as there are only four *pceA* sequences known (Neumann et al. 1998; Regeard et al. 2004; Kimoto et al. 2010; Buttet et al. 2013; Hug et al. 2013) but several studies with primer pairs derived from the nucleotide sequence information for *S. multivorans* PceA showed positive results in PCE-contaminated habitats and/or enrichment cultures obtained from such environments (see Table 10.1). Screening of organohalide-contaminated environments with 16S rRNA gene specific primers revealed numerous habitats, where *Sulfurospirillum* species closely related to *S. halorespirans* or *S. multivorans* are present, often occurring together with *Dehalococcoides*, *Desulfitobacterium*, or *Dehalobacter* species (see Table 10.1). This observation might on the one hand be based on the ability of the corresponding *Sulfurospirillum* species to respire organohalides, on the other hand *Sulfurospirillum* might be involved in a syntrophic relationship with *Dehalobacter* and *Dehalococcoides*, either providing corrinoid derivatives or hydrogen produced via fermentation of carboxylic acids. The produced hydrogen could then serve as electron donor for obligate organohalide-respiring organisms. The occurence of *Sulfurospirillum* species in arsenate-containing habitats is also striking (see Table 10.1). Obviously, many *Sulfurospirillum* spp. are not only able to gain energy from arsenate reduction, but probably can also detoxify arsenic species, enabling them to thrive in environments with high arsenic concentrations. The finding of *Sulfurospirillum* spp. in several chlorophenol-containing habitats (Li et al. 2014) led to the conclusion that a few *Sulfurospirillum* species might be involved in the dechlorination of aromatic organohalides. However, later it was found out, that a δ-proteobacterium was responsible for dechlorination of 4-chlorophenol (Suzuki et al. 2014).

Table 10.1 Habitats of *Sulfurospirillum* spp.

Habitat	Country	Isolate or most similar isolate	References
Isolates			
Pond sediment	Germany	*S. deleyianum*	Wolfe and Penning (1977), Schumacher et al. (1992)
Anaerobic wastewater digester	The Netherlands	*Sulfurospirillum* species DSM 806	Laanbroek et al. (1977)
Selenate-contaminated freshwater marsh	US	*S. barnesii*	Oremland et al. (1994), Stolz et al. (1999)
Arsenic contaminated sediments	US	*S. arsenophilum*	Ahmann et al. (1994), Stolz et al. (1999)
Activated sludge	Germany	*S. multivorans*	Scholz-Muramatsu et al. (1995)
Marine surface sediment	France	*S. arcachonense*	Finster et al. (1997)
Alvinella pompejana deep sea ridge tubeworm	Pacific Ocean	*Sulfurospirillum* species AM-N	Campbell et al. (2001)
Deep sea sediment	Pacific Ocean	*Sulfurospirillum* species 18.1	Campbell et al. (2001)
Groundwater	US	*Sulfurospirillum* species JPD-1	Pietari (2002)
Upflow packed bed bio-reactor inoculated with produced water from an oil field	Canada	*Sulfurospirillum* species NO2B *Sulfurospirillum* species NO3A	Hubert et al. (2003)
Contaminated soil	The Netherlands	*S. halorespirans*	Luijten et al. (2003)
Saale river sediment enrichment culture with TCPP, *Dehalococcoides*	Germany	*Sulfurospirillum* species EK7	Ballerstedt et al. (2004)
Marine sediment	North Sea	*S. carboxydovorans*	Jensen and Finster (2005)
Underground crude oil storage cavity	Japan	*S. cavolei*	Kodama et al. (2007)
Groundwater with high arsenic concentration	US	*Sulfurospirillum* species NP4	MacRae et al. (2007)
Oil field	Canada	*Sulfurospirillum* species C6 *Sulfurospirillum* species KW	Hubert and Voordouw (2007)
Thiopaq reactor	The Netherlands	*S. alkalitolerans*	Sorokin et al. (2013)
Malaysian giant turtle (stool sample)	The Netherlands	*Sulfurospirillum* species 11S05485-2	Gilbert et al. (2014)
Environmental samples and enrichment cultures—PceA sequences			
Saale river sediment	Germany	*S. multivorans* (95–96 %, 152 AA fragment) CAC37186-91	von Wintzingerode et al. (2001)

(continued)

Table 10.1 (continued)

Habitat	Country	Isolate or most similar isolate	References
Groundwater sample from industrial site	Japan	*S. halorespirans* (94 %) BAJ09319	Kimoto et al. (2010)
PCE-contaminated groundwater	Netherlands	*S. halorespirans* (PceA-TCE—93 %) AGW23615; *S. multivorans* pceA (PceA-DCE—97 %) AGW23613	Maillard et al. (2011), Buttet et al. (2013)
Industrial site contaminated with PCE and TCE	France	*S. multivorans* (96 and 98 %, 157 AA fragments) ADJ40004-5	Dugat-Bony et al. (2011)
Environmental samples and enrichment cultures—16S rRNA gene sequences			
TCE contaminated groundwater	US	*S. multivorans*, *S. halorespirans* (97 %) AY667251	Macbeth et al. (2004)
Subsurface microbial community near a deep sea hydrothermal field	Indian ocean	*Sulfurospirillum* sp. AM-N (98 %) AB113186	Takai et al. (2004)
Pelican Lake oilfield Sedimentary Basin, 400 m depth	Canada	*S. multivorans* (99 %) AY570611	Grabowski et al. (2005)
TCE-contaminated groundwater (KB1 culture, with Dehalococcoides)	Canada	*S. multivorans* (98 %) AY780560	Duhamel and Edwards (2006)
Peach lye wastewater	South Africa	*S. halorespirans*, *S. arsenophilum*, (98 %) (577 bp) DQ191240-1	Keyser et al. (2007)
Industrial wastewater	Greece	n.a.	Mamais et al. (2007), Panagiotakis et al. (2014)
Arsenic contaminated fine-grained sediment, 9 m below surface	Cambodia	*Sulfurospirillum* sp. 11S05485-2 (100 %) 400 bp EF014932	Lear et al. (2007)
Contaminated sediment (Spittelwasser)	Germany	n.a.	Bunge et al. (2007)
Oil field	Argentina	*S. alkalitolerans* (99 %) EU628152-4	Grigoryan et al. (2008)
Anaerobic sludge thickener of a wastewater treatment plant	China	*Sulfurospirillum* sp. 11S05485-2 (99 %) JX944560	Chen et al. (2008), Liu et al. (2013)
PCE-contaminated aquifer	Italy	*S. deleyianum* (98 %) EF644515	Rossetti et al. (2008)
Arsenic polluted aquifer sediments	India	*Sulfurospirillum* sp. NP4 (99 %) n.a.	Hery et al. (2008)

(continued)

Table 10.1 (continued)

Habitat	Country	Isolate or most similar isolate	References
Ebro/Danube river sediment; DCA dechlorinating enrichment culture	Spain, Hungary	n.a.	van der Zaan et al. (2009)
Septic tank effluent waste water	US	*S. multivorans*, *S. halorespirans* (99 %) EU403615	Tomaras et al. (2009)
Mesothermic oil field	Canada	n.a.	Shartau et al. (2010)
Endophyte of *Phragmites australis* roots grown on wetlands to treat wastewater	China	*S. multivorans and S. halorespirans* (99 %) HMI142815, GUI78864 (720 bp)	Li et al. (2010, 2011)
Contaminated groundwater	Germany	*S. halorespirans* (97–98 %) HE652870, HE652873, HE652875-6	Cichocka et al. (2010), Kaufhold et al. (2013)
Microbial *Beggiatoa* mats from a fish farm	Chile	*S. carboxydovorans* (93 %) 700 bp fragments) fj875183-5	Aranda et al. (2010)
Arsenate-containing Kafni Glacier soil, Himalaya, 3500 m high	India	Clone KF-356 *S. arsenophilum* (97 %), KF-220, KF-7 *S. multivorans* (98 %) EF445480-3	Srinivas et al. (2011)
Pond sediment, PCE/H_2-fed enrichment culture	US	*S. deleyianum* (97 %) GQ377111	Dong et al. (2011)
Athabasca oil sands formation waters, Muskeg River oil sands mine	Canada	*S. multivorans* (98 %) JF789598 (700 bp)	Hubert et al. (2012)
Phototrophic hydrogen-producing bacterial consortium from pig slaughterhouse wastewater	Brazil	*S. cavolei*, (92 %, 100 %) JQ437239, JQ437246 (~450 bp)	Lazaro et al. (2012)
Uncontaminated paddy soil enrichment culture (with *Dehalobacter* and *Geobacter*)	Japan	*S. multivorans* (99 %) JQ781057 *S. multivorans* (98 %) AB700601	Zhang et al. (2012, 2013)
Subglacial waters	Iceland	n.a.	Marteinsson et al. (2013)
Contaminated groundwater	Hungary	*S. halorespirans* (96 %) JQ436571	Meszaros et al. (2013)
Manganate and TCE contaminated groundwater	US	n.a. (phylochip analysis)	Sercu et al. (2013)
Arako River sediment, contaminated with TCP (*Dehalobacter*) and 4-CP degrading enrichment culture	Japan	*S. multivorans* (99 %) AB721403, AB755789	Li et al. (2013, 2014)

(continued)

Table 10.1 (continued)

Habitat	Country	Isolate or most similar isolate	References
Biocathode with bacteria from a wastewater basin	USA	n.a.	Marshall et al. (2013)

Amino acid sequence identities or nucleotide sequence identities to the closest relative isolate is given as percent values in brackets, followed by the Genbank accession numbers of the sequence. *n.a.* No NCBI genbank entry is available

10.2.3 General Physiology

Nearly all *Sulfurospirillum* spp. are, like most free-living ε-proteobacteria, facultative microaerophilic species growing with 1–5% oxygen in the headspace as electron acceptor. They are oxidase-positive and, with the exception of *S. deleyianum*, also catalase-positive (Schumacher et al. 1992; Jensen and Finster 2005). Electron donors used in general are formate, hydrogen, sulfide, and organic acids such as pyruvate, lactate, or TCA cycle intermediates such as succinate, citrate, and malate (see Table 10.2). Pyruvate or fumarate is fermented by most Sulfurospirilla. Electron acceptors for anaerobic respiration used by all *Sulfurospirillum* species so far are nitrate (except *S. arcachonense*), fumarate and sulfur (polysulfide). Sulfate is not reduced in the energy metabolism by any *Sulfurospirillum* species described so far. All species described until now grow organotrophically and lithotrophically; acetate is needed as a carbon source in the latter case. Substrates in general not used by *Sulfurospirillum* spp. are sugars and fatty acids. Amino acids may be used by some strains: alanine was reported to be an electron donor for *S. carboxydovorans*, aspartate for *S. deleyianum*, while *S. arcachonense* has been shown to utilize alanine and glutamate. Alcohols are rarely used; glycerol has been shown to be used by *S. carboxydovorans* and *S. multivorans*. A high H_2 affinity of below 0.1 nM was described for *S. arsenophilum* and the PCE-respiring *S. multivorans* and *S. halorespirans* (Luijten et al. 2004b; Heimann et al. 2007). Menaquinone 6 was identified as the major quinone in *Sulfurospirillum* spp. Minor amounts (10–20 %) of thermoplasmaquinone 6 were also found. In *S. deleyianum*, small amounts of menaquinone 5 were detected. *Sulfurospirillum* cells contain cytochromes *b* and *c* (Schumacher et al. 1992; Scholz-Muramatsu et al. 1995; Finster et al. 1997; Ballerstedt et al. 2004). Ubiquinones were not found in *Sulfurospirillum* spp.

Table 10.2 Main physiological features of *Sulfurospirillum* isolates

	Sulfurospirillum											
	multivorans[1]	*halorespirans*[2]	*arsenophilum*[3]	*barnesii*[3]	*deleyianum*[4]	*cavolei*[5]	*carboxydovorans*[6]	*arcachonense*[7]	*AM-N*[8]	*alkalitolerans*[9]	*NP4*[10]	*EK7*[11]
Electron acceptors												
PCE	+	+	ND	ND	–	–	+	ND	ND	ND	ND	–
Fumarate	+	+	+	+	+	ND	ND	ND	ND	+	+	+
Sulfur/polysulfide	+	+	+	+	+	+	+	+	+	+	+	+
Thiosulfate	+[13]	–	+	+	+	+	+	–	–	+	+	+
Sulfite	ND	–	ND	–	+	+	+	–	–	–	+	–
DMSO	+[13]	ND	ND	–	+	+	+	–	ND	–	ND	ND
TMAO	+[13]	ND	ND	+	–[13]	ND	ND	–	ND	–	ND	ND
Nitrate	+	+	+	+	+	+	+	–	ND	+	+	+
Nitrite	ND	+	+	+	+	–	+	–	ND	+	–	ND
Arsenate	+[14]	+	+	+	+[14]	+	+	–	ND	+	+	+
Selenate	+[14]	+	+[14]	+	–[14]	–	+	–	ND	–	+	+
Manganate	–[14]	+[14]	+[14]	+[14]	+[14]	ND	ND	ND	ND	–	+	ND
Electron donors												
Lactate	+	+	+	+	–	+	+	+	ND	+	+	+
Pyruvate	+	+	+	+	+	+	+	+	–	+	+	+
Succinate	ND	ND	ND	ND	+	+	+	+	ND	ND	–	+
Malate	ND	ND	ND	ND	+	+	+	+	ND	–	ND	ND
Glycerol	+	ND	ND	ND	–	ND	+	–	ND	ND	ND	ND
Formate	+	+	+	+	+	+	+	+	+	+	+	+
Sulfide	+	ND	ND	ND	+	ND	ND	ND	ND	+	ND	+

(continued)

Table 10.2 (continued)

	Sulfurospirillum											
	multi-vorans[1]	*halorespi-rans*[2]	*arsenophi-lum*[3]	*barnesii*[3]	*deley-ianum*[4]	*cav-olei*[5]	*carboxydo-vorans*[6]	*arcachon-ense*[7]	*AM-N*[8]	*alkalitoler-ans*[9]	*NP4*[10]	*EK7*[11]
Hydrogen	+	+	+	+	+	+	+	+	–	+	ND	+
Fermentation												
Pyruvate	+	+	+[15]	ND	+	+	–	–	–	+	+	ND
Fumarate	+[15]	+	+	+	+	+	+	+	+	+	+	+

ND not determined
[1]Scholz-Muramatsu et al. (1995), [2]Luijten et al. (2003), [3]Stolz et al. (1999), [4]Schumacher et al. (1992), [5]Kodama et al. (2007), [6]Jensen and Finster (2005), [7]Finster et al. (1997), [8]Campbell et al. (2001), [9]Sorokin et al. (2013), [10]MacRae et al. (2007), [11]Ballerstedt et al. (2004), [12]Neumann et al. (1994), [13]Goris et al. (2014), [14]Luijten et al. (2004a), [15](unpublished data)

10.3 Phylogeny and Diversity

10.3.1 Phylogeny

The genus *Sulfurospirillum* is placed into the order *Campylobacterales* of the ε-proteobacteria. The closest relative genera are *Arcobacter* and *Campylobacter*, two clades, which contain pathogenic and free-living species. Three branches of *Sulfurospirillum* can be separated due to a 16S rRNA gene-based phylogenetic tree (Fig. 10.2): The marine clade (with *S. arcachonense*, *S. carboxydovorans,* and *Sulfurospirillum* sp. AM-N), the terrestrial clade (including *S. multivorans*) and a third clade with *S. alkalitolerans* being the sole taxonomically named member. The phylogenetic affiliation of the latter clade to *Sulfurospirillum* has to be confimed through further studies, as the 16S rRNA genes of *S. alkalitolerans* and close relatives seem to be only distantly related to those of other Sulfurospirilla (around 90 % sequence identity). Two further isolates, designated NO2B and NO3A, were enriched in a bioreactor inoculated with produced water from a Canadian oil field; they are placed in that preliminary *Sulfurospirillum* clade as well and originally described to be most closely related to *Campylobacter lari* (Hubert et al. 2003). Several clones from enrichment cultures, mostly found in areas contaminated with oil (Hubert and Voordouw 2007; Grigoryan et al. 2008) are also placed into this clade.

10.3.2 Diversity

Morphology, motility and many growth characteristics are very similar for all of the described *Sulfurospirillum* species. Deviant from the general characteristics described in 10.2.3 are a temperature optimum of 20 °C for *S. arsenophilum* and of 42 °C for *Sulfurospirillum* strain AM-N, the alkaline pH optimum of about 8.5 for *S. alkalitolerans* and the halophilic growth in the presence of 2 % NaCl for the marine species.

The most obvious physiological diversity is found in the use of different electron donors and electron acceptors which support growth via anaerobic respiration in the different *Sulfurospirillum* spp. (see also Table 10.2).

Sulfur and arsenate, as well as nitrate and fumarate, are suitable electron acceptors for most Sulfurospirilla, while sulfite and selenate are supporting growth in fewer species (Table 10.2). The product of nitrate reduction is nitrite, which is ammonified by most *Sulfurospirillum* species. The nitrite reductase is a periplasmic, membrane-bound dimeric multihaem enzyme. The crystal structure of this cytochrome *c* nitrate reductase of *S. deleyianum* was solved (Einsle et al. 1999). Manganate, trimethylamine N-oxide (TMAO), and dimethyl sulfoxide (DMSO) were tested for only half of the species described so far, so that a comparison is not suitable. Tetrathionate supported slow growth of *S. multivorans*, but not that of any

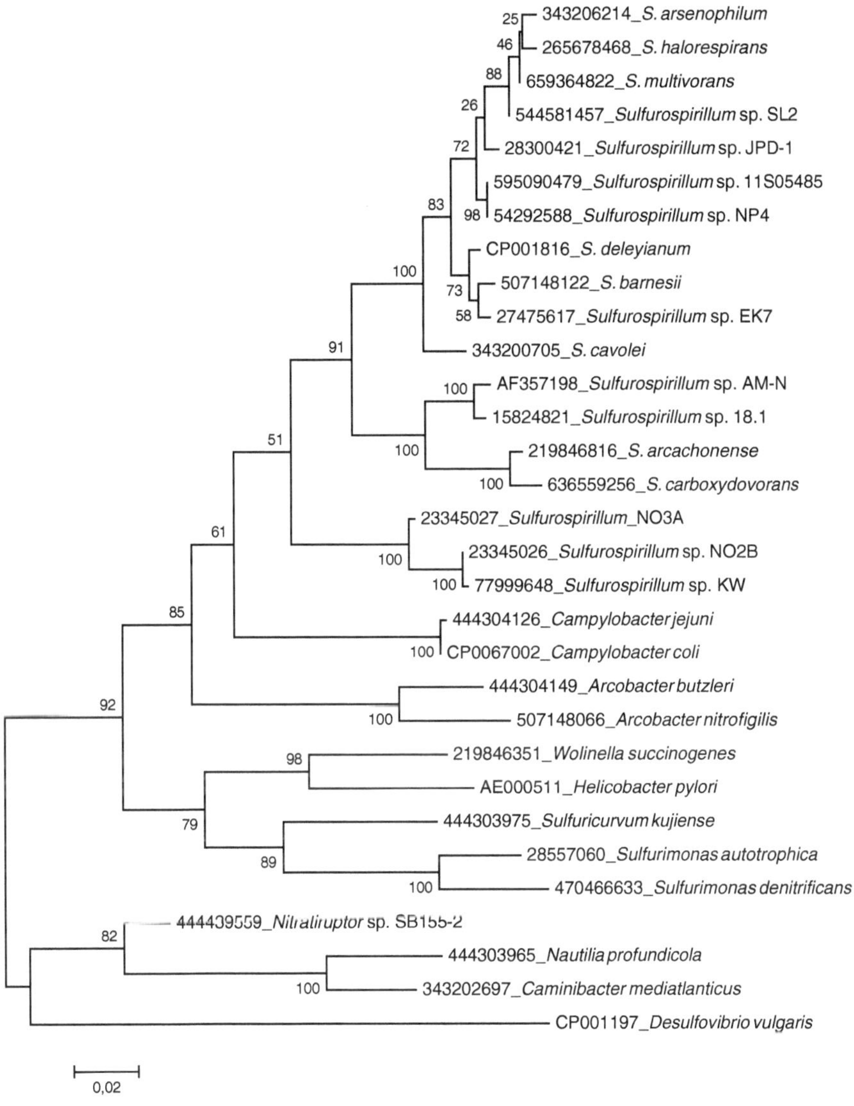

Fig. 10.2 Phylogenetic tree of *Sulfurospirillum* spp. Selected ε-proteobacterial 16S rRNA gene sequences were derived from NCBI Genbank, GI identifiers are given for each sequence. The following 16S rRNA genes were chosen if a complete genome sequence was available: *Desulfovibrio vulgaris* DvMF_R0027 (used as outgroup), *S. deleyianum* Sdel_R0037, *Helicobacter pylori* 26695 HP_r07, *Campylobacter coli* 15-537360 N149_0029 KJ081211. The tree was generated with MEGA 6.0 using the maximum likelihood algorithm with 500 bootstraps and standard settings. Sequences were trimmed to the size of the shortest available sequence (resulting in 1223 bp included) and aligned with Muscle. Bootstrap values of lower than 70 % are considered as low, those nodes should be taken with care.

other *Sulfurospirillum* species tested so far (including *S. deleyianum* and *S. barnesii* (Goris et al. 2014)). Ferric iron (Fe^{3+}) is reduced indirectly by *Sulfurospirillum* species abiotically by metabolically produced extracellular sulfur species, but is most likely not used as sole electron acceptor for anaerobic respiration and growth (Luijten et al. 2004a; Straub and Schink 2004; Lohmayer et al. 2014).

S. carboxydovorans is the only *Sulfurospirillum* species in which carbon monoxide oxidation was observed so far. CO supported slow growth (doubling time of ~50 h) with elemental sulfur, DMSO, or thiosulfate as electron acceptors (Jensen and Finster 2005). Hydrogen is produced upon CO oxidation and was observed to be "recycled" later presumably by an uptake hydrogenase. A CO dehydrogenase gene cluster was found to be present in the genome of *Sulfurospirillum* strain SCADC binned from metagenomic data (Tan and Foght 2014) (see 10.3.4.1).

The only physiological trait by which *S. halorespirans* can be distinguished from *S. multivorans* is the usage of thiosulfate as electron acceptor in the latter and the reduction of manganate in the former bacterium (Luijten et al. 2003, 2004a; Goris et al. 2014).

10.3.3 Organohalide Respiration and Reductive Dehalogenation Characteristics

Two *Sulfurospirillum* isolates are hitherto unambiguously described as dehalogenating species, i.e. *S. multivorans* and *S. halorespirans*. *S. carboxydovorans* was described as PCE dehalogenating (Jensen and Finster 2005), but this could not be reproduced in studies conducted in our laboratory with the strain ordered from the German Collection of Microorganisms and Cell cultures (DSMZ) (unpublished results). Another strain described as PCE-dechlorinating is the psychrotrophic *Sulfurospirillum* sp. strain JPD-1, described in a PhD thesis (Pietari 2002). This strain is deposited in the American Type Culture Collection (ATCC) under the non-confirmed taxonomic name *S. tacomaensis*. The presence of a complete *pceA* gene is not always correlating to *in vivo* dechlorination, as demonstrated for *S. multivorans* strain N. This organism does not dechlorinate any chlorinated ethenes, despite harboring a complete *pceA* gene; the organism is also unable to synthesize corrinoids *de novo*. The ability to dechlorinate cannot be restored by the addition of external vitamin B_{12} and the PceA apoenzyme is not synthesized in the cells (Siebert et al. 2002). The crystal structure of PceA from *S. multivorans* was solved recently (Bommer et al. 2014). A highly enriched culture containing two dehalogenating *Sulfurospirillum* strains was described recently. One of these strains dechlorinates PCE to TCE exclusively, whereas the other one dechlorinates PCE to cDCE. Both strains contain identical 16S rRNA gene sequences and therefore seem to be two strains of the same species. They can be distinguished by their different amino acid sequences of PceA (93 % identity) (Buttet et al. 2013).

S. halorespirans and *S. multivorans* are very similar in their overall dehalogenation characteristics; their PceA amino acid sequence identity is about 92 %. Both organisms dechlorinate PCE primarily to cDCE (Luijten et al. 2003) and *S. multivorans* grows with either PCE or TCE as electron acceptor up to concentrations of at least about 300 μM (Neumann et al. 1994). For organohalide respiration, pyruvate or lactate seem to be the best electron donors with respect to the growth rate. Slower growth is observed with formate or hydrogen as electron donor when acetate is used as carbon source. Vitamin B_{12} is not required for growth and there is no difference between growth in the absence or presence of B_{12} with PCE as the electron acceptor. *S. multivorans* is able to produce a unique corrinoid, norpseudovitamin B_{12}, the cofactor of the reductive dehalogenase PceA, *de novo* (Kräutler et al. 2003). Small amounts of yeast extract (0.05–0.2%) have a positive effect on growth. Possibly due to the low activity of PceA with chlorinated propenes (Neumann et al. 2002) or 3,5-dichlorophenol (unpublished results) as substrates, growth was not observed with these compounds. PceA debrominates several brominated ethenes (tribromoethene and dibromoethene which are reduced to vinyl bromide) at high activities. Dibromoethene (DBE) supported growth in a concentration up to about 150 μM in the medium, while higher DBE concentrations inhibited growth (Ye et al. 2010). Tribromoethene (TBE), while being a substrate for PceA, was toxic for *S. multivorans* in the concentration range used for DBE, so that growth could not be monitored with TBE.

Chlorinated methanes were shown to inhibit PCE respiration of *S. multivorans* (Eisenbeis et al. 1997), whereas fumarate and sulfur lowered PCE dechlorination rates, possibly due to competing electron acceptors. Nitrate had no direct negative effect on PCE dechlorination (Neumann et al. 1994).

When *S. multivorans* was grown with electron acceptors other than chlorinated ethenes, reductive dechlorination was slowly down-regulated until it ceases about 100 generations later. This downregulation was not due to a gene loss or other genetic changes in the reductive dehalogenase region. PCE-dependent growth was inducible again within a few days (John et al. 2009). A similar behavior of downregulation was also shown for *S. halorespirans* (unpublished results). While initially PceA was believed to be located in the cytoplasm (Neumann et al. 1994; Miller et al. 1996), it was later shown that PceA was membrane-associated and periplasmically oriented (John et al. 2006). This is in-line with the amino acid sequence bearing a TAT signal motif (Neumann et al. 1998). However, when *S. multivorans* is grown for several generations in the absence of PCE, the major part of the PceA fraction was localized in the cytoplasm, indicating regulation of the enzyme's translocation by its substrate (John et al. 2006).

Table 10.3 Comparison of available *Sulfurospirillum* genomes (as of 2014–12)

	Sulfurospirillum					
	multivorans	*deleyianum*	*barnesii*	*arcachonense*	*AM-N*	*SCADC*
Genome size (kb)	3.18	2.31	2.51	>2.66	>2.26	>2.66
GC content (%)	41	39	39	30	32	41.6
CRISPR	1	1	1	–	1	1
rRNA operons	2	3	2	3	3	2
tRNAs	45	43	41	33	33	39
Genes						
rdh genes	2	–	–	–	–	–
nap cluster	+	+	+	–	+	+
nrf cluster	+	+	+	–	+	+
nos cluster	(+)	–	–	–	+	–
nor cluster	–	–	–	–	+	–
aio cluster	+	–	+	–	–	+
arr cluster	+	–	+	–	–	–
CODH	–	–	–	–	–	+
SOX	–	–	–	–	–	+
fdh clusters (#)	4	3	4	3	2	4
Lactate dehydrogenase	+	+	+	+	–	–
ATP CL gene	+	–	–	–	–	–
Complex I clusters	2	1	2	2	2	2
nif clusters	2	–	1	1	–	1
SOD (Fe)	+	+	+	–	–	+
SOD (Cu–Zn)	+	–	–	–	–	–
SOD (Mn)	–	+	+	–	–	+
Catalase	+	–	+	–	–	+
Corrinoid biosynthesis cluster	+(nor)	–	+	+	–	–
Hydrogenase genes						
MBH	+	+	+	+	+	+
HupSL	+	–	–	–	+	+
CooH	+	+	+	–	–	+
Hyf	+	+	+	+	+	+

Brackets indicate an incomplete cluster. A number indicates presence of more than one cluster. *rdh* Reductive dehalogenase homologous gene, *nap* nitrate reductase, *nrf* ammonifying nitrite reductase, *nos* nitrous oxide reductase, *nor* cytoplasmic nitrite reductase, *aioA* arsenite oxidase, *arr* arsenate reductase, *CODH* carbon monoxide dehydrogenase, *SOX* sulfur oxidation system, *fdh* formate dehydrogenase, *ATP CL* ATP citrate lyase, *nif* nitrogenase, *SOD* superoxide dismutase, *MBH* membrane-bound hydrogenase, *HupSL* cytoplasmic hydrogen uptake hydrogenase, *CooH* carbon monoxide hydrogenase like complex, *Hyf E. coli* hydrogenase 4-like complex. (*nor*) norpseudovitamin B_{12} biosynthesis cluster. For details and genome accession numbers see text

10.3.4 Genomic Features

10.3.4.1 General Genomic Features

Currently (as of December 2014), there are six different complete or near complete *Sulfurospirillum* genome sequences deposited in the publicly available databases (NCBI Genbank and Portal of the Department of Energy Joint Genome Institute—JGI). Two additional partial genome sequences from deep sea *Sulfurospirillum* spp. derived from single cell sequencing projects are available at the JGI server (*Sulfurospirillum* SCGC AAA036-N09 and SCGC AAA036-P08). The three completely sequenced genomes belong to *S. deleyianum*, *S. barnesii* (Genbank accession number CP003333), and *S. multivorans* (Sikorski et al. 2010; Goris et al. 2014). A draft genome of the marine isolate *S. arcachonense* is available through genbank (RefSeq accession number NZ_JFBL00000000), as is a genome from an environmental sample constructed of the metagenome data of an alkene degrading community enriched from an oil sands tailings pond in Canada (SCADC) (Tan and Foght 2014). A draft genome of the marine *Sulfurospirillum* strain AM-N, living in a symbiotic relationship with a tubeworm, is available through the JGI database (JGI IMG object ID 2502171155). Currently, we are working on assembly of the genomes of *S. multivorans* strain N, *S. halorespirans*, and *Sulfurospirillum* enrichment clone SL2-TCE (unpublished results).

The genomes of *Sulfurospirillum* species vary greatly in size (2.3–3.2 Mbp). Together with the *Arcobacter* spp. genomes (2.2–3.5 Mbp), *Sulfurospirillum* species harbor the largest chromosomes among the ε-proteobacteria. The GC content of the terrestrial species ranges from 38 to 41 %, while that of the marine species is considerably lower (between 30 and 32 %). Plasmids were not found in any of the sequenced *Sulfurospirillum* spp. Usually, two or three rRNA operons are found and 41–45 tRNAs (39 for the draft genome of *Sulfurospirillum* strain SCADC and 33 for the draft genomes of the marine species *S. arcachonense* and *Sulfurospirillum* strain AM-N). One CRISPR locus is mostly found. In the *Sulfurospirillum* genomes, a remarkable variety of catabolic enzymes are encoded with several hydrogenases and formate dehydrogenases and often many other electron donor and acceptor enzyme systems, the latter being mostly molybdopterin oxidoreductases. The marine species in general contain less genes coding for oxidoreductases participating in energy metabolism (Table 10.3).

Interestingly, several of the sequenced *Sulfurospirillum* genomes harbor corrinoid biosynthesis gene clusters, a feature normally very rarely found among ε-proteobacteria. Up to 2009 no corrinoid biosynthesis gene clusters were found in ε-proteobacteria (Zhang et al. 2009). In the organohalide-respiring *S. multivorans,* the corrinoid (norpseudovitamin B_{12}) biosynthesis gene cluster is located directly downstream of the reductive dehalogenase gene cluster and exhibits similarities to the fusobacterial corrinoid biosynthesis cluster (Goris et al. 2014). In the other species carrying a corrinoid biosynthesis cluster (*S. barnesii* and *S. arcachonense*), the corresponding genes are not related to those of the *S. multivorans* (or fusobacterial) cluster, but instead to that of *Vibrio* and *Shewanella* species (γ-proteobacteria). The corrinoid produced by *S. barnesii* and presumably *S. arcachonense* was identified as pseudovitamin B_{12} (Goris et al. 2014). Besides the reductive dehalogenase, two other enzymes are putatively corrinoid-dependent, namely, ethanolamine ammonia lyase and glutamate mutase, but their requirement for norpseudovitamin B_{12} has not yet been determined. Ribonucleotide reductase and methionine synthase, reported to be often dependent on corrinoids, are found to be encoded in non-corrinoid-dependent forms on the genome of *S. multivorans* and the other *Sufurospirillum* spp. genomes. *S. barnesii* contains the gene encoding the ethanolamine ammonia lyase, *S. arcachonense* additionally the gene for glutamate mutase, while only the latter is found in the genome of *Sulfurospirillum* strain SCADC. Thus, the distribution of corrinoid-dependent enzymes correlates in most *Sulfurospirillum* spp. with the presence of a corrinoid biosynthesis cluster, with *Sulfurospirillum* strain SCADC being the exception. No corrinoid uptake system was identified in *Sulfurospirillum* strain SCADC, which raises the question whether the glutamate mutase is not functional in this *Sulfurospirillum* species or whether the enzyme might function without a corrinoid as cofactor. *Sulfurospirillum* strain SCADC encodes an anaerobic carbon monoxide dehydrogenase (CooS), clustered with genes encoding a putative hydrogen-evolving hydrogenase. This complex might function as a catabolic enzyme similar to that described for *Carboxydothermus hydrogenoformans* (Soboh et al. 2002). Since this physiological trait was described for *S. carboxydovorans*, it is feasible that *Sulfurospirillum* strain SCADC is also able to oxidize CO and use it for energy conservation. Genes encoding enzymes usually involved in the reductive acetyl-CoA pathway (Wood-Ljungdahl pathway) were not found in the genome of *Sulfurospirillum* species SCADC in a genome-wide analysis conducted in our laboratory (unpublished results), despite being mentioned in the corresponding genome announcement (Tan and Foght 2014). Carbon dioxide fixation is not a physiological feature found in *Sulfurospirillum* spp. (see 10.2.3). Therefore, genes for an autotrophic lifestyle are not found in general in this taxon. A gene encoding the ATP citrate lyase is found solely in *Sulfurospirillum multivorans* (Goris et al. 2014). This enzyme was reported to be involved in CO_2 fixation via the reductive TCA cycle in other ε-proteobacteria (Hügler et al. 2005). While the other key enzymes necessary for a reductive TCA cycle (2-oxoglutarate:ferredoxin oxidoreductase and malate:quinone oxidoreductase) are in general found in the ε-proteobacterial version of the TCA cycle, ATP citrate lyase is usually found only in some autotrophic, mainly marine ε-proteobacteria and *S. multivorans*. In the latter organism, a reductive TCA cycle does not seem to be operative, since all attempts to grow the bacterium autotrophically failed so far.

Recently (January to March 2015), five additional *Sulfurospirillum* spp. genomes became available through the Genbank database. Three of them belong to *S. cavolei* strains based on 16S rRNA gene comparison and genomic distance calculation: a draft genome of *S. cavolei* NBRC 109482 from the Japanese NBRC culture collection (Genbank accession no. BBQE00000000), a draft genome reconstructed from a metagenome of a microbial electrosynthetic system (*S. cavolei*

MES (Ross et al. 2015)) and a complete sequence of *S. cavolei* UCH003 (Genbank accession AP014724) isolated from trichloroethene-contaminated groundwater in Japan. An additional *Sulfurospirillum* species was isolated from this contaminated site, designated *Sulfurospirillum* UCH001 (based on 16S rRNA sequence similarity closely related to *Sulfurospirillum* sp. NP4 and *Sulfurospirillum* sp. 11S05485-2). The genome is completely sequenced (Genbank accession AP014723). Additionally, the genome of *S. arsenophilum* strain NBRC 109478 from the NBRC culture collection was sequenced and is available as a draft genome (Genbank accession BBQF00000000). All five genomes are midsized compared to the other *Sulfurospirillum* spp. genomes (around 2.6–2.8 Mbp) and have in general very similar features as the genomes described above. None of the newly sequenced genomes contain any reductive dehalogenase genes or sequence similarities to the OHR region, which is surprising for the *Sulfurospirillum* spp. isolated from trichloroethene-contaminated sites. In addition, none of these five genome sequences contain the genetic information for producing corrinoids.

10.3.5 The OHR Region

Comparison of *Sulfurospirillum* spp. genomes revealed the genomic region which encodes the complete organohalide respiratory apparatus, including gene products responsible for maturation and regulation. One region, located roughly opposite of the origin of replication in the S. multivorans genome, is found in no other non-dechlorinating *Sulfurospirillum* and harbors the PCE reductive dehalogenase gene. Therefore, this region is termed organohalide respiration (OHR) region (Fig. 10.3). The same region is found at a similar position in the as yet unpublished genomes of the organohalide-respiring *S. halorespirans* and *Sulfurospirillum* species SL2. In all of these genomes, the cluster seems to be nearly 100 % conserved in terms of synteny and nucleotide sequence identity. In the *Sulfurospirillum* genomes

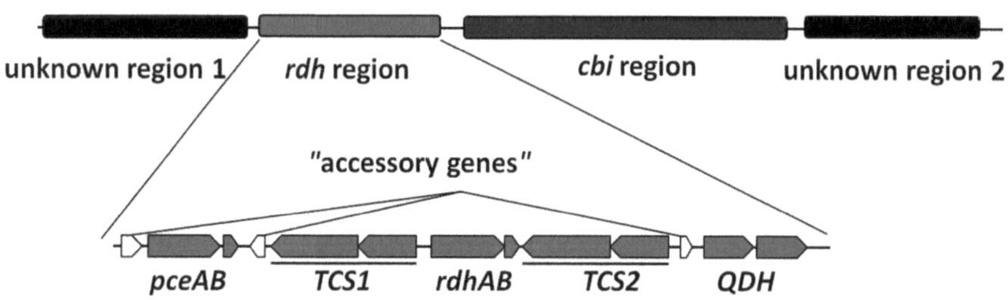

Fig. 10.3 Schematic overview of the complete OHR region of *Sulfurospirillum* spp. The *rdh* region with the quinol dehydrogenase genes downstream is shown in more detail at the *bottom*. *cbi*: corrinoid biosynthesis, TCS: two-component regulatory system, QDH: quinol dehydrogenase genes

containing the OHR region, genes encoding for the PCE reductive dehalogenase cluster together with genes encoding a two-component regulatory system, a second reductive dehalogenase gene cluster, a putative quinol dehydrogenase and the corrinoid biosynthesis gene cluster. Additionally, several genes are found which might play an accessory role in PCE respiration (Goris et al. 2014, Goris et al. 2015). These are (1) a gene encoding a peroxidase family protein, (2) an IscU/NifU-like protein, and (3) a small membrane protein (12 kDa, 3 predicted TM helices) not bearing similarities to the PceB protein. All of these proteins are not encoded in any of the OHRB outside the *Sulfurospirillum* genus. Furthermore, the OHR region of *Sulfurospirillum* spp. is unique among OHRB in that it is unusually large (around 50 kb) and that the functionally related genes of maturation and PCE respiration are colocalized. The putative quinol dehydrogenase gene products are similar to the NapGH/NosGH-type quinol dehydrogenases involved in nitrate and nitrous oxide reduction in *Wolinella succinogenes* (Kern and Simon 2008). They are composed of one periplasmic subunit, harboring four [4Fe–4S] cluster binding motifs and one membrane integral subunit, which contains two [4Fe–4S] cluster motifs. This is the first time that the involvement of a quinol dehydrogenase is described for an organohalide respiratory chain (Goris et al. 2014). While dedicated quinol dehydrogenase genes are not found in any other organohalide-respiring bacteria, PceC/RdhC of *Desulfitobacterium* and *Dehalobacter* spp. bear restricted similarities of the N-terminus of the gene product to the NapGH type quinol dehydrogenases.

The pceA gene products of Sulfurospirilla (non-maturated forms are 55–56 kDa in size) harbor a TAT signal and binding motifs for two [4Fe–4S] clusters and encode a periplasmic reductive dehalogenase. It is probably membrane-bound through its putative membrane anchor PceB, encoded by the gene PceB*pceA* (Neumann et al. 1998). is predicted to contain only two transmembrane helices, as opposed to PceB of *Desulfitobacterium*, *Dehalobacter*, and *Dehalococcoides mccartyi* which contains three transmembrane helices. Both genes, *pceA* and *pceB*, are phylogenetically clearly distinct from other reductive dehalogenase genes with an amino acid sequence identity of only 31 % for PceA and less than 33 % for PceB to sequences from other bacterial genera. An exception is an *rdhA* gene from *Desulfomonile tiedjei* with 39 % similarity to PceA. A second *rdh* cluster is found downstream of the *pceA* cluster with the same gene order in *S. multivorans*. Its deduced amino acid sequence has about 70 % identity to PceA and the function is not known. This second *rdhA* gene is 100% conserved in the other genomes of organohalide-respiring *Sulfurospirillum* spp., indicating an important, still unknown function.

Up- and downstream of the OHR region, several genes are located, which do not seem to play a role in organohalide respiration, although located on the OHR region. The genes do not have orthologues in non-OHR region containing genomes and they are conserved in OHR *Sulfurospirillum* species. A schematic overview of the arrangement of the OHR region is given in Fig. 10.3.

Acknowledgements This work was supported by the German Research Foundation (DFG), Research Unit FOR 1530.

References

Ahmann D, Roberts AL, Krumholz LR, Morel FM (1994) Microbe grows by reducing arsenic. Nat 371:750. doi:10.1038/371750a0

Aranda C, Paredes J, Valenzuela C, Lam P, Guillou L (2010) 16S rRNA gene-based molecular analysis of mat-forming and accompanying bacteria covering organically-enriched marine sediments underlying a salmon farm in Southern Chile (Calbuco Island). Gayana 74:125–135

Ballerstedt H et al (2004) Properties of a trichlorodibenzo-p-dioxin-dechlorinating mixed culture with a *Dehalococcoides* as putative dechlorinating species. FEMS Microbiol Ecol 47:223–234. doi:10.1016/S0168-6496(03)00282-4

Bommer M, Kunze C, Fesseler J, Schubert T, Diekert G, Dobbek H (2014) Structural basis for organohalide respiration. Science 346:455–458. doi:10.1126/science.1258118

Bunge M et al (2007) Biological activity in a heavily organohalogen-contaminated river sediment. Environ Sci Pollut Res 14:3–10. doi:10.1065/espr2006.03.298

Buttet G, Holliger C, Maillard J (2013) Functional genotyping of *Sulfurospirillum* spp. in mixed cultures allowed the identification of a new tetrachloroethene reductive dehalogenase. Appl Environ Microbiol 79:6941–6947. doi:10.1128/AEM.02312-13

Campbell B, Jeanthon C, Kostka J, Luther G, Cary S (2001) Growth and phylogenetic properties of novel bacteria belonging to the epsilon subdivision of the proteobacteria enriched from *Alvinella pompejana* and deep-sea hydrothermal vents. Appl Environ Microbiol 67:4566–4572. doi:10.1128/AEM.67.10.4566-4572.2001

Chen C, Ren N, Wang A, Yu Z, Lee DJ (2008) Microbial community of granules in expanded granular sludge bed reactor for simultaneous biological removal of sulfate, nitrate and lactate. Appl Microbiol Biotechnol 79:1071–1077. doi:10.1007/s00253-008-1503-5

Cichocka D, Nikolausz M, Haest PJ, Nijenhuis I (2010) Tetrachloroethene conversion to ethene by a *Dehalococcoides*-containing enrichment culture from Bitterfeld. FEMS Microbiol Ecol 72:297–310. doi:10.1111/j.1574-6941.2010.00845.x

Dong Y, Butler EC, Philp RP, Krumholz LR (2011) Impacts of microbial community composition on isotope fractionation during reductive dechlorination of tetrachloroethylene. Biodegradation 22:431–444. doi:10.1007/s10532-010-9416-2

Dugat-Bony E et al (2011) HiSpOD: probe design for functional DNA microarrays. Bioinformatics 27:641–648. doi:10.1093/bioinformatics/btq712

Duhamel M, Edwards EA (2006) Microbial composition of chlorinated ethene-degrading cultures dominated by *Dehalococcoides*. FEMS Microbiol Ecol 58:538–549. doi:10.1111/j.1574-6941.2006.00191.x

Einsle O et al (1999) Structure of cytochrome c nitrite reductase. Nature 400:476–480. doi:10.1038/22802

Eisenbeis M, BauerKreisel P, ScholzMuramatsu H (1997) Studies on the dechlorination of tetrachloroethene to *cis*-1,2-dichloroethene by *Dehalospirillum multivorans* in biofilms. Water Sci Technol 36:191–198. doi:10.1016/S0273-1223(97)00352-1

Eisenmann E, Beuerle J, Sulger K, Kroneck P, Schumacher W (1995) Lithotrophic growth of *Sulfurospirillum deleyianum* with sulfide as electron donor coupled to respiratory reduction of nitrate to ammonia. Arch Microbiol 164:180–185. doi:10.1007/BF02529969

Finster K, Liesack W, Tindall B (1997) *Sulfurospirillum arcachonense* sp. nov., a new-microaerophilic sulfur-reducing bacterium. Int J Syst Bacteriol 47:1212–1217

Gevertz D, Telang AJ, Voordouw G, Jenneman GE (2000) Isolation and characterization of strains CVO and FWKO B, two novel nitrate-reducing, sulfide-oxidizing bacteria isolated from oil field brine. Appl Environ Microbiol 66:2491–2501

Gilbert MJ et al (2014) Occurrence, diversity, and host association of intestinal *Campylobacter*, *Arcobacter*, and *Helicobacter* in reptiles. PLoS ONE 9:e101599. doi:10.1371/journal.pone.0101599

Goris T et al (2014) Insights into organohalide respiration and the versatile catabolism of *Sulfurospirillum multivorans* gained from comparative genomics and physiological studies. Environ Microbiol 16:3562–3580 doi:10.1111/1462-2920.12589

Goris T et al (2015) Proteomics of the organohalide-respiring Epsilonproteobacterium *Sulfurospirillum multivorans* adapted to tetrachloroethene and other energy substrates. Sci Rep 5:13794. doi:10.1038/srep13794

Grabowski A, Nercessian O, Fayolle F, Blanchet D, Jeanthon C (2005) Microbial diversity in production waters of a low-temperature biodegraded oil reservoir. FEMS Microbiol Ecol 54:427–443. doi:10.1016/j.femsec.2005.05.007

Grigoryan AA et al (2008) Competitive oxidation of volatile fatty acids by sulfate- and nitrate-reducing bacteria from an oil field in Argentina. Appl Environ Microbiol 74:4324–4335. doi:10.1128/AEM.00419-08

Heimann AC et al (2007) Hydrogen thresholds and steady-state concentrations associated with microbial arsenate respiration. Environ Sci Technol 41:2311–2317

Hery M, Gault A, Rowland H, Lear G, Polya D, Lloyd J (2008) Molecular and cultivation-dependent analysis of metal-reducing bacteria implicated in arsenic mobilisation in south-east asian aquifers. Appl Geochem 23:3215–3223. doi:10.1016/j.apgeochem.2008.07.003

Hubert C, Voordouw G (2007) Oil field souring control by nitrate-reducing *Sulfurospirillum* spp. that outcompete sulfate-reducing bacteria for organic electron donors. Appl Environ Microbiol 73:2644–2652. doi:10.1128/AEM.02332-06

Hubert C, Nemati M, Jenneman G, Voordouw G (2003) Containment of biogenic sulfide production in continuous up-flow packed-bed bioreactors with nitrate or nitrite. Biotechnol Prog 19:338–345. doi:10.1021/bp020128f

Hubert CR et al (2012) Massive dominance of Epsilonproteobacteria in formation waters from a Canadian oil sands reservoir containing severely biodegraded oil. Environ Microbiol 14:387–404. doi:10.1111/j.1462-2920.2011.02521.x

Hug LA et al (2013) Overview of organohalide-respiring bacteria and a proposal for a classification system for reductive dehalogenases. Philos Trans R Soc Lond B Biol Sci 368:20120322. doi:10.1098/rstb.2012.0322

Hügler M, Wirsen CO, Fuchs G, Taylor CD, Sievert SM (2005) Evidence for autotrophic CO2 fixation via the reductive tricarboxylic acid cycle by members of the epsilon subdivision of proteobacteria. J Bacteriol 187:3020–3027. doi:10.1128/JB.187.9.3020-3027.2005

Jensen A, Finster K (2005) Isolation and characterization of *Sulfurospirillum carboxydovorans* sp nov., a new microaerophilic carbon monoxide oxidizing epsilon proteobacterium. Antonie Van Leeuwenhoek Int J Gen Mol Microbiol 87:339–353. doi:10.1007/s10482-004-6839-y

John M, Schmitz R, Westermann M, Richter W, Diekert G (2006) Growth substrate dependent localization of tetrachloroethene reductive dehalogenase in *Sulfurospirillum multivorans*. Arch Microbiol 186:99–106. doi:10.1007/s00203-006-0125-5

John M, Rubick R, Schmitz RP, Rakoczy J, Schubert T, Diekert G (2009) Retentive memory of bacteria: long-term regulation of dehalorespiration in *Sulfurospirillum multivorans*. J Bacteriol 191:1650–1655. doi:10.1128/JB.00597-08

Kaufhold T, Schmidt M, Cichocka D, Nikolausz M, Nijenhuis I (2013) Dehalogenation of diverse halogenated substrates by a highly enriched *Dehalococcoides*-containing culture derived from the contaminated mega-site in Bitterfeld. FEMS Microbiol Ecol 83:176–188. doi:10.1111/j.1574-6941.2012.01462.x

Kern M, Simon J (2008) Characterization of the NapGH quinol dehydrogenase complex involved in *Wolinella succinogenes* nitrate respiration. Mol Microbiol 69:1137–1152. doi:10.1111/j.1365-2958.2008.06361.x

Keyser M, Britz T, Witthuhn R (2007) Fingerprinting and identification of bacteria present in UASB granules used to treat winery, brewery, distillery or peach-lye canning wastewater. S Afr Enol Vitic 28:69–79

Kimoto H et al (2010) Cloning of a novel dehalogenase from environmental DNA. Biosci Biotechnol Biochem 74:1290–1292. doi:10.1271/bbb.100027

Kodama Y, Ha L, Watanabe K (2007) *Sulfurospirillum cavolei* sp nov, a facultatively anaerobic sulfur-reducing bacterium isolated from an underground crude oil storage cavity. Int J Syst Evol Microbiol 57:827–831. doi:10.1099/ijs.0.64823-0

Kräutler B et al (2003) The cofactor of tetrachloroethene reductive dehalogenase of *Dehalospirillum multivorans* is norpseudo-B(12), a new type of a natural corrinoid. Helv Chim Acta 86:3698–3716

Laanbroek H, Kingma W, Veldkamp H (1977) Isolation of an aspartate-fermenting, free-living *Campylobacter* species. FEMS Microbiol Lett 1:99–102. doi:10.1016/0378-1097(77)90010-6

Lazaro C, Vich D, Hirasawa J, Varesche M (2012) Hydrogen production and consumption of organic acids by a phototrophic microbial consortium. Int J Hydrogen Energy 37:11691–11700. doi:10.1016/j.ijhydene.2012.05.088

Lear G, Song B, Gault AG, Polya DA, Lloyd JR (2007) Molecular analysis of arsenate-reducing bacteria within Cambodian sediments following amendment with acetate. Appl Environ Microbiol 73:1041–1048. doi:10.1128/AEM.01654-06

Li YH, Zhu JN, Zhai ZH, Zhang Q (2010) Endophytic bacterial diversity in roots of *Phragmites australis* in constructed Beijing Cuihu Wetland (China). FEMS Microbiol Lett 309:84–93. doi:10.1111/j.1574-6968.2010.02015.x

Li YH, Liu QF, Liu Y, Zhu JN, Zhang Q (2011) Endophytic bacterial diversity in roots of *Typha angustifolia L.* in the constructed Beijing Cuihu Wetland (China). Res Microbiol 162:124–131. doi:10.1016/j.resmic.2010.09.021

Li Z, Suzuki D, Zhang C, Yoshida N, Yang S, Katayama A (2013) Involvement of Dehalobacter strains in the anaerobic dechlorination of 2,4,6-trichlorophenol. J Biosci Bioeng 116:602–609. doi:10.1016/j.jbiosc.2013.05.009

Li Z et al (2014) Anaerobic 4-chlorophenol mineralization in an enriched culture under iron-reducing conditions. J Biosci Bioeng. doi:10.1016/j.jbiosc.2014.04.007

Liu L, Tsyganova O, Lee D, Chang J, Wang A, Ren N (2013) Double-chamber microbial fuel cells started up under room and low temperatures. Int J Hydrogen Energy 38:15574–15579. doi:10.1016/j.ijhydene.2013.02.090

Lohmayer R, Kappler A, Lösekann-Behrens T, Planer-Friedrich B (2014) Sulfur species as redox partners and electron shuttles for ferrihydrite reduction by *Sulfurospirillum deleyianum*. Appl Environ Microbiol 80:3141–3149. doi:10.1128/AEM.04220-13

Luijten ML et al (2003) Description of *Sulfurospirillum halorespirans* sp. nov., an anaerobic, tetrachloroethene-respiring bacterium, and transfer of *Dehalospirillum multivorans* to the genus *Sulfurospirillum* as *Sulfurospirillum multivorans* comb. nov. Int J Syst Evol Microbiol 53:787–793

Luijten M et al (2004a) Anaerobic reduction and oxidation of quinone moieties and the reduction of oxidized metals by halorespiring and related organisms. FEMS Microbiol Ecol 49:145–150. doi:10.1016/j.femsec.2004.01.015

Luijten ML, Roelofsen W, Langenhoff AA, Schraa G, Stams AJ (2004b) Hydrogen threshold concentrations in pure cultures of halorespiring bacteria and at a site polluted with chlorinated ethenes. Environ Microbiol 6:646–650. doi:10.1111/j.1462-2920.2004.00608.x

Macbeth TW, Cummings DE, Spring S, Petzke LM, Sorenson KS (2004) Molecular characterization of a dechlorinating community resulting from in situ biostimulation in a trichloroethene-contaminated deep, fractured basalt aquifer and comparison to a derivative laboratory culture. Appl Environ Microbiol 70:7329–7341. doi:10.1128/AEM.70.12.7329-7341.2004

MacRae J, Lavine I, McCaffery K, Ricupero K (2007) Isolation and characterization of NP4, arsenate-reducing *Sulfurospirillum*, from Maine groundwater. J Environ Eng ASCE 133:81–88. doi:10.1061/(ASCE)0733-9372(2007)133:1(81)

Maillard J et al (2011) Reductive dechlorination of tetrachloroethene by a stepwise catalysis of different organohalide respiring bacteria and reductive dehalogenases. Biodegradation. doi:10.1007/s10532-011-9454-4

Mamais D et al (2007) Optimization of nitrogen removal and start-up of Psyttalia sewage treatment works. Environ Technol 28:129–136. doi:10.1080/09593332808618779

Marshall CW, Ross DE, Fichot EB, Norman RS, May HD (2013) Long-term operation of microbial electrosynthesis systems improves acetate production by autotrophic microbiomes. Environ Sci Technol 47:6023–6029. doi:10.1021/es400341b

Marteinsson VT et al (2013) Microbial communities in the subglacial waters of the Vatnajökull ice cap, Iceland. ISME J 7:427–437. doi:10.1038/ismej.2012.97

Meszaros E, Sipos R, Pal R, Romsics C, Marialigeti K (2013) Stimulation of trichloroethene biodegradation in anaerobic three-phase microcosms. Int Biodeterior Biodegrad 84:126–133. doi:10.1016/j.ibiod.2012.08.006

Middeldorp P et al (1998) Stimulation of reductive dechlorination for in situ bioremediation of a soil contaminated with chlorinated ethenes. Water Sci Technol 37:105–110. doi:10.1016/S0273-1223(98)00240-6

Miller E, Wohlfarth G, Diekert G (1996) Studies on tetrachloroethene respiration in *Dehalospirillum multivorans*. Arch Microbiol 166:379–387. doi:10.1007/BF01682983

Neumann A, Scholz-Muramatsu H, Diekert G (1994) Tetrachloroethene metabolism of *Dehalospirillum multivorans*. Arch Microbiol 162:295–301

Neumann A, Wohlfarth G, Diekert G (1998) Tetrachloroethene dehalogenase from *Dehalospirillum multivorans*: cloning, sequencing of the encoding genes, and expression of the *pceA* gene in *Escherichia coli*. J Bacteriol 180:4140–4145

Neumann A, Seibert A, Trescher T, Reinhardt S, Wohlfarth G, Diekert G (2002) Tetrachloroethene reductive dehalogenase of *Dehalospirillum multivorans*: substrate specificity of the native enzyme and its corrinoid cofactor. Arch Microbiol 177:420–426. doi:10.1007/s00203-002-0409-3

Oremland RS et al (1994) Isolation, growth, and metabolism of an obligately anaerobic, selenate-respiring bacterium, strain SES-3. Appl Environ Microbiol 60:3011–3019

Panagiotakis I, Mamais D, Pantazidou M, Rossetti S, Aulenta F, Tandoi V (2014) Predominance of *Dehalococcoides* in the presence of different sulfate concentrations. Water Air Soil Pollut 225:1–14. doi:10.1007/s11270-013-1785-9

Pietari J (2002) Characterization of tetrachloroethene dechlorinating cultures and isolation of a novel tetrachloroethene to cis-1,2-dichloroethene halorespiring bacterium. University of Washington, Washington

Regeard C, Maillard J, Holliger C (2004) Development of degenerate and specific PCR primers for the detection and isolation of known and putative chloroethene reductive dehalogenase genes. J Microbiol Methods 56:107–118

Ross DE, Marshall CW, May HD, Norman RS (2015) Draft genome sequence of *Sulfurospirillum* sp. strain MES, reconstructed from the metagenome of a microbial electrosynthesis system. Genome Announc 3(1):e01336-14. doi:10.1128/genomeA.01336-14

Rossetti S, Aulenta F, Majone M, Crocetti G, Tandoi V (2008) Structure analysis and performance of a microbial community from a contaminated aquifer involved in the complete reductive dechlorination of 1,1,2,2-tetrachloroethane to ethene. Biotechnol Bioeng 100(2):240–249. doi:10.1002/bit.21776

Scholz-Muramatsu H, Neumann A, Messmer M, Moore E, Dieker G (1995) Isolation and characterization of *Dehalospirillum multivorans* gen. nov., sp. nov., a tetrachloroethene-utilizing, strictly anaerobic bacterium. Arch Microbiol 163:48–56. doi:10.1007/BF00262203

Schumacher W, Kroneck P (1992) Anaerobic energy metabolism of the sulfur-reducing bacterium "Spirillum" 5175 during dissimilatory nitrate reduction to ammonia. Arch Microbiol 157:464–470. doi:10.1007/BF00249106

Schumacher W, Kroneck P, Pfennig N (1992) Comparative systematic study on "Spirillum" 5175, *Campylobacter* and *Wolinella* species. Arch Microbiol 158:287–293. doi:10.1007/BF00245247

Sercu B et al (2013) The influence of in situ chemical oxidation on microbial community composition in groundwater contaminated with chlorinated solvents. Microb Ecol 65:39–49. doi:10.1007/s00248-012-0092-0

Shartau SL et al (2010) Ammonium concentrations in produced waters from a mesothermic oil field subjected to nitrate injection decrease through formation of denitrifying biomass and anammox activity. Appl Environ Microbiol 76:4977–4987. doi:10.1128/AEM.00596-10

Siebert A, Neumann A, Schubert T, Diekert G (2002) A non-dechlorinating strain of *Dehalospirillum multivorans*: evidence for a key role of the corrinoid cofactor in the synthesis of an active tetrachloroethene dehalogenase. Arch Microbiol 178:443–449. doi:10.1007/s00203-002-0473-8

Sikorski J et al (2010) Complete genome sequence of *Sulfurospirillum deleyianum* type strain (5175). Stand Genomic Sci 2:149–157. doi:10.4056/sigs.671209

Soboh B, Linder D, Hedderich R (2002) Purification and catalytic properties of a CO-oxidizing: H2-evolving enzyme complex from *Carboxydothermus hydrogenoformans*. Eur J Biochem 269:5712–5721 3282 [pii]

Sorokin D, Tourova T, Muyzer G (2013) Isolation and characterization of two novel alkalitolerant sulfidogens from a Thiopaq bioreactor, *Desulfonatronum alkalitolerans* sp nov., and *Sulfurospirillum alkalitolerans* sp nov. Extremophiles 17:535–543. doi:10.1007/s00792-013-0538-4

Srinivas TN et al (2011) Comparison of bacterial diversity in proglacial soil from Kafni Glacier, Himalayan Mountain ranges, India, with the bacterial diversity of other glaciers in the world. Extremophiles 15:673–690. doi:10.1007/s00792-011-0398-8

Stolz JF, Gugliuzza T, Blum JS, Oremland R, Murillo FM (1997) Differential cytochrome content and reductase activity in *Geospirillum barnesii* strain SeS3. Arch Microbiol 167:1–5

Stolz J, Ellis D, Blum J, Ahmann D, Lovley D, Oremland R (1999) *Sulfurospirillum barnesii* sp nov and *Sulfurospirillum arsenophilum* sp nov., new members of the *Sulfurospirillum* clade of the epsilon proteobacteria. Int J Syst Bacteriol 49:1177–1180

Straub KL, Schink B (2004) Ferrihydrite-dependent growth of *Sulfurospirillum deleyianum* through electron transfer via sulfur cycling. Appl Environ Microbiol 70:5744–5749. doi:10.1128/AEM.70.10.5744-5749.2004

Suzuki D, Li Z, Cui X, Zhang C, Katayama A (2014) Reclassification of *Desulfobacterium anilini* as *Desulfatiglans anilini* comb. nov. within *Desulfatiglans* gen. nov., and description of a 4-chlorophenol-degrading sulfate-reducing bacterium, *Desulfatiglans parachlorophenolica* sp. nov. Int J Syst Evol Microbiol 64:3081–3086. doi:10.1099/ijs.0.064360-0

Takai K et al (2004) Geochemical and microbiological evidence for a hydrogen-based, hyperthermophilic subsurface lithoautotrophic microbial ecosystem (HyperSLiME) beneath an active deep-sea hydrothermal field. Extremophiles 8:269–282. doi:10.1007/s00792-004-0386-3

Tan B, Foght J (2014) Draft genome sequences of campylobacterales (epsilonproteobacteria) obtained from methanogenic oil sands tailings pond metagenomes. Genome Announc 2(5):e01034-14. doi:10.1128/genomeA.01034-14

Tomaras J, Sahl JW, Siegrist RL, Spear JR (2009) Microbial diversity of septic tank effluent and a soil biomat. Appl Environ Microbiol 75:3348–3351. doi:10.1128/AEM.00560-08

van der Zaan B, de Weert J, Rijnaarts H, de Vos WM, Smidt H, Gerritse J (2009) Degradation of 1,2-dichloroethane by microbial communities from river sediment at various redox conditions. Water Res 43:3207–3216. doi:10.1016/j.watres.2009.04.042

von Wintzingerode F, Schlötelburg C, Hauck R, Hegemann W, Göbel UB (2001) Development of primers for amplifying genes encoding CprA- and PceA-like reductive dehalogenases in anaerobic microbial consortia, dechlorinating trichlorobenzene and 1,2-dichloropropane. FEMS Microbiol Ecol 35:189–196

Wolfe RS, Penning N (1977) Reduction of sulfur by spirillum 5175 and syntrophism with Chlorobium. Appl Environ Microbiol 33:427–433

Ye L, Schilhabel A, Bartram S, Boland W, Diekert G (2010) Reductive dehalogenation of brominated ethenes by *Sulfurospirillum multivorans* and *Desulfitobacterium hafniense* PCE-S. Environ Microbiol 12:501–509. doi:10.1111/j.1462-2920.2009.02093.x

Zhang Y, Rodionov DA, Gelfand MS, Gladyshev VN (2009) Comparative genomic analyses of nickel, cobalt and vitamin B12 utilization. BMC Genom 10:78. doi:10.1186/1471-2164-10-78

Zhang C, Suzuki D, Li Z, Ye L, Katayama A (2012) Polyphasic characterization of two microbial consortia with wide dechlorination spectra for chlorophenols. J Biosci Bioeng 114:512–517. doi:10.1016/j.jbiosc.2012.05.025

Zhang C, Li Z, Suzuki D, Ye L, Yoshida N, Katayama A (2013) A humin-dependent Dehalobacter species is involved in reductive debromination of tetrabromobisphenol A. Chemosphere 92:1343–1348. doi:10.1016/j.chemosphere.2013.05.051

Chapter 11
Organohalide-Respiring *Deltaproteobacteria*

Robert A. Sanford, Janamejaya Chowdhary and Frank E. Löffler

Abstract Organohalide respiration was first discovered in the deltaproteobacterium *Desulfomonile tiedjei*, which used 3-chlorobenzoate as the respiratory electron acceptor. Since this breakthrough discovery, the organohalide-respiring phenotype was demonstrated in 6 out of the 21 currently published families of the class *Deltaproteobacteria*. A survey of 208 available deltaproteobacterial genome sequences identified putative reductive dehalogenase genes in about 10 % of the genomes, suggesting that the ability to perform reductive dechlorination is not rare among the *Deltaproteobacteria*. For example, free-living *Geobacter lovleyi* strains dechlorinate the priority pollutants tetrachloroethene and trichloroethene in freshwater aquifers whereas the sponge-associated species *Desulfoluna spongiiphila* uses bromo- and iodophenols as electron acceptors in marine environments. Organohalide-respiring *Deltaproteobacteria* inhabit diverse habitats where they fulfill key functions in the global cycling of halogens, and have relevant roles in bioremediation applications.

R.A. Sanford
Department of Geology, University of Illinois at Urbana-Champaign, Urbana, IL 61801, USA

J. Chowdhary · F.E. Löffler
Biosciences Division, Oak Ridge National Laboratory, Oak Ridge, TN 37831, USA

F.E. Löffler
Joint Institute for Biological Sciences (JIBS), University of Tennessee and Oak Ridge National Laboratory (UT-ORNL), Oak Ridge, TN 37831, USA

F.E. Löffler (✉)
Center for Environmental Biotechnology, Department of Microbiology, Department of Civil and Environmental Engineering, University of Tennessee, Knoxville, TN 37996, USA
e-mail: frank.loeffler@utk.edu

L. Adrian and F.E. Löffler (eds.), *Organohalide-Respiring Bacteria*,
DOI 10.1007/978-3-662-49875-0_11

11.1 Introduction

Organohalide respiration (respiratory reductive dehalogenation) supports growth in a number of bacterial phyla. Within the phylum *Proteobacteria*, the class *Deltaproteobacteria* comprises the greatest diversity of organohalide-respiring taxa, which occur in aquatic, marine, subsurface groundwater, and soil ecosystems. Organohalide-respiring *Deltaproteobacteria* use a variety of chlorinated compounds as electron acceptors, including chlorinated phenols, chlorinated benzoates, and chlorinated ethenes. Some brominated and iodinated compounds can also be reductively dehalogenated but growth with brominated and iodinated electron acceptors has received much less attention. Based on thermodynamic considerations, fluorinated hydrocarbons could also support organohalide respiration (Parsons et al. 2008) but to date no evidence exists that the reductive cleavage of a carbon–fluorine bond can be linked with energy conservation. In the presence of suitable chloroorganic compounds, organohalide respiration leads to growth yields that exceed those seen with sulfate reduction and iron reduction (Bethke et al. 2011), indicating that organohalide-respiring bacteria compete with non-dechlorinators for electron donors in anoxic environments.

Although much attention currently focuses on organohalide-respiring *Chloroflexi*, in particular *Dehalococcoides mccartyi* (Löffler et al. 2013), the *Deltaproteobacteria* deserve some special mention because the first organism demonstrated to respire with a chlorinated compound belongs to this class. Reductive dechlorination at the time (early 1980s) was viewed mainly as a co-metabolic process linked to CO_2/H_2 reductive acetogenesis and methanogenesis. The isolation of *Desulfomonile tiedjei* strain DCB-1 was a breakthrough discovery because it demonstrated that halogenated compounds, in this case 3-chlorobenzoate, can serve as electron acceptors and fuel microbial energy metabolism (Shelton and Tiedje 1984). Originally, strain DCB-1 was part of a consortium shown to produce methane, carbon dioxide, and inorganic chloride as final products from 3-chlorobenzoate (Suflita et al. 1982). Shelton and Tiedje (1984) demonstrated that this activity required a consortium containing strain DCB-1, which reductively dechlorinated 3-chlorobenzoate to benzoate, a benzoate-oxidizing bacterium that produced H_2, and two H_2-consuming methanogens. Subsequent experiments with strain DCB-1 confirmed the theoretical analysis that reductive dechlorination reactions are exergonic and chlorinated compounds act as respiratory electron acceptors (Dolfing and Tiedje 1987). This work set the stage for the many publications that followed since it represented a fundamental shift in the recognized compounds that could be used as respiratory terminal electron acceptors by bacteria.

Phylogenetically, dehalogenating taxa within the *Deltaproteobacteria* are very diverse as they occur in five orders and six families within this class (Fig. 11.1). Interestingly, a dichotomy exists within the *Deltaproteobacteria* in terms of halogenated electron acceptors utilized. Only members of the order *Desulfuromonadales* (i.e., *Geobacter*, *Desulfuromonas*) respire certain aliphatic chlorinated hydrocarbons, whereas the known organohalide-respiring taxa

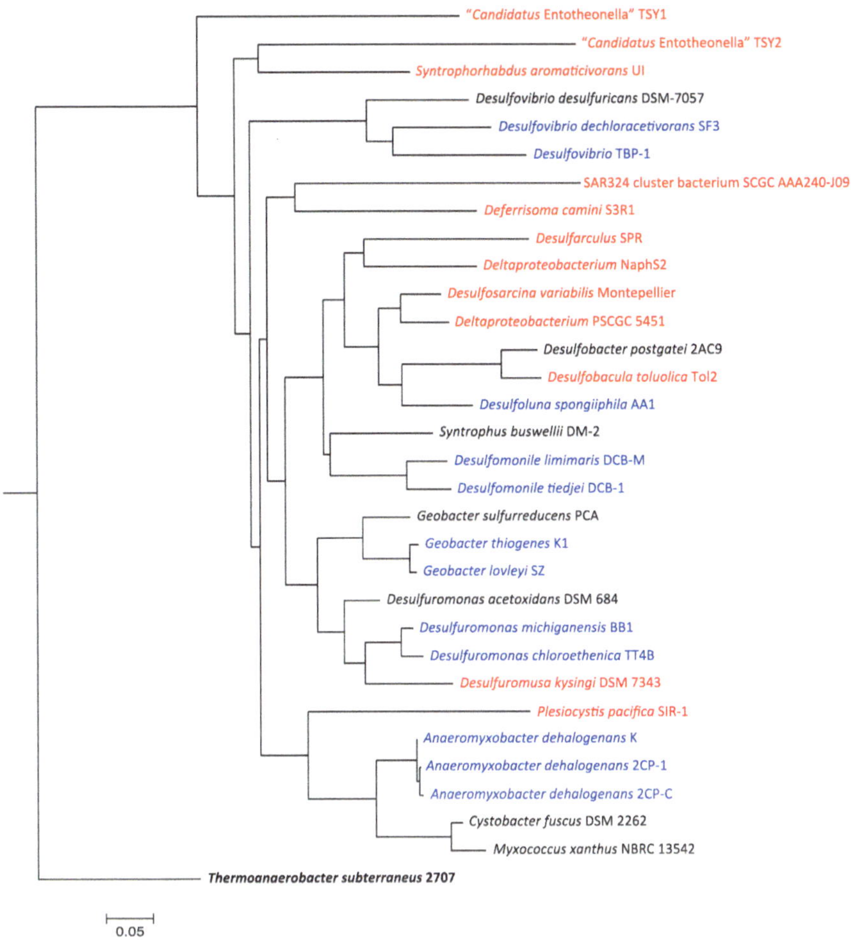

Fig. 11.1 16S rRNA gene sequence-based phylogenetic relationships of taxa within the class *Deltaproteobacteria* known to grow via organohalide respiration (*blue*), demonstrated to harbor putative reductive dehalogenase genes (but growth via organohalide respiration has not been demonstrated) (*red*), or do not possess reductive dehalogenation capabilities (*black*). Sequence alignments were performed with MUSCLE (Edgar 2004). The phylogenetic tree was constructed with RAxML (Stamatakis 2014) in Geneious (Kearse et al. 2012) and visualized with MEGA6 (Tamura et al. 2013)

of the orders *Myxococcales* (i.e., *Anaeromyxobacter*) *Desulfovibrionales* (i.e., *Desulfovibrio*), *Desulfobacterales* (i.e., *Desulfoluna*), and *Syntrophobacterales* (i.e., *Desulfomonile*) respire halogenated aromatic compounds (Table 11.1). The *Deltaproteobacteria* demonstrated to respire halogenated organic compounds are listed in Table 11.1. A shared feature of the organohalide-respiring *Deltaproteobacteria* is some degree of metabolic versatility and other (i.e., non-chlorinated) electron acceptors (e.g., nitrate, fumarate, sulfate, ferric iron) or fermentation substrates (e.g., pyruvate) support growth.

Table 11.1 *Deltaproteobacteria* isolates shown to utilize halogenated compound as electron acceptors

Organism (family)	Halogenated substrate(s)[e]	Electron donor(s)	Source	References
Anaeromyxobacter dehalogenans[a] (*Cystobacteraceae*)	2-CP; 2,6-DCP; 2,5-DCP; 2-BrP	Acetate, H_2, succinate, pyruvate, formate	Soil, compost, aquifers, river sediments	Sanford et al. (2002)
Desulfoluna spongiiphila strain AA1 (*Desulfobacteraceae*)	2-BrP; 4-BrP; 2,4-DBrP; 2,6-DBrP; 2,4,6-TBrP; 2-IP; 3-IP; 2-Br-4-FP; 3,5-DBr-4-OHBA	Propionate, 1 actate, pyruvate, succinate, citrate, benzoate, glucose	Marine sponge	Ahn et al. (2009)
Desulfomonile tiedjei strain DCB-1 (*Syntrophaceae*)	3-CBA; 3,5-DCBA	Pyruvate, acetate, benzoate	Sewage sludge	DeWeerd et al. (1990)
Desulfomonile limimaris[b] (*Syntrophaceae*)	3-CBA; 3-BrBA; 2,3-DBrBA; 2,5-DBrBA; 3,5-DCBA	Lactate, pyruvate, formate, H_2	Marine sediments	Sun et al. (2001)
Desulfovibrio dechloroacetivorans strain SF3 (*Desulfovibrionaceae*)	2-CP; 2,6-DCP	Acetate, lactate, pyruvate, propionate	Marine sediment	Sun et al. (2000)
Desulfovibrio sp. strain TBP-1 (*Desulfovibrionaceae*)	2-BrP; 4-BrP; 2,4-DBrP; 2,6-DBrP; 2,4,6-TBrP	Lactate, pyruvate, formate, H_2	Estuarine sediment	Boyle et al. (1999)
Desulfuromonas chloroethenica (*Desulfuromonadaceae*)	PCE; TCE	Acetate, pyruvate	Freshwater sediments	Krumholz (1997)
Desulfuromonas michiganensis[c] (*Desulfuromonadaceae*)	PCE; TCE	Acetate, fumarate, malate, succinate, pyruvate, lactate	River sediment, contaminated aquifer	Sung et al. (2003)
Geobacter lovleyi (*Geobacteriaceae*)	PCE; TCE	Acetate, pyruvate, H_2	Contaminated freshwater sediment, aquifer	Sung et al. (2006), Wagner et al. (2012)
Geobacter thiogenes strain K1[d] (*Geobacteriaceae*)	TCAc	Acetate, acetoin	Contaminated soil	De Wever et al. (2000)

[a]Strains 2CP-1, 2CP-C, K, R, DCP-18, and FRCW
[b]Strains DCB-M and DCB-F
[c]Strains BB1 and BRS1
[d]Formerly named "Trichlorobacter thiogenes"
[e]Halogenated substrates: 2-CP, 2-chlorophenol; 2,6-DCP, 2,6-dichlorophenol; 2,5-DCP, 2,5-dichlorophenol; 2-BrP, 2-bromophenol; 3-BrP, 3-bromophenol; 4-BrP, 4-bromophenol; 2,4-DBrP, 2,4-dibromophenol; 2,6-DBrP, 2,6-dibromophenol; 2,4,6-TBrP, 2,4,6-tribromophenol; 2-IP, 2-iodophenol; 3-IP, 3-iodophenol; 3-CBA, 3-chlorobenzoate; 3,5-CBA, 3,5-dichlorobenzoate; 3-BrBA, 3-bromobenzoate; 2,3-DBrBA, 2,3-dibromobenzoate; 2,5-DBrBA, 2,5-dibromobenzoate; 3,5-DBr-4-OHBA, 3,5-dibromo-4-hydroxybenzoate; 2-Br-4-FP, 2-bromo-4-fluorophenol; PCE, tetrachloroethene; TCE, trichloroethene; TCAc, trichloroacetate
NA—not available

11.2 Characteristics of Studied Organohalide-Respiring *Deltaproteobacteria*

This section provides details about the isolation and physiology of organohalide-respiring taxa within the class *Deltaproteobacteria.*

11.2.1 Anaeromyxobacter dehalogenans

This bacterial species represents the first member of the *Myxococcales* that was shown to grow via any type of anaerobic metabolism (Cole et al. 1994; Sanford et al. 2002). Although the initial isolates were obtained based on their ability to respire ortho-substituted halophenols, subsequent studies demonstrated that strains of this species share versatile metabolisms. In addition to respiratory reductive dehalogenation, *Anaeromyxobacter dehalogenans* strains respire oxygen, nitrate, nitrite, nitrous oxide, and fumarate, as well as oxidized metal species including ferric iron, MnO_2, and hexavalent uranium. Michigan stream sediment and tropical surface soil served as the sources for isolating the first *A. dehalogenans* isolates, strains 2CP-1 and 2CP-C, respectively (Cole et al. 1994; Sanford et al. 2002). Additional dehalogenating isolates were obtained from compost (strains 2CP-3 and 2CP-5) (Sanford et al. 2002), Michigan pond sediment (stain 2CP-2), Korean river sediment (strain K) (Genome accession number: CP001131), agricultural soils (strains R and DCP-18) (unpublished, Sanford et al.), and uranium-contaminated groundwater (strain FRCW) (Thomas et al. 2010). Community structure analyses performed with a variety of environmental samples revealed that *Anaeromyxobacter*-related bacteria are common inhabitants of agricultural and unmanaged soils, freshwater sediments, and aquifers (Treude et al. 2003; Thomas et al. 2010; Kudo et al. 2013). Phylogenetically, all dehalogenating isolates share >99.8 % sequence identity for their respective 16S rRNA genes. *Anaeromyxobacter* sp. strain FAc12 shares 99.5 % 16S rRNA gene sequence similarity to *A. dehalogenans* strain 2CP-1, although the ability of strain FAc12 to respire chlorophenols has not been demonstrated (Treude et al. 2003). Genomes from four dehalogenating strains have been sequenced (strains 2CP-1, 2CP-C, K and FRCW). Not surprisingly, many genomic similarities existed, but a few features clearly distinguished individual strains. For example, only strain K has the genetic potential to use sulfate as a source of sulfur for growth, and growth experiments confirmed this ability (Sanford, unpublished data).

A. dehalogenans strains use different reduced carbon compounds, including acetate, as electron donors for growth when reducing a number of electron acceptors, including chlorinated organic compounds. This metabolic versatility allowed direct comparisons of growth yields with different electron acceptors. *A. dehalogenans* strains partition about 65 % of the electrons available from acetate oxidation for respiratory reductive dechlorination with chlorinated phenols and energy generation ($f_e = 0.65$), and 45 % of the electron are directed toward biomass synthesis ($f_s = 0.35$) (Löffler et al. 1999). Growth yields with 2-chlorophenol as the electron acceptor were 2.9 g of cells (dry weight) per mole 2-CP dechlorinated to phenol (Table 11.2) (Cole et al. 1994). Comparative growth studies showed that different *A. dehalogenans* strains attained similar growth yields with nitrate and chlorophenol respiration (Sanford et al. 2002), while less biomass was produced with ferric iron (as ferric citrate) as an electron acceptor ($f_s = 0.20$) (He and

Table 11.2 Comparison of cell yields and f_e values for *Deltaproteobacteria* isolates demonstrated to grow with halogenated substrates as electron acceptors

Organism	Halogenated substrate(s)	Yield (mg protein/mmol electron acceptor)	f_e^a	References
Anaeromyxobacter dehalogenans strain 2CP-1	2-CP	2.9	0.66	Sanford et al. (2002)
Desulfoluna spongiiphila strain AA1	2-BrP	NA	NA	Ahn et al. (2009)
Desulfomonile tiedjei strain DCB-1	3-CBA	1.9	0.75	Sun et al. (2001)
Desulfomonile limimaris strain DCB-M	3-CBA	1.7	0.77	Sun et al. (2001)
Desulfovibrio dechloroacetivorans strain SF3	2-CP	0.92[b]	0.86	Sun et al. (2000)
Desulfovibrio sp. strain TBP-1	2-BrP	1.42	0.80	Boyle et al. (1999)
Desulfuromonas chloroethenica strain TT4B	PCE, TCE	0.15	0.97	Krumholz (1997)
Desulfuromonas michiganensis strain BB1	PCE, TCE	0.39-2.42[b,c]	0.70-0.94[c]	Sung et al. (2003)
Geobacter lovleyi strain SZ	PCE, TCE	NA	NA	Sung et al. (2006)
Geobacter thiogenes strain K1	TCAc	0.50[b]	0.92	De Wever et al. (2000)

2-CP, 2-chlorophenol; 2-BrP, 2-bromophenol; 3-CBA, 3-chlorobenzoate; PCE, tetrachloroethene; TCE, trichloroethene; TCAc, trichloroacetate

[a] f_e signifies the fraction of electrons generated in electron donor oxidation and used for electron acceptor reduction for the purpose of energy conservation. Mathematically, f_e equals the moles of electrons used for electron acceptor reduction relative to the total moles of electron donor consumed. The f_e values were reported in the literature or, if no fe data were provided, calculated from yield data presented in the respective references.

[b] Values estimated from f_e data

[c] *Desulfuromonas michiganensis* strain BB1 uncouples growth from respiration once stationary phase has been reached. Depending on the growth state, the yield and f_e values will vary, which explain why ranges rather than exact values are provided. The low yield reflects the final measurement made in stationary phase cultures and the f_e values are correspondingly higher. The *lower* f_e value corresponds to the measured value for this organism during the exponential growth phase, which corresponds to the higher yield measurement

NA—not available

Sanford 2003). For growth to occur with chlorophenols, their concentration could not exceed 200 μM, but growth was not significantly inhibited by up to 8 mM phenol (the end product of reductive dechlorination) (He and Sanford 2004a, b).

11.2.2 *Desulfoluna spongiiphila* Strain AA1

Strain AA1 was isolated from the marine sponge *Aplysina aerophoba* collected in the Mediterranean Sea, and represents the only animal-associated organohalide-respiring bacterium in the *Deltaproteobacteria* isolated to date (Ahn et al. 2009). This organism primarily respires brominated phenols and is unable to use the chlorinated analogs as electron acceptors (Table 11.1). The preference for brominated compounds was reported to be due to the generation of brominated tyrosine derivatives by the host sponge, which have structural similarity to bromophenols (Ahn et al. 2003). Strain AA1 also is one of the few organisms shown to reduce iodophenols. *Desulfoluna spongiiphila* also uses sulfate, sulfite, and thiosulfate as electron acceptors for growth. Lactate, pyruvate, succinate, citrate, benzoate, and glucose serve as electron donors and are partially oxidized to acetate, a common feature among the sulfate-reducing bacteria.

11.2.3 *Desulfomonile tiedjei* Strain DCB-1

Strain DCB-1 (**De**Chlorinating **B**acterium **#1**) was isolated from an anaerobic consortium enriched from sewage sludge that exhibited complete mineralization of 3-chlorobenzoate. Strain DCB-1 is aptly named since it was the first such organism identified and thoroughly characterized (Suflita et al. 1982; Shelton and Tiedje 1984; DeWeerd et al. 1990; Mohn and Tiedje 1991). As the first isolate capable to fulfill its energy requirement from a single reductive dechlorination reaction, the organism was studied in detail. Free energy calculations explicitly showed that reductive dechlorination of 3-chlorobenzoate releases considerable energy, with a $\Delta G^{\circ\prime}$ of -125 kJ per dechlorination reaction (Dolfing and Tiedje 1987; Dolfing and Novak 2015). Subsequent efforts demonstrated the generation of a proton-motive force and ATP generation associated with the reductive dechlorination of 3-chlorobenzoate (Mohn and Tiedje 1991). This study unequivocally established that 3-chlorobenzoate reductive dehalogenation is a respiratory process and that most of the ATP generation was not associated with substrate-level phosphorylation. Growth yields are shown in Table 11.2 and are comparable to those observed with other organohalide-respiring organisms. Strain DCB-1 was not able to dehalogenate 3-fluorobenzoate but this compound (0.1 mM) induced the 3-chlorobenzoate reductive dehalogenase and facilitated biomass production for enzymatic studies (Cole et al. 1995; Ni et al. 1995).

D. tiedjei cells have a characteristic morphology and grow as long rods with a unique collar-like structure (Mohn et al. 1990). For axenic growth, this bacterium requires the vitamins thiamine, nicotinamide, and 1,4-naphthoquinone. The original culture medium contained rumen fluid, which apparently served as a source for these critical vitamins. Pyruvate, benzoate, and H_2 all serve as electron donors when sulfate or thiosulfate are supplied as electron acceptors. Organohalide respiration of meta-substituted chlorobenzoates occurs best when pyruvate is provided as an electron donor; however, acetate can also be used as a source of reducing equivalents. Pyruvate also supports fermentative growth with generation of acetate and lactate. Strain DCB-1 shows the ability to assimilate CO_2 into biomass and can also demethylate methyl-substituted benzoates for use as a carbon source (Stevens et al. 1988). Initial results suggested that strain DCB-1 was able to reduce nitrate (5 mM) to nitrite although no growth data were presented (Shelton and Tiedje 1984). A subsequent study with 20 mM nitrate showed no enhanced growth with pyruvate as an electron donor; however, no nitrate reduction data were shown and high nitrate concentrations were not inhibitory to pyruvate fermentation (Stevens et al. 1988). An analysis of the recently completed genome sequence for this organism suggests that it does possess respiratory nitrate reductase genes and would likely be able to use this substrate as an electron acceptor as originally reported. Interestingly, the genome also indicates that strain DCB-1 possesses a nitrous oxide reductase (*nosZ*) gene even though it apparently lacks the *nir* and *nor* gene clusters of the canonical denitrification pathway (Sanford et al. 2012). Spores have not been observed in old strain DCB-1 cultures; however, viable inocula could be obtained from liquid cultures grown in basal salts medium amended with 10 mM pyruvate and 1 mM 3-chlorobenzoate following a 14-year starvation period (Löffler, unpublished data).

11.2.4 *Desulfomonile limimaris* Strains DCB-M and DCB-F

Marine sediments served as the source material for 3-chlorobenzoate-dechlorinating enrichment cultures that eventually yielded additional *Desulfomonile* isolates (Sun et al. 2001). Although morphologically similar to strain DCB-1, possessing the same collar structure, the marine strains have key phylogenetic and physiological differences. The 16S rRNA genes of strain DCB-M and strain DCB-1 shared only 93 % similarity, suggesting that these two strains represent different species. Strain DCB-F shares 99 % sequence identity with the 16S rRNA gene of strain DCB-M suggesting these isolates belong to a single species. Physiological data support this phylogenetic relationship and both *Desulfomonile limimaris* strains require NaCl for growth and thrive at salt concentrations about half the strength of seawater. In contrast, elevated salt concentrations impair growth of strain DCB-1. Benzoate, lactate, formate, H_2, butyrate, propionate, and pyruvate all serve as electron donors; however, pyruvate fermentation alone does not sustain growth as was observed with strain DCB-1. In contrast to strain DCB-1, the *D. limimaris* strains cannot oxidize acetate coupled with the reduction of meta-substituted chlorobenzoates. Sulfate, sulfite, thiosulfate, nitrate (5 mM), fumarate, and halobenzoates all serve as electron acceptors for *D. limimaris*, with the best growth occurring with sulfite and thiosulfate. Organohalide respiration of 3-chlorobenzoate generates similar biomass yields in both strain DCB-M and strain DCB-1, and 1.7 g and 1.9 g protein per mol of benzoate produced were generated, respectively (Dolfing and Tiedje 1987; Sun et al. 2001).

11.2.5 *Desulfovibrio dechloroacetivorans* Strain SF3

This bacterium was enriched from marine sediments collected in San Francisco Bay using 2-chlorophenol as electron acceptor (Sun et al. 2000). Strain SF3 shows a relatively narrow range of haloorganic substrates and only uses ortho-substituted chlorophenols as an electron acceptors, similar to what has been observed for *A. dehalogenans*. As expected for members of the *Desulfovibrio* genus, strain SF3 has the typical sulfate-reducing phenotype and couples the incomplete oxidation of lactate and pyruvate to the reduction of sulfate, sulfite, and thiosulfate (Odum and Singleton 1993). Nitrate (5 mM) and fumarate also support growth with these electron donors, and are reduced to nitrite and succinate, respectively. Interestingly, additional electron donors including acetate, fumarate, propionate, alanine, and ethanol support organohalide respiration. The ability for this *Desulfovibrio* strain to completely oxidize acetate when growing with 2-chlorophenol is unusual and represents a deviation from the generally accepted physiological description of the *Desulfovibrio* genus (Voordouw 1995). Sulfate-reducing *Desulfovibrio* typically incompletely oxidize organic electron donors to acetate and CO_2. As a marine sediment isolate, strain SF3 requires NaCl between 0.016 and 2.5 % in the growth medium; however, KCl concentrations above 0.32 % are inhibitory (Sun et al. 2000). The reported reaction stoichiometry with acetate as an electron donor and 2-chlorophenol as an electron acceptor indicated that 86 % of the electrons were used for organohalide respiration (i.e., $f_e = 0.86$) (Table 11.2) (Sun et al. 2000). Thus, organohalide respiration with 2-chlorophenol as electron acceptor and acetate as the electron donor in strain SF3 is energetically less efficient than in *A. dehalogenans* strain 2CP-C.

11.2.6 *Desulfovibrio sp.* Strain TBP-1

Strain TBP-1 was obtained from estuary sediments taken from the Arthur Kill tidal strait near the New York/New Jersey harbor. The original anoxic enrichment cultures received 2,4,6-tribromophenol as the sole added electron acceptor (Boyle et al. 1999). With acetate plus H_2 added as carbon source and electron donor, respectively, isolate strain TBP-1 grew by using 2,4,6-tribromophenol reduction to phenol as a terminal electron-accepting process. As expected for a member of the *Desulfovibrio* genus, strain TBP-1 coupled the incomplete oxidation of organic electron donors such as lactate, pyruvate, and fumarate to sulfate reduction. In addition to sulfate, strain TBP-1 utilized sulfite, sulfur, and thiosulfate as electron acceptors. No growth was observed with nitrate (20 mM), fumarate, or acrylate as electron acceptors. For organohalide respiration, strain TBP-1 exclusively used ortho- and meta-substituted bromophenols, which were stoichiometrically reduced to phenol. Protein growth yields for organohalide respiration were somewhat lower than those reported for *Desulfomonile* strains with 1.42 g protein per mole 4-bromophenol reduced (Table 11.2). On a per electron basis, the cell protein yield for sulfate reduction was higher than that for reductive debromination (Boyle et al. 1999). This lower growth yield suggests that strain TBP-1 is not as energetically efficient as some of the other organohalide-respiring bacteria including members of the *Anaeromyxobacter* and *Desulfomonile* genera. Since strain TBP-1 is able to use H_2 as an electron donor, fermentative metabolism does not explain its ability to grow. It has been suggested that electron bifurcation may be involved with energy coupling with organohalide respiration when H_2 is an electron donor, which could explain why the apparent yield is lower for some bacteria when using haloorganic compounds as an electron acceptor (Buckel and Thauer 2013).

11.2.7 Desulfuromonas chloroethenica Strain TT4B

Strain TT4B was enriched and ultimately isolated from anoxic sediment collected from a small stream in Massachusetts that was contaminated with industrial solvents including trichloroethene (TCE) (Krumholz et al. 1996). This nonmotile organism uses tetrachloroethene (PCE) and TCE as electron acceptors for growth generating stoichiometric amounts of *cis*-1,2-dichloroethene (cDCE) (Krumholz 1997). No other chloroorganic compounds were dechlorinated by this strain. Fumarate, polysulfide and ferric iron (as nitrilotriacetate) also serve as respiratory electron acceptors with acetate or pyruvate provided as electron donors. Phylogenetically and physiologically, strain TT4B is very similar to *Desulfuromonas acetexigens* although this related organism has not been demonstrated to respire haloorganic compounds (Sung et al. 2003). Strain TT4B only produced 0.15 g of protein per mole chloride liberated with acetate as the electron donor, which is a much lower yield compared to other organohalide-respiring bacteria (Table 11.2). It is possible that this strain cannot efficiently conserve the free energy change associated with PCE-to-cDCE reductive dechlorination; however, an alternative explanation is that growth became uncoupled from respiration once stationary phase had been reached. This phenomenon was observed in *D. michiganensis* strain BB1 and is discussed below. Uncoupling of reductive dechlorination from energy conservation would lead to an underestimate of the true growth yield. *D. chloroethenica* strain TT4B was deposited in the American Type Culture Collection (accession no. ATCC 700295) but the strain was apparently lost and is no longer available.

11.2.8 *Desulfuromonas michiganensis* Strains BB1 and BRS1

Two PCE-to-cDCE dechlorinating isolates were obtained from enrichments amended with PCE as electron acceptor and acetate as electron donor. Strain BB1 was derived from pristine river sediment and strain BRS1 from PCE-impacted aquifer material at the Bachman Road site in Oscoda, Michigan (Sung et al. 2003). Phylogenetic analysis determined that both isolates belonged to the genus *Desulfuromonas*. Based on phenotypic differences (e.g., motility, substrate utilization) to the closest relative *Desulfuromonas chloroethenica* strain TT4B, the new isolates were grouped in the new species *Desulfuromonas michiganensis* (Sung et al. 2003). Lactate, acetate, pyruvate, succinate, malate, and fumarate served as electron donors for PCE-to-cDCE reductive dechlorination. Alternate electron acceptors included fumarate, malate, ferric citrate and sulfur. Strains BB1 and BRS1 require sulfide as a sulfur source for growth and neither strain can use H_2 as an electron donor. The addition of 1 mM sulfite to the medium completely inhibited dechlorination activity. As with *D. chloroethenica*, the energy efficiency of organohalide respiration as indicated by growth yield was low compared to other bacteria. Strain BB1 only generated 0.39 g of protein per mole of chloride; however, the organism uncouples growth from respiration once stationary phase of growth has been reached, which makes yield determination challenging (Table 11.2) (Sung et al. 2003). The fraction of electrons from the electron donor, f_e, used for energy generation ranges from 0.6 to 0.7 during exponential growth, which is similar to f_e values determined for other organohalide-respiring bacteria. In the stationary growth phase, f_e approached the theoretical maximum of 1.0 as the organism continued to dechlorinate PCE but no further growth occurred. This example demonstrates that yield data can vary considerably depending on the time of analysis (e.g., exponential growth phase versus stationary growth phase), and some caution is warranted when interpreting growth yield data obtained for organohalide-respiring bacteria. One additional point to make about strain BB1 is that a non-dechlorinating variant strain with identical phylogeny and phenotype, except for the ability to utilize PCE and TCE as electron acceptors, was obtained from strain BB1 cultures transferred repeatedly with fumarate as electron acceptor and in the absence of PCE (Sung et al. 2003). Apparently, the ability to respire PCE and TCE can be lost in the absence of the chlorinated electron acceptor(s) when rapid growth with an alternate electron acceptor (i.e., fumarate) occurs under laboratory conditions. This raises the question of stability of the reductive dechlorination trait in *D. michiganensis* under in situ conditions. Repeated microcosm setup with the same Pere Marquette River sediment materials, however, consistently yielded PCE-to-cDCE-dechlorinating enrichment cultures and *D. michiganensis* 16S rRNA gene-targeted PCR assays demonstrated the presence of this type of organism (Sung et al. 2003). These observations suggest that the PCE/TCE-to-cDCE reductive dechlorination trait is stable under in situ conditions; however, this observation serves as a cautionary notice for making any inferences about dechlorination potential or activity based on 16S rRNA gene-based analysis alone.

11.2.9 *Geobacter lovleyi* Strain SZ

Su-Zi Creek sediment in South Korea served as the source for the enrichment and isolation of this PCE-to-cDCE-dechlorinating organism (Sung et al. 2006). Members of the *Geobacter* genus are distributed in anoxic freshwater sediments and aquifers, and have versatile anaerobic metabolisms (Lovley 1993; Mahadevan et al. 2011). Similar to known *Geobacter* strains, *Geobacter lovleyi* used both soluble and insoluble ferric iron as an electron acceptor for growth. Nitrate (reduced to ammonium), fumarate, malate, elemental sulfur, hexavalent uranium, and MnO_2 also served as electron acceptors for strain SZ. Only PCE and TCE were used for organohalide respiration yielding stoichiometric amounts of cDCE and inorganic chloride as end products. Acetate served as an electron donor as did H_2 as long as small amounts of acetate or lactate were supplied as a source of carbon. H_2 consumption threshold concentrations were lowest with PCE provided as an electron acceptor compared to ferric iron (as ferric citrate) and nitrate. These H_2 consumption threshold measurements suggested that organohalide respiration in this organism was energetically more efficient than dissimilatory iron reduction or dissimilatory nitrate reduction to ammonium. When acetate threshold concentrations (i.e., the observed acetate concentration when respiration ceased in the presence of electron acceptor) were measured, a consistent residual concentration of 3.0 nM was observed with PCE, and similar acetate consumption thresholds were determined for ferric citrate and nitrate. These values were considerably lower than the 69 nM acetate threshold reported for 2-chlorophenol-respiring *A. dehalogenans* cells (He and Sanford 2004b), suggesting a possible energetic cost to using chlorophenols versus PCE as an electron acceptor. In contrast to H_2 consumption threshold concentrations, only a few studies have reported acetate threshold concentrations; however, the studies with *G. lovleyi* suggest that acetate utilization is also constrained by the same thermodynamic principles as H_2. Phylogenetically, the strain SZ 16S rRNA gene sequence is 98.4 % similar to that found in *Geobacter thiogenes*, another organohalide-respiring member of this genus (see following section). PCR assays specific for the *G. lovleyi* 16S rRNA gene have been designed and detected this type of dechlorinator in the commercial bioaugmentation culture KB-1 (Amos et al. 2007b). Further, *G. lovleyi* 16S rRNA gene sequences were detected in PCE/TCE-dechlorinating microcosms established with soil collected from the PCE/TCE-impacted East Gate Disposal Yard in Fort Lewis, WA (Amos et al. 2007a; Fletcher et al. 2011). Subsequent efforts identified *G. lovleyi* strain KB-1, a close relative of strain SZ, as the key PCE-to-cDCE dechlorinator in the consortium KB-1, and additional *G. lovleyi* strains were isolated from Fort Lewis microcosms (Wagner et al. 2012). Community structure analyses and the application of the *G. lovleyi* 16S rRNA gene-targeted PCR assay suggested that this type of organism is often found at sites impacted with chlorinated ethenes (Amos et al. 2007a; Wagner et al. 2012). Interestingly, strain SZ and strain KB-1 harbor plasmid pSZ77, which carries several genes required for de novo corrinoid biosynthesis. The reductive dehalogenase enzyme systems require coenzyme B_{12} and *G. lovleyi* growth with PCE in vitamin B_{12}-free medium demonstrated this organism's ability of de novo corrinoid biosynthesis (Yan et al. 2012). Attempts to cure the plasmid by repeated transfers in medium amended with vitamin B_{12} were not successful suggesting that this plasmid is stable in *G. lovleyi* (Wagner et al. 2012).

11.2.10 Geobacter thiogenes Strain K1 (Formerly "Trichlorobacter thiogenes")

Strain K1 was isolated from trichloroacetate (TCAc)-fed enrichment cultures seeded with contaminated subsurface soil collected in Michigan. The original isolate, designated "Trichlorobacter thiogenes", grew in reduced basal salts medium amended with acetate (2.5 mM) and TCAc (1 mM) (De Wever et al. 2000). This strain stoichiometrically reduced TCAc to dichloroacetic acid and required acetate concentrations above 0.1 mM to maintain this activity (Table 11.2). In the initial characterization, sulfur and fumarate were identified as alternative electron acceptors for growth, while no other halogenated compounds, including PCE, served as electron acceptors. Observations of elemental sulfur appearing in cultures grown with TCAc or fumarate led to the discovery that these electron-accepting processes (i.e., reductive dechlorination and fumarate reduction) were linked to sulfur cycling between sulfide (a reductant added to the medium) and sulfur. Apparently, sulfide acts as the direct electron donor for TCAc reductive dechlorination and is thereby oxidized to elemental sulfur (S°). Sulfur subsequently acts as the physiological electron acceptor for the organism with acetate as an electron donor. Therefore, strain K1 does not use TCAc as respiratory electron acceptor and is not an organohalide-respiring bacterium sensu stricto. Since sulfur reduction to sulfide does not provide as much energy as reductive dechlorination, the use of this cycle for TCAc reductive dechlorination may explain why higher acetate concentrations (>0.1 mM) than observed for acetate-oxidizing organohalide-respiring organisms were required to maintain the dechlorinating activity. Recently, this type of sulfur-mediated electron shuttling was implicated in bacterial iron reduction as well (Flynn et al. 2014).

Since its isolation, strain K1 has been reclassified as *Geobacter thiogenes* due to its phylogenetic and physiological similarities to other members of the *Geobacter* genus (Nevin et al. 2007). Strain K1 was shown to grow with ferric iron as an electron acceptor; however, sulfur-mediated electron shuttling was not demonstrated. Based on the recent findings exploring ferric iron reduction under alkaline conditions (Flynn et al. 2014), it would not be surprising if sulfur-mediated electron shuttling is involved in ferric iron reduction in strain K1 cultures.

11.3 Ecophysiology

Organohalide-respiring *Deltaproteobacteria* occur in a variety of habitats including contaminated and uncontaminated soil, freshwater sediments, aquifers, and marine/estuarine environments. The ability to gain energy from reductive dechlorination reactions confers an obvious advantage to organisms inhabiting contaminated environments. It is less obvious why organohalide-respiring *Deltaproteobacteria* were isolated from pristine environments not impacted by anthropogenic chloroorganic chemicals. Interestingly, in such uncontaminated habitats, natural halogenated compounds have been noted, which could explain the occurrence of the organohalide-respiring phenotype in pristine environments (Gribble 1994, 2000; Gordon 2004) (see Chap. 2). For example the sponge *Aplysina aerophoba* produces a brominated tyrosine derivative and has been shown to harbor reductively dehalogenating bacteria (Ahn et al. 2003). Fungi, which are abundant in soil habitats are known to produce substantial amounts of chlorinated metabolites that could potentially support organohalide respiration in the soil environment (de Jong et al. 1994; Ortiz-Bermudez et al. 2007; Garvie et al. 2015). Coupled biotic–abiotic soil processes can lead to the formation of chlorinated aliphatic compounds, including vinyl chloride (Keppler et al. 2002). These examples demonstrate that niches for organohalide-respiring microorganisms exist in environments not impacted by anthropogenic contamination, and the organohalide-respiring phenotype very likely predates the industrial revolution, which marks the begin of anthropogenic releases of chlorinated chemicals into the environment.

The reductive dehalogenase enzyme systems, which act as terminal oxidoreductases in the respiratory reductive dechlorination process, require a corrinoid co-factor for activity. De novo corrinoid synthesis is a metabolically expensive process requiring more than 20 gene products (Escalante-Semerena 2007). Strictly organohalide-respiring *Chloroflexi* like *D. mccartyi* do not possess the biochemical machinery for corrin ring biosynthesis, even though their energy metabolism hinges on reductive dechlorination reactions catalyzed by corrinoid-dependent reductive dehalogenases (Löffler et al. 2013). A recent analysis of sequenced genomes suggested that about 35 and 25 % of *Bacteria* and *Archaea*, respectively, have the genetic potential for de novo corrinoid biosynthesis, and can potentially provide this essential co-factor (unpublished data); however, *D. mccartyi* strains have specific corrinoid co-factor requirements. For example, the non-dechlorinator *Geobacter sulfurreducens* produces a corrinoid that does not support reductive dechlorination and growth of corrinoid auxotrophic *D. mccartyi* strains. In contrast, the PCE-respiring species *G. lovleyi* generates a corrinoid that enables *D. mccartyi* activity and growth (Yan et al. 2012, 2013). This observation suggests that the reductive dehalogenase enzyme systems of the PCE-respiring bacterium *G. lovleyi* and those of organohalide-respiring *D. mccartyi* strains are functional using the same corrinoid co-factor. Therefore, the cooccurrence of corrinoid auxotrophic, organohalide-respiring *Chloroflexi* and corrinoid-producing,

PCE/TCE-respiring *Deltaproteobacteria* in aquifers impacted with PCE and/or TCE (Amos et al. 2007b) or bioaugmentation cultures derived from PCE and/or TCE-impacted aquifers is not surprising (Wagner et al. 2012). Defined co-culture studies with a *Dhc* strain and a methanogen or an acetogen pure culture demonstrated that dimethylbenzimidazole (i.e., the lower base attached to the corrin ring in biologically functional corrinoids) supports *Dhc* activity; however, the exact corrinoid co-factor requirement of organohalide-respiring *Chloroflexi* has yet to be elucidated (Yan et al. 2013).

11.4 Reductive Dehalogenases

The discovery of reductive dechlorination as a respiratory process sparked considerable interest in the reductive dehalogenase enzyme systems that cleave the carbon–chlorine bond and replace the chlorine substituent with a hydrogen atom. Because *D. tiedjei* was the first available organohalide-respiring isolate, initial efforts focused on the 3-chlorobenzoate reductive dehalogenase of strain DCB-1. The reductive dehalogenase was partially purified and the preliminary characterization suggested a heme-containing heterodimer consisting of subunits with molecular weights of 64 and 37 kilodalton (Ni et al. 1995). All bacterial reductive dehalogenase enzyme systems that have been subsequently characterized appear to consist of a single polypeptide chain with molecular weights ranging between 41.8 (Kim et al. 2012) and 117.9 kD (Chen et al. 2013), and possess a corrinoid rather than a heme co-factor (Smidt and de Vos 2004; Bommer et al. 2014; Payne et al. 2015). An exception to the common monomeric structure appears to be the PceA reductive dehalogenase from *Sulfurospirillum multivorans*, which forms a functional homodimer in the crystalline state (PDB:4UQU) (Bommer et al. 2014). This homodimer formation was not observed in vivo (Neumann et al. 1995) suggesting that the absence of the membrane-anchoring B protein and/or the periplasmic membrane triggers homodimerization of PceA. Further experimental studies of reductive dehalogenase dimerization in, perhaps, surfactant vesicles or similar assemblies could test this possibility.

Unfortunately, the 3-chlorobenzoate reductive dehalogenase of strain DCB-1 remains the only such enzyme system biochemically characterized from a member of the class *Deltaproteobacteria.* This is unfortunate because the widespread distribution of organohalide-respiring *Deltaproteobacteria* implies relevant roles in global chlorine cycling. Further, PCE-to-cDCE dechlorinating *Deltaproteobacteria* play key roles for initiating PCE and TCE detoxification and contribute to enhanced dissolution of free-phase PCE and TCE in contaminated aquifers (Amos et al. 2007a). Hence, both from an ecological and applied environmental perspective, more information about the reductive dehalogenase enzyme systems of members of the *Deltaproteobacteria* is desirable.

11.5 Insights from Genome Analyses

A total of 208 (as of May 2015) *Deltaproteobacteria* genome sequences are publically available from the Joint Genome Institute, including those of a few organohalide-respiring *Deltaproteobacteria* (Table 11.3). Interestingly, the genomes of 20 *Deltaproteobacteria*, some of which have not been recognized to perform reductive dehalogenation, possess one or multiple reductive dehalogenase gene operons (Table 11.3). Although gene function and the ability of the respective hosts to grow via organohalide respiration remains to be determined, the observation that ~10 % of the sequenced *Deltaproteobacteria* genomes contain one or multiple reductive dehalogenase gene indicates that the ability to perform reductive dehalogenation is not uncommon among the *Deltaproteobacteria*. Of the 20 putative reductive dehalogenase gene-containing genomes, seven genomes have putative membrane anchor B protein-encoding genes suggesting an organohalide-respiring phenotype. A total of 32 putative reductive dehalogenase genes were identified, of which five encode proteins with the TAT-signal peptide, and 15 have a neighboring gene encoding a short transmembrane protein, presumably membrane-anchoring B proteins.

The reductive dehalogenase gene operon structures of the organohalide-respiring *Deltaproteobacteria G. lovleyi* strain SZ (PCE → cDCE), *A. dehalogenans* strain 2CP-C (2-chlorophenol → phenol), and *D. tiedjei* strain DCB-1 (3-chlorobenzoate → benzoate) are compared in Fig. 11.2. The *pceA-pce*B open reading frames are duplicated on the strain SZ genome (Wagner et al. 2012). The strain 2CP-C operon, presumably encoding the 2-chlorophenol reductive dehalogenase, comprises two nonidentical reductive dehalogenase genes, and the corresponding protein sequences share 26 % identity (unpublished data). The genome analysis of strain DCB-1 identified three putative reductive dehalogenase genes organized into two operons (Fig. 11.2) (unpublished data). Two nonidentical genes encoding proteins with 34 % identity are organized in one operon, suggestive of gene duplication and divergence. A third reductive dehalogenase gene present in a second operon encodes a protein sharing 31 and 33 % identity, respectively, with the aforementioned gene products. All reductive dehalogenase genes of strain DCB-1 have neighboring genes that encode proteins with 54 and 61 residues, respectively, in one operon and 72 residues in the second operon. These putative membrane-anchoring proteins are predicted to have two transmembrane helices. Similarly, the reductive dehalogenase gene operon of strain 2CP-C encodes two transmembrane proteins with 55 and 83 residues, which may serve as reductive dehalogenase membrane anchors; however, their role has yet to be experimentally confirmed.

The reductive dehalogenase gene operon neighborhood in *G. lovleyi* strain SZ contains integrase and transposase gene mobile genetic elements, which imply potential horizontal gene transfer events. No mobile genetic elements were found upstream and downstream of the reductive dehalogenase genes of strain 2CP-C and strain DCB-1. The differences in reductive dehalogenase gene neighborhoods suggest that the acquisition of the *pceA-pce*B by *G. lovleyi* strain SZ

Table 11.3 *Deltaproteobacteria* with (putative) reductive dehalogenase (RDase) genes based on genome sequence analysis

Organism (family)	Genome accession #	# of RDase Genes (with associated B gene)	Source[a]	References
Anaeromyxobacter dehalogenans strain 2CP-1 (*Cystobacteraceae*)	CP001359	2 (2)	Sediment (Iso)	Sanford et al. (2002)
Anaeromyxobacter dehalogenans strain 2CP-C (*Cystobacteraceae*)	CP000251	2 (2)	Soil (Iso)	Thomas et al. (2008)
Anaeromyxobacter dehalogenans strain K (*Cystobacteraceae*)	CP001131	2 (2)	Soil (Iso)	Unpublished
"*Candidatus* Entotheonella sp." strain TSY1 (Unclassified)	AZHW01000000	2 (0)	Marine sponge (MG)	Wilson et al. (2014)
"*Candidatus* Entotheonella sp. " strain TSY2 (Unclassified)	AZHX01000001	2 (0)	Marine sponge (MG)	Wilson et al. (2014)
Deferrisoma camini strain S3R1 (*Deferrisoma*)	JAFN01000000	1 (0)	Hydrothermal vent (Iso)	Slobodkina et al. (2012)
Deltaproteobacterium NaphS2 (Unclassified)	ADZZ00000000	1 (0)	Marine sediments (Iso)	DiDonato et al. (2010)
Deltaproteobacterium PSCGC 5419 (Unclassified)	PRJNA187123	1 (0)	Coastal sediments (SC)	Unpublished
Deltaproteobacterium PSCGC 5451 (Unclassified)	PRJNA187122	1 (0)	Coastal sediments (SC)	Unpublished
Desulfarculus species SPR (*Desulfarculaceae*)	AZAC01000000	1 (0)	Marine sediments (Iso)	Unpublished
Desulfobacular toluolica strain Tol2 (*Desulfobacteraceae*)	NC_018645	3 (2)	Marine sediment (Iso)	Wöhlbrand et al. (2013)
Desulfomonile tiedjei strain DCB-1 (*Syntrophaceae*)	CP003360	3 (3)	Sewage sludge (Iso)	Unpublished
Desulfosarcina variabilis strain Montpellier (*Desulfobacteraceae*)	PRJNA50095	1 (0)	Marine sediments (Iso)	Unpublished

(continued)

Table 11.3 (continued)

Organism (family)	Genome accession #	# of RDase Genes (with associated B gene)	Source[a]	References
Desulfuromusa kysingii strain DSM 7344 (*Desulfuromonadaceae*)	PRJNA262212	2 (2)	Marine sediment (Iso)	Unpublished
Geobacter lovleyi strain SZ (*Geobacteriaceae*)	CP001089	2 (2)	Contaminated freshwater sediment, aquifer (Iso)	Wagner et al. (2012)
Plesiocystis pacifica strain SIR-1 (*Myxococcales*)	PRJNA19341	2 (0)	Sea grass/beach (Iso)	Unpublished
SAR324 cluster bacterium SCGC AAA240-J09 (Unclassified)	AFIA00000000	1 (0)	Seawater (SC)	Swan et al. (2011)
SAR324 cluster bacterium SCGC AAA008-E19 (Unclassified)	A8E19_0001	1 (0)	Marine (SC)	Unpublished
SAR324 cluster bacterium SCGC AAA240-O02 (Unclassified)	A0O02_0001	1 (0)	Marine (SC)	Unpublished
Syntrophorhabdus aromaticivorans strain UI (*Syntrophobacterales*)	PRJNA63195	1 (0)	Sludge (Iso)	Nobu et al. (2014)

[a]Iso, isolate available; MG, genome assembled from metagenomic datasets; SC, genome obtained from single cell genome sequencing

occurred more recently, or that these reductive dehalogenase genes have distinct evolutionary histories within the *Deltaproteobacteria*.

11.6 Enrichment and Isolation of Organohalide-Respiring *Deltaproteobacteria*

The known *Deltaproteobacteria* harboring reductive dehalogenase genes are versaphiles and can use different substrates to support growth. While this feature may facilitate the isolation of novel *Deltaproteobacteria*, enrichment in the absence of the halogenated (e.g., chlorinated or brominated) electron acceptor may select for strains that lack the organohalide-respiring phenotype. Several organohalide-respiring *Deltaproteobacteria* form colonies on solid surfaces (i.e., agar plates), but obtaining colonies with a halogenated hydrocarbon as electron acceptor can be challenging. Obstacles include technical difficulties supplying the halogenated electron acceptor, in particular volatile compounds, in sufficient quantities while

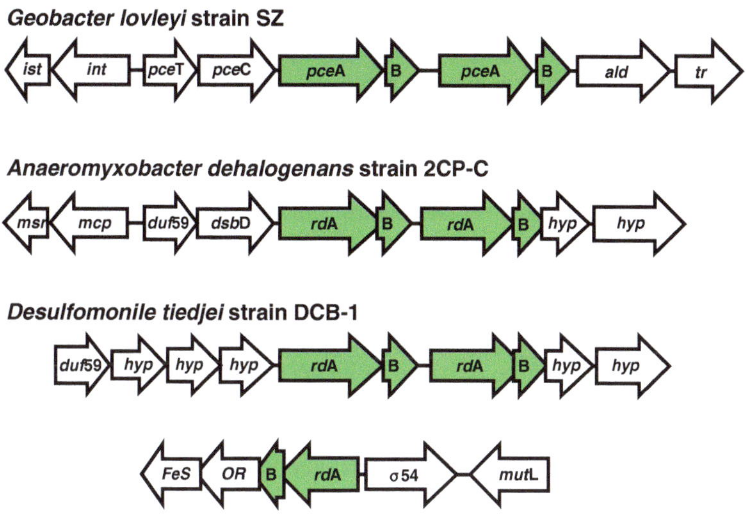

Fig. 11.2 Local reductive dehalogenase gene neighborhoods in *Geobacter lovleyi* strain SZ, *Anaeromyxobacter dehalogenans* strain 2CP-C and *Desulfomonile tiedjei* strain DCB1. Inferred gene functions are abbreviated as follows: *pceA*—tetrachloroethene reductive dehalogenase, *pceB*—putative membrane anchor protein subunit, *pceT*—peptidylproly *cis-trans* isomerase, *pceC*—FMN-binding and polyferredoxin-containing protein, *int*—integrase, *ist*—ATPase, *tr*—transposase, *ald*—aldehyde dehydrogenase, *rd*A—reductive dehalogenase A protein, B—reductive dehalogenase membrane anchor protein, *hyp*—hypothetical protein, *dsbD*—cytochrome c biogenesis protein, *duf59*—protein with domain of unknown function, *mcp*—methyl-accepting chemotaxis sensory transducer, *msr*—methionine-R-sulfoxide reductase, OR—NADH: ubiquinone oxidoreductase chain-1 like protein, σ54—Sigma54 specific transcriptional regulator, and mutL—DNA mismatch repair protein

avoiding toxicity effects. In most cases, pure cultures maintained in liquid medium would not tolerate high concentrations of the halogenated electron acceptor. For example, chlorophenols should be added to the medium in the 100–200 μM concentration range for sustaining activity and sometimes even lower. Any use of higher concentrations in a liquid medium may result in inhibition (Sanford et al. 2002). Aqueous phase PCE concentrations should not exceed 0.5 mM to avoid toxic effects (Amos et al. 2007a). The two constraints on the electron acceptor concentrations provided in the medium are toxicity (i.e., the upper concentration limit) and detection/quantification limits of the analytical procedure for monitoring electron acceptor consumption (i.e., the lower concentration limit). Dehalogenation activity (i.e., substrate consumption and product formation) in liquid cultures may be detected but due to the low electron acceptor concentrations, growth cannot be measured with usual analytical tools such as turbidity (i.e., optical density) measurements or even protein analysis. To quantify actual biomass increases linked to organohalide respiration, two approaches have been successfully applied. First, the sequential refeeding of cultures once the halogenated electron acceptor has been consumed leads to cumulative growth, which can more easily be measured. The reductive dechlorination products (e.g., 2-chlorophenol → phenol; PCE → cDCE) are generally less toxic to the dechlorinating organism than the parent compounds. Alternate approaches to demonstrate growth include the sensitive measurement of cell increases with quantitative PCR (He et al. 2003), the determination of ^{14}C assimilated into biomass from a radiolabeled carbon source (He and Sanford 2003), or traditional microscopic cell counts (Hatt et al. 2013). Obviously, appropriate control cultures (e.g., cultures receiving the organic dehalogenation product but not the halogenated electron acceptor) should accompany these experiments to conclusively link growth with reductive dehalogenation activity. The versatile nature of *Deltaproteobacteria* metabolism enables alternate approaches to obtain isolates. For example, nontoxic electron acceptors such as fumarate can be provided in higher concentrations, and the resulting colonies can be tested individually in liquid medium for dehalogenating activity (Sung et al. 2003, 2006). For organohalide-respiring bacteria that do not form colonies on solid media, the dilution-to-extinction principle and soft-agar shake cultures have been used to obtain isolates (Löffler et al. 2005). Many organohalide-respiring *Deltaproteobacteria* use acetate as electron donor and medium amended with acetate as the sole electron donor and the halogenated compound as electron acceptor generates selective enrichment conditions. These methodologies can lead to rapid success in enriching and isolating organohalide-respiring *Deltaproteobacteria*, and expand our understanding of the distribution and relevance of this metabolic trait among members of this ubiquitously distributed bacterial class.

References

Ahn Y-B, Kerkhof LJ, Häggblom MM (2009) *Desulfoluna spongiiphila* sp. nov., a dehalogenating bacterium in the *Desulfobacteraceae* from the marine sponge *Aplysina aerophoba*. Int J Syst Evol Microbiol 59:2133–2139

Ahn YB, Rhee SK, Fennell DE, Kerkhof LJ, Hentschel U, Häggblom MM (2003) Reductive dehalogenation of brominated phenolic compounds by microorganisms associated with the marine sponge *Aplysina aerophoba*. Appl Environ Microbiol 69:4159–4166

Amos BK, Christ JA, Abriola LM, Pennell KD, Löffler F (2007a) Experimental evaluation and mathematical modeling of microbially enhanced tetrachloroethene (PCE) dissolution. Environ Sci Technol 41:963–970

Amos BK, Sung Y, Fletcher KE, Gentry TJ, Wu W-M, Criddle CS et al (2007b) Detection and quantification of *Geobacter lovleyi* strain SZ: implications for bioremediation at tetrachloroethene- and uranium-impacted sites. Appl Environ Microbiol 73:6898–6904

Bethke CM, Sanford RA, Kirk MF, Jin Q, Flynn TM (2011) The thermodynamic ladder in geomicrobiology. Am J Sci 311:183–210

Bommer M, Kunze C, Fesseler J, Schubert T, Diekert G, Dobbek H (2014) Structural basis for organohalide respiration. Science 346:455–458

Boyle AW, Phelps CD, Young LY (1999) Isolation from estuarine sediments of a *Desulfovibrio* strain which can grow on lactate coupled to the reductive dehalogenation of 2,4,6-tribromophenol. Appl Environ Microbiol 65:1133–1140

Buckel W, Thauer RK (2013) Energy conservation via electron bifurcating ferredoxin reduction and proton/Na(+) translocating ferredoxin oxidation. Biochim Biophys Acta 1827:94–113

Chen K, Huang L, Xu C, Liu X, He J, Zinder SH et al (2013) Molecular characterization of the enzymes involved in the degradation of a brominated aromatic herbicide. Mol Microbiol 89:1121–1139

Cole JR, Cascarelli AL, Mohn W, Tiedje J (1994) Isolation and characterization of a novel bacterium growing via reductive dehalogenation of 2-chlorophenol. Appl Environ Microbiol 60:3536–3542

Cole JR, Fathepure BZ, Tiedje J (1995) Tetrachloroethene and 3-chlorobenzoate dechlorination activities are co-induced in *Desulfomonile tiedjei* DCB-1. Biodegradation 6:167–172

de Jong E, Field JA, Spinnler H-E, Wijnberg JBPA, de Bont JAM (1994) Significant biogenesis of chlorinated aromatics by fungi in natural environments. Appl Environ Microbiol 60:264–270

De Wever H, Cole JR, Fettig MR, Hogan DA, Tiedje J (2000) Reductive dehalogenation of trichloroacetic acid by *Trichlorobacter thiogenes* gen. nov., sp. nov. Appl Environ Microbiol 66:2297–2301

DeWeerd KA, Mandelco L, Tanner RS, Woese CR, Suflita J (1990) *Desulfomonile tiedjei* gen. nov. and sp. nov., a novel anaerobic, dehalogenating, sulfate-reducing bacterium. Arch Microbiol 154:23–30

DiDonato RJ, Young ND, Butler JE, Chin K-J, Hixson KK, Mouser P et al (2010) Genome sequence of the deltaproteobacterial strain NaphS2 and analysis of differential gene expression during anaerobic growth on naphthalene. PLoS ONE 5:e14072

Dolfing J, Novak I (2015) The Gibbs free energy of formation of halogenated benzenes, benzoates and phenols and their potential role as electron acceptors in anaerobic environments. Biodegradation 26:15–27

Dolfing J, Tiedje J (1987) Growth yield increase linked to reductive dechlorination in a defined 3-chlorobenzoate degrading methanogenic coculture. Arch Microbiol 149:102–105

Edgar RC (2004) MUSCLE: multiple sequence alignment with high accuracy and high throughput. Nucleic Acids Res 32:1792–1797

Escalante-Semerena JC (2007) Conversion of cobinamide into adenosylcobamide in bacteria and archaea. J Bacteriol 189:4555–4560

Fletcher KE, Costanza J, Pennell KD, Löffler FE (2011) Electron donor availability for microbial reductive processes following thermal treatment. Water Res 45:6625–6636

Flynn TM, O'Loughlin EJ, Mishra B, DiChristina TJ, Kemner KM (2014) Sulfur-mediated electron shuttling during bacterial iron reduction. Science 344:1039–1042

Garvie LAJ, Wilkens B, Groy TL, Glaeser JA (2015) Substantial production of drosophilin A methyl ether (tetrachloro-1,4-dimethoxybenzene) by the lignicolous basidiomycete *Phellinus badius* in the heartwood of mesquite (*Prosopis juliflora*) trees. Sci Nat 102:18

Gordon WG (2004) Amazing organohalogens: although best known as synthetic toxicants, thousands of halogen compounds are, in fact, part of our natural enviornment. Am Sci 92:342–349

Gribble GW (1994) The natural production of chlorinated compounds. Environ Sci Technol 28:310A–319A

Gribble GW (2000) The natural production of organobromine compounds. Environ Sci Pollut Res 7:37–47

Hatt JK, Ritalahti KM, Ogles DM, Lebrón CA, Löffler F (2013) Design and application of an internal amplification control to improve *Dehalococcoides mccartyi* 16S rRNA gene enumeration by qPCR. Environ Sci Technol 47:11131–11138

He JZ, Ritalahti KM, Yang KL, Koenigsberg SS, Löffler F (2003) Detoxification of vinyl chloride to ethene coupled to growth of an anaerobic bacterium. Nature 424:62–65

He Q, Sanford RA (2003) Characterization of Fe(III) reduction by chlororespiring *Anaeromyxobacter dehalogenans*. Appl Environ Microbiol 69:2712–2718

He Q, Sanford RA (2004a) Acetate threshold concentrations suggest varying energy requirements during anaerobic respiration by *Anaeromyxobacter dehalogenans*. Appl Environ Microbiol 70:6940–6943

He Q, Sanford RA (2004b) The generation of high biomass from chlororespiring bacteria using a continuous fed-batch bioreactor. Appl Microbiol Biotechnol 65:377–382

Kearse M, Moir R, Wilson A, Stones-Havas S, Cheung M, Sturrock S et al (2012) Geneious basic: an integrated and extendable desktop software platform for the organization and analysis of sequence data. Bioinformatics (Oxford, England) 28:1647–1649

Keppler F, Borchers R, Pracht J, Rheinberger S, Schöler HF (2002) Natural formation of vinyl chloride in the terrestrial environment. Environ Sci Technol 36:2479–2483

Kim S-H, Harzman C, Davis J, Hutcheson R, Broderick J, Marsh T et al (2012) Genome sequence of *Desulfitobacterium hafniense* DCB-2, a Gram-positive anaerobe capable of dehalogenation and metal reduction. BMC Microbiol 12:21

Krumholz L (1997) *Desulfuromonas chloroethenica* sp. nov. uses tetrachloroethylene and trichloroethylene as electron acceptors. Int J Syst Bacteriol 47:1262–1263

Krumholz L, Sharp R, Fishbain SS (1996) A freshwater anaerobe coupling acetate oxidation to tetrachloroethylene dehalogenation. Appl Environ Microbiol 62:4108–4113

Kudo K, Yamaguchi N, Makino T, Ohtsuka T, Kimura K, Dong DT et al (2013) Release of arsenic from soil by a novel dissimilatory arsenate-reducing bacterium, *Anaeromyxobacter* sp. strain PSR-1. Appl Environ Microbiol 79:4635–4642

Löffler F, Ritalahti KM, Zinder SH (2013) *Dehalococcoides* and reductive dechlorination of chlorinated solvents. In: Stroo HF (ed) Bioaugmentation for groundwater remediation. Springer, New York, pp 39–88

Löffler F, Tiedje J, Sanford RA (1999) Fraction of electrons consumed in electron acceptor reduction and hydrogen thresholds as indicators of halorespiratory physiology. Appl Environ Microbiol 65:4049–4056

Löffler FE, Sanford RA, Ritalahti KM (2005) Enrichment, cultivation, and detection of reductively dechlorinating bacteria. In: Jared RL (ed) Methods in Enzymology. Academic Press, New York, pp 77–111

Lovley DR (1993) Dissimilatory metal reduction. Annu Rev Microbiol 47:263–290

Mahadevan R, Palsson BO, Lovley DR (2011) In situ to in silico and back: elucidating the physiology and ecology of *Geobacter* spp. using genome-scale modelling. Nat Rev Microbiol 9:39–50

Mohn W, Linkfield TG, Pankratz HS, Tiedje J (1990) Involvement of a collar structure in polar growth and cell division of strain DCB-1. Appl Environ Microbiol 56:1206–1211

Mohn W, Tiedje J (1991) Evidence for chemiosmotic coupling of reductive dechlorination and ATP synthesis in *Desulfomonile tiedjei*. Arch Microbiol 157:1–6

Neumann A, Wohlfarth G, Diekert G (1995) Properties of tetrachloroethene and trichloroethene dehalogenase of *Dehalospirillum multivorans*. Arch Microbiol 163:276–281

Ni S, Fredrickson JK, Xun L (1995) Purification and characterization of a novel 3-chlorobenzoate-reductive dehalogenase from the cytoplasmic membrane of *Desulfomonile tiedjei* DCB-1. J Bacteriol 177:5135–5139

Nobu MK, Narihiro T, Hideyuki T, Qiu YL, Sekiguchi Y, Woyke T et al (2014) The genome of *Syntrophorhabdus aromaticivorans* strain UI provides new insights for syntrophic aromatic compound metabolism and electron flow. Environ Microbiol 17(12):4861–4872

Odum JM, Singleton RJE (1993) The sulphate-reducing bacteria: contemporary perspectives. Springer-Verlag, New York. ISBN 0-387-97865-97868

Ortiz-Bermudez P, Hirth KC, Srebotnik E, Hammel KE (2007) Chlorination of lignin by ubiquitous fungi has a likely role in global organochlorine production. Proc Natl Acad Sci 104:3895–3900

Parsons JR, Saez M, Dolfing J, de Voogt P (2008) Biodegradation of perfluorinated compounds. Rev Environ Contam Toxicol 196:53–71

Payne KAP, Quezada CP, Fisher K, Dunstan MS, Collins FA, Sjuts H et al (2015) Reductive dehalogenase structure suggests a mechanism for B_{12}-dependent dehalogenation. Nature 517:513–516

Sanford RA, Cole JR, Tiedje J (2002) Characterization and description of *Anaeromyxobacter dehalogenans* gen. nov., sp. nov., an aryl-halorespiring facultative anaerobic myxobacterium. Appl Environ Microbiol 68:893–900

Sanford RA, Wagner DD, Wu Q, Chee-Sanford JC, Thomas SH, Cruz-García C et al (2012) Unexpected nondenitrifier nitrous oxide reductase gene diversity and abundance in soils. Proc Natl Acad Sci 109:19709–19714

Shelton DR, Tiedje J (1984) Isolation and partial characterization of bacteria in an anaerobic consortium that mineralizes 3-chlorobenzoic acid. Appl Environ Microbiol 48:840–848

Slobodkina GB, Reysenbach AL, Panteleeva AN, Kostrikina NA, Wagner ID, Bonch-Osmolovskaya EA et al (2012) *Deferrisoma camini* gen. nov., sp. nov., a moderately thermophilic, dissimilatory iron(III)-reducing bacterium from a deep-sea hydrothermal vent that forms a distinct phylogenetic branch in the *Deltaproteobacteria*. Int J Syst Evol Microbiol 62:2463–2468

Smidt H, de Vos WM (2004) Anaerobic microbial dehalogenation. Annu Rev Microbiol 58:43–73

Stamatakis A (2014) RAxML version 8: a tool for phylogenetic analysis and post-analysis of large phylogenies. Bioinformatics (Oxford, England) 30:1312–1313

Stevens TO, Linkfield TG, Tiedje J (1988) Physiological characterization of strain DCB-1, a unique dehalogenating sulfidogenic bacterium. Appl Environ Microbiol 54:2938–2943

Suflita J, Horowitz A, Shelton DR, Tiedje J (1982) Dehalogenation: a novel pathway for the anaerobic biodegradation of haloaromatic compounds. Science 218:1115–1117

Sun B, Cole JR, Sanford RA, Tiedje J (2000) Isolation and characterization of *Desulfovibrio dechloracetivorans* sp. nov., a marine dechlorinating bacterium growing by coupling the oxidation of acetate to the reductive dechlorination of 2-chlorophenol. Appl Environ Microbiol 66:2408–2413

Sun B, Cole JR, Tiedje J (2001) *Desulfomonile limimaris* sp. nov., an anaerobic dehalogenating bacterium from marine sediments. Int J Syst Evol Microbiol 51:365–371

Sung Y, Fletcher K, Ritalahti KM, Apkarian RP, Ramos-Hernández N, Sanford RA et al (2006) *Geobacter lovleyi* sp. nov. strain SZ, a novel metal-reducing and tetrachloroethene-dechlorinating bacterium. Appl Environ Microbiol 72:2775–2782

Sung Y, Ritalahti KM, Sanford RA, Urbance JW, Flynn SJ, Tiedje J et al (2003) Characterization of two tetrachloroethene-reducing, acetate-oxidizing anaerobic bacteria and their description as *Desulfuromonas michiganensis* sp. nov. Appl Environ Microbiol 69:2964–2974

Swan BK, Martinez-Garcia M, Preston CM, Sczyrba A, Woyke T, Lamy D et al (2011) Potential for chemolithoautotrophy among ubiquitous bacteria lineages in the dark ocean. Science 333:1296–1300

Tamura K, Stecher G, Peterson D, Filipski A, Kumar S (2013) MEGA6: molecular evolutionary genetics analysis version 6.0. Mol Biol Evol 30:2725–2729

Thomas SH, Sanford RA, Amos BK, Leigh MB, Cardenas E, Löffler F (2010) Unique ecophysiology among U(VI)-reducing bacteria as revealed by evaluation of oxygen metabolism in *Anaeromyxobacter dehalogenans* strain 2CP-C. Appl Environ Microbiol 76:176–183

Thomas SH, Wagner RD, Arakaki AK, Skolnick J, Kirby JR, Shimkets LJ et al (2008) The mosaic genome of *Anaeromyxobacter dehalogenans* strain 2CP-C suggests an aerobic common ancestor to the delta-proteobacteria. PLoS ONE 3:e2103

Treude N, Rosencrantz D, Liesack W, Schnell S (2003) Strain FAc12, a dissimilatory iron-reducing member of the *Anaeromyxobacter* subgroup of *Myxococcales*. FEMS Microbiol Ecol 44:261–269

Voordouw G (1995) The genus *Desulfovibrio*—the centennial. Appl Environ Microbiol 61:2813–2819

Wagner DD, Hug LA, Hatt JK, Spitzmiller MR, Padilla-Crespo E, Ritalahti KM et al (2012) Genomic determinants of organohalide-respiration in *Geobacter lovleyi*, an unusual member of the Geobacteraceae. BMC Genom 13:200

Wilson MC, Mori T, Rückert C, Uria AR, Helf MJ, Takada K et al (2014) An environmental bacterial taxon with a large and distinct metabolic repertoire. Nature 506:58–62

Wöhlbrand L, Jacob JH, Kube M, Mussmann M, Jarling R, Beck A et al (2013) Complete genome, catabolic sub-proteomes and key-metabolites of *Desulfobacula toluolica* Tol2, a marine, aromatic compound-degrading, sulfate-reducing bacterium. Environ Microbiol 15:1334–1355

Yan J, Im J, Yang Y, Löffler F (2013) Guided cobalamin biosynthesis supports *Dehalococcoides mccartyi* reductive dechlorination activity. Philos Trans R Soc B Biol Sci 368:20120320

Yan J, Ritalahti KM, Wagner DD, Löffler F (2012) Unexpected specificity of interspecies cobamide transfer from *Geobacter* spp. to organohalide-respiring *Dehalococcoides mccartyi* strains. Appl Environ Microbiol 78:6630–6636

Chapter 12
Comparative Physiology of Organohalide-Respiring Bacteria

Koshlan Mayer-Blackwell, Holly Sewell, Maeva Fincker and Alfred M. Spormann

Abstract The potential for reductive organohalide respiration is relatively widespread among bacteria. In this chapter, we highlight metabolic differences between facultative and obligate organohalide-respiring bacteria. In addition, we compare the genomic architecture and evolution of the bacteria that comprise the obligate organohalide respiring *Dehalococcoides*, *Dehalobacter*, and *Dehalogenimonas* genera. Major unresolved questions remain about the necessary and sufficient enzymes for energy conservation coupled to reductive dehalogenation in these microorganisms. Although comparative physiology among these three genera reveals considerable metabolic and eco-physiological diversity consistent with their unique phylogeny, these microorganisms share similar genomic signatures, suggestive of convergent adaptive niche specialization to catabolism of naturally occurring organohalidc compounds.

12.1 Physiological Differences Between Obligate and Facultative Organohalide-Respiring Bacteria

The potential for reductive organohalide respiration is widespread among anaerobic and facultative anaerobic bacteria based on genomic evidence as well as experimental data. Included are the bacterial phyla Proteobacteria, Chloroflexi, and

K. Mayer-Blackwell · H. Sewell · M. Fincker · A.M. Spormann (✉)
Civil and Environmental Engineering, Stanford University,
318 Campus Drive, Stanford, CA 94305, USA
e-mail: spormann@stanford.edu

A.M. Spormann
Geological and Environmental Sciences, and Chemical Engineering,
Stanford University, 318 Campus Drive, Stanford, CA 94305, USA

L. Adrian and F.E. Löffler (eds.), *Organohalide-Respiring Bacteria*,
DOI 10.1007/978-3-662-49875-0_12

Firmicutes. So far, microorganisms in the genera *Anaeromyxobacter, Comamonas, Geobacter, Dehalobacter, Dehalococcoides, Dehalogenimonas, Desulfitobacterium, Desulfoluna, Desulfomonile, Desulfovibrio, Desulfuromonas, Shewanella,* and *Sulfurospirillum* have experimentally evidenced reductive dehalogenation (Christiansen and Ahring 1996; Krumholz 1997; Holliger et al. 1998; Sun et al. 2001; Sanford et al. 2002; Sung et al. 2003; Ahn et al. 2009; Moe et al. 2009; Lohner and Spormann 2013; Chen et al. 2013). Additionally, microorganisms of many other genera contain genes encoding putative reductive dehalogenases (Hug et al. 2013).

The exact physiological role of the majority of predicted reductive dehalogenases in the respective microorganisms is unknown. However, in well-studied cases, reductive dehalogenases enable the use of halogenated (chlorinated or brominated) organic compounds as catabolic electron acceptors in microbial energy metabolism. Based on this function, it is useful to distinguish between obligate and facultative organohalide respiring microorganisms. Obligate organohalide-respiring bacteria such as *Dehalococcoides, Dehalogenimonas*, and *Dehalobacter* spp. are highly niche-specialized to organohalide respiration, whereas well-studied *Desulfitobacterium, Geobacter, Sulfurospirillum, Desulfovibrio,* and *Anaeromyxobacter* spp. are metabolically versatile and use such electron acceptors as NO_3^-, Fe(III), Mn(IV), DMSO, or oxidized sulfur compounds (Fig. 12.1). The niche specialization of obligate and facultative organohalide-respiring bacteria is also reflected in the spectrum of electron donors used: facultative organohalide-respiring bacteria can use diverse electron donors, typically the organic end products of primary fermenters, as well as H_2 and formate. The obligate organohalide-respiring bacteria are in nearly all studied cases restricted to H_2 (Fig. 12.1).

The physiological role of organohalide respiration in obligate and facultative organohalide-respiring bacteria can be viewed in context of respiration as a common mode of energy conservation. While most catabolic reactions involve oxidation and reduction reactions, not all electron transfer reactions in catabolic pathways are coupled to energy conservation. 'Respiration' signifies energy conservation coupled to redox reactions at the cytoplasmic membrane, wherein exergonic redox reactions are coupled to the endergonic generation of a chemiosmotic proton or sodium gradient required to drive the phosphorylation of ADP via a membrane-bound ATPase. A particularly well-studied example involves the oxidation of NADH by Complex I in aerobic bacteria, where the reduction of membrane-integrated quinones is coupled to proton translocation (Trumpower 1990; Friedrich and Scheide 2000; Friedrich 2001; Efremov et al. 2010). Additional protons may be translocated during the subsequent quinol oxidation, where electrons are transferred directly, or via a cytochrome c, to molecular oxygen as a terminal electron acceptor. Interestingly, the degree to which energy is conserved during quinol oxidation is mechanistically variable and under selection (Poole and Ingledew 1987; Anraku and Gennis 1987). One example is the anaerobic catabolism of *Shewanella oneidensis* MR-1. When this microbe grows anaerobically with lactate as electron donor, lactate is oxidized to acetate with acetyl-phosphate as the substrate level phosphorylation intermediate (Hau and Gralnick 2007). The reducing equivalents generated are transferred to

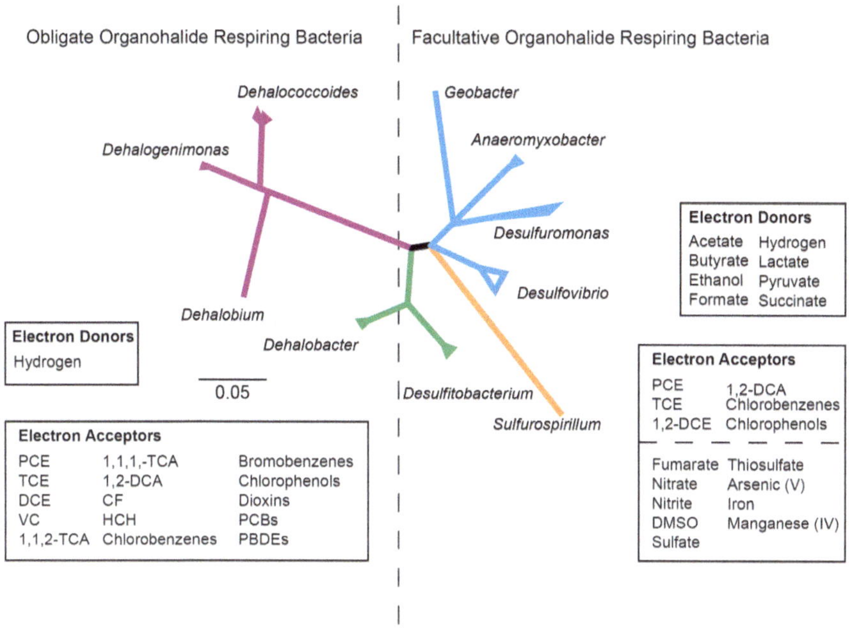

Fig. 12.1 Phylogenetic tree of major organohalide respiring genera based on 16S rRNA gene sequences and commonly used electron donors and acceptors. Multiple sequence alignment was generated using the SILVA aligner (Pruesse et al. 2007, 2012). Reference bar represents divergence. Obligate organohalide respiring microorganisms have shown capacity to use brominated benzenes (Wagner et al. 2012a) and polybrominated diphenyl ethers (He et al. 2006) as electron acceptors in addition to many chlorinated hydrocarbons. The figure is modified after Maphosa et al. (2010). Abbreviations are as follows: *PCE*—tetrachloroethene, *TCE*—trichloroethene, *DCE*—dichloroethene, *VC*—vinyl choride, *TCA*—trichloroethane, *DCA*—dichloroethane, *CF*—chloroform, *HCH*—hexachlorocyclohexane, *PCB*—polychlorinated biphenol, *PBDE*—polybrominated diphenyl ether, *DMSO*—dimethyl sulfoxide

quinones, and quinol oxidation occurs without energy conservation. Here, electrons are transferred from the quinol pool to terminal Fe(III), NO_3^-, NO_2^-, or DMSO reductases in the periplasm and outer membrane via the electron transfer protein CymA without apparent generation of proton motive force (Myers and Myers 2000; Hunt et al. 2010).

This strategy observed in *Shewanella* can be considered as "facilitated fermentation," as it represents a catabolic metabolism that neither meets the classical definitions of fermentation nor respiration. In facilitated fermentation, an organic electron donor is coupled to the reduction of an electron acceptor; however, energy is not conserved via electron transport-coupled phosphorylation. Instead, microorganisms utilizing facilitated fermentation conserve energy in the oxidative pathway, primarily or exclusively, via substrate level phosphorylation. Despite being thermodynamically inefficient, facilitated fermentation could be beneficial to the microorganism in some environments where it enables more rapid organic substrate uptake and oxidation (Kreft and Bonhoeffer 2005).

12.1.1 Function of Reductive Dehalogenases in Energy Metabolism

Do some organohalide-respiring bacteria perform reductive dehalogenation in the context of facilitated fermentation? If so, theory would predict that molar growth yield for those microorganisms should be less than that observed under fully coupled respiratory conditions. One suggestive example is the facultative organohalide respiring bacterium *Desulfovibrio* strain TBP-1, isolated from estuarine sediments in New Jersey, USA (Boyle et al. 1999). Thermodynamics predicts that more Gibbs free energy is available when coupling lactate oxidation to 4-bromophenol versus sulfate reduction (Thauer et al. 1977; Dolfing and Novak 2014). During lactate utilization, strain TBP-1 manifests growth yield of 0.7 and 1.1 mg protein/mmol electron equivalent using 4-bromophenol or sulfate as a terminal electron acceptor, respectively (Boyle et al. 1999). Thus, the lower growth yield using 4-bromophenol as electron acceptor implies that fewer, if any, electron transfer steps are coupled to cation translocation during organohalide "respiration" in *Desulfovibrio* TBP-1. In this case, and perhaps in other chemoorganotrophic facultative organohalide-respiring bacteria, the primary function of reductive dehalogenases may be to balance internal pools of reducing equivalents and to increase catabolic rate. Determining whether reductive dehalogenase-dependent facilitated fermentation is common will require more studies that directly compare carefully determined molar growth yields of chemoorganotrophic facultative organohalide-respiring bacteria utilizing both halogenated and nonhalogenated electron acceptors. Early studies of *Desulfitobacterium dehalogenans* suggested that it may be capable of both dechlorination-linked "respiration" with hydrogen and formate, but employ a facilitated fermentation-type mechanism when grown on lactate or pyruvate (Mackiewicz and Wiegel 1998; van de Pas et al. 2001). Studies in obligate organohalide-respiring bacteria have not shown evidence of a higher molar growth yield (cell mass per mol halide released) compared with that achieved by facultative organohalide-respiring bacteria (Table 12.1). In general, caution should be exercised when comparing molar growth yields across organohalide-respiring bacteria, as precipitates of inorganic compounds that may form in the medium and may vary by medium formulation confound protein or cellular weight quantification. As a result, most molar growth yields of obligate facultative organohalide-respiring bacteria are inferred indirectly from quantitative PCR derived cell density estimates (Löffler et al. 2013).

In microorganisms where the facilitated fermentation hypothesis holds, reduction of a halogenated substrate would provide the greatest advantage if reductive dehalogenation occurs at the same or higher rate as substrate level phosphorylation, which determines the maximum rate of ATP synthesis. Since fermentative substrate level phosphorylation pathways can be rapid, a metabolism based on facilitated fermentation coupled to organohalide reduction might exert evolutionary selection on the kinetic parameters of reductive dehalogenases. With current data, however, it is difficult to determine whether there are systematic

Table 12.1 Growth yields in selected organohalide-respiring bacteria

Organism	Electron donor	Electron acceptor	Yield		Unit	Source
			Cells by qPCR	Dry weight		
Desuflomonile tiedjei DCB-1	Formate	3,5-dichlorobenzoate		2.8	g dry cell weight/Cl^-	Mohn and Tiedje (1990)
Desuflomonile tiedjei DCB-1	Hydrogen	3,5-dichlorobenzoate		1.8	g dry cell weight/Cl^-	Mohn and Tiedje (1990)
Desulfitobacterium dehalogenans	Pyruvate	None (fermenative)		14	g dry cell weight/ pyruvate	van de Pas et al. (2001)
Desulfitobacterium dehalogenans	Pyruvate	3-chloro-4-hydroxypheny-lacetate		13.1	g dry cell weight/Cl^-	van de Pas et al. (2001)
Desulfitobacterium dehalogenans	Lactate	3-chloro-4-hydroxypheny-lacetate		5.5	g dry cell weight/Cl^-	van de Pas et al. (2001)
Desulfitobacterium dehalogenans	Formate	3-chloro-4-hydroxypheny-lacetate		3.1	g dry cell weight/Cl^-	van de Pas et al. (2001)
Desulfitobacterium dehalogenans	Hydrogen	3-chloro-4-hydroxypheny-lacetate		3.2	g dry cell weight/Cl^-	van de Pas et al. (2001)
Desulfitobacterium dehalogenans	Formate	3-chloro-4-hydroxypheny-lacetate		1.2–5.7	g dry cell weight/eeq	Mackiewicz and Wiegel (1998)
Desulfovibrio strain TBP-1	Lactate	Sulfate		2.2	g dry weight/mol 2 eeq	Boyle et al. (1999)
Desulfovibrio strain TBP-1	Lactate	4-bromophenol		1.4	g dry weight/mol C^-	Boyle et al. (1999)
Sulfurospirillum multivorans	Hydrogen	Tetrachloroethene		1.4	g dry weight/Cl^-	Gabriele Diekert (Personal communication)
Dehalobacter restrictus	Hydrogen	Trichloroethene		3.3–5.6	g dry weight/Cl^-	Christof Hollinger (Personal communication)

(continued)

Table 12.1 (continued)

Organism	Electron donor	Electron acceptor	Yield		Unit	Source
			Cells by qPCR	Dry weight		
Dehalococcoides mccartyi 195	Hydrogen	Trichloroethene	2.96×10^8	3.8[a]	g dry weight/Cl^-	Maymó-Gatell et al. (1997)
Dehalococcoides mccartyi GT	Hydrogen	Trichloroethene	3.1×10^8	3.8*	g dry weight/Cl^-	He et al. (2003)
Dehalococcoides mccartyi GT	Hydrogen	Vinyl chloride	3.1×10^8	3[a]	g dry weight/Cl^-	Sung et al. (2006)
Dehalococcoides mccartyi CBDB1	Hydrogen	Tetrachloroethene	1.3×10^8	1.6[a]	g dry weight/Cl^-	Sung et al. (2006)
Dehalococcoides mccartyi CBDB1	Hydrogen	2,3-dichlorophenol	$7.6 * 10^7$	0.9[a]	g dry weight/Cl^-	Adrian et al. (2007)
Dehalococcoides mccartyi BAV1	Hydrogen	Vinyl chloride	6.3×10^7	0.76[a]	g dry weight/Cl^-	Marco-Urrea et al. (2011)
Dehalogenimonas lykanthroporepellens	Hydrogen	Chloropropanes		None reported		Moe et al. (2009)
Geobacter lovleyi	Acetate	Tetrachloroethene		None reported		Sung et al. (2006)

[a]Estimates require a conversion factor from cell per mL. We adopt the convention in *Dehalococcoides* sp. description (Löffler et al. 2013)

kinetic differences in reductive dehalogenases present in facultative versus obligate organohalide-respiring bacteria. The few reductive dehalogenases that have been characterized reveal a range of kinetic parameters. For instance, the reported K_m of PceA in *Sulfurospirillum multivorans* is 200 μM (Neumann et al. 1996), whereas the reported K_m of a PceA isolated from *Desulfitobacterium* PCE-S was 10 μM (Miller et al. 1998), similar to the K_m of vinyl chloride reductase from *Dehalococcoides* strain VS (Müller et al. 2004; Parthasarathy et al. 2015). Following recent successes in heterologous expression of active reductive dehalogenases (Sjuts et al. 2012; Mac Nelly et al. 2014; Bommer et al. 2014; Payne et al. 2015; Parthasarathy et al. 2015) should accelerate the elucidation of kinetic diversity among reductive dehalogenases from both facultative and obligate organohalide-respiring bacteria. Systematic differences in kinetics and growth yields across bacterial phyla would support the hypothesis that organohalide "respiration" has considerable mechanistic variability.

Another proposed physiological function of organohalide reduction not associated with energy conservation involves the dehalogenation of organohalogen compounds for the purpose of making them amenable to degradation as purely organic compounds via pathways of central metabolism. In this capacity, reductive dehalogenation would enable catabolic access to a carbon source not available to microorganisms lacking reductive dehalogenases (Lohner and Spormann 2013).

12.1.2 Are Menaquinones Universal Electron Carriers in Organohalide Respiration?

The identity of essential electron transfer components and mechanisms involved in energy conservation coupled to reductive dehalogenation has been a central but unresolved issue in organohalide respiration research. Based on our current understanding, some constraints appear to apply broadly. Reductive dehalogenases with activity toward aliphatic or aromatic halogenated hydrocarbons studied to date are membrane-associated. [Some in *Dehalogenimonas* may prove to be an exception (Siddaramappa et al. 2012).] There is consensus—based on biochemical and predictive bioinformatics—that the reductive dehalogenases are associated with the outer side of the cytoplasmic membrane (Suyama et al. 2002; Hölscher et al. 2004). There is also strong evidence that in many microorganisms reductive dehalogenation may be quinone-dependent. However, insufficient evidence exists to decide whether quinones are an essential feature of organohalide respiration in general, particularly in slow-growing, obligate organohalide-respiring bacteria.

Quinones are membrane-bound electron carriers with a key role in electron transport phosphorylation. Menaquinones have a well-documented role in electron transport and energy conservation, e.g., of sulfate and Fe(III) reducing bacteria, and facultative organohalide-respiring bacteria *D. dehalogenans* and *S. multivorans* have been experimentally shown to utilize menaquinones during organohalide reduction (Goris et al. 2014; Kruse et al. 2015). Furthermore, menaquinone is essential in some facultative organohalide-respiring bacteria. For instance, *Desulfitobacterium dichloroeliminans* strain DCA1 is auxotrophic for menaquinone and requires the amendment of vitamin K_2 in the medium for growth on lactate and 1,2-dichloroethane (De Wildeman et al. 2004).

However, a quinone-independent mode of reductive dehalogenation in *Dehalococcoides* and *Dehalogenimonas*-type microorganisms is predicted based on the absence of genes for *de novo* quinone biosynthesis as well as the presence of only a few annotated genes with quinone/quinol function. Moreover, growth of axenic strains has been observed in medium devoid of vitamin K_2 (Schipp et al. 2013). One contrasting study provided experimental data for two *Dehalococcoides mccartyi* strains, FL2 and BAV1, which reports an unusually high level of ubiquinone, which is atypical in anaerobic respiration and which was present at a 20-fold higher concentration than observed in aerobically grown *Escherichia coli*; menaquinone (MK-5) was also detected in the lipid analysis (White et al. 2005). No follow-up studies were published to resolve this interesting observation.

Therefore, the absence of conclusive support for the presence and involvement of menaquinone in *Dehalococcoides* raises the possibility of a different, "quinone-independent" mechanism of energy conservation and perhaps distinct electron transfer mechanisms in some reductive dehalogenases. If in the H_2-utilizing *Dehalococcoides* energy is conserved via a Complex I type mechanism, as has been proposed in *Sulfurospirillum* or *Desulfitobacterium* spp. (Goris et al. 2014; Kruse et al. 2015), then a so far unidentified electron carrier, functioning as the quinone equivalent, is expected to be present and transfer electrons to the reductive dehalogenase (Fig. 12.2). Interestingly, the Nuo-type (Complex 1) operon in *Dehalococcoides* is significantly different from that in *Desulfitobacterium hafniense* strains Y51 and DCB2. Complex I in *Desulfitobacterium* is predicted to contain 14 subunits and complete NADH binding (N), proton translocating (P), and quinone-reduction (Q) modules. In contrast, the operon encoding for Complex I in *Dehalococcoides* lacks the genes for the Nuo N-module (*nuoE-G*), leaving only 11 subunits (Seshadri et al. 2005; Nonaka et al. 2006). However, the module comprised of NuoE, NuoF, and NuoG is frequently missing in eubacteria and archaea. In fact, Complex I is hypothesized to have arisen through the combined evolution of separate functional modules (Friedrich 2001).

Dehalobacter spp, which all evidence suggests use menaquinone, also contain an 11 subunit Nuo-type complex. However, the 11-subunit Nuo complex is functionally and structurally similar to energy-conserving [NiFe]-hydrogenase complexes, which couple the oxidation of ferredoxin to H_2 formation with the generation of a proton motive force independent of a quinone pool (Buckel and Thauer 2013). As of yet, no specific partner protein has been identified to functionally substitute for the N-module, and it is hypothesized that the 11-subunit Complex 1 may instead interact with various electron donor and acceptor proteins (Moparthi and Hagerhall 2011). Alternatively, the absent classical NADH binding module could be complemented by subunits from other oxidoreductases distal to

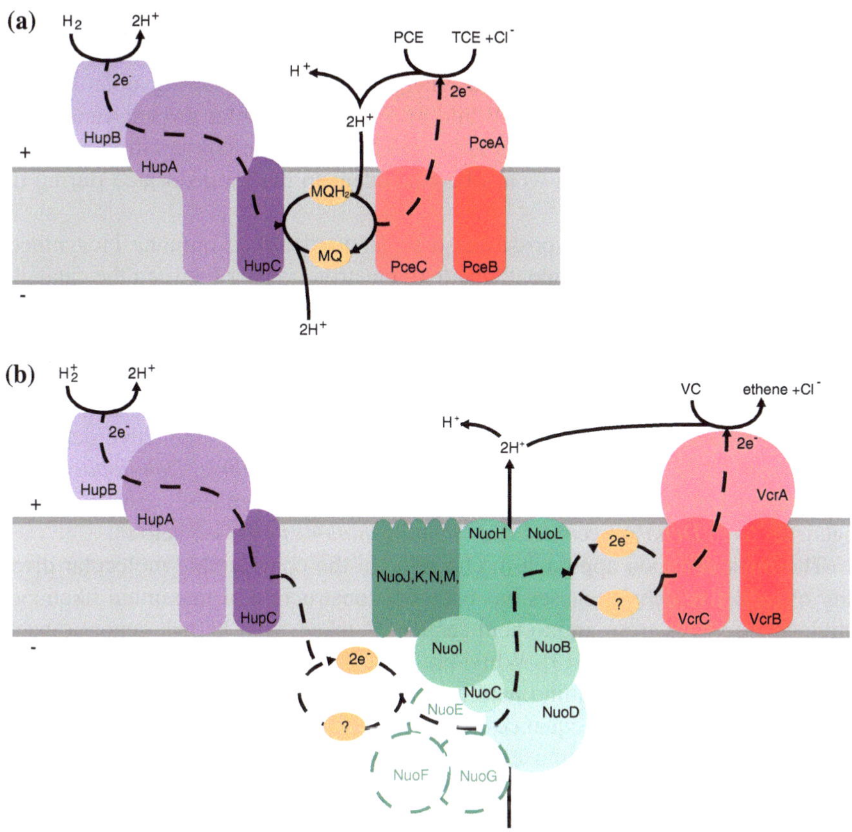

Fig. 12.2 Proposed models for energy conservation in quinone-dependent and in quinone-independent reductive dehalogenation with H_2 as electron donor. **a** hypothesized electron transfer process in *Dehalobacter* spp. (see chapter on *Dehalobacter*). **b** Possible role of Nuo-type respiratory complex in electron transfer and energy conservation mechanism in *Dehalococcoides mccartyi*. The NADH accepting subunits NuoEFG are not found in the Nuo operon in either *Dehalococcoides mccartyi* or *Dehalobacter* spp., but they are found in *Desulfitobacterium* spp. (*dashed lines* represent missing subunits not found in the operon of *Dehalococcoides mccartyi* or *Dehalobacter* spp.) Currently, biochemical evidence is lacking to determine whether the Nuo-type Complex I is in fact involved in *Dehalococcoides* respiratory energy conservation

the Nuo operon (Seshadri et al. 2005). In *D. mccartyi,* it is unclear whether the Nuo-type Complex I is essential in respiratory energy conservation.

For the "quinone-dependent" dehalogenating bacteria—growing on lactate, pyruvate, or even H_2—energy can be conserved via a Complex I type reduction of quinones, where the membrane-bound reductive dehalogenase serves to oxidize quinol. During the quinol oxidation via reductive dehalogenases, energy may or may not be conserved, while the reduction of quinones is most likely coupled to electron transport phosphorylation via a Complex I type mechanism. If energy were conserved at both steps, the expected growth yield would be closer to the theoretical thermodynamic maximum; however, currently, there is neither experimental nor theoretical evidence that energy is conserved during quinol oxidation and chloroethene reduction. While data is sparse, molar growth yields from *Dehalobacter* grown on hydrogen are consistent with the conservation of 1 ATP per chloride released, which would be equivalent to $3H^+$ translocated during the energy-conserving process.

Unless *Dehalococcoides* possesses a so-far unidentified quinone biosynthetic pathway, a "quinone-independent" mode of electron transfer between the catabolic uptake hydrogenase (HupL) and the reductive dehalogenase is likely. From a biochemical point of view, it can be hypothesized that a different path for electron uptake to a halogenated substrate exists in "quinone-independent" reductive dehalogenases. An underlying structural difference may explain the significant divide in amino acid sequence between those "quinone-dependent" reductive dehalogenases found in the fast growing *Sulfurospirillum, Desulfovibrio, Dehalobacter,* and *Desulfitobacterium* spp., and the assumed "quinone-independent" reductive dehalogenases as in *Dehalococcoides* and *Dehalogenimonas* spp (see below).

The most common approach to characterize the considerable molecular diversity of reductive dehalogenases has been the construction of maximum likelihood phylogenetic trees from amino acid sequences taken from known genomic backgrounds (see Hug et al. 2013). While phylogenetic trees emphasize well relations between closely related amino acid sequences, their interpretation is more difficult for distantly related sequences. Figure 12.3 displays an amino acid sequence similarity network graph as a complementary approach, where nodes represent unique reductive dehalogenase sequences, and edges represent homology between sequences where pairwise alignments share greater than 40 % identity. Edge length approximates evolutionary divergence encoded in bitscore of each pairwise alignment (Weston et al. 2004; Atkinson et al. 2009).

This network shows a bifurcation between almost all Chloroflexi and non-Chloroflexi reductive dehalogenases. Despite the common acquisition of these genes via horizontal gene transfer in the Choroflexi, there are no high similarity alignments between Chloroflexi and non-Chloroflexi reductive dehalogenase amino acid sequences. By contrast, there are multiple high similarity (>90 % at the amino acid level) homologs between *Desulfitobacterium* spp. and *Dehalobacter* spp., with possible cross-phylum lateral gene transfer to *Geobacter lovleyi* (Wagner et al. 2012b). As hypothesized above, barriers to gene transfer could be mechanistic at the protein level where "quinone-dependent" oxidoreductases are incompatible with the physiological environment and biochemical constraints amenable to "quinone-independent" reductive dehalogenases. Alternatively, the lack of shared nucleotide-level homologs with high pairwise sequence identity could be due to genetic barriers to recombination confronting mobile genetic elements or phage specialized to a Chloroflexi genomic background, which is a well-recognized dynamic in the maintenance of bacterial diversity (Lawrence 1997, 1999).

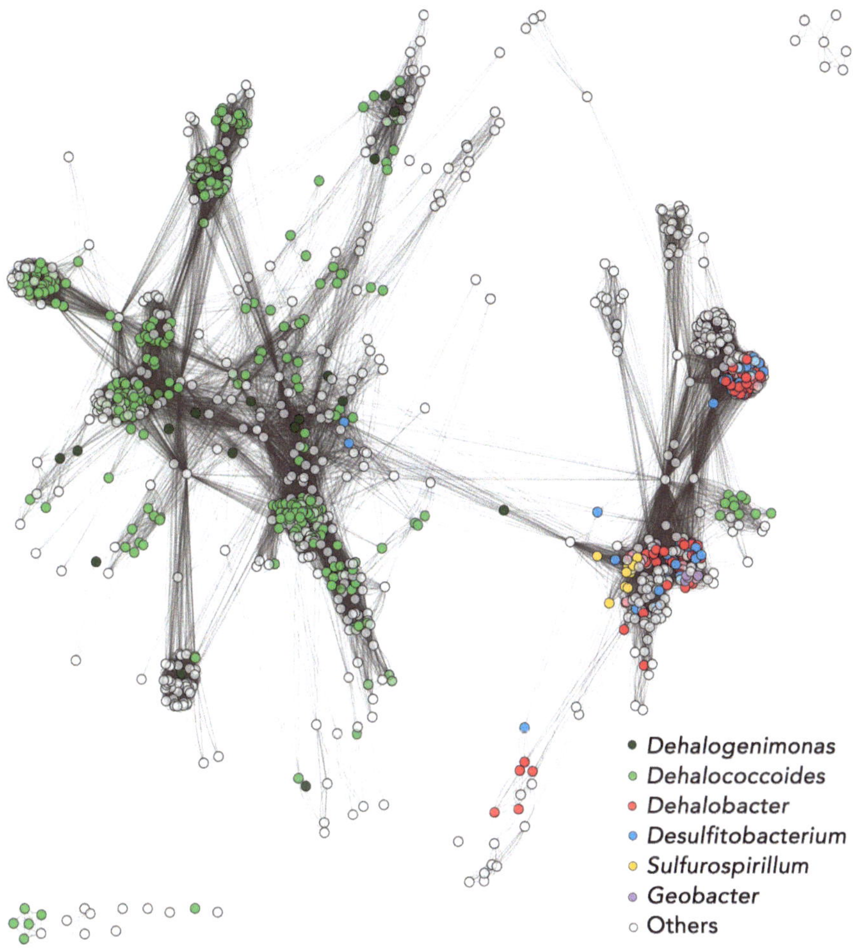

Fig. 12.3 Reductive dehalogenase amino acid sequence similarity network displaying as nodes full and near-full length sequences in the reductive dehalogenase protein family (IPR012832 InterPro v51.0, updated April 2015) with TAT signal peptides removed. Edges represent pairwise percent identity (greater than 40 % at the amino acid level) spanning a blastp alignment of at least 80 % the length of the shorter protein sequence. The graph is arranged in an edge-weighted layout by the software Cytoscape where distances between nodes approximate evolutionary divergence. Node color corresponds to the genus of the host organism in which each reductive dehalogenase was identified. Environmental sequences are categorized as other. A network bifurcation exists between Chloroflexi and non-Chloroflexi protein sequences

12.2 Comparative Genomics and Evolution of Obligate Organohalide-Respiring Bacteria

We follow our discussion of comparative physiology by examining what has been learned from comparative genomics by focusing on the obligate organohalide-respiring bacteria that constitute the *Dehalococcoides*, *Dehalobacter*, and *Dehalogenimonas* genera.

12.2.1 The High Plasticity Region Concept in Dehalococcoides mccartyi

Significant information about gene content and gene synteny was revealed with the completion of the *D. mccartyi* 195 genome (Seshadri et al. 2005) and subsequent comparison to that of the *D. mccartyi* CBDB1 genome (Kube et al. 2005). Kube and colleagues noted the striking feature that "whereas the positions and orientations of more than 90 % of the genes not associated with *rdh* loci are conserved," the positions of the *rdh* gene loci demonstrate remarkably "high plasticity," with orthologous *rdh* between the genomes scattered in different positions. Analysis of nucleotide k-mer frequency by Regeard et al. (2005), sequencing of highly identical *tceA* genes from 12 geographically distinct environments by Krajmalnik-Brown et al. (2007), and identification of unusual codon usage bias in *vcrA* and a site-specific mechanism of integration at the *ssrA* locus by McMurdie et al. (2007, 2011) strengthened the notion that reductive dehalogenases in *Dehalococcoides* are subject to significant horizontal gene transfer via genomic islands (Regeard et al. 2005; Krajmalnik Brown et al. 2007; McMurdie et al. 2007).

The further comparison of genomes of *D. mccartyi* CBDB1 and *D. mccartyi* 195 with the vinyl chloride respiring strains *D. mccartyi* VS and BAV1 (McMurdie et al. 2009) substantiated the high-plasticity region concept. The high frequency of repeated elements, loci of site-specific integration, and genes associated with recombination appear to maintain the variability of these regions, within an otherwise stable genomic core. Moreover, the clustering of reductive dehalogenases to these regions could facilitate co-expression of multiple *rdhA* as has been observed under electron acceptor limited conditions (Johnson et al. 2008). Indeed, the genomes of all eleven *D. mccartyi* strains sequenced by 2015 share this distinct genomic architecture, with two high plasticity regions near the origin of replication (see Fig. 12.4).

McMurdie et al. (2009) compared the *D. mccartyi* genomes to those of *Canditatus Pelagibacter*, the smallest known free-living bacterium, noting that evolutionary pressure for genomic streamlining has been somewhat weaker in *D. mccartyi* based on the presence of significantly more transposons and pseudogenes in *D. mccartyi* chromosomes. It is hypothesized that in the case of *D. mccartyi*, evolutionary selection acted toward maintaining *rdhA* diversity, presumably enabling

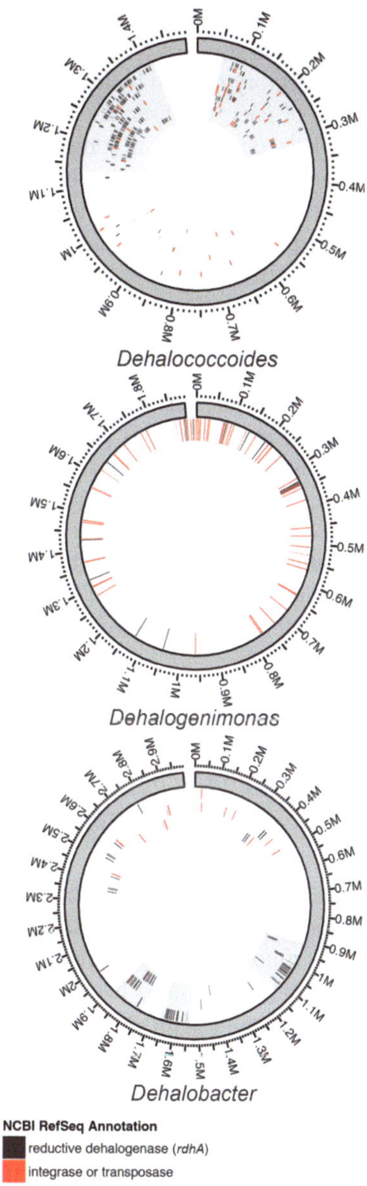

Fig. 12.4 Genomic positions of annotated reductive dehalogenase (*black*), integrase and transposase (*red*) genes in finished genomes of *Dehalococcoides, Dehalogenimonas, and Dehalobacter* spp. Shaded regions highlight areas of high reductive dehalogenase concentration. Notably, all eleven *Dehalococcoides mccartyi* strains share a distinct genomic architecture, with two reductive dehalogenase-enriched high plasticity regions near the origin of replication. Reductive dehalogenases in *Dehalobacter* spp. are also spatially concentrated. The genome of *Dehalogenimonas lykanthroporepellens* is highly enriched with transposon-annotated genes, compared to the other genera

opportunistic adaptation to changing niches defined by diverse halogenated substrates. The small size of the *D. mccartyi* genomes (~1.4 Mbp) reflects an evolutionary trajectory marked by irreversible commitment to organohalide respiration as sole mode of energy conservation and cofactor auxotrophy. In contrast to *Pelagibacter*, which is considered a completely free-living microbe, *D. mccartyi*, may *in situ* depend on a proximal microbial community providing not only the catabolic H_2 but also essential cofactors, including corrinoids or precursors. Whether cofactor cross-feeding is an artifact of laboratory culturing in growth medium containing corrinoids is still a subject of debate, especially as comparative genomics is supported by metagenomic efforts. For instance, shotgun metagenomic sequencing of a microbial consortium from the Alameda Naval Air Station (California, USA) identified 9 biosynthesis genes associated with cobalamin biosynthesis on a *Dehalococcoides* associated scaffold, which had not been previously observed in pure culture genomes (Brisson et al. 2012). A genomic propensity for rapid gene loss (e.g., of corrinoid biosynthetic genes) is consistent with the observed genome streamlining as well as the recent acquisition of genomic islands for anthropogenic chloroethene reduction (McMurdie et al. 2011). Much recent and ongoing work is targeted at understanding possible fitness and kinetic difference among closely related coexisting strains of *D. mccartyi* (Morris et al. 2007; Lee et al. 2011; Hug et al. 2011; Heavner et al. 2013; Marshall et al. 2014; Mayer-Blackwell et al. 2014).

12.2.2 Comparative Genomics Between *Dehalococcoides mccartyi* and *Dehalogenimonas lykanthroporepellens*

With the completion of the *Dehalogenimonas lykanthroporepellens* BL-DC-9 genome sequence, genomic insights from a closely related *Dehalococcoides* "out-group" microorganism became available. Strain BL-DC-9 is another deep-branching Chloroflexi bacterium capable of respiring a variety of chloropropanes (Siddaramappa et al. 2012; Mukherjee et al. 2014).

The chromosome of *D. lykanthroporepellens* is 200 kB larger than that of the largest of *D. mccartyi* genome sequenced to date (1.47 MB), and it contains significantly different genomic content. It has approximately 700 genes absent in any *D. mccartyi* strain. Siddaramappa et al. (2012) also evaluated the positions of shared genes between *D. mccartyi* strains and *D. lykanthroporepellens* BL-DC-9 and hypothesized that the lack of synteny suggested a divergent evolution. McMurdie et al. (2011) used the 432 core orthologous protein encoding genes shared between *D. mccartyi* and *D. lykanthroporepellens* to estimate 1.2–34 million years as a lower bound on the time of divergence from a most recent common ancestor, depending on assumption about the microorganisms' doubling time in nature.

Many of the unique genes with no homolog between the *Dehalococcoides* and *Dehalogenimonas* genera (determined by a 20 % amino identity cut-off) encode

Table 12.2 Genomic features of selected organohalide respiring genera

Genus	Finished genomes	Phylum	Genome size (MB)	GC %	Annotated genes	tRNA	*rdhA* genes
Dehalococcoides	11	Chloroflexi	1.34–1.47	47–49	1436–1642	46–47	17–43
Dehalogenimonas	1	Chloroflexi	1.7	55	1771	47	18
Dehalobacter	3	Firmicutes	2.9–3.1	44–45	2908–3040	51–52	16–21
Desulfitobacterium	6	Firmicutes	3.2–5.7	42–48	3152–5489	59–75	1–7
Sulfurospirillum	1	Proteobacteria (epsilon)	3.2	40	3285	45	2
Anaeromyxobacter	1	Proteobacteria (delta)	5	74	4524	49	1

Genomic data was tabulated from the finished version of genomes available at integrated microbial genomics resource at the joint genome institute. To ensure standard criteria for annotation of a reductive dehalogase major subunit genes (*rdhA*), the number of annotated reductive dehalogenases per genome is based on genes containing and reductive dehalogenase_domain (IPR028894) using InterPro v51.0 (Jones et al. 2014)

"endonucleases/methylases, heterodisulfide reductases, acetyltransferases, kinases, phosphatases, and dehalogenases." Notably, the *D. lykanthroporepellens* genome contains genes predicted to encode for the biosynthesis of osmoprotectants, which are absent in *D. mccartyi* and which may allow *D. lykanthroporepellens* to occupy a broader ecological range allowing growth in environments with fluctuating salinity (Siddaramappa et al. 2012). Interestingly in this microorganism, many of the strain-specific genes were found in the genomic neighborhood of prophage and mobile elements. While genes encoding transposases are colocated in *D. lykanthroporepellens* with many *rdhA* genes, they do not appear to be associated with *Dehalococcoides*-type genomic islands and are not concentrated into high plasticity regions as was observed in *D. mccartyi* (Fig. 12.4). Furthermore, the lack of cognate *rdhB* genes associated with 6 of 25 *rdhA*, raises speculation that reductive dechlorination in *D. lykanthroporepellens* may occur in the cytoplasm (Siddaramappa et al. 2012). Further members of the *Dehalogenimonas* genus have been isolated (Bowman et al. 2013), and the identification of their genomic sequences should be a high priority within the organohalide respiration research community (Table 12.2).

12.2.3 Comparative Genomics Between Dehalococcoides mccartyi and Dehalobacter spp.

While the comparison of *D. mccartyi* to *D. lykanthroporepellens* is informative about divergent evolution, comparing *D. mccartyi* strains to recently sequenced *Dehalobacter* spp. offers a view on convergent evolution between members of distinct bacterial Phyla to occupy remarkably similar and narrow metabolic niches.

Upon inspection of the *Dehalobacter restrictus* genome in 2013, Rupakula et al. noted many features present in *Dehalobacter* absent in *Dehalococcoides*: (i) chemotaxis and flagellar machinery, (ii) complete menaquinone biosynthesis genes starting from chorismate, (iii) near complete cobalamin biosynthesis operon, (iv) a complete Wood–Ljungdahl pathway, and (v) a dominance of *cprK*-type rather than *marR* mode of gene regulation. Nevertheless, *D. restrictus,* along with newly sequenced *Dehalobacter* CF and *Dehalobacter* 11-DCA genomes (Tang et al. 2012), share a number of features with *D. mccartyi.*

Foremost, both are characterized by relatively small genomes, with *Dehalobacter* taking an intermediate position between the streamlined *D. mccartyi* genome and that of the metabolically versatile and more closely related *Desulfitobacterium* (Rupakula et al. 2013) (Fig. 12.4). Since the *Dehalobacter* and *Desulfitobacterium* genera are phylogenetically more closely related, an inspection of those metabolic features absent in *Dehalobacter* spp. may suggest important steps towards genome reduction and expansion of the *rdhA* repertoire associated with a transition to an obligate organohalide respiring lifestyle.

Dehalobacter compared to *Desulfitobacterium* show a strong specialization to hydrogen metabolism. [Fermentation of dichloromethane may also be possible in some strains (Justicia-Leon et al. 2012; Lee et al. 2012).] To this end, the *D. restrictus* genome contains 8 hydrogenases (including membrane-bound HupL, Ech-type, Hyc-types) as well as an 11 subunit Nuo-type respiratory Complex 1. But perhaps most notably, *D. restrictus*, like *D. mccartyi.* lacks functional capacity for B_{12} biosynthesis despite the cofactor's crucial role in the strains' reductive dehalogenases. *D. restrictus'* loss of B_{12} biosynthetic capacity is much more recent than for *D. mccartyi* as evidenced by a near intact cobalamin biosynthesis operon with a single frame shift mutation in the gene *cbiH* (Rupakula et al. 2013), implying that some *Dehalobacter* spp. may still synthesize cobalamin *de novo* rather than scavenging and modifying the cofactor from a supporting microbial community or an investigator in the laboratory. The possibility that the *cbiH* mutation observed in *D. restrictus* was simply acquired during laboratory cultivation in the presence of a vitamin solution cannot be ruled out; however, global proteomic survey showed that the upper portion of the cobalamin biosynthesis operon were not expressed in the presence of vitamin B_{12}, suggesting that *D. restrictus* regulates portions of the pathway separately to avoid the biosynthetic costs, while retaining the capacity for cofactor modification (see more about the post-translational regulation of cobalamin biosynthesis in the *Dehalobacter* chapter of this book). This may well represent a first evolutionary step toward full cobalamin auxotrophy via gene loss as observed in the *Dehalococcoides* genus. Growth of various *Dehalobacter* containing consortia that have not been previously subjected to strong population bottlenecking—in a chemostat system with varied vitamin supplement regimes—would be a good candidate system to test hypotheses concerning genomic reduction.

12.3 Conclusion

Comparative physiology of organohalide reducing bacteria reveals: (i) Microorganisms use halogenated electron acceptors for a variety of metabolic strategies, not all of which may fit under the classical definition of respiration. (ii) Selective pressures act at the level of the genomes of obligate organohalide-respiring bacteria, reflecting the degree to which organohalide respiration defines the niche of these microorganisms. (iii) The common facultative versus obligate classification scheme used to understand organohalide respiring microorganisms is useful; however, the mechanism of energy conservation associated with reductive dehalogenation, particularly in the *D. mccartyi* and *Dehalogenimonas* spp., remains an unresolved question. Further mechanistic studies of catabolic electron flow is needed to fully understand—on a biochemical level—what unites and distinguishes "organohalide respiring" microorganisms.

Literature

Adrian L, Hansen SK, Fung JM, Görisch H, Zinder SH (2007) Growth of *Dehalococcoides* strains with chlorophenols as electron acceptors. Environ Sci Technol 41:2318–2323. doi:10.1021/es062076m

Ahn Y-B, Kerkhof LJ, Häggblom MM (2009) *Desulfoluna spongiiphila* sp. nov., a dehalogenating bacterium in the Desulfobacteraceae from the marine sponge *Aplysina aerophoba*. Int J Syst Evol Microbiol 59:2133–2139. doi:10.1099/ijs.0.005884-0

Anraku Y, Gennis RB (1987) The aerobic respiratory chain of *Escherichia coli*. Trends Biochem Sci 12:262–266. doi:10.1016/0968-0004(87)90131-9

Atkinson HJ, Morris JH, Ferrin TE, Babbitt PC (2009) Using sequence similarity networks for visualization of relationships across diverse protein superfamilies. PLoS ONE 4:e4345. doi:10.1371/journal.pone.0004345

Bommer M, Kunze C, Fesseler J, Schubert T, Diekert G, Dobbek H (2014) Structural basis for organohalide respiration. Science 346:455–458. doi:10.1126/science.1258118

Bowman KS, Nobre MF, da Costa MS, Rainey FA, Moe WM (2013) *Dehalogenimonas alkenigignens* sp. nov., a chlorinated-alkane-dehalogenating bacterium isolated from groundwater. Int J Syst Evol Microbiol 63:1492–1498. doi:10.1099/ijs.0.045054-0

Boyle AW, Phelps CD, Young LY (1999) Isolation from estuarine sediments of a *Desulfovibrio* strain which can grow on lactate coupled to the reductive dehalogenation of 2,4,6-tribromophenol. Appl Environ Microbiol 65:1133–1140

Brisson VL, West KA, Lee PK, Tringe SG, Brodie EL, Alvarez-Cohen L (2012) Metagenomic analysis of a stable trichloroethene-degrading microbial community. ISME J 6:1702–1714. doi:10.1038/ismej.2012.15

Buckel W, Thauer RK (2013) Energy conservation via electron bifurcating ferredoxin reduction and proton/Na^+ translocating ferredoxin oxidation. Biochimica et Biophysica Acta (BBA)—Bioenergetics 1827:94–113. doi:10.1016/j.bbabio.2012.07.002

Chen K, Huang L, Xu C, Liu X, He J, Zinder SH, Li S, Jiang J (2013) Molecular characterization of the enzymes involved in the degradation of a brominated aromatic herbicide. Mol Microbiol 89:1121–1139. doi:10.1111/mmi.12332

Christiansen N, Ahring B (1996) *Desulfitobacterium hafniense* sp. nov., an anaerobic, reductively dechlorinating bacterium. Int J Syst Evol Microbiol 46:442–448. doi:10.1099/00207713-46-2-442

De Wildeman S, Linthout G, Van Langenhove H, Verstraete W (2004) Complete lab-scale detoxification of groundwater containing 1,2-dichloroethane. Appl Microbiol Biotechnol 63:609–612. doi:10.1007/s00253-003-1363-y

Dolfing J, Novak I (2014) The Gibbs free energy of formation of halogenated benzenes, benzoates and phenols and their potential role as electron acceptors in anaerobic environments. Biodegradation 26:15–27. doi:10.1007/s10532-014-9710-5

Efremov RG, Baradaran R, Sazanov LA (2010) The architecture of respiratory complex I. Nature 465:441–445. doi:10.1038/nature09066

Friedrich T (2001) Complex I: a chimaera of a redox and conformation-driven proton pump? J Bioenerg Biomembr 33:169–177. doi:10.1023/A:1010722717257

Friedrich T, Scheide D (2000) The respiratory complex I of bacteria, archaea and eukarya and its module common with membrane-bound multisubunit hydrogenases. FEBS Lett 479:1–5. doi:10.1016/S0014-5793(00)01867-6

Goris T, Schubert T, Gadkari J, Wubet T, Tarkka M, Buscot F, Adrian L, Diekert G (2014) Insights into organohalide respiration and the versatile catabolism of *Sulfurospirillum multivorans* gained from comparative genomics and physiological studies. Environ Microbiol 16:3562–3580. doi:10.1111/1462-2920.12589

Hau HH, Gralnick JA (2007) Ecology and biotechnology of the genus *Shewanella*. 61:237–258. doi: 10.1146/annurev.micro.61.080706.093257

He J, Ritalahti KM, Yang K-L, Koenigsberg SS, Löffler FE (2003) Detoxification of vinyl chloride to ethene coupled to growth of an anaerobic bacterium. Nature 424:62–65. doi:10.1038/nature01717

He J, Robrock KR, Alvarez-Cohen L (2006) Microbial reductive debromination of polybrominated diphenyl ethers (PBDEs). Environ Sci Technol 40:4429–4434. doi:10.1021/es052508d

Heavner GLW, Rowe AR, Mansfeldt CB, Pan JK, Gossett JM, Richardson RE (2013) Molecular biomarker-based biokinetic modeling of a PCE-dechlorinating and methanogenic mixed culture. Environ Sci Technol 47:3724–3733. doi:10.1021/es303517s

Holliger C, Hahn D, Harmsen H, Ludwig W, Schumacher W, Tindall B, Vazquez F, Weiss N, Zehnder AJB (1998) *Dehalobacter restrictus* gen. nov. and sp. nov., a strictly anaerobic bacterium that reductively dechlorinates tetra- and trichloroethene in an anaerobic respiration. Arch Microbiol 169:313–321. doi:10.1007/s002030050577

Hölscher T, Krajmalnik Brown R, Ritalahti KM, von Wintzingerode F, Görisch H, Löffler FE, Adrian L (2004) Multiple nonidentical reductive-dehalogenase-homologous genes are common in *Dehalococcoides*. Appl Environ Microbiol 70:5290–5297. doi:10.1128/AEM.70.9.5290-5297.2004

Hug LA, Salehi M, Nuin P, Tillier ER, Edwards EA (2011) Design and verification of a pangenome microarray oligonucleotide probe set for *Dehalococcoides* spp. Appl Environ Microbiol 77:5361–5369. doi:10.1128/AEM.00063-11

Hug LA, Maphosa F, Leys D, Löffler FE, Smidt H, Edwards EA, Adrian L (2013) Overview of organohalide-respiring bacteria and a proposal for a classification system for reductive dehalogenases. Philos Trans R Soc Lond B Biol Sci 368:20120322. doi:10.1098/rstb.2012.0322

Hunt KA, Flynn JM, Naranjo B, Shikhare ID, Gralnick JA (2010) Substrate-level phosphorylation is the primary source of energy conservation during anaerobic respiration of *Shewanella oneidensis* strain MR-1. J Bacteriol 192:3345–3351. doi:10.1128/JB.00090-10

Johnson DR, Brodie EL, Hubbard AE, Andersen GL, Zinder SH, Alvarez-Cohen L (2008) Temporal transcriptomic microarray analysis of "*Dehalococcoides ethenogenes*" strain 195 during the transition into stationary phase. Appl Environ Microbiol 74:2864–2872. doi:10.1128/AEM.02208-07

Jones P, Binns D, Chang H-Y, Fraser M, Li W, McAnulla C, McWilliam H, Maslen J, Mitchell A, Nuka G, Pesseat S, Quinn AF, Sangrador-Vegas A, Scheremetjew M, Yong S-Y, Lopez R, Hunter S (2014) InterProScan 5: genome-scale protein function classification. Bioinformatics 30:1236–1240. doi:10.1093/bioinformatics/btu031

Justicia-Leon SD, Ritalahti KM, Mack EE, Löffler FE (2012) Dichloromethane fermentation by a *Dehalobacter* sp. in an enrichment culture derived from pristine river sediment. Appl Environ Microbiol 78:1288–1291. doi:10.1128/AEM.07325-11

Krajmalnik Brown R, Sung Y, Ritalahti KM, Saunders FM, Löffler FE (2007) Environmental distribution of the trichloroethene reductive dehalogenase gene (tceA) suggests lateral gene transfer among *Dehalococcoides*. FEMS Microbiol Ecol 59:206–214. doi:10.1111/j.1574-6941.2006.00243.x

Kreft J-U, Bonhoeffer S (2005) The evolution of groups of cooperating bacteria and the growth rate versus yield trade-off. Microbiology 151:637–641. doi:10.1099/mic.0.27415-0

Krumholz LR (1997) Desulfuromonas chloroethenica sp. nov. uses tetrachloroethylene and trichloroethylene as electron acceptors. Int J Syst Evol Microbiol 47:1262–1263. doi:10.1099/00207713-47-4-1262

Kruse T, van de Pas BA, Atteia A, Krab K, Hagen WR, Goodwin L, Chain P, Boeren S, Maphosa F, Schraa G, de Vos WM, van der Oost J, Smidt H, Stams AJM (2015) Genomic, proteomic, and biochemical analysis of the organohalide respiratory pathway in *Desulfitobacterium dehalogenans*. J Bacteriol 197:893–904. doi:10.1128/JB.02370-14

Kube M, Beck A, Zinder SH, Kuhl H, Reinhardt R, Adrian L (2005) Genome sequence of the chlorinated compoundlrespiring bacterium *Dehalococcoides* species strain CBDB1. Nat Biotechnol 23:1269–1273. doi:10.1038/nbt1131

Lawrence JG (1997) Selfish operons and speciation by gene transfer. Trends Microbiol 5:355–359. doi:10.1016/S0966-842X(97)01110-4

Lawrence JG (1999) Gene transfer, speciation, and the evolution of bacterial genomes. Curr Opin Microbiol 2:519–523. doi:10.1016/S1369-5274(99)00010-7

Lee PKH, Cheng D, Hu P, West KA, Dick GJ, Brodie EL, Andersen GL, Zinder SH, He J, Alvarez-Cohen L (2011) Comparative genomics of two newly isolated *Dehalococcoides* strains and an enrichment using a genus microarray. ISME J 5:1014–1024. doi:10.1038/ismej.2010.202

Lee M, Low A, Zemb O, Koenig J, Michaelsen A, Manefield M (2012) Complete chloroform dechlorination by organochlorine respiration and fermentation. Environ Microbiol 14:883–894. doi:10.1111/j.1462-2920.2011.02656.x

Löffler FE, Yan J, Ritalahti KM, Adrian L, Edwards EA, Konstantinidis KT, Müller JA, Fullerton H, Zinder SH, Spormann AM (2013) *Dehalococcoides mccartyi* gen. nov., sp. nov., obligately organohalide-respiring anaerobic bacteria relevant to halogen cycling and bioremediation, belong to a novel bacterial class, Dehalococcoidia classis nov., order Dehalococcoidales ord. nov. and family Dehalococcoidaceae fam. nov., within the phylum Chloroflexi. Int J Syst Evol Microbiol 63:625–635. doi:10.1099/ijs.0.034926-0

Lohner ST, Spormann AM (2013) Identification of a reductive tetrachloroethene dehalogenase in *Shewanella sediminis*. Philos Trans R Soc Lond B Biol Sci 368:20120326. doi:10.1098/rstb.2012.0326

Mac Nelly A, Kai M, Svatoš A, Diekert G, Schubert T (2014) Functional heterologous production of reductive dehalogenases from *Desulfitobacterium hafniense* strains. Appl Environ Microbiol 80:4313–4322. doi:10.1128/AEM.00881-14

Mackiewicz M, Wiegel J (1998) Comparison of energy and growth yields for *Desulfitobacterium dehalogenans* during utilization of chlorophenol and various traditional electron acceptors. Appl Environ Microbiol 64:352–355

Maphosa F, De Vos WM, Smidt H (2010) Exploiting the ecogenomics toolbox for environmental diagnostics of organohalide-respiring bacteria. Trends Biotech 28(6):308–316. doi:10.1016/j.tibtech.2010.03.005

Marco-Urrea E, Nijenhuis I, Adrian L (2011) Transformation and carbon isotope fractionation of tetra- and trichloroethene to trans-dichloroethene by *Dehalococcoides* sp. strain CBDB1. Environ Sci Technol 45:1555–1562. doi:10.1021/es1023459

Marshall IPG, Azizian MF, Semprini L, Spormann AM (2014) Inferring community dynamics of organohalide-respiring bacteria in chemostats by covariance of rdhA gene abundance. FEMS Microbiol Ecol 87:428–440. doi:10.1111/1574-6941.12235

Mayer-Blackwell K, Azizian MF, Machak C, Vitale E, Carpani G, de Ferra F, Semprini L, Spormann AM (2014) Nanoliter qPCR platform for highly parallel, quantitative assessment of reductive dehalogenase genes and populations of dehalogenating microorganisms in complex environments. Environ Sci Technol 48:9659–9667. doi:10.1021/es500918w

Maymó-Gatell X, Chien Y-T, Gossett JM, Zinder SH (1997) Isolation of a bacterium that reductively dechlorinates tetrachloroethene to ethene. Science 276:1568–1571. doi:10.1126/science.276.5318.1568

McMurdie PJ, Behrens SF, Holmes S, Spormann AM (2007) Unusual codon bias in vinyl chloride reductase genes of *Dehalococcoides* species. Appl Environ Microbiol 73:2744–2747. doi:10.1128/AEM.02768-06

McMurdie PJ, Behrens SF, Müller JA, Göke J, Ritalahti KM, Wagner R, Goltsman E, Lapidus A, Holmes S, Löffler FE, Spormann AM (2009) Localized plasticity in the streamlined genomes of vinyl chloride respiring *Dehalococcoides*. PLoS Genet 5:e1000714. doi:10.1371/journal.pgen.1000714

McMurdie PJ, Hug LA, Edwards EA, Holmes S, Spormann AM (2011) Site-specific mobilization of vinyl chloride respiration islands by a mechanism common in *Dehalococcoides*. BMC Genom 12:287. doi:10.1186/1471-2164-12-287

Miller E, Wohlfarth G, Diekert G (1998) Purification and characterization of the tetrachloroethene reductive dehalogenase of strain PCE-S. Arch Microbiol 169:497–502. doi:10.1007/s002030050602

Moe WM, Yan J, Nobre MF, da Costa MS, Rainey FA (2009) *Dehalogenimonas lykanthroporepellens* gen. nov., sp. nov., a reductively dehalogenating bacterium isolated from chlorinated solvent-contaminated groundwater. Int J Syst Evol Microbiol 59:2692–2697. doi:10.1099/ijs.0.011502-0

Mohn WW, Tiedje JM (1990) Strain DCB-1 conserves energy for growth from reductive dechlorination coupled to formate oxidation. Arch Microbiol 153:267–271. doi:10.1007/BF00249080

Moparthi VK, Hägerhäll C (2011) The evolution of respiratory chain complex I from a smaller last common ancestor consisting of 11 protein subunits. J of Mol Evol, 72(5-6):484–497. doi:10.1007/s00239-011-9447-2

Morris RM, Fung JM, Rahm BG, Zhang S, Freedman DL, Zinder SH, Richardson RE (2007) Comparative proteomics of *Dehalococcoides* spp. reveals strain-specific peptides associated with activity. Appl Environ Microbiol 73:320–326. doi:10.1128/AEM.02129-06

Mukherjee K, Bowman KS, Rainey FA, Siddaramappa S, Challacombe JF, Moe WM (2014) *Dehalogenimonas lykanthroporepellens* BL-DC-9T simultaneously transcribes many rdhA genes during organohalide respiration with 1,2-DCA, 1,2-DCP, and 1,2,3-TCP as electron acceptors. FEMS Microbiol Lett 354:111–118. doi:10.1111/1574-6968.12434

Müller JA, Rosner BM, von Abendroth G, Meshulam-Simon G, McCarty PL, Spormann AM (2004) Molecular identification of the catabolic vinyl chloride reductase from *Dehalococcoides* sp. strain VS and its environmental distribution. Appl Environ Microbiol 70:4880–4888. doi:10.1128/AEM.70.8.4880-4888.2004

Myers JM, Myers CR (2000) Role of the tetraheme cytochrome CymA in anaerobic electron transport in cells of *Shewanella putrefaciens* MR-1 with normal levels of menaquinone. J Bacteriol 182:67–75. doi:10.1128/JB.182.1.67-75.2000

Neumann A, Wohlfarth G, Diekert G (1996) Purification and characterization of tetrachloroethene reductive dehalogenase from *Dehalospirillum multivorans*. J Biol Chem 271:16515–16519

Nonaka H, Keresztes G, Shinoda Y, Ikenaga Y, Abe M, Naito K, Inatomi K, Furukawa K, Inui M, Yukawa H (2006) Complete genome sequence of the dehalorespiring bacterium *Desulfitobacterium hafniense* Y51 and comparison with *Dehalococcoides ethenogenes* 195. J Bacteriol 188:2262–2274. doi:10.1128/JB.188.6.2262-2274.2006

Parthasarathy A, Stich TA, Lohner ST, Lesnefsky A, Britt RD, Spormann AM (2015) Biochemical and EPR-spectroscopic investigation into heterologously expressed vinyl chloride reductive dehalogenase (VcrA) from *Dehalococcoides mccartyi* strain VS. J Am Chem Soc 137:3525–3532. doi:10.1021/ja511653d

Payne KAP, Quezada CP, Fisher K, Dunstan MS, Collins FA, Sjuts H, Levy C, Hay S, Rigby SEJ, Leys D (2015) Reductive dehalogenase structure suggests a mechanism for B12-dependent dehalogenation. Nature 517:513–516. doi:10.1038/nature13901

Poole RK, Ingledew WJ (1987) Pathways of electrons to oxygen: *Escherichia coli* and *Salmonella typhimurium:* cellular and molecular biology. Am Soc Microbiol (Washington, DC)

Pruesse E, Quast C, Knittel K, Fuchs BM, Ludwig W, Peplies J, Glöckner FO (2007) SILVA: a comprehensive online resource for quality checked and aligned ribosomal RNA sequence data compatible with ARB. Nucl Acids Res 35:7188–7196. doi:10.1093/nar/gkm864

Pruesse E, Peplies J, Glöckner FO (2012) SINA: accurate high-throughput multiple sequence alignment of ribosomal RNA genes. Bioinformatics 28:1823–1829. doi:10.1093/bioinformatics/bts252

Regeard C, Maillard J, Dufraigne C, Deschavanne P, Holliger C (2005) Indications for acquisition of reductive dehalogenase genes through horizontal gene transfer by *Dehalococcoides* ethenogenes strain 195. Appl Environ Microbiol 71:2955–2961. doi:10.1128/AEM.71.6.2955-2961.2005

Rupakula A, Kruse T, Boeren S, Holliger C, Smidt H, Maillard J (2013) The restricted metabolism of the obligate organohalide respiring bacterium *Dehalobacter restrictus*: lessons from tiered functional genomics. Philos Trans R Soci Lond B Biol Sci 368:20120325. doi:10.1098/rstb.2012.0325

Sanford RA, Cole JR, Tiedje JM (2002) Characterization and description of *Anaeromyxobacter dehalogenans* gen. nov., sp. nov., an Aryl-halorespiring facultative anaerobic myxobacterium. Appl Environ Microbiol 68:893–900. doi:10.1128/AEM.68.2.893-900.2002

Schipp CJ, Marco-Urrea E, Kublik A, Seifert J, Adrian L (2013) Organic cofactors in the metabolism of *Dehalococcoides mccartyi* strains. Philos Trans R Soc Lond B Biol Sci 368:20120321. doi:10.1098/rstb.2012.0321

Seshadri R, Adrian L, Fouts DE, Eisen JA, Phillippy AM, Methe BA, Ward NL, Nelson WC, Deboy RT, Khouri HM, Kolonay JF, Dodson RJ, Daugherty SC, Brinkac LM, Sullivan SA, Madupu R, Nelson KE, Kang KH, Impraim M, Tran K, Robinson JM, Forberger HA, Fraser CM, Zinder SH, Heidelberg JF (2005) Genome sequence of the PCE-dechlorinating bacterium *Dehalococcoides ethenogenes*. Science 307:105–108. doi:10.1126/science.1102226

Siddaramappa S, Challacombe JF, Delano SF, Green LD, Daligault H, Bruce D, Detter C, Tapia R, Han S, Goodwin L, Han J, Woyke T, Pitluck S, Pennacchio L, Nolan M, Land M, Chang Y-J, Kyrpides NC, Ovchinnikova G, Hauser L, Lapidus A, Yan J, Bowman KS, da Costa MS, Rainey FA, Moe WM (2012) Complete genome sequence of *Dehalogenimonas lykanthroporepellens* type strain (BL-DC-9T) and comparison to "*Dehalococcoides*" strains. Stand Genomic Sci 6:251–264. doi:10.4056/sigs.2806097

Sjuts H, Fisher K, Dunstan MS, Rigby SE, Leys D (2012) Heterologous expression, purification and cofactor reconstitution of the reductive dehalogenase PceA from *Dehalobacter restrictus*. Protein Expr Purif 85:224–229. doi:10.1016/j.pep.2012.08.007

Sun B, Cole JR, Tiedje JM (2001) *Desulfomonile limimaris* sp. nov., an anaerobic dehalogenating bacterium from marine sediments. Int J Syst Evol Microbiol 51:365–371. doi:10.1099/00207713-51-2-365

Sung Y, Ritalahti KM, Sanford RA, Urbance JW, Flynn SJ, Tiedje JM, Löffler FE (2003) Characterization of two tetrachloroethene-reducing, acetate-oxidizing anaerobic bacteria and their description as *Desulfuromonas michiganensis* sp. nov. Appl Environ Microbiol 69:2964–2974. doi:10.1128/AEM.69.5.2964-2974.2003

Sung Y, Ritalahti KM, Apkarian RP, Löffler FE (2006) Quantitative PCR confirms purity of strain GT, a novel trichloroethene-to-ethene-respiring *Dehalococcoides* isolate. Appl Environ Microbiol 72:1980–1987. doi:10.1128/AEM.72.3.1980-1987.2006

Suyama A, Yamashita M, Yoshino S, Furukawa K (2002) Molecular characterization of the PceA reductive dehalogenase of *Desulfitobacterium* sp. strain Y51. J Bacteriol 184:3419–3425. doi:10.1128/JB.184.13.3419-3425.2002

Tang S, Gong Y, Edwards EA (2012) Semi-automatic in silico gap closure enabled de novo assembly of two *Dehalobacter* genomes from metagenomic data. PLoS ONE 7:e52038. doi:10.1371/journal.pone.0052038

Thauer RK, Jungermann K, Decker K (1977) Energy conservation in chemotrophic anaerobic bacteria. Bacteriol Rev 41:100. doi:10.1038/ismej.2014.24

Trumpower B (1990) The protonmotive Q cycle. Translocation 265:11409–11412

van de Pas BA, Jansen S, Dijkema C, Schraa G, de Vos WM, Stams AJM (2001) Energy yield of respiration on chloroaromatic compounds in *Desulfitobacterium dehalogenans*. Appl Environ Microbiol 67:3958–3963. doi:10.1128/AEM.67.9.3958-3963.2001

Wagner A, Cooper M, Ferdi S, Seifert J, Adrian L (2012a) Growth of *Dehalococcoides mccartyi* strain CBDB1 by reductive dehalogenation of brominated benzenes to benzene. Environ Sci Technol 46:8960–8968. doi:10.1021/es3003519

Wagner DD, Hug LA, Hatt JK, Spitzmiller MR, Padilla-Crespo E, Ritalahti KM, Edwards EA, Konstantinidis KT, Löffler FE (2012b) Genomic determinants of organohalide-respiration in *Geobacter lovleyi*, an unusual member of the Geobacteraceae. BMC Genom 13:200. doi:10.1186/1471-2164-13-200

Weston J, Elisseeff A, Zhou D, Leslie CS, Noble WS (2004) Protein ranking: from local to global structure in the protein similarity network. PNAS 101:6559–6563. doi:10.1073/pnas.0308067101

White DC, Geyer R, Peacock AD, Hedrick DB, Koenigsberg SS, Sung Y, He J, Löffler FE (2005) Phospholipid furan fatty acids and ubiquinone-8: lipid biomarkers that may protect *Dehalococcoides* strains from free radicals. Appl Environ Microbiol 71:8426–8433. doi:10.1128/AEM.71.12.8426-8433.2005

Part III
Ecology of Organohalide-Respiring Bacteria

Chapter 13
Electron Acceptor Interactions Between Organohalide-Respiring Bacteria: Cross-Feeding, Competition, and Inhibition

Kai Wei, Ariel Grostern, Winnie W.M. Chan, Ruth E. Richardson and Elizabeth A. Edwards

Abstract Because of the stepwise, progressive nature of reductive dehalogenation reactions, polyhalogenated parent electron acceptors and their corresponding intermediary dehalogenation products are almost always simultaneously present in the environments where these processes occur. Moreover, a wide variety of polyhalogenated industrial chemicals find their way into the environment, frequently at the same manufacturing or processing facility, resulting in complex mixtures of pollutants in the subsurface. Therefore, cross-feeding, competition, and inhibition are inevitable in these systems and their magnitude or impact must be quantified to better predict and promote the rate and extent of detoxification. Numerical simulations of reactive transport that incorporate fitted parameters describing these processes provide useful tools to evaluate scenarios. Direct experimental evidence of inhibition or competition using defined enzyme and microbial assays provides a more mechanistic understanding of these effects. Combining carefully executed, well-designed experiments with modeling ultimately provides the most useful data for fundamental understanding as well as decision-making in the context of remediation.

K. Wei · A. Grostern · W.W.M. Chan · E.A. Edwards (✉)
Department of Chemical Engineering and Applied Chemistry, University of Toronto, Toronto, ON, Canada
e-mail: elizabeth.edwards@utoronto.ca

R.E. Richardson
Department of Civil and Environmental Engineering, Cornell University Ithaca, 14853 New York, USA
e-mail: RER26@cornell.edu

L. Adrian and F.E. Löffler (eds.), *Organohalide-Respiring Bacteria*,
DOI 10.1007/978-3-662-49875-0_13

13.1 Introduction

Many natural and industrialized environments contain a varied assortment of halogenated organic compounds. Naturally occurring organohalides are highly diverse but are present at relatively low concentrations (see Chap. 2). In contrast, chlorinated solvents and other halogenated chemicals produced at industrial scale and inadvertently released into the environment have concentrations that range over many orders of magnitude up to aqueous solubilities and even beyond as they form non aqueous phase liquids (NAPL) in the subsurface (see Chap. 22). Moreover, contaminated sites often contain mixtures of contaminants and these mixtures do frequently confound cleanup efforts. How does this diversity of potential halogenated electron acceptors influence the resident or bioaugmented microbial communities, particularly the organohalide reducing bacteria (OHRB), and the resulting rates of dehalogenation? And importantly, what are the best strategies for enhancing bioremediation at sites with multiple co-contaminants? This Chapter describes observations and on-going efforts to understand and quantify competition and inhibition in anaerobic dehalogenating microbial communities.

13.2 Lessons from Field Observations and Enrichment Cultures

OHRB are now used commonly for remediation of chlorinated solvent contamination (e.g., Chap. 22). Over the past 30 years, research has revealed that reductive dechlorination of halogenated organics proceeds through sequential two electron-equivalent steps, and that pathways can converge, creating critical bottlenecks where inhibition is particularly noticeable. Much of the work towards understanding interactions and inhibition has been driven by the need to predict and model these effects in the context of remediation at sites with high concentrations of solvents (DNAPL sites) or where multiple contaminants co-exist. The most common halogenated organic contaminants in groundwater are the chlorinated ethenes (e.g., perchloroethene—PCE and trichloroethene—TCE), chlorinated ethanes (trichloroethanes—TCA and dichloroethanes—DCA) and chloromethanes (chloroform—CF, and dichloromethane—DCM) and thus to date the majority of work has been focused on these co-contaminants. A summary of the major dechlorination pathways for simple chlorinated aliphatic compounds and microbes that have been found to catalyze these interconnecting reactions is illustrated in Fig. 13.1 highlighting for example the important role that *Dehalococcoides* plays in removing the bottleneck through *cis*-dichloroethene (cDCE) and vinyl chloride (VC) to nontoxic ethene. This is one of the prime reasons why bioaugmentation with *Dehalococcoides*-containing cultures is often highly effective. While this funnel toward VC is well known for the chlorinated alkanes and alkenes, many other contaminants are also released into the environment as mixtures, with their own

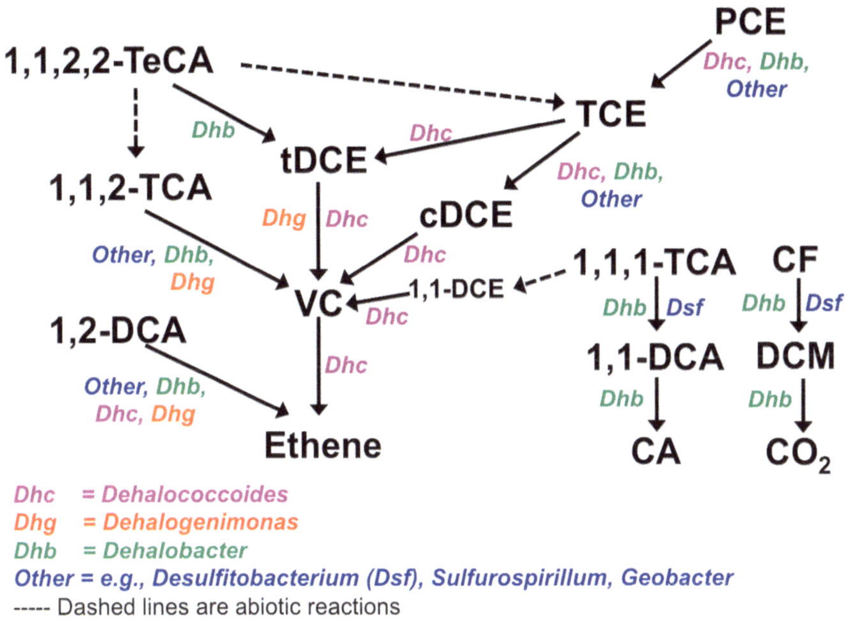

Fig. 13.1 Overview of common chlorinated solvent dechlorination pathways and organisms responsible. Chlorinated ethenes, chlorinated ethanes, and chlorinated methanes make up the largest fraction of the organic halogenated contaminants found at contaminated sites, and frequently co occur. Figure 13.1 illustrates the bottleneck at the level of cDCE and VC for many dechlorination pathways. For example, chloroform and 1,1,1-TCA inhibit VC dechlorination, consequently affecting dechlorination of all upstream compounds. To date, cDCE and VC dechlorination have been observed only within members of the *Dehalococcoidia*, while many of the other dechlorination steps can be catalyzed by several different genera. Abbreviations: *TeCA* Tetrachloroethene; *PCE* Perchloroethene; *TCE* Trichloroethene; *DCE* Dichloroethene; *VC* Vinyl Chloride; *TCA* Trichloroethane; *DCA* Dichloroethane; *CF* Chloroform; *DCM* Dichloromethane

patterns of dechlorination and cross-inhibition. Prime examples include the polychlorinated biphenyls (PCBs), polychlorinated benzenes, polychlorinated phenols, polychlorinated alkanes, and brominated flame retardants. Research is on-going for many of these contaminants and others to further the understanding of dehalogenation patterns and to help promote detoxification, although quantifying inhibition is challenging, as described further below.

13.2.1 *Simulations of Field and Laboratory Observations*

Various models of microbial growth and biodegradation with or without transport have been used extensively to simulate the fate of contaminants and their daughter products observed in laboratory studies and at field sites, ranging from very simple models to extraordinarily complex multidimensional and multiphase models (Chambon et al. 2010, 2013; Sleep and Sykes 1993; Fennell and Gossett 1998; Smatlak et al. 1996). The United States Geological Survey (USGS) provides resources for some widely used commercial models (e.g. MODPATH and MT3D transport models). Statistician George Box stated aptly that "all models are wrong, but some are useful". Many contaminant fate and transport models are tremendously useful for evaluating scenarios, establishing rate-limiting steps, and discovering knowledge-gaps in different contexts. In these models, certain assumptions about microbially catalyzed reactions have to be made. Often they are modeled using first order decay constants, sometimes Monod kinetic expressions are included, perhaps incorporating fermentation reactions, and possibly with additional thermodynamic constraints. As the models increase in complexity, considerably more knowledge on the responsible microbial populations and their abundance, reaction pathways and kinetics is needed. Simple first order kinetic models of the stepwise reductive dechlorination rarely fit experimental data well. Slower than predicted rates of dechlorination of lesser chlorinated daughter products, an observation practitioners and site owners like to call "cDCE- or VC-stall", lead to investigations into causes and mechanisms for inhibition of these key dechlorination steps.

Substrate inhibition is manifested when specific rates of reaction decrease at higher substrate concentration. In the context of sequential reductive dechlorination, substrate inhibition is sometimes referred to as "self-inhibition" when a compound and any of its daughter products cause inhibition of any dehalogenation step. The frequently modeled mechanism assumes a single enzyme (or organism) and substrates and dechlorinated products competing for the active site. As such, it is very difficult to distinguish between specific mechanisms of inhibition (substrate inhibition versus competitive inhibition) because at any given time, the parent compound as well as several dehalogenated products may be present. PCE and TCE and their daughter products cDCE and VC have been the subject of the most research. The more chlorinated ethenes have been found to competitively inhibit the reductive dechlorination of the less chlorinated ethenes and, to a lesser extent, vice versa (Cupples et al. 2004; Yu et al. 2005; Yu and Semprini 2004). This type of competitive inhibition, as well as substrate inhibition, have been incorporated into some contaminant fate and transport models used for evaluating remediation alternatives and effectiveness (Chambon et al. 2010; Chen et al. 2013; Christ and Abriola 2007; Haest et al. 2010, 2012; Heavner et al. 2013; Huang and Becker 2011; Lai and Becker 2013; Lee et al. 2004; Mendoza-Sanchez and Cunningham 2012; Pon et al. 2003; Popat and Deshusses 2011; Sabalowsky and Semprini 2010; Yu et al. 2005). These models typically assume either a single aggregate dechlorinating population, or perhaps two distinct populations, one for more chlorinated and another for less chlorinated ethenes. Neither reflects the true situation because microbial populations are diverse assemblages of subpopulations with different substrate preferences that often vary spatially and temporally. A thorough review of kinetic models describing reductive dechlorination of chlorinated ethenes was recently published (Chambon et al. 2013). Inhibition constants (K_i)

for self-inhibition among chlorinated ethenes were reviewed and summarized, and generally fall within the range between 3–10 μM (0.5–15 mg/L), although one of the many challenges noted in this review is that the range of published kinetic and inhibition constants used to describe these phenomena varies over four orders of magnitude in some cases, meaning that the fundamental underlying mechanisms are still poorly understood, not independent, very complex, experimentally difficult to analyze, and dynamic. Experiments to quantify the magnitude of inhibition are difficult to control as the concentration of substrates and products are always changing during sequential stepwise dechlorination reactions. The enzymes and microbes catalyzing these dechlorination reactions vary from site to site, and enzymes from different microbes have different inherent kinetic and inhibition properties that affect rates and extent of dechlorination. Despite these challenges, models have proven very useful to predict the most likely ranges of outcomes, to evaluate scenarios, and especially to point towards knowledge gaps that need better understanding or more data.

Cross-inhibition occurs when different series of halogenated substrates acted on by distinct microbial populations interact, such as the observed inhibition of chlorinated ethene dechlorination by chlorinated ethanes. Contaminated sites often contain mixtures of different types of chlorinated solvents as well as other contaminants and these mixtures do frequently confound cleanup efforts. An important example is the co-contamination of sites with TCE and 1,1,1-trichloroethane (1,1,1-TCA) (Scheutz et al. 2011), which occurs at ~20 % of sites on the U.S. Environmental Protection Agency's National Priorities List (Grostern and Edwards 2006a). 1,1,1-TCA was introduced as a less toxic alternative to TCE (Doherty 2000) but was subsequently found to be a potent inhibitor of methanogenesis (Adamson and Parkin 2000). Chloroform (CF) is another frequent co-contaminant with chlorinated ethenes. 1,1,1-TCA and CF have specifically been identified to inhibit VC dechlorination resulting in the buildup of the toxic intermediates cDCE and VC (Chan et al. 2011; Chung and Rittmann 2008; Duchesneau et al. 2007; Duhamel et al. 2002; Grostern and Edwards 2006a; Scheutz et al. 2011; Yu et al. 2005). The reverse is also true: VC itself inhibits dechlorination of chlorinated ethanes and methanes (Grostern et al. 2009). A review of sites with comingled halogenated contaminants vividly illustrates the challenges for cleanup (Scheutz et al. 2011).

13.2.2 *Microbial Diversity, Competition, Inhibition, and Reductive Dehalogenases*

Competition and inhibition drive evolution through natural selection. These processes result in highly diverse microbial populations with different patterns of resource use that occupy different niches. We appreciate now more than ever the diversity of microbes that can dehalogenate, as well as the diversity of reductive dehalogenases within these microbes that mediate these reactions, as covered in the many chapters of this book. Moreover, these organohalide-respiring

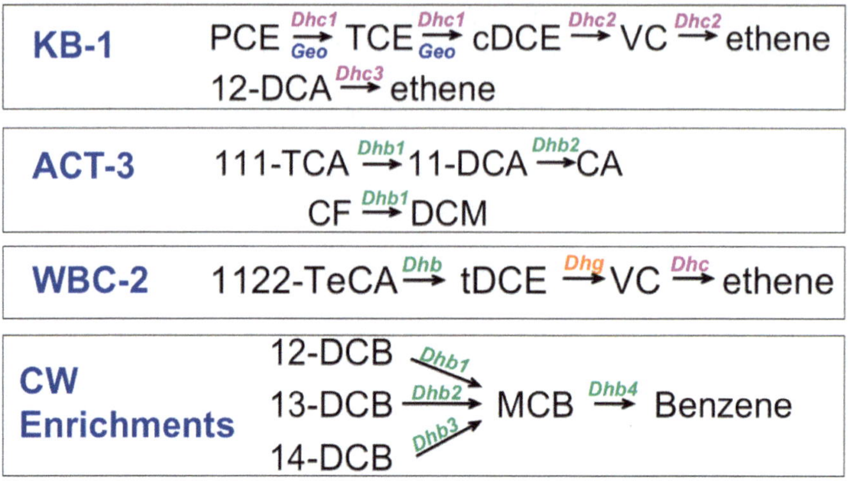

Fig. 13.2 Substrate specialization in OHRB. Enrichment cultures (referred to as KB-1, ACT-3, WBC-2, and CW) that dechlorinate PCE, chloroform and 1,1,1-TCA, 1,1,2,2-tetrachloroethene, and chlorinated benzenes have been developed and in many cases successfully deployed for remediation. Recent research is revealing that within these dechlorinating enrichment cultures, different OHRB specialize to individual steps in dechlorination. In some cases, very close relatives carry out distinct steps, such as two nearly identical strains of *Dehalobacter* (*Dhb*) in the case of 1,1,1-TCA and 1,1-DCA dechlorination in the ACT-3 enrichment culture, or very different microbes working together, in the case of *Dehalobacter*, *Dehalogenimonas,* and *Dehalococcoides* to bring about transformation of 1,1,2,2-TeCA to ethene in the WBC-2 enrichment culture. See Fig. 13.1 for definition of chemical names. Nomenclature *Dhb1*, *Dhb2* refers to presence of distinct strains of *Dehalobacter* specialized for the dechlorination step indicated

bacteria are critically dependent on fermenting and acetogenic microbes within their community to provide essential electron donors, co-factors, and other nutrients for growth (Chap. 14). It is now becoming clear that OHRB are often extreme "niche-specialists", with rather pronounced strain and enzyme specialization observed under some conditions (Fig. 13.2). For example, in the studies of chlorobenzene dechlorination in enrichment cultures from the Chambers Works (CW) site (Nelson et al. 2011), *Dehalobacter* was found to be responsible for observed dechlorination, but different, yet closely related strains, were specialized for each of the three dichlorobenzene isomers (Nelson et al. 2014) and yet a fourth strain for the substrate monochlorobenzene (MCB) (Fig. 13.2). Moreover, some strains could not dechlorinate the other's substrate. The same has been observed in the ACT-3 enrichment culture that dechlorinates 1,1,1-TCA and 1,1-dichloroethane (1,1-DCA) (Tang and Edwards 2013; Tang et al. 2012) with yet again different *Dehalobacter* strains (Fig. 13.2). Chloroform and 1,1,1-TCA (methyl chloroform) are dechlorinated by the same *Dehalobacter* strain, while 1,1-DCA is dechlorinated by a closely related but distinct strain (Tang and Edwards 2013). The challenge is in detecting this strain differentiation, as often these organisms have identical 16S rRNA sequences—the difference lies primarily in the complement of reductive dehalogenase genes that each organism harbors. The evolutionary and ecological implications of this specialization are fascinating to consider and may also provide a mechanism for the community to overcome competition for and inhibition by closely related electron acceptors. It is important to therefore develop investigative tools to explore the evolution of microbial communities in the presence of multiple competing substrates. These pressures will select for a different microbial populations compared to experiments with fewer substrates, but will likely develop more interesting models for understanding adaptation and inhibition in the context of evolution, and certainly provide a means for generating more robust inocula for bioaugmentation.

At the heart of the processes involved in competition and inhibition are the diverse reductive dehalogenases that catalyze the different reactions. Different OHRB harbor and express a different array of enzymes and thus have different substrate and inhibition profiles. Selection conditions established at field sites and in enrichment cultures will therefore influence microbial populations and associated kinetic and inhibition properties. Rudimentary understanding of the relationship between enzyme, microbe, and specific activity has already greatly improved bioremediation outcomes, and further understanding will further advance the technology. In particular, the recent elucidation of the three-dimensional structures of two reductive dehalogenases (see Chap. 20) is very exciting. Briefly, a tetrachloroethene dehalogenase, PceA, produced by *Sulfurospirillum multivorans* was purified and structurally characterized by Bommer et al. (2014) and a catabolic (nonrespiratory) reductive dehalogenase was characterized by Payne et al. (2015), with striking similarity to the structure of PceA. These structures revealed a "letterbox-like" insertion unit or substrate channel that must strongly contribute to substrate preferences of the enzyme. Enzymes with similar substrates appear to share "letterbox" substrate channel sequences. These structures, together with recent advances in the ability to heterologous express reductive dehalogenases in more tractable hosts (MacNelly et al. 2014; Parthasarathy et al. 2015) now provide the opportunity to examine these enzymes at a more mechanistic level, and perhaps will eventually enable the design proteins and microbes that are less susceptible to inhibition by competing or co-occurring substrates. Characterization of single enzymes is essential for fundamental understanding and to inform more complex systems, and may eventually lead to rationally designed enzymes to deal with problematic pollutants. At the same time, approaches to characterize activity and inhibition in complex natural and engineered mixed microbial dechlorinating communities are also needed, as described below.

13.3 Experimental Approaches to Quantify Inhibition in Microbial Communities

Inhibition observed in anaerobic dehalogenating microbial communities in the lab and in the field results from multiple factors, including direct inhibition exerted by the inhibitor on OHRB, general toxicity to essential bacterial cell functions, competition and survival among different bacterial species, and suboptimum site parameters like pH, redox, or salinity. For example, halogenated compounds are relatively hydrophobic and impact multiple cellular functions, and may not simply interact directly or specifically with reductive dehalogenase enzymes. Chloroform is a well-known inhibitor of methanogens and other C1-metabolizing organisms that are not OHRB (Yu and Smith 2000), possibly as a result of nonspecific binding to the many porphyrin-like co-factors and methyltransferases involved in metabolism. OHRB and reductive dechlorination can also be inhibited by other electron acceptors, such as oxygen, nitrate, and various forms of sulfur and iron, as well as detergents and solvents used to *enhance* remediation and factors such as pH or salinity (Wei and Finneran 2011; McGuire and Hughes 2003; Paul et al. 2013; Paul and Smolders 2014; Townsend and Suflita 1997; Yeh et al. 1999). In order to specifically classify the mechanisms for inhibition of reductive dechlorination in complex systems, we shall restrict our analysis here to the effect of the simultaneous presence of multiple distinct halogenated organics on the growth of OHRB and on dehalogenation kinetics, as seen in the context of groundwater remediation.

13.3.1 Deconvoluting the Mechanisms of Inhibition

In the context of OHRB living in a mixed anaerobic community of fermenters, acetogens and methanogens in their environment, we can envisage three general levels of inhibition, ranging from effects directly on the catalytic reductive dehalogenase enzyme, to effects on the OHRB more generally or effects on other microbes on which the OHRB depend:

(1) Enzyme level: Inhibition at the level of the membrane-bound reductive dehalogenase enzyme itself. For example the inhibitor can bind to the enzyme at the active site in the classical notion of competitive inhibition or can specifically bind elsewhere on the enzyme or enzyme–substrate complex, resulting in decreased activity.
(2) Organism level: Inhibition at the level of the OHRB itself, where the inhibitor interferes with or binds to critical cellular components other than reductive dehalogenases, resulting in impaired growth and dehalogenating activity. For example inhibitors may interact with vital metabolic enzymes or electron transport components, or impair membrane function via hydrophobic interactions.
(3) Community level: at this level, the inhibitor interferes with other microbes in the community responsible for key processes of electron donor fermentation and provision of hydrogen and other essential nutrients or co-factors to the OHRB (and possibly vice versa).

13.3.2 Challenges in Experimental Design to Deconvolute Inhibitory Effects

The study of inhibition is challenging because typically, multiple superimposed mechanisms underlie an experimental observation. Several strategies must be used to tease apart nonspecific from specific interactions, and to create controlled conditions and quantifiable, transferable results. This is quite a challenging task in the context of mixed anaerobic microbial cultures and multiple substrates. Inhibition can be competitive or not, and it can be reversible or not, depending on the nature of the interaction. In all cases, the effect will be very much dependent on the concentration of the inhibiting compound and concentration of microbial populations. Thus, experiments need to be carefully designed to tease apart confounding effects as much as possible while maintaining sufficient complexity to be useful to extend to field scale.

There are many approaches to investigate inhibitory effects, often requiring many experiments at many different concentrations and environmental conditions. The frustration expressed by Chambon et al. (2013) regarding the range of kinetic constants reported in the literature reflects this challenge. In order to minimize the number of uncontrolled variables in such experiments, a systematic evaluation of an inhibitor is a sound yet feasible approach, starting with experiments in cell free extracts (CFE) to interrogate putative inhibitor at the level of enzymes, followed by resting cell assays that look at additional effects on the whole organism while uncoupled from growth, and then topped off by experiments with growing microbial communities. The advantage of such an approach is that the range of conditions to be tested in the more complex growth experiments can be narrowed down in earlier experiments. This approach is illustrated in Sects. 13.4–13.6.

13.4 Inhibition of Reductive Dehalogenases (Enzyme Level)

The classical approach to determine inhibitor effects at the enzyme level is to perform enzyme kinetic assays on purified enzyme preparations measuring initial rates of reaction as a function of initial substrate concentration to derive Michaelis-Menten parameters. Such experiments carried out in the presence of increasing concentrations of a specific inhibitor then provide information on the potency of the inhibitor. Kinetic data can be fit to models of competitive, uncompetitive, noncompetitive, or Haldane inhibition to ascertain nature of interactions between the inhibitor and substrate. From data of initial rate (V_0) plotted against initial substrate concentration $[S]$ for increasing inhibitor concentration $[I]$, the kinetic parameters and inhibition constants can be extracted using any number of mathematical tools, though some software offer built-in features to extract enzyme kinetics (e.g. SigmaPlot® or JMP®). The most appropriate model (competitive, noncompetitive, and uncompetitive) for a given data set is selected based on statistical criteria (Hurvich and Tsai 1989).

Such classical enzyme assays can be adapted to characterizing the dehalogenases in OHRB with careful consideration of their oxygen sensitivity and the volatile nature of many organohalide substrates (Loffler et al. 1996; Rosner et al. 1997). Because reductive dehalogenases are not easily purified nor heterologously expressed, dehalogenase enzyme assays have often been applied to crude cell extracts, typically prepared from mixed cultures. This approach works well despite uncertainties in true active protein concentrations, provided that the same batch of crude extract is used for an entire series of concentrations of substrate and inhibitor. Enzyme assays for reductive dechlorination typically use titanium citrate-reduced methyl viologen as the artificial electron donor under anaerobic conditions and halogenated organic as the electron acceptor. As mentioned before, the analytical requirements to determine velocity are also a little tricky owing to the fact that the substrates are typically volatile and hydrophobic. These single substrate dechlorination assays have been adapted to explore the inhibitory effects of different halogenated compounds in mixed cultures (Chan et al. 2011; Grostern et al. 2009). Briefly, to perform tests for inhibition, cells from active dechlorinating cultures are collected by centrifugation, disrupted by sonication in lysis buffer, and the resulting CFE are dispensed into aliquots and stored frozen at −80 °C until use. All manipulations are carried out without exposure to oxygen. Crude protein extracts in assay buffer containing electron-accepting substrates, inhibitors, and reduced methyl viologen as electron donor are incubated anaerobically for 1–3 h and then sampled to determine the extent and rate of dechlorination normalized to total protein concentration. The data obtained from each substrate/inhibitor combination are then fit to enzyme kinetic models to quantify and categorize the nature of the inhibition. What results are estimates of kinetic and model parameters, including the inhibition constant that defines the strength of the inhibition. This approach is illustrated with examples below.

Chlorinated ethenes (e.g., PCE and TCE) are frequently found in groundwater in combination with chlorinated ethanes (e.g., 1,1,1-TCA) and methanes (e.g., CF) at industrial sites, and dechlorination rates are negatively impacted in these situations, thus inhibition by these co-contaminants has been the focus of several investigations. Inhibition was observed to act in both directions: chlorinated ethenes inhibited the reductive dechlorination of chlorinated alkanes, and vice versa. These examples are described below, in Sects. 13.4.1 and 13.4.2.

13.4.1 Inhibition of Chloroalkene Reductive Dehalogenases by Chlorinated Alkanes

The inhibitory effects exerted by chlorinated alkanes on the reductive dechlorination of chlorinated ethenes were investigated in mixed cultures enriched on PCE or TCE. Cell-free extract from three chloroethene-dechlorinating enrichment cultures, including KB-1 (Duhamel et al. 2002), OW (Daprato et al. 2007), and Biodechlor Inoculum (BDI) (Ritalahti et al. 2006) were amended with

TCE, cDCE, or VC and challenged with 1,1,1-TCA, 1,1-DCA or CF to determine the extent of the inhibitory effects (Chan et al. 2011). These three cultures contain *Dehalococcoides* and express the vinyl chloride reductase, VcrA, but were enriched from geographically different contaminated sites using different electron donors to provide energy (hydrogen) and carbon (acetate) to the dechlorinating organisms. All three of these mixed cultures include additional dechlorinating genera beyond *Dehalococcoides*, such as *Geobacter* (KB-1 and BDI) and *Dehalobacter* (OW). However, none of these enrichment cultures dechlorinates the suspected inhibiting compounds, 1,1,1-TCA, or CF.

An example of the data that is obtained from kinetic assays is shown in Fig. 13.3 depicting (A) the effect of 1,1,1-TCA on VC dechlorination in whole cell suspension (CS) assays, as well as (B) the effect of TCE on CF dechlorination in cell-free extract assays. The inhibition constant, K_i, can be thought of as that concentration of inhibitor that results in a 50 % decrease in dechlorination rate relative to the uninhibited case: the lower the K_i, the stronger the influence of the inhibitor. In a study comparing the effect of 1,1,1-TCA on VC dechlorination in three distinct mixed dechlorinating cultures, a very similar response was observed despite different rates of dechlorination. The key findings from this study were that 1,1,1-TCA strongly inhibited VC dechlorination in cell-free extract assays, suggesting that 1,1,1-TCA specifically interacts with the VC reductase associated with VC-to-ethene reductive dechlorination, i.e., the protein VcrA common to the three cultures (Table 13.1). An inhibition constant, K_i, of around 2.0 μM (270 μg/L) 1,1,1-TCA was calculated from these experimental data (Table 13.1). The reductive dehalogenases involved in cDCE and TCE dechlorination were also inhibited by 1,1,1-TCA, but to a lesser extent. In sharp contrast, 1,1-DCA had no pronounced inhibitory effects (i.e. large K_i) on any chlorinated ethene reductive dehalogenases (Table 13.1), indicating that removal of 1,1,1-TCA via reductive dechlorination to 1,1-DCA is a viable strategy to relieve inhibition (Grostern et al. 2009). Interestingly, 1,1,1-TCA was less inhibitory to the TCE reductive dehalogenases in consortia BDI and OW suggesting that the reductive dehalogenases in the TCE to cDCE dechlorinating microbes (*Geobacter* and *Dehalobacter*) present in these cultures may be slightly less inhibited by 1,1,1-TCA than *Dehalococcoides*—which was actually confirmed using a *Geobacter* enrichment from KB-1 (Chan et al. 2011).

13.4.2 Inhibition of Chloroalkane Reductive Dehalogenases by Chlorinated Alkenes

Inhibition was also found to occur in the reverse direction, where chlorinated ethenes inhibit the reductive dechlorination of chlorinated alkanes. These experiments were conducted using the *Dehalobacter*-containing enrichment culture referred to as ACT-3 (Fig. 13.2) that dechlorinates 1,1,1-TCA, 1,1-DCA, and CF, but none of the chlorinated ethenes (Grostern et al. 2009). Cell-free extracts of ACT-3 were amended with the substrates 1,1,1-TCA, 1,1-DCA or CF, and

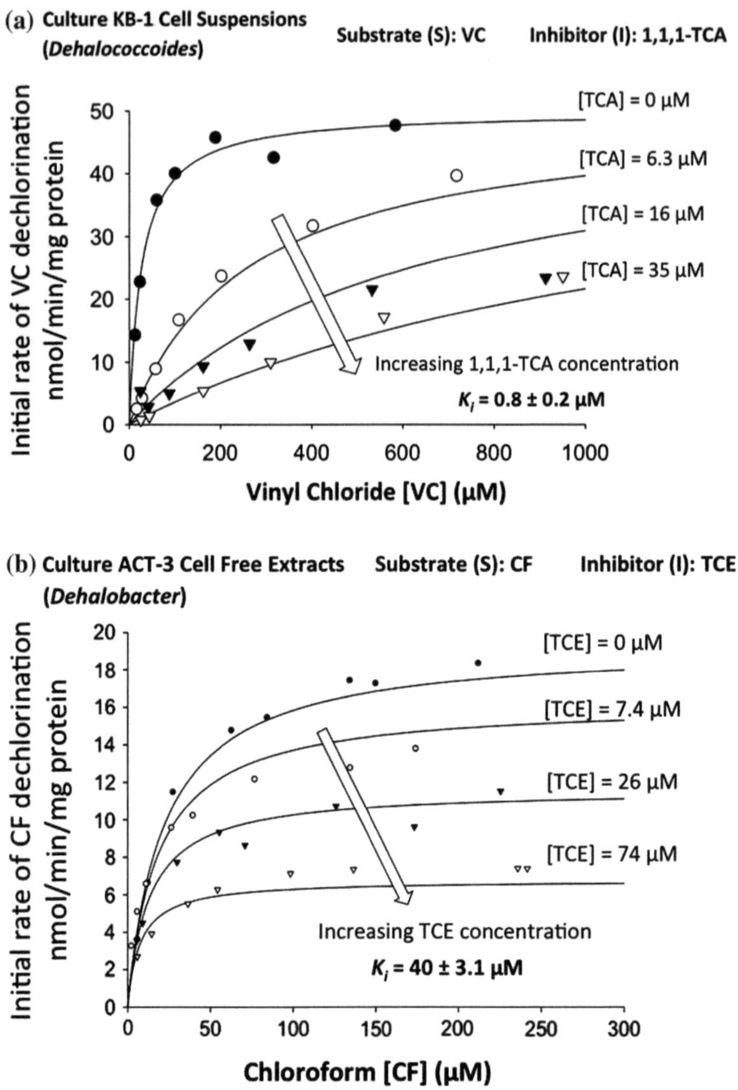

Fig. 13.3 Michaelis Menten kinetics for two distinct enrichment cultures in the presence of inhibitors. *Panel A* Kinetics of vinyl chloride (*VC*) dechlorination in cell suspensions of *Dehalococcoides*-containing culture KB-1 in the presence of increasing concentrations of the inhibitor 1,1,1-TCA. *Panel B* Kinetics of chloroform (*CF*) dechlorination in cell free extracts of *Dehalobacter*-containing culture ACT-3 in the presence of increasing concentrations of the inhibitor TCE. *Symbols* are experimental data and *solid lines* represent the best fit to each data set based on nonlinear regression to a competitive model (panel A) or uncompetitive model inhibition model (panel B). Adapted from Chan et al. (2011) and Wei (2012)

Table 13.1 Ranked inhibition constants for chloroethene dechlorination inhibited by chlorinated ethanes and methanes

				$K_I \pm 95\%$CI	
Culture	**Substrate**	**Inhibitor**	**Preparation**	μM	μg/L (ppb)
Cell Suspensions (Competitive Model):					
OW	VC	1,1,1-TCA	CS	0.2 ± 0.1	33 ± 11
BDI	VC	1,1,1-TCA	CS	0.4 ± 0.1	58 ± 13
KB-1	VC	1,1,1-TCA	CS[a]	0.8 ± 0.2	100 ± 27
Cell Suspensions (Non-Competitive Model):					
KB-1	TCE	CF	CS	6.8 ± 0.3	800 ± 38
KB-1	VC	CF	CS	59 ± 4.9	7000 ± 585
Cell-Free Extracts (Non-Competitive Model):					
KB-1	TCE	1,1,1-TCA	CFE	1.5 ± 0.6	210 ± 74
KB-1	TCE	CF	CFE	2.0 ± 0.1	240 ± 13
KB-1	VC	1,1,1-TCA	CFE	2.0 ± 0.3	270 ± 40
OW	VC	1,1,1-TCA	CFE	2.0 ± 0.4	270 ± 50
BDI	VC	1,1,1-TCA	CFE	2.1 ± 0.4	280 ± 50
KB-1/Dhc	TCE	1,1,1-TCA	CFE	2.2 ± 0.6	300 ± 75
KB-1	VC	CF	CFE	4.2 ± 0.2	500 ± 26
KB-1/Geo	TCE	1,1,1-TCA	CFE	5.1 ± 2	690 ± 260
BDI	cDCE	1,1,1-TCA	CFE	5.5 ± 0.8	730 ± 110
KB-1	cDCE	CF	CFE	11 ± 0.9	1300 ± 110
KB-1	cDCE	1,1,1-TCA	CFE	19 ± 4	2,500 ± 600
OW	TCE	1,1,1-TCA	CFE	40 ± 9	5,300 ± 1,200
BDI	TCE	1,1,1-TCA	CFE	43 ± 17	5,800 ± 2,300
OW	cDCE	1,1,1-TCA	CFE	86 ± 17	11,500± 2,300
OW	VC	1,1-DCA	CFE	104 ± 24	10,300± 2,400
BDI	cDCE	1,1-DCA	CFE	110 ± 18	11,000± 1,800
OW	cDCE	1,1-DCA	CFE	130 ± 57	13,000± 5,600
BDI	VC	1,1-DCA	CFE	162 ± 39	16,000± 3,700
KB-1	VC	1,1-DCA	CFE	224 ± 111	29,700±11,000
KB-1	cDCE	1,1-DCA	CFE	830 ± 280	82,000±28,000
KB-1	TCE	1,1-DCA	CFE	No inhibition	No inhibition

Data compiled from Chan et al. (2011) for TCA and DCA and Wei (2012) for CF. The noncompetitive model was typically the best fit for all cell-free extract experiments

CFE Cell-free extract

CS Whole (resting) cell suspension

KB-1/Dhc (Highly enriched *Dehalococcoides* subculture of KB-1)

KB-1/Geo (*Geobacter* enrichment from KB-1 dechlorinating PCE to cDCE only)

Shaded rows at the bottom of the table highlight high inhibition constants with 1,1-DCA, and thus little to no inhibition by this compound

[a]Data set plotted in Fig. 13.3 panel A

challenged with individual chlorinated ethenes to determine the extent to which TCE, cDCE or VC inhibited reductive dechlorination (Grostern et al. 2009). In particular, VC was found to profoundly inhibit chloroform reductive dechlorination. The inhibition constant K_i was estimated to be as low as ~0.6 μM (40 μg/L)

Table 13.2 Kinetic parameters (V_{max}, K_m and K_i) for 1,1,1-TCA, 1,1-DCA and CF reductive dechlorination in cell-free extracts and resting whole cell suspensions of a *Dehalobacter* enrichment culture in the presence of chlorinated ethenes

Substrate	Inhibitor	Preparation	V_{max} (nmol/min/mg)	K_m (μM)	K_i (μM)	K_i (μg/L)
1,1,1-TCA	TCE	CFE	102 ± 7	42 ± 6	42 ± 6	5500
1,1,1-TCA	cDCE	CFE	86 ± 11	34 ± 10	126 ± 38	12,000
1,1,1-TCA	VC	CFE	73 ± 8	33 ± 8	35 ± 8	1600
1,1-DCA	TCE	CFE	63 ± 6	46 ± 64	No inhibition	
1,1-DCA	cDCE	CFE	44 ± 5	289 ± 68	No inhibition	
1,1-DCA	VC	CFE	53 ± 3	396 ± 42	No inhibition	
CF	TCE	CFE[a]	19 ± 0.63	22 ± 2.4	40 ± 3.1	5300
CF	cDCE	CFE	23 ± 1.0	17 ± 3.4	6.7 ± 0.70	650
CF	VC	CFE	24 ± 1.0	23 ± 4.0	0.56 ± 0.052	35
CF	VC	CS	9.4 ± 0.12	1.5 ± 0.11	8.4 ± 1.1	530
1,1,1-TCA	TCE	CS	4.5 ± 0.3	14 ± 3.0	242 ± 124	32,000
1,1,1-TCA	cDCE	CS	5.0 ± 0.3	11 ± 3.0	872 ± 498	85,000
1,1,1-TCA	VC	CS	3.0 ± 0.3	13 ± 4.0	228 ± 167	14,000
1,1-DCA	TCE	CS	2.2 ± 0.2	87 ± 30	No inhibition	
1,1-DCA	cDCE	CS	2.2 ± 0.1	147 ± 15	189 ± 27	18,000
1,1-DCA	VC	CS	5.6 ± 0.4	192 ± 26	83 ± 19	5200

Data compiled from Grostern et al. (2009) for substrates 1,1,1-TCA and 1,1-DCA and Wei (2012) for CF

The best fit for all data was to an uncompetitive model

Error values represent 95 % confidence intervals

CFE Cell-free extract

CS Whole cell suspension

[a]Data set plotted in Fig. 13.3 panel B

in cell-free extracts (Table 13.2). The strong inhibition of CF dechlorination by VC is consistent with the problematic persistence of CF at contaminated sites. Moreover, although the same enzyme catalyzes both 1,1,1-TCA and CF dechlorination, 1,1,1-TCA dechlorination was significantly less inhibited by VC and cDCE (Table 13.2) than CF. These observations are consistent with unusual isotope fractionation behavior with these same reactions (Chan et al. 2012) and suggests a rather strong interaction between 1,1,1-TCA and the enzyme. In contrast, none of the chlorinated ethenes inhibited 1,1-DCA dechlorination in cell-free extract assays. This is consistent with the proven existence of two distinct *Dehalobacter* strains in the ACT-3 culture, one that dechlorinates 1,1,1-TCA and the other 1,1-DCA. One strain produces a reductive dehalogenase (CfrA) that dechlorinates 1,1,1-TCA and CF, but not 1,1-DCA, while the other strain produces a reductive dehalogenase (DcrA) that dechlorinates 1,1-DCA but not 1,1,1-TCA nor CF (Tang and Edwards 2013). These enzymes are very similar at the sequence-level, and thus kinetic observations point to ways that one can begin to investigate mechanisms of substrate interactions and protein structure in more detail.

13.5 Inhibition in Whole Cell Assays

Once inhibition is quantified in cell-free extracts, the next step is to extend the analysis in order to determine if the inhibiting substance might have an effect on critical cell components other than reductive dehalogenases. For this, whole CS assays can be used. These assays are similar to CFE assays except that concentrated, intact cells are used instead of disrupted cells. Hydrogen or another physiological electron donor is provided instead of reduced methyl viologen and the assay is run in growth medium, not lysis buffer. The reaction is initiated with addition of the electron acceptor and is terminated 1–3 h later, before any significant cell growth has occurred, hence these assays are sometimes called "resting cell assays". Cell growth can be assumed to be negligible over a period of 3 h or less because OHRB tend to have relatively long doubling times on the order of 0.8–3 days (Grostern et al. 2010; Loffler et al. 2013; He et al. 2007). This assay will detect any inhibition on dehalogenases as in CFE assays, but inhibition will be modulated by the presence of other cellular components, such as intact cytoplasmic and outer membranes, that might alter the rate-determining step. Since these assays require a functional electron transport chain, inhibition of hydrogenases may also be detected in CS assays. One of the advantages of whole CS assays is that the kinetic constants are more relevant for use in models of microbial growth and dechlorination since the rate constants can be normalized to the number of dechlorinating organisms present, and the inhibition and half-saturation constants are completely transferrable to models involving growing cells. The next two sections (13.5.1 and 13.5.2) illustrate some of the features of whole cell assays.

13.5.1 Cross-Inhibitory Effects Examined in Whole Cell Suspension Assays

The cross-inhibitory effects of chlorinated ethenes and alkanes described in Sects. 13.4.1 and 13.4.2 were also investigated resting CS assays (Grostern et al. 2009; Chan et al. 2011; Wei 2012). The key findings in CS assays as well as the comparison with the results obtained from cell-free extract assays are summarized in this section.

Whole CS assays were used to determine the inhibitory effects of chlorinated ethanes on TCE sequential reductive dechlorination. The smallest most potent inhibition constants, ranging from 0.3 or 0.8 μM (40–100 μg/L) were measured for 1,1,1-TCA inhibiting VC dechlorination in these assays. As the inhibition constants were similar between CS assays and cell-free extract (CFE) assays, the data suggest that 1,1,1-TCA acts directly on the reductive dehalogenase enzyme system and does not exert a general toxic effect of *Dehalococcoides* cells (Chan et al. 2011). When all experiments and associated uncertainties were considered, the inhibition constant for 1,1,1-TCA on VC ranged between 30–270 μg/L. This range represents the threshold where VC dechlorination rates will be significantly impacted relative to conditions without 1,1,1-TCA, providing direct guidance to site managers (Chan et al. 2011).

Whole CS assays were also used to examine the inhibitory effects of chlorinated alkenes on the reductive dechlorination of chlorinated ethanes by a *Dehalobacter*-containing culture, ACT-3. A comparison of the cell-free extract and whole CS data revealed that inhibition was less pronounced in whole CSs compared to cell-free extracts for both CF and 1,1,1-TCA, as shown by higher K_i values by an order of magnitude in whole CSs (Table 13.2). For example, the K_i measured for VC inhibition of CF dechlorination was ~8 μM (500 μg/L) versus 0.6 μM (40 μg/L) in cell-free extracts. These findings suggest that when the reductive dehalogenase activity is assayed in intact cells of *Dehalobacter*, the membrane offers some form of protection against inhibition. In these experiments, the rate of the dehalogenation reaction on a total protein basis (V_0) was much faster in cell-free extracts (where methyl viologen is electron donor) than in CS assays (Table 13.2), suggesting that electron transport from H_2 and not dechlorination is rate-limiting in whole cell assays. This observation has also been reported by others (Nijenhuis and Zinder 2005), and can certainly confound interpretation of results. In whole cell assay, VC was much less inhibitory to 1,1-DCA dechlorination compared to 1,1,1-TCA or CF, with a K_i ~ 80 μM or 8 mg/L (Table 13.2). Although relatively high, a K_i of 8 mg/L is nonetheless a concentration relevant to some DNAPL sites and certainly relevant to enrichment cultures where inhibition was indeed observed (Grostern and Edwards 2006a). As illustrated, these whole cell inhibition constants are a useful guide to determine when inhibition needs to be considered and when it is appropriate to omit.

13.5.2 Inhibition Model Type and Fit and Underlying Mechanisms

The kinetic model that best fit the data presented in the examples (Tables 13.1 and 13.2) was not the same. In experiments with chlorinated ethene-dechlorinating cultures (*Dehalococcoides*-dominated), the noncompetitive model fit best, except for in assays for 1,1,1-TCA inhibition with whole cell where the competitive model fit better (Table 13.1). Perhaps different rate-limiting steps or interactions were interrogated in cell free versus whole cell assays. The ability to truly discriminate between different models is a function of the number of experiments conducted and reproducibility, and thus one should be cautious in over-interpreting the data. Nevertheless, noncompetitive inhibition assumes that the inhibitor binds somewhere other than at the active site, which is a likely interaction in the light of the recent 3D structure of a reductive dehalogenase, suggesting a substrate-tailored opening into a cavity containing the active site (see Chap. 20). It may be that 1,1,1-TCA and CF, being nonplanar, occlude this opening in chlorethene reductases.

In all of the experiments conducted with the *Dehalobacter*-containing ACT-3 culture, the best fit of the data was to the uncompetitive model, which assumes that the inhibitor binds only to the enzyme–substrate complex (Table 13.2). The differences in inhibitory mechanism suggested by model fit may simply reflect experimental error and not reflect a governing underlying principle. But perhaps they may point to real mechanistic differences and avenues for future research. The diversity of reductive dehalogenase sequences and associated proteins provides ample possibility for differences in protein interactions. It is clear that each enzyme will have different kinetic constants and different interactions with potential inhibitors. A better understanding of the active enzymes in a given system, and their structures, will ultimately improve our fundamental understanding of inhibition and how to represent these phenomena in models.

13.6 Inhibition in Actively Growing Microbial Cultures and Communities

Ultimately, the need is to quantify and predict the effects of inhibitors during active culture growth as it occurs in laboratory batch and continuous experiments, packed flowing columns, and especially in engineered or natural systems in the field (Schaefer et al. 2008). The approach outlined herein of starting from CFE and resting CS assays to deconvolute simultaneous effects can substantially narrow down the experimental conditions and concentration ranges that then need to be tested in more complex realistic situations. The examples provided illustrating the specific effects between chlorinated ethanes, methanes, and ethenes have revealed interesting patterns of cross-inhibition relevant to both the scientific understanding of the underlying microorganisms and their respective reductive dehalogenases, as well as to the application of dechlorinating cultures for bioaugmentation. There are certainly other experimental approaches to control variables and extract meaningful kinetic data from mixed cultures, such as the batch multiequilibration method demonstrated by (Yu et al. 2005), which also minimizes growth. As it has become easier to also monitor microbial populations and proteins in cultures and field samples, these approaches will be informed with the knowledge of the actual concentration of OHRB and the identity of the specific proteins expressed resulting in more accurate, specific, and transferrable kinetic parameters and half-saturation and inhibition constants for use in models and hypothesis testing. Researchers should be encouraged to collect such information for their systems. The various models described in Sect. 13.2.1 already are extensively used because they provide useful frameworks to integrate complex simultaneous phenomena and to improve conceptual understanding of site data and remediation performance. Understanding the molecular basis for inhibition is improving significantly with new developments in microbial community analysis and protein characterization, and this knowledge will certainly translate to improved predictive ability and utility of modeling efforts.

In this chapter, we have so far only surveyed data for chlorinated ethenes, chloroform, and two chlorinated ethanes. Needless to say, many more interactions are possible among the myriad of chemicals that are present in the environment and at contaminated sites. Many investigations have revealed other complex patterns of inhibition that are often difficult to understand. In a study with chlorinated ethanes with halogens on vicinal carbons, such as 1,1,2-TCA and 1,2-DCA that tend to be dechlorinated via dihaloelimination (see Fig. 13.1), 1,2-DCA was not dechlorinated by *Dehalogenimonas* until 1,1,2-TCA reached low concentrations (Dillehay et al. 2014). Lai and Becker (2013) used a dual Monod model to predict population abundance and survival of two PCE-dechlorinating genera by incorporating PCE and TCE inhibition on VC dechlorination by *Dehalococcoides*, and of VC inhibition of PCE and TCE dechlorination by *Dehalobacter* in the coculture.

Considering more complex molecules beyond chlorinated aliphatics, such as chlorinated and brominated aromatic compounds (He et al. 2006), polychlorinated biphenyls (PCBs) (Demirtepe et al. 2015), halogenated pesticides, dioxins (Häggblom and Bossert 2003), and many more described in this book, the possible combinations of substrates and inhibitors become truly daunting. Complex patterns of dechlorination have been observed with chlorinated benzenes and toluenes (Nelson et al. 2014) and PCBs (Bedard 2008; Demirtepe et al. 2015), underlying multiple strains and species that carry out distinct reactions. Nevertheless, these mixed microbial populations can and eventually dechlorinate these compounds. Knowledge of these interactions can help to promote more productive pathways. A key factor in the transformation of some of the more chlorinated hydrophobic compounds is that their concentrations are typically limited by the low aqueous solubility of the parent compound and thus may not reach critical thresholds for inhibition, meaning that slow but steady dechlorination can persist. So while on the one hand low aqueous concentrations means slower dehalogenation, it may also permit a greater diversity of reactions and ultimately more complete dehalogenation. In these slower systems with lower concentrations of substrates, competition for nutrients beyond the halogenated electron acceptor plays a major role, as introduced in the next section.

13.7 Cross-Feeding and Competition in Anaerobic Dehalogenating Microbial Communities

In the context of microbial ecology, intraspecies and interspecies competition exists within anaerobic dehalogenating microbial communities, which may contribute to observed inhibitory effects on reductive dechlorination at contaminated sites and in other environments (Chap. 14). In many of the described anaerobic dehalogenating microbial communities, more than one OHRB population is stably maintained. Fermenting and acetogenic bacteria are also critical for providing essential electron donors, co-factors, and nutrients. Moreover, other groups of anaerobes, such as iron-reducing and sulfate-reducing bacteria, have been found in anaerobic dehalogenating microbial communities when the appropriate electron acceptor is available. The effect of competition on OHRB and their dehalogenation performance may result from competition between different species of OHRB for the same halogenated electron acceptor, or from competition between OHRB and other anaerobic bacteria for electron donor due to the presence of alternative electron acceptors, such as sulfate, nitrate, iron, and others.

Anaerobic dehalogenating microbial communities grown on different halogenated compound classes typically comprise obligate and facultative OHRB growing together in stable enrichment cultures. Combinations include *Desulfitobacterium* and *Dehalococcoides* (Rouzeau-Szynalski et al. 2011; Ise et al. 2011; Yang et al. 2005), *Dehalococcoides* and *Dehalobacter* (Grostern and Edwards 2006b; Daprato et al. 2007), *Dehalococcoides* and *Geobacter* (Duhamel et al. 2002; Ziv-El et al. 2012; Ritalahti et al. 2006), *Sulfurospirillum* and *Dehalococcoides* (Maillard et al. 2011; He et al. 2006), and often more than two genera: *Dehalococcoides, Desulfuromonas, Desulfitobacterium* (Ballerstedt et al. 2004), *Dehalogenimonas*, *Dehalobacter* and *Dehalococcoides* (Manchester et al. 2012), and *Dehalococcoides*, *Dehalobacter* and *Desulfitobacterium* (Vandermeeren et al. 2014). Even when only a single OHRB genus is present, multiple distinct strains are often observed. This has been shown in cultures dechlorinating chloroalkenes/alkanes (Duhamel et al. 2002; Duhamel and Edwards 2006; Daprato et al. 2007; Futagami et al. 2011; Ritalahti et al. 2006; Brisson et al. 2012; Vainberg et al. 2009; Tang and Edwards 2013), chlorobenzenes (Nelson et al. 2011, 2014) and PCBs (Watts et al. 2001; Fagervold et al. 2005).

Certain Monod kinetic-based models have explored competition between different OHRB genera (Becker 2006; Becker and Seagren 2009; Lai and Becker 2013). For example, Becker and Seagren (2009) modeled competition between *Dehalococcoides* and *Desulfuromonas michiganensis* and the implications for PCE DNAPL bioenhancement. They concluded that the potential for bioenhancement of DNAPL was better at low flow velocity. Chen et al. (2013) also explored and modeled enhanced DNAPL dissolution in a column setting. *Geobacter* and *Dehalococcoides* were the modeled populations. Enhanced dissolution showed high sensitivity to *Dehalococcoides* kinetic constants for PCE and cDCE as well as *Dehalococcoides* growth. In a dual Monod model for *Dehalobacter* and *Dehalococcoides* grown on PCE in coculture with non-limiting H_2 levels, Lai and Becker (2013) conclude that accurate models required that both competition and inhibition should be considered.

When multiple electron acceptors are simultaneously present, OHRB will complete with other anaerobic respiring bacteria for resources, primarily electron donor, and the availability of electron donor generally governs the order of terminal electron acceptor utilization. When electron donor is limiting, the most energy-yielding electron-accepting processes and associated organisms will be favored thermodynamically (Fennell and Gossett 1998; Smatlak et al. 1996). When electron donor is in excess, multiple electron-accepting processes will occur simultaneously. In most laboratory enrichment cultures, the only external electron acceptor added is the organohalide of interest though CO_2 as bicarbonate is typically present in the medium, and donor is often in excess, thus methanogenesis, acetogenesis, and dechlorination occur simultaneously. Then, when alternate terminal electron acceptors are added, the microbial community tends to change significantly. Iron reducers, sulfate reducers, and even denitrifiers have been reported in OHRB communities when the appropriate electron acceptor is added. Wei and Finneran (2013) reported methanogenesis, iron reduction, and sulfate reduction are simultaneous with OHRB. Higher acetate levels did not speed up reductive dechlorination but instead increased methanogenesis. In membrane bioreactors TCE, 1,1,1-TCA, and CF were reductively dehalogenated even in the presence of nitrate and sulfate reduction (Zhang et al. 2010). Aulenta et al. (2007) maintained cultures on a variety of electron donors with concurrent reduction of a mixture of electron acceptors [chlorinated ethenes, nitrate, sulfate, and Fe(III)] but they reported that over 99 % of the reducing equivalents were channeled to alternate electron acceptors rather than organohalides. In others reports, the generated sulfide and/or the competition with sulfate reducers for electron donors diminished reductive dechlorination when sulfate was provided. For example, parallel PCE-dechlorinating chemostats were established and sulfate was fed to one (at 1 mM) (Berggren et al. 2013). Reductive dechlorination efficiency decreased following the onset of complete sulfate reduction—with VC and cDCE rather than ethene as the major end product. In the trichlorobenzene-dechlorinating enrichment culture that ultimately yielded *Dehalococcoides* strain CBDB1, sulfate addition clearly inhibited reductive dechlorination (Adrian et al. 1998). The ecological principles of competition for resources are equally applicable in OHRB communities. In the typically resource limited natural environments, thermodynamic thresholds for growth are critical, and in this context OHRB do well, because the energy from dehalogenation is typically greater than that from methanogenesis, acetogenesis, or sulfate reduction, but growth is generally limited by low concentrations of electron acceptor. In artificial situations of contaminated sites or bioreactors, conditions can be tailored or created to channel electrons from available donors to dehalogenation. Much effort has been focused on the value of different donors in this regard, owing to these donors being the driver of cost for remediation. Models that consider the whole microbial community often incorporate thermodynamic thresholds for competing electron-accepting processes, as reviewed in (Häggblom and Bossert 2003; Chambon et al. 2013).

13.8 Summary

Inhibition and competition within OHRB communities are clearly a complex multilayered subject. Nevertheless, enzyme assays carried out with crude cell extracts confirm that some halogenated compounds interact and inhibit certain reductive dehalogenases in a quantifiable and reproducible manner, indicating that inhibition is often a direct result of a specific interaction with a dehalogenase protein complex. A systematic approach starting from known enzymes and cultures to obtain meaningful data may preclude some of the frustrations among experimentalists and modelers alike in attempts to reconcile data to find adequate model parameters for meaningful simulations. An on-going dialog between experimentalists and modelers is needed to obtain approximations that do not over simplify dominant rate-limiting steps. Moreover, too many studies invoke inhibition without any consideration of the concentrations at which these effects are measured.

Fundamentally, the inhibition constants (K_i) reflect the affinity of inhibitors for their target and provide the inhibitor concentration that causes substantial inhibition (typically 50 %) of the process. While the values of K_i differ depending on the model type (e.g., competitive, noncompetitive, uncompetitive or Haldane or other), in the context of field work and the errors associated with the many assumptions and data that enter a model, they provide an adequate approximate measure of relative potency and a place to begin, regardless of model. The many K_i values reported in the literature for chlorinated solvents, many summarized in recent publications (Lai and Becker 2013; Chambon et al. 2013) and others illustrated herein (Chan et al. 2011; Grostern et al. 2009; Wei 2012) provide useful guidelines for practitioners to decide whether co-contaminant concentrations in a particular situation are likely or not to affect the microbial reductive dechlorination of the target contaminants.

Inhibition constants, just like K_m, V_{max}, pH, T optima, co-factors, and substrate ranges are intrinsic properties particular to each enzyme and reflect the underlying structural properties of the protein. These properties should be measured as part of the data sets required for more complete protein characterization. As more and more reductive dehalogenases are identified, purified, and characterized more fully, patterns among these interactions will become more clear and more robust inhibition constants or models of activity will be available, leading to increased insight into the mechanism of the reactions catalyzed by these enzymes, as well as to more practically useful models of dehalogenation.

Similarly, now that we can monitor complete microbial communities using rapid and inexpensive sequencing tools, we are beginning to see more accurate analysis of inhibition and competition effects in mixed communities, and better modeling of microbial population dynamics and function. This window into the community is opening ever-wider and revealing new levels of complexity but also monitoring tools to track and predict governing patterns of resource allocation and utilization. The analysis of microbiome function is not only being driven by environmental processes as in past decades, but now has received enormous attention from human health biology and agriculture as well. The next decade of research is bound to reveal remarkable dynamics within these communities that contribute significantly to their function and evolution, and to the health of the planet and all its inhabitants.

Acknowledgements The authors acknowledge the contributions of the many students, postdocs, and industrial and academic collaborators who have contributed to the research and insights into reductive dehalogenation and anaerobic microbial processes over the years.

References

Adamson DT, Parkin GF (2000) Impact of mixtures of chlorinated aliphatic hydrocarbons on a high-rate, tetraehloroethene-dechlorinating enrichment culture. Environ Sci Technol 34(10):1959–1965

Adrian L, Manz W, Szewzyk U, Görisch H (1998) Physiological characterization of a bacterial consortium reductively dechlorinating 1,2,3- and 1,2,4-trichlorobenzene. Appl Environ Microbiol 64(2):496–503

Aulenta F, Pera A, Rossetti S, Papini MP, Majone M (2007) Relevance of side reactions in anaerobic reductive dechlorination microcosms amended with different electron donors. Water Res 41(1):27–38

Ballerstedt H, Hantke J, Bunge M, Werner B, Gerritse J, Andreesen JR, Lechner U (2004) Properties of a trichlorodibenzo-p-dioxin-dechlorinating mixed culture with a *Dehalococcoides* as putative dechlorinating species. FEMS Microbiol Ecol 47(2):223–234

Becker JG (2006) A modeling study and implications of competition between *Dehalococcoides ethenogenes* and other tetrachloroethene-respiring bacteria. Environ Sci Technol 40(14):4473–4480

Becker JG, Seagren EA (2009) Modeling the effects of microbial competition and hydrodynamics on the dissolution and detoxification of dense nonaqueous phase liquid contaminants. Environ Sci Technol 43(3):870–877

Bedard DL (2008) A case study for microbial biodegradation: anaerobic bacterial reductive dechlorination of polychlorinated biphenyls—from sediment to defined medium. Annu Rev Microbiol 62(1):253–270

Berggren DRV, Marshall IPG, Azizian MF, Spormann AM, Semprini L (2013) Effects of sulfate reduction on the bacterial community and kinetic parameters of a dechlorinating culture under chemostat growth conditions. Environ Sci Technol 47(4):1879–1886

Bommer M, Kunze C, Fesseler J, Schubert T, Diekert G, Dobbek H (2014) Structural basis for organohalide respiration. Science 346(6208):455–458

Brisson VL, West KA, Lee PKH, Tringe SG, Brodie EL, Alvarez-Cohen L (2012) Metagenomic analysis of a stable trichloroethene-degrading microbial community. ISME J 6(9):1702–1714

Chambon JC, Bjerg PL, Scheutz C, Baelum J, Jakobsen R, Binning PJ (2013) Review of reactive kinetic models describing reductive dechlorination of chlorinated ethenes in soil and groundwater. Biotechnol Bioeng 110(1):1–23

Chambon JC, Broholm MM, Binning PJ, Bjerg PL (2010) Modeling multi-component transport and enhanced anaerobic dechlorination processes in a single fracture-clay matrix system. J Contam Hydrol 112(1–4):77–90

Chan CCH, Mundle SOC, Eckert T, Liang X, Tang S, Lacrampe-Couloume G, Edwards EA, Sherwood Lollar B (2012) Large carbon isotope fractionation during biodegradation of chloroform by *Dehalobacter* cultures. Environ Sci Technol 46:10154–10160

Chan WWM, Grostern A, Loffler FE, Edwards EA (2011) Quantifying the effects of 1,1,1-trichloroethane and 1,1-dichloroethane on chlorinated ethene reductive dehalogenases. Environ Sci Technol 45(22):9693–9702

Chen MJ, Abriola LM, Amos BK, Suchomel EJ, Pennell KD, Löffler FE, Christ JA (2013) Microbially enhanced dissolution and reductive dechlorination of PCE by a mixed culture: model validation and sensitivity analysis. J Contam Hydrol 151:117–130

Christ JA, Abriola LM (2007) Modeling metabolic reductive dechlorination in dense non-aqueous phase liquid source-zones. Adv Water Resour 30(6–7):1547–1561

Chung J, Rittmann BE (2008) Simultaneous bio-reduction of trichloroethene, trichloroethane, and chloroform using a hydrogen-based membrane biofilm reactor. Water Sci Technol 58(3):495–501

Cupples AM, Spormann AM, McCarty PL (2004) Comparative evaluation of chloroethene dechlorination to ethene by *Dehalococcoides*-like microorganisms. Environ Sci Technol 38(18):4768–4774

Daprato RC, Löffler FE, Hughes JB (2007) Comparative analysis of three tetrachloroethene to ethene halorespiring consortia suggests functional redundancy. Environ Sci Technol 41(7):2261–2269

Demirtepe H, Kjellerup B, Sowers KR, Imamoglu I (2015) Evaluation of PCB dechlorination pathways in anaerobic sediment microcosms using an anaerobic dechlorination model. J Hazard Mater 296:120–127

Dillehay JL, Bowman KS, Yan J, Rainey FA, Moe WM (2014) Substrate interactions in dehalogenation of 1,2-dichloroethane, 1,2-dichloropropane, and 1,1,2-trichloroethane mixtures by *Dehalogenimonas* spp. Biodegradation 25(2):301–312

Doherty RE (2000) A history of the production and use of carbon tetrachloride, tetrachloroethylene, trichloroethylene and 1,1,1-trichloroethane in the united states: part 2–trichloroethylene and 1,1,1-trichloroethane. Environ Forens 1(2):83–93

Duchesneau MN, Workman R, Baddour FR, Dennis P (2007) Combined *Dehalobacter* and *Dehalococcoides* bioaugmentation for bioremediation of 1,1,1-trichloroethane and chlorinatedethenes. Paper presented at the 9th international in situ and on-site bioremediation symposium, Baltimore, Maryland

Duhamel M, Edwards EA (2006) Microbial composition of chlorinated ethene-degrading cultures dominated by *Dehalococcoides*. FEMS Microbiol Ecol 58(3):538–549

Duhamel M, Wehr SD, Yu L, Rizvi H, Seepersad D, Dworatzek S, Cox EE, Edwards EA (2002) Comparison of anaerobic dechlorinating enrichment cultures maintained on tetrachloroethene, trichloroethene, *cis*-dichloroethene and vinyl chloride. Water Res 36(17):4193–4202

Fagervold SK, Watts JEM, May HD, Sowers KR (2005) Sequential reductive dechlorination of meta-chlorinated polychlorinated biphenyl congeners in sediment microcosms by two different chloroflexi phylotypes. Appl Environ Microbiol 71(12):8085–8090

Fennell DE, Gossett JM (1998) Modeling the production of and competition for hydrogen in a dechlorinating culture. Becker, Jennifer G, Research Support, US Govt, Non-PHS United States. Environ Sci Technol 40(14):4473–4480 (2006 July 15); 32(16):2450–2460

Futagami T, Okamoto F, Hashimoto H, Fukuzawa K, Higashi K, Nazir K, Wada E, Suyama A, Takegawa K, Goto M, Nakamura K, Furukawa K (2011) Enrichment and characterization of a trichloroethene-dechlorinating consortium containing multiple "*Dehalococcoides*" strains. Biosci Biotechnol Biochem 75(7):1268–1274

Grostern A, Chan WWM, Edwards EA (2009) 1,1,1-trichloroethane and 1,1-dichloroethane reductive dechlorination kinetics and co-contaminant effects in a *Dehalobacter*-containing mixed culture. Environ Sci Technol 43(17):6799–6807

Grostern A, Duhamel M, Dworatzek S, Edwards EA (2010) Chloroform respiration to dichloromethane by a *Dehalobacter* population. Environ Microbiol 12(4):1053–1060

Grostern A, Edwards EA (2006a) A 1,1,1-trichloroethane-degrading anaerobic mixed microbial culture enhances biotransformation of mixtures of chlorinated ethenes and ethanes. Appl Environ Microbiol 72(12):7849–7856

Grostern A, Edwards EA (2006b) Growth of *Dehalobacter* and *Dehalococcoides* spp. during degradation of chlorinated ethanes. Appl Environ Microbiol 72(1):428–436

Haest PJ, Springael D, Seuntjens P, Smolders E (2012) Self-inhibition can limit biologically enhanced TCE dissolution from a TCE DNAPL. Chemosphere 89(11):1369–1375

Haest PJ, Springael D, Smolders E (2010) Dechlorination kinetics of TCE at toxic TCE concentrations: Assessment of different models. Water Res 44(1):331–339

Häggblom MM, Bossert ID (2003) Dehalogenation: microbial processes and environmental applications. Springer, Berlin

He J, Holmes VF, Lee PK, Alvarez-Cohen L (2007) Influence of vitamin B12 and cocultures on the growth of *Dehalococcoides* isolates in defined medium. Appl Environ Microbiol 73(9):2847–2853

He JZ, Robrock KR, Alvarez-Cohen L (2006) Microbial reductive debromination of polybrominated diphenyl ethers (PBDEs). Environ Sci Technol 40(14):4429–4434

Heavner GLW, Rowe AR, Mansfeldt CB, Pan JK, Gossett JM, Richardson RE (2013) Molecular biomarker-based biokinetic modeling of a PCE-dechlorinating and methanogenic mixed culture. Environ Sci Technol 47(8):3724–3733

Huang D, Becker JG (2011) Dehalorespiration model that incorporates the self-inhibition and biomass inactivation effects of high tetrachloroethene concentrations. Environ Sci Technol 45(3):1093–1099

Hurvich CM, Tsai C-L (1989) Regression and time series model selection in small samples. Biometrika 76:297–307

Ise K, Suto K, Inoue C (2011) Microbial diversity and changes in the distribution of dehalogenase genes during dechlorination with different concentrations of *cis*-DCE. Environ Sci Technol 45(12):5339–5345

Lai YJ, Becker JG (2013) Compounded effects of chlorinated ethene inhibition on ecological interactions and population abundance in a *Dehalococcoides–Dehalobacter* coculture. Environ Sci Technol 47(3):1518–1525

Lee IS, Bae JH, Yang YR, McCarty PL (2004) Simulated and experimental evaluation of factors affecting the rate and extent of reductive dehalogenation of chloroethenes with glucose. J Contam Hydrol 74(1–4):313–331

Löffler FE, Sanford RA, Tiedje JM (1996) Initial characterization of a reductive dehalogenase from *Desulfitobacterium chlororespirans* Co23. Appl Environ Microbiol 62(10):3809–3813

Löffler FE, Yan J, Ritalahti KM, Adrian L, Edwards EA, Konstantinidis KT, Müller JA, Fullerton H, Zinder SH, Spormann AM (2013) *Dehalococcoides mccartyi gen. nov., sp. nov.*, obligately organohalide-respiring anaerobic bacteria relevant to halogen cycling and bioremediation, belong to a novel bacterial class, *Dehalococcoidia classis nov.*, order *Dehalococcoidales ord. nov.* and family *Dehalococcoidaceae fam. nov.*, within the phylum Chloroflexi. Int J Syst Evol Microbiol 63(Pt 2):625–635

MacNelly A, Kai M, Svatos M, Diekert G, Schubert T (2014) Functional heterologous production of reductive dehalogenases from *Desulfitobacterium hafniense* strains. App Envron Microbiol 80(14):4313–4322

Maillard J, Charnay M-P, Regeard C, Rohrbach-Brandt E, Rouzeau-Szynalski K, Rossi P, Holliger C (2011) Reductive dechlorination of tetrachloroethene by a stepwise catalysis of different organohalide respiring bacteria and reductive dehalogenases. Biodegradation 22(5):949–960

Manchester MJ, Hug LA, Zarek M, Zila A, Edwards EA (2012) Discovery of a trans-Dichloroethene-respiring *Dehalogenimonas* species in the 1,1,2,2-tetrachloroethane-dechlorinating WBC-2 consortium. Appl Environ Microbiol 78(15):5280–5287

McGuire T, Hughes JB (2003) Effects of surfactants on the dechlorination of chlorinated ethenes. Environ Toxicol Chem 22(11):2630–2638

Mendoza-Sanchez I, Cunningham J (2012) Efficient algorithms for modeling the transport and biodegradation of chlorinated ethenes in groundwater. Transp Porous Media 92(1):165–185

Nelson JL, Fung JM, Cadillo-Quiroz H, Cheng X, Zinder SH (2011) A role for *Dehalobacter* spp. in the reductive dehalogenation of dichlorobenzenes and monochlorobenzene. Environ Sci Technol 45(16):6806–6813

Nelson JL, Jiang J, Zinder SH (2014) Dehalogenation of chlorobenzenes, dichlorotoluenes, and tetrachloroethene by three *Dehalobacter* spp. Environ Sci Technol 48(7):3776–3782

Nijenhuis I, Zinder SH (2005) Characterization of hydrogenase and reductive dehalogenase activities of *Dehalococcoides ethenogenes* strain 195. Appl Environ Microbiol 71(3):1664–1667

Parthasarathy A, Stich TA, Lohner ST, Lesnefsky A, Britt RD, Spormann A (2015) Heterologously expressed vinyl chloride reductive dehalogenase (VcrA) from *Dehalococcoides mccartyi* strain VS. J Am Chem Soc 137:3525–3532

Paul L, Herrmann S, Koch CB, Philips J, Smolders E (2013) Inhibition of microbial trichloroethylene dechorination by Fe (III) reduction depends on Fe mineralogy: A batch study using the bioaugmentation culture KB-1. Water Res 47(7):2543–2554

Paul L, Smolders E (2014) Inhibition of iron (III) minerals and acidification on the reductive dechlorination of trichloroethylene. Chemosphere 111:471–477

Payne KA, Quezada CP, Fisher K, Dunstan MS, Collins FA, Sjuts H, Levy C, Hay S, Rigby SE, Leys D (2015) Reductive dehalogenase structure suggests a mechanism for B12-dependent dehalogenation. Nature 517(7535):513–516

Pon G, Hyman MR, Semprini L (2003) Acetylene inhibition of trichloroethene and vinyl chloride reductive dechlorination. Environ Sci Technol 37(14):3181–3188

Popat SC, Deshusses MA (2011) Kinetics and inhibition of reductive dechlorination of trichloroethene, *cis*-1,2-dichloroethene and vinyl chloride in a continuously fed anaerobic biofilm reactor. Environ Sci Technol 45(4):1569–1578

Ritalahti KM, Amos BK, Sung Y, Wu QZ, Koenigsberg SS, Löffler FE (2006) Quantitative PCR targeting 16S rRNA and reductive dehalogenase genes simultaneously monitors multiple *Dehalococcoides* strains. Appl Environ Microbiol 72(4):2765–2774

Rosner BM, McCarty PL, Spormann AM (1997) In vitro studies on reductive vinyl chloride dehalogenation by an anaerobic mixed culture. Appl Environ Microbiol 63(11):4139–4144

Rouzeau-Szynalski K, Maillard J, Holliger C (2011) Frequent concomitant presence of *Desulfitobacterium* spp. and "*Dehalococcoides*" spp. in chloroethene-dechlorinating microbial communities. Appl Microbiol Biotechnol 90(1):361–368

Sabalowsky AR, Semprini L (2010) Trichloroethene and *cis*-1,2-dichloroethene concentration-dependent toxicity model simulates anaerobic dechlorination at high concentrations: I. Batch-Fed Reactors. Biotechnol Bioeng 107(3):529–539

Schaefer CE, Condee CW, Vainberg S, Steffan RJ (2008) Bioaugmentation for chlorinated ethenes using *Dehalococcoides* sp.: comparison between batch and column experiments. Chemosphere 75:141–148

Scheutz C, Durant ND, Hansen MH, Bjerg PL (2011) Natural and enhanced anaerobic degradation of 1,1,1-trichloroethane and its degradation products in the subsurface—a critical review. Water Res 45(9):2701–2723

Sleep BE, Sykes JF (1993) Compositional simulation of groundwater contamination by organic compounds: 1. model development and verification. Water Resour Res 29(6):1697–1708

Smatlak CR, Gossett JM, Zinder SH (1996) Comparative kinetics of hydrogen utilization for reductive dechlorination of tetrachloroethene and methanogenesis in an anaerobic enrichment culture. Environ Sci Technol 30(9):2850–2858

Tang S, Edwards EA (2013) Identification of *Dehalobacter* reductive dehalogenases that catalyse dechlorination of chloroform, 1, 1, 1-trichloroethane and 1, 1-dichloroethane. Philos Trans R Soc B: Biol Sci 368 (1616)

Tang S, Gong Y, Edwards EA (2012) Semi-automatic in silico gap closure enabled de novo assembly of two *Dehalobacter* genomes from metagenomic data. PLoS ONE 7(12):e52038

Townsend GT, Suflita JM (1997) Influence of sulfur oxyanions on reductive dehalogenation activities in *Desulfomonile tiedjei*. Appl Environmental Microbiology 63(9):3594–3599

Vainberg S, Condee CW, Steffan RJ (2009) Large-scale production of bacterial consortia for remediation of chlorinated solvent-contaminated groundwater. J Ind Microbiol Biotechnol 36(9):1189–1197

Vandermeeren P, Herrmann S, Cichocka D, Busschaert P, Lievens B, Richnow HH, Springael D (2014) Diversity of dechlorination pathways and organohalide respiring bacteria in chlorobenzene dechlorinating enrichment cultures originating from river sludge. Biodegradation 25(5):757–776

Watts JE, Wu Q, Schreier SB, May HD, Sowers KR (2001) Comparative analysis of polychlorinated biphenyl-dechlorinating communities in enrichment cultures using three different molecular screening techniques. Environ Microbiol 3(11):710–719

Wei K (2012) Substrates and substrate interactions in anaerobic dechlorinating cultures. Master's Thesis, University of Toronto, Toronto, ON

Wei N, Finneran KT (2011) Influence of ferric iron on complete dechlorination of trichloroethylene (TCE) to ethene: Fe(III) reduction does not always inhibit complete dechlorination. Environ Sci Technol 45(17):7422–7430

Wei N, Finneran KT (2013) Low and high acetate amendments are equally as effective at promoting complete dechlorination of trichloroethylene (TCE). Biodegradation 24(3):413–425

Yang YR, Pesaro M, Sigler W, Zeyer J (2005) Identification of microorganisms involved in reductive dehalogenation of chlorinated ethenes in an anaerobic microbial community. Water Res 39(16):3954–3966

Yeh DH, Pennell KD, Pavlostathis SG (1999) Effect of tween surfactants on methanogenesis and microbial reductive dechlorination of hexachlorobenzene. Environ Toxicol Chem 18(7):1408–1416

Yu S, Semprini L (2004) Kinetics and modeling of reductive dechlorination at high PCE and TCE concentrations. Biotechnol Bioeng 88(4):451–464

Yu SH, Dolan ME, Semprini L (2005) Kinetics and inhibition of reductive dechlorination of chlorinated ethylenes by two different mixed cultures. Environ Sci Technol 39(1):195–205

Yu ZT, Smith GB (2000) Inhibition of methanogenesis by C-1- and C-2-polychlorinated aliphatic hydrocarbons. Environ Toxicol Chem 19(9):2212–2217

Zhang H, Ziv-El M, Rittmann BE, Krajmalnik-Brown R (2010) Effect of dechlorination and sulfate reduction on the microbial community structure in denitrifying membrane-biofilm reactors. Environ Sci Technol 44(13):5159–5164

Ziv-El M, Popat SC, Parameswaran P, Kang DW, Polasko A, Halden RU, Rittmann BE, Krajmalnik-Brown R (2012) Using electron balances and molecular techniques to assess trichoroethene-induced shifts to a dechlorinating microbial community. Biotechnol Bioeng 109(9):2230–2239

Chapter 14
Organohalide-Respiring Bacteria as Members of Microbial Communities: Catabolic Food Webs and Biochemical Interactions

Ruth E. Richardson

Abstract Organohalide-respiring bacteria (OHRB) have been isolated from a wide range of anoxic environments worldwide and can easily be enriched in the laboratory. Obligate OHRB generally thrive best in mixed communities as part of anaerobic food webs that typically involve interspecies hydrogen (H_2) transfer from fermenters to OHRB, and often OHRB compete for H_2 with hydrogenotrophic methanogens. In laboratory enrichments, the community composition of the non-OHRB fraction of the communities is dependent on which electron donor is used for enrichment as well as other factors (e.g., the concentrations of organohalide substrate). In addition to catabolic food webs, other biochemical interactions in these communities include provision of key cofactors (e.g., corrinoids), relief of toxicity due to reactive oxygen species, as well as the organohalides themselves. Multiple OHRB often coexist stably in enrichment cultures and environmental communities. This diversity in OHRB populations creates complex interactions among different OHRB—with the partially dehalogenated end product of one population serving as substrate for other populations. Recent broad surveys of bacterial and archaeal community structure at sites undergoing in situ bioremediation are confirming that fermenters, methanogens, and OHRB are all stimulated by enhanced bioremediation efforts but that aerobes including methanotrophs and organohalide-oxidizing aerobes are also stimulated—especially in downgradient plume regions. The chapter will also discuss roles of OHRB populations in pristine environments including soils and sediments where they dehalogenate naturally produced halogenated organic matter and may compete with sulfate reducers and iron reducers when appropriate electron acceptors are available.

R.E. Richardson (✉)
School of Civil and Environmental Engineering, Cornell University, Ithaca, NY, USA
e-mail: rer26@cornell.edu

L. Adrian and F.E. Löffler (eds.), *Organohalide-Respiring Bacteria*,
DOI 10.1007/978-3-662-49875-0_14

Abbreviations

Dhc	*Dehalococcoides*
OHRB	Organohalide-respiring bacteria
Dhb	*Dehalobacter*
RDase	Reductive dehalogenase
PCE	Tetrachloroethene
TCE	Trichloroethene
cis-DCE	*cis*-1,2-dichloroethene
VC	Vinyl chloride
DNAPL	Dense nonaqueous phase liquid
PCBs	Polychlorinated biphenyls

14.1 Introduction

Enrichment cultures with organohalide-respiring capabilities have led to isolation of various Organohalide-respiring bacteria (OHRB). Pure culture studies have enabled metabolic characterization of the OHRB including electron donor range characterization and range of organohalides used as electron acceptors. Though pure culture work is essential for such fundamental characterizations, the parent enrichment culture is often easier to maintain than the derived pure cultures. Companies selling bioaugmentation cultures maintain their dechlorinating cultures as consortia. Comparisons of volume-normalized respiration rates of various organohalide enrichment cultures and pure cultures show that rates vary over more than four orders of magnitude and, in most cases, organohalide dehalogenation rates by obligate OHRB are faster in mixed enrichment cultures than in pure cultures (see Table 14.1 and references therein). Interspecies food webs exist in these enrichments—both among different OHRB populations as well as between non-OHRB and OHRB. The phylogenetic and metabolic groups that comprise the non-OHRB populations vary as a function of several factors including electron donor fed, incubation conditions, and type of organohalide substrate as well as the availability of other types of electron acceptors like sulfate or iron. Though electron donor does influence community structure, even after enrichment for years on a single electron donor, organohalide-respiring communities retain diverse fermentation capabilities.

In pristine settings, OHRB are members of a natural organohalide biogeochemical cycle. Sediments and soils harbor natural, albeit small, populations of OHRB, which is the basis of monitored natural attenuation in contaminated aquifers. At such contaminated field sites, local geochemistry and specific bioremediation enhancement strategies impact the structure of in situ organohalide-respiring communities. Figure 14.1 presents a schematic of general microbial food web interactions in organohalide-respiring communities. Included are those interactions common in stable laboratory enrichment cultures (solid lines in Fig. 14.1)

Table 14.1 Representative respiration rates by various organohalide-respiring cultures normalized per liter of culture

Organohalide substrate	Culture name[a]	Type of culture[b]	Key OHRB population(s)	Noted non-OHRB populations	Electron donor	Organohalide respiration rate μmole *X*/L/h (*X* = halogen)	Reference(s)
TCE	DehaloR^2	E	*Dhc*, *Geobacter*	*Acetobacterium*, *Clostridium*, *Spirochaetes*	Lactate and Methanol	130	Delgado et al. (2014), Ziv-El et al. (2011)
PCE	D2[a]	E	*Dhc*	*Syntrophomonas*, other Clostridiaceae *Methanospirillum*, *Methanosaeta*	Butyrate	82	Mansfeldt et al. (2014), Rowe et al. (2008)
TCE	KB-1-U of Toronto[a]	E	*Dhc* (>1 strain) *Geobacter*	*Acetobacterium*, *Sporomusa*, *Spirochaetes*, *Bacteroidetes*, *Methanomethylovorans*, *Methanomicrobiales*, *Methanosaeta*, and *Methanosarcina*	Methanol	9.2	Duhamel et al. (2002), Duhamel and Edwards (2006)
TCE	*Dhc* strain 195 cocultures	C	*Dhc*	*Desulfovibrio*	Lactate	9.2	Men et al. (2012)
TCE	*Dhc* strain 195	P	*Dhc*	None	H_2	3.2	Men et al. (2012)
PCE	*Dhc* strain 195	P	*Dhc*	None	H_2	120	Maymó-Gatell et al. (1997)
TCE	*Dhc* strain FL2	P	*Dhc*	None	H_2	2.3	He et al. (2005)
PCE	*Sulfurospirillum multivorans*	P	*Sulfurospirillum*	None	H_2	45	Scholz-Muramatsu et al. (1995)
PCE	*Desulfitobacterium* strain PCE1	P	*Desulfitobacterium*	None	Formate	5.6	Gerritse et al. (1996)
1,1,1-Trichloroethane	ACT-3 (aka MS)[a]	E	*Dhb* (>1 strain)	*Bacteroides*, *Desulfovibrio*, *Clostridium*, *Sedimentibacter*	Methanol, Ethanol, and Lactate	0.87 (2.4 for chloroform in subculture CF)	Grostern and Edwards (2006a), Tang et al. (2012)

(continued)

Table 14.1 (continued)

Organohalide substrate	Culture name[a]	Type of culture[b]	Key OHRB population(s)	Noted non-OHRB populations	Electron donor	Organohalide respiration rate µmole *X*/L/h (*X* = halogen)	Reference(s)
1,1,2,2-Tetrachloroethane	WBC-2	E	*Dhb*, *Dhc*, *Dehalogenimonas*	*Acetobacterium, Clostridium, Anaeromusa, Desulfovibrio*	Ethanol and Lactate	1.7	Manchester et al. (2012), Jones et al. (2006)
1,2,4-Trichlorobenzene	Unnamed	E	*Dhc*	*Desulfovibrionaceae* and gammaproteobacteria	Pyruvate	0.96	Adrian et al. (1998), (2000a), (personal communication[c])
Dichlorobenzenes	Unnamed	E	*Dhb* (>1 strain)	unspecified methanogens and bacterial fermenters	Yeast extract	3.3	(Fung et al. (2009)
1,2-Dichlorobenzene	CW 12DCB	E	*Dhb* (>1 strain)	Firmicutes, Bacteroidetes	Butyrate	58	Nelson et al. (2011)
Trichlorobenzenes	*Dhc* Strain CBDB1	P	*Dhc*	N/A	H_2	0.06–0.09	Adrian et al. (2000b)
β-Hexachlorohexane	E1[a]	E	*Dhb*	*Sedimentibacter*	H_2 or lactate	0.45	van Doesburg et al. (2005), Maphosa et al. (2012)
Arochlor 1260 PCBs	JN	E	*Dhc*	*Thauera, Geobacter, Pseudomonas*, various Clostridiales, Bacteroidetes	H_2	0.83	Bedard et al. (2006)
Dioxins	Unnamed	E	*Desulfitobacterium, Desulfuromonas* and *Dhc*	*Acetobacterium*	Pyruvate and fumarate	0.23	Ballerstedt et al. (2004)
Polybrominated diphenyl ethers	GY	E	*Dhc*	*Desulfovibrio*	H_2	0.0062	Lee et al. (2011b)

Representatives are included for enrichment cultures grown on chlorinated ethenes, chlorinated ethanes, chlorinated benzenes, PCBs, polychlorinated dioxins, polybrominated biphenyl ethers, and hexachlorohexane. Rates for some pure cultures and axenic cocultures are provided for comparison

[a]Culture food web depicted in Fig. 14.2

[b]Type of culture: Enrichment (E), Pure (P), or Axenic coculture (C)

[c]Published rate of 0.1 µmoles/L/hr (Adrian et al., 2000a) has increased 10× since the original publication

as well as additional processes involved in natural environments such as soils and sediments (dashed lines). The main metabolic food webs involve flows of reducing power (electron donor) and carbon.

14.2 Catabolic Food Webs in Organohalide-Respiring Microbial Communities

Catabolic interactions among populations in organohalide-respiring communities can broadly be divided into organohalide food webs among OHRB and electron donor flow through the communities.

14.2.1 Organohalide Food webs

As discussed in Chap. 13, it is common to find more than one OHRB population in organohalide-respiring communities. Individual OHRB catalyze only a subset of possible reductive dechlorination steps and utilize organohalide molecules with specific dehalogenation patterns. This leads to a multitude of potential webs of inter-reliant populations of OHRB. The resultant cross-feeding creates an organohalide food web among OHRB populations. Examples of cultures fed a single organohalide molecule but sustaining-interdependent OHRB populations include those respiring chlorobenzenes (Fung et al. 2009; Nelson et al. 2011, 2014), chlorinated ethenes (Duhamel and Edwards 2006; Ritalahti et al. 2006), Polychlorinated biphenyls (PCBs) (Fagervold et al. 2007; Watts et al. 2001; Wu et al. 2002), and chlorinated ethanes (Manchester et al. 2012). Some of these examples are discussed in Chap. 13.

14.2.2 Electron Donor and Carbon Flow

In natural environments and laboratory enrichments alike, fermentable organic substrates are usually the electron donor source for organohalide-respiring communities. In enrichment cultures, organic electron donors include methanol, ethanol, lactate, butyrate, propionate, benzoate, acetate, or glucose. Some of the lower molecular weight organic acid anions (e.g., lactate or acetate) can be used directly by some OHRB. Primary fermentation reactions produce a range of simple organic acids (e.g., acetic, propionic, butyric, lactic, formic), alcohols, carbon dioxide, and H_2. Secondary fermentation reactions further catabolize organic intermediates into acetate, H_2, and carbon dioxide (Group 1 in Fig. 14.1). For higher molecular weight polymeric substrates (e.g., polylactate), hydrolysis and subsequent fermentation are critical to provide OHRB with both carbon and reducing power.

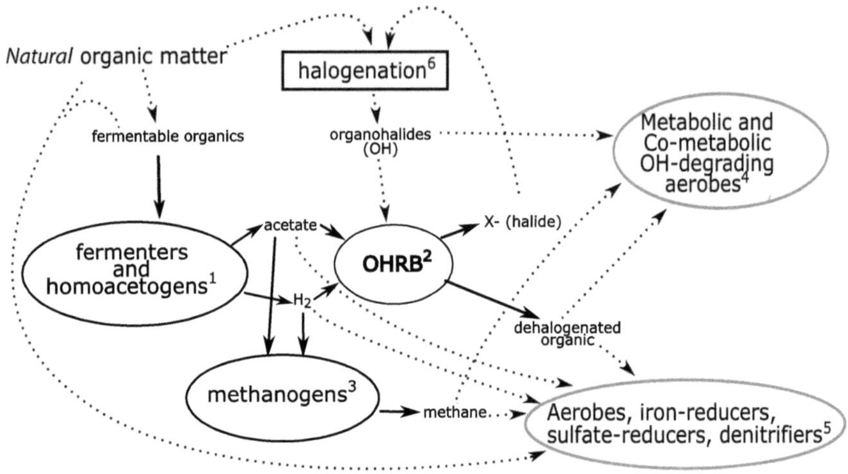

Fig. 14.1 Schematic of catabolic interactions in OHRB-containing communities. *Dotted lines* represent processes not usually sustained in laboratory enrichments but important in natural environments including pristine locations. *Solid lines* represent processes that are important in both environmental settings and in lab cultures. *Ovals* represent microbial physiological groups and the *box* represents halogenation reactions. Other important biochemical interactions such as cofactor sharing, organohalide detoxification, antagonism, and oxygen detoxification are not presented but are discussed in the text. *Numbers* denote key physiological groups of microorganisms or enzyme classes. *1* fermenters and homoacetogens. Fermenters include primary fermenters that produce a range of organic intermediates, H_2, and carbon dioxide and secondary fermenters that further degrade organic intermediates to acetate, H_2, and carbon dioxide (details not shown). Common fermenters in OHRB communities include various Firmicutes, Bacteroidetes, and Proteobacteria. Homoacetogens generate acetate from one-carbon compounds and can also produce H_2 from substrates such as methanol. Commonly found genera of homoacetogens in enrichment cultures include *Acetobacterium, Sporomusa*, and *Sedimentibacter. 2* OHRB populations including both obligate and facultative OHRB. Multiple OHRB genera and strains of individual species are commonly found even in well-established enrichment cultures. *3* methanogens with both hydrogenotrophic and acetoclastic physiologies. A wide range of both types of methanogens are observed in OHRB communities. *4* aerobes that can catabolize partially dehalogenated organohalide compounds either to conserve energy (e.g., *Polaromonas* sp. strain JS666 and various *Nocardiodes* strains) or in cometabolic processes (e.g., the methanotroph *Methylosinus* which utilizes methane as a primary substrate). *5* physiologies that compete with OHRB and methanogens for electron donor (especially in the form of H_2 or acetate) include aerobes, denitrifiers, iron reducers (such as *Geobacter, Anaeromyxobacter*, and *Shewanella*), and sulfate reducers (various delta-Proteobacteria and Firmicutes). These groups may also utilize the end products of organohalide respiration and methanogenesis as electron donors. *6* enzymatic and abiotic formation of organohalides from halide ions and natural organic matter

Molecular hydrogen (H_2) is an important intermediate in fermenting communities (Fig. 14.1). All isolated obligate OHRB and many facultative OHRB use H_2 as an electron donor. Given that the standard free energy changes ($\Delta G^{o'}$) associated with organohalide respiration reactions are large and negative, the processes are thermodynamically favorable even at very low H_2 levels [e.g., <1 nM for *Dhc*;

(Löffler et al. 2013)]. In mixed cultures, H_2 is usually produced via fermentation reactions of organic substrates. Commonly found fermenters include *Desulfovibrio* and various Clostridiales and Bacteroidetes (Men et al. 2012; Grostern and Edwards 2006a; Tang et al. 2012; Nelson et al. 2011; Lee et al. 2011b; Grostern and Edwards 2009; Adrian et al. 1998). Interspecies H_2 transfer is a common phenomenon and in some cases the process is syntrophic—thereby beneficial to both the fermenter and the H_2 consumer. Though fermentation of substrates like lactate, ethanol, and sugars is thermodynamically favorable even at high levels of H_2, other organic substrates such as propionate and butyrate (with fermentation $\Delta G^{o'}$ values greater than zero) can be fermented only when H_2 is kept at low levels by a hydrogenotrophic organism—such as OHRB and hydrogenotrophic methanogens (Groups 2 and 3 in Fig. 14.1, respectively) (Fennell et al. 1997; Mathai et al. 2015; Morris et al. 2013; Schink and Stams 2013). The following equations summarize the comparison of H_2-generating lactate and butyrate fermentation reactions (He et al. 2002).

$$\text{lactate}^- + 2\text{H}_2\text{O} \rightarrow \text{acetate}^- + \text{HCO}_3^- + \text{H}^+ + 2\text{H}_2 \quad \Delta G^{0'} = -3.96\frac{\text{kJ}}{\text{reaction}} \tag{14.1}$$

$$\text{butyrate}^- + 2\text{H}_2\text{O} \rightarrow 2\text{acetate}^- + \text{H}^+ + 2\text{H}_2 \quad \Delta G^{0'} = +48.30\frac{\text{kJ}}{\text{reaction}} \tag{14.2}$$

H_2 in organohalide-respiring communities can also be generated by homoacetogens [e.g., certain Firmicutes, Spirochaeta (Dickert and Wohlfarth 1994)], specific acetoclastic methanogens such as *Methanosarcina* (Heimann et al. 2006; Kulkarni et al. 2009; Zinder and Anguish 1992; Lovley and Ferry 1985), and acetate-oxidizing *Geobacter* (Cord-Ruwisch et al. 1998; Löffler and Sanford 2005).

Acetate is a key product of primary and secondary fermentation processes. Even those obligate OHRB that rely upon H_2 for reducing power require acetate for cell growth (*Dhc*, some *Dhb*, *Dehalogenimonas,* and *Dehalobium* strain DF1). Homoacetogens can generate acetate from H_2/CO_2 or reduced one-carbon (C1) compounds such as methanol. As mentioned above, some acetoclastic methanogens can release H_2. In fact, acetate can be used as the sole electron donor for complete dechlorination of chlorinated ethenes to ethene in defined cocultures and mixed communities (Heimann et al. 2006; He et al. 2002). In subcultures of a trichloroethene (TCE)- and methanol-enriched community [KB–1; (Duhamel and Edwards 2007)] that were evolved to use just acetate as electron donor, *Methanosarcina* produced H_2 from acetate as a side reaction to methanogenesis (Heimann et al. 2006). Another acetoclastic methanogen, *Methanosaeta,* could not generate H_2 under these conditions. Both of these methanogens were detected in the parent culture (Duhamel and Edwards 2006). The produced H_2 drove organohalide respiration by *Dhc* strains (Heimann et al. 2006).

Methane is a natural end point for catabolized organic carbon under anaerobic conditions. Various genera of acetoclastic and hydrogenotrophic methanogens (Group 3 in Fig. 14.1) are found in enrichment cultures and environmental communities performing organohalide respiration. Acetoclastic methanogenesis generates CO_2 and methane while hydrogenotrophic methanogenesis consumes H_2 and CO_2 to generate methane and water. Though methanogens are commonly found in these communities, some enrichment cultures lack methanogenic populations. High concentrations of chlorinated solvents inhibit methanogens (Yu and Smith 2000; DiStefano et al. 1991) and some researchers diminish methanogens intentionally using 2-bromoethane sulfonate (BES)—an inhibitor of methanogenesis. The fact that some OHRB have lower H_2 thresholds than hydrogenotrophic methanogens has led to successful strategies to preferentially stimulate organohalide respiration over methanogenesis (Fennell et al. 1997; Yang and McCarty 1998). Other strategies to enrich OHRB over methanogens include limiting inorganic carbon (CO_2) for hydrogenotrophic methanogens by lowering bicarbonate levels in the medium. Delgado et al. (2014) succeeded in lowering both methanogenesis and homoacetogenesis by lowering bicarbonate levels sixfold from 30 to 5 mM.

In environmental settings (and in enrichment cultures), microbes that use alternate electron acceptors (e.g., iron, oxygen, sulfate, nitrate) can also compete for H_2 and acetate when appropriate electron acceptors are present (Group 4 in Fig. 14.1). Some of these populations also directly consume higher molecular weight carbon in soils/sediments, thereby competing more with fermenters than with OHRB, but the net effect on OHRB is one of competition for electron donor and carbon. The facultative OHRB have physiological diversity and they may be involved in multiple biogeochemical processes including organohalide respiration, sulfate reduction, fermentation, denitrification, and iron reduction. For some of these genera (e.g., *Geobacter*, *Desulfuromonas*, *Desulfovibrio*), the majority of isolates are not capable of organohalide respiration. Some evidence of in situ competition for H_2 at natural attenuation sites between iron reducers and Vinyl chloride (VC) respirers has been observed in phylogenetic surveys (Shani et al. 2013).

Carbon monoxide and formate are partially oxidized C1 compounds that usually play smaller roles in catabolic food webs as electron donors, but some evidence suggests that these may be used by non-OHRB community members such as some methanogens and acetogens (Oelgeschläger and Rother 2008; Sieber et al. 2014). Carbon monoxide has been shown to be released by *Dhc* strain 195 by enzymes of its Wood–Ljungdahl pathway (Zhuang et al. 2014). The role carbon monoxide plays in food webs in OHRB communities is currently unclear. In *Dhc*, the most highly expressed membrane-bound oxidoreductase (besides the RDases) is annotated as a formate dehydrogenase ("Fdh"). However, formate is not produced nor utilized by the *Dhc* and the true function of "Fdh" remains unknown (Maymó-Gatell et al. 1997; Morris et al. 2006, 2007; Werner et al. 2009). *Dehalobium* isolate DF-1 and some *Dehalobacter* isolates were reported to use formate as an electron donor (Wu et al. 2002; Sun et al. 2002). Various methanogens can also utilize carbon monoxide and formate directly (Zinder and Anguish 1992).

There is a possibility that direct electron transfer and/or use of electron shuttle mediators could also be involved in interspecies transfer of reducing power. Aulenta et al. (2007) showed that electrodes poised at low redox potential could provide reducing equivalents to OHRB (presumably *Dhc*) as long as the artificial electron shuttle methyl viologen was present (Aulenta et al. 2007).

14.2.3 Commonly Observed Groups of Non-OHRB Involved in Metabolic Food Webs

Though the various phylogenetically characterized enrichment cultures contain different specific populations, there are some phylogenetic lineages that appear more commonly than others. These include delta-Proteobacteria (especially nondechlorinating strains of *Desulfovibrio* and *Desulfuromonas*), gammaproteobacteria (*Citrobacter*), methanogenic euryarchaeota (various acetoclastic and hydrogenotrophic families), Clostridiales (including *Clostridium*; *Sedimentibacter*; *Eubacterium*; *Syntrophomonas*; *Acetobacterium*), Selenomonadales (*Sporomusa* and *Pelosinus*), *Bacteroidetes*, *Spirochaeta*, *Chloroflexi*, and *Nitrospira* (Duhamel and Edwards 2006; Hug et al. 2012). Table 14.1 presents key populations in several representative organohalide-enriched communities. Many of these populations are fermenters, hydrogenotrophic methanogens, or homoacetogens, while others are more cryptic in physiology (e.g., *Nitrospira*, non-*Dehalococcoidales* members of the *Chloroflexi*).

Of the phylogenetic groups listed above, some are more commonly found than others. Most notably, *Desulfovibrio* is a very common member of OHRB communities. Most observed *Desulfovibrio* spp. cannot perform organohalide dechlorination and instead act as fermenters in organohalide-respiring communities (Sun et al. 2000; Zhang et al. 2010; Grostern and Edwards 2006a). Some researchers have shown that *Desulfovibrio* strains form stable cocultures with *Dhc* strain 195 and in some cases the coculture has faster dechlorination rates than *Dhc* pure culture controls (Table 14.1) (Men et al. 2012; He et al. 2007; Wu et al. 2002). Firmicutes with fermentative and even homoacetogenic capabilities are also commonly found (e.g., *Sporomusa, Pelosinus, Clostridium, Sedimentibacter, Eubacterium, Syntrophomonas, and Acetobacterium*) (Group "1" in Fig. 14.1). *Sedimentibacter* strains have been found in OHRB communities grown on a range of organohalide compounds (Cheng et al. 2010; van Doesburg et al. 2005; Maphosa et al. 2010; Oba et al. 2014; Vandermeeren et al. 2014) including a notable coculture that fully dechlorinates β-hexachlorohexane to benzene (Culture "E1" in Table 14.1 and Fig. 14.2e). Homoacetogens (e.g., *Acetobacterium, Sporomusa, Spirochaeta*) are commonly observed in cultures-fed reduced one-carbon compounds (e.g., methanol) or H_2 plus carbon dioxide. The specific types of spirochetes present in these OHRB communities show some overlap (Gu et al. 2004).

14.3 Other Biochemical Interactions in OHRB Communities

In addition to direct catabolic interactions where products from one community member are used as catabolic substrates for others (sometimes syntrophically), there are additional biochemical interactions at play in these communities—both direct and indirect. Direct provision of cofactors and nutrients and indirect detoxifying reactions are discussed below.

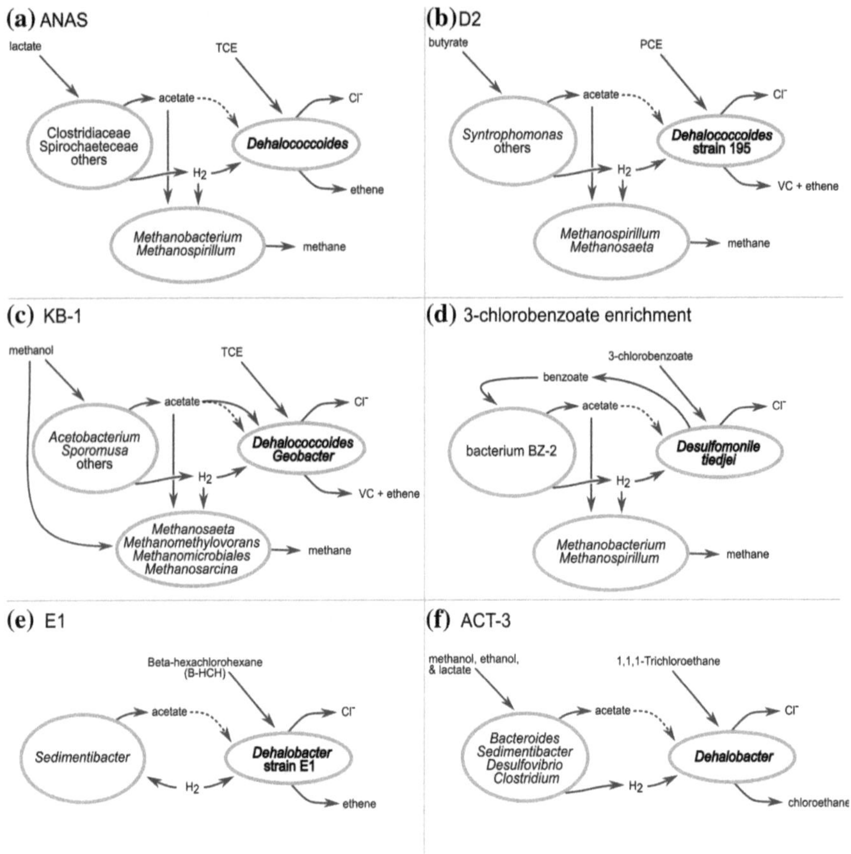

Fig. 14.2 Examples of metabolic food webs in organohalide-respiring enrichment cultures. Key phylogenetic groups carrying out different catabolic processes are shown for each culture. *Dashed lines* represent uptake of acetate as a carbon source. "Others" indicates that other small populations may also contribute to the process. **a** ANAS culture grown on lactate and TCE (Richardson et al. 2002; Brisson et al. 2012); **b** KB-1 culture grown on methanol and TCE (Duhamel et al. 2002; Hug et al. 2012; Duhamel and Edwards 2007); **c** D2 culture grown on butyrate and PCE (Fennell and Gossett 1998; Rowe et al. 2008); **d** *Desulfomonile tiedjei* consortium grown on 3-chlorobenzoate (as both organohalide and sole carbon source) (Dolfing and Tiedje 1991; Mohn and Tiedje 1992; Shelton and Tiedje 1984); **e** E1 culture grown on H_2/CO_2 and β-hexachlorohexane (van Doesburg et al. 2005; Maphosa et al. 2012), **f** ACT-3 culture grown on 1,1,1-trichloroethane and a mix of methanol, ethanol, and lactate (Grostern et al. 2009; Tang et al. 2012)

14.3.1 Cofactors and Nutrients

The most focused-upon cofactors in these OHRB communities are corrinoids. Functional RDase enzymes in OHRB require a corrinoid cofactor (see Chaps. 17 and 19) but the vast majority of the obligate OHRB with sequenced genomes do not contain complete biosynthetic pathways for corrinoids on their genomes (with two exceptions: a *Dhc* strain in the ANAS mixed culture (Brisson et al. 2012) and *Dehalobacter restrictus* PER-K23 (Rupakula et al. 2013)). Recent studies suggest that PER-K23 has a deletion in its *cpiH* gene rendering the pathway nonfunctional (Rupakula et al. 2015). However, other *Dhb* strains with fully functional biosynthetic pathways are emerging (Tang et al. 2012). The interspecies transfer of corrinoids has been studied in enrichment cultures as well as axenic cocultures (Men et al. 2012, 2014; Yi et al. 2012; Johnson et al. 2009; Yan et al. 2012, 2013). Genera with strains that can fill the role of corrinoid biosynthesis in organohalide-respiring communities include *Geobacter* (Yan et al. 2012; Wagner et al. 2012), *Acetobacterium* (Ziv-El et al. 2012), *Sporomusa* or *Methanosarcina barkeri*, and *Pelosinus* when 5′,6′-dimethylbenzimidazole (DMB) is provided (Yan et al. 2013). Comparing metagenomes across three chlorinated ethene-dechlorinating communities, corrinoid biosynthetic pathways were well represented in the non-organohalide-respiring members of these communities (Hug et al. 2012; Brisson et al. 2012). Other as-yet-undetermined cofactors/micronutrients will likely be elucidated in the near future with additional metagenome-sequencing projects and advances in high-resolution metabolomic profiling. A recent analysis summarizes the current state of knowledge regarding *Dhc*'s organic cofactors (Schipp et al. 2013).

Nitrogen for biomass growth is usually provided to enrichment cultures as fixed nitrogen. However, at least one OHRB (*Dhc* strain 195) can fix nitrogen (N_2) and thus can help provide the community with a key anabolic nutrient requirement (Lee et al. 2009, 2012b). In *Dhc*, all amino acids can be synthesized de novo despite the fact that some pathways appeared incomplete following genome annotation (Seshadri et al. 2005; Schipp et al. 2013; Tang et al. 2009; Kube et al. 2005; Marco-Urrea et al. 2012). Recent work has shown a novel approach to methionine biosynthesis, which involves enzymes of the Wood–Ljungdahl pathway (Zhuang et al. 2014). Despite the capacity for de novo synthesis, some amino acids are presumed to be taken up by ABC-type transporters rather than synthesized (Zhuang et al. 2011). *Dehalobacter restrictus* PERK23 was shown to be auxotrophic for some amino acids (Holliger et al. 1998).

14.3.2 Toxins and Inhibitors

O_2 scavenging Another benefit of growth in mixed culture for the obligate OHRB is that of O_2 scavenging when the community is exposed to oxygen. Organohalide-respiring communities may contain organisms that use O_2 for respiration as well as organisms that possess detoxification pathways for dealing with reactive oxygen species. Comparing the metagenomes of three chlorinated ethene-respiring communities, many orthologs of superoxide dismutase were found from multiple genera in each community (Hug et al. 2012). One of the facultative OHRB, *Anaeromyxobacter dehalogenans*, is actually a microaerophile (Thomas et al.

2008). The ability to deplete oxygen in cultures is reliant upon sufficient reduced substrates (which can even include dead biomass).

Toxic organohalide levels Organohalide compounds can be highly toxic to many organisms. Methanogens have been shown to be inhibited by Tetrachloroethene (PCE), TCE, 1,1,1-trichloroethane, and chloroform—with chloroform being inhibitory at concentrations as low as 0.8 μM (DiStefano et al. 1991; Men et al. 2013; Yu and Smith 2000). Even some of the OHRB themselves find organohalide substrates toxic and/or inhibitory at higher concentrations and there are several examples of inhibition by nonsubstrate organohalides—e.g., the inhibition of chlorinated ethene-respiring *Dhc* by 1,1,1-trichloroethane or chloroform (Grostern and Edwards 2006a) as well as competitive inhibition among substrates such as TCE and *cis*-1,2-dichloroethene (*cis*-DCE) by the TCE RDase, TceA (Magnuson et al. 2000). In some competitive inhibition cases, the inhibition constants can have values as low as 0.2 μM (see Chap. 13 Tables and references therein). Futagami et al. (2006) showed that chloroform and carbon tetrachloride inhibited PCE dechlorination in the facultative OHRB *Desulfitobacterium hafniense* strain Y51 and even favored the growth of non-organohalide respiration mutants of strain Y51, which had lost the ability to express the PCE-RDase PceA.
Destruction of organohalides by OHRB can thereby benefit the entire community. Even in situations where concentrations are not toxic to cells, organohalides can cause competitive inhibition and limit the activities of RDases. Examples of such competitive inhibition are discussed in Chap. 13. Many of the existing biokinetic models have had to incorporate inhibition of organohalide respiration rates at high organohalide concentrations as well as competitive inhibition by substrates acted upon by the same RDases (Lai and Becker 2013; Haest et al. 2010a; Yu and Semprini 2004; Yu et al. 2005; Heavner et al. 2013). This further complicates the interplay among OHRB populations in mixed cultures. Reductive dechlorination by one OHRB can alleviate (or amplify) inhibition of other OHRB. For example, Grostern and Edwards combined two cultures—culture KB-1 enriched on chlorinated ethenes that finds 1,1,1-trichloroethane inhibitory and culture ACT-3/MS enriched on chlorinated ethanes that finds VC inhibitory (Grostern and Edwards 2006a). Dechlorination of 1,1,1-trichloroethane to 1,1-dichloroethane by a *Dhb* population relieved toxicity for the *Dhc* strains, which then dechlorinated the chlorinated ethenes to ethene. After the chlorinated ethenes were fully converted to ethene, dechlorination of 1,1-dichloroethane to chloroethane by a distinct *Dhb* population proceeded.

14.4 Variations in Community Ecology and Composition

Community composition in enrichment cultures can vary for a number of reasons including different inoculum, electron donor, organohalide type, organohalide concentration, and incubation conditions (temperature, pH, micronutrients, alternate electron acceptors). In this section, trends for some of these variables are presented.

14.4.1 Electron Donor and Electron Acceptor/Electron Donor Ratio

A wide range of electron donors have been used to enrich and sustain organohalide-respiring communities. Biochemical and thermodynamic characteristics of the electron donors can significantly affect community structure. Electron donor amount relative to stoichiometric organohalide demands plays a large factor in the species richness and the fraction of biomass comprising OHRB populations. Methanogens (both acetoclastic and/or hydrogenotrophic) can thrive when electron donor is added in excess relative to the amount needed to stoichiometrically reduce the organohalide provided. On the other extreme is the electron donor-limited situation where endogenous decay of biomass is the source of reducing equivalents and very little methanogenesis occurs (Sleep et al. 2005). Methanol as the enrichment electron donor directly supports homoacetogens (e.g., *Acetobacterium* or *Sporomusa*) and various methanogens (Aulenta et al. 2002, 2005; Daprato et al. 2007; Duhamel and Edwards 2007). Acetate as the only electron donor can support acetoclastic methanogens, some of which (e.g., the *Methanosarcina* strain in consortium KB-1) can release small amounts of H_2 for use by obligate OHRB (Heimann et al. 2006). H_2 as the electron donor can support methanogens as well as homoacetogens, if carbon dioxide is also available.

In short-term incubations of subcultures of a TCE-to-ethene-dechlorinating consortium on lactate, methanol, propionate, and butyrate community stability was observed (Freeborn et al. 2005). Phylogenetic profiles showed only minor shifts in community structure, consistent with the evidence from the culture's metagenome that genes for multiple fermentation pathways persisted even after decade of incubation with lactate (Brisson et al. 2012; Hug et al. 2012). It is likely that the same phylogenetic groups persisted by upregulating the appropriate catabolic pathways needed for the new electron donor. Hug et al. (2012) compared metagenomes from three stable PCE/TCE-respiring OHRB communities (Hug et al. 2012). These were the consortia ANAS (Richardson et al. 2002; Brisson et al. 2012), D2 (Rowe et al. 2008), and KB-1 (Duhamel and Edwards 2006) (Table 14.1 and Fig. 14.2a–c). The authors found that multiple groups in the community can metabolize the electron donor used to maintain each culture (lactate, butyrate, or methanol, respectively). Interestingly, even after years of selection on one electron donor, each community's metagenome still contained genes for metabolism of a wide range of organic electron donors (Hug et al. 2012).

With respect to thermodynamic considerations, some electron donors can be used to selectively support OHRB over methanogens or homoacetogens. While substrates such as sugars, lactate, ethanol, and methanol can be fermented to release high levels of H_2, the fermentation of others, such as butyrate and propionate, is only thermodynamically feasible if H_2 levels are kept low (See Eqs. 14.1 and 14.2). In studies by Fennell et al. (1997) with the *Dhc*-containing Cornell enrichment culture (later named D2; Fig. 14.2b and Table 14.1), higher H_2 levels and methanogenesis rates were observed in cultures-fed ethanol or lactate compared to propionate or butyrate. Hydrogenases are important enzymes in these communities—particularly in obligately hydrogenotrophic microorganisms such as many of the obligate OHRB. Several metagenomic sequencing projects found multiple hydrogenase genes present on the genomes of different community members (Brisson et al. 2012; Hug et al. 2012; Maphosa et al. 2012; Tang et al. 2012). Transcript levels for hydrogenases have been shown to correlate with overall respiration rate in both *Dhc* and hydrogenotrophic methanogens (Rowe et al. 2012, 2013; Heavner 2013; Rahm and Richardson 2008). With advances in high-throughput transcriptome analysis methods (e.g., microarrays and high-throughput sequencing technologies), hundreds to thousands of unique hydrogenase transcripts can be monitored simultaneously. Marshall et al. (2012) have designed a tiling oligonucleotide DNA microarray ("Hydrogenase Chip") to characterize H_2-producing and H_2-consuming microbes in microbial communities including an OHRB community. Their microarray analysis of transcript levels in PCE-dechlorinating consortia showed strong correlation with rates of PCE respiration.

14.4.2 Organohalide Class

OHRB communities have been enriched with different types of organohalides, including chloroalkanes, chloromethanes, chloroalkenes, chlorinated monoaromatics (benzenes, phenols, benzoates, phthalates), PCBs, halogenated diethers, halogenated dioxins, and halogenated furans. Within any one of these classes of organohalide compounds, specific organohalides differ in the degree of halogenation, type of halogen, and in the specific arrangement of the halogen atom(s) on the carbon backbone. Overall, there is no apparent trend in the types of non-OHRB found in communities solely as a function of class of chlorinated organohalide. However, some trends are beginning to emerge with respect to the genera of OHRB found as a function of organohalide class, degree of halogenation, and specific halogenation patterns. Various OHRB genera contribute to organohalide respiration of chlorinated ethenes, chlorinated ethanes, and chlorinated monoaromatics. Only *Dhb* were proposed to respire chloroform (Lee et al. 2012a; Grostern et al. 2010; Justicia-Leon et al. 2014) and 1,1,1-trichloroethane (Grostern and Edwards 2006a; Grostern et al. 2009). Recently, however, another member of the *Peptococcaceae* family, *Desulfitobacterium* sp. strain PR, was isolated and shown to dechlorinate 1,1,1-trichloroethane and chloroform (Ding et al. 2014). In general, for cultures enriched on chlorinated ethanes, *Dhb*, *Dhc, Dehalogenimonas,* or *Desulfitobacterium* have been identified as the OHRB populations (Grostern and Edwards 2006a, b, 2009; Grostern et al. 2010; Justicia-Leon et al. 2012; van Doesburg et al. 2005; Marzorati et al. 2007; De Wildeman et al. 2003). *Dehalococcoidia* (especially *Dhc* and *Dehalobium*) are usually found in communities enriched on dioxins, furans, and polychlorinated biphenyls (Ahn et al. 2007; Ballerstedt et al. 2004; Bunge and Lechner 2009; Liu et al. 2014; Yoshida et al. 2005; Cutter et al. 2001; Fagervold et al. 2007; Holoman et al. 1998; Watts et al.

2001; Wu et al. 2002; Bedard et al. 2007; Yan et al. 2006). *Dehalogenimonas* spp. seems to specialize in chlorinated ethanes or chlorinated ethenes (Bowman et al. 2013; Moe et al. 2009; Manchester et al. 2012).

With polybrominated diphenyl ethers (PBDE) as the growth substrate, the same OHRB genera are seen as with chlorinated organohalides—particularly *Dhc* and *Dhb* (He et al. 2006; Lee et al. 2011b). Recently, a community enriched with lactate and triiodophenol was described as having no typical OHRB genera. Instead the researchers found an *Aneaerolinea* strain within the *Chloroflexi* and RDases with only 75 % identity to known *Dhb* and *Dhc* RDases, suggesting that deiodinating strains and their RDases may be specialized (Oba et al. 2014). The non-OHRB populations (*Clostridium* and *Sedimentibacter*) matched genera typically found in organochloride-respiring communities.

Some studies have done head-to-head comparisons of short-term incubations on different specific organohalides. The KB-1 culture (Table 14.1 and Fig. 14.2c), originally enriched on PCE, was later divided into four subcultures maintained on PCE, TCE, *cis*-DCE, and VC. The most notable change was that the culture maintained on VC retained only a subset of the original culture's *Dhc* strains and also lost the *Geobacter* population that can only respire PCE or TCE to *cis*-DCE (Duhamel et al. 2002; Duhamel and Edwards 2006). Yang et al. (2005) set up different enrichments with variations in electron donor (H_2 versus acetate) and organohalide (PCE versus *cis*-DCE versus none). They found significant *Desulfitobacterium* populations only in cultures-fed PCE while *Dehalococcoides* were found in cultures-fed either PCE or *cis*-DCE. A homoacetogenic Firmicute (*Sporomusa*) was present in all cultures that received H_2 as the electron donor. Gu et al. (2004) compared communities with TCE versus *cis*-DCE as the organohalide and observed the dominant phylogenetic groups (*Dhc* and Clostridiales) in both enrichments but a spirochete occurred only in *the cis*-DCE enrichment. Though there was no clear reason for the spirochete's presence in only the *cis*-DCE enrichment, it is notable that the spirochete's 16S rRNA gene sequence was very similar to one found in a chloropropanes-respiring community (GenBank Accession # AJ306787) and another found in a dioxin-respiring community (Yoshida et al. 2005)—suggesting that particular spirochete lineages may be preferentially enriched in organohalide-respiring communities.

14.5 Case Studies in Community Ecology: Simple Mixed Communities

Stable cocultures containing obligate OHRB and one or two other populations have been both enriched and assembled from pure cultures (Men et al. 2012, 2014; Yan et al. 2012; Shelton and Tiedje 1984; van Doesburg et al. 2005). In these cocultures, the non-OHRB often ferments the added electron donor to provide the obligate OHRB with both its carbon source and its electron donor. In cultures enriched with H_2/CO_2, homoacetogens may grow that generate acetate for obligate OHRB to use as a carbon source. Additionally, the non-OHRB populations have the ability to biosynthesize key cofactors needed by the OHRB (e.g., corrinoid cofactors required for RDase function). Studies with an axenic community containing *Desulfomonile tiedjei*, a *Methanospirillum* strain, and bacterium BZ-2 showed a very interesting food web when fed 3-chlorobenzoate: *D. tiedjei* dechlorinated 3-chlorobenzoate to benzoate, which was fermented by BZ-2 to H_2, acetate, and CO_2. These products were used by both *D. tiedjei* and the methanogen (Fig. 14.2d). BZ-2 and *Methanospirillum* were also proposed to synthesize important cofactors used by the dechlorinator (Shelton and Tiedje 1984; Mohn and Tiedje 1992). The metagenome of a stable coculture ("E1") of *Dhb* and a *Sedimentibacter* that dechlorinates hexachlorocyclohexane to benzene (Fig. 14.2e) confirmed the presence of corrinoid synthesis genes in the *Sedimentibacter* and a lack of the synthesis pathway in the *Dhb* population (Maphosa et al. 2012; van Doesburg et al. 2005). Other stable cocultures include *Dhb* and *Acetobacterium* with 1,2-dichloroethane (Grostern and Edwards 2009), *Dhb, Dhc,* and *Acetobacterium* with 1,1,2-trichloroethane (Grostern and Edwards 2009), *Dhc* and *Desulfovibrio* with PBDEs (Lee et al. 2011b), and *Dhc* and a spirochete with TCE (He et al. 2005). Stable assembled cocultures include *Dhc* and *Desulfovibrio* with or without *Methanobacterium* (Men et al. 2012, 2014; Lee et al. 2011b) and *Geobacter* and *Dhc* with chlorinated ethenes (Yan et al. 2012).

14.6 Organohalide Respiration Rates in OHRB Communities and Modeling of Community Food Webs

In terms of dechlorination rates observed in organohalide-respiring cultures, volume-normalized organohalide respiration rates vary widely (Löffler et al. 2013; Adrian et al. 2000b; Schaefer et al. 2009; Cupples et al. 2003; Mansfeldt et al. 2014; Maymó-Gatell et al. 1997; Gerritse et al. 1996). In pure cultures, obligate OHRB generally grow to lower densities than in the enrichment cultures from which they were isolated. For example, chlorinated ethene-respiring *Dhc* isolates rarely grow above 1E8 cells/mL while multiple researchers reported mixed cultures containing 1E9 cells/mL *Dhc*. This higher cell density translates into faster overall culture respiration rates.

Table 14.1 presents a collection of organohalide respiration rates and microbial populations from representative cultures enriched on various organohalide compounds. With respect to rates of organohalide respiration in mixed communities, it is difficult to directly compare rate values as different researchers normalize rates with different units (e.g., per genome copy, per g dry soil/sediment, per L culture, or per mg total protein). In comparing rates here, reported rates were converted to common units of µmoles Cl^- released per L of culture per hour (µmoles Cl^-/L/h). The fastest rates of organohalide respiration have been reported for cultures respiring chlorinated ethenes. For example, rates of approximately 100 μmole Cl^-/L/h have been reported in cultures respiring chlorinated ethenes and containing *Dhc* populations near 1E9 cells/mL (Rowe et al. 2008; Delgado et al. 2014; Ziv-El et al. 2011; Duhamel et al. 2002; Yu et al. 2005; Maymó-Gatell et al. 1997; Mansfeldt et al. 2014). The fastest reported culture to fully dechlorinate TCE to ethene was grown in a chemostat with a three-day retention time and achieved 130 μmole Cl^-/L/h (Delgado et al. 2014). The original report on the isolation of *Dhc* strain 195 presented data suggesting that pure cultures can reach high density (1E9 cells/mL) and rates (120 μmoles Cl^-/L/h) approaching those of mixed cultures (Maymó-Gatell et al. 1997). *Sulfurospirillum* is a pure culture with high rates of dechlorination. For example, *Sulfurospirillum multivorans* was shown to dechlorinate PCE to TCE at a rate of 45 μmoles Cl^-/L/h (Scholz-Muramatsu et al. 1995). However, in most studies, pure culture rates are much lower than rates reported for mixed cultures or defined cocultures.

Organohalide respiration rates for other organohalide classes are generally lower than for chlorinated ethenes. Chlorinated ethane-respiring cultures show rates on the order of 1 μmole Cl^-/L/h (Grostern and Edwards 2006a; Manchester et al. 2012). Chlorobenzene-enriched cultures have shown rates up to 58 μmoles Cl^-/L/h (Nelson et al. 2011). In communities respiring high molecular weight organohalides with low aqueous solubility (e.g., polychlorinated dioxins, polychlorinated furans, PCBs), organohalide respiration rates are more likely to be limited by mass transfer rates than for higher solubility organohalide compounds. Cultures respiring trichlorodibenzo-p-dioxin were reported to have rates of 0.22 μmole Cl^-/L/h (Ballerstedt et al. 2004) and culture GY respired PBDE with rates near 0.006 μmole Cl^-/L/h (Lee et al. 2011b). Overall, reported rates range by five orders of magnitude across the different organohalide classes.

Effective models for community behavior require accounting for important physiological groups that impact OHRB activity. The production and consumption of H_2 and, to a lesser extent, acetate are central processes in these OHRB communities. Several models based on Monod kinetics exist for chloroethene-respiring OHRB populations growing in mixed cultures (Heavner 2013; Yu and Semprini 2004; Yu et al. 2005; Haest et al. 2010a, b; Baelum et al. 2013; Becker 2006; Becker and Seagren 2009; Lai and Becker 2013; Fennell and Gossett 1998; Chen et al. 2013; Schneidewind et al. 2014; Maphosa et al. 2010). Chambon et al. (2013) summarizes models of organohalide respiration in a recent review. Many of these existing modeling efforts consider the activity of only the OHRB. The concentrations of H_2 and acetate are usually assumed to be constant (Schneidewind et al. 2014). A few models have also biokinetically modeled the activities of non-OHRB community members including the effects of fermenters and methanogens upon H_2 and acetate availability (Fennell and Gossett 1998; Heavner et al. 2013). The use of DNA, RNA, and protein biomarkers to quantify different populations and their key enzymes is allowing development and validation of community models in ways not previously possible and should improve predictive models in the near future.

14.7 Community Structure in Contaminated Aquifer/ Sediment Environments

As discussed in Chap. 22, successful bioremediation at contaminated sites hinges upon OHRB activities being realized in situ. In a remediation context, enhanced bioremediation efforts historically involved adding electron donor in excess of what would be needed to completely reduce site organohalide concentrations—to ensure complete dechlorination. Electron donor additions stimulate many microbes in addition to OHRB—including fermenters, methanogens, and others physiological groups supported by site geochemistry (e.g., sulfate reducers, iron reducers). At natural attenuation sites, electron donor may be limiting and may be derived from hydrocarbon co-contaminants at the site (e.g., petroleum products) or decay of endogenous biomass. Compared to their relatively high abundance in laboratory cultures, OHRB are a much smaller percentage of the biomass at contaminated field sites. This is true even in dense nonaqueous phase liquid (DNAPL) source zones where organohalide levels are high and/or when OHRB cultures are bioaugmented into the aquifer (commonly at 100–10,000 fold dilutions of inoculum into aquifer porewater) (Major et al. 2002; Baelum et al. 2013; Tas et al. 2009; Ellis et al. 2000; Lendvay et al. 2003).

Several surveys of in situ biomass have been conducted from materials retrieved from field sites undergoing enhanced bioremediation (e.g., biostimulation alone or combined with bioaugmentation) or natural attenuation. Many studies have assayed biomass specifically for known OHRB via methods targeting their 16S rRNA genes and/or their RDase genes. The assays include qualitative/semiquantitative methods (e.g., PCR amplification and clone library analysis) as well as quantitative assays by qPCR and/or competitive PCR (Matturro et al. 2013; Hatt and Loffler 2012; Kjellerup et al. 2008; Hendrickson et al. 2002; Pérez-de-Mora et al. 2014; Lu et al. 2006; Rahm et al. 2006; Lee et al. 2008). In general, these studies show that native OHRB are often present in situ, can be biostimulated by electron donor additions, and can even persist following bioaugmentation with other OHRB cultures (Pérez-de-Mora et al. 2014; Baelum et al. 2014). Other researchers have utilized the benefits of degenerate primers for either ribosomal RNA genes of the *Dhc*-like group of the Chloroflexi (Lowe et al. 2002) or RDase genes (Tas et al. 2009, 2011; Dugat-Bony et al. 2012; Lee et al. 2008). Dugat-Bony et al. (2012) have introduced a DechlorArray microarray with probes to 92 distinct RDase genes covering a range of available RDase diversity. Using this array on groundwater biomass from four chlorinated ethene-impacted sites undergoing biostimulation, the authors found that there was great temporal and spatial variability of diverse OHRB genera. The VC RDases, *bvcA,* and *vcrA* were below detection at a location stalled at *cis*-DCE but present where dechlorination past *cis*-DCE occurred. Tas et al. (2009) examined biomass from a PCE-contaminated aquifer using functional gene arrays including probes for 153 RDase genes as well as probes for many other functional genes. They found diverse RDase genes of the

Dhc type in the DNA pool and they also witnessed high variability across time and location.

General community structures at contaminated sites have been examined using methods such as high-throughput sequencing of 16S rRNA gene amplicons (Kotik et al. 2013; Lee et al. 2011a), phylogenetic microarrays (Conrad et al. 2010; Lee et al. 2012c; Nemir et al. 2010), and 16S rRNA gene clone libraries (Rahm et al. 2006; Dojka et al. 1998; Bowman et al. 2006; Macbeth et al. 2004; Lee et al. 2012c). Macbeth et al. (2004) surveyed biomass from a TCE-contaminated deep-fractured basalt aquifer that had been biostimulated. The Firmicute *Acetobacterium* was the most abundant sequence in the small clone library (32/93 clones in bacterial library). Other phylogenetic groups present included *Sphingobacteria, Bacteroides, Spirochaetes, Mollicutes, Proteobacteria,* and candidate divisions OP11 and OP3. Bowman et al. (2006) studied acidic groundwater near a DNAPL source zone (mostly chlorinated ethanes) using both general and *Dhc*-specific 16S rRNA gene-targeted primers to establish clone libraries. In the library established with general 16S rRNA gene primers, Firmicutes sequences were dominant (62 % of the library) and no *Dhc* sequences were detected. However, *Dhc*-specific PCR did confirm presence of *Dhc* populations. Rahm et al. (2006) had similar observations at the Seal Beach site—that *Dhc* were not observed in clone libraries created with degenerate primers, but direct PCR with *Dhc* primers was positive. Direct comparisons between results from clone libraries and phylogenetic microarrays found that clone libraries would need to be very large to enable detection of *Dhc* at their low in situ populations (Lee et al. 2012c). It is notable that recommended densities for in situ bioremediation are above 1E4/mL (1E7/L) (Lu et al. 2006). This may be a very small fraction of the community overall and detection limits may be high in community survey methods (e.g., clone libraries, microarrays, and high-throughput sequencing of 16S rRNA gene amplicons). However, some studies have found OHRB in clone libraries created using bacterial 16S rRNA gene-targeted primers or general 16S rRNA gene-targeted primers (Dojka et al. 1998; Lowe et al. 2002). Lowe et al. (2002) found obligate OHRB 16S rRNA gene sequences in clone libraries established with DNA obtained from locations impacted with chlorinated ethanes but not from pristine locations—specifically *Dehalobacter*, not *Dehalococcoidales*.

Remarkable advances in sequencing are enabling much more thorough investigations of microbial community changes at field sites using PCR amplicon sequencing and even metagenome sequencing. Such studies are improving resolution of community structure. With increased numbers of samples, and appropriate metadata, microbial responses to contamination and remediation practices can now be traced. Studies utilizing phylogenetic microarrays or 454 pyrosequencing on chlorinated ethene plume biomass agreed in observing a wide range of physiological types in the plume communities—with obligately anaerobic and aerobic organisms both present in samples (Conrad et al. 2010; Kotik et al. 2013). Conrad et al. (2010) performed a transect along a two-kilometer-long plume and found that methane produced in the active biostimulation area (likely by *Methanosarcinaceae* populations) was stimulating methanotrophs (particularly

Methylosinus) further down the plume. These methanotrophs can degrade both methane and chlorinated ethenes (Group 4 in Fig. 14.1), the latter via cometabolic reactions catalyzed by their methane monooxygenase. In work by Kotik et al. (2013), bacterial rRNA gene amplicons were analyzed by 454 pyrosequencing. Though *Albidiferax* (beta Proteobacteria) were predominant in all locations, the two most highly contaminated sites did have several putative facultative OHRB genera (*Anaeromyxobacter*, *Desulfuromonas*, *Desulfovibrio,* and *Geobacter*), though it is not known if the strains contained genes for organohalide respiration. These populations may be respiring with sulfate or iron and some can grow fermentatively. Even though the deep sequencing method had a detection level of 0.25 % of total amplicons, no obligate OHRB sequences were detected. Instead, the community structure suggested that *cis*-DCE and VC may be acted upon by the aerobic chlorinated ethene degraders detected by their analyses. These included *Polaromonas*, the only known aerobe that grows with *cis*-DCE as its sole carbon and energy source. They also found possible aerobic cometabolizers of chlorinated ethenes (*Burkholderia* and *Methylobacter*). Lee et al. (2012c) used phylogenetic microarrays to track groundwater communities during biostimulation and bioaugmentation for TCE bioremediation. Archaea grew over time—consistent with a two order of magnitude increase in methane at the site. *Dhc* were detected by their microarray method.

It is clear that biostimulation efforts at contaminated groundwater sites stimulate a wide range of microbial populations directly and indirectly—including various anaerobes (e.g., OHRB, methanogens, fermenters) and aerobes (methanotrophs, aerobic organohalide-mineralizing populations, cometabolic organohalide degraders).

14.8 Roles of OHRB in Natural Systems

Though releases of pure organohalide compounds into the environment creates an unnatural situation, organohalides in general are not solely anthropogenic—they are produced naturally and can support native populations of OHRB (see Chap. 2). Figure 14.1 includes several processes that are proposed to play important roles in natural organohalide cycles.

The naturally occurring organohalides are diverse (Leri et al. 2007). Gribble (2003) reported that more than 3800 individual organohalides are present in natural systems and most of these are chlorinated and brominated organohalides, though fluorinated and iodinated organohalides do exist. A variety of studies have clearly demonstrated the existence of natural organohalide pools in weathered plant material, soil, and sediment (freshwater and marine) (Asplund and Grimvall 1991; Gribble 1994, 2003; Leri et al. 2007). Organohalides are also found associated with living animals (e.g., marine sponges produce brominated organics intentionally) (Ahn et al. 2003).

With respect to the abundance of organohalides, in several studies of soils the organic chloride pool was actually larger than the inorganic chloride pool (Oberg et al. 2005; Rohlenova et al. 2009; Redon et al. 2011). Chlorinated organics were shown to correlate strongly with the prevalence of organic matter content of soils (Gustavsson et al. 2012). In humus layers, nearly 100 % of the soil chlorine atoms were tied up as chlorinated organics (Redon et al. 2011). Absolute levels of chlorinated organics across 51 soils in France ranged from 34 to 689 mg/kg soil (Redon et al. 2011).

Biogeochemical cycling of natural organohalides involves both formation and destruction. The OHRB likely play a major role in organohalide destruction. Formation of organohalides occurs by both abiotic and biotic mechanisms. Several recent studies have concluded that in organic soils, the biotic route predominates (Aeppli et al. 2013; Bastviken et al. 2007, 2009; Clarke et al. 2009; Rohlenova et al. 2009). Biological halogenation reactions can occur by different enzymatic classes including haloperoxidases and flavin dehydrogenases (Reaction 6 in Fig. 14.1) (Aeppli et al. 2013; Krzmarzick et al. 2012; Bengtson et al. 2009). Leri et al. (2007) used microscopic techniques to visualize hotspots of chlorinated organics and colocalization of fungi suggests that they play a substantial role in chlorinated organics formation. The reasons for organohalide production by biota (fungi and bacteria and animals) include antagonism, defense (against predation), and signaling (Bengtson et al. 2009; Clarke et al. 2009).

The production of less halogenated and nonhalogenated organic matter by OHRB detoxifies antagonistic compounds and also provides cross-feeding opportunities. The cross-feeding takes two forms in native soils/sediments. As discussed earlier (and in Chap. 13), *successive reductive dehalogenation can set up an OHRB food web*. Additionally, these partially dehalogenated organohalides can be oxidized by organisms utilizing alternative electron acceptors such as O_2, iron, sulfate, and nitrate (Groups labeled "4" and "5" in Fig. 14.1). The aerobes include those that cometabolically degrade organohalides by methane-, toluene-, and alkene-oxygenase enzyme systems (Arp et al. 2001; Mattes et al. 2010) as well as those that can use partially dehalogenated organics as sole carbon and energy sources such as specific *Polaromonas* and *Nocardiodes* strains (Coleman et al. 2002a, b; Jennings et al. 2009). Less chlorinated ethenes and aromatics can drive aerobic catabolism even at extremely low O_2 levels (Gossett 2010). The chloride released by OHRB can be recycled back to be used by halogenating enzymes (Reaction 6 in Fig. 14.1) (Aeppli et al. 2013; Bengtson et al. 2009; Krzmarzick et al. 2012).

14.9 Summary

The increasing speed and power of phylogenetic and metagenomic profiling methods is enabling studies that shed light on the ecology of OHRB communities from simple two-member cocultures to highly diverse communities. We see trends in the reactions mediated by non-OHRB populations including the provision of cofactors, carbon, and reducing power (especially H_2) to OHRB. Many OHRB (especially the obligate OHRB) utilize H_2 and acetate in the presence of suitable organohalide electron acceptors. In OHRB communities, the OHRB populations often compete with methanogens for H_2 and acetate. In fact, many OHRB communities and their food webs resemble anaerobic digester communities in their non-OHRB populations, although clustering of metagenomic sequencing raw reads did suggest distinct composition among the OHRB communities versus anaerobic methanogenic digester communities (Hug et al. 2012). Euryarcheaota, Proteobacteria, Bacteroidetes, Firmicutes, and Spirocheata are the most commonly found non-OHRB groups in organohalide-respiring communities. Another common observation is that multiple OHRB usually coexist in enrichment cultures even after decades of enrichment—presumably due to cross-feeding or niche specialization. In pristine, carbon-rich settings, a natural organohalide cycle exists where organohalides are produced biotically for signaling, defense, and antagonism purposes. The various organohalide molecules are dehalogenated by OHRB and the organic products of organohalide respiration support respirers of other terminal electron acceptors (e.g., oxygen, iron and sulfate). Organisms that use these alternate electron acceptors can also directly compete with OHRB for H_2 and acetate. Significant mysteries remain, such as the full suite of cofactors, nutrients, and signaling molecules transferred among different community members as well as the ecological reason for the general lack of full dehalogenation capabilities on individual OHRB genomes. Further meta-omic studies will help find answers to these remaining questions.

Acknowledgments The author acknowledges the collective work of many researchers worldwide and the public and private funding agencies that have made this area of research possible. Thanks to Elizabeth Edwards and Cresten Mansfeldt for input on the chapter content.

References

Adrian L, Manz W, Szewzyk U, Gorisch H (1998) Physiological characterization of a bacterial consortium reductively dechlorinating 1,2,3- and 1,2,4-trichlorobenzene. Appl Environ Microbiol 64(2):496–503

Adrian L, Szewzyk U, Gorisch H (2000a) Bacterial growth based on reductive dechlorination of trichlorobenzenes. Biodegradation 11(1):73–81

Adrian L, Szewzyk U, Wecke J, Gorisch H (2000b) Bacterial dehalorespiration with chlorinated benzenes. Nature 408(6812):580–583

Aeppli C, Bastviken D, Andersson P, Gustafsson Ö (2013) Chlorine isotope effects and composition of naturally produced organochlorines from chloroperoxidases, flavin-dependent halogenases, and in forest soil. Environ Sci Technol 47(13):6864–6871. doi:10.1021/es3037669

Ahn YB, Rhee SK, Fennell DE, Kerkhof LJ, Hentschel U, Häggblom MM (2003) Reductive dehalogenation of brominated phenolic compounds by microorganisms associated with the marine sponge *Aplysina aerophoba*. Appl Environ Microbiol 69(7):4159–4166

Ahn Y-B, Haggblom MM, Kerkhof LJ (2007) Comparison of anaerobic microbial communities from Estuarine sediments amended with halogenated compounds to enhance dechlorination of 1,2,3,4-tetrachlorodibenzo-p-dioxin. FEMS Microbiol Ecol 61(2):362–371. doi:10.1111/j.1574-6941.2007.00342.x

Arp DJ, Yeager CM, Hyman MR (2001) Molecular and cellular fundamentals of aerobic cometabolism of trichloroethylene. Biodegradation 12(2):81–103. doi:10.1023/a:1012089908518

Asplund G, Grimvall A (1991) Organohalogens in nature. Environ Sci Technol 25(8):1346–1350

Aulenta F, Majone M, Verbo P, Tandoi V (2002) Complete dechlorination of tetrachloroethene to ethene in presence of methanogenesis and acetogenesis by an anaerobic sediment microcosm. Biodegradation 13(6):411–424

Aulenta F, Gossett JM, Papini MP, Rossetti S, Majone M (2005) Comparative study of methanol, butyrate, and hydrogen as electron donors for long-term dechlorination of tetrachloroethene in mixed anerobic cultures. Biotechnol Bioeng 91(6):743–753

Aulenta F, Catervi A, Majone M, Panero S, Reale P, Rossetti S (2007) Electron transfer from a solid-state electrode assisted by methyl viologen sustains efficient microbial reductive dechlorination of TCE. Environ Sci Technol 41(7):2554–2559

Baelum J, Chambon JC, Scheutz C, Binning PJ, Laier T, Bjerg PL, Jacobsen CS (2013) A conceptual model linking functional gene expression and reductive dechlorination rates of chlorinated ethenes in clay rich groundwater sediment. Water Res 47(7):2467–2478. doi:10.1016/j.watres.2013.02.016

Baelum J, Scheutz C, Chambon JC, Jensen CM, Brochmann RP, Dennis P, Laier T, Broholm MM, Bjerg PL, Binning PJ, Jacobsen CS (2014) The impact of bioaugmentation on dechlorination kinetics and on microbial dechlorinating communities in subsurface clay till. Environ Pollut 186:149–157. doi:10.1016/j.envpol.2013.11.013

Ballerstedt H, Hantke J, Bunge M, Werner B, Gerritse J, Andreesen JR, Lechner U (2004) Properties of a trichlorodibenzo-p-dioxin-dechlorinating mixed culture with a *Dehalococcoides* as putative dechlorinating species. FEMS Microbiol Ecol 47(2):223–234

Bastviken D, Thomsen F, Svensson T, Karlsson S, Sanden P, Shaw G, Matucha M, Oberg G (2007) Chloride retention in forest soil by microbial uptake and by natural chlorination of organic matter. Geochim Cosmochim Acta 71(13):3182–3192. doi:10.1016/j.gca.2007.04.028

Bastviken D, Svensson T, Karlsson S, Sanden P, Oberg G (2009) Temperature sensitivity indicates that chlorination of organic matter in forest soil is primarily biotic. Environ Sci Technol 43(10):3569–3573. doi:10.1021/es8035779

Becker JG (2006) A modeling study and implications of competition between *Dehalococcoides ethenogenes* and other tetrachloroethene-respiring bacteria. Environ Sci Technol 40(14):4473–4480

Becker JG, Seagren EA (2009) Modeling the effects of microbial competition and hydrodynamics on the dissolution and detoxification of dense nonaqueous phase liquid contaminants. Environ Sci Technol 43(3):870–877. doi:10.1021/es801616f

Bedard DL, Bailey JJ, Reiss BL, Jerzak GV (2006) Development and characterization of stable sediment-free anaerobic bacterial enrichment cultures that dechlorinate aroclor 1260. Appl Environ Microbiol 72(4):2460–2470

Bedard DL, Ritalahti KA, Löffler FE (2007) The *Dehalococcoides* population in sediment-free mixed cultures metabolically dechlorinates the commercial polychlorinated biphenyl mixture aroclor 1260. Appl Environ Microbiol 73(8):2513–2521. doi:10.1128/aem.02909-06

Bengtson P, Bastviken D, de Boer W, Oberg G (2009) Possible role of reactive chlorine in microbial antagonism and organic matter chlorination in terrestrial environments. Environ Microbiol 11(6):1330–1339. doi:10.1111/j.1462-2920.2009.01915.x

Bowman KS, Moe WM, Rash BA, Bae HS, Rainey FA (2006) Bacterial diversity of an acidic Louisiana groundwater contaminated by dense nonaqueous-phase liquid containing chloroethanes and other solvents. FEMS Microbiol Ecol 58(1):120–133

Bowman KS, Nobre MF, da Costa MS, Rainey FA, Moe WM (2013) *Dehalogenimonas alkenigignens* sp nov., a chlorinated-alkane-dehalogenating bacterium isolated from groundwater. Int J Syst Evol Microbiol 63:1492–1498. doi:10.1099/ijs.0.045054-0

Brisson VL, West KA, Lee PKH, Tringe SG, Brodie EL, Alvarez-Cohen L (2012) Metagenomic analysis of a stable trichloroethene-degrading microbial community. ISME J 6(9):1702–1714. doi:10.1038/ismej.2012.15

Bunge M, Lechner U (2009) Anaerobic reductive dehalogenation of polychlorinated dioxins. Appl Microbiol Biotechnol 84(3):429–444. doi:10.1007/s00253-009-2084-7

Chambon JC, Bjerg PL, Scheutz C, Baelum J, Jakobsen R, Binning PJ (2013) Review of reactive kinetic models describing reductive dechlorination of chlorinated ethenes in soil and groundwater. Biotechnol Bioeng 110(1):1–23. doi:10.1002/bit.24714

Chen MJ, Abriola LM, Amos BK, Suchomel EJ, Pennell KD, Löffler FE, Christ JA (2013) Microbially enhanced dissolution and reductive dechlorination of PCE by a mixed culture: Model validation and sensitivity analysis. J Contam Hydrol 151:117–130. doi:10.1016/j.jconhyd.2013.05.005

Cheng D, Chow WL, He J (2010) A *Dehalococcoides*-containing co-culture that dechlorinates tetrachloroethene to trans-1,2-dichloroethene. ISME J 4(1). doi:10.1038/ismej.2009.90

Clarke N, Fuksova K, Gryndler M, Lachmanova Z, Liste HH, Rohlenova J, Schroll R, Schroder P, Matucha M (2009) The formation and fate of chlorinated organic substances in temperate and boreal forest soils. Environ Sci Pollut Res 16(2):127–143. doi:10.1007/s11356-008-0090-4

Coleman NV, Mattes TE, Gossett JM, Spain JC (2002a) Biodegradation of *cis*-dichloroethene as the sole carbon source by a beta-proteobacterium. Appl Environ Microbiol 68(6):2726–2730

Coleman NV, Mattes TE, Gossett JM, Spain JC (2002b) Phylogenetic and kinetic diversity of aerobic vinyl chloride-assimilating bacteria from contaminated sites. Appl Environ Microbiol 68(12):6162–6171

Conrad ME, Brodie EL, Radtke CW, Bill M, Delwiche ME, Lee MH, Swift DL, Colwell FS (2010) Field evidence for co-metabolism of trichloroethene stimulated by addition of electron donor to groundwater. Environ Sci Technol 44(12):4697–4704. doi:10.1021/es903535j

Cord-Ruwisch R, Lovley DR, Schink B (1998) Growth of *Geobacter sulfurreducens* with acetate in syntrophic cooperation with hydrogen-oxidizing anaerobic partners. Appl Environ Microbiol 64(6):2232–2236

Cupples AM, Spormann AM, McCarty PL (2003) Growth of a *Dehalococcoides*-like microorganism on vinyl chloride and *cis*-dichloroethene as electron acceptors as determined by competitive PCR. Appl Environ Microbiol 69(2):953–959

Cutter LA, Watts JE, Sowers KR, May HD (2001) Identification of a microorganism that links its growth to the reductive dechlorination of 2,3,5,6-chlorobiphenyl. Environ Microbiol 3(11):699–709

Daprato RC, Löffler FE, Hughes JB (2007) Comparative analysis of three tetrachloroethene to ethene halorespiring consortia suggests functional redundancy. Environ Sci Technol 41(7):2261–2269

De Wildeman S, Diekert G, Van Langenhove H, Verstraete W (2003) Stereoselective microbial dehalorespiration with vicinal dichlorinated alkanes. Appl Environ Microbiol 69(9):5643–5647

Delgado AG, Fajardo-Williams D, Popat SC, Torres CI, Krajmalnik-Brown R (2014) Successful operation of continuous reactors at short retention times results in high-density, fast-rate *Dehalococcoides* dechlorinating cultures. Appl Microbiol Biotechnol 98(6):2729–2737. doi:10.1007/s00253-013-5263-5

Diekert G, Wohlfarth G (1994) Metabolism of homoacetogens. Antonie Van Leeuwenhoek 66(1–3):209–221. doi:10.1007/BF00871640

Ding C, Zhao S, He J (2014) A *Desulfitobacterium* sp. strain PR reductively dechlorinates both 1,1,1-trichloroethane and chloroform: strain PR dechlorinates TCA and chloroform. Environ Microbiol 16(11):3387–3397. doi:10.1111/1462-2920.12387

DiStefano TD, Gossett JM, Zinder SH (1991) Reductive dechlorination of high concentrations of tetrachloroethene to ethene by an anaerobic enrichment culture in the absence of methanogenesis. Appl Environ Microbiol 57(8):2287–2292

Dojka MA, Hugenholtz P, Haack SK, Pace NR (1998) Microbial diversity in a hydrocarbon- and chlorinated-solvent-contaminated aquifer undergoing intrinsic bioremediation. Appl Environ Microbiol 64(10):3869–3877

Dolfing J, Tiedje JM (1991) Kinetics of two complementary hydrogen sink reactions in a defined 3-chlorobenzoate degrading methanogenic co-culture. FEMS Microbiol Lett 86(1):25–32. doi:10.1111/j.1574-6968.1991.tb04792.x

Dugat-Bony E, Biderre-Petit C, Jaziri F, David MM, Denonfoux J, Lyon DY, Richard J-Y, Curvers C, Boucher D, Vogel TM, Peyretaillade E, Peyret P (2012) In situ TCE degradation mediated by complex dehalorespiring communities during biostimulation processes. Microb Biotechnol 5(5):642–653. doi:10.1111/j.1751-7915.2012.00339.x

Duhamel M, Edwards EA (2006) Microbial composition of chlorinated ethene-degrading cultures dominated by *Dehalococcoides*. FEMS Microbiol Ecol 58(3):538–549

Duhamel M, Edwards EA (2007) Growth and yields of dechlorinators, acetogens, and methanogens during reductive dechlorination of chlorinated ethenes and dihaloelimination of 1,2-dichloroethane. Environ Sci Technol 41(7):2303–2310

Duhamel M, Wehr SD, Yu L, Rizvi H, Seepersad D, Dworatzek S, Cox EE, Edwards EA (2002) Comparison of anaerobic dechlorinating enrichment cultures maintained on tetrachloroethene, trichloroethene, *cis*-dichloroethene and vinyl chloride. Water Res 36(17):4193–4202

Ellis DE, Lutz EJ, Odom JM, Buchanan RJ, Bartlett CL, Lee MD, Harkness MR, Deweerd KA (2000) Bioaugmentation for accelerated in situ anaerobic bioremediation. Environ Sci Technol 34(11):2254–2260

Fagervold SK, May HD, Sowers KR (2007) Microbial reductive dechlorination of aroclor 1260 in Baltimore Harbor sediment microcosms is catalyzed by three phylotypes within the phylum *Chloroflexi*. Appl Environ Microbiol 73(9):3009–3018. doi:10.1128/Aem.02958-06

Fennell DE, Gossett JM (1998) Modeling the production of and competition for hydrogen in a dechlorinating culture. Environ Sci Technol 32(16):2450–2460

Fennell DE, Gossett JM, Zinder SH (1997) Comparison of butyric kid, ethanol, lactic acid, and propionic acid as hydrogen donors for the reductive dechlorination of tetrachloroethene. Environ Sci Technol 31(3):918–926

Freeborn RA, West KA, Bhupathiraju VK, Chauhan S, Rahm BG, Richardson RE, Alvarez-Cohen L (2005) Phylogenetic analysis of TCE-dechlorinating consortia enriched on a variety of electron donors. Environ Sci Technol 39(21):8358–8368

Fung JM, Weisenstein BP, Mack EE, Vidumsky JE, Ei TA, Zinder SH (2009) Reductive dehalogenation of dichlorobenzenes and monochlorobenzene to benzene in microcosms. Environ Sci Technol 43(7):2302–2307. doi:10.1021/es802131d

Futagami T, Yamaguchi T, Nakayama SI, Goto M, Furukawa K (2006) Effects of chloromethanes on growth of and deletion of the pce gene cluster in dehalorespiring *Desulfitobacterium hafniense* strain Y51. Appl Environ Microbiol 72(9):5998–6003. doi:10.1128/aem.00979-06

Gerritse J, Renard V, Pedro Gomes TM, Lawson PA, Collins MD, Gottschal JC (1996) *Desulfitobacterium* sp. strain PCE1, an anaerobic bacterium that can grow by reductive dechlorination of tetrachloroethene or *ortho*-chlorinated phenols. Arch Microbiol 165(2):132–140

Gossett JM (2010) Sustained aerobic oxidation of vinyl chloride at low oxygen concentrations. Environ Sci Technol 44(4):1405–1411. doi:10.1021/es9033974

Gribble GW (1994) The natural production of chlorinated compounds. Environ Sci Technol 28(7):A310–A319

Gribble GW (2003) The diversity of naturally produced organohalogens. Chemosphere 52(2):289–297. doi:10.1016/S0045-6535(03)00207-8

Grostern A, Edwards EA (2006a) A 1,1,1-trichloroethane-degrading anaerobic mixed microbial culture enhances biotransformation of mixtures of chlorinated ethenes and ethanes. Appl Environ Microbiol 72(12):7849–7856

Grostern A, Edwards EA (2006b) Growth of *Dehalobacter* and *Dehalococcoides* spp. during degradation of chlorinated ethanes. Appl Environ Microbiol 72(1):428–436. doi:10.1128/aem.72.1.428-436.2006

Grostern A, Edwards EA (2009) Characterization of a *Dehalobacter* coculture that dechlorinates 1,2-dichloroethane to ethene and identification of the putative reductive dehalogenase gene. Appl Environ Microbiol 75(9):2684–2693

Grostern A, Chan WW, Edwards EA (2009) 1,1,1-trichloroethane and 1,1-dichloroethane reductive dechlorination kinetics and co-contaminant effects in a *Dehalobacter*-containing mixed culture. Environ Sci Technol 43(17):6799–6807

Grostern A, Duhamel M, Dworatzek S, Edwards EA (2010) Chloroform respiration to dichloromethane by a *Dehalobacter* population. Environ Microbiol 12(4):1053–1060. doi:10.1111/j.1462-2920.2009.02150.x

Gu AZ, Hedlund BP, Staley JT, Strand SE, Stensel HD (2004) Analysis and comparison of the microbial community structures of two enrichment cultures capable of reductively dechlorinating TCE and *cis*-DCE. Environ Microbiol 6(1):45

Gustavsson M, Karlsson S, Öberg G, Sandén P, Svensson T, Valinia S, Thiry Y, Bastviken D (2012) Organic matter chlorination rates in different boreal soils: the role of soil organic matter content. Environ Sci Technol 46(3):1504–1510. doi:10.1021/es203191r

Haest PJ, Springael D, Smolders E (2010a) Dechlorination kinetics of TCE at toxic TCE concentrations: assessment of different models. Water Res 44(1):331–339. doi:10.1016/j.watres.2009.09.033

Haest PJ, Springael D, Smolders E (2010b) Modelling reactive CAH transport using batch experiment degradation kinetics. Water Res 44(9):2981–2989. doi:10.1016/j.watres.2010.02.031

Hatt JK, Loffler FE (2012) Quantitative real-time PCR (qPCR) detection chemistries affect enumeration of the *Dehalococcoides* 16S rRNA gene in groundwater. J Microbiol Methods 88(2):263–270. doi:10.1016/j.mimet.2011.12.005

He J, Sung Y, Dollhopf ME, Fathepure BZ, Tiedje JM, Löffler FE (2002) Acetate versus hydrogen as direct electron donors to stimulate the microbial reductive dechlorination process at chloroethene-contaminated sites. Environ Sci Technol 36(18):3945–3952

He J, Sung Y, Krajmalnik-Brown R, Ritalahti KM, Löffler FE (2005) Isolation and characterization of *Dehalococcoides* sp. strain FL2, a trichloroethene (TCE)- and 1,2-dichloroethene-respiring anaerobe. Environ Microbiol 7(9):1442–1450

He JZ, Robrock KR, Alvarez-Cohen L (2006) Microbial reductive debromination of polybrominated diphenyl ethers (PBDEs). Environ Sci Technol 40(14):4429–4434

He J, Holmes VF, Lee PK, Alvarez-Cohen L (2007) Influence of vitamin B12 and cocultures on the growth of *Dehalococcoides* isolates in defined medium. Appl Environ Microbiol 73(9):2847–2853

Heavner GLW (2013) Biokinetic modeling, laboratory examination and field analysis of DNA, RNA and protein as robust molecular biomarkers of chloroethene reductive dechlorination in *Dehalococcoides mccartyi*. Ph.D. Disseration, Cornell University

Heavner GLW, Rowe AR, Mansfeldt CB, Pan JK, Gossett JM, Richardson RE (2013) Molecular biomarker-based biokinetic modeling of a PCE-dechlorinating and methanogenic mixed culture. Environ Sci Technol 47(8):3724–3733. doi:10.1021/es303517s

Heimann AC, Batstone DJ, Jakobsen R (2006) *Methanosarcina* spp. drive vinyl chloride dechlorination via interspecies hydrogen transfer. Appl Environ Microbiol 72(4):2942–2949

Hendrickson ER, Payne JA, Young RM, Starr MG, Perry MP, Fahnestock S, Ellis DE, Ebersole RC (2002) Molecular analysis of *Dehalococcoides* 16S ribosomal DNA from chloroethene-contaminated sites throughout North America and Europe. Appl Environ Microbiol 68(2):485–495

Holliger C, Hahn D, Harmsen H, Ludwig W, Schumacher W, Tindall B, Vazquez F, Weiss N, Zehnder AJB (1998) *Dehalobacter restrictus* gen. nov. and sp. nov., a strictly anaerobic bacterium that reductively dechlorinates tetra- and trichloroethene in an anaerobic respiration. Arch Microbiol 169(4):313–321. doi:10.1007/s002030050577

Holoman TRP, Elberson MA, Cutter LA, May HD, Sowers KR (1998) Characterization of a defined 2,3,5,6-tetrachlorobiphenyl-*ortho*-dechlorinating microbial community by comparative sequence analysis of genes coding for 16S rRNA. Appl Environ Microbiol 64(9):3359–3367

Hug LA, Beiko RG, Rowe AR, Richardson RE, Edwards EA (2012) Comparative metagenomics of three *Dehalococcoides*-containing enrichment cultures: the role of the non-dechlorinating community. BMC Genomics 13. doi:10.1186/1471-2164-13-327

Jennings LK, Chartrand MMG, Lacrampe-Couloume G, Lollar BS, Spain JC, Gossett JM (2009) Proteomic and transcriptomic analyses reveal genes upregulated by *cis*-dichloroethene in *Polaromonas* sp strain JS666. Appl Environ Microbiol 75(11):3733–3744. doi:10.1128/aem.00031-09

Johnson DR, Nemir A, Andersen GL, Zinder SH, Alvarez-Cohen L (2009) Transcriptomic microarray analysis of corrinoid responsive genes in *Dehalococcoides ethenogenes* strain 195. FEMS Microbiol Lett 294(2):198–206

Jones EJP, Voytek MA, Lorah MM, Kirshtein JD (2006) Characterization of a microbial consortium capable of rapid and simultaneous dechlorination of 1,1,2,2-tetrachloroethane and chlorinated ethane and ethene intermediates. Bioremediat J 10(4):153–168. doi:10.1080/10889860601021399

Justicia-Leon SD, Ritalahti KM, Mack EE, Löffler FE (2012) Dichloromethane fermentation by a *Dehalobacter* sp in an enrichment culture derived from pristine river sediment. Appl Environ Microbiol 78(4):1288–1291. doi:10.1128/aem.07325-11

Justicia-Leon SD, Higgins S, Mack EE, Griffiths DR, Tang S, Edwards EA, Löffler FE (2014) Bioaugmentation with distinct *Dehalobacter* strains achieves chloroform detoxification in microcosms. Environ Sci Technol 48:1851–1858. doi:10.1021/es403582f

Kjellerup BV, Sun X, Ghosh U, May HD, Sowers KR (2008) Site-specific microbial communities in three PCB-impacted sediments are associated with different in situ dechlorinating activities. Environ Microbiol 10(5):1296–1309. doi:10.1111/j.1462-2920.2007.01543.x

Kotik M, Davidova A, Voriskova J, Baldrian P (2013) Bacterial communities in tetrachloroethene-polluted groundwaters: a case study. Sci Total Environ 454–455:517–527. doi:10.1016/j.scitotenv.2013.02.082

Krzmarzick MJ, Crary BB, Harding JJ, Oyerinde OO, Leri AC, Myneni SCB, Novak PJ (2012) Natural niche for organohalide-respiring chloroflexi. Appl Environ Microbiol 78(2):393–401. doi:10.1128/aem.06510-11

Kube M, Beck A, Zinder SH, Kuhl H, Reinhardt R, Adrian L (2005) Genome sequence of the chlorinated compound respiring bacterium *Dehalococcoides* species strain CBDB1. Nat Biotechnol 23(10):1269–1273

Kulkarni G, Kridelbaugh DM, Guss AM, Metcalf WW (2009) Hydrogen is a preferred intermediate in the energy-conserving electron transport chain of *Methanosarcina barkeri*. Proc Natl Acad Sci USA 106(37):15915–15920. doi:10.1073/pnas.0905914106

Lai YJ, Becker JG (2013) Compounded effects of chlorinated ethene inhibition on ecological interactions and population abundance in a *Dehalococcoides*—*Dehalobacter* coculture. Environ Sci Technol 47(3):1518–1525. doi:10.1021/es3034582

Lee PKH, Macbeth TW, Sorenson KS, Deeb RA, Alvarez-Cohen L (2008) Quantifying genes and transcripts to assess the in situ physiology of "*Dehalococcoides*" spp. in a trichloroethene-contaminated groundwater site. Appl Environ Microbiol 74(9):2728–2739. doi:10.1128/aem.02199-07

Lee PKH, He JZ, Zinder SH, Alvarez-Cohen L (2009) Evidence for nitrogen fixation by "*Dehalococcoides ethenogenes*" strain 195. Appl Environ Microbiol 75(23):7551–7555. doi:10.1128/aem.01886-09

Lee J, Lee TK, Löffler FE, Park J (2011a) Characterization of microbial community structure and population dynamics of tetrachloroethene-dechlorinating tidal mudflat communities. Biodegradation 22(4):687–698. doi:10.1007/s10532-010-9429-x

Lee LK, Ding C, Yang KL, He JZ (2011b) Complete debromination of tetra- and penta-brominated diphenyl ethers by a coculture consisting of *Dehalococcoides* and *Desulfovibrio* species. Environ Sci Technol 45(19):8475–8482. doi:10.1021/es201559g

Lee M, Low A, Zemb O, Koenig J, Michaelsen A, Manefield M (2012a) Complete chloroform dechlorination by organochlorine respiration and fermentation. Environ Microbiol 14(4):883–894. doi:10.1111/j.1462-2920.2011.02656.x

Lee PKH, Dill BD, Louie TS, Shah M, VerBerkmoes NC, Andersen GL, Zinder SH, Alvarez-Cohen L (2012b) Global transcriptomic and proteomic responses of *Dehalococcoides ethenogenes* strain 195 to fixed nitrogen limitation. Appl Environ Microbiol 78(5):1424–1436. doi:10.1128/aem.06792-11

Lee PKH, Warnecke F, Brodie EL, Macbeth TW, Conrad ME, Andersen GL, Alvarez-Cohen L (2012c) Phylogenetic microarray analysis of a microbial community performing reductive dechlorination at a TCE-contaminated site. Environ Sci Technol 46(2):1044–1054. doi:10.1021/es203005k

Lendvay JM, Löffler FE, Dollhopf M, Aiello MR, Daniels G, Fathepure BZ, Gebhard M, Heine R, Helton R, Shi J, Krajmalnik-Brown R, Major CL, Barcelona MJ, Petrovskis E, Hickey R, Tiedje JM, Adriaens P (2003) Bioreactive barriers: a comparison of bioaugmentation and biostimulation for chlorinated solvent remediation. Environ Sci Technol 37(7):1422–1431

Leri AC, Marcus MA, Myneni SCB (2007) X-ray spectromicroscopic investigation of natural organochlorine distribution in weathering plant material. Geochim Cosmochim Acta 71(23):5834–5846. doi:10.1016/j.gca.2007.09.001

Liu H, Park JW, Haggblom MM (2014) Enriching for microbial reductive dechlorination of polychlorinated dibenzo-p-dioxins and dibenzofurans. Environ Pollut 184:222–230. doi:10.1016/j.envpol.2013.08.019

Löffler FE, Yan JY, Ritalahti KM, Adrian L, Edwards EA, Konstantinidis KT, Mueller JA, Fullerton H, Zinder SH, Spormann AM (2013) *Dehalococcoides mccartyi* gen. nov., sp. nov., obligate organohalide-respiring anaerobic bacteria, relevant to halogen cycling and bioremediation, belong to a novel bacterial class, Dehalococcoidetes classis nov., within the phylum Chloroflexi. Int J Syst Evol Microbiol 63:625–635. doi:10.1099/ijs.0.034926-0

Löffler FE, Sanford RA (2005) Analysis of trace hydrogen metabolism. In: Methods in enzymology, vol 397. Academic Press, New York, pp 222–237. doi:http://dx.doi.org/10.1016/S0076-6879(05)97013-4

Lovley DR, Ferry JG (1985) Production and consumption of H_2 during growth of *Methanosarcina* spp. on acetate. Appl Environ Microbiol 49(1):247–249

Lowe M, Madsen EL, Schindler K, Smith C, Emrich S, Robb F, Halden RU (2002) Geochemistry and microbial diversity of a trichloroethene-contaminated Superfund site undergoing intrinsic in situ reductive dechlorination. FEMS Microbiol Ecol 40(2):123–134

Lu XX, Wilson JT, Kampbell DH (2006) Relationship between *Dehalococcoides* DNA in ground water and rates of reductive dechlorination at field scale. Water Res 40(16):3131–3140

Macbeth TW, Cummings DE, Spring S, Petzke LM, Sorenson KS (2004) Molecular characterization of a dechlorinating community resulting from in situ biostimulation in a trichloroethene-contaminated deep, fractured basalt aquifer and comparison to a derivative laboratory culture. Appl Environ Microbiol 70(12):7329–7341

Magnuson JK, Romine MF, Burris DR, Kingsley MT (2000) Trichloroethene reductive dehalogenase from *Dehalococcoides ethenogenes*: sequence of tceA and substrate range characterization. Appl Environ Microbiol 66(12):5141–5147

Major DW, McMaster ML, Cox EE, Edwards EA, Dworatzek SM, Hendrickson ER, Starr MG, Payne JA, Buonamici LW (2002) Field demonstration of successful bioaugmentation to achieve dechlorination of tetrachloroethene to ethene. Environ Sci Technol 36(23):5106–5116

Manchester MJ, Hug LA, Zarek M, Zila A, Edwards EA (2012) Discovery of a trans-dichloroethene-respiring *Dehalogenimonas* species in the 1,1,2,2-tetrachloroethane-dechlorinating WBC-2 consortium. Appl Environ Microbiol 78(15):5280–5287. doi:10.1128/AEM.00384-12

Mansfeldt CB, Rowe AR, Heavner GLW, Zinder SH, Richardson RE (2014) Meta-analyses of *Dehalococcoides mccartyi* strain 195 transcriptomic profiles identify a respiration rate-related gene expression transition point and interoperon recruitment of a key oxidoreductase subunit. Appl Environ Microbiol 80(19):6062–6072. doi:10.1128/AEM.02130-14

Maphosa F, Smidt H, De Vos WM, Roling WFM (2010) Microbial community- and metabolite dynamics of an anoxic dechlorinating bioreactor. Environ Sci Technol 44(13):4884–4890. doi:10.1021/es903721s

Maphosa F, van Passel MWJ, de Vos WM, Smidt H (2012) Metagenome analysis reveals yet unexplored reductive dechlorinating potential of *Dehalobacter* sp E1 growing in co-culture with *Sedimentibacter* sp. Environ Microbiol Rep 4(6):604–616. doi:10.1111/j.1758-2229.2012.00376.x

Marco-Urrea E, Seifert J, von Bergen M, Adrian L (2012) Stable isotope peptide mass spectrometry to decipher Amino Acid Metabolism in *Dehalococcoides* Strain CBDB1. J Bacteriol 194:4169–4177. doi:10.1128/jb.00049-12

Marshall IPG, Berggren DRV, Azizian MF, Burow LC, Semprini L, Spormann AM (2012) The hydrogenase chip: a tiling oligonucleotide DNA microarray technique for characterizing hydrogen-producing and -consuming microbes in microbial communities. ISME J 6(4):814–826. doi:10.1038/ismej.2011.136

Marzorati M, de Ferra F, Van Raemdonck H, Borin S, Allifranchini E, Carpani G, Serbolisca L, Verstraete W, Boon N, Daffonchio D (2007) A novel reductive dehalogenase, identified in a contaminated groundwater enrichment culture and in *Desulfitobacterium dichloroeliminans* strain DCA1, is linked to dehalogenation of 1,2-dichloroethane. Appl Environ Microbiol 73(9):2990–2999. doi:10.1128/AEM.02748-06

Mathai PP, Zitomer DH, Maki JS (2015) Quantitative detection of syntrophic fatty acid-degrading bacterial communities in methanogenic environments. Microbiology 161:1189–1197. doi:10.1099/mic.0.000085

Mattes TE, Alexander AK, Coleman NV (2010) Aerobic biodegradation of the chloroethenes: pathways, enzymes, ecology, and evolution. FEMS Microbiol Rev 34(4):445–475. doi:10.1111/j.1574-6976.2010.00210.x

Matturro B, Heavner GL, Richardson RE, Rossetti S (2013) Quantitative estimation of *Dehalococcoides* mccartyi at laboratory and field scale: comparative study between CARD-FISH and real time PCR. J Microbiol Methods 93(2):127–133. doi:10.1016/j.mimet.2013.02.011

Maymó-Gatell X, Chien Y, Gossett JM, Zinder SH (1997) Isolation of a bacterium that reductively dechlorinates tetrachloroethene to ethene. Science 276(5318):1568–1571

Men YJ, Feil H, VerBerkmoes NC, Shah MB, Johnson DR, Lee PKH, West KA, Zinder SH, Andersen GL, Alvarez-Cohen L (2012) Sustainable syntrophic growth of *Dehalococcoides ethenogenes* strain 195 with *Desulfovibrio vulgaris* Hildenborough and *Methanobacterium congolense*: global transcriptomic and proteomic analyses. ISME J 6(2):410–421. doi:10.1038/ismej.2011.111

Men Y, Lee PKH, Harding KC, Alvarez-Cohen L (2013) Characterization of four TCE-dechlorinating microbial enrichments grown with different cobalamin stress and methanogenic conditions. Appl Microbiol Biotechnol 97(14):6439–6450. doi:10.1007/s00253-013-4896-8

Men YJ, Seth EC, Yi S, Allen RH, Taga ME, Alvarez-Cohen L (2014) Sustainable growth of *Dehalococcoides mccartyi* 195 by corrinoid salvaging and remodeling in defined lactate-fermenting consortia. Appl Environ Microbiol 80(7):2133–2141. doi:10.1128/aem.03477-13

Moe WM, Yan J, Nobre MF, da Costa MS, Rainey FA (2009) *Dehalogenimonas lykanthroporepellens* gen. nov., sp nov., a reductively dehalogenating bacterium isolated from chlorinated solvent-contaminated groundwater. Int J Syst Evol Microbiol 59:2692–2697. doi:10.1099/ijs.0.011502-0

Mohn WW, Tiedje JM (1992) Microbial reductive dehalogenation. Microbiol Rev 56(3):482–507

Morris RM, Sowell S, Barofsky D, Zinder S, Richardson R (2006) Transcription and mass-spectroscopic proteomic studies of electron transport oxidoreductases in *Dehalococcoides ethenogenes*. Environ Microbiol 8(9):1499–1509

Morris RM, Fung JM, Rahm BG, Zhang S, Freedman DL, Zinder SH, Richardson RE (2007) Comparative proteomics of *Dehalococcoides* spp. reveals strain-specific peptides associated with activity. Appl Environ Microbiol 73(1):320–326. doi:10.1128/AEM.02129-06

Morris BEL, Henneberger R, Huber H, Moissl-Eichinger C (2013) Microbial syntrophy: interaction for the common good. FEMS Microbiol Rev 37:384–406. doi:10.1111/1574-6976.12019

Nelson JL, Fung JM, Cadillo-Quiroz H, Cheng X, Zinder SH (2011) A role for *Dehalobacter* spp. in the reductive dehalogenation of dichlorobenzenes and monochlorobenzene. Environ Sci Technol 45(16):6806–6813. doi:10.1021/es200480k

Nelson JL, Jiang J, Zinder SH (2014) Dehalogenation of chlorobenzenes, dichlorotoluenes, and tetrachloroethene by three *Dehalobacter* spp. Environ Sci Technol 48(7):3776–3782. doi:10.1021/es4044769

Nemir A, David MM, Perrussel R, Sapkota A, Simonet P, Monier J-M, Vogel TM (2010) Comparative phylogenetic microarray analysis of microbial communities in TCE-contaminated soils. Chemosphere 80(5):600–607. doi:10.1016/j.chemosphere.2010.03.036

Oba Y, Futagami T, Amachi S (2014) Enrichment of a microbial consortium capable of reductive deiodination of 2,4,6-triiodophenol. J Biosci Bioeng 117(3):310–317. doi:10.1016/j.jbiosc.2013.08.011

Oberg G, Holm M, Sanden P, Svensson T, Parikka M (2005) The role of organic-matter-bound chlorine in the chlorine cycle: a case study of the Stubbetorp catchment, Sweden. Biogeochemistry 75(2):241–269. doi:10.1007/s10533-004-7259-9

Oelgeschläger E, Rother M (2008) Carbon monoxide-dependent energy metabolism in anaerobic bacteria and archaea. Arch Microbiol 190:257–269. doi:10.1007/s00203-008-0382-6

Pérez-de-Mora A, Zila A, McMaster ML, Edwards EA (2014) Bioremediation of chlorinated ethenes in fractured bedrock and associated changes in dechlorinating and nondechlorinating microbial populations. Environ Sci Technol 48(10):5770–5779. doi:10.1021/es404122y

Rahm BG, Richardson RE (2008) Correlation of respiratory gene expression levels and pseudo-steady-state PCE respiration rates in *Dehalococcoides* ethenogenes. Environ Sci Technol 42(2):416–421. doi:10.1021/es071455s

Rahm BG, Chauhan S, Holmes VF, Macbeth TW, Sorenson KS Jr, Alvarez-Cohen L (2006) Molecular characterization of microbial populations at two sites with differing reductive dechlorination abilities. Biodegradation 17(6):523–534. doi:10.1007/s10532-005-9023-9

Redon P-O, Abdelouas A, Bastviken D, Cecchini S, Nicolas M, Thiry Y (2011) Chloride and organic chlorine in forest soils: storage, residence times, and influence of ecological conditions. Environ Sci Technol 45(17):7202–7208. doi:10.1021/es2011918

Richardson RE, Bhupathiraju VK, Song DL, Goulet TA, Alvarez-Cohen L (2002) Phylogenetic characterization of microbial communities that reductively dechlorinate TCE based upon a combination of molecular techniques. Environ Sci Technol 36(12):2652–2662

Ritalahti KM, Amos BK, Sung Y, Wu QZ, Koenigsberg SS, Löffler FE (2006) Quantitative PCR targeting 16S rRNA and reductive dehalogenase genes simultaneously monitors multiple *Dehalococcoides* strains. Appl Environ Microbiol 72(4):2765–2774. doi:10.1128/aem.72.4.2765-27774.2006

Rohlenova J, Gryndler M, Forczek ST, Fuksova K, Handova V, Matucha M (2009) Microbial chlorination of organic matter in forest soil: investigation using Cl-36-chloride and its methodology. Environ Sci Technol 43(10):3652–3655. doi:10.1021/es803300f

Rowe AR, Lazar BJ, Morris RM, Richardson RE (2008) Characterization of the community structure of a dechlorinating mixed culture and comparisons of gene expression in planktonic and biofloc-associated "*Dehalococcoides*" and *Methanospirillum* species. Appl Environ Microbiol 74(21):6709–6719. doi:10.1128/aem.00445-08

Rowe AR, Heavner GL, Mansfeldt CB, Werner JJ, Richardson RE (2012) Relating chloroethene respiration rates in *Dehalococcoides* to protein and mRNA biomarkers. Environ Sci Technol 46(17):9388–9397. doi:10.1021/es300996c

Rowe AR, Mansfeldt CB, Heavner GL, Richardson RE (2013) *Methanospirillum* respiratory mRNA biomarkers correlate with hydrogenotrophic methanogenesis rate during growth and competition for hydrogen in an organochlorine-respiring mixed culture. Environ Sci Technol 47(1):372–381. doi:10.1021/es303061y

Rupakula A, Kruse T, Boeren S, Holliger C, Smidt H, Maillard J (2013) The restricted metabolism of the obligate organohalide respiring bacterium *Dehalobacter restrictus*: lessons from tiered functional genomics. Philos Trans R Soc Lond Ser B Biol Sci 368(1616). doi:10.1098/rstb.2012.0325

Rupakula A, Lu Y, Kruse T, Boeren S, Holliger C, Smidt H, Maillard J (2015) Functional genomics of corrinoid starvation in the organohalide-respiring bacterium *Dehalobacter restrictus* strain PER-K23. Frontiers in Microbiology 5. doi:10.3389/fmicb.2014.00761

Schaefer CE, Condee CW, Vainberg S, Steffan RJ (2009) Bioaugmentation for chlorinated ethenes using *Dehalococcoides* sp.: comparison between batch and column experiments. Chemosphere 75(2):141–148. doi:10.1016/j.chemosphere.2008.12.041

Schink B, Stams AJM (2013) Syntrophism among prokaryotes. In: Rosenberg E, DeLong EF, Lory S, Stackebrandt E, Thompson F (eds) The Prokaryotes. Springer, Berlin, pp 471–493

Schipp CJ, Marco-Urrea E, Kublik A, Seifert J, Adrian L (2013) Organic cofactors in the metabolism of *Dehalococcoides mccartyi* strains. Philos Trans R Soc B-Biol Sci 368(1616). doi:10.1098/rstb.2012.0321

Schneidewind U, Haest PJ, Atashgahi S, Maphosa F, Hamonts K, Maesen M, Calderer M, Seuntjens P, Smidt H, Springael D, Dejonghe W (2014) Kinetics of dechlorination by *Dehalococcoides mccartyi* using different carbon sources. J Contam Hydrol 157:25–36. doi:10.1016/j.jconhyd.2013.10.006

Scholz-Muramatsu H, Neumann A, Messmer M, Moore E, Diekert G (1995) Isolation and characterization of *Dehalospirillum multivorans* gen. nov., sp. nov., a tetrachloroethene-utilizing, strictly anaerobic bacterium. Arch Microbiol 163(1):48–56. doi:10.1007/BF00262203

Seshadri R, Adrian L, Fouts DE, Eisen JA, Phillippy AM, Methe BA, Ward NL, Nelson WC, Deboy RT, Khouri HM, Kolonay JF, Dodson RJ, Daugherty SC, Brinkac LM, Sullivan SA, Madupu R, Nelson KT, Kang KH, Impraim M, Tran K, Robinson JM, Forberger HA, Fraser CM, Zinder SH, Heidelberg JF (2005) Genome sequence of the PCE-dechlorinating bacterium *Dehalococcoides ethenogenes*. Science 307(5706):105–108

Shani N, Rossi P, Holliger C (2013) Correlations between environmental variables and bacterial community structures suggest Fe(III) and vinyl chloride reduction as antagonistic terminal electron-accepting processes. Environ Sci Technol 47(13):6836–6845. doi:10.1021/es304017s

Shelton DR, Tiedje JM (1984) Isolation and partial characterization of bacteria in an anaerobic consortium that mineralizes 3-chlorobenzoic acid. Appl Environ Microbiol 48(4):840–848

Sieber JR, Le HM, McInerney MJ (2014) The importance of hydrogen and formate transfer for syntrophic fatty, aromatic and alicyclic metabolism: importance of interspecies hydrogen and formate transfer. Environ Microbiol 16:177–188. doi:10.1111/1462-2920.12269

Sleep BE, Brown AJ, Lollar BS (2005) Long-term tetrachlorethene degradation sustained by endogenous cell decay. J Environ Eng Sci 4(1):11–17

Sun B, Cole JR, Sanford RA, Tiedje JM (2000) Isolation and characterization of *Desulfovibrio dechloracetivorans* sp. nov., a marine dechlorinating bacterium growing by coupling the oxidation of acetate to the reductive dechlorination of 2-chlorophenol. Appl Environ Microbiol 66(6):2408–2413

Sun BL, Griffin BM, Ayala-del-Rio HL, Hashsham SA, Tiedje JM (2002) Microbial dehalorespiration with 1,1,1-trichloroethane. Science 298(5595):1023–1025

Tang YJJ, Yi S, Zhuang WQ, Zinder SH, Keasling JD, Alvarez-Cohen L (2009) Investigation of carbon metabolism in "*Dehalococcoides ethenogenes*" strain 195 by use of isotopomer and transcriptomic analyses. J Bacteriol 191(16):5224–5231. doi:10.1128/jb.00085-09

Tang S, Gong Y, Edwards EA (2012) Semi-automatic in silico gap closure enabled de novo assembly of two *Dehalobacter* genomes from metagenomic data. PLoS ONE 7(12):e52038. doi:10.1371/journal.pone.0052038

Tas N, van Eekert MHA, Schraa G, Zhou J, de Vos WM, Smidt H (2009) Tracking functional guilds: "*Dehalococcoides*" spp. in European river basins contaminated with hexachlorobenzene. Appl Environ Microbiol 75(14):4696–4704. doi:10.1128/AEM.02829-08

Tas N, van Eekert MHA, Wagner A, Schraa G, de Vos WM, Smidt H (2011) Role of "*Dehalococcoides*" spp. in the anaerobic transformation of hexachlorobenzene in European rivers. Appl Environ Microbiol 77(13):4437–4445. doi:10.1128/aem.01940-10

Thomas SH, Wagner RD, Arakaki AK, Skolnick J, Kirby JR, Shimkets LJ, Sanford RA, Loffler FE (2008) The mosaic genome of *Anaeromyxobacter dehalogenans* strain 2CP-C suggests an aerobic common ancestor to the delta-Proteobacteria. Plos One 3(5). doi:10.1371/journal.pone.0002103

van Doesburg W, van Eekert MHA, Middeldorp PJM, Balk M, Schraa G, Stams AJM (2005) Reductive dechlorination of beta-hexachlorocyclohexane (beta-HCH) by a *Dehalobacter* species in coculture with a *Sedimentibacter* sp. FEMS Microbiol Ecol 54(1):87–95. doi:10.1016/j.femsec.2005.03.003

Vandermeeren P, Herrmann S, Cichocka D, Busschaert P, Lievens B, Richnow HH, Springael D (2014) Diversity of dechlorination pathways and organohalide respiring bacteria in chlorobenzene dechlorinating enrichment cultures originating from river sludge. Biodegradation 25(5):757–776. doi:10.1007/s10532-014-9697-y

Wagner DD, Hug LA, Hatt JK, Spitzmiller MR, Padilla-Crespo E, Ritalahti KM, Edwards EA, Konstantinidis KT, Löffler FE (2012) Genomic determinants of organohalide-respiration in *Geobacter lovleyi*, an unusual member of the *Geobacteraceae*. BMC Genomics 13. doi:10.1186/1471-2164-13-200

Watts JE, Wu Q, Schreier SB, May HD, Sowers KR (2001) Comparative analysis of polychlorinated biphenyl-dechlorinating communities in enrichment cultures using three different molecular screening techniques. Environ Microbiol 3(11):710–719

Werner JJ, Ptak AC, Rahm BG, Zhang S, Richardson RE (2009) Absolute quantification of *Dehalococcoides* proteins: enzyme bioindicators of chlorinated ethene dehalorespiration. Environ Microbiol 11(10):2687–2697. doi:10.1111/j.1462-2920.2009.01996.x

Wu Q, Watts JEM, Sowers KR, May HD (2002) Identification of a bacterium that specifically catalyzes the reductive dechlorination of polychlorinated biphenyls with doubly flanked chlorines. Appl Environ Microbiol 68(2):807–812

Yan T, LaPara TM, Novak PJ (2006) The reductive dechlorination of 2,3,4,5-tetrachlorobiphenyl in three different sediment cultures: evidence for the involvement of phylogenetically similar *Dehalococcoides*-like bacterial populations. FEMS Microbiol Ecol 55(2):248–261

Yan J, Ritalahti KM, Wagner DD, Löffler FE (2012) Unexpected specificity of interspecies cobamide transfer from *Geobacter* spp. to organohalide-respiring *Dehalococcoides mccartyi* strains. Appl Environ Microbiol 78(18):6630–6636. doi:10.1128/aem.01535-12

Yan J, Im J, Yang Y, Löffler FE (2013) Guided cobalamin biosynthesis supports *Dehalococcoides mccartyi* reductive dechlorination activity. Philos Trans R Soc B-Biol Sci 368(1616). doi:10.1098/rstb.2012.0320

Yang YR, McCarty PL (1998) Competition for hydrogen within a chlorinated solvent dehalogenating anaerobic mixed culture. Environ Sci Technol 32(22):3591–3597

Yang YR, Pesaro M, Sigler W, Zeyer J (2005) Identification of microorganisms involved in reductive dehalogenation of chlorinated ethenes in an anaerobic microbial community. Water Res 39(16):3954–3966

Yi S, Seth EC, Men YJ, Stabler SP, Allen RH, Alvarez-Cohen L, Taga ME (2012) Versatility in corrinoid salvaging and remodeling pathways supports corrinoid-dependent metabolism in *Dehalococcoides mccartyi*. Appl Environ Microbiol 78(21):7745–7752. doi:10.1128/aem.02150-12

Yoshida N, Takahashi N, Hiraishi A (2005) Phylogenetic characterization of a polychlorinated-dioxin-dechlorinating microbial community by use of microcosm studies. Appl Environ Microbiol 71(8):4325–4334

Yu S, Semprini L (2004) Kinetics and modeling of reductive dechlorination at high PCE and TCE concentrations. Biotechnol Bioeng 88(4):451–464. doi:10.1002/bit.20260

Yu ZT, Smith GB (2000) Inhibition of methanogenesis by C-1- and C-2-polychlorinated aliphatic hydrocarbons. Environ Toxicol Chem 19(9):2212–2217

Yu SH, Dolan ME, Semprini L (2005) Kinetics and inhibition of reductive dechlorination of chlorinated ethylenes by two different mixed cultures. Environ Sci Technol 39(1):195–205. doi:10.1021/es0496773

Zhang H, Ziv-El M, Rittmann BE, Krajmalnik-Brown R (2010) Effect of dechlorination and sulfate reduction on the microbial community structure in denitrifying membrane-biofilm reactors. Environ Sci Technol 44(13):5159–5164. doi:10.1021/es100695n

Zhuang WQ, Yi S, Feng XY, Zinder SH, Tang YJJ, Alvarez-Cohen L (2011) Selective utilization of exogenous amino acids by *Dehalococcoides ethenogenes* strain 195 and its effects on growth and dechlorination activity. Appl Environ Microbiol 77(21):7797–7803. doi:10.1128/aem.05676-11

Zhuang WQ, Yi S, Bill M, Brisson VL, Feng XY, Men YJ, Conrad ME, Tang YJJ, Alvarez-Cohen L (2014) Incomplete Wood-Ljungdahl pathway facilitates one-carbon metabolism in organohalide-respiring *Dehalococcoides mccartyi*. Proc Natl Acad Sci USA 111(17):6419–6424. doi:10.1073/pnas.1321542111

Zinder SH, Anguish T (1992) Carbon monoxide, hydrogen, and formate metabolism during methanogenesis from acetate by thermophilic cultures of *Methanosarcina* and *Methanothrix* strains. Grostern, Ariel Edwards, Elizabeth A, Research Support, Non-US Govt Research Support, US Govt, Non-PHS United States. Appl Environ Microbiol 75(9):2684–2693 (2009 May); 58(10):3323–3329 (Epub 6 Mar 2009)

Ziv-El M, Delgado AG, Yao Y, Kang DW, Nelson KG, Halden RU, Krajmalnik-Brown R (2011) Development and characterization of DehaloR2, a novel anaerobic microbial consortium performing rapid dechlorination of TCE to ethene. Appl Microbiol Biotechnol 92(5):1063–1071. doi:10.1007/s00253-011-3388-y

Ziv El M, Popat SC, Parameswaran P, Kang DW, Polasko A, Halden RU, Rittmann BE, Krajmalnik-Brown R (2012) Using electron balances and molecular techniques to assess trichoroethene-induced shifts to a dechlorinating microbial community. Biotechnol Bioeng 109(9):2230–2239. doi:10.1002/Bit.24504

Part IV
Genomics and Regulation of Organohalide-Respiring Bacteria

Chapter 15
Comparative Genomics and Transcriptomics of Organohalide-Respiring Bacteria and Regulation of *rdh* Gene Transcription

Thomas Kruse, Hauke Smidt and Ute Lechner

Abstract Comparison of the genomes of organohalide-respiring bacteria has improved our understanding of the genetic background of the organohalide respiration process. In this chapter the remarkable differences between obligate and facultative organohalide-respiring bacteria in the number of reductive dehalogenase-encoding genes and the numbers and types of accessory genes are discussed in relation to different lifestyles and evolutionary aspects. Furthermore, the putative function of accessory genes is discussed and a unifying nomenclature is proposed. The genomes also reflect distinct mechanisms for the synthesis or acquisition of the corrinoid cofactors of reductive dehalogenases, which are well in accord with the observed growth requirements of the respective organohalide-respiring bacteria. The value of microarray-based comparative genomics, transcriptomics, and quantitative transcription analyses for understanding the physiology and environmental significance of organohalide respiration is discussed. The reductive dehalogenase genes are in general associated with genes encoding transcriptional regulators, which are likely involved in sensing the halogenated electron acceptors. The role of two types of regulators in transcriptional regulation of organohalide respiration has been investigated. A multiple antibiotic resistance regulator (MarR)-type regulator was shown to regulate negatively the transcription of reductive dehalogenase

T. Kruse · H. Smidt
Laboratory of Microbiology, Wageningen University, 6703 WB Wageningen, The Netherlands
e-mail: thomas@theviking.dk

H. Smidt
e-mail: hauke.smidt@wur.nl

U. Lechner (✉)
Institute of Biology/Microbiology, Martin Luther University Halle-Wittenberg, 06120 Halle, Germany
e-mail: ute.lechner@mikrobiologie.uni-halle.de

L. Adrian and F.E. Löffler (eds.), *Organohalide-Respiring Bacteria*,
DOI 10.1007/978-3-662-49875-0_15

genes in *Dehalococcoides mccartyi*. In *Desulfitobacterium hafniense*, the cAMP receptor protein/fumarate and nitrate reduction (CRP/FNR) regulator, CprK, activates transcription of reductive dehalogenase genes. The molecular mechanism of how *ortho*-chlorophenols act as effectors has been elucidated and how, through the induction of structural changes, they lead to DNA binding of the regulator.

15.1 General Genomic Features of OHRB

The full genome sequences of two organohalide-respiring bacteria (OHRB) were published in 2005, just 10 years after the publication of the first bacterial genome (Kube et al. 2005; Seshadri et al. 2005; Fleischmann et al. 1995). These two genome sequences were obtained for *Dehalococcoides mccartyi* strains CBDB1 and 195^T (previously *D. ethenogenes* 195^T, here and throughout this chapter ^{T} indicates a type strain) both of which belong to the *Dehalococcoides* genus, whose members are known as obligate OHRB using hydrogen as sole electron donor (Löffler et al. 2013). In 2006 the first genome of a facultative OHRB, *Desulfitobacterium hafniense* Y51, was published (Nonaka et al. 2006). Currently (March 2015) 23 genomes of OHRB have been formally published, and many more are listed in the Genomes Online database (GOLD; https://gold.jgi-psf.org/) as planned or ongoing projects (Table 15.1) (Pagani et al. 2012). With the rapid advances in sequencing technology and the trend towards publishing the full genome sequence, together with the physiological characterisation of new strains, it is expected that these numbers will increase in the future.

The availability of full genome sequences has boosted our understanding of the physiology and evolution of OHRB. The genome size gives a crude indication of an organism's metabolic capacity. Organisms with a diverse metabolism therefore generally have larger genomes than organisms inhabiting more restricted ecological niches. In line with this assumption, the obligate organohalide-respiring *D. mccartyi* and *Dehalogenimonas lykanthroporepellens* have very small genomes with an average size of 1.4 Mbp, which is among the smallest among free-living bacteria (Table 15.1) (Kube et al. 2005), whereas *Dehalobacter* spp. have an intermediate genome size of 3 Mbp, which is rather large for an organism with such a restricted metabolism (Table 15.1). It has been shown that at least some *Dehalobacter* strains are able to ferment dichloromethane, which suggests this genus has a more extended metabolic repertoire than previously assumed (Lee et al. 2012a; Justicia-Leon et al. 2012). In contrast, facultative OHRB in general have larger genomes ranging from 3.2 to 6.5 Mbp (Table 15.1). For example, the genomes of *Desulfitobacterium* spp. encode the largest numbers (53–57) of molybdopterin oxidoreductases (protein family (pfam) 01568) of any organism with a sequenced genome, highlighting their versatile metabolism. The function of these proteins is to a large extent unknown. However, an example of a well-characterised member of this group is the catalytic subunit FdhA of the membrane bound, outward-facing formate dehydrogenase FdhABC (for details see Chap. 10) (Kim et al. 2012; Nonaka et al. 2006; Kruse et al. 2015).

Table 15.1 Basic genome characteristics of organohalide-respiring bacteria for which the genome sequences are available

Organism	Genome size (Mbp)	GC %	Genes	RDH	Closed	Accession	Year of publication	Reference
Dehalococcoides mccartyi 195^{T}	1.5	49	1647	17	Yes	NC_002936	2005	Seshadri et al. (2005)
Dehalococcoides mccartyi CBDB1	1.4	47	1517	32	Yes	NC_007356	2005	Kube et al. (2005)
Dehalococcoides mccartyi VS	1.4	47	1489	36	Yes	NC_013552	2009	McMurdie et al. (2009)
Dehalococcoides mccartyi BAV1	1.3	47	1443	11	Yes	NC_009455	2009	McMurdie et al. (2009)
Dehalococcoides mccartyi BTF08	1.5	47	1580	20	Yes	CP004080	2013	Pöritz et al. (2013)
Dehalococcoides mccartyi DCMB5	1.4	47	1526	23	Yes	CP004079	2013	Pöritz et al. (2013)
Dehalococcoides mccartyi CG1	1.5	46.9	1640	35	Yes	CP006949	2014	Wang et al. (2014a)
Dehalococcoides mccartyi CG4	1.4	48.7	1510	15	Yes	CP006950	2014	Wang et al. (2014a)
Dehalococcoides mccartyi CG5	1.4	47.2	1490	26	Yes	CP006951	2014	Wang et al. (2014a)
Dehalococcoides mccartyi SG1	1.4 (23 kbp)[a]	47	1486	28	Yes	JPRE01000000	2014	Wang et al. (2014b)
Dehalococcoides mccartyi GT	1.4	47	1483	20	Yes	NC_013890		Unpublished, JGI
Dehalococcoides mccartyi GY50	1.4	47	1591	25	Yes	NC_022964		Unpublished, National University of Singapore

(continued)

Table 15.1 (continued)

Organism	Genome size (Mbp)	GC %	Genes	RDH	Closed	Accession	Year of publication	Reference
Dehalogenimonas lykanthroporepellens	1.7	55	1771	25	Yes	Nc_014314	2012	Siddaramappa et al. (2012)
Dehalobacter sp E1	2.6	45.1	2587	10	No	CANE01000001-CANE01000102	2012	Maphosa et al. (2012)
Dehalobacter sp CF	3.1	44.3	3040	17	Yes	CP003870	2012	Tang et al. (2012)
Dehalobacter sp 11DCA	3.1	44.6	3038	17	Yes	CP003869	2012	Tang et al. (2012)
Dehalobacter restrictus PER-K23^T	2.9	44.6	2908	25	Yes	CP007033	2013	Kruse et al. (2013)
Dehalobacter sp UNSWDHB	3.2	44.9	3041	17	No	AUUR00000000	2013	Deshpande et al. (2013)
Desulfitobacterium hafniense Y51	5.7	47	5208	1^b	Yes	NC_007907	2006	Nonaka et al. (2006)
Desulfitobacterium hafniense DCB-2^T	5.3	48	5042	7^b	Yes	CP001336	2012	Kim et al. (2012)
Desulfitobacterium dehalogenans JW/IU-DC1^T	4.3	45	4252	6^b	Yes	NC_018017	2015	Kruse et al. (2015)
Desulfitobacterium hafniense PCE-S	5.7	47.3	5456	2^b	No	LK996017-LK996040^c	2015	Goris et al. (2015)
Desulfitobacterium hafniense TCP-A	5	47	4877	5^b	No	KB900390:KB900391		Kruse et al. (in preparation)
Desulfitobacterium hafniense PCP-1	5.6	48	5358	7^b	No	KB902317:KB902361		Kruse et al. (in preparation)

(continued)

Table 15.1 (continued)

Organism	Genome size (Mbp)	GC %	Genes	RDH	Closed	Accession	Year of publication	Reference
Desulfitobacterium hafniense TCE-1	5.7	47	5570	1[b]	No	Gp0007974[d]		Kruse et al. (in preparation)
Desulfitobacterium hafniense DP7	5.2	48	5456	0[b]	No	JH414432:JH414494		Kruse et al. (in preparation)
Desulfitobacterium hafniense LBE	5.5	47	5341	2[b]	Yes	Gp0007973		Kruse et al. (in preparation)
Desulfitobacterium sp. PCE-1	4.2	45	4123	6[b]	Yes	Gp0006289[d]		Kruse et al. (in preparation)
Desulfitobacterium dichloroeliminans LMG[T]	3.6	44	3537	1[b]	Yes	NC_019903		Kruse et al. (in preparation)
Desulfitobacterium metallireducens[T]	3.2	42	3152	0	Yes	NZ_AGJB00000000		Kruse et al. (in preparation)
Geobacter lovley SZ[T]	3.9 (77kbp)[a]	54.7	3777	2	no	NC_010815	2012	Wagner et al. (2012)
Shewanella sediminis HAW-EB3[T]	5.5	46.1	4666	5	Yes	NC_009831	2007	Unpublished, JGI
Sulfurospirillum multivorans[T]	3.2	40.9	3301	2	Yes	CP007201		Goris et al. (2014)
Anaeromyxobacter dehalogenans 2CP-C	5	75	4421	2	Yes	CP000251	2008	Thomas et al. (2008)
Anaeromyxobacter dehalogenans 2CP-1[T]	5	75	4540	2	yes	CP001359		Unpublished, JGI
Desulfomonile tiedjei DCB-1[T]	6.5 (27kbp)[a]	50.1	5628	1	Yes	CP003360		Unpublished, JGI

[a]Plasmid, [b]not counting the noncanonical 2.4.6-trichlorophenol reductase, CrdA (Boyer et al. 2003), [c]available through the European Molecular Biology Laboratory (EMBL), [d]Genome online database (Gold) ID. [T]Type strain

Furthermore, genome sequencing of obligate OHRB such as *D. mccartyi*, *D. lykanthroporepellens*, and *Dehalobacter* spp. has revealed large number of genes encoding reductive dehalogenase homologues (Rdh), ranging from 11 to 36 in number (Table 15.1, and references therein). Generally, facultative OHRB carry a lower number of reductive dehalogenase-encoding genes, ranging from one to seven for currently available OHRB genomes. These *rdhAB* genes are scattered throughout the genomes (Table 15.1, and references therein). Notably, some strains of the genus *Desulfitobacterium* do not carry *rdhAB* genes in their genomes. A possible explanation could be that the genes encoding reductive dehalogenases have been lost from the genome prior to isolation or genome sequencing. For example, genes encoding reductive dehalogenases were absent in the genome of *Desulfitobacterium metallireducens*^{T} although this strain was previously reported to be an OHRB (Finneran et al. 2002).

Dehalococcoides mccartyi strains have highly conserved genomes and the core genome, i.e., genes present in all members of the group (strains 195^T; CBDB1; BAV1, and VS), has been reported to comprise 1029–1118 genes, corresponding to 63–77 % of the genome (McMurdie et al. 2009; Ahsanul Islam et al. 2010). The inter-strain variability is mainly found in two genomic regions located at either side of the origin of replication, often referred to as 'high plasticity regions' (Kube et al. 2005; McMurdie et al. 2009). Most of the *rdh* genes are located in these regions, leading to the speculation that the 'high plasticity regions' serve as evolutionary hotspots with frequent horizontal gene transfers allowing rapid evolution of *D. mccartyi* strains while at the same time preserving core metabolic functions (McMurdie et al. 2009). Interestingly, 21 out of 25 *rdh* genes encoded in the genome of *Dehalobacter restrictus* PER-K23^T are also located in two genomic regions. The majority of the *rdh* genes from these regions were not found in the genome of the closely related *Dehalobacter* sp. E1, although the genomes showed an otherwise high degree of similarity (Maphosa et al. 2012; Kruse et al. 2013). *Dehalococcoides* and *Dehalobacter* belong to the Chloroflexi and Firmicutes phyla, respectively, indicating that evolution favors this genomic organization for metabolically restricted OHRB, and it is generally accepted that horizontal gene transfer was crucial in the spreading of *rdh* genes (McMurdie et al. 2009; Regeard et al. 2005; Nonaka et al. 2006; Kruse et al. 2013).

15.2 Comparative Genomics and Functionality of Corrinoid Biosynthesis

The availability of full genome sequences has paved the way for further research on OHRB. Microarrays have been designed for both *D. mccartyi* and *D. hafniense* DCB-2^T for comparative genomics and transcriptomics (Mansfeldt et al. 2014; Kim et al. 2012; Johnson et al. 2009; McMurdie et al. 2009; West et al. 2008; Lee et al. 2011; Hug et al. 2011). The first OHRB microarray was designed based on the genome of *D. mccartyi* 195^T. This microarray was used to analyze gDNA obtained from a microbial consortium, ANAS, containing various *D. mccartyi* strains able to degrade PCE completely to ethene (West et al. 2008). DNA from the ANAS culture hybridized with 87 % of the probes. The majority of the probes without a match were for genes located in predicted insertion elements or the previously mentioned 'high plasticity regions', revealing for the first time that the presence of 'high plasticity regions' is a general characteristic of *D. mccartyi* and not merely a feature of the 195^T and CBDB1 strains (West et al. 2008). The detection range of *D. mccartyi* microarrays was later expanded by adding probes from additional strains as more genomes became available (Lee et al. 2011) or by designing probes targeting consensus sequences conserved between strains of *D. mccartyi* in order to construct a pan-genome microarray targeting both characterized and uncharacterized *D. mccartyi* strains (Hug et al. 2011). Finally a shotgun metagenome microarray has been constructed, based on sequences from the KB-1 enrichment culture containing several *D. mccartyi* strains (Waller et al. 2012). Microarrays have been used to gain insight into the global response of OHRB to external factors such as oxygen stress, the presence of symbionts, and variations in the levels of nutrients or corrinoids (Kim et al. 2012; Men et al. 2012, 2013, 2014; Johnson et al. 2009). Another interesting finding using these microarrays was increased transcription of prophage genes in the late growth phase (Johnson et al. 2008) or after starving the culture by omitting chlorinated compounds from the medium (Waller et al. 2012). Although no *rdhA* genes were found within the prophage region, the *tceA* gene, encoding an RDase that dechlorinates trichlorethene to dichloroethene, was shown to be located in close proximity to a prophage in the KB-1 culture and in strain BTF08 (Pöritz et al. 2013). Thus, it cannot be excluded that mispackaging of phage particles occasionally leads to transfer of *tceA* together with the phage (Waller et al. 2012). It has been speculated that phages might act as shuttles for horizontal transfer of *rdh* genes in both *D. mccartyi* (Waller et al. 2012; McMurdie et al. 2011; Pöritz et al. 2013) and *Sulfurospirillum multivorans* (Goris et al. 2014), although experimental evidence for this hypothesis is still lacking.

In some cases the presence or absence of specific genes has led to directly testable hypotheses. For example, genes encoding a protein with high similarity to PceA, catalysing the degradation of tetrachloroethene and trichloroethene to dichloroethene, was discovered in the genome of *Dehalobacter* sp. E1 and *Shewanella sediminis* (Maphosa 2010; Lohner and Spormann 2013; Maphosa et al. 2012). Similarly, the predicted minimum gene set for nitrogen fixation was identified in the genome of *D. hafniense* DCB-2^T (Kim et al. 2012; Dos Santos et al. 2012). The functionality of both was confirmed experimentally, highlighting the strength of experiments based on genome analysis (Maphosa 2010; Kim et al. 2012; Lohner and Spormann 2013). On the other hand, analysis of the genome of *D. mccartyi* 195^T suggested incomplete synthesis pathways for some amino acids (Seshadri et al. 2005). However, the experimental findings demonstrated that complete amino acid synthesis machineries are present in *D. mccartyi* 195^T, indicating the existence of new biosynthetic pathways (Tang et al. 2009).

Another important result of comparative genomics of OHRB was the identification of differences in the acquisition of the corrinoid cofactor. The RdhA catalytic subunit of reductive dehalogenases requires incorporation of a corrinoid cofactor in the active site (Banerjee and Ragsdale 2003). The facultative OHRB, *Geobacter lovleyi*, *Desulfitobacterium* spp. and *Sulfurospirillum* spp. all encode the full gene set for de novo synthesis of corrinoids (Kim et al. 2012; Goris et al. 2014; Wagner et al. 2012; Nonaka et al. 2006; Kruse et al. 2015). The latter has been shown to synthesize norpseudo B_{12}, a novel type of corrinoid (Kräutler et al. 2003). In contrast, the obligate OHRB, *D. mccartyi, D. lykanthroporepellens*, *Dehalobacter restrictus* and *Dehalobacter* sp. E1, all encode incomplete de novo corrinoid synthesis pathways (Johnson et al. 2009; Maphosa et al. 2012; Kruse et al. 2013; Yan et al. 2012; Zhang et al. 2009; Siddaramappa et al. 2012). The presence of a seemingly intact corrinoid synthesis pathway has been observed in the genome of four *Dehalobacter* spp. strains (Rupakula et al. 2015). It has not yet been determined, however, whether these strains are capable of de novo corrinoid synthesis.

This may seem counterintuitive considering the essential role of corrinoids for the metabolism of obligate OHRB. A possible explanation could be the energy cost associated with de novo corrinoid synthesis, a process involving more than 30 enzymatic steps (Moore et al. 2013). Organohalides are present in limited quantities in pristine natural environments, probably favoring a salvaging strategy rather than de novo synthesis of corrinoids for obligate OHRB (Gribble 1998). Two studies investigated the effect of cultivating the corrinoid prototrophic *D. hafniense* TCE1 or Y51 in the presence or absence of corrinoids. It was found that the absence of corrinoids led to either a prolonged lag phase or faster loss of the transposon encoding the corrinoid-containing RDase, PceA, (Choudhary et al. 2013; Reinhold et al. 2012). Both *D. mccartyi* and *Dehalobacter* spp. encode the prokaryotic *btuBFCD* corrinoid uptake system belonging to the ABC transporter family (Rupakula et al. 2013; Zhang et al. 2009; Maphosa et al. 2012; Borths et al. 2002). Interspecies transfer of corrinoids between different symbionts and *D. mccartyi* has been demonstrated (Yan et al. 2012, 2013; Men et al. 2014).

The nature of the lower ligand is essential for corrinoids utilized by *D. mccartyi* as 5′,6′-dimethylbenzimidazole (DMB) and some other ligands promoted growth, whereas corrinoids containing, e.g., phenolic lower ligands did not (Yi et al. 2012). *D. mccartyi* can substitute the lower ligands to yield the suitable form of corrinoids if the correct lower ligand is present in the medium (Yi et al. 2012; Men et al. 2014). In fact it has been shown that remodeling of corrinoids is intimately linked to salvage (Yi et al. 2012). The *btuFCD* genes encoding the corrinoid transporter, are located next to *cbiZ*, which codes for the amidohydrolase responsible for cleaving the lower ligand in both *D. mccartyi* and *Dehalobacter* spp. (Yi et al. 2012; Woodson and Escalante-Semerena 2004; Men et al. 2014; Rupakula et al. 2015). A recent study indicated remodeling of the phenolic ligand-containing corrinoids provided by the cocultivated *Pelosinus fermentans* R7 to *D. mccartyi* 195^T in the presence of added DMB. Based on a microarray analysis, up-regulation of *btuFCD* expression was confirmed, whereas an uncharacterised operon encoding a putative Fe^{3+}/cobalamin transporter lacking an associated *cbiZ* gene was strongly down-regulated in cocultures supplemented with DMB compared to cocultures treated with B_{12} (Men et al. 2014).

Table 15.2 Accessory genes found to be associated with *rdhAB* genes from four well-characterized genera of organohalide-respiring bacteria and the elucidated or predicted function of the encoded proteins

Nomenclature	First traceable use of name	Verified or predicted function	*Desulfitobacterium*	*Dehalococcoides*	*Dehalobacter*	*Sulfurospirillum*
rdhA	Neumann et al. (1998)	Reductive dehalogenase, catalytic subunit	*cprA*, Dhaf_0737, 219537062	*vcrA*, DhcVS_1291, 270154553	*pceA*, DEHRE_12145, 570739860	*pceA*, SMUL_1531, 584602139
rdhB	Neumann et al. (1998)	Reductive dehalogenase, membrane anchor	*cprB*, Dhaf_0736, 219537061	*vcrB*, DhcVS_1290, 270154552	*pceB*, DEHRE_12140, 570739859	*pceB*, SMUL_1532, 584602140
rdhC	Smidt et al. (2000)	Putative function in regulation or electron transport	*cprC*, Dhaf_0738, 219537063	*vcrC*, DhcVS_1289, 270154551	*pceC*, DEHRE_12135, 570739858	Not observed
rdhC (use discouraged, changed to *rdhS*)	Kube et al. (2005)	Sensory histidine kinase, part of two-component regulatory system	see *rdhS* for examples			
rdhD (use discouraged, changed to *rdhP*)	Kube et al. (2005)	DNA-binding response regulator, part of two-component regulatory system	see *rdhP* for examples			
rdhD (use discouraged, changed to *rdhE*)	Smidt et al. (2000)	Chaperonine 60 protein	*cprD*, Dhaf_0735, 219537060	Not observed	DEHRE_03960, 570738474	Not observed
rdhE	Smidt et al. (2000)	Chaperonine 60 protein	*cprE*, Dhaf_0740, 219537065	Not observed	DEHRE_03955, 570738473	Not observed
rdhF	Kube et al. (2005)	CbiZ like protein, potentially involved in modification or insertion of corrinoids	Not observed	cbdb_A87, 73748025	Not observed	Not observed

(continued)

Table 15.2 (continued)

Nomenclature	First traceable use of name	Verified or predicted function	*Desulfitobacterium*	*Dehalococcoides*	*Dehalobacter*	*Sulfurospirillum*
rdhG	Kube et al. (2005)	hypothetical protein, putatively involved in insertion of metal cofactors	Not observed	cbdb_A97, 73748036	Not observed	Not observed
rdhH	Kube et al. (2005)	hypothetical protein, containing a DUF71 domain, putatively involved in corrinoid modification or insertion	not observed	cbdbA89, 3623418	Not observed	Not observed
rdhI	Kube et al (2005)	radical SAM domain protein, putatively involved in cofactor modification or insertion	Not observed	cbdb_A93, 73748032	Not observed	Not observed
rdhJ (previously named *rdhI*)	McMurdie et al. (2009)	Vitamin B_{12} binding radical SAM domain protein	not observed	cbdb_A92, 73748031	Not observed	Not observed
rdhK	Smidt et al. (2000)	CRP/FNR type regulator	*cprK*, Dhaf_0733, 219537058	cbdbA262, 3624107[b]	DEHRE_03985, 570740417	Not observed
rdhM[a]		NapG like membrane-bound menaquinone dehydrogenase	Desde_4093, 390526623	Not observed	Not observed	SMUL_1541, 584602149
rdhN[a]		NapH like membrane-bound menaquinone dehydrogenase	Desde_4092, 390526622	Not observed	Not observed	SMUL_1542, 84602150
rdhO[a]		Methyl-accepting chemotaxis protein	Dhaf_0686, 219537011		Not observed	Not observed

(continued)

Table 15.2 (continued)

Nomenclature	First traceable use of name	Verified or predicted function	*Desulfitobacterium*	*Dehalococcoides*	*Dehalobacter*	*Sulfurospirillum*
rdhP		DNA-binding response regulator, part of two-component regulatory system	Desde_4097, 13116968	DET0170, 3230558	Not observed	SMUL_1539, 584602147
rdhR	Kube et al. (2005)	MarR type transcriptional regulator	Not observed	cbdb_A1625, 73749322	Not observed	Not observed
rdhS		Sensory histidine kinase, part of two-component regulatory system	Desde_4098, 13119763	DET0171, 3230557	Not observed	SMUL_1538, 584602146
rdhT	Smidt et al. (2000)	Trigger factor	*cprT*, Dhaf_0732, 219537057	Not observed	DEHRE_12130, 570739857	Not observed
rdhZ	Smidt et al. (2000)	Putative chaperone, pfam 11068	*cprZ*, Dhaf_0734, 219537059	Not observed	DEHRE_03965, 570738475	Not observed

Columns from left to right: (1) currently used nomenclature; proposed changes of previously published names are indicated between brackets; [a]indicates that this gene had not previously been assigned an *rdh* prefix; (2) first traceable use of suffix for genes encoding reductive dehalogenase or associated genes; (3) verified or predicted function; (4–7) representative examples from *Desulfitobacterium*, *Dehalococcoides*, *Dehalobacter* and *Sulfurospirillum* are shown. Gene names are given when the substrate of the encoded proteins is known, followed by locus tag and NCBI gene identity (GI) number, [b]unclear if associated with organohalide respiration

Based on these observations, it is tempting to speculate on the presence of two salvage systems, one of which would be dedicated to corrinoid remodeling and a second system would be responsible for uptake without adjustment of the lower ligand. To what extent remodeling of corrinoids takes place in *Dehalogenimonas* spp. and *Dehalobacter* spp. has still not been resolved. It is noteworthy, however, that the genomes of *D. mccartyi* encode several *cbiZ* homologues associated with BtuFCD transporters and reductive dehalogenases (Table 15.2). This will be discussed in more detail together with other genes associated with *rdhA* later in this chapter.

15.3 Organization of *rdh* Gene Clusters

With few exceptions, the *rdhA* genes encoding the catalytic subunits of reductive dehalogenases are adjacent to a small gene, *rdhB*, coding for a hydrophobic protein believed to act as a membrane anchor for the RdhA (van de Pas et al. 1999; Neumann et al. 1998; Magnuson et al. 2000). This minimal *rdhAB* gene cluster is frequently accompanied by a variable set of accessory genes, for most of which the exact function is still unknown. Some of these have been shown to encode proteins regulating expression of the *rdhAB* genes (Wagner et al. 2013; Gábor et al. 2008; Pop et al. 2004) or are chaperones probably assisting the folding of RdhA (Maillard et al. 2011; Mac Nelly et al. 2014; Morita et al. 2009). Traditionally, the nomenclature of these accessory genes has followed the name of the reductive dehalogenase subunit, followed by a letter specific for genes predicted to be homologues of previously described genes. Rather confusingly in some cases, however, different letters have been assigned to accessory genes predicted to encode proteins with similar functions, both within the same *rdh* gene cluster or between different organisms. Similarly, the same letter has been used for genes for which the gene product seems to have different functions (Table 15.2) (Kube et al. 2005; McMurdie et al. 2009; Pop et al. 2004; Smidt et al. 2000). As more genomes of OHRB become available it will be necessary to develop an uniform nomenclature for these accessory *rdh* genes.

In the following paragraphs we will give a short overview of accessory genes identified in *D. mccartyi, Dehalobacter* spp., *Desulfitobacterium* spp. and *Sulfurospirillum* spp. comprising the four most thoroughly studied genera of OHRB. We also propose names for *rdh*-associated genes that have not yet been named as such. The known or proposed function of Rdh proteins is schematically shown in Fig. 15.1. Figure 15.2 shows exemplarily the organization of *rdh* gene clusters in Firmicutes and Chloroflexi.

The ***rdhA*** gene encodes the catalytic subunit of reductive dehalogenases, catalysing the removal of one or more halogens from the organohalide substrate. Reductive dehalogenases contain a twin arginine translocation (TAT) signal sequence, two Fe/S clusters and a corrinoid cofactor, although the type of the incorporated corrinoid may vary (for details see Chap. 17) (Magnuson et al. 2000; Palmer and Berks 2012; Kräutler et al. 2003; Yi et al. 2012; Keller et al. 2013; Yan et al. 2012).

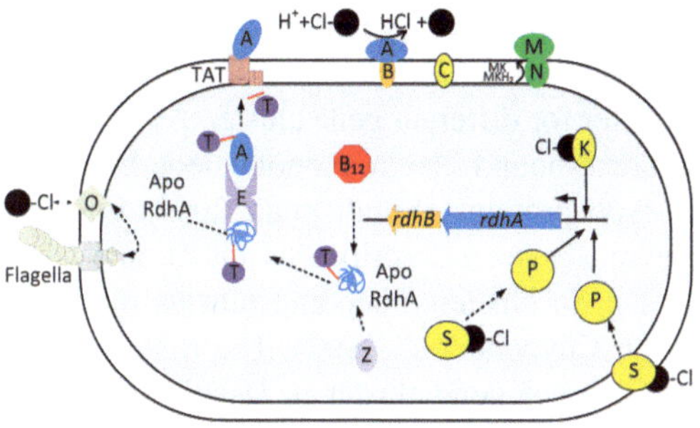

Fig. 15.1 Schematic representation of proteins encoded by accessory genes found in *rdh* gene clusters. *Left* Consensus cell representing *Desulfitobacterium* spp., *Dehalobacter* spp. and *Sulfurospirillum* spp., *right Dehalococcoides mccartyi. Note* not all *rdhAB* accessory genes are found in a single organism (see Table 15.2). *Letters inside symbols* indicate the protein name without the Rdh prefix. *Curved arrows* chemical reactions; *dashed arrows* direction of movement or interaction; *bent arrow* transcriptional start site; *straight solid lines* interaction of transcriptional regulators with *rdhA* promoters (*arrow* indicates activation, an inverse T repression); *blue* reductive dehalogenase; *orange* membrane anchor; *red bar* TAT signal sequence; *purple* chaperones; *red* corrinoids; *light red* corrinoid interacting protein, *gray* metal cofactor binding protein; *green* electron transferring complex; *yellow* transcriptional regulators; *black circles* chlorinated compounds; *light green* chemotaxis and motility. See main text or Table 15.2 for details on function of individual proteins

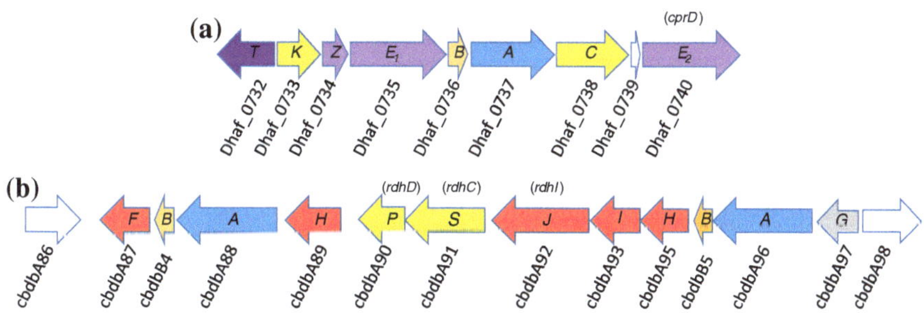

Fig. 15.2 Examples of the organization of *rdh* gene clusters in Firmicutes (**a**) and Chloroflexi (**b**). **a** The *rdh* gene cluster of *Desulfitobacterium hafniense* strain DCB-2^T and *Desulfitobacterium dehalogenans* is depicted, which encodes the chlorophenol reductive dehalogenase CprA (Smidt et al. 2000; Kim et al. 2012). **b** A detail of 'high plasticity region' 1 of the genome of *Dehalococcoides mccartyi* strain CBDB1 is shown. The *rdhAB* and accessory genes are given without the *rdh* prefix and colored corresponding to the respective proteins shown in Fig. 15.1. When a new gene designation is proposed the originally reported name is given in brackets (for details see the text). *Empty arrows* unknown function. The locus tags are indicated

The ***rdhB*** gene encodes a small membrane protein that is believed to function as a membrane anchor for RdhA (van de Pas et al. 1999; Neumann et al. 1998; Magnuson et al. 2000). As mentioned already, the *rdhA* gene is usually accompanied by an *rdhB* gene; however, *D. lykanthroporepellens* BL-DC-9T is a recently described exception where only six of twenty five *rdhAs* have a cognate *rdhB* (Magnuson et al. 2000; Siddaramappa et al. 2012). Furthermore, the order of *rdhA* and *rdhB* (i.e., *rdhBA* vs. *rdhAB)* can differ for different gene clusters.

The ***rdhC*** gene encodes a membrane-bound flavin mononucleotide-binding protein resembling NosR- and NirI-type proteins (Smidt et al. 2000; Futagami et al. 2008). These proteins have previously been described as membrane-bound proteins controlling the expression of gene clusters encoding nitrous oxide and nitrite reductases (Saunders et al. 1999; Cuypers et al. 1992). The first described *rdhC* gene was *cprC*, which is part of the *cpr* gene cluster in *Desulfitobacterium dehalogenans* JW/IU-DC1^T (Smidt et al. 2000). Based on amino acid similarity, it was speculated that RdhC might have a similar function to NosR- and NirI-type proteins (Smidt et al. 2000; Futagami et al. 2008; Maillard et al. 2005). However, it was later shown that NosR likely plays a role in electron transfer (Wunsch and Zumft 2005), for a recent review see Simon and Klotz (2013). It has been recently proposed that RdhC has a function in electron transport to RdhA (Futagami et al. 2014; Prat 2009), although there is still no experimental evidence to support this proposition.

rdhD see *rdhE* below

The ***rdhE*** gene encodes a protein with a high level of amino acid sequence similarity to the 60 kD subunit of the well-characterized *Escherichia coli* GroEL/GroES two-component chaperonins (Smidt et al. 2000). This complex assists in protein folding and the mechanism has been described in detail in excellent reviews (Horwich et al. 2007; Thirumalai and Lorimer 2001). A nomenclature where the large subunit is named *cpn60* and the small subunit *cpn10*, has been proposed, with multiple *cpn60*s being numbered by their order of appearance in the genome (Lund 2009; Coates et al. 1993). The *cpr* gene cluster of *Desulfitobacterium* spp. encoding a chlorophenol RDase has been reported to harbor two *cpn60* homologues, designated *cprD* and *cprE* (Gábor et al. 2006; Smidt et al. 2000). We suggest to rename these to $cprE_1$ and $cprE_2$, following the nomenclature suggested by (Lund 2009). Consequently *cpn60* homologues found in *rdh* gene clusters should be named *rdhE* with numbers in subscript if there are multiple *rdhE* homologues, unless the substrate for the associated *rdhA* is known.

The ***rdhF*** gene encodes a protein resembling CbiZ, an amidohydrolase involved in salvaging cobinamide from the environment (Yi et al. 2012; Kube et al. 2005). In bacteria, *cbiZ* genes are often found in the proximity of genes encoding corrinoid transporters (BtuFCD) (Gray et al. 2008). It has been suggested that *cbiZ* homologues located distant from genes related to synthesis or salvage of cobinamide play a role in processes other than cobalamin salvage (Woodson and Escalante-Semerena 2004; Gray et al. 2008; de Crécy-Lagard et al. 2012). A role in remodeling the lower ligand of cobalamin has been demonstrated in *Rhodobacter sphaeroides* (Gray and Escalante-Semerena 2009). Although no experimental evidence for the function of *rdhF* exists, it is possible that it plays a role in ensuring incorporation of the correct form of cobalamin into RDases.

The ***rdhG*** gene encodes a hypothetical protein without any signal sequence or transmembrane helices, suggesting a cytoplasmic location. This protein belongs to pfam03692. The function of this protein is not yet known but it encodes a conserved CxxCxxCC domain, which has been speculated to be involved in zinc or iron chelation (Finn et al. 2014). Reductive dehalogenases contain two *N*-terminal Fe/S clusters (Hug et al. 2013; Magnuson et al. 2000), hence it is possible that this protein plays a role in maturation of RDases. One part of RdhG contains a zinc-dependent carboxypeptidase-like domain leading to the proposal that it might have protease activity (Kube et al. 2005).

The ***rdhH*** gene encodes a hypothetical protein without any signal sequence or transmembrane helices, strongly suggesting a cytoplasmic location. This protein contains a domain of unknown function 71, a recent comparative genomics study using the published genome sequences of bacteria and archaea suggested that it originates from archaea and may have a function similar to CbiZ in modifying the lower ligand in cobalamins (de Crécy-Lagard et al. 2012).

The ***rdhI*** gene encodes a hypothetical protein without any signal sequences or transmembrane helices, strongly suggesting a cytoplasmic location. This protein contains a predicted radical SAM domain, and therefore might have a role in incorporation or modification of cofactors (Kube et al. 2005).

RdhJ. In the literature this gene has also been referred to as *rdhI* (McMurdie et al. 2009). Here we propose to rename this gene *rdhJ* to distinguish these two nonhomologous proteins. The *rdhJ* gene encodes a hypothetical protein without any signal sequence or transmembrane helices, strongly suggesting a cytoplasmic location. This protein contains both a predicted *N*-terminal vitamin B_{12}-binding domain and a *C*-terminal radical SAM domain, again suggesting a role in the modification or incorporation of cofactors (McMurdie et al. 2009; Kube et al. 2005).

The ***rdhK*** gene encodes a CRP/FNR (cAMP receptor protein/fumarate and nitrate reduction)-type transcriptional regulator. Among these, CprK from *D. hafniense* DCB-2^T has been characterized in detail (see Sect. 15.5), and shown to act as a transcriptional activator of *cprA*, which encodes a chlorophenol RDase found in several strains of *Desulfitobacterium* spp. (Smidt et al. 2000; Kim et al. 2012; Villemur et al. 2002, 2006). CprA catalyses the degradation of 3-chloro-4-hydroxyphenylacetic acid to 4-hydroxyphenylacetic acid (van de Pas et al. 1999; Gerritse et al. 1999; Christiansen and Ahring 1996).

The ***rdhM*** and ***rdhN*** genes encode two proteins resembling NapG and NapH, respectively. NapH is a membrane-bound Fe/S protein that receives electrons from the menaquinol pool. Electrons are channeled from NapH to the periplasmic NapG, from which the electrons are transferred to the periplasmic NapA, catalyzing the reduction of nitrate to nitrite (Simon and Klotz 2013). RdhN is an Fe/S protein predicted to contain several transmembrane regions, whereas RdhM contains both a signal sequence and sequences with homology to Fe/S clusters suggesting that, like NapG, it is a periplasmic protein involved in electron transfer. In the newly published genome of *Sulfurospirillum multivorans* a pair of *rdhMN* genes was found in the same gene cluster as the genes encoding the PCE dehalogenase PceA and another RDase for which the substrate is not known (Goris et al.

2014). Expression of these genes was only detected when PCE was used as electron acceptor, but not when fumarate was used (Goris et al. 2014). These findings, and the similarity to NapGH, make it plausible that RdhMN facilitate electron transport from membrane-embedded menaquinones to RDases attached to the outside of the cell membrane.

The ***rdhO*** gene encodes a protein predicted to function as a membrane-bound methyl-accepting chemotaxis protein. These act as membrane-bound receptors able to detect extracellular factors like the presence of suitable electron acceptors. Information is channeled from membrane-bound methyl-accepting chemotaxis protein via the Che signaling pathway to the flagellar motor enabling the cell to alter the direction of swimming (for reviews see Wadhams and Armitage 2004; Suzuki et al. 2010; Mascher et al. 2006).

The ***rdhP*** gene was originally named *rdhD* (Kube et al. 2005). We propose to rename it *rdhP* according to the predicted function of the gene product as the **p**hosphoryl group-accepting DNA-binding response regulator of a two-component signal transduction system. It probably regulates gene expression upon phosphorylation or dephosphorylation by the associated sensory histidine kinase RdhS.

The ***rdhR*** gene encodes a transcriptional regulator of the MarR-type. It was recently demonstrated that expression of some reductive dehalogenases in *D. mccartyi* is indeed regulated by RdhR proteins, as will be discussed in detail later in this chapter (Wagner et al. 2013).

The rdhS gene was originally named *rdhC* (Kube et al. 2005). To distinguish it from the above-mentioned *cprC* we here propose to rename it *rdhS*. The *rdhS* gene is predicted to encode a sensory histidine kinase that is part of a two-component signal transduction system. The histidine kinases possess diverse signal input domains often located outside the cell allowing sensing of extracellular signals, although it should be noted that, in *Dehalococcoides*, the RdhS proteins are cytoplasmic. In two-component signal transduction systems, a conserved autophosphorylation and phosphotransfer pathway links the input signals to output responses mediated by the phosphorylated response regulator (Gao and Stock 2009). RdhS is proposed to sense specific organohalides, but it may also recognize other types of stimuli such as the redox status of the cell (Kube et al. 2005).

The ***rdhT*** gene encodes a trigger factor-like protein, predicted to be involved in folding and maturation of RdhA (Smidt et al. 2000). Full-length trigger factors contain three domains: an *N*-terminal ribosome-binding domain; a peptidyl-prolyl *cis/trans* isomerase domain; and a *C*-terminal domain. The trigger factor binds to the large ribosomal subunits near the exit channel and chaperones the folding of newly synthesized proteins (reviewed in (Kim et al. 2013). However, the *rdhT* gene does not encode an *N*-terminal ribosome-binding domain, showing that it most likely does not function as a classical trigger factor. It was shown that PceT from *D. hafniense* Y51 and TCE1 binds specifically to the TAT signal sequence, which may delay the translocation of PceA thereby increasing the chance of correct folding and incorporation of all cofactors (Morita et al. 2009; Maillard et al. 2011). Furthermore, the peptidyl-prolyl *cis/trans* isomerase activity of PceT was confirmed, which supports the suggestion that PceT and likely RdhTs in general act as dedicated chaperones for the correct folding of RdhAs (Morita et al. 2009; Maillard et al. 2011). In agreement with this, co-expression of *pceA* and *pceT* from *D. hafniense* Y51 or *rdhA$_3$* and *rdhT$_3$* from *D. hafniense* DCB-2^T in *Shimwellia blattae* led to increased amounts of soluble reductive dehalogenase compared to when *pceA* or *rdhA$_3$* were expressed alone (Mac Nelly et al. 2014). The PceA encoding gene cluster in *D. hafniense* Y51 and TCE1 is constitutively expressed (Reinhold et al. 2012; Prat et al. 2011). Cultivation of strain Y51 in medium devoid of PCE led to the formation of intracellular protein aggregates, consisting of apo-PceA with bound PceT and CobT, a protein involved in the later stage of corrinoid synthesis (Reinhold et al. 2012; Claas et al. 2010; Trzebiatowski et al. 1994). This may suggest posttranslational regulation, but the role, if any, of RdhT in this process still warrants further studies.

The ***rdhZ*** gene encodes a small protein with a size of approximately 135 amino acids belonging to pfam11068 (Finn et al. 2014). This protein does not contain any signal sequence or membrane helices suggesting a cytoplasmic localisation. Based on analysis of the crystal structure of a member of the same pfam, obtained from the non-organohalide-respiring bacterium *Synechococcus* sp. strain WH8102, a function as a molecular chaperone has been suggested (http://pfam.xfam.org/structure/4dci). The location of *rdhZ* next to the chaperone-encoding *rdhE* gene in both *Desulfitobacterium* spp. and *Dehalobacter restrictus* lends support to this speculation.

15.4 Transcription of *rdh* Genes

The transcription of *rdh* genes has been studied to elucidate the molecular mechanisms underlying the perception of, and response to, organohalides. This work has also aimed to identify the RdhA enzyme catalysing the degradation of a specific organohalide compound and to develop RNA-based tools for the prediction of dehalogenation activity. Here, the current knowledge in the field of *rdh* gene transcription and regulation is summarized.

As already mentioned most known OHRBs possess several *rdhA* genes in their genome. The function of some of these has been elucidated using traditional biochemistry (for a review see Chap. 17). Until recently, the lack of a system for functional heterologous expression of RDases (Mac Nelly et al. 2014) has prevented detailed biochemical analysis of the majority of RdhAs. Instead, transcription analyses can provide first hints to the substrate range of an RdhA when gene expression is induced by a specific substrate. Transcriptional analyses included reverse transcription of total RNA, the amplification of *rdhA* fragments using degenerate primers, followed by cloning and sequencing. This approach identified BvcA, the vinyl chloride (VC) RDase of *D. mccartyi* strain BAV1 (Krajmalnik-Brown et al. 2004), DcpA, the 1,2-dichloropropene-dichloro-eliminating RDase of *D. lykanthroporepellens* and of nonidentified *D. mccartyi* strains in two enrichment cultures (Padilla-Crespo et al. 2014), and RdhA1, responsible for dichloro-elimination of 1,2-dichloroethane by a *Dehalobacter* strain (Grostern and Edwards 2009). The involvement of DceA6 (MbrA) in the dechlorination of PCE to *trans*-DCE by *D. mccartyi* MB was also supported by transcription analyses (Chow et al. 2010). Transcriptional analyses also gave hints to an extended substrate range of the PCE reductive dehalogenase PceA of *D. mccartyi* 195^T. The *pceA* gene was one of the most highly up-regulated *rdhA* genes in strain 195^T, not only during growth with PCE but also during growth with 2,3-dichlorophenol, suggesting that both compounds are substrates of PceA (Fung et al. 2007).

Simultaneous transcription of multiple *rdhA* genes has also been observed, such as in one of the first transcriptional studies conducted on the *Dehalococcoides*-containing enrichment culture KB1 (Waller et al. 2005). This appears logical, if reductive dechlorination of a compound involves a series of dehalogenation steps. Indeed, dissection of the transcriptional response of the four *rdhA* genes *cprA2* to *cprA5* in *D. hafniense* PCP-1 during the course of 2,4,6-trichlorophenol (TCP) dechlorination revealed that transcription of *cprA3*, which encodes an *ortho*-dechlorinating RDase (Bisaillon et al. 2010) (and of *cprA2* with an unknown function) preceded the induction of *cprA5* (Bisaillon et al. 2011). This probably reflected the turnover of 2,4,6-TCP to the transiently formed intermediate 2,4-dichlorophenol, which is a substrate of CprA5 (Thibodeau et al. 2004). This expression pattern points to a transcriptional regulation by the successive substrates and dechlorination intermediates in *Desulfitobacterium*. Remarkably, *rdhA* transcription in obligate OHRB such as *Dehalobacter*, *Dehalococcoides,* and *Dehalogenimonas* seems to be controlled by gene- and substrate-specific mechanisms as well as higher levels of regulation. Although the catabolic genes were transcribed to the highest levels in the presence of their putative substrate, transcripts of many additional *rdhA* genes were also detected. In *Dehalobacter restrictus*, transcripts of almost all 25 rdhA genes were identified during growth on PCE (Rupakula et al. 2013), although the *rdhA24* transcript encoding PceA was most abundant. In three representatives of the Chloroflexi, *D. mccartyi* strains 195^T and CBDB1 and *D. lykanthroporepellens* BL-DC-9^T, a similar general up-regulation of almost all *rdhA* genes was observed. In strain 195^T, a total of 17, 13 and 9 *rdhA* transcripts were detected in the presence of PCE, TCE, or 2,3-DCP, respectively, however, in levels varying over five orders of magnitude (Fung et al. 2007). Among those, the *rdhA* gene DET0162 exhibited one of the highest transcript levels despite containing a point mutation probably leading to a nonfunctional protein (Fung et al. 2007; Rahm et al. 2006). For strain CBDB1, transcription of all 32 *rdhA* genes was induced during growth on 1,2,3- or 1,2,4-trichlorobenzene (Wagner et al. 2009), and a similar response was observed during growth with 2,3-dichlorodibenzo-*p*-dioxin (Wagner et al. 2013), however, with a distinct up-regulation of the chlorobenzene RDase-encoding *cbrA* to the highest number of transcripts observed. Recently, the transcription of all 25 *rdhA* genes present in the genome of strain BL-DC-9^T was analyzed during dichloro-elimination of 1,2-dichloroethane, 1,2-dichloropropane and 1,2,3-trichloropropane and indicated transcription of 19 *rdhA* genes (Mukherjee et al. 2014). Among these were 13 full-length and, similar to DET0162 in *D. mccartyi* 195^T, six incomplete *rdhA* genes lacking a predicted *N*-terminal TAT leader sequence or lacking iron–sulphur cluster-binding motifs in the *C*-terminus (Mukherjee et al. 2014). A comparative analysis of microarray–based transcriptomic studies of *D. mccartyi* 195^T and *Dehalococcoides* strains in the KB-1 enrichment culture also indicated the multitude of transcribed *rdhA* genes under a variety of conditions (Islam et al. 2014). This apparently gratuitous expression might reflect a fundamental strategy in the obligate OHRBs for the adaptation to newly emerging halogenated compounds in their environment by ensuring the continuous presence of low levels of various *rdhA* transcripts. It is conceivable that posttranscriptional or posttranslational regulation mechanisms of RDase activity exist in OHRB, which is strongly supported by a recent comparison of mRNA and protein abundances per cell (Rowe et al. 2015).

The expression of *rdhA* genes showed considerable differences depending on the growth phase and variations in other environmental parameters. Initial transcriptional studies quantified the up-regulation of specific functional *rdhA* genes (e.g., *tceA* and *vcrA*) in the *D. mccartyi*-containing mixed cultures ANAS and D2 (Johnson et al. 2005; Lee et al. 2006; Rahm et al. 2006). Within the first hours of exposure, a several fold increase in gene expression was observed, reaching a plateau lasting beyond the decline of active dechlorination and followed by a slow mRNA decay, which leveled off at low basal transcript numbers (Lee et al. 2006). A similar initial response and long-lasting stability of the functional gene transcript *cbrA* to trichlorobenzene exposure was reported for *D. mccartyi* CBDB1 (Wagner et al. 2009). To explore the suitability of functional *rdhA* transcripts as bio-indicators for predicting in situ dehalogenation activity, the relation of transcript copy numbers with growth or dehalogenation activity was analyzed. The pseudo-steady-state respiration rates in mixed cultures containing *D. mccartyi* 195^T continuously fed with PCE or TCE were positively correlated with the *tceA* transcript abundance (Rahm and Richardson 2008; Rowe et al. 2012). Further, microarray-derived transcriptomic profiles suggested a close interrelationship between the global transcriptional regulation and the organohalide respiration status of the cells (Mansfeldt et al. 2014). DNA microarray studies of *D. mccartyi* 195^T growing on TCE demonstrated a major shift in global gene expression during transition into the stationary phase with a large number of genes being up- or down-regulated (Johnson et al. 2008). Several *rdhA* transcripts showed differential expression dynamics across the time-course of the experiment. The *tceA* gene was highly expressed throughout the experiment. Another group of *rdhA*s (DET0173, DET0180, DET1534 and DET1545) showed strong up-regulation during the transition from the exponential to the stationary phase. Interestingly, DET0180 is part of a conserved syntenic region present in all sequenced *D. mccartyi* genomes (Hug et al. 2013). The delayed transcriptional induction of DET0173 and DET1545 orthologs in a mixed culture was repeatedly observed (West et al. 2013). DET1545 possesses orthologs in most *D. mccartyi* genomes and is localized within a 12 kb DNA region with high gene synteny close to the 3′ end of 'high plasticity region' 2 (McMurdie et al. 2009). DET1545 and its orthologs lack a directly associated regulatory gene, however, its transcriptional up-regulation was positively correlated with starvation-related stress such as low respiration rates (Rowe et al. 2012;

Mansfeldt et al. 2014), consumption of the electron acceptor (Rahm et al. 2006), absence of a chlorinated compound (Waller et al. 2012; Islam et al. 2014) or fixed nitrogen limitation (Lee et al. 2012b). This suggests an important role of the RDase in the survival strategy of *D. mccartyi*.

Transcripts of functional *rdhA* genes can be used as indicators of ongoing in situ reductive dechlorination. For instance, during biostimulation and bioaugmentation studies in a TCE-contaminated groundwater site (Lee et al. 2008) and in microcosms (Baelum et al. 2013) the expression of *vcrA* and *bvcA* was correlated with the observed dechlorination of VC to ethene. In a PCE-dechlorinating flow column (Behrens et al. 2008) the transcript abundance of *bvcA*, *tceA*, and *vcrA* correlated with the zone of most active dechlorination. Also, in PCE-amended microcosms enriched from Yangtze River sediment the dynamics of transcript copy numbers of *tceA* and *vcrA* correlated with the observed turnover of TCE and VC (Kranzioch et al. 2014). However, transcription analyses also revealed some inconsistencies. Although a complete conversion of TCE (Lee et al. 2008; Baelum et al. 2013) or PCE (Behrens et al. 2008; Kranzioch et al. 2014) to ethene occurred, transcripts of *pceA* or *tceA* encoding proteins catalysing the conversion of PCE or TCE to DCE were absent or inconsistently detected. This might suggest the presence of other OHRB such as *Desulfitobacterium* or *Dehalobacter* (Kranzioch et al. 2014) or *Eubacterium* (Behrens et al. 2008), which compete successfully for the primary steps of PCE or TCE dechlorination, while their genes were probably not targeted by the primers used. Another possible reason for the discrepancy in the results might be due to the expression of *rdhA*s in *D. mccartyi* with an as yet unrecognized function. For instance, PCE-to-*trans*-DCE dechlorination by an enrichment culture (Futamata et al. 2009) or *D. mccartyi* strain MB (Cheng and He 2009), containing neither *pceA* nor *tceA* pointed to the existence of other RDases with this specific function. Furthermore, transcript abundance did not always correlate with dechlorination activity. For instance, oxygen-amended, dechlorination-inactive *D. mccartyi* cells still formed *tceA* and *vcrA* transcripts (Amos et al. 2008). An increase in *rdhA* gene expression was also observed at elevated temperatures in cultures with an impaired dechlorination activity suggesting that this up-regulation was a general stress response (Fletcher et al. 2011). These observations highlight the necessity for a deeper understanding of composition and function of the dehalogenating community to allow an appropriate biomarker selection.

15.5 Regulation of *rdh* Gene Transcription

Despite the multitude of observed *rdh* transcripts from a broad range of reported OHRBs, relatively little is known about the regulatory mechanisms controlling their synthesis. From the genome sequences available it appears that these mechanisms might differ according to OHRB phylogeny rather than the organism's status as facultative or obligate OHRB. In members of the Firmicutes, i.e., *Desulfitobacterium* and *Dehalobacter,* the canonical *rdh* operon structure (Fig. 15.2) is well conserved and often associated upstream or downstream with *rdhK,* which encodes a member of the CRP/FNR family of transcriptional regulators (Gábor et al. 2008; Fletcher et al. 2011; Rupakula et al. 2013) (Table 15.2, Fig. 15.2). Its function in the *ortho*-chlorophenol-dependent regulation of organohalide respiration has been elucidated in detail for CprK (see below). In the Epsilonproteobacterium Sulfurospirillum *multivorans*, the genes encoding PceAB are closely associated with two *rdhPS* pairs encoding two component signal transduction system response regulators and histidine kinases (Table 15.2) (Goris et al. 2014). The RdhS homologs found in *Sulfurospirillum multivorans* are predicted to contain several transmembrane helices and thus are most likely located in the cell membrane (Goris et al. 2014). For *Sulfurospirillum multivorans*, induction of dechlorination in response to PCE or TCE concomitant with the formation of active PceA was demonstrated (John et al. 2009), suggesting an involvement of these two-component signal transduction systems in the regulation of organohalide respiration. Contrary to this observation, the *Sulfurospirillum multivorans pceA* mRNA and the PceA protein were present over many generations in the absence of PCE, suggesting that an additional, still unknown signal might contribute to this kind of long-term 'memory' (John et al. 2009). In the *Dehalococcoidia* class of the Chloroflexi, the *rdhA* genes are mostly associated with genes encoding either two-component signal transduction system or MarR regulators. Analysis of the deduced amino acid sequences of the two-component signal transduction systems suggests a cytoplasmic localization for the histidinc kinasc components, as no transmembrane helices are predicted, raising the question whether the chlorinated compound itself or another signal such as the redox status of the cell is the recognized signal (Kube et al. 2005). The MarR-type regulator family mediates cellular responses to changing environmental conditions, such as antibiotic or peroxide stress or adaptation to the catabolism of aromatic compounds (Wilkinson and Grove 2006). It is characterized by a winged helix-turn-helix (HTH) motif mediating contact between the dimeric regulator and short palindromic sites in the target DNA. A high number of MarR-type regulators is encoded in *rdh* clusters in *D. mccartyi* genomes (Wagner et al. 2013). Recently, one representative was also detected in an *rdh* cluster in *Dehalobacter* sp. E1 (Maphosa et al. 2012). Two further types of putative transcriptional regulators have been predicted in *D. mccartyi.* In most *D. mccartyi* genomes a single RdhK is encoded, however, this is not closely associated with *rdhA* genes. The *vcrC* gene encoding a member of the NirI/NosZ family (RdhC) is part of the *vcrABC* operon, which is essential for VC respiration in strain VS. Notably, members of the NosI/NosR family are also encoded in the canonical *rdhTKZE$_1$BACE$_2$* gene cluster of the Firmicutes (CprC; Table 15.2, Fig. 15.2) (Smidt et al. 2000).

The role of one MarR-type regulator, CbdbA1625, in the transcriptional control of the *rdhA* gene *cbdbA1624* in *D. mccartyi* CBDB1 has been studied in detail. The *rdhA* gene *cbdbA1624* is divergently oriented to the MarR-encoding gene and was specifically up-regulated in the presence of 1,2,4-trichlorobenzene (Wagner et al. 2009, 2013). To determine whether CbdbA1625 exerts a regulatory function, it was heterologously produced and the interaction with the PCR-amplified intergenic region was studied in vitro. Electrophoretic mobility shift assays pointed to two binding sites within a perfect 40 bp palindrome, which contained short inverted repeats in its half sites typical of MarR binding motifs. This region also contained the overlapping transcriptional start sites and the −10 promoter regions of the *rdhA* and the MarR-encoding gene (*marR*). Transcriptional promoter-*lacZ* fusions introduced as single copy into the *E. coli* chromosome were used to study the in vivo interaction of CbdbA1625 (produced heterologously from a pBAD vector) with promoters of the *rdhA* genes cbdbA1624 (and cbdbA1453 as a negative control) and the *marR* gene. The resulting β-galactosidase activities of the *marR* (Wagner et al. 2013) and *rdhA* (cbdbA1624) (Lydia Krasper, personal communication) promoters were significantly reduced upon synthesis of CbdbA1625, whereas the activity of the control promoter was not impaired, indicating specificity of interaction and negative autoregulation (Fig. 15.1). MarR regulators are known to be released from their target promoters by allosteric binding of a ligand (Wilkinson and Grove 2006); however, the signals acting on MarR regulators in *D. mccartyi* are still unknown.

In summary, the MarR-dependent repression of *rdhA* gene transcription may be the rule rather than the exception in *D. mccartyi,* ensuring specific *rdhA* transcription when halogenated compounds become available.

Several studies have provided molecular insight into *rdhA* gene regulation by CprK, a CRP-FNR-family regulator from *Desulfitobacterium* spp. This family of regulators is characterized by a length of 230–250 amino acids, an *N*-terminally located nucleotide (effector)-binding domain and a *C*-terminal HTH DNA-binding motif. The family responds to a broad spectrum of signals leading to activation of gene transcription. To accomplish the regulatory function the *N*-terminal domain binds allosteric effector molecules, after which a signal is transmitted to the DNA-binding domain (Körner et al. 2003). CprK is encoded in the *cprTKZE$_1$-BACE$_2$* gene cluster (Fig. 15.2, Table 15.2) in *Desulfitobacterium dehalogenans* JW/IU-DC1^T and *D. hafniense* DCB-2^T including the gene encoding the *ortho*-chlorophenol reductive dehalogenase CprA (van de Pas et al. 1999; Smidt et al. 2000; Kim et al. 2012). Transcription of this gene cluster was initiated from three promoters in the presence of 3-chloro-4-hydroxyphenylacetic acid, whereas *cprK* was constitutively expressed at a low level. Electrophoretic mobility shift assays, DNase footprinting studies and promoter-*lacZ* fusion experiments using heterologously expressed CprK from the two *Desulfitobacterium* spp. strains mentioned above indicated that CprK acts as a transcriptional activator of the *cpr* gene cluster in the presence of 3-chloro-4-hydroxyphenylacetic acid (Pop et al. 2004; Gábor et al. 2006). 3-Chloro-4-hydroxyphenylacetic acid binds with high affinity to CprK, which promotes its specific interaction with a dehalo-box motif upstream of the -10 and -35 promoter regions of several *cpr* genes. CprK is inactivated by oxygen in vitro (Pop et al. 2004) due to the formation of an intermolecular disulphide bridge (Cys11 and Cys200); however, a physiological role in redox-sensing seems unlikely (Gábor et al. 2006). The availability of the genome sequence of *D. hafniense* DCB-2^T allowed a comparative analysis of a number of CprK proteins (Gábor et al. 2008). Genes encoding five CprK-type proteins encoded in the genome are clustered with *cpr* genes. The *cprK1* gene is part of the cluster encoding the biochemically characterized *ortho*-chlorophenol RDase CprA1 of *D. hafniense* (Christiansen et al. 1998). Eleven dehalo-box motifs were identified within the five *cpr* gene clusters. They consisted of a 5 bp imperfect inverted repeat with 4 nt spacing and exhibited TTAGT-N_4-ACTAA as consensus sequence. Interaction studies of three heterologously produced CprK proteins with the corresponding dehalo-box motifs in the presence and absence of different chlorinated compounds revealed partly overlapping functions. Two CprK proteins were specifically activated by 3-chloro-4-hydroxyphenylacetic acid and other *ortho*-halogenated phenols, and another one by *meta*-chlorinated phenols. Thus, a range of chlorophenols can be sensed by *D. hafniense* DCB-2^T in a cooperative manner. (Gábor et al. 2008).

The crystal structures of oxidized *D. hafniense* CprK1 with bound 3-chloro-4-hydroxyphenylacetic acid and of reduced, ligand-free CprK from *Desulfitobacterium dehalogenans* JW/IU-$DC1^T$ allowed the first mechanistic insight and identified allosteric conformational changes induced by ligand binding (Joyce et al. 2006). Both the phenolic OH-group and the *ortho*-chlorine were required for tight binding, in which a conserved lysine residue (Lys133) in the central α-helix interacts with the OH-group. Reduced and oxidized protein dimerizes without forming a disulphide bond. Based on results of DNA-binding assays and macromolecular native mass spectrometry, in combination with limited proteolysis, a dynamic model for the activation process was proposed (Mazon et al. 2007). This was refined by a complete structural description of the redox-dependent and allosteric molecular arrangements of CprK1 (Levy et al. 2008). In the absence of the ligand, the dimeric CprK is flexible. Binding of the ligand leads to a concerted reorganization of both monomers that moves the *C*-terminal DNA-binding domains into positions compatible with binding to the palindromic dehalo-box motif sequence (Levy et al. 2008). This motion is induced by interactions of the phenolate with Lys133 as well as docking of the chloride atom into a hydrophobic pocket provided by the central α-helix connecting the *N*- and *C*-terminal domains. As a consequence, three amino acids of the *N*-terminal effector domain interact with the DNA-binding domain and thus stabilize the DNA-binding conformation. Nucleotide-specific contacts are formed between the HTH motif recognition helix and the DNA that lead to the observed specificity for the AT-rich dehalo-box motif. The physical presence of the bulky halogen atom as well as its inherent electronegativity lowering the pK_a of the phenol are necessary for the formation of the Lys133-phenolate salt bridge, resulting in high affinity binding (Kemp et al. 2013). Therefore, neither the end product of dechlorination, 4-hydroxyphenylacetic acid, nor 3-chlorophenylacetic acid lacking either the chlorine substituent or the hydroxyl group are recognized by CprK. Lys133 is strictly conserved within the CprK family suggesting that its function might be limited to sensing phenolic compounds in the environment. Indeed, *pceA* genes in *Desulfitobacterium* are often part of the canonical transposon-encoded *pceABCT* gene cluster lacking a *cprK* ortholog and are constitutively expressed

(Duret et al. 2012; Suyama et al. 2002; Peng et al. 2012), perhaps reflecting an evolutionary adaptation to high anthropogenic concentrations of PCE and TCE in the environment.

These first studies of regulators in OHRB revealed the existence of completely different mechanisms of transcriptional regulation in two phylogenetically distant bacteria. It can be expected that the investigation of other types of regulators such as two-component system members will further expand our understanding of how bacteria sense and respond to halogenated compounds as electron acceptors for organohalide respiration.

Acknowledgement This work was funded by the German Research Foundation (DFG research unit FOR 1530). Work of HS and TK was financially supported by the EcoLinc Project of the Netherlands Genomics Initiative, the BE-BASIC foundation through projects F07.001.05 and F08.004.01, as well as the European Community program FP7 (grants KBBE-211684; BACSIN, and KBBE-222625; METAEXPLORE). We are grateful to Gary Sawers for comments on the manuscript.

References

Ahsanul Islam M, Edwards EA, Mahadevan R (2010) Characterizing the metabolism of *Dehalococcoides* with a constraint-based model. PLoS Comput Biol 6(8). doi:10.1371/journal.pcbi.1000887

Amos BK, Ritalahti KM, Cruz-Garcia C, Padilla-Crespo E, Löffler FE (2008) Oxygen effect on *Dehalococcoides* viability and biomarker quantification. Environ Sci Technol 42(15):5718–5726

Baelum J, Chambon JC, Scheutz C, Binning PJ, Laier T, Bjerg PL, Jacobsen CS (2013) A conceptual model linking functional gene expression and reductive dechlorination rates of chlorinated ethenes in clay rich groundwater sediment. Water Res 47(7):2467–2478. doi:10.1016/j.watres.2013.02.016

Banerjee R, Ragsdale SW (2003) The many faces of vitamin B12: catalysis by cobalamin-dependent enzymes. Annu Rev Biochem 72:209–247

Behrens S, Azizian MF, McMurdie PJ, Sabalowsky A, Dolan ME, Semprini L, Spormann AM (2008) Monitoring abundance and expression of "*Dehalococcoides*" species chloroethene-reductive dehalogenases in a tetrachloroethene-dechlorinating flow column. Appl Environ Microbiol 74(18):5695–5703. doi:10.1128/AEM.00926-08

Bisaillon A, Beaudet R, Lépine F, Deziel E, Villemur R (2010) Identification and characterization of a novel CprA reductive dehalogenase specific to highly chlorinated phenols from *Desulfitobacterium hafniense* strain PCP-1. Appl Environ Microbiol 76(22):7536–7540. doi:10.1128/AEM.01362-10

Bisaillon A, Beaudet R, Lépine F, Villemur R (2011) Quantitative analysis of the relative transcript levels of four chlorophenol reductive dehalogenase genes in *Desulfitobacterium hafniense* PCP-1 exposed to chlorophenols. Appl Environ Microbiol 77(17):6261–6264. doi:10.1128/AEM.00390-11

Borths EL, Locher KP, Lee AT, Rees DC (2002) The structure of *Escherichia coli* BtuF and binding to its cognate ATP binding cassette transporter. Proc Natl Acad Sci USA 99(26):16642–16647. doi:10.1073/pnas.262659699

Boyer A, Pagé-Bélanger R, Saucier M, Villemur R, Lépine F, Juteau P, Beaudet R (2003) Purification, cloning and sequencing of an enzyme mediating the reductive dechlorination of 2,4,6-trichlorophenol from *Desulfitobacterium frappieri* PCP-1. Biochem J 373(1):297–303

Cheng D, He JZ (2009) Isolation and characterization of "*Dehalococcoides*" sp. strain MB, which dechlorinates tetrachloroethene to *trans*-1,2-dichloroethene. Appl Environ Microbiol 75(18):5910–5918. doi:10.1128/Aem.00767-09

Choudhary PK, Duret A, Rohrbach-Brandt E, Holliger C, Sigel RKO, Maillard J (2013) Diversity of cobalamin riboswitches in the corrinoid-producing organohalide respirer *Desulfitobacterium hafniense*. J Bacteriol 195(22):5186–5195. doi:10.1128/jb.00730-13

Chow WL, Cheng D, Wang S, He J (2010) Identification and transcriptional analysis of *trans*-DCE-producing reductive dehalogenases in *Dehalococcoides* species. ISME J 4(8):1020–1030. doi:10.1038/ismej.2010.27

Christiansen N, Ahring BK (1996) *Desulfitobacterium hafniense* sp. nov., an anaerobic, reductively dechlorinating bacterium. Int J Syst Bacteriol 46(2):442–448

Christiansen N, Ahring BK, Wohlfarth G, Diekert G (1998) Purification and characterization of the 3-chloro-4-hydroxy-phenylacetate reductive dehalogenase of *Desulfitobacterium hafniense*. FEBS Lett 436(2):159–162

Claas KR, Parrish JR, Maggio-Hall LA, Escalante-Semerena JC (2010) Functional analysis of the nicotinate mononucleotide: 5,6-dimethylbenzimidazole phosphoribosyltransferase (CobT) enzyme, involved in the late steps of coenzyme B12 biosynthesis in *Salmonella enterica*. J Bacteriol 192(1):145–154

Coates ARM, Shinnick TM, Ellis RJ (1993) Chaperonin nomenclature. Mol Microbiol 8(4):787

Cuypers H, Viebrock-Sambale A, Zumft WG (1992) NosR, a membrane-bound regulatory component necessary for expression of nitrous oxide reductase in denitrifying *Pseudomonas stutzeri*. J Bacteriol 174(16):5332–5339

de Crécy-Lagard V, Forouhar F, Brochier-Armanet C, Tong L, Hunt JF (2012) Comparative genomic analysis of the DUF71/COG2102 family predicts roles in diphthamide biosynthesis and B12 salvage. Biol Direct 7:32

Deshpande NP, Wong YK, Mansfield M, Wilkins MR, Lee M (2013) Genome sequence of *Dehalobacter* UNSWDHB, a chloroform-dechlorinating bacterium. Genome Announc 1(5):e00720–e00713. doi:10.1128/genomeA.00720-13

Dos Santos PC, Fang Z, Mason SW, Setubal JC, Dixon R (2012) Distribution of nitrogen fixation and nitrogenase-like sequences amongst microbial genomes. BMC Genomics 13(1):162

Duret A, Holliger C, Maillard J (2012) The physiological opportunism of *Desulfitobacterium hafniense* strain TCE1 towards organohalide respiration with tetrachloroethene. Appl Environ Microbiol 78(17):6121–6127. doi:10.1128/AEM.01221-12

Finn RD, Bateman A, Clements J, Coggill P, Eberhardt RY, Eddy SR, Heger A, Hetherington K, Holm L, Mistry J, Sonnhammer ELL, Tate J, Punta M (2014) Pfam: the protein families database. Nucleic Acids Res 42(D1):D222–D230. doi:10.1093/nar/gkt1223

Finneran KT, Forbush HM, VanPraagh CV, Lovley DR (2002) *Desulfitobacterium metallireducens* sp. nov., an anaerobic bacterium that couples growth to the reduction of metals and humic acids as well as chlorinated compounds. Int J Syst Evol Microbiol 52(Pt 6):1929–1935

Fleischmann RD, Adams MD, White O, Clayton RA, Kirkness EF, Kerlavage AR, Bult CJ, Tomb JF, Dougherty BA, Merrick JM, McKenney K, Sutton G, FitzHugh W, Fields C, Gocayne JD, Scott J, Shirley R, Liu LI, Venter JC (1995) Whole-genome random sequencing and assembly of *Haemophilus influenzae* Rd. Science 269(5223):496–521

Fletcher KE, Costanza J, Cruz-Garcia C, Ramaswamy NS, Pennell KD, Löffler FE (2011) Effects of elevated temperature on *Dehalococcoides* dechlorination performance and DNA and RNA biomarker abundance. Environ Sci Technol 45(2):712–718. doi:10.1021/es1023477

Fung JM, Morris RM, Adrian L, Zinder SH (2007) Expression of reductive dehalogenase genes in *Dehalococcoides ethenogenes* strain 195 growing on tetrachloroethene, trichloroethene, or 2,3-dichlorophenol. Appl Environ Microbiol 73(14):4439–4445. doi:10.1128/AEM.00215-07

Futagami T, Goto M, Furukawa K (2008) Biochemical and genetic bases of dehalorespiration. Chem Rec 8(1):1–12

Futagami T, Goto M, Furukawa K (2014) Genetic systems of organohalide-respiring bacteria. In: Nojiri H, Tsuda M, Fukuda M, Kamagata Y (eds) Biodegradative bacteria. How bacteria degrade, survive, adapt and evolve. Springer, Tokyo, pp 59–81. doi:10.1007/978-4-431-54520-0

Futamata H, Kaiya S, Sugawara M, Hiraishi A (2009) Phylogenetic and transcriptional analyses of a tetrachloroethene-dechlorinating "*Dehalococcoides*" enrichment culture TUT2264 and its reductive-dehalogenase genes. Microbes Environ 24(4):330–337

Gábor K, Verissimo CS, Cyran BC, Ter Horst P, Meijer NP, Smidt H, de Vos WM, van der Oost J (2006) Characterization of CprK1, a CRP/FNR-type transcriptional regulator of halorespiration from *Desulfitobacterium hafniense*. J Bacteriol 188(7):2604–2613. doi:10.1128/JB.188.7.2604-2613.2006

Gábor K, Hailesellasse Sene K, Smidt H, de Vos WM, van der Oost J (2008) Divergent roles of CprK paralogues from *Desulfitobacterium hafniense* in activating gene expression. Microbiology 154(Pt 12):3686–3696. doi:10.1099/mic.0.2008/021584-0

Gao R, Stock AM (2009) Biological insights from structures of two-component proteins. Annu Rev Microbiol 63:133–154

Gerritse J, Drzyzga O, Kloetstra G, Keijmel M, Wiersum LP, Hutson R, Collins MD, Gottschal JC (1999) Influence of different electron donors and acceptors on dehalorespiration of tetrachloroethene by *Desulfitobacterium frappieri* TCE1. Appl Environ Microbiol 65(12):5212–5221

Goris T, Schubert T, Gadkari J, Wubet T, Tarkka M, Buscot F, Adrian L, Diekert G (2014) Insights into organohalide respiration and the versatile catabolism of *Sulfurospirillum multivorans* gained from comparative genomics and physiological studies. Environ Microbiol 16(11):3562–3580. doi:10.1111/1462-2920.12589

Goris T, Hornung B, Kruse T, Reinhold A, Westermann M, Schaap PJ, Smidt H, Diekert G (2015) Draft genome sequence and characterization of *Desulfitobacterium hafniense* PCE-S. Stand Genomic Sci 10:15. doi:10.1186/1944-3277-10-15

Gray MJ, Tavares NK, Escalante-Semerena JC (2008) The genome of *Rhodobacter sphaeroides* strain 2.4.1 encodes functional cobinamide salvaging systems of archaeal and bacterial origins. Mol Microbiol 70(4):824–836

Gray MJ, Escalante-Semerena JC (2009) The cobinamide amidohydrolase (cobyric acid-forming) CbiZ enzyme: a critical activity of the cobamide remodelling system of *Rhodobacter sphaeroides*. Mol Microbiol 74(5):1198–1210. doi:10.1111/j.1365-2958.2009.06928.x

Gribble GW (1998) Naturally occurring organohalogen compounds. Acc Chem Res 31(3):141–152

Grostern A, Edwards EA (2009) Characterization of a *Dehalobacter* coculture that dechlorinates 1,2-dichloroethane to ethene and identification of the putative reductive dehalogenase gene. Appl Environ Microbiol 75(9):2684–2693. doi:10.1128/AEM.02037-08

Horwich AL, Fenton WA, Chapman E, Farr GW (2007) Two families of chaperonin: Physiology and mechanism. Annu Rev Cell Dev Biol 23:115–145

Hug LA, Salehi M, Nuin P, Tillier ER, Edwards EA (2011) Design and verification of a pangenome microarray oligonucleotide probe set for *Dehalococcoides* spp. Appl Environ Microbiol 77(15):5361–5369. doi:10.1128/AEM.00063-11

Hug LA, Maphosa F, Leys D, Löffler FE, Smidt H, Edwards EA, Adrian L (2013) Overview of organohalide-respiring bacteria and a proposal for a classification system for reductive dehalogenases. Philos Trans R Soc Lond B Biol Sci 368(1616):20120322. doi:10.1098/rstb.2012.0322

Islam MA, Waller AS, Hug LA, Provart NJ, Edwards EA, Mahadevan R (2014) New insights into *Dehalococcoides mccartyi* metabolism from a reconstructed metabolic network-based systems-level analysis of *D. mccartyi* transcriptomes. PLoS ONE 9(4):e94808. doi:10.1371/journal.pone.0094808

John M, Rubick R, Schmitz RP, Rakoczy J, Schubert T, Diekert G (2009) Retentive memory of bacteria: long-term regulation of dehalorespiration in *Sulfurospirillum multivorans*. J Bacteriol 191(5):1650–1655. doi:10.1128/JB.00597-08

Johnson DR, Lee PK, Holmes VF, Fortin AC, Alvarez-Cohen L (2005) Transcriptional expression of the *tceA* gene in a *Dehalococcoides*-containing microbial enrichment. Appl Environ Microbiol 71(11):7145–7151. doi:10.1128/AEM.71.11.7145-7151.2005

Johnson DR, Brodie EL, Hubbard AE, Andersen GL, Zinder SH, Alvarez-Cohen L (2008) Temporal transcriptomic microarray analysis of "*Dehalococcoides ethenogenes*" strain 195 during the transition into stationary phase. Appl Environ Microbiol 74(9):2864–2872

Johnson DR, Nemir A, Andersen GL, Zinder SH, Alvarez-Cohen L (2009) Transcriptomic microarray analysis of corrinoid responsive genes in *Dehalococcoides ethenogenes* strain 195. FEMS Microbiol Lett 294(2):198–206

Joyce MG, Levy C, Gábor K, Pop SM, Biehl BD, Doukov TI, Ryter JM, Mazon H, Smidt H, van den Heuvel RH, Ragsdale SW, van der Oost J, Leys D (2006) CprK crystal structures reveal mechanism for transcriptional control of halorespiration. J Biol Chem 281(38):28318–28325. doi:10.1074/jbc.M602654200

Justicia-Leon SD, Ritalahti KM, Mack EE, Löffler FE (2012) Dichloromethane fermentation by a *Dehalobacter* sp. in an enrichment culture derived from pristine river sediment. Appl Environ Microbiol 78(4):1288–1291. doi:10.1128/AEM.07325-11

Keller S, Ruetz M, Kunze C, Kräutler B, Diekert G, Schubert T (2013) Exogenous 5,6-dimethylbenzimidazole caused production of a non-functional tetrachloroethene reductive dehalogenase in *Sulfurospirillum multivorans*. Environ Microbiol 16(11):3361–3369. doi:10.1111/1462-2920.12268

Kemp LR, Dunstan MS, Fisher K, Warwicker J, Leys D (2013) The transcriptional regulator CprK detects chlorination by combining direct and indirect readout mechanisms. Philos Trans R Soc Lond B Biol Sci 368(1616):20120323. doi:10.1098/rstb.2012.0323

Kim S-H, Harzman C, Davis J, Hutcheson R, Broderick J, Marsh T, Tiedje J (2012) Genome sequence of *Desulfitobacterium hafniense* DCB-2, a Gram-positive anaerobe capable of dehalogenation and metal reduction. BMC Microbiol 12(1):21. doi:10.1186/1471-2180-12-21

Kim YE, Hipp MS, Bracher A, Hayer-Hartl M, Ulrich Hartl F (2013) Molecular chaperone functions in protein folding and proteostasis. Annu Rev Biochem 82:323–355

Körner H, Sofia HJ, Zumft WG (2003) Phylogeny of the bacterial superfamily of Crp-Fnr transcription regulators: exploiting the metabolic spectrum by controlling alternative gene programs. FEMS Microbiol Rev 27(5):559–592

Krajmalnik-Brown R, Hölscher T, Thomson IN, Saunders FM, Ritalahti KM, Löffler FE (2004) Genetic identification of a putative vinyl chloride reductase in *Dehalococcoides* sp. strain BAV1. Appl Environ Microbiol 70(10):6347–6351. doi:10.1128/AEM.70.10.6347-6351.2004

Kranzioch I, Ganz S, Tiehm A (2014) Chloroethene degradation and expression of *Dehalococcoides* dehalogenase genes in cultures originating from Yangtze sediments. Environ Sci Pollut Res Int 20(10):7046–7056. doi:10.1007/s11356-014-3574-4

Kräutler B, Fieber W, Ostermann S, Fasching M, Ongania KH, Gruber K, Kratky C, Mikl C (2003) The cofactor of tetrachloroethene reductive dehalogenase of *Dehalospirillum multivorans* is norpseudo-B-12, a new type of a natural corrinoid. Helv Chim Acta 86(11):3698–3716

Kruse T, Maillard J, Holliger C, Goodwin L, Woyke T, Teshima H, Bruce D, Detter C, Tapia R, Han C, Huntemann M, Wei C-L, Han J, Chen A, Kyrpides N, Szeto E, Markowitz V, Ivanova N, Pagani I, Pati A, Pitluck S, Nolan M, Holliger C, Smidt H (2013) Complete genome sequence of *Dehalobacter restrictus* PER-K23. Stand Genomic Sci 8(3):375–388

Kruse T, van de Pas BA, Atteia A, Krab K, Hagen WR, Goodwin L, Chain P, Boeren S, Maphosa F, Schraa G, de Vos WM, van der Oost J, Smidt H, Stams AJ (2015) Genomic, proteomic and biochemical analysis of the organohalide respiratory pathway in *Desulfitobacterium dehalogenans*. J Bacteriol 197(5):893–904. doi:10.1128/JB.02370-14

Kube M, Beck A, Zinder SH, Kuhl H, Reinhardt R, Adrian L (2005) Genome sequence of the chlorinated compound-respiring bacterium *Dehalococcoides* species strain CBDB1. Nat Biotechnol 23(10):1269–1273. doi:10.1038/nbt1131

Lee PK, Johnson DR, Holmes VF, He J, Alvarez-Cohen L (2006) Reductive dehalogenase gene expression as a biomarker for physiological activity of *Dehalococcoides* spp. Appl Environ Microbiol 72(9):6161–6168. doi:10.1128/AEM.01070-06

Lee PK, Macbeth TW, Sorenson KS Jr, Deeb RA, Alvarez-Cohen L (2008) Quantifying genes and transcripts to assess the in situ physiology of "*Dehalococcoides*" spp. in a trichloroethene-contaminated groundwater site. Appl Environ Microbiol 74(9):2728–2739. doi:10.1128/AEM.02199-07

Lee PKH, Cheng D, Hu P, West KA, Dick GJ, Brodie EL, Andersen GL, Zinder SH, He J, Alvarez-Cohen L (2011) Comparative genomics of two newly isolated *Dehalococcoides* strains and an enrichment using a genus microarray. ISME J 5(6):1014–1024. doi:10.1038/ismej.2010.202

Lee M, Low A, Zemb O, Koenig J, Michaelsen A, Manefield M (2012a) Complete chloroform dechlorination by organochlorine respiration and fermentation. Environ Microbiol 14(4):883–894. doi:10.1111/j.1462-2920.2011.02656.x

Lee PK, Dill BD, Louie TS, Shah M, Verberkmoes NC, Andersen GL, Zinder SH, Alvarez-Cohen L (2012b) Global transcriptomic and proteomic responses of *Dehalococcoides ethenogenes* strain 195 to fixed nitrogen limitation. Appl Environ Microbiol 78(5):1424–1436. doi:10.1128/AEM.06792-11

Levy C, Pike K, Heyes DJ, Joyce MG, Gábor K, Smidt H, van der Oost J, Leys D (2008) Molecular basis of halorespiration control by CprK, a CRP-FNR type transcriptional regulator. Mol Microbiol 70(1):151–167. doi:10.1111/j.1365-2958.2008.06399.x

Löffler FE, Yan J, Ritalahti KM, Adrian L, Edwards EA, Konstantinidis KT, Müller JA, Fullerton H, Zinder SH, Spormann AM (2013) *Dehalococcoides mccartyi* gen. nov., sp. nov., obligately organohalide-respiring anaerobic bacteria relevant to halogen cycling and bioremediation, belong to a novel bacterial class, *Dehalococcoidia* classis nov., order *Dehalococcoidales* ord. nov. and family *Dehalococcoidaceae* fam. nov., within the phylum *Chloroflexi*. Int J Syst Evol Microbiol 63(PART2):625–635

Lohner ST, Spormann AM (2013) Identification of a reductive tetrachloroethene dehalogenase in *Shewanella sediminis*. Philos Trans R Soc Lond B Biol Sci 368(1616):20120326. doi:10.1098/rstb.2012.0326

Lund PA (2009) Multiple chaperonins in bacteria—why so many? FEMS Microbiol Rev 33(4):785–800

Mac Nelly A, Kai M, Svatoš A, Diekert G, Schubert T (2014) Functional heterologous production of reductive dehalogenases from *Desulfitobacterium hafniense* strains. Appl Environ Microbiol 80(14):4313–4322. doi:10.1128/aem.00881-14

Magnuson JK, Romine MF, Burris DR, Kingsley MT (2000) Trichloroethene reductive dehalogenase from *Dehalococcoides ethenogenes*: sequence of *tceA* and substrate range characterization. Appl Environ Microbiol 66(12):5141–5147

Maillard J, Regeard C, Holliger C (2005) Isolation and characterization of Tn-Dha1, a transposon containing the tetrachloroethene reductive dehalogenase of *Desulfitobacterium hafniense* strain TCE1. Environ Microbiol 7(1):107–117

Maillard J, Genevaux P, Holliger C (2011) Redundancy and specificity of multiple trigger factor chaperones in Desulfitobacteria. Microbiology 157(8):2410–2421

Mansfeldt CB, Rowe AR, Heavner GL, Zinder SH, Richardson RE (2014) Meta-analyses of *Dehalococcoides mccartyi* strain 195 transcriptomic profiles identify a respiration rate-related gene expression transition point and interoperon recruitment of a key oxidoreductase subunit. Appl Environ Microbiol 80(19):6062–6072. doi:10.1128/AEM.02130-14

Maphosa F (2010) Chasing organohalide respirers: ecogenomics approaches to assess the bioremediation capacity of soils. PhD thesis, Wageningen University, Wageningen

Maphosa F, Van Passel MWJ, De Vos WM, Smidt H (2012) Metagenome analysis reveals yet unexplored reductive dechlorinating potential of *Dehalobacter* sp. E1 growing in co-culture with *Sedimentibacter* sp. Environ Microbiol Rep 4(6):604–616. doi:10.1111/j.1758-2229.2012.00376.x

Mascher T, Helmann JD, Unden G (2006) Stimulus perception in bacterial signal-transducing histidine kinases. Microbiol Mol Biol Rev 70(4):910–938

Mazon H, Gábor K, Leys D, Heck AJ, van der Oost J, van den Heuvel RH (2007) Transcriptional activation by CprK1 is regulated by protein structural changes induced by effector binding and redox state. J Biol Chem 282(15):11281–11290. doi:10.1074/jbc.M611177200

McMurdie PJ, Behrens SF, Müller JA, Goke J, Ritalahti KM, Wagner R, Goltsman E, Lapidus A, Holmes S, Löffler FE, Spormann AM (2009) Localized plasticity in the streamlined genomes of vinyl chloride respiring *Dehalococcoides*. PLoS Genet 5(11):e1000714. doi:10.1371/journal.pgen.1000714

McMurdie PJ, Hug LA, Edwards EA, Holmes S, Spormann AM (2011) Site-specific mobilization of vinyl chloride respiration islands by a mechanism common in *Dehalococcoides*. BMC Genomics 12:287. doi:10.1186/1471-2164-12-287

Men Y, Feil H, Verberkmoes NC, Shah MB, Johnson DR, Lee PK, West KA, Zinder SH, Andersen GL, Alvarez-Cohen L (2012) Sustainable syntrophic growth of *Dehalococcoides ethenogenes* strain 195 with *Desulfovibrio vulgaris* Hildenborough and *Methanobacterium congolense*: global transcriptomic and proteomic analyses. ISME J 6(2):410–421. doi:10.1038/ismej.2011.111

Men Y, Lee PKH, Harding KC, Alvarez-Cohen L (2013) Characterization of four TCE-dechlorinating microbial enrichments grown with different cobalamin stress and methanogenic conditions. Appl Microbiol Biotechnol 97(14):6439–6450

Men Y, Seth EC, Yi S, Allen RH, Taga ME, Alvarez-Cohen L (2014) Sustainable growth of *Dehalococcoides mccartyi* 195 by corrinoid salvaging and remodeling in defined lactate-fermenting consortia. Appl Environ Microbiol 80(7):2133–2141

Moore SJ, Lawrence AD, Biedendieck R, Deery E, Frank S, Howard MJ, Rigby SEJ, Warren MJ (2013) Elucidation of the anaerobic pathway for the corrin component of cobalamin (vitamin B12). Proc Natl Acad Sci USA 110(37):14906–14911. doi:10.1073/pnas.1308098110

Morita Y, Futagami T, Goto M, Furukawa K (2009) Functional characterization of the trigger factor protein PceT of tetrachloroethene-dechlorinating *Desulfitobacterium hafniense* Y51. Appl Microbiol Biotechnol 83(4):775–781

Mukherjee K, Bowman KS, Rainey FA, Siddaramappa S, Challacombe JF, Moe WM (2014) *Dehalogenimonas lykanthroporepellens* BL-DC-9^T simultaneously transcribes many *rdhA* genes during organohalide respiration with 1,2-DCA, 1,2-DCP, and 1,2,3-TCP as electron acceptors. FEMS Microbiol Lett 354(2):111–118. doi:10.1111/1574-6968.12434

Neumann A, Wohlfarth G, Diekert G (1998) Tetrachloroethene dehalogenase from *Dehalospirillum multivorans*: cloning, sequencing of the encoding genes, and expression of the *pceA* gene in *Escherichia coli*. J Bacteriol 180(16):4140–4145

Nonaka H, Keresztes G, Shinoda Y, Ikenaga Y, Abe M, Naito K, Inatomi K, Furukawa K, Inui M, Yukawa H (2006) Complete genome sequence of the dehalorespiring bacterium *Desulfitobacterium hafniense* Y51 and comparison with *Dehalococcoides ethenogenes* 195. J Bacteriol 188(6):2262–2274

Padilla-Crespo E, Yan J, Swift C, Wagner DD, Chourey K, Hettich RL, Ritalahti KM, Löffler FE (2014) Identification and environmental distribution of *dcpA*, which encodes the reductive dehalogenase catalyzing the dichloroelimination of 1,2-dichloropropane to propene in organohalide-respiring *Chloroflexi*. Appl Environ Microbiol 80(3):808–818. doi:10.1128/AEM.02927-13

Pagani I, Liolios K, Jansson J, Chen IMA, Smirnova T, Nosrat B, Markowitz VM, Kyrpides NC (2012) The Genomes OnLine Database (GOLD) v. 4: status of genomic and metagenomic projects and their associated metadata. Nucleic Acids Res 40(D1):D571–D579

Palmer T, Berks BC (2012) The twin-arginine translocation (Tat) protein export pathway. Nature Rev Microbiol 10(7):483–496

Peng X, Yamamoto S, Vertes AA, Keresztes G, Inatomi K, Inui M, Yukawa H (2012) Global transcriptome analysis of the tetrachloroethene-dechlorinating bacterium *Desulfitobacterium hafniense* Y51 in the presence of various electron donors and terminal electron acceptors. J Ind Microbiol Biotechnol 39(2):255–268. doi:10.1007/s10295-011-1023-7

Pop SM, Kolarik RJ, Ragsdale SW (2004) Regulation of anaerobic dehalorespiration by the transcriptional activator CprK. J Biol Chem 279(48):49910–49918. doi:10.1074/jbc.M409435200

Pöritz M, Goris T, Wubet T, Tarkka MT, Buscot F, Nijenhuis I, Lechner U, Adrian L (2013) Genome sequences of two dehalogenation specialists—*Dehalococcoides mccartyi* strains BTF08 and DCMB5 enriched from the highly polluted Bitterfeld region. FEMS Microbiol Lett 343(2):101–104. doi:10.1111/1574-6968.12160

Prat L (2009) Identification and characterization of proteins supporting dehalorespiration in *Desulfitobacterium hafniense* strain TCE1. Ph.D. thesis, École Polytechnique Fédérale de Lausanne, Lausanne

Prat L, Maillard J, Grimaud R, Holliger C (2011) Physiological adaptation of *Desulfitobacterium hafniense* strain TCE1 to tetrachloroethene respiration. Appl Environ Microbiol 77(11):3853–3859

Rahm BG, Richardson RE (2008) Correlation of respiratory gene expression levels and pseudo-steady-state PCE respiration rates in *Dehalococcoides ethenogenes*. Environ Sci Technol 42(2):416–421

Rahm BG, Morris RM, Richardson RE (2006) Temporal expression of respiratory genes in an enrichment culture containing *Dehalococcoides ethenogenes*. Appl Environ Microbiol 72(8):5486–5491. doi:10.1128/AEM.00855-06

Regeard C, Maillard J, Dufraigne C, Deschavanne P, Holliger C (2005) Indications for acquisition of reductive dehalogenase genes through horizontal gene transfer by *Dehalococcoides ethenogenes* strain 195. Appl Environ Microbiol 71(6):2955–2961

Reinhold A, Westermann M, Seifert J, von Bergen M, Schubert T, Diekert G (2012) Impact of Vitamin B12 on formation of the tetrachloroethene reductive dehalogenase in *Desulfitobacterium hafniense* strain Y51. Appl Environ Microbiol 78(22):8025–8032

Rowe AR, Heavner GL, Mansfeldt CB, Werner JJ, Richardson RE (2012) Relating chloroethene respiration rates in *Dehalococcoides* to protein and mRNA biomarkers. Environ Sci Technol 46(17):9388–9397. doi:10.1021/es300996c

Rowe AR, Mansfeldt CB, Heavner GL, Richardson RE (2015) Relating mRNA and protein biomarker levels in a *Dehalococcoides* and *Methanospirillum*-containing community. Appl Microbiol Biotechnol 99(5):2313–2327. doi:10.1007/s00253-014-6220-7

Rupakula A, Kruse T, Boeren S, Holliger C, Smidt H, Maillard J (2013) The restricted metabolism of the obligate organohalide respiring bacterium *Dehalobacter restrictus*: lessons from tiered functional genomics. Philos Trans R Soc Lond B Biol Sci 368(1616):20120325. doi:10.1098/rstb.2012.0325

Rupakula A, Lu Y, Kruse T, Boeren S, Holliger C, Smidt H, Maillard J (2015) Functional genomics of corrinoid starvation in the organohalide-respiring bacterium *Dehalobacter restrictus* strain PER-K23. Front Microbiol 5:751. doi:10.3389/fmicb.2014.00751

Saunders NFW, Houben ENG, Koefoed S, De Weert S, Reijnders WNM, Westerhoff HV, De Boer APN, Van Spanning RJM (1999) Transcription regulation of the *nir* gene cluster encoding nitrite reductase of *Paracoccus denitrificans* involves NNR and NirI, a novel type of membrane protein. Mol Microbiol 34(1):24–36

Seshadri R, Adrian L, Fouts DE, Eisen JA, Phillippy AM, Methe BA, Ward NL, Nelson WC, Deboy RT, Khouri HM, Kolonay JF, Dodson RJ, Daugherty SC, Brinkac LM, Sullivan SA, Madupu R, Nelson KT, Kang KH, Impraim M, Tran K, Robinson JM, Forberger HA, Fraser CM, Zinder SH, Heidelberg JF (2005) Genome sequence of the PCE-dechlorinating bacterium *Dehalococcoides ethenogenes*. Science 307(5706):105–108

Siddaramappa S, Challacombe JF, Delano SF, Green LD, Daligault H, Bruce D, Detter C, Tapia R, Han S, Goodwin L, Han J, Woyke T, Pitluck S, Pennacchio L, Nolan M, Land M, Chang YJ, Kyrpides NC, Ovchinnikova G, Hauser L, Lapidus A, Yan J, Bowman KS, da Costa MS, Rainey FA, Moe WM (2012) Complete genome sequence of *Dehalogenimonas lykanthroporepellens* type strain (BL-DC-9T) and comparison to "*Dehalococcoides*" strains. Stand Genomic Sci 6(2):251–264. doi:10.4056/sigs.2806097

Simon J, Klotz MG (2013) Diversity and evolution of bioenergetic systems involved in microbial nitrogen compound transformations. Biochim Biophys Acta 1827(2):114–135. doi:10.1016/j.bbabio.2012.07.005

Smidt H, van Leest M, van der Oost J, de Vos WM (2000) Transcriptional regulation of the *cpr* gene cluster in *ortho*-chlorophenol-respiring *Desulfitobacterium dehalogenans*. J Bacteriol 182(20):5683–5691

Suyama A, Yamashita M, Yoshino S, Furukawa K (2002) Molecular characterization of the PceA reductive dehalogenase of *Desulfitobacterium* sp. strain Y51. J Bacteriol 184(13):3419–3425

Suzuki D, Irieda H, Homma M, Kawagishi I, Sudo Y (2010) Phototactic and chemotactic signal transduction by transmembrane receptors and transducers in microorganisms. Sensors 10(4):4010–4039

Tang YJ, Yi S, Zhuang W-Q, Zinder SH, Keasling JD, Alvarez-Cohen L (2009) Investigation of carbon metabolism in "*Dehalococcoides ethenogenes*" strain 195 by use of isotopomer and transcriptomic analyses. J Bacteriol 191(16):5224–5231

Tang S, Gong Y, Edwards EA (2012) Semi-automatic *in silico* gap closure enabled de novo assembly of two *Dehalobacter* genomes from metagenomic data. PLoS ONE 7(12):e52038

Thibodeau J, Gauthier A, Duguay M, Villemur R, Lépine F, Juteau P, Beaudet R (2004) Purification, cloning, and sequencing of a 3,5-dichlorophenol reductive dehalogenase from *Desulfitobacterium frappieri* PCP-1. Appl Environ Microbiol 70(8):4532–4537. doi:10.1128/AEM.70.8.4532-4537.2004

Thirumalai D, Lorimer GH (2001) Chaperonin-mediated protein folding. Annu Rev Biophys Biomol Struct 30:245–269

Thomas SH, Wagner RD, Arakaki AK, Skolnick J, Kirby JR, Shimkets LJ, Sanford RA, Löffler FE (2008) The mosaic genome of *Anaeromyxobacter dehalogenans* strain 2CP-C suggests an aerobic common ancestor to the delta-proteobacteria. PLoS ONE 3(5):e2103. doi:10.1371/journal.pone.0002103

Trzebiatowski JR, O'Toole GA, Escalante-Semerena JC (1994) The *cobT* gene of *Salmonella typhimurium* encodes the NaMN: 5,6-dimethylbenzimidazole phosphoribosyltransferase responsible for the synthesis of N1-(5-phospho-α-D-ribosyl)-5,6-dimethylbenzimidazole, an intermediate in the synthesis of the nucleotide loop of cobalamin. J Bacteriol 176(12):3568–3575

van de Pas BA, Smidt H, Hagen WR, van der Oost J, Schraa G, Stams AJ, de Vos WM (1999) Purification and molecular characterization of *ortho*-chlorophenol reductive dehalogenase, a key enzyme of halorespiration in *Desulfitobacterium dehalogenans*. J Biol Chem 274(29):20287–22092

Villemur R, Saucier M, Gauthier A, Beaudet R (2002) Occurrence of several genes encoding putative reductive dehalogenases in *Desulfitobacterium hafniense/frappieri* and *Dehalococcoides ethenogenes*. Can J Microbiol 48(8):697–706

Villemur R, Lanthier M, Beaudet R, Lepine F (2006) The *Desulfitobacterium* genus. FEMS Microbiol Rev 30(5):706–733

Wadhams GH, Armitage JP (2004) Making sense of it all: bacterial chemotaxis. Nature Rev Mol Cell Biol 5(12):1024–1037

Wagner A, Adrian L, Kleinsteuber S, Andreesen JR, Lechner U (2009) Transcription analysis of genes encoding homologues of reductive dehalogenases in "*Dehalococcoides*" sp. strain CBDB1 by using terminal restriction fragment length polymorphism and quantitative PCR. Appl Environ Microbiol 75(7):1876–1884

Wagner DD, Hug LA, Hatt JK, Spitzmiller MR, Padilla-Crespo E, Ritalahti KM, Edwards EA, Konstantinidis KT, Löffler FE (2012) Genomic determinants of organohalide-respiration in *Geobacter lovleyi*, an unusual member of the *Geobacteraceae*. BMC Genomics 13(1):200. doi:10.1186/1471-2164-13-200

Wagner A, Segler L, Kleinsteuber S, Sawers G, Smidt H, Lechner U (2013) Regulation of reductive dehalogenase gene transcription in *Dehalococcoides mccartyi*. Philos Trans R Soc Lond B Biol Sci 368(1616):20120317. doi:10.1098/rstb.2012.0317

Waller AS, Krajmalnik-Brown R, Löffler FE, Edwards EA (2005) Multiple reductive-dehalogenase-homologous genes are simultaneously transcribed during dechlorination by *Dehalococcoides*-containing cultures. Appl Environ Microbiol 71(12):8257–8264. doi:10.1128/Aem.71.12.8257-8264.2005

Waller AS, Hug LA, Mo K, Radford DR, Maxwell KL, Edwards EA (2012) Transcriptional analysis of a *Dehalococcoides*-containing microbial consortium reveals prophage activation. Appl Environ Microbiol 78(4):1178–1186. doi:10.1128/AEM.06416-11

Wang S, Chng KR, Wilm A, Zhao S, Yang KL, Nagarajan N, He J (2014a) Genomic characterization of three unique *Dehalococcoides* that respire on persistent polychlorinated biphenyls. Proc Natl Acad Sci USA 111(33):12103–12108

Wang S, Chng KR, Wu C, Wilm A, Nagarajan N, He J (2014b) Draft genome sequence of polychlorinated biphenyl-dechlorinating *Dehalococcoides mccartyi* strain SG1, which carries a circular putative plasmid. Genome Announc 2(5):e00901–e00914. doi:10.1128/genomeA.00901-14

West KA, Johnson DR, Hu P, DeSantis TZ, Brodie EL, Lee PK, Feil H, Andersen GL, Zinder SH, Alvarez-Cohen L (2008) Comparative genomics of "*Dehalococcoides ethenogenes*" 195 and an enrichment culture containing unsequenced "*Dehalococcoides*" strains. Appl Environ Microbiol 74(11):3533–3540. doi:10.1128/AEM.01835-07

West KA, Lee PK, Johnson DR, Zinder SH, Alvarez-Cohen L (2013) Global gene expression of *Dehalococcoides* within a robust dynamic TCE-dechlorinating community under conditions of periodic substrate supply. Biotechnol Bioeng 110(5):1333–1341. doi:10.1002/bit.24819

Wilkinson SP, Grove A (2006) Ligand-responsive transcriptional regulation by members of the MarR family of winged helix proteins. Curr Issues Mol Biol 8(1):51–62

Woodson JD, Escalante-Semerena JC (2004) CbiZ, an amidohydrolase enzyme required for salvaging the coenzyme B12 precursor cobinamide in archaea. Proc Natl Acad Sci USA 101(10):3591–3596

Wunsch P, Zumft WG (2005) Functional domains of NosR, a novel transmembrane iron-sulfur flavoprotein necessary for nitrous oxide respiration. J Bacteriol 187(6):1992–2001

Yan J, Ritalahti KM, Wagner DD, Löffler FE (2012) Unexpected specificity of interspecies cobamide transfer from *Geobacter* spp. to organohalide-respiring *Dehalococcoides mccartyi* strains. Appl Environ Microbiol 78(18):6630–6636

Yan J, Im J, Yang Y, Löffler FE (2013) Guided cobalamin biosynthesis supports *Dehalococcoides mccartyi* reductive dechlorination activity. Philos Trans R Soc Lond B Biol Sci 368(1616):20120321. doi:10.1098/rstb.2012.0320

Yi S, Seth EC, Men YJ, Stabler SP, Allen RH, Alvarez-Cohen L, Taga ME (2012) Versatility in corrinoid salvaging and remodeling pathways supports corrinoid-dependent metabolism in *Dehalococcoides mccartyi*. Appl Environ Microbiol 78(21):7745–7752

Zhang Y, Rodionov D, Gelfand M, Gladyshev V (2009) Comparative genomic analyses of nickel, cobalt and vitamin B12 utilization. BMC Genomics 10(1):78

Chapter 16
Diversity, Evolution, and Environmental Distribution of Reductive Dehalogenase Genes

Laura A. Hug

Abstract Reductive dehalogenases, the active enzymes in organohalide respiration, are a diverse protein family with low sequence similarity and a punctuated distribution across the tree of life. The diversity, environmental distribution, and evolution of the reductive dehalogenases remain open questions. This chapter describes reductive dehalogenase sequence similarity and domain architecture, highlighting why predicting substrate specificity from sequence similarity is unreliable for these enzymes. Common in contaminated environments, but also identified in soda lakes, ocean sediment, and even as part of the human microbiome, the global distribution of reductive dehalogenases is broad and continually expanding. A map view of current locations where reductive dehalogenases have been detected provides compelling evidence for the ubiquity of these proteins in the environment, in keeping with predictions of an ancient origin for the group. The evolutionary history of the reductive dehalogenases includes vertical inheritance alongside a myriad of mechanisms for lateral transfer, including integrases, circularizing transposable elements, and, possibly, phage-mediated transfer. The reductive dehalogenases remain incompletely characterized from the perspectives of sequence diversity, substrate specificities, global distribution, and modes of inheritance.

16.1 Introduction

The first reductive dehalogenase was purified in 1995 (Ni et al. 1995), with the first gene sequence determined in 1998 (Neumann et al. 1998). Since these initial studies, the number, sequence diversity, and taxonomic distribution of reductive dehalogenase homologous sequences (*rdhs*) have all increased substantially.

L.A. Hug (✉)
Department of Biology, University of Waterloo, Waterloo, ON, Canada
e-mail: laura.hug@uwaterloo.ca

L. Adrian and F.E. Löffler (eds.), *Organohalide-Respiring Bacteria*,
DOI 10.1007/978-3-662-49875-0_16

Reductive dehalogenase genes comprise an operon containing *rdhA*, the gene for the enzyme, *rdhB*, a gene encoding a predicted membrane-integral protein putatively anchoring the RdhA subunit to the membrane, and other associated genes (see Chap. 15 for a full discussion of the operonic structures and functions therein). RdhAs are the critical proteins in organohalide respiration, a metabolism that is now recognized as a significant component of the global halogen cycle (Krzmarzick et al. 2012) and routinely leveraged as an industrial tool for bioremediation (Peale et al. 2008; Maphosa et al. 2012).

The advent of high-throughput sequencing has contributed to the rapid increase in the number of unique *rdh* genes identified, in conjunction with better resolution of the global distribution of this gene family. *rdh* gene sequence diversity has increased with each novel genome, metagenome, and environmental screen conducted; any existing constraints on the gene family's diversity and environmental distribution remain unknown. As novel genome sequences populate the organohalide-respiring clades on the bacterial tree of life, the origin and mode of inheritance of *rdh* genes remain unclear. This chapter aims to explore these open questions, and situate them within the current knowledge space. First the sequence diversity of the *rdhA* and *rdhB* gene families is described, with an examination of RdhA domain structures, a comparison of the RdhA and RdhB families to the sequence diversity of other defined protein families, and a summary of a recently proposed classification system for RdhAs that leverages this sequence diversity. Next, the environmental distribution of *rdhA* genes is examined, summarizing the contaminated sites, human-designed bioreactors and wastewater treatment plants, and pristine locations from which *rdhA* genes have been identified, describing the distribution of these genes across the globe. Finally, the evolution of organohalide respiration is explored, from the number and genomic context of *rdhA* genes within different phyla to the predicted modes of inheritance for these genes. The implications of the evolutionary history of this gene family are discussed in relation to using *rdhA* genes as biomarkers for remediation efforts.

16.2 Sequence Diversity

16.2.1 Protein Family Sequence Diversity and Domain Organization

Until very recently, no three-dimensional structure for a reductive dehalogenase had been determined, meaning that the definition of this enzyme family and assignment of novel sequences have been based on sequence similarity alone. As a result, the RdhA family is recognized by the HMM-based databases of homologous proteins PFAM and TIGRFAM (PF13486, TIGR02486), but is not included in PROSITE or SMART, which require structural information. The PFAM and TIGRFAM models are similar in length, spanning 308 and 314 amino acids respectively of what is typically a ~500 amino acid protein. The PFAM model

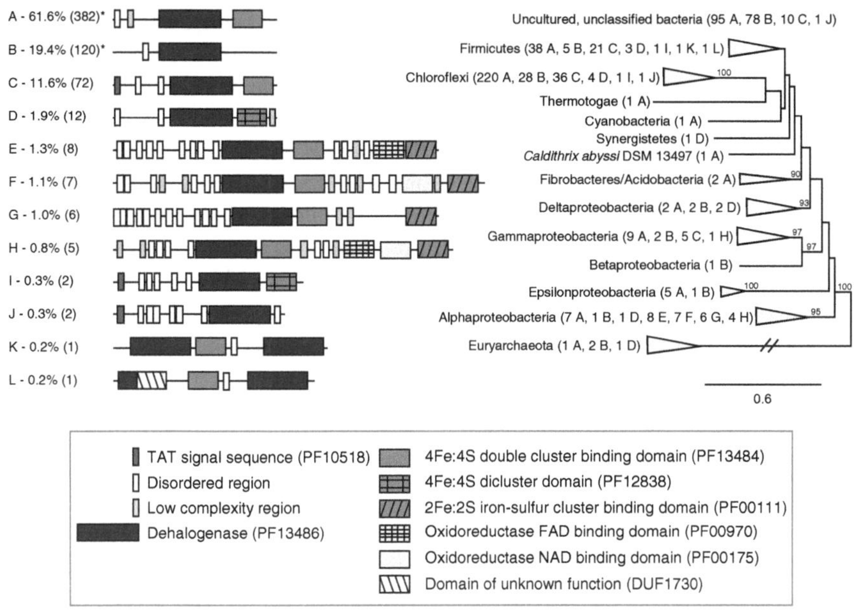

Fig. 16.1 RdhA domain organizations and their phylogenetic affiliations. The number of sequences for each domain architecture is noted at *left*. Protein architectures were adapted from the reductive dehalogenase PFAM PF13486 and are not to scale (pfam.sanger.ac.uk). The phylogeny was generated with the 16S rRNA genes of all organisms with an RdhA in PF13486. rRNA sequences were mined from NCBI (ncbi.nlm.nih.gov) and the JCVI (moore.jcvi.org) and aligned with Muscle (Edgar 2004). The tree was generated using PhyML under the HKY85 model of evolution with 100 bootstrap resamplings (Guindon and Gascuel 2003), with bootstrap support values above 50 % reported. *For architectures A and B, predicted TAT signal sequence domains are present on many full-length sequences, but were not annotated by PFAM based on low sequence similarity to the domain HMM. Similarly for architecture B, predicted 4Fe:4S double cluster binding domains are present on certain sequences but do not pass the PFAM domain identification scoring threshold

gives a relatively complete picture of the sequence diversity within this family. It was built with a seed alignment of 89 RdhAs, with a full dataset of 620 protein sequences (accessed pfam.sanger.ac.uk/family/PF13486 in June 2014). The average identity for the full alignment is 24 % at the amino acid level, with an average 54.6 % coverage of a given protein sequence by this model. Both the PFAM and TIGRFAM models cover only the N-terminal domain of RdhAs, and do not include C-terminal 4Fe–4S cluster binding domains or TAT signal sequences, features typically associated with RdhAs (Fig. 16.1). The C-terminal portion of the RdhA proteins is typically more highly conserved, containing a relatively tightly constrained 4Fe–4S double cluster binding domain (PF13484, average identity of 42 %). For an estimate of the full-length sequence divergence across RdhAs, a near-full-length alignment of a curated set of RdhA sequences, lacking only the N-terminal TAT signal sequence, had an average identity of 29.6 % (Hug et al. 2013b).

There is currently no PFAM for the RdhB protein family. The *rdhB* genes encode small proteins of ~90 amino acids with predicted trans-membrane regions, and can be located up or downstream of the *rdhA* gene. In an analysis of 337 RdhB sequences identified by keyword search in the NCBI database, the sequence identity for the full alignment was 29.8 %, ranging from 3.2 to 100 % pairwise identity. The level of sequence conservation across the RdhA and RdhB families is low, but is in-line with other described protein families: the top 20 largest PFAMs have an average identity of 35 % (range 15–86 %), while the haloacid dehalogenase PFAMs (PF12710, PF13419, PF00702, PF08282, PF13344) have an average identity of 22 % (range 16–28 %). The sequence divergence within the reductive dehalogenases is thus well within the expected range for a protein family.

From the 620 proteins included in the PFAM RdhA protein family, 12 different domain architectures have been described from twelve bacterial and one archaeal phyla (Fig. 16.1). The dominant architecture (61.6 % of all sequences) contains an N-terminal unstructured region, followed by a region of low complexity, the reductive dehalogenase domain, and the 4Fe–4S cluster binding domain. The four least common architectures comprise only 1 % of the RdhA-containing proteins (I-L in Fig. 16.1), and are found in genes from known dechlorinating organisms within the genera *Dehalococcoides* and *Desulfitobacterium*, whose genomes contain multiple reductive dehalogenases (Seshadri et al. 2005; Kube et al. 2005; Nonaka et al. 2006; Maphosa et al. 2010). These rare conformations may represent functional variants for this enzyme class, or may belong to pseudogenes whose loss of activity was unimportant to organisms with multiple functional RdhAs remaining. The four structural variants containing a C-terminal 2Fe:2S iron–sulfur cluster binding domains (E-H in Fig. 16.1) and an oxidoreductase domain downstream of the dehalogenase domain (E, F, H) are found exclusively in Proteobacteria, from taxonomic groups that, until recently, had no documented organohalide respiration activity. The presence of the oxidoreductase domains and second Fe:S binding domain may confer a distinct, non-dehalogenating functionality to these proteins, however, the recent characterization of a reductive dehalogenase with a C-terminal oxidoreductase and 2Fe:S binding cluster from the Betaproteobacterium *Comomonas* sp. strain 7D-2 suggests these architectures may well belong to functional RdhAs (Chen et al. 2013). The *Comomonas* sp. strain 7D-2 reductive dehalogenase is active under aerobic conditions, a somewhat unexpected finding given this enzymatic function is canonically anaerobic. While some RdhAs have been shown to be oxygen tolerant (Neumann et al. 1994; Löffler et al. 1996), native aerobic activity within an organism had not previously been shown. The *Comomonas* sp. strain 7D-2 protein shares 49 % sequence identity to the nearest RdhA, from the Alphaproteobacterium *Rhodobacteraceae* bacterium HIMB11, but has 91 % sequence identity with its best hit from the NCBI database, a gene annotated as NAD-binding oxidoreductase from *Hydrogenophaga* sp. strain PBC. The top BLAST hits for the *Comomonas* sp. strain 7D-2 RdhA are all annotated as NAD-binding oxidoreductases despite containing conserved reductive dehalogenase domains, indicating this RdhA sequence type has been known for some time, but has been missed during automated annotation of genomes. As heterologous expression of active reductive dehalogenases is now possible (Mac Nelly et al. 2014), these putatively mis-annotated genes would be interesting targets for functional classification. The identification of an instance of reductive dechlorination in aerobic conditions from noncanonical reductive dehalogenase sequences underscores the sequence space and diversity that remain unexplored for this biochemical function.

16.2.2 Prediction of Substrate Specificity from Sequence Similarity

The combined substrate specificities of the reductive dehalogenases encoded on an organism's genome determine the range of halogenated electron acceptors that organism can respire. In turn, this substrate specificity defines which organisms can be utilized for targeted bioremediation of contaminated sites. As biochemical characterization of RdhAs lags behind the identification of novel *rdhA* gene sequences, there is considerable interest in being able to apply sequence similarity to predict substrate specificity. This is not a straightforward proposition, as RdhA sequence similarity does not always correspond to shared substrate specificities and vice versa. As an example, the two closely related *Dehalobacter* strains CF and DCA encode the *cfrA* and *dcrA* genes respectively. *cfrA* and *dcrA* are syntenic within the two genomes, share 95 % sequence identity at the amino acid level, and are tightly grouped within a global reductive dehalogenase tree (Hug et al. 2013b; Tang and Edwards 2013). Despite these strong relationships, the CfrA and DcrA proteins have been biochemically proven to dechlorinate chloroform/1,1,1-trichloroethane and 1,1-dichloroethane respectively; nonoverlapping substrate ranges (Tang and Edwards 2013). Similarly, the *Desulfitobacterium dichloroeliminans* dichloroethane-dehalogenating enzyme DcaA shares 90 % amino acid identity with the *Dehalobacter restrictus* strain PER-K23 tetrachloroethene (PCE)-dechlorinating enzyme PceA, a relatively high level of conservation within this protein family. Despite this, the two enzymes do not share a common substrate (Marzorati et al. 2007). As an example of the reverse scenario, the five proteins biochemically proven to share specificity for tetrachloroethene come from different genera (*Shewanella*, *Dehalococcoides*, *Dehalobacter*, *Sulfurospirillum*, and *Desulfitobacterium* (Neumann et al. 1996; Magnuson et al. 1998; Maillard et al. 2003; Tsukagoshi et al. 2006; Lohner and Spormann 2013)), are broadly distributed on an evolutionary tree of reductive dehalogenases, and share at maximum 29.8 % pairwise identity (range 20.9–29.8 %) (Hug et al. 2013b). From this, it is clear that prediction of putative substrates for novel reductive dehalogenases based on sequence similarity cannot be considered reliable.

16.2.3 Classification of Reductive Dehalogenases into Ortholog Groups

The increase in reductive dehalogenase sequences from next-generation sequencing of genomes and metagenomes as well as primer-based environmental surveys has lead to a recognized need for organized nomenclature within the gene family. A characterization of the RdhA family proposed a flexible classification system for defining reductive dehalogenase orthologs, allowing rapid comparison of novel sequences to the existing RdhA diversity (Hug et al. 2013b). The proposed system defined reductive dehalogenase ortholog groups (RD_OGs) as groups of RdhAs with pairwise amino acid percent identity of 90 % or greater between all members of a group, along with consistent monophyly for the group within phylogenetic analyses (Hug et al. 2013b). The ortholog groups provide a universal framework for comparison of novel reductive dehalogenase sequences, while allowing definition of new RD_OGs once closely related homologs to existing sequences are identified from new environments or organisms. The system requires a certain level of curation before an RdhA sequence can be included: it is recommended that the organism whose genome encodes the *rdhA* be known, with at least a 16S rRNA gene associated with the *rdhA* sequence. Initially, 46 RD_OGs were defined, which contained a total of 176 RdhA sequences from the 264 included in the analysis. The maximum number of RdhAs in a given ortholog group was 10, with a minimum, by definition, of 2. The ortholog groups were numbered based on the date of first publication of the earliest described sequence in the group, with new RD_OGs to be added sequentially.

The classification system was subsequently applied in the description of two novel *Dehalococcoides mccartyi* genomes, strains BTF08 and DCMB5 (Löffler et al. 2012; Pöritz et al. 2013). Of the 20 and 23 RdhA predicted from these genomes, 15 and 12, respectively, could be assigned to an existing RD_OG based on ≥90 % percent identity (PID) to the ortholog group members. Of the remaining 16 RdhA, 2 from strain BTF08 and 3 from strain DCMB5 form new RD_OGs with RdhAs from other *D. mccartyi* strains' genomes. Notably, all five of the shared RdhA present in both strains' genomes were classifiable within the existing RD_OG system, identifying these as broadly distributed orthologs within the *Dehalococcoides* genus (Pöritz et al. 2013). It is intended for the classification system to be updated with each subsequent isolated genome and well-curated metagenome, so that it remains current and continues to provide a mechanism for capturing and organizing the sequence diversity of the RdhA family. The current version of the classification system is publicly available for download and/or amendment here: https://docs.google.com/folderview?id=0BwCzK8wzlz8ON1o2Z3FTbHFPYXc.

16.2.4 Discovery of Novel Reductive Dehalogenases

Environmental sequencing, in the form of metagenomics or primer-based analyses, provides an opportunity to examine the RdhA family beyond the documented diversity of reductive dehalogenases from isolates or organisms in culture. To date, novel environmental RdhA sequences have largely been identified through PCR-directed studies targeting orthologs of biochemically characterized reductive dehalogenases (Magnuson et al. 2000; von Wintzingerode et al. 2001; Rhee et al. 2003; Regeard et al. 2004; Hölscher et al. 2004; Krajmalnik-Brown et al. 2004; Futagami et al. 2009). Such screens are powerful for amplification of low-abundance genes from complex environments, but they rely on specificity of primers, lowering the novelty of the sequences that can be amplified and sequenced. PCR screens have identified highly diverse *rdhA* genes from marine subsurface sediment samples with only 33–64 % percent identities to previously sequenced genes (Futagami et al. 2009). PCR screens have also led to sequencing of *rdhA* genes from estuarine sediment enrichment cultures with 21–34 % percent identity to the characterized proteins against which the primers were designed (Rhee et al. 2003). Moving beyond primer sets designed from the biochemically characterized enzymes, a PCR primer suite covering the known diversity of the *rdhA* gene family has been developed (Hug and Edwards 2013). The primer suite comprises 44 different primer sets specifically designed with degeneracies to allow amplification of less conserved gene sequences. Illumina sequencing and assembly of the amplified gene products from six different samples, four environmental sites and two enrichment cultures, led to identification of nearly 800 *rdhA* gene sequences. Two-thirds of the identified RdhA shared >90 % amino acid identity to existing database sequences, and many of the proteins were broadly distributed, both among the six samples surveyed and within sequenced genomes (Hug and Edwards 2013). The remaining third of the sequences were more novel, though 15 % had close homologs within the six samples surveyed. In total, the study resulted in 241 new RdhA ortholog types, substantially increasing the sequence space RdhAs occupy.

Metagenomic surveys for reductive dehalogenase genes are less common, though *rdhA* genes have been reported from some shotgun sequencing efforts (Hug et al. 2013a; Kawai et al. 2014). Metagenome sequencing allows highly novel sequences to be identified, as it does not suffer from the amplification specificity of PCR screens. However, metagenomics does require that the organisms whose genomes encode the *rdhA* be at sufficiently high abundance in the environment to be tractable for assembly and gene prediction. Metagenomic sequencing of marine sediments identified highly abundant and diverse *rdhA* genes, with a greater sequence diversity than had previously been reported from a PCR screen of the same environment (Futagami et al. 2009; Kawai et al. 2014). Similarly, an aquifer sediment metagenome from a heavy metal impacted, but not chlorinated compound contaminated, site reported 82 *rdhA* genes, with weak similarity to *rdhA* from the *Chloroflexi* and *Firmicutes* (Hug et al. 2013a). These environmental screens have led to a broader picture of reductive dehalogenase sequence diversity than is currently captured in the PFAM database.

A perfect example of the novelty, diversity, and substrate functionality that remains to be determined for RdhAs is described in a recent examination of enrichment cultures selecting for dechlorination of chlorinated xanthones, analogs of naturally occurring halo organics (Krzmarzick et al. 2014). The microcosms show enrichment of organisms from a novel Firmicute clade, related to but distinct from the *Dehalobacter* and *Desulfitobacterium* genera. Despite a wide variety of published primer sets tested, at present no *rdhA* genes have been identified from these cultures (Krzmarzick et al. 2014).

16.3 Environmental Distribution

Initial research into organohalide respiration and *rdh* genes leveraged existing environmental contamination as an enrichment factor for the dehalogenation activity of interest. The vast majority of organohalide-respiring organisms have been isolated or enriched from contaminated sites around the world, largely from contaminated soils and sediments (e.g., Duhamel and Edwards 2006; Marzorati et al. 2007; West et al. 2008), but also from wastewater treatment facilities (Fennell et al. 1997; Kim et al. 2012). Isolation of organohalide-respiring organisms from uncontaminated sites has also been accomplished, but is significantly more rare (Sung et al. 2006; Duhamel and Edwards 2006; Justicia-Leon et al. 2012). In contrast, the majority of organisms whose genomes encode a predicted reductive dehalogenase but who have not been shown to conduct organohalide respiration have been isolated from putatively pristine environments (e.g., Hafenbradl et al. 1996; DiDonato et al. 2010; Sattley and Blankenship 2010). These pristine environments include diverse soils, freshwater and marine sediments, and varying depths in ocean waters, and are from sites located across the globe (Fig. 16.2). *RdhA* genes have additionally been identified from isolated organisms that are pathogens or syntrophs of macroscopic organisms, including *Pseudovibrio* sp. strain JE062, isolated from black band diseased coral off the coast of Florida, U.S.A. (Bondarev et al. 2013), and *Phaeobacter gallaeciensis* strain 2.10, isolated from the marine algae *Ulva lactuca* off the coast of Australia (Thole et al. 2012). *RdhA* genes have even been identified from bacteria colonizing humans, specifically in the genomes of two *Clostridium difficile* strains (Brouwer et al. 2011) and *Lachnospiraceae* bacterium 2_1_58FAA (NCBI accession: GCA_000218465).

Beyond isolated organisms, both primer-based (Magnuson et al. 2000; von Wintzingerode et al. 2001; Rhee et al. 2003; Regeard et al. 2004; Hölscher et al. 2004; Futagami et al. 2009) and metagenomic surveys (Hug et al. 2013a; Kawai et al. 2014) have expanded the environmental distribution of documented *RdhA* genes. *RdhA* genes have been identified from harsh environments like the soda lakes in Mongolia (Sorokin et al. 2008), hot spring microbial mat communities in Iceland (Sattley and Blankenship 2010), and even from trester, the solid

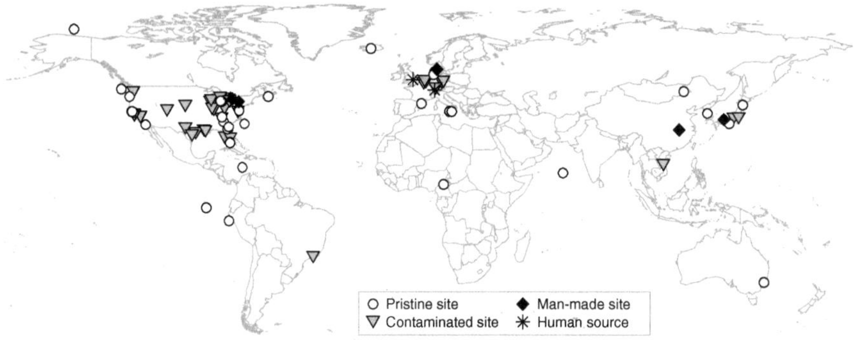

Fig. 16.2 Global distribution of reductive dehalogenase genes, as reported from enrichment cultures, isolate genomes, PCR-based screens, and metagenomic datasets. Sites are labeled as pristine (*white circles*) if no specific halogenated contamination was listed for the site, while contaminated sites (*gray triangles*) had measured halogenated compound contamination. Man-made sites (*black diamonds*) refer to industrial wastewater sewage treatment plants and mines, while human-sourced *rdh* genes (*asterisks*) derived from human subjects, with the location at which the sample was taken plotted on the map

wastes derived from the wine making process (Padilla-Crespo et al. 2014). The distribution of known sites harboring reductive dehalogenases is uneven, with a skew toward regions where research groups studying organohalide respiration are located (e.g., North America (west coast and eastern seaboard) as well as Germany) (Fig. 16.2). Despite this skew, what has emerged from the combination of enrichment culture studies and environmental screens is a global distribution of reductive dehalogenases that suggests these genes are near-ubiquitous across the planet (Fig. 16.2), or at least that there are no currently identifiable constraints to their distribution.

There are a number of cases where reductive dehalogenases have been sought in a given environment, but have not been detected. Some have intriguing implications for the ubiquity of this gene family; of six ocean sediment sites assayed with PCR-based methods, only one had 100 % amplification of *rdhA* genes across its depth profile, while the others showed variable amplification success, often preferentially from shallower depths (Futagami et al. 2009). However, all six sites did have at least one depth from which *rdhA* genes could be detected, albeit only after DNA amplification of the samples, a preliminary finding that hints at the abundance of this gene family in the marine subsurface. The reductive dehalogenase gene *dcpA*, implicated in the dichloroelimination of 1,2-dichloropropene to propene, was identified in the contaminated groundwater plume at Ft. Pierce, FL, but was not detected in pristine sediment samples upstream of the plume from the same site (Padilla-Crespo et al. 2014). Similarly, degenerate primer PCR amplification of DNA from groundwater from a contaminated site in West Louisiana yielded over 150 unique *rdhA* genes, while a sediment sample from the site did not result in a successful amplification of any *rdhA* (Hug and Edwards 2013).

From the same study, pristine sediment from Toronto, ON and contaminated sediments from sites near Genoa, Italy and Priolo, Italy did not show amplification of *rdhA* genes in any of the 44 degenerate PCR reactions assayed (Hug and Edwards 2013). Taken together, it is unclear whether sediment samples contain inhibitory compounds that prevent PCR amplification of *rdhA*s, or whether the planktonic fraction of subsurface microbial communities genuinely harbors the majority of the reductive dehalogenase-bearing organisms, with *rdhA* genes falling below the PCR detection limit in sediment-derived DNA extracts. In contrast, a metagenome survey of heavy metal-contaminated aquifer sediment from Rifle, CO did identify *rdhA* genes from each of the three depths surveyed (Hug et al. 2013a). Punctuated detection of *rdhA*s across environmental sites is not limited to groundwater/sediment comparisons. In a PCR-based assay of a variety of groundwater samples from Moffett Field in California, the vinyl chloride reductase *vcrA* was only detected in the groundwater samples undergoing reduction of vinyl chloride to ethene (Müller et al. 2004). Enrichment of *rdhAB*s in an environment is tied to the growth profile of the organism whose genome encodes the genes. It has been shown that enrichment of microbial consortia on specific chlorinated compounds causes loss of organohalide-respiring organisms and their associated *rdhAB* genes when no longer relevant to the dechlorination process at hand (Grostern and Edwards 2006; Krajmalnik-Brown et al. 2007). This likely holds true at environmental sites as well, with the contamination profile or native halogenated substrates at the site dictating the *rdhA* genes that are abundant and thus detected. In the future, with deep sequencing costs consistently dropping and bioinformatics tools for processing large sequence datasets improving, the *rdhA* gene distribution is predicted to expand, both as lower abundance genes become tractable for assembly and annotation, and as more environments are surveyed. In future, it will be interesting to see if constraints to this distribution emerge, and if so, what the mechanism(s) behind those constraints might be.

16.4 Evolution

The origins of organohalide respiration and reductive dehalogenases are unknown but are predicted to be ancient based on phylogenetic depth (Hug et al. 2013b), molecular dating (McMurdie et al. 2011), and distribution of *rdhA* genes deep in the sub-seafloor (Kawai et al. 2014). The gene family is present across the prokaryotic tree of life in a punctuated distribution: present in only a few taxa from phyla or classes whose members do not otherwise contain homologs to these genes. The *Chloroflexi* are a good example of this, with reductive dehalogenases present in *Dehalococcoidia* species but not in any other members of the phylum, including relatively closely related members of the same class (Hug et al. 2013a; Wasmund et al. 2014; Kaster et al. 2014). This punctuated distribution makes it difficult to discern whether reductive dehalogenases represent an ancient enzyme family with a strong pattern of gene loss through prokaryotic evolution, or a more taxonomically limited adaptation that has subsequently spread through gene duplication and gene transfer. There is evidence for each of these modes of inheritance, suggesting all were factors in the present-day distribution of reductive dehalogenases.

In a global tree of RdhA proteins, evidence for vertical transmission was confined to broad phylogenetic affiliations; proteins from the same phyla occasionally clustered together, including one large cluster of *Dehalococcoidia*-associated RdhA (Hug et al. 2013b). However, this cluster did not contain all *Dehalococcoidia* RdhA sequences: a smaller clade placed outside this radiation with all non-*Chloroflexi*-derived sequences. RdhAs from Proteobacteria did not show evidence of phylum- or class-level monophyly. There are a number of possible explanations for the lack of correlation between the RdhA tree and the organism phylogeny. Many organohalide-respiring organisms encode multiple reductive dehalogenases on their genomes, which may be the result of gene duplication. In this case, the global RdhA tree contains paralog groups and would not be expected to match a 16S rRNA gene phylogeny. However, if gene duplication and punctuated loss were sufficient to describe RdhA evolution, one would expect the organism relationships within RdhA paralog groups to mimic the organismal 16S rRNA phylogeny. This is manifestly not the case for the RdhA tree. Many *Dehalococcoides* RdhA clades do not support the evolutionary relationships defined by the organisms' 16S rRNA genes (Hug et al. 2013b). The deltaproteobacterial RdhAs show evidence for gene duplication within the *Anaeromyxobacter*, with each of two clusters resolving the same relationships, but the two RdhAs from *Geobacter lovleyi* strain SZ are not clustered with or near these other deltaproteobacterial RdhAs. In order to fully explain the RdhA protein tree, a mixture of vertical inheritance, gene duplication and gene loss, and lateral gene transfer (LGT) must be invoked.

The existing evidence for lateral gene transfer of *rdhA* genes suggests this process is occurring within multiple genera, via several different mechanisms (Rhee et al. 2003; Maillard et al. 2005; Krajmalnik-Brown et al. 2007; McMurdie et al. 2007, 2011). An examination of the vinyl chloride reductases *vcrA* and *bvcA* from *D. mccartyi* strains identified a codon bias in both genes strongly suggestive of an evolutionary origin for these genes outside of the *Dehalococcoides* (McMurdie et al. 2007). Comparative genomics revealed that the *vcrABC* operon is contained within a genomic island that was horizontally acquired (Fig. 16.3c) (McMurdie et al. 2009). Further investigation revealed the genomic island's mechanism of mobility, a site-specific integrase (DsiB), and that such elements are common in the high plasticity regions of *D. mccartyi* genomes; bordered by direct repeats, the elements contain a conserved integration module and a variable 'cargo' region, often including reductive dehalogenases (McMurdie et al. 2011). The *Dehalococcoides* trichloroethene reductase gene *tceA* has also been implicated in LGT. In a PCR-based survey, *tceA* genes with unexpectedly high sequence identities (>95 %) were identified in twelve enrichment cultures derived from geographically distinct sites within the United States (Krajmalnik-Brown et al. 2007). The high percent identity of the *tceAB* regions suggested a recent

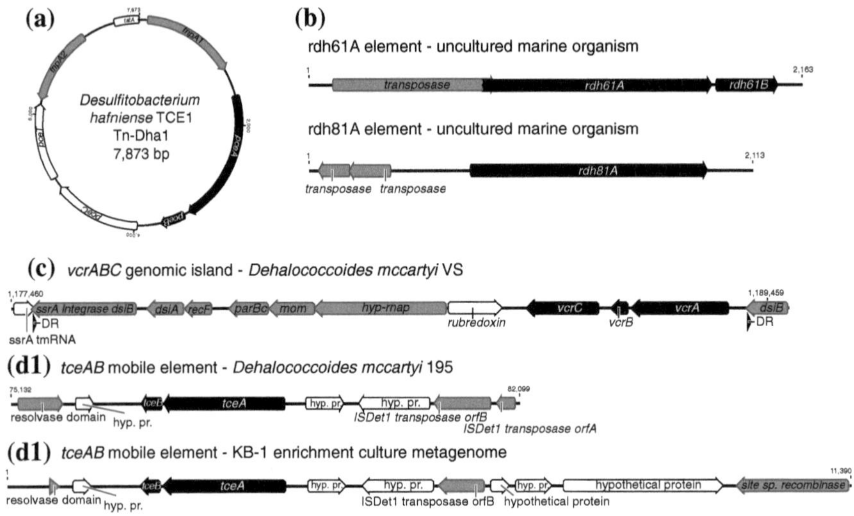

Fig. 16.3 Hypothesized mechanisms of reductive dehalogenase lateral gene transfer. Genes associated with mobile elements are colored gray, *rdhAB* genes in *black*, and other genes in *white*. A: circularizing transposable element carrying the *pceABCT* cluster from *Desulfitobacterium hafniense* TCE1 (Maillard et al. 2005). B: *tceA* genes identified from a sulfidogenic 2-bromophenol-degrading consortium with associated transposase genes (Rhee et al. 2003). C: *vcrABC*-bearing genomic island from *D. mccartyi* VS (McMurdie et al. 2011). The island is defined by direct repeats (DR) within site-specific recombinases (*ssrA*). D: conserved predicted transposable element carrying *tceAB* genes identified in *D. mccartyi* 195's genome (D1) (Seshadri et al. 2005) and the KB-1 enrichment culture metagenome (D2) (Waller et al. 2012). Abbreviations: *DR* direct repeat; dsiA = *Dehalococcoides* ssrA-specific integrase A; dsiB = PinR and serine recombinase domain-containing protein; hyp. pr. = hypothetical protein; hyp-rnap = putative DNA-directed RNA polymerase; mom = Mom homolog, methylation protein; parBc = ParBC domain-containing protein; recF = recombinase F homolog; tnpA1/tnpA2 = putative transposases

dissemination within *Dehalococcoides* populations, corroborated by the presence of predicted transposase genes adjacent to the *tceAB* operon (Krajmalnik-Brown et al. 2007). In *D. mccartyi* 195's genome, the *tceAB* genes are flanked by a resolvase domain protein and two transposases (Fig. 16.3, D1), while a syntenic *tceA* genomic region in the KB-1 enrichment culture metagenome has a truncated resolvase domain protein and only one transposase (Fig. 16.3, d2) (Waller et al. 2012). The synteny and high sequence identity between these two *tceAB* regions, coupled to the absence of synteny on the chromosomes outside this region suggest this region represents a shared mobile element. In a similar vein, two *rdhA*-containing regions amplified from a sulfidogenic 2-bromophenol-degrading enrichment consortium carry predicted transposases adjacent to the *rdhA*s (Fig. 16.3b) (Rhee et al. 2003), making LGT a plausible mechanism of acquisition for these genes. Perhaps the most compelling evidence for LGT of *rdh* genes was the identification of a circularizing transposon from the genome of *D. hafniense* strain TCE1 encoding the tetrachloroethene reductive dehalogenase *pceA* and its associated *pceBCT* genes (Fig. 16.3a) (Maillard et al. 2005). In addition to confirmation of a circularized form of this predicted transposon, the *D. hafniense* TCE1 *pceABCT* gene cluster shares 100 % sequence identity with those from *D. restrictus* and *Desulfitobacterium* strains PCE-S and Y51, further indication of a more recent acquisition than via vertical inheritance. Interestingly, while both *Dehalobacter* and *Desulfitobacterium* strains carry the *pceABCT* cluster, only the *Desulfitobacterium* strains' are contained within a circularizing transposon (Maillard et al. 2005).

The evolutionary origin of *rdh* genes is a complex question, one that will likely require complete genomes from many more representatives to be untangled. The apparent role of LGT in *rdh* gene dissemination has implications for bioremediation efforts, as correlations between *rdhA* and organism abundances will be affected if closely related genes are distributed across multiple genera or phyla.

16.5 Summary

The RdhA and B enzyme families have high sequence divergence, and are distributed across the prokaryotic tree of life. The protein families' sequence divergences are in-line with other defined protein families, though some RdhA enzymes contain domain organizations that may not all be conducive for dehalogenation. In the historic absence of a protein structure, the RdhA family has been defined by sequence similarity; a classification system leveraging this sequence diversity has been developed. Globally, there are no known constraints for the *rdh* gene distribution. *rdhA* genes have been identified from most natural environments, including regions with harsh akaline conditions or extreme temperatures as well as from bacteria associated with human and other eukaryotic hosts. The origins of organohalide respiration are not well understood, but evidence exists that both vertical inheritance and lateral gene transfer have contributed to the distribution of *rdhA* genes across taxonomic groups. Multiple mechanisms of LGT have been proposed, including transposon-, insertion element-, and phage-mediated gene movement. As more environments are examined, novel RdhA with new substrate profiles will undoubtedly be discovered, continuing to expand the global distribution, sequence diversity, and known functionality of this protein family.

References

Bondarev V, Richter M, Romano S et al (2013) The genus *Pseudovibrio* contains metabolically versatile bacteria adapted for symbiosis. Environ Microbiol 15:2095–2113. doi:10.1111/1462-2920.12123

Brouwer MSM, Warburton PJ, Roberts AP et al (2011) Genetic organisation, mobility and predicted functions of genes on integrated, mobile genetic elements in sequenced strains of *Clostridium difficile*. PLoS ONE 6:e23014. doi:10.1371/journal.pone.0023014

Chen K, Huang L, Xu C et al (2013) Molecular characterization of the enzymes involved in the degradation of a brominated aromatic herbicide. Mol Microbiol 89:1121–1139. doi:10.1111/mmi.12332

DiDonato RJ, Young ND, Butler JE et al (2010) Genome sequence of the deltaproteobacterial strain NaphS2 and analysis of differential gene expression during anaerobic growth on naphthalene. PLoS ONE 5:e14072. doi:10.1371/journal.pone.0014072

Duhamel M, Edwards EA (2006) Microbial composition of chlorinated ethene-degrading cultures dominated by *Dehalococcoides*. FEMS Microbiol Ecol 58:538–549. doi:10.1111/j.1574-6941.2006.00191.x

Edgar RC (2004) MUSCLE: multiple sequence alignment with high accuracy and high throughput. Nucleic Acids Res 32:1792–1797. doi:10.1093/nar/gkh340

Fennell DE, Gossett JM, Zinder SH (1997) Comparison of butyric acid, ethanol, lactic acid, and propionic acid as hydrogen donors for the reductive dechlorination of tetrachloroethene. Environ Sci Technol 31:918–926. doi:10.1021/es960756r

Futagami T, Morono Y, Terada T et al (2009) Dehalogenation activities and distribution of reductive dehalogenase homologous genes in marine subsurface sediments. Appl Environ Microbiol 75:6905–6909. doi:10.1128/AEM.01124-09

Grostern A, Edwards EA (2006) Growth of *Dehalobacter* and *Dehalococcoides* spp. during degradation of chlorinated ethanes. Appl Environ Microbiol 72:428–436. doi:10.1128/AEM.72.1.428-436.2006

Guindon S, Gascuel O (2003) A simple, fast, and accurate algorithm to estimate large phylogenies by maximum likelihood. Syst Biol 52:696–704

Hafenbradl D, Keller M, Dirmeier R et al (1996) *Ferroglobus placidus* gen. nov., sp. nov., A novel hyperthermophilic archaeum that oxidizes Fe2+ at neutral pH under anoxic conditions. Arch Microbiol 166:308–314

Hölscher T, Krajmalnik-Brown R, Ritalahti KM et al (2004) Multiple nonidentical reductive-dehalogenase-homologous genes are common in *Dehalococcoides*. Appl Environ Microbiol 70:5290–5297. doi:10.1128/AEM.70.9.5290-5297.2004

Hug LA, Edwards EA (2013) Diversity of reductive dehalogenase genes from environmental samples and enrichment cultures identified with degenerate primer PCR screens. Front Microbiol 4:341. doi:10.3389/fmicb.2013.00341

Hug LA, Castelle CJ, Wrighton KC et al (2013a) Community genomic analyses constrain the distribution of metabolic traits across the Chloroflexi phylum and indicate roles in sediment carbon cycling. Microbiome 1:22. doi:10.1186/2049-2618-1-22

Hug LA, Maphosa F, Leys D et al (2013b) Overview of organohalide-respiring bacteria and a proposal for a classification system for reductive dehalogenases. Philos Trans R Soc Lond B Biol Sci 368:20120322. doi:10.1098/rstb.2012.0322

Justicia-Leon SD, Ritalahti KM, Mack EE, Löffler FE (2012) Dichloromethane fermentation by a *Dehalobacter* sp. in an enrichment culture derived from pristine river sediment. Appl Environ Microbiol 78:1288–1291

Kaster A-K, Mayer-Blackwell K, Pasarelli B, Spormann AM (2014) Single cell genomic study of *Dehalococcoidetes* species from deep-sea sediments of the Peruvian Margin. ISME J 8:1831–1842. doi:10.1038/ismej.2014.24

Kawai M, Futagami T, Toyoda A et al (2014) High frequency of phylogenetically diverse reductive dehalogenase-homologous genes in deep subseafloor sedimentary metagenomes. Front Microbiol 5:80. doi:10.3389/fmicb.2014.00080

Kim S-H, Harzman C, Davis JK et al (2012) Genome sequence of *Desulfitobacterium hafniense* DCB-2, a Gram-positive anaerobe capable of dehalogenation and metal reduction. BMC Microbiol 12:21

Krajmalnik-Brown R, Hölscher T, Thomson IN et al (2004) Genetic identification of a putative vinyl chloride reductase in *Dehalococcoides* sp. strain BAV1. Appl Environ Microbiol 70:6347–6351. doi:10.1128/AEM.70.10.6347-6351.2004

Krajmalnik-Brown R, Sung Y, Ritalahti KM et al (2007) Environmental distribution of the trichloroethene reductive dehalogenase gene (tceA) suggests lateral gene transfer among *Dehalococcoides*. FEMS Microbiol Ecol 59:206–214. doi:10.1111/j.1574-6941.2006.00243.x

Krzmarzick MJ, Crary BB, Harding JJ et al (2012) Natural niche for organohalide-respiring Chloroflexi. Appl Environ Microbiol 78:393–401. doi:10.1128/AEM.06510-11

Krzmarzick MJ, Miller HR, Yan T, Novak PJ (2014) Novel Firmicutes group implicated in the dechlorination of two chlorinated xanthones, analogues of natural organochlorines. Appl Environ Microbiol 80:1210–1218. doi:10.1128/AEM.03472-13

Kube M, Beck A, Zinder SH et al (2005) Genome sequence of the chlorinated compound-respiring bacterium *Dehalococcoides* species strain CBDB1. Nat Biotechnol 23:1269–1273. doi:10.1038/nbt1131

Löffler FE, Sanford RA, Tiedje JM (1996) Initial characterization of a reductive dehalogenase from *Desulfitobacterium chlororespirans* Co23. Appl Environ Microbiol 62:3809–3813

Löffler FE, Yan J, Ritalahti KM et al (2012) *Dehalococcoides mccartyi* gen. nov., sp. nov., obligate organohalide-respiring anaerobic bacteria, relevant to halogen cycling and bioremediation, belong to a novel bacterial class, *Dehalococcoidetes* classis nov., within the phylum Chloroflexi. Int J Syst Evol Microbiol 63:625–635. doi:10.1099/ijs.0.034926-0

Lohner ST, Spormann AM (2013) Identification of a reductive tetrachloroethene dehalogenase in *Shewanella sediminis*. Philos Trans R Soc Lond B Biol Sci 368:20120326. doi:10.1098/rstb.2012.0326

Mac Nelly A, Kai M, Svatoš A et al (2014) Functional heterologous production of reductive dehalogenases from *Desulfitobacterium hafniense* strains. Appl Environ Microbiol 80:4313–4322. doi:10.1128/AEM.00881-14

Magnuson JK, Stern RV, Gossett JM et al (1998) Reductive dechlorination of tetrachloroethene to ethene by a two-component enzyme pathway. Appl Environ Microbiol 64:1270–1275

Magnuson JK, Romine MF, Burris DR, Kingsley MT (2000) Trichloroethene reductive dehalogenase from *Dehalococcoides ethenogenes*: sequence of *tceA* and substrate range characterization. Appl Environ Microbiol 66:5141–5147

Maillard J, Schumacher W, Vazquez F et al (2003) Characterization of the corrinoid iron-sulfur protein tetrachloroethene reductive dehalogenase of *Dehalobacter restrictus*. Appl Environ Microbiol 69:4628–4638

Maillard J, Regeard C, Holliger C (2005) Isolation and characterization of Tn-Dha1, a transposon containing the tetrachloroethene reductive dehalogenase of *Desulfitobacterium hafniense* strain TCE1. Environ Microbiol 7:107–117. doi:10.1111/j.1462-2920.2004.00671.x

Maphosa F, de Vos WM, Smidt H (2010) Exploiting the ecogenomics toolbox for environmental diagnostics of organohalide-respiring bacteria. Trends Biotechnol 28:308–316. doi:10.1016/j.tibtech.2010.03.005

Maphosa F, Lieten SH, Dinkla I et al (2012) Ecogenomics of microbial communities in bioremediation of chlorinated contaminated sites. Front Microbiol 3:351. doi:10.3389/fmicb.2012.00351

Marzorati M, de Ferra F, Van Raemdonck H et al (2007) A novel reductive dehalogenase, identified in a contaminated groundwater enrichment culture and in *Desulfitobacterium dichloroeliminans* strain DCA1, is linked to dehalogenation of 1,2-dichloroethane. Appl Environ Microbiol 73:2990–2999. doi:10.1128/AEM.02748-06

McMurdie PJ, Behrens SF, Holmes S, Spormann AM (2007) Unusual codon bias in vinyl chloride reductase genes of *Dehalococcoides* species. Appl Environ Microbiol 73:2744–2747. doi:10.1128/AEM.02768-06

McMurdie PJ, Behrens SF, Müller JA et al (2009) Localized plasticity in the streamlined genomes of vinyl chloride respiring *Dehalococcoides*. PLoS Genet 5:e1000714. doi:10.1371/journal.pgen.1000714

McMurdie PJ, Hug LA, Edwards EA et al (2011) Site-specific mobilization of vinyl chloride respiration islands by a mechanism common in *Dehalococcoides*. BMC Genom 12:287. doi:10.1186/1471-2164-12-287

Müller JA, Rosner BM, Von Abendroth G et al (2004) Molecular identification of the catabolic vinyl chloride reductase from *Dehalococcoides* sp. strain VS and its environmental distribution. Appl Environ Microbiol 70:4880–4888. doi:10.1128/AEM.70.8.4880-4888.2004

Neumann A, Scholz-Muramatsu H, Diekert G (1994) Tetrachloroethene metabolism of *Dehalospirillum multivorans*. Arch Microbiol 162:295–301

Neumann A, Wohlfarth G, Diekert G (1996) Purification and characterization of tetrachloroethene reductive dehalogenase from *Dehalospirillum multivorans*. J Biol Chem 271:16515–16519

Neumann A, Wohlfarth G, Diekert G (1998) Tetrachloroethene dehalogenase from *Dehalospirillum multivorans*: cloning, sequencing of the encoding genes, and expression of the *pceA* gene in *Escherichia coli*. J Bacteriol 180:4140–4145

Ni S, Fredrickson JK, Xun L (1995) Purification and characterization of a novel 3-chlorobenzoate-reductive dehalogenase from the cytoplasmic membrane of *Desulfomonile tiedjei* DCB-1. J Bacteriol 177:5135–5139

Nonaka H, Keresztes G, Shinoda Y et al (2006) Complete genome sequence of the dehalorespiring bacterium *Desulfitobacterium hafniense* Y51 and comparison with *Dehalococcoides ethenogenes* 195. J Bacteriol 188:2262–2274. doi:10.1128/JB.188.6.2262-2274.2006

Padilla-Crespo E, Yan J, Swift C et al (2014) Identification and environmental distribution of *dcpA*, which encodes the reductive dehalogenase catalyzing the dichloroelimination of 1,2-dichloropropane to propene in organohalide-respiring Chloroflexi. Appl Environ Microbiol 80:808–818. doi:10.1128/AEM.02927-13

Peale JGD, Bakkom E, Lakhwala F et al (2008) TCE plume remediation via ISCR-enhanced bioremediation utilizing EHC® and KB-1®. Remediat J 18:19–31. doi:10.1002/rem.20179

Pöritz M, Goris T, Wubet T et al (2013) Genome sequences of two dehalogenation specialists—*Dehalococcoides mccartyi* strains BTF08 and DCMB5 enriched from the highly polluted Bitterfeld region. FEMS Microbiol Lett 343:101–104. doi:10.1111/1574-6968.12160

Regeard C, Maillard J, Holliger C (2004) Development of degenerate and specific PCR primers for the detection and isolation of known and putative chloroethene reductive dehalogenase genes. J Microbiol Methods 56:107–118

Rhee S-K, Fennell DE, Häggblom MM, Kerkhof LJ (2003) Detection by PCR of reductive dehalogenase motifs in a sulfidogenic 2-bromophenol-degrading consortium enriched from estuarine sediment. FEMS Microbiol Ecol 43:317–324. doi:10.1111/j.1574-6941.2003.tb01072.x

Sattley WM, Blankenship RE (2010) Insights into heliobacterial photosynthesis and physiology from the genome of *Heliobacterium modesticaldum*. Photosynth Res 104:113–122. doi:10.1007/s11120-010-9529-9

Seshadri R, Adrian L, Fouts DE et al (2005) Genome sequence of the PCE-dechlorinating bacterium *Dehalococcoides ethenogenes*. Science 307:105–108. doi:10.1126/science.1102226

Sorokin DY, Tourova TP, Mussmann M, Muyzer G (2008) *Dethiobacter alkaliphilus* gen. nov. sp. nov., and *Desulfurivibrio alkaliphilus* gen. nov. sp. nov.: two novel representatives of reductive sulfur cycle from soda lakes. Extremophiles 12:431–439. doi:10.1007/s00792-008-0148-8

Sung Y, Fletcher KE, Ritalahti KM et al (2006) *Geobacter lovleyi* sp. nov. strain SZ, a novel metal-reducing and tetrachloroethene-dechlorinating bacterium. Appl Environ Microbiol 72:2775–2782. doi:10.1128/AEM.72.4.2775-2782.2006

Tang S, Edwards EA (2013) Identification of *Dehalobacter* reductive dehalogenases that catalyse dechlorination of chloroform, 1,1,1-trichloroethane and 1,1-dichloroethane. Philos Trans R Soc Lond B Biol Sci 368:20120318. doi:10.1098/rstb.2012.0318

Thole S, Kalhoefer D, Voget S et al (2012) *Phaeobacter gallaeciensis* genomes from globally opposite locations reveal high similarity of adaptation to surface life. ISME J 6:2229–2244. doi:10.1038/ismej.2012.62

Tsukagoshi N, Ezaki S, Uenaka T et al (2006) Isolation and transcriptional analysis of novel tetrachloroethene reductive dehalogenase gene from *Desulfitobacterium* sp. strain KBC1. Appl Microbiol Biotechnol 69:543–553. doi:10.1007/s00253-005-0022-x

Von Wintzingerode F, Schlötelburg C, Hauck R et al (2001) Development of primers for amplifying genes encoding CprA- and PceA-like reductive dehalogenases in anaerobic microbial consortia, dechlorinating trichlorobenzene and 1,2-dichloropropane. FEMS Microbiol Ecol 35:189–196

Waller AS, Hug LA, Mo K et al (2012) Transcriptional analysis of a *Dehalococcoides*-containing microbial consortium reveals prophage activation. Appl Environ Microbiol 78:1178–1186. doi:10.1128/AEM.06416-11

Wasmund K, Schreiber L, Lloyd KG et al (2014) Genome sequencing of a single cell of the widely distributed marine subsurface *Dehalococcoidia*, phylum Chloroflexi. ISME J 8:383–397. doi:10.1038/ismej.2013.143

West KA, Johnson DR, Hu P et al (2008) Comparative genomics of "*Dehalococcoides ethenogenes*" 195 and an enrichment culture containing unsequenced "*Dehalococcoides*" strains. Appl Environ Microbiol 74:3533–3540. doi:10.1128/AEM.01835-07

Part V
Biochemistry of Organohalide-Respiring Bacteria

Chapter 17
Comparative Biochemistry of Organohalide Respiration

Torsten Schubert and Gabriele Diekert

Abstract Corrinoid-containing reductive dehalogenases (RDases) play a key role in the energy metabolism of anaerobic organohalide-respiring bacteria (OHRB). In such microorganisms the reductive dehalogenation of organohalides catalyzed by RDases is coupled to ATP synthesis via electron transport phosphorylation. The overview presented here summarizes the actual knowledge about the biochemical properties and catalytic mechanism(s) of these enzymes found in bacteria of various phylogenetic affiliation. Furthermore, based on recent findings the multistep biosynthesis of the membrane-associated RDases and the achievements in functional heterologous production of these corrinoid-containing iron–sulfur proteins are described. Up to date, little is known about the composition of the organohalide respiratory chain in OHRB and the interaction of the RDases with other electron-transferring components in the cytoplasmic membrane. In this summary, actual models of different organohalide respiratory chains are included.

17.1 Introduction

Alkyl and aryl halides occur as intermediates of the natural halogen cycle, which is driven by geological or biological processes in the environment (Gribble 2003). In addition, halogenated organic compounds are produced in huge amounts in industry for diverse applications (Euro 2014). Extensive use of chlorinated pesticides, which started in the first half of the last century, was followed by detection of these organohalides as frequent contaminants of soil or groundwater (see Part I, Chap. 2). Polyhalogenated hydrocarbons in particular exhibit a high persistence against aerobic

T. Schubert (✉) · G. Diekert (✉)
Department of Applied and Ecological Microbiology, Institute of Microbiology, Friedrich Schiller University, Philosophenweg 12, D-07743 Jena, Germany
e-mail: torsten.schubert@uni-jena.de

G. Diekert
e-mail: gabriele.diekert@uni-jena.de

L. Adrian and F.E. Löffler (eds.), *Organohalide-Respiring Bacteria*,
DOI 10.1007/978-3-662-49875-0_17

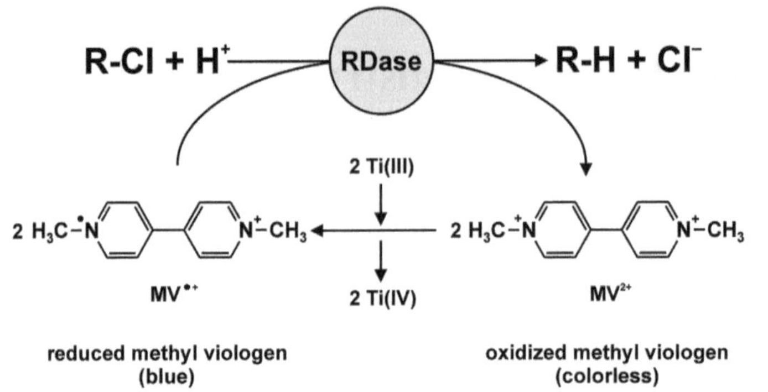

Fig. 17.1 Spectrophotometric RDase enzyme assay. Ti(III), Ti(IV) = different oxidation states of citrate-complexed titanium (Zehnder and Wuhrmann 1976); *RDase* reductive dehalogenase; *MV* methyl viologen

biodegradation, which makes investigations toward an anaerobic conversion and detoxification worthwhile. The biological degradation of organohalides under anaerobic conditions has been observed in the late 1960s by the work of McCarty and coworkers (Hill and McCarty 1967; see Part II, Chap. 4). Enrichment of dehalogenating microbial communities ensued, leading to the discovery of the anaerobic reductive dehalogenation in the early 1980s (Bouwer and McCarty 1983). Several attempts were made to isolate reductively dehalogenating anaerobes from different, mostly contaminated habitats. The first isolated organism was the sulfate-reducing deltaproteobacterium *Desulfomonile tiedjei* (DeWeerd et al. 1990). The organism was shown to reductively dehalogenate 3-chlorobenzoate to benzoate under anaerobic conditions. Using reduced methyl viologen ($MV^{\bullet+}$) as artificial electron donor ($E^{\circ\prime} = -446$ mV, $\varepsilon_{578\mathrm{nm}} = 9.7\ \mathrm{mM}^{-1} \times \mathrm{cm}^{-1}$; for details on the electrochemistry of viologens see Bird and Kuhn 1981), enzymatic reductive dehalogenation of 3-chlorobenzoate has been measured in cell extracts by the change in the absorption upon oxidation of $MV^{\bullet+}$ (DeWeerd and Suflita 1990). Later on, the microbial reductive dehalogenation of the polyhalogenated organohalide tetrachloroethene (PCE) was monitored via spectrophotometry in cell extracts of the epsilonproteobacterium *Sulfurospirillum multivorans* (formerly: *Dehalospirillum multivorans*) (Neumann et al. 1994) (Fig. 17.1). The photometric assay for visualizing the dehalogenating enzyme activity allowed for the first enrichment and purification of a reductive dehalogenase enzyme (RDase).

D. tiedjei couples the reductive dehalogenation of 3-chlorobenzoate to energy conversion via electron transport phosphorylation, an observation which pointed to a function of the RDase as terminal reductase in a novel type of anaerobic respiratory chain (Mohn and Tiedje 1990) (see 'RDases as terminal reductases in organohalide respiratory chains'). The organism reductively dehalogenates 3-chlorobenzoate as electron acceptor with formate as electron donor. The use of formate as sole source of reducing equivalents excludes energy conservation via substrate-level phosphorylation in *D. tiedjei*, something that would be possible with

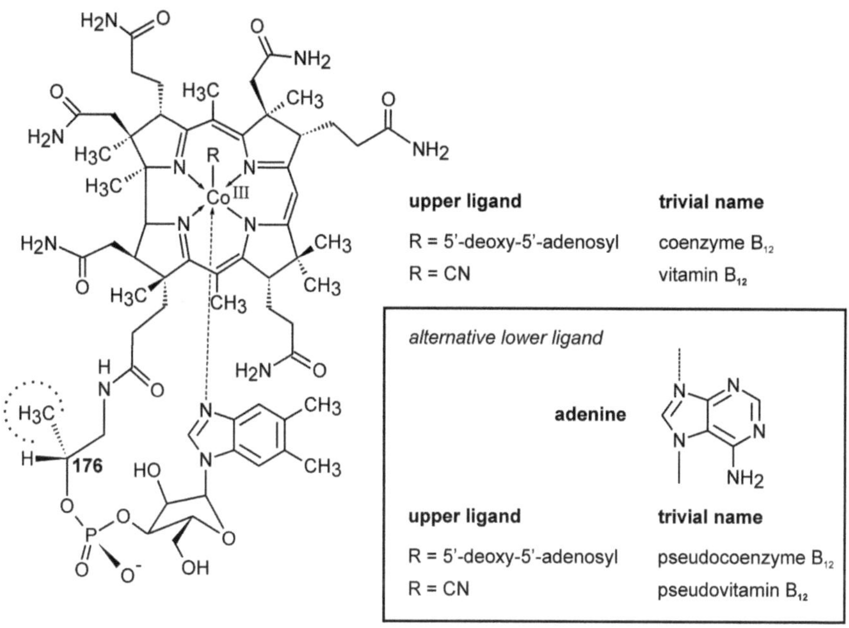

Fig. 17.2 Structure of base-on cobamides ('complete' corrinoids containing an upper and a lower ligand). Coβ-5′-deoxy-5′-adenosyl-Coα-5,6-dimethylbenzimidazolyl-cobamide = coenzyme B_{12}, Coβ-cyano-Coα-5,6-dimethylbenzimidazolyl-cobamide = vitamin B_{12}, Coβ 5′ deoxy-5′-adenosyl-Coα-adeninyl-cobamide = pseudocoenzyme B_{12}, Coβ-cyano-Coα-adeninyl-cobamide = pseudovitamin B_{12}. The presence of the cyano group (–CN) as upper ligand is a result of the corrinoid extraction procedure (Stupperich et al. 1986). Methyl group 177 (*dotted circle*) is replaced by a hydrogen substituent in norcobamides, e.g., norpseudovitamin B_{12} (Kräutler et al. 2003)

more complex organic compounds such as pyruvate as alternative electron donor. Hence, the reductive dehalogenation does not serve just as electron sink in formate-grown cells but rather as the terminal reaction in an electron transport chain that is coupled to energy conservation via a chemiosmotic mechanism (organohalide respiration; *obsolete synonyms*: dehalorespiration, halorespiration).

The first RDase purified was the heterodimeric 3-chlorobenzoate RDase of *D. tiedjei* (Ni et al. 1995). Based on the absorption spectrum of the enzyme-containing fraction, the authors concluded the presence of a heme cofactor in the RDase. However, stimulated from earlier studies on the reaction of alkyl halides with reduced corrinoids (Wood et al. 1968; Krone et al. 1989a) (for details on the corrinoid structure and nomenclature see Fig. 17.2 and see Part V, Chap. 19), the involvement of cobalt-containing heterocyclic organic macrocycles in the microbial reductive dehalogenation was proposed. The inhibition of the reductive dehalogenation by 1-iodopropane (propyl iodide) in cell extracts of the tetrachloroethene-dehalogenating bacterium *S. multivorans* supported this assumption (Neumann et al. 1995). The inhibition by propyl iodide was reversed by illumination. This result was interpreted as cleavage of the light-sensitive cobalt–carbon sigma bond (Kennedy et al. 1969) of the propylated corrinoid cofactor. The presence of a strong reducing agent was a prerequisite for the inhibitory effect of propyl iodide, which pointed to a reduction of the central cobalt ion in the corrinoid prior to the reaction with the alkylating agent (Neumann et al. 1995). No effect of propyl iodide has been observed when the 3-chlorobenzoate reduction in cell extracts of *D. tiedjei* was tested in the presence of the inhibitor (Louie and Mohn 1999). This result was interpreted as proof that a corrinoid does not play a role in this particular reductive dechlorination. Nevertheless, the involvement of corrinoids as metal-containing cofactors in RDase-catalyzed reactions was unambiguously proven when the PCE reductive dehalogenase of *S. multivorans* was purified and characterized (Neumann et al. 1996) and the structure of its norpseudo-B_{12} cofactor was elucidated (Kräutler et al. 2003).

Three formal oxidation states are known for the cobalt atom in the corrinoid structure, namely [Co^{III}], [Co^{II}], and [Co^{I}] (Lexa and Saveant 1983). A decrease in the oxidation state is usually accompanied by changes in Co ligation. Only artificial electron donors with a standard redox potential below −360 mV were effective in supporting the PCE reductive dehalogenation reaction in cell extracts of *S. multivorans* (Miller et al. 1996). Reduced pyridine nucleotides (e.g., NADH) or flavins (e.g., $FADH_2$) could not replace reduced methyl viologen ($MV^{\bullet+}$) in the enzymatic assay (Neumann et al. 1995). The necessity of a low potential electron donor for the enzymatic reductive dehalogenation indicated the involvement of the superreduced corrinoid, i.e., the [Co^{I}] oxidation state. The formation of the [Co^{I}] from the [Co^{II}] state in free corrinoids proceeds at a very low standard redox potential (e.g., cyanocobalamin in water, $E°$ [Co^{II}]/[Co^{I}] ≤ -800 mV at physiological pH and 22 °C), while the [Co^{III}] form is reduced at more positive values (e.g., cyanocobalamin in water, $E°$ [Co^{III}]/[Co^{II}] ~ 0 mV at physiological pH and 22 °C) (Lexa and Saveant 1983). The generation of the low potential [Co^{I}] in the terminal reductase of the organohalide respiratory chain poses a thermodynamical problem of OHRB, when 'high' potential electron donors, such as quinones are part of the respiratory chain (see 'RDases as terminal reductases in organohalide respiratory chains'). The thermodynamical problem is even greater assuming that the electrons are transferred to the corrinoid via iron–sulfur clusters in the RDase, since the midpoint redox potential of these clusters may even be lower (see 'Biochemical characteristics of RDases'). Recently, cytoplasmic non-respiratory RDases were identified in aerobes that are functionally linked to NAD(P)H-dependent oxidoreductases (Chen et al. 2013; Payne et al. 2015). The electrons derived from NAD(P)H oxidation might be transferred via a bound FMN or ferredoxin to such RDases, which also contain corrinoid cofactors in the active site. A transfer of reducing equivalents from NADH ($E^{\circ\prime} = -320$ mV) to [Co^{II}] is more feasible than from quinones as for the respiratory RDases. In the future, the differences in the electron transfer pathways leading to respiratory or non-respiratory RDases have to be unraveled for a better understanding of their function in bacterial energy metabolism.

On the basis of the initial findings listed above a variety of RDase genes (see Part IV, Chap. 16) and gene products were described, the role and characteristics of which were subject to several reviews (Wohlfarth and Diekert 1997; Holliger et al. 1998b;

Smidt and de Vos 2004; Futagami et al. 2008). The aim of this chapter is to summarize the recent knowledge about the biochemistry of corrinoid-containing RDase enzymes, which serve on the one hand as terminal reductases in the energy metabolism of organohalide-respiring bacteria (OHRB) or on the other hand appeared to have another function in non-organohalide respirers.

17.2 Biochemical Characteristics of RDases

The RDases (e.g., tetrachloroethene reductive dehalogenase PceA—EC 1.97.1.8) catalyze the reductive dehalogenation of aliphatic and aromatic organohalides either by hydrogenolysis or dihaloelimination (Fig. 17.3).

The number of RDases encoded in the genomes of OHRB ranges from less than 10 in the metabolic versatile OHRB, e.g., *Sulfurospirillum* spp. and *Desulfitobacterium* spp., up to more than 30 in the OHRB strictly depending on this respiratory metabolism such as *Dehalococcoides mccartyi* and *Dehalobacter* spp. (Hug et al. 2013; see Part IV, Chap. 16). All RDases purified and characterized so far (Table 1) (with one exception described by Ni et al. 1995; see above) are iron–sulfur proteins that harbor a corrinoid cofactor. Recently, the 3D-structure of the PceA enzyme of *S. multivorans* was elucidated and displayed the corrinoid cofactor in the base-off conformation non-covalently attached to the enzyme at the active site (Bommer et al. 2014; see Part V, Chap. 20). This mode of B_{12}-binding was also found in the structure of the soluble cytoplasmic *ortho*-dibromophenol RDase of the marine alphaproteobacterium *Nitratireductor pacificus* pht-3B (Payne et al. 2015). The latter enzyme was obtained from heterologous production in the non-dechlorinating, but corrinoid-producing *Bacillus megaterium*. In *N. pacificus* the enzyme does not appear to have a respiratory function. The corrinoid cofactor of the RDases analyzed so far is bound in the base-off conformation and contains cobalt in the [Co^{II}] oxidation state, as it was shown by electron paramagnetic resonance spectroscopy (EPR) for different purified enzymes in their 'as isolated' state (Schumacher et al. 1997; van de Pas et al. 1999; Kräutler et al. 2003; Payne et al. 2015). In an actual study, the heterologously produced and reconstituted vinyl chloride (VC) RDase (VcrA) from *D. mccartyi* VS was analyzed by EPR and shown to harbor a corrinoid cofactor in the base-on conformation (Parthasarathy et al. 2015), a result that

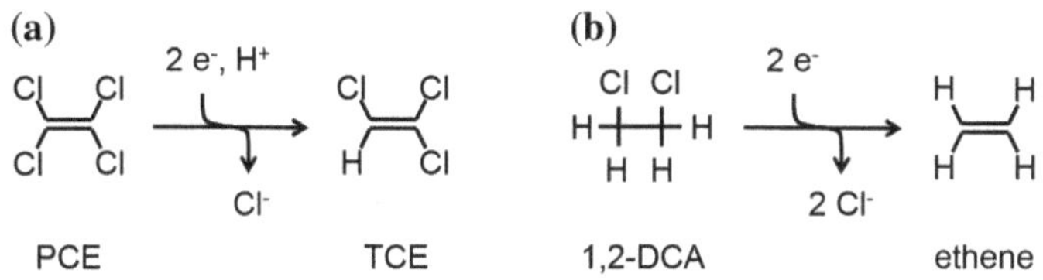

Fig. 17.3 Examples for reductive dehalogenation reactions. **a** Hydrogenolysis of tetrachloroethene (PCE) to trichloroethene (TCE). **b** Dihaloelimination of 1,2-dichloroethane (1,2-DCA) to ethene

Table 1 Biochemical properties of purified RDases

RDase	Organism	Specific RDase activity (nkat/mg)		Apparent K_m (μM)	Cofactor(s)	Reference(s)
		crude extract	purified enzyme			
Respiratory RDases						
3-Chlorobenzoate RDase	*Desulfomonile tiedjei*	3	307	n.d.	heme, Fe/S ?	Ni et al. 1995
Tetrachloroethene RDase	*Sulfurospirillum multivorans*	25	2640	200 (PCE), 240 (TCE)	norpseudo-B_{12}, 2 × [4Fe-4S]	Neumann et al. 1996, Kräutler et al. 2003, Bommer et al. 2014
Tetrachloroethene RDase	*Desulfitobacterium hafniense* PCE-S	4	650	10 (PCE), 4 (TCE)	corrinoid; 8 Fe, 10 S[c]	Miller et al. 1998
ClOHPA RDase	*Desulfitobacterium hafniense* DCB-2	9	103	n.d.	corrinoid; 12 Fe, 13 S[c]	Christiansen et al. 1998
Tetrachloroethene RDase	*Dehalococcoides mccartyi* 195	5[a]	342	n.d.	corrinoid[d], Fe/S ?	Magnuson et al. 1998
Trichloroethene RDase	*Dehalococcoides mccartyi* 195	9[a]	202	n.d.	corrinoid[d], Fe/S ?	Magnuson et al. 1998
ClOHPA RDase	*Desulfitobacterium dehalogenans*	5	465	20 (ClOHPA)	corrinoid, 1 × [4Fe-4S], 1 × [3Fe-4S]	van de Pas et al. 1999
Tetrachloroethene RDase	*Dehalobacter restrictus* PER-K23	10	227	20 (PCE)	corrinoid, 2 × [4Fe-4S]	Schumacher et al. 1997, Maillard et al. 2003
ClOHPA RDase	*Desulfitobacterium hafniense* PCE1	2	68	n.d.	corrinoid[d], Fe/S ?	van de Pas et al. 2001
Tetrachloroethene RDase	*Desulfitobacterium hafniense* PCE1	5	92	n.d.	corrinoid[d], Fe/S ?	van de Pas et al. 2001
Tetrachloroethene RDase	*Desulfitobacterium hafniense* TCE1	2	167	n.d.	corrinoid ?, Fe/S ?	van de Pas et al. 2001

(continued)

Table 1 (continued)

RDase	Organism	Specific RDase activity (nkat/mg)		Apparent K_m (μM)	Cofactor(s)	Reference(s)
		crude extract	purified enzyme			
ClOHPA RDase	*Desulfitobacterium chlororespirans*	1.4	257	310 (ClOHPA)	corrinoid, 8 Fe, 8 S[c]	Krasotkina et al. 2001
Tetrachloroethene RDase	*Desulfitobacterium hafniense* Y51	0.8	2	106 (PCE), 535 (TCE)	corrinoid[d], Fe/S ?	Suyama et al. 2002
3, 5-Dichlorophenol RDase	*Desulfitobacterium hafniense* PCP-1	0.5[b]	7	49.3 (3, 5-DCP)	corrinoid[d], Fe/S ?	Thibodeau et al. 2004
Vinyl Chloride RDase	*Dehalococcoides mccartyi* VS	4	17	n.d.	corrinoid ?, Fe/S ?	Müller et al. 2004
Vinyl Chloride RDase	*Dehalococcoides mccartyi* VS (heterologously produced/ reconstituted)	-	37.5	13 (VC), 45 (1,1-DCE), 1300 (1,2-DCA)	B_{12}, 2 x [4Fe-4S]	Parthasarathy et al. 2015
Non-respiratory RDases						
Bromoxynil RDase	*Comamonas* sp. 7D-2	0.8	27	38 (DBHB), 149 (BHB)	no corrinoid ?[e], Fe/S ?	Chen et al. 2013
ortho-Dibromophenol RDase	*Nitratireductor pacificus* (heterologously produced)	n.d.	approx. 2 (DBHB)	approx. 25 (DBHB)	B_{12}, 2 × [4Fe-4S]	Payne et al. 2015

n. d. not determined, *TCE* trichloroethene, *ClOHPA* 3-chloro-4-hydroxy-phenylacetate, *3, 5-DCP* 3, 5-dichlorophenol, *VC* vinyl chloride, *1, 1-DCE* 1,1-dichloroethene, *1,2-DCA* 1,2-dichloroethane, *DBHB* 3, 5-dibromo-4-hydroxybenzoate, *BHB* 3-bromo-4-hydroxybenzoate

[a]cell membranes, [b]solubilized membrane fraction, [c]determination of iron and acid-labile sulfide, [d]light-sensitive inhibition with iodoalkanes, [e]no light-sensitive inhibition with iodoalkanes

differs from what was reported for other RDases. The cobalt of the corrinoid cofactor in the reconstituted VcrA was in the [Co^{II}] oxidation state, which is consistent with the data obtained before. The paramagnetic [Co^{II}] is detectable by EPR, while the diamagnetic [Co^{I}] and [Co^{III}] oxidation states are EPR-silent. The midpoint redox potential of the [Co^{II}]/[Co^{I}] couple of the PceA-bound norpseudo-B_{12} cofactor in *S. multivorans* was −380 mV (versus the standard hydrogen electrode, SHE) at pH 7.5 as determined by EPR (Kräutler et al. 2003). This redox potential of the cofactor was about 100 mV more positive than the midpoint potential of the [Co^{II}]/[Co^{I}] couple of the extracted and purified norpseudovitamin B_{12} (−480 mV vs. SHE, pH 7.0), which indicates an influence of the protein environment on the redox potential. In earlier experiments, a similar midpoint redox potential of the [Co^{II}]/[Co^{I}] couple was reported for PceA of *Dehalobacter restrictus* (-350 ± 20 mV, pH 8.0; Schumacher et al. 1997) and CprA of *Desulfitobacterium dehalogenans* (−370 mV, pH 7.8; van de Pas et al. 1999). The structure of the corrinoid cofactor in PceA of *D. restrictus* has not been elucidated so far, however, the corrinoid-auxotrophic organism is expected to 'recruit' vitamin B_{12} added to the growth medium (Holliger et al. 1998a; Rupakula et al. 2013). The corrinoid cofactor bound to CprA of *D. dehalogenans* has also not been structurally characterized up to date. The oxidation of [Co^{II}] at positive redox potentials, i.e., the formation of the [Co^{III}] oxidation state, has never been observed for any RDase tested (e.g., potentials > 150 mV were applied to PceA of *D. restrictus*; Schumacher et al. 1997). The [Co^{III}]/[Co^{II}] redox couple of the enzyme-bound corrinoid seems to have an unusually high redox potential. Whether the [Co^{III}] state plays a role in the catalytic cycle of the enzymes remains unclear (see 'Catalytic mechanism of RDases'). The electron transfer to the active site buried inside the RDase proceeds most probably via iron–sulfur clusters bound to the enzyme. Conserved cysteine residues residing in the C-terminal half of all RDase proteins bind two [4Fe–4S] clusters or alternatively one [4Fe–4S] and one [3Fe–4S] cluster; the latter was only reported for the *ortho*-chlorophenol reductive dehalogenase (CprA) of *D. dehalogenans* (van de Pas et al. 1999). In redox titration experiments coupled with EPR both [4Fe–4S] clusters of PceA of *D. restrictus* showed a midpoint redox potential of about −480 mV at pH 8.0 (Schumacher et al. 1997). A comparably low value (−440 mV at pH 7.8) was also measured for the [4Fe–4S] cluster of CprA of *D. dehalogenans*, however, the [3Fe–4S] cluster of this enzyme displayed a positive midpoint redox potential (+70 mV at pH 7.8; van de Pas et al. 1999).

Most of the respiratory RDase enzymes were reported to be oxygen-sensitive with a half-life time in air ranging from approximately one hour to several hours depending on the experimental conditions (see references listed in Table 1). The 3-chlorobenzoate reductase of *D. tiedjei* was purified under air and did not seem to be affected (Ni et al. 1995). An insensitivity toward oxygen was also observed for the recently described non-respiratory RDases of the aerobic *Comamonas* sp. 7D-2 (Chen et al. 2013) and the marine *N. pacificus* (Payne et al. 2015), raising the question how these enzymes manage to avoid oxidative damage when compared to the respiratory RDases. The pH optima measured for the purified RDases are in between pH 7.2 and 8.2. The molecular mass of respiratory RDases ranges from 46 to 65 kDa with respect to the protein monomer. However, the structural analysis of PceA of *S. multivorans* suggested a homodimeric form of this respiratory RDase under native conditions (Bommer et al. 2014). The mature respiratory RDases are located in the exoplasm of the bacterial cells (periplasm in Gram-negative bacteria) and are associated to the cytoplasmic membrane (Nijenhuis and Zinder 2005; John et al. 2006; Reinhold et al. 2012). The formation of oligomeric states by the respiratory RDases inside the cell prior to the membrane translocation was not verified so far. The membrane association of the enzyme is most probably based on the interaction with the putative membrane-integral B protein (Neumann et al. 1998), a variant of which is encoded in almost every RDase gene cluster. Up to date, the affinity of the RDases to the surface of the membrane or to an anchor protein within this membrane (e.g., the B protein) has not been quantified. The mature form of the PCE reductive dehalogenase (PceA) in *D. hafniense* Y51 was described as soluble protein present in the exoplasm of the cell (Suyama et al. 2002), although the *pceB* gene product was assumed to be present in the organism. This result pointed to a weak or only transient interaction of the RDase with the membrane or membrane-integral proteins such as PceB. The mature PceA enzyme of *S. multivorans* was shown to be membrane associated in whole cells by freeze-fracture replica immunogold labeling (John et al. 2006). Nevertheless, the enzyme was efficiently sheared from the cytoplasmic membrane upon cell lysis, which implied a loose membrane attachment. This fact allowed for the purification of the mature PceA from the soluble fraction rather than the membranes of this organism (Neumann et al. 1996). These results indicate that the membrane association of at least some of the respiratory RDases is not very tight. However, prior to the purification of several RDases a solubilization of the enzyme from the membrane was indispensable. In most cases Triton X-100 (0.1–1 % [w/v]) was applied as detergent for extraction of the enzyme from the membrane fraction.

The RDases can be divided into enzymes converting aliphatic halogenated organic compounds and RDases dehalogenating aromatic organohalides. Overlapping substrate spectra, which means the conversion of both types of substrates by a single enzyme, have been rarely described. The conversion of aliphatic and aromatic organohalides was suggested for the 3-chlorobenzoate RDase of *D. tiedjei*, for which the dehalogenation of PCE has been detected in cell extracts containing the 3-chlorobenzoate RDase (Townsend and Suflita 1996). However, the conversion of PCE has never been proven with the purified enzyme. In general, RDases, which convert aromatic organohalides such as the *ortho*-chlorophenol RDase (CprA) of *D. dehalogenans* (van de Pas et al. 1999) did not dehalogenate chlorinated ethenes at significant rates. In contrast, chloroaromatic compounds such as chlorobenzoates or chlorophenols were not converted by PCE and TCE reductive dehalogenases (Neumann et al. 1996; Miller et al. 1998; Maillard et al. 2003). PceA of *S. multivorans* produces *cis*-1,2-dichloroethene (*c*DCE) via TCE from PCE and also dehalogenates brominated ethenes and chloropropenes but does not convert 3-chlorobenzoate (Neumann et al. 1996; Ye et al. 2010). Overlapping substrate spectra between RDases converting either aliphatic or aromatic organohalides cannot be ruled out completely, but there is a clear preference among the purified enzymes so far.

The exclusive formation of *c*DCE instead of *trans*-1,2-dichloroethene (*t*DCE) or 1,1-dichloroethene (1,1-DCE) from PCE or TCE dechlorination was reported for the purified PCE reductive dehalogenases of *D. hafniense* strains PCE-S, PCE1, and Y51 as well as *D. restrictus* PER-K23 and *D. mccartyi* 195 (Miller et al. 1998; Magnuson et al. 1998; van de Pas et al. 2001a, b; Suyama et al. 2002; Maillard et al. 2003). The predominant formation of *c*DCE from PCE or TCE reductive dechlorination was also observed for the abiotic reaction with free vitamin B_{12} (Glod et al. 1996) and might be explained by the comparably high stability of the respective *cis*-1,2-dichloroethen-1-yl radical upon TCE dechlorination (Nonnenberg et al. 2002). However, *Dehalococcoides*-containing mixed cultures have been described, which convert PCE into *trans*-1,2-dichloroethene (*t*DCE) rather than *c*DCE (Griffin et al. 2004; Cheng et al. 2010). The formation of *t*DCE from PCE was also observed for the pure strain *D. mccartyi* CBDB1 (Marco-Urrea et al. 2011). The RDases responsible for this conversion seem to overcome the mechanistic constraints of protein-free corrinoids forming preferentially *c*DCE upon TCE reductive dehalogenation. Complete dechlorination of PCE to ethene has so far not been reported for PCE reductive dehalogenases. Production of ethene was only observed for TCE, DCE, and VC-converting RDases, which did not dehalogenate PCE (Magnuson et al. 1998; Müller et al. 2004).

A distinct group of RDases is represented by those enzymes catalyzing the dihaloelimination of vicinal dichlorinated alkanes, e g., the 1,2-dichloroethane RDase (DcaA) of *D. dichloroeliminans* strain DCA1 (De Wildeman et al. 2003), which produces ethene from the substrate. Neither VC nor monochloroethane was formed as a free intermediate indicating conversion of the substrate by a single enzyme. Although DcaA or a similar RDase enzyme has not been purified and characterized so far, the involvement of a corrinoid cofactor was deduced from propyl iodide inhibition studies (De Wildeman et al. 2003). The conversion of chlorinated ethenes or phenols was not observed in cell extracts of *D. dichloroeliminans* containing DcaA.

Besides the inhibition by propyl iodide, which is most probably based on alkylation of the cobalt ion in the corrinoid cofactor (see 'Introduction'), RDase-mediated reductive dehalogenation in cell extracts was inhibited by cyanide, sulfite, nitrite, and nitrous oxide (Neumann et al. 1995; Suyama et al. 2002). Nitrous oxide was reported to oxidize the superreduced [Co^{I}] state of corrinoids (Schrauzer et al. 1971). Cyanide, sulfite, and nitrite ions were shown to efficiently replace the natural ligands of the central cobalt ion in corrinoids (Firth et al. 1969). Binding of an artificial ligand would have been reversed upon reduction of the corrinoid cofactor, however, the cyanide-mediated inhibition of PceA of *S. multivorans* was not reversed in the presence of titanium citrate (Neumann et al. 1995). Thus, an impact of cyanide on another part of the protein was also considered. The PceA enzyme of *S. multivorans* was efficiently inhibited by chloromethanes (50 % inhibition with either 0.8 μM CH_3Cl, 50 μM CH_2Cl_2, 25 μM $CHCl_3$, or 100 μM CCl_4) (Neumann et al. 1996). It is not known, whether this sensitivity toward chloromethanes is a general property of RDases. However, since *Dehalobacter* spp. appeared to convert chloroform (CCl_3) (Lee et al. 2012) and *Desulfitobacterium* sp. strain PR was shown to reductively dechlorinate chloromethanes by an RDase (Ding et al. 2014), a general inhibitory effect is unlikely.

The purification of RDases is often hampered by the low cell yield obtained from the cultivation of reductively dehalogenating bacteria, especially of those strictly depending on organohalides as electron acceptors, e.g., *D. mccartyi* (Löffler et al. 2013). Furthermore, the assignment of a substrate spectrum to a specific RDase may be hindered by the concomitant expression of several RDase genes in a single reductively dehalogenating organism. Functional heterologous production of respiratory RDases was accomplished recently (Mac Nelly et al. 2014; see 'Heterologous production of RDases'), but it remains complicated and challenging and was not applied to *D. mccartyi* RDases up to date. Nevertheless, besides the purified RDases several enzymes were partially characterized in cell extracts with respect to their substrate range via blue-native polyacrylamide gel electrophoresis (BN-PAGE) coupled to the measurement of RDase activity in gel slices and the identification of the proteins involved via tryptic digestion, liquid chromatography of the peptides, and tandem mass spectrometry (LC-MS/MS) (Adrian et al. 2007). RDases, which were produced in the presence of TCE, *c*DCE, *t*DCE, VC, or 1,2-dichloroethane by the KB-1 consortium (Duhamel et al. 2002), have been characterized using this approach (Tang et al. 2013). The KB-1 consortium was shown before to contain mainly reductively dehalogenating *Dehalococcoides* spp. and *Geobacter* spp. (Duhamel and Edwards 2006) and was applied for successful bioremediation of a tetrachloroethene-contaminated field site (Major et al. 2002). In the consortium, the RDase BvcA of *D. mccartyi* BAV1 was shown to dehalogenate VC, all dichloroethene isomers, TCE, and even 1,2-dichloroethane to ethene, while PCE was not converted (Tang et al. 2013). Hence, the BvcA enzyme seems to combine both the ability for reductive hydrogenolysis and dihaloelimination (Fig. 17.3) making it an interesting target for mechanistic studies. Recently, the heterologously produced and reconstituted vinyl chloride RDase of *D. mccartyi* VS was shown to catalyze on the one hand the reductive hydrogenolysis of vinyl chloride or 1,1-dichloroethene and on the other hand the dihaloelimination of 1,2-dichloroethane (Parthasarathy et al. 2015). This study with a purified RDase proved the combination of both catalytic mechanisms in a single enzyme.

17.3 Catalytic Mechanism of RDases

For catalysis of the reductive dehalogenation, the RDases contain two types of metal cofactors which were proposed to be essential for catalysis (see 'Biochemical characteristics of RDases'). From the structural analysis of PceA of *S. multivorans* (Bommer et al. 2014) and the *ortho*-bromophenol RDase of *N. pacificus* (Payne et al. 2015) a consecutive electron transfer via the 'wire' of two [4Fe-4S] clusters to the corrinoid cofactor at the active site was deduced. A previously envisaged splitted electron flow within the enzyme, i.e., each electron is transferred separately via a single iron–sulfur cluster to the cobalt (Neumann et al. 1996), cannot be excluded, but it seems rather unlikely taken the recent structural data into account (see Part V, Chap. 20). The physiological electron donor, which transfers the electrons to the membrane-associated RDase enzyme in the organohalide respiratory chains, is unknown so far and might differ among the OHRB (see 'RDases as terminal reductases in organohalide respiratory chains').

Abiotic reductive dehalogenation especially of polyhalogenated organohalides, e.g., tetrachlorethene or hexachlorobenzene, was shown earlier for porphinoid macrocyles containing the transition metals cobalt (B_{12}), nickel (factor F_{430}), or iron (hematin) in the reduced state (Krone et al. 1989a, b; Gantzer and Wackett 1991). Such a reactivity toward organohalides has not been observed for iron–sulfur clusters or copper-containing cofactors, pointing toward the aforementioned porphinoid metal complexes as suitable cofactors for enzymatic reductive dehalogenation. All RDase enzymes, purified and characterized so far were shown to harbor a cobalt-containing porphyrin-derived corrinoid cofactor (see 'Biochemical characteristics of RDases'), with a single exception: the 3-chlorobenzoate RDase of *D. tiedjei* (Ni et al. 1995).

The reaction rate of the PceA-mediated reductive dehalogenation of PCE was found to be considerably higher in comparison to the abiotic PCE dechlorination by the extracted and purified norpseudo-B_{12} cofactor of *S. multivorans* (Neumann et al. 2002), which raised the question for similarities or differences in the reaction mechanism between abiotic and enzymatic reductive dehalogenation. From the analysis of dual-element isotope effects during the reductive dehalogenation of PCE by PceA of *S. multivorans* in comparison to the extracted norpseudo-B_{12} cofactor, mechanistic differences were deduced (Renpenning et al. 2014; see Part V, Chap. 18). However, such differences were not observed for the TCE conversion. Studies on the corrinoid-mediated abiotic reductive dehalogenation of PCE and TCE (summarized in Kliegman and McNeill 2008) and indirect biochemical evidences (Schumacher et al. 1997; Schmitz et al. 2007; Ye et al. 2010) allowed for the deduction of a tentative reaction scheme including two possible pathways (Fig. 17.4). Nevertheless, the complete catalytic cycle of the corrinoid-dependent reductive dehalogenation remains enigmatic up to date.

The superreduced corrinoid ([Co^I]), which was identified as the reactive species initially attacking the organohalide in abiotic (Wood et al. 1968) and enzymatic (Miller et al. 1996; Schumacher et al. 1997, van de Pas et al. 1999; Parthasarathy et al. 2015) reductive dehalogenation, can either act as reductant in an electron transfer mechanism (Fig. 17.4, pathway A) or as strong nucleophile attacking the electron-deficient carbon backbone of the substrate (Fig. 17.4, pathway B). The electron-transfer mechanism in reductive dechlorination of PCE (Fig. 17.4, pathway A) is expected to lead to the formation of a trichlorovinyl radical via an intermediate carbanion radical upon elimination of a chlorine substituent. After reduction and protonation of the trichlorovinyl radical the final product TCE is formed. The second electron transfer step might proceed via the cobalt ion in the center of the corrinoid cofactor or the proximal [4Fe–4S] cluster, which is located in close proximity to the corrinoid cofactor and the substrate binding site of the

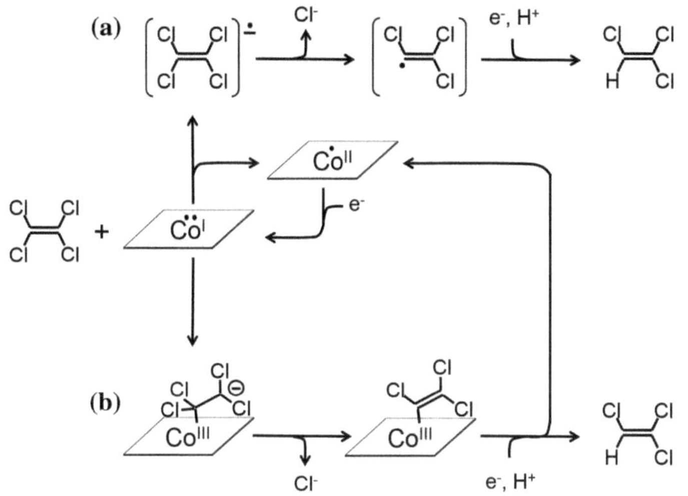

Fig. 17.4 Possible reaction mechanisms of tetrachloroethene reductive dehalogenation catalyzed by RDase enzymes via **a** external electron transfer or **b** corrinoid alkylation

RDase (Bommer et al. 2014; see Part V, Chap. 20). When upon reaction with PCE the superreduced [Co^{I}] is oxidized, [Co^{II}] is formed (Schumacher et al. 1997). The formation of an alkylated [Co^{III}]-state through combination of [Co^{II}] and the trichlorovinyl radical has not been proven, but cannot be ultimately excluded. The second possible pathway, in which [Co^{I}] reacts as a nucleophile (Fig. 17.4, pathway B), leads to the formation of a Co–C bond and the elimination of one chloride anion from PCE, probably via the formation of a cobalt–tetrachloroethyl complex as an intermediate. The Co–C bond in the resulting cobalt–trichlorovinyl complex might then undergo homolytic cleavage yielding a trichlorovinyl radical and a [Co^{II}] or the cobalt trichlorovinyl complex directly abstracts a hydrogen atom and releases TCE.

An indication for the formation of a substrate radical as intermediate, either via pathway A or B (Fig. 17.4), was provided by the detection of adducts formed from methyl viologen radicals ($MV^{\bullet+}$) and chloropropenyl radicals upon the reductive dechlorination of chloropropenes by the purified PceA enzyme of *S. multivorans* (Fig. 17.4, Schmitz et al. 2007). In the RDase enzyme assay, the $MV^{\bullet+}$ radical is formed by a single electron transfer to MV^{2+} from titanium citrate used as primary artificial electron donor (Fig. 17.1). When PCE or TCE were applied as substrates in the assay, both compounds were converted stoichiometrically to *c*DCE and the methyl viologen oxidized in the course of the reaction could be completely recovered by reduction. In contrast, when 2,3-dichloropropene, *trans*-1,3-dichloropropene, and 1,1,3-trichloropropene were dehalogenated by PceA, the amount of the dechlorination product formed was lower than the amount of substrate consumed in the assay and only a part of the $MV^{\bullet+}$ oxidized during the reaction could be re-reduced. This was explained by adduct formation from $MV^{\bullet+}$ and chloropropenyl radical intermediates produced in the catalytic cycle. Adducts were not observed when PCE was applied as substrate, which might be caused either by a different reaction mechanism or by a tighter binding of the short-lived trichlorovinyl radical to the active site. Another indication for the transient formation of a substrate radical in the catalytic cycle of PceA is the production of all three isomers of dibromoethene from tribromoethene (Ye et al. 2010). While purified PceA of *S. multivorans* converts PCE via TCE exclusively to *c*DCE, the enzyme forms vinyl bromide via *cis*-1,2-dibromoethene, *trans*-1,2-dibromoethene, or 1,1-dibromoethene upon reductive dehalogenation of tribromoethene. The formation of all product isomers was interpreted as indication for an external electron transfer mechanism. Taken all these indirect evidences into account, the involvement of a substrate radical during enzymatic reductive dehalogenation of PCE is feasible, even though not yet proven.

The PCE, TCE, and tribromoethene turnover rate of PceA was increased in the presence of ammonium ions (4 mM) by a factor of about two (Neumann et al. 1996; Ye et al. 2010). Such a positive effect has not been observed for conversion of all three dibromoethene isomers by the same enzyme (Ye et al. 2010). When the purified 3-chloro-4-hydroxy-phenylacetate RDase of *D. hafniense* DCB-2 was tested in the presence of ammonium sulfate, an increase of the enzyme activity of about 40 % was achieved (Christiansen et al. 1998). Such an effect was not seen for the PCE and TCE reductive dehalogenation catalyzed by the purified PceA of *D. hafniense* PCE-S (Miller et al. 1998). The effect of ammonium ions on the reactivity of RDases remains mysterious so far, however, the differential effects observed for PceA of *S. multivorans* converting either polyhalogenated ethenes such as PCE, TCE, and tribromoethene or less halogenated ethenes (i.e., the dibromoethene isomers) might imply mechanistic differences in the conversion of different substrates.

Recently obtained structural data support the involvement of an external electron transfer rather than a nucleophilic attack by [Co^I] in the reaction mechanism. The formation of a cobalt–carbon bond during catalytic turnover of PCE by PceA of *S. multivorans* (Fig. 17.4b) seems unlikely given the steric constraints imposed by the active site (Bommer et al. 2014). Similar structural restrictions became obvious when the *ortho*-bromophenol RDase of *N. pacificus* was structurally analyzed in complex with its substrate 3,5-dibromo-4-hydroxybenzoic acid (Payne et al. 2015). From the positioning of the substrate and the corrinoid cofactor in the active site, the formation of a transient cobalt–carbon bond between both was excluded. Using EPR spectroscopy, evidence became available for a direct interaction of the cobalt ion with the halogen substituent in the substrate (Payne et al. 2015). Based on these findings a novel reaction mechanism was proposed, which includes a removal of the halogen substituent by the formation of a cobalt–halogen bond (see Part V, Chap. 20). In order to prove the applicability of this tentative mechanism for the huge diversity of organohalides, such experiments have to be conducted and extended with other RDases displaying different substrate preferences.

17.4 RDases as Terminal Reductases in Organohalide Respiratory Chains

Up to date, genes encoding RDase enzymes were detected in several different phyla of the bacterial kingdom; a single RDase gene was identified in *Ferroglobus placidus*, a hyperthermophilic archaeon (Hug et al. 2013). Different roles may be assigned to RDases in the bacterial catabolism: (i) terminal reductase in the organohalide respiratory chain, (ii) electron sink under fermentative growth conditions, (iii) reductive dehalogenation as preceding step of complete degradation of aromatic organohalides, and iv) detoxification. The use of an RDase enzyme as terminal reductase of a respiratory chain is the unifying property of OHRB (see Part II, Chap. 5). Numerous biochemical studies on organohalide respiration accompanied by genome sequencing projects implied an unforeseen diversity in the composition and topology of the electron transfer pathways present in the phylogenetically diverse organohalide respirers. OHRB are able to use different electron donors for organohalide respiration, such as hydrogen, formate, pyruvate, and lactate. When hydrogen or formate is used, energy conservation is strictly depending on electron transport phosphorylation (ETP). The oxidation of pyruvate or other complex organic substrates might be also coupled to ETP via organohalide respiration, however, it would also allow for energy conservation via substrate level phosphorylation. Hence, organohalide-dependent or -stimulated growth with such complex organic electron donors is not a clear proof for energy conservation via organohalide respiration in reductively dehalogenating bacteria; organohalides might as well just serve as electron sink for fermentation processes. The involvement of a chemiosmotic mechanism has to be proven carefully in such cases.

Early studies with *D. tiedjei* showed that formate oxidation was stimulated by the reductive dehalogenation of 3-chlorobenzoate in this organism (Mohn and Tiedje 1990). This stimulation led to an increase in the intracellular ATP level, which pointed to a coupling of formate oxidation and reductive dehalogenation to energy conservation via a chemiosmotic mechanism (Mohn and Tiedje 1991). By measuring the acidification of the medium during H_2-driven reductive dehalogenation of 3-chlorobenzoate, an H^+/e^--ratio of approx. 1 was determined in *D. tiedjei*. The amount of electrons channeled into the respiratory chain was calculated from the conversion of the halogenated substrate. A comparable H^+/e^--ratio was obtained for the PCE-dehalogenating *D. restrictus* (Schumacher and Holliger 1996). Up to date, no such data are available for *Dehalococooides* spp., which strictly rely on organohalide respiration (see Part II, Chap. 6). For the release of a halide ion from the substrate two electrons are needed, hence, per organohalide converted two protons are liberated to the exoplasm. The liberation of protons (i.e., acidification) is based on proton translocation across the membrane either by proton pumps or redox loops or on proton release upon substrate conversion by exoplasmic oxidases such as hydrogenase or formate dehydrogenase. Taken into account that 3–4 protons are required for synthesis of a single ATP from ADP plus P_i (Weber and Senior 2003; Mayer and Müller 2014; Silverstein 2014), less than one ATP is formed per halide ion released in organohalide respiration. From the dehalogenation of PCE via TCE to *c*DCE driven by hydrogen oxidation a Gibbs free energy ($\Delta G^{\circ\prime}$) of 189 kJ per mol H_2 is gained (Holliger et al. 1998b). This energy would allow for the formation of almost 2.5 ATP, when 70–80 kJ/mol are assumed to be necessary to form one ATP from ADP and P_i in vivo (Thauer et al. 1977; Schink and Friedrich 1994). However, the low growth yields of different PCE-dechlorinating OHRB with hydrogen as electron donor implied a lower value than one ATP per halide ion removed (Scholz-Muramatsu et al. 1995; Holliger et al. 1998a; Maymó-Gatell et al. 1997). The latter observation is in line with the H^+/e^--ratio measured for the *D. restrictus* PCE respiratory chain.

Enzyme activity measurements, cell fractionation, inhibition studies, and quinone identification in combination with proteomic and transcriptomic approaches provided first hints on the components possibly involved in the electron transfer chains of the various OHRB. In several studies, the electron-donating enzymes, i.e., hydrogenase and formate dehydrogenase, were shown to be membrane associated and facing the outside of the cytoplasmic membrane (Miller et al. 1996, 1997; Schumacher and Holliger 1996; Louie and Mohn 1999; van de Pas et al. 2001a, b; Nijenhuis and Zinder 2005). The participation of quinones as electron shuttle in the membrane was tested by inhibition experiments with the quinone analog 2-*n*-heptyl-4-hydroxyquinoline N-oxide (HQNO). In cell suspensions of *D. restrictus* PCE respiration was significantly inhibited in the presence of HQNO (10 nmol/mg protein) (Schumacher and Holliger 1996). With *D. tiedjei* cells an inhibitory effect of HQNO (150 nmol or 1.5 μmol/mg protein) on the organohalide respiration with 3-chlorobenzoate was observed as well (Louie and Mohn 1999). No effect of the inhibitor (10 nmol/mg protein) on the PCE respiration in *S. multivorans* has been observed (Miller et al. 1996). However, more recent results obtained with cell suspensions applying higher concentrations of HQNO (80, 240, or 320 nmol/mg protein) showed a clear inhibition of PCE respiration in this organism (Krauter 2006). In order to prove a specific effect of HQNO on quinol-dependent respiratory chains in *S. multivorans*, the fumarate respiration was tested as a control and was shown to be efficiently inhibited by HQNO. The redox difference spectra of membrane fractions obtained from *D. restrictus* cells reduced with hydrogen and subsequently oxidized with PCE also indicated the involvement of menaquinone in the electron transfer between hydrogen oxidation and PCE reduction (Schumacher and Holliger 1996). Moreover, in *D. restrictus* cell suspensions the PCE dechlorination could be driven by reduced 2,3-dimethyl-1,4-naphthoquinone ($DMNH_2$), a menaquinone analog. These results present evidence for an involvement of quinones in the organohalide respiratory chains of *D. restrictus*, *D. tiedjei*, and *S. multivorans*. This conclusion is supported by the finding that the pathway for quinone biosynthesis was found to be encoded in the genomes of *D. restrictus* and *S. multivorans* (Rupakula et al. 2013; Goris et al. 2014) and by extraction of menaquinones from the membrane fraction (Scholz-Muramatsu et al. 1995; Holliger et al. 1998a). A quinone-like compound with an unknown structure has been extracted from *D. tiedjei* (Louie and Mohn 1999). The presence of menaquinone and the biosynthetic pathway, however, may also be based on the essential role of this compound in other anaerobic respiratory processes in these organisms.

D. mccartyi is not capable of a respiration other than that of organohalides. Hence, the absence or presence of quinones and/or the genes encoding the pathway for quinone biosynthesis would be an important indication for or against the involvement of quinones in organohalide respiration mediated by these organisms. So far, quinones have been extracted only from *D. mccartyi* strains BAV1 and FL2 (White et al. 2005), although their source is unclear. A complete quinone biosynthesis pathway is not encoded in any of the *D. mccartyi* genomes sequenced up to date (Schipp et al. 2013) and quinones could not be detected by mass spectrometric analysis of membrane extracts of *D. mccartyi* CBDB1 (L. Adrian, personal communication). No effect of the quinone analog HQNO on the H_2-driven organohalide respiration of 1,2,3-trichlorobenzene (1,2,3-TCB) by *D. mccartyi* CBDB1 was found when the cells (5×10^7 cells/ml) were tested for conversion of the substrate in the presence of 1 μM HQNO (Jayachandran et al. 2004). This result implies a quinone-independent electron transfer. In addition, $DMNH_2$ did not function as artificial electron donor for 1,2,3-TCB reductive dechlorination in the organism (Jayachandran et al. 2004). This observation supported the hypothesis of a quinone-independent electron transfer in the organohalide respiratory chain of the latter organism and led to the assumption that quinone-dependent and quinone-independent organohalide respiratory chains may occur among the different OHRB (Fig. 17.5).

So far, biochemical data on the RDases are only available for the apparently quinone-dependent organohalide respiratory chains. From the structural information and from biochemical analyses of respiratory RDases, no indication for a direct interaction of the membrane-associated enzymes with the menaquinone pool in the cytoplasmic membrane has been obtained. Hence, the involvement of an additional component for quinol oxidation and transfer of the electrons to the terminal reductase is expected (Fig. 17.5a). The RDases are attached to the outer face of the cytoplasmic membrane most probably via the small membrane-integral B protein (Neumann et al. 1998). The presence of the *B* gene is a general feature identified in almost all RDase gene clusters (see Part IV, Chap. 15). Hydrophobicity plots allowed for the prediction of either two or in most cases three transmembrane helices in B proteins (Neumann et al. 1998; van de Pas et al. 1999). Since classical sequence motifs for cofactor or metal binding are absent in the B proteins, electron-conducting metal centers such as iron–sulfur clusters or heme groups are not expected to be present. Hence, a direct involvement of the B proteins in electron transfer seems to be rather unlikely. Most RDase gene clusters do not include genes encoding putative membrane-associated or membrane-integral electron transfer proteins. However, the C protein, a gene product that might exert such a function, is encoded in a small number of RDase gene clusters (e.g., *D. hafniense* strains TCE1 and Y51 (Maillard et al. 2005; Nonaka et al. 2006) and *Geobacter lovleyi* (Wagner et al. 2012)). Although the C proteins show sequence similarity to membrane-integral regulatory proteins of the NirI/NosR-type (Saunders et al. 1999; Wunsch and Zumft 2005), an involvement in the electron transfer to the RDase cannot be excluded. A peripheral, non-membrane-integral putative FMN-binding domain in the N-terminal half of the C proteins might

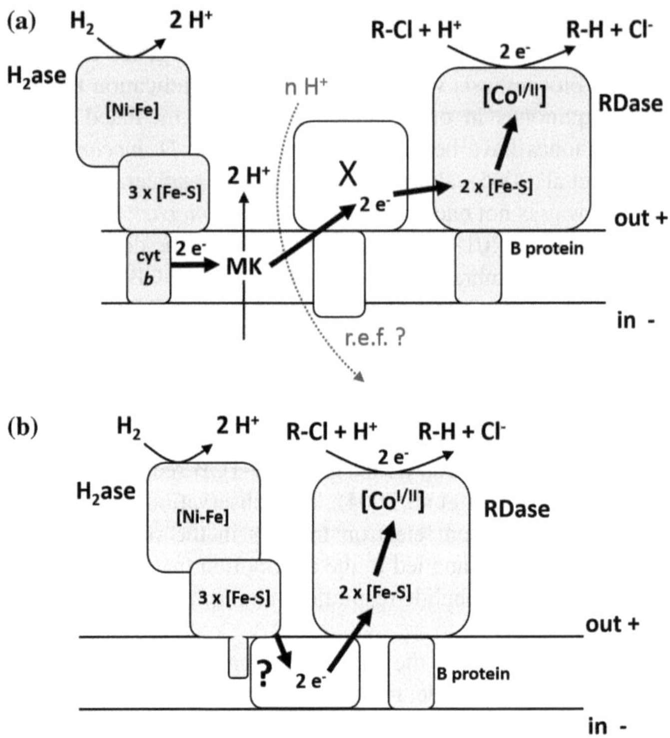

Fig. 17.5 Tentative schemes for the composition of quinone-dependent (**a**) and quinone-independent (**b**) organohalide respiratory chains. *MK* menaquinone, *X* quinol dehydrogenase, *r.e.f.* proton motive force-driven reverse electron flow (indications for r.e.f. only available for *S. multivorans*, Miller et al. 1996)

harbor a flavin cofactor possibly involved in the electron transfer chain. The membrane-integral C-terminal part displays similarities to the membrane protein NapH, which is a subunit of the quinol dehydrogenase NapGH involved in electron transfer to the periplasmic nitrate reductase NapA (Kern and Simon 2008). This similarity might imply a similar function, however, cysteine-containing motifs binding two [4Fe–4S] clusters in NapH are absent in PceC. The involvement of a quinol dehydrogenase in the organohalide respiratory chain was also discussed for *S. multivorans* and *D. tiedjei*, since genes encoding such an enzyme are colocating with the RDase structural genes in these organisms (Goris et al. 2014). A strong indication for a functional coupling of the NapGH-like gene products encoded in close proximity to the *pceAB* genes in *S. multivorans* was obtained from co-regulation of the gene expression of both gene clusters.

The redox couples PCE/TCE and TCE/*c*DCE have a positive standard redox potential of 580 and 540 mV, respectively (Vogel et al. 1987), a fact that makes PCE or TCE suitable electron acceptors for microbial respiration. Aromatic halogenated organic compounds such as chlorinated phenols, benzenes, or benzoates also exhibit positive standard redox potential, which range from 300 to 500 mV (Dolfing and Novak 2014). The MK/MKH_2 couple has a standard redox potential versus SHE of −74 mV at pH 7.0 (Thauer et al. 1977), therefore, the electron transfer to the halogenated substrate appears thermodynamically feasible. However, the [Co^I] state of the corrinoid cofactor in the RDase with a low midpoint redox potential (≤−350 mV) was identified as the reactive species attacking the organohalide and the iron–sulfur clusters as possible electron donors for the corrinoid reduction exhibit an even lower midpoint redox potential (see 'Biochemical characteristics of RDases'). The high potential difference between the menaquinone pool and the cofactors of the RDase might be overcome by a reverse electron flow (r.e.f.), in which the electrochemical proton potential drives a thermodynamically unfavorable electron transfer reaction. When suspensions of *S. multivorans* cells were tested for PCE reduction coupled to hydrogen or formate oxidation in the presence of protonophores such as carbonyl cyanide-*p*-trifluoromethoxyphenyl hydrazone (FCCP; 15 nmol/mg protein), reductive dehalogenation was completely inhibited (Miller et al. 1996). Fumarate respiration was not affected under these conditions. In addition, PCE reduction was not observed in cell-free extracts of the organism or in cells treated with mild detergents. From these results, the requirement of the proton gradient for the organohalide respiration in *S. multivorans* became obvious and the involvement of a reverse electron flow to overcome the thermodynamical barrier between the midpoint redox potential of the quinone and the metal cofactors of the RDase was proposed. The dependence on the proton motive force does not seem to be a general property of organohalide respiratory chains. Since H_2-reduced cells of *D. restrictus* efficiently dehalogenated PCE or TCE in the presence of the uncoupling agent carbonyl cyanide-*m*-chlorophenyl hydrazone (CCCP; 15 nmol/mg protein), the involvement of a reverse electron flow in this quinone-dependent organohalide respiratory chain is doubtful (Schumacher and Holliger 1996). Whether this difference reflects a different composition of the organohalide respiratory chain has to be investigated in the future.

Only little is known about the composition of the electron transfer chain in the organohalide-respiring *D. mccartyi*. Since quinones do not appear to be involved in electron transfer, a direct interaction of electron-donating and electron-accepting oxidoreductases is feasible (Fig. 17.5b). From the measurement of the protein abundance in diverse *D. mccartyi* strains initial results were obtained about the types of electron-donating enzymes involved in the organohalide respiration (Adrian et al. 2007; Morris et al. 2007). *D. mccartyi* is able to use hydrogen as sole electron donor, but not formate (Löffler et al. 2013). Among the highly abundant proteins in *D. mccartyi* cells, a membrane-bound uptake [NiFe] hydrogenase was identified and a putative formate dehydrogenase (Fdh). Since *D. mccartyi* lacks a formate metabolism, the latter enzyme appears to convert a different but not yet identified substrate. An essential cysteine or selenocysteine, which is present in the active site of formate dehydrogenases of other organisms, is replaced by a serine in the *D. mccartyi* enzyme. The role of this oxidoreductase in the *D. mccartyi* organohalide respiratory chain has to be unraveled in further studies.

17.5 Assembly of RDases

The physiologically active form of the respiratory RDases is located at the outer face of the cytoplasmic membrane (Nijenhuis and Zinder 2005; John et al. 2006; Reinhold et al. 2012), most probably attached to the membrane-integral B protein. The precursor protein of the RDase is produced inside the cell with a Tat (twin arginine translocation) signal peptide at its N-terminus (Fig. 17.6). This signal peptide is required for the recognition of the precursor by the Tat export machinery in the membrane (Palmer and Berks 2012). Except for the conserved Tat consensus sequence, including the essential twin arginine, a high sequence variability is observed among the RDase signal peptides, which might reflect an adaptation to the different Tat translocases in the phylogenetically diverse OHRB. The membrane-integral Tat translocase is known to export folded, in most cases cofactor-containing proteins across the cytoplasmic membrane. The biosynthesis of the metal-cofactors of the RDases, i.e., two iron–sulfur clusters (Fontecave and Ollagnier-de-Choudens 2008) and the corrinoid cofactor (Warren et al. 2002; Moore et al. 2013), is mediated in the cytoplasm of the cell. Each type of cofactor is produced by a specific set of enzymes and is transferred, most probably in a fully assembled form, into the RDase apoprotein. Some OHRB are not able to synthesize corrinoids de novo (see Part V, Chap. 19). These organisms are dependent on corrinoid salvaging from the environment (e.g., *D. mccartyi*; Löffler et al. 2013), which is initiated by the ABC transport system BtuCDF (Hvorup et al. 2007; Korkhov et al. 2014). Natural cobamides ('complete' corrinoids containing

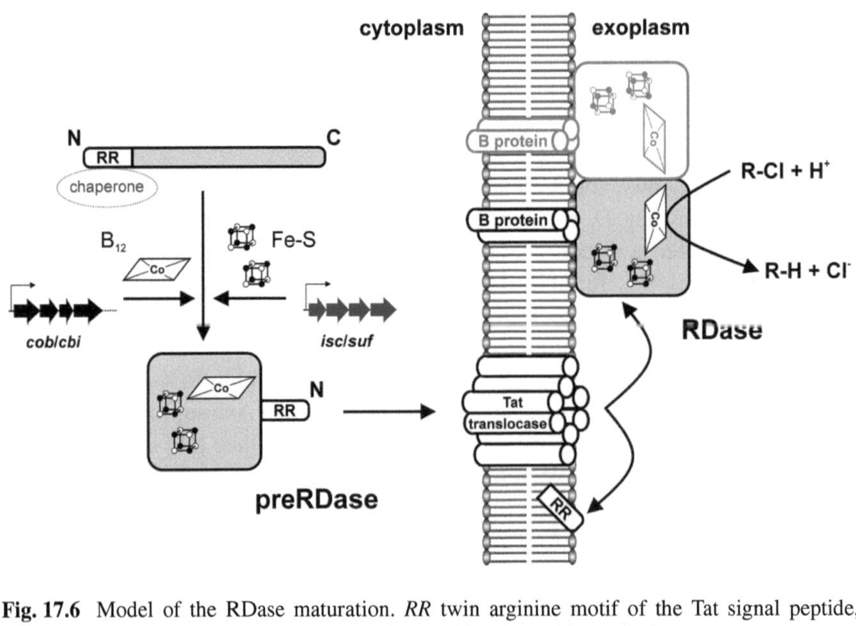

Fig. 17.6 Model of the RDase maturation. *RR* twin arginine motif of the Tat signal peptide, *cob/cbi* corrinoid biosynthesis genes, *isc/suf* iron sulfur cluster biosynthesis genes

an upper and a lower ligand) display structural diversity in the nucleotide loop, which harbors the cobamide's lower ligand base (Fig. 17.2) (Crofts et al. 2013). Recently, evidence became available that among RDase enzymes specificities toward structurally different corrinoid cofactors exist, especially with respect to the type of lower ligand base present in the nucleotide loop substructure (Yan et al. 2012; Yi et al. 2012; Keller et al. 2014).

Either RDases contain two [4Fe–4S] clusters or one [4Fe–4S] cluster and one [3Fe–4S] cluster (Table 1). In general, three biosynthetic machineries were described for the formation of iron–sulfur clusters (summarized in Fontecave and Ollagnier-de-Choudens 2008). The *nif* system is involved in nitrogenase maturation (Jacobson et al. 1989). The *isc* (iron–sulfur-cluster formation)-system (Zheng et al. 1998) seems to fulfill the role of a housekeeping biosynthetic pathway, since it was found in many organisms. The also frequently detected *suf* (sulfur mobilization)-system (Takahashi and Tokumoto 2002) is required for iron–sulfur cluster biosynthesis under unfavorable conditions (e.g., high level of reactive oxygen species or limitations in iron supply) (Nachin et al. 2003; Outten et al. 2004). All three pathways display an overlapping functionality, which allow for mutual complementation, provided that more than one pathway is present in a single organism. The pre-assembly of the iron–sulfur clusters on scaffold proteins (IscU, SufU, NifU) is a common theme in all three biosynthetic pathways. The incorporation of the clusters into the target apoprotein requires a close contact with the proteins involved in their biosynthesis. Such a protein–protein interaction is expected to occur during maturation of the RDase precursor in the cytoplasm but was not experimentally proven so far. A classical scaffold protein, which serves as 'backbone' for cofactor assembly, seems to be absent in the corrinoid biosynthetic pathway (Warren et al. 2002). However, the adenosyltransferase CobA is present, which is on the one hand responsible for the adenosylation of the central cobalt during the late steps of corrinoid production and on the other hand was shown to deliver cobamides to the target enzymes (Padovani et al. 2008). If this mechanism applies to RDases has not yet been tested.

In general, a proper incorporation of the cofactors was found to be a prerequisite for correct folding and subsequent Tat-dependent export of metal cofactor-containing redox enzymes (Sargent 2007). Impairment of cofactor biosynthesis usually leads to an enrichment of the enzyme precursor in the cytoplasm and is often accompanied by degradation of the unfolded apoprotein. When the PCE-dehalogenating *S. multivorans* was isolated, a non-dechlorinating variant of the organism has been obtained (Siebert et al. 2002). The so-called *S. multivorans* strain N lacks the corrinoid cofactor of the PCE reductive dehalogenase, which makes the organism unable to grow with PCE. The precursor of PceA (prePceA) has not been detected in crude extracts of strain N, although the *pceA* gene is intact. The *pceA* transcript was produced, albeit its level was reduced in comparison to the PCE-dechlorinating *S. multivorans* isolate.

A negative effect on the Tat-dependent translocation of PceA in *S. multivorans* has been observed when the organism was cultivated for a few generations with fumarate as terminal electron acceptor instead of PCE (John et al. 2006). Under these conditions, the precursor of the enzyme, prePceA, accumulated in the cytoplasm. The PceA activity in such fumarate-grown cells was comparable to cells cultivated with PCE. Hence, the maturation of the enzyme, i.e., the cofactor acquisition, was not affected. Only the membrane export was hindered. The mechanism underlying this unique phenomenon is still unknown. Comparable growth experiments with the PCE-converting *D. hafniense* Y51 resulted in an accumulation and aggregation of prePceA in the cytoplasm during subcultivation of the organism on fumarate instead of PCE (Reinhold et al. 2012). A significant decrease in the corrinoid cofactor production was monitored concomitantly, which might be responsible for the aggregate formation of catalytically inactive prePceA inside the cells. Besides prePceA, the intracellular protein aggregates contained PceT, which is a PceA-specific chaperone (see below), and CobT (nicotinate–nucleotide dimethylbenzimidazole phosphoribosyltransferase) (Crofts et al. 2013), which is an enzyme of the corrinoid biosynthetic pathway. The co-aggregation of these proteins in significant amounts together with prePceA points to an impaired maturation of the RDase in fumarate-grown cells, which leaves an intermediate protein complex behind.

The maturation of RDases is a multistep process including cofactor incorporation and protein folding. In case of the respiratory RDases, the biosynthesis is completed by the Tat-dependent membrane export of the catalytically active enzyme. The involvement of general molecular chaperones such as the Trigger Factor, DnaK, or GroEL (Kim et al. 2013; Castanié-Cornet et al. 2014) in the biosynthesis of RDases has not been proven up to date. In addition, substrate specific chaperones such as redox enzyme maturation proteins (REMPs) (Turner et al. 2004) do not seem to be associated with most of the RDases. However, a few RDase gene clusters include open reading frames encoding putative folding helper proteins. The *cpr* gene cluster of *D. dehalogenans* harbors the accessory genes *cprD* and *cprE*, the gene products of which show the highest sequence similarity to chaperones of the GroEL type (Smidt et al. 2000). A role of the respective gene products in the maturation of the *ortho*-chlorophenol reductive dehalogenase (CprA) could not yet been assigned to either of the putative chaperones. Also encoded in the *cpr* gene region is CprT, a putative trigger factor like chaperone. CprT displays amino acid sequence similarity to the trigger factor protein, which is a ribosome-associated general chaperone acting on nascent protein chains (Kim et al. 2013). Out of the three protein domains present in the trigger factor structure, i.e., the N-terminal domain responsible for ribosome-binding, the peptidyl–prolyl *cis–trans* isomerase domain, and the C-terminal domain, only two are encoded in the *cprT* gene. The N-terminal-domain of trigger factor is absent in CprT. RDase specific trigger factor-like chaperones of the CprT type are only found in a small number of RDase gene clusters including the clusters that encode the PCE reductive dehalogenase (PceA) of *D. hafniense* strains TCE1 and Y51 (Maillard et al. 2005; Nonaka et al. 2006) and of *G. lovleyi* (Wagner et al. 2012) as well as the chlorophenol RDase (RdhA3) of *D. hafniense* DCB-2 (Kim et al. 2012).

The role of the PceT protein of *D. hafniense* strains Y51 and TCE1 in RDase biosynthesis was investigated in more detail (Morita et al. 2009; Maillard et al. 2011). The peptidyl–prolyl *cis–trans* isomerase domain of PceT was proven to be functional and an interaction of the chaperone with prePceA, the Tat signal peptide bearing precursor of the PCE reductive dehalogenase, was demonstrated. Maillard et al. (2011) reported evidence for binding of PceT to the Tat signal peptide of prePceA and a positive effect on the solubility of the enzyme's precursor in the heterologous expression host *Escherichia coli*. Recently, the functional heterologous production of *Desulfitobacterium* RDases together with CprT/PceT-like chaperones was accomplished (Mac Nelly et al. 2014; see 'Heterologous production of RDases'), which benefited from those earlier observations. Since the Tat signal peptide is required for the recognition of prePceA by the Tat translocase, the release of PceT from the signal peptide prior to prePceA export across the membrane is proposed. Structural data for PceA of *S. multivorans* showed the mature enzyme in a homodimeric form (Bommer et al. 2014). It is unclear, if the respiratory RDases obtain their oligomeric state already inside the cells, which might also be assisted by chaperones. Recently, RDase enzymes were described, which are lacking the Tat signal peptide and are therefore predicted to be located in the cytoplasm of the cell (Chen et al. 2013; Payne et al. 2015). The 3D-structure of the *ortho*-bromophenol RDase of *N. pacificus* showed a monomeric enzyme containing a corrinoid and two [4Fe-4S] clusters as metal cofactors. Hence, it is likely that all RDase apoproteins, including the non-respiratory RDases, need accessory proteins for cofactor acquisition and folding.

17.6 Heterologous Production of RDases

Over the last 20 years several hundred RDase gene sequences were deposited in databases (Hug et al. 2013). RDase enzymes are present in bacteria belonging to diverse phyla and expected to play a role in the global halogen cycle at various habitats (e.g., forest soil, fresh water sediments, and marine subseafloor). However, only a little number of RDases has been purified and biochemically characterized. A systematic analysis of RDases from a microbial community or a certain organism is often hampered by difficulties in the cultivation and isolation of OHRB. The growth yields of OHRB on halogenated substrates are usually low, which makes the production of sufficient biomass for enzyme purification laborious. This applies especially to the metabolically restricted OHRB, which cannot be cultivated with alternative electron acceptors such as nitrate or fumarate to obtain more biomass. In addition, in most OHRB more than one RDase gene is present and the expression profiles of the respective genes in the presence of different halogenated substrates often overlap. Hence, the assignment of a substrate spectrum to a specific RDase is almost impossible when crude extract rather than purified enzyme is used in RDase activity measurements.

During the last two decades efforts were made to produce RDases heterologously using *E. coli* as standard expression host (Neumann et al. 1998; Suyama et al. 2002; Kimoto et al. 2010; Sjuts et al. 2012). Such experiments resulted in the production of catalytically inactive apoprotein, arrested in cytoplasmic protein aggregates (i.e., inclusion bodies). *E. coli* is not able to synthesize corrinoids de novo (Blattner et al. 1997), which was discussed as the reason for the formation of inactive RDase protein in this organism (Neumann et al. 1998). Recent publications reported the corrinoid-producing Gram-negative gammaproteobacterium *Shimwellia blattae*, formerly *Escherichia blattae* (Burgess et al. 1973), and the Gram-positive *Bacillus megaterium* (Payne et al. 2015) as suitable hosts for functional heterologous production of RDases. Mac Nelly et al. (2014) accomplished the production of respiratory RDases of two *Desulfitobacterium* strains, namely the PCE reductive dehalogenase (PceA) of *D. hafniense* Y51 and the chlorophenol RDase (RdhA3) of *D. hafniense* DCB-2. The RDases were produced with the N-terminal Tat signal peptide to ensure correct maturation and folding of the protein. However, both RDases appeared to be not exported in *S. blattae*, since the Tat signal peptide was not cleaved off. The RDase activity in crude extract of the *S. blattae* production strains was stimulated by the coproduction of the dedicated chaperones PceT or RdhT, respectively. Actually, the formation of catalytically active RdhA3 was strictly dependent on the presence of the folding helper protein. Up to date no RDase seemingly lacking a dedicated chaperone was tested in this system. The formation of active RDase in the *S. blattae* production strains was also increased by raising the intracellular corrinoid level. This was achieved by the addition of exogenous corrinoid (i.e., hydroxocobalamin) and 5,6-dimethylbenzimidazole, the precursor of the corrinoid's lower ligand base, to the growth medium and by cultivation of *S. blattae* on glycerol as growth substrate. For the conversion of glycerol a corrinoid-dependent glycerol dehydratase is required in *S. blattae*, the corrinoid demand of which might stimulate the corrinoid production by the organism (Andres et al. 2004). The cultivation of *S. blattae* for RDase production was performed under anaerobic conditions. This was dispensable when the oxygen-insensitive non-respiratory *ortho*-bromophenol RDase of *N. pacificus* was heterologously produced in *B. megaterium* cultivated in complex medium (Payne et al. 2015). Based on this achieved progress, the functional heterologous production might pave the way for a better understanding of RDase function and reaction mechanism in organohalide respiration and for an in-depth analysis of the biosynthesis of these exceptional enzymes. Recently, the vinyl chloride RDase (VcrA) from *D. mccartyi* VS was functionally reconstituted by the incorporation of a corrinoid cofactor and two iron–sulfur clusters into the heterologously produced VcrA apoprotein (Parthasarathy et al. 2015). In the future, this technique might allow for the biochemical analysis of different RDases from *D. mccartyi* strains, which were not functionally produced so far by heterologous expression of the respective genes.

Acknowledgements This work was supported by the DFG Research Unit FOR1530 and the DFG grants DI314/12-2 and SCHU2605/1-1.

References

Adrian L, Rahnenführer J, Gobom J, Hölscher T (2007) Identification of a chlorobenzene reductive dehalogenase in *Dehalococcoides* sp. strain CBDB1. Appl Environ Microbiol 73:7717–7724

Andres S, Wiezer A, Bendfeldt H, Waschkowitz T, Toeche-Mittler C, Daniel R (2004) Insights into the genome of the enteric bacterium *Escherichia blattae*: cobalamin (B_{12}) biosynthesis, B_{12}-dependent reactions, and inactivation of the gene region encoding B_{12}-dependent glycerol dehydratase by a new Mu-like prophage. J Mol Microbiol Biotechnol 8:150–168

Bird CL, Kuhn AT (1981) Electrochemistry of the viologens. Chem Soc Rev 10:49–82

Blattner FR, Plunkett G 3rd, Bloch CA, Perna NT, Burland V, Riley M, Collado-Vides J, Glasner JD, Rode CK, Mayhew GF, Gregor J, Davis NW, Kirkpatrick HA, Goeden MA, Rose DJ, Mau B, Shao Y (1997) The complete genome sequence of *Escherichia coli* K-12. Science 277:1453–1462

Bommer M, Kunze C, Fesseler J, Schubert T, Diekert G, Dobbek H (2014) Structural basis for organohalide respiration. Science 346:455–458

Bouwer EJ, McCarty PL (1983) Transformations of 1- and 2-carbon halogenated aliphatic organic compounds under methanogenic conditions. Appl Environ Microbiol 45:1286–1294

Burgess NR, McDermott SN, Whiting J (1973) Aerobic bacteria occurring in the hind-gut of the cockroach, *Blatta orientalis*. J Hyg (Lond) 71:1–7

Castanié-Cornet MP, Bruel N, Genevaux P (2014) Chaperone networking facilitates protein targeting to the bacterial cytoplasmic membrane. Biochim Biophys Acta 1843:1442–1456

Chen K, Huang L, Xu C, Liu X, He J, Zinder SH, Li S, Jiang J (2013) Molecular characterization of the enzymes involved in the degradation of a brominated aromatic herbicide. Mol Microbiol 89:1121–1139

Cheng D, Chow WL, He J (2010) A *Dehalococcoides*-containing co-culture that dechlorinates tetrachloroethene to *trans*-1,2-dichloroethene. ISME J 4:88–97

Christiansen N, Ahring BK, Wohlfarth G, Diekert G (1998) Purification and characterization of the 3-chloro-4-hydroxy-phenylacetate reductive dehalogenase of *Desulfitobacterium hafniense*. FEBS Lett 436:159–162

Crofts TS, Seth EC, Hazra AB, Taga ME (2013) Cobamide structure depends on both lower ligand availability and CobT substrate specificity. Chem Biol 20:1265–1274

De Wildeman S, Diekert G, Van Langenhove H, Verstraete W (2003) Stereoselective microbial dehalorespiration with vicinal dichlorinated alkanes. Appl Environ Microbiol 69:5643–5647

DeWeerd KA, Suflita JM (1990) Anaerobic aryl reductive dehalogenation of halobenzoates by cell extracts of "*Desulfomonile tiedjei*". Appl Environ Microbiol 56:2999–3005

DeWeerd KA, Mandelco L, Tanner RS, Woese CR, Suflita JM (1990) *Desulfomonile tiedjei* gen. nov. and sp. nov., a novel anaerobic, dehalogenating, sulfate-reducing bacterium. Arch Microbiol 154:23–30

Ding C, Zhao S, He J (2014) A *Desulfitobacterium* sp. strain PR reductively dechlorinates both 1, 1, 1-trichloroethane and chloroform. Environ Microbiol 16:3387–3397

Dolfing J, Novak I (2014) The Gibbs free energy of formation of halogenated benzenes, benzoates and phenols and their potential role as electron acceptors in anaerobic environments. Biodegradation 26:15–27

Duhamel M, Edwards EA (2006) Microbial composition of chlorinated ethene-degrading cultures dominated by *Dehalococcoides*. FEMS Microbiol Ecol 58:538–549

Duhamel M, Wehr SD, Yu L, Rizvi H, Seepersad D, Dworatzek S, Cox EE, Edwards EA (2002) Comparison of anaerobic dechlorinating enrichment cultures maintained on tetrachloroethene, trichloroethene, *cis*-dichloroethene and vinyl chloride. Water Res 36:4193–4202

Euro C (2014) Chlorine Industry Review 2013–2014. http://www.eurochlor.org/

Firth RA, Hill HAO, Pratt JM, Thorp RG, Williams, RJP (1969) The chemistry of vitamin B_{12}. Part XI. Some further formation constants. J Chem Soc A 381–386

Fontecave M, Ollagnier-de-Choudens S (2008) Iron–sulfur cluster biosynthesis in bacteria: mechanisms of cluster assembly and transfer. Arch Biochem Biophys 474:226–237

Futagami T, Goto M, Furukawa K (2008) Biochemical and genetic bases of dehalorespiration. Chem Rec 8:1–12

Gantzer CJ, Wackett LP (1991) Reductive dechlorination catalyzed by bacterial transition-metal coenzymes. Environ Sci Technol 25:715–722

Glod G, Angst W, Holliger C, Schwarzenbach RP (1996) Corrinoid-mediated reduction of tetrachloroethene, trichloroethene, and trichlorofluoroethene in homogeneous aqueous solution: reaction kinetics and reaction mechanisms. Environ Sci Technol 31:253–260

Goris T, Schubert T, Gadkari J, Wubet T, Tarkka M, Buscot F, Adrian L, Diekert G (2014) Insights into organohalide respiration and the versatile catabolism of *Sulfurospirillum multivorans* gained from comparative genomics and physiological studies. Environ Microbiol 16:3562–3580

Gribble GW (2003) The diversity of naturally produced organohalogens. Chemosphere 52:289–297

Griffin BM, Tiedje JM, Löffler FE (2004) Anaerobic microbial reductive dechlorination of tetrachloroethene to predominately *trans*-1,2-dichloroethene. Environ Sci Technol 38:4300–4303

Hill DW, McCarty PL (1967) Anaerobic degradation of selected chlorinated hydrocarbon pesticides. J Water Pollut Control Fed 39:1259–1277

Holliger C, Hahn D, Harmsen H, Ludwig W, Schumacher W, Tindall B, Vazquez F, Weiss N, Zehnder AJ (1998a) *Dehalobacter restrictus* gen. nov. and sp. nov., a strictly anaerobic bacterium that reductively dechlorinates tetra- and trichloroethene in an anaerobic respiration. Arch Microbiol 169:313–321

Holliger C, Wohlfarth G, Diekert G (1998b) Reductive dechlorination in the energy metabolism of anaerobic bacteria. FEMS Microbiol Rev 22:383–398

Hug LA, Maphosa F, Leys D, Löffler FE, Smidt H, Edwards EA, Adrian L (2013) Overview of organohalide-respiring bacteria and a proposal for a classification system for reductive dehalogenases. Philos Trans R Soc Lond B Biol Sci 368:20120322

Hvorup RN, Goetz BA, Niederer M, Hollenstein K, Perozo E, Locher KP (2007) Asymmetry in the structure of the ABC transporter-binding protein complex BtuCD-BtuF. Science 317:1387–1390

Jacobson MR, Cash VL, Weiss MC, Laird NF, Newton WE, Dean DR (1989) Biochemical and genetic analysis of the *nifUSVWZM* cluster from *Azotobacter vinelandii*. Mol Gen Genet 219:49–57

Jayachandran G, Görisch H, Adrian L (2004) Studies on hydrogenase activity and chlorobenzene respiration in *Dehalococcoides* sp. strain CBDB1. Arch Microbiol 182:498–504

John M, Schmitz RP, Westermann M, Richter W, Diekert G (2006) Growth substrate dependent localization of tetrachloroethene reductive dehalogenase in *Sulfurospirillum multivorans*. Arch Microbiol 186:99–106

Keller S, Ruetz M, Kunze C, Kräutler B, Diekert G, Schubert T (2014) Exogenous 5,6-dimethylbenzimidazole caused production of a non-functional tetrachloroethene reductive dehalogenase in *Sulfurospirillum multivorans*. Environ Microbiol 16:3361–3369

Kennedy FS, Buckman T, Wood JM (1969) Carbenoid intermediates from the photolysis of haloalkylcobalamins. Biochim Biophys Acta 177:661–663

Kern M, Simon J (2008) Characterization of the NapGH quinol dehydrogenase complex involved in *Wolinella succinogenes* nitrate respiration. Mol Microbiol 69:1137–1152

Kim SH, Harzman C, Davis JK, Hutcheson R, Broderick JB, Marsh TL, Tiedje JM (2012) Genome sequence of *Desulfitobacterium hafniense* DCB-2, a Gram-positive anaerobe capable of dehalogenation and metal reduction. BMC Microbiol 8(12):21

Kim YE, Hipp MS, Bracher A, Hayer-Hartl M, Hartl FU (2013) Molecular chaperone functions in protein folding and proteostasis. Annu Rev Biochem 82:323–355

Kimoto H, Suye S, Makishima H, Arai J, Yamaguchi S, Fujii Y, Yoshioka T, Taketo A (2010) Cloning of a novel dehalogenase from environmental DNA. Biosci Biotechnol Biochem 74:1290–1292

Kliegman S, McNeill K (2008) Dechlorination of chloroethylenes by cob (I) alamin and cobalamin model complexes. Dalton Trans 32:4191–4201

Korkhov VM, Mireku SA, Veprintsev DB, Locher KP (2014) Structure of AMP-PNP-bound BtuCD and mechanism of ATP-powered vitamin B_{12} transport by BtuCD-F. Nat Struct Mol Biol. doi:10.1038/nsmb.2918

Krasotkina J, Walters T, Maruya KA, Ragsdale SW (2001) Characterization of the B_{12}- and iron–sulfur-containing reductive dehalogenase from *Desulfitobacterium chlororespirans*. J Biol Chem 276:40991–40997

Krauter H (2006) Untersuchungen zur Beteiligung membranständiger redoxaktiver Komponenten bei der Dehalorespiration in *Sulfurospirillum multivorans*. Diploma thesis. Friedrich Schiller University Jena

Kräutler B, Fieber W, Ostermann S, Fasching M, Ongania KH, Gruber K, Kratky C, Mikl C, Siebert A, Diekert G (2003) The cofactor of tetrachloroethene reductive dehalogenase of *Dehalospirillum multivorans* is Norpseudo-B_{12}, a new type of a natural corrinoid. Helv Chim Acta 86:3698–3716

Krone UE, Thauer RK, Hogenkamp HPC (1989a) Reductive dehalogenation of chlorinated C_1-hydrocarbons mediated by corrinoids. Biochemistry 28:4908–4914

Krone UE, Laufer K, Thauer RK, Hogenkamp HP (1989b) Coenzyme F_{430} as a possible catalyst for the reductive dehalogenation of chlorinated C_1 hydrocarbons in methanogenic bacteria. Biochemistry 28:10061–10065

Lee M, Low A, Zemb O, Koenig J, Michaelsen A, Manefield M (2012) Complete chloroform dechlorination by organochlorine respiration and fermentation. Environ Microbiol 14:883–894

Lexa D, Saveant JM (1983) The electrochemistry of vitamin B_{12}. Acc Chem Res 16:235–243

Löffler FE, Yan J, Ritalahti KM, Adrian L, Edwards EA, Konstantinidis KT, Müller JA, Fullerton H, Zinder SH, Spormann AM (2013) *Dehalococcoides mccartyi* gen. nov., sp. nov., obligately organohalide-respiring anaerobic bacteria relevant to halogen cycling and bioremediation, belong to a novel bacterial class, *Dehalococcoidia* classis nov., order *Dehalococcoidales* ord. nov. and family *Dehalococcoidaceae* fam. nov., within the phylum *Chloroflexi*. Int J Syst Evol Microbiol 63:625–635

Louie TM, Mohn WW (1999) Evidence for a chemiosmotic model of dehalorespiration in *Desulfomonile tiedjei* DCB-1. J Bacteriol 181:40–46

Mac Nelly A, Kai M, Svatoš A, Diekert G, Schubert T (2014) Functional heterologous production of reductive dehalogenases from *Desulfitobacterium hafniense* strains. Appl Environ Microbiol 80:4313–4322

Magnuson JK, Stern RV, Gossett JM, Zinder SH, Burris DR (1998) Reductive dechlorination of tetrachloroethene to ethene by a two-component enzyme pathway. Appl Environ Microbiol 64:1270–1275

Maillard J, Schumacher W, Vazquez F, Regeard C, Hagen WR, Holliger C (2003) Characterization of the corrinoid iron–sulfur protein tetrachloroethene reductive dehalogenase of *Dehalobacter restrictus*. Appl Environ Microbiol 69:4628–4638

Maillard J, Regeard C, Holliger C (2005) Isolation and characterization of Tn-Dha1, a transposon containing the tetrachloroethene reductive dehalogenase of *Desulfitobacterium hafniense* strain TCE1. Environ Microbiol 7:107–117

Maillard J, Genevaux P, Holliger C (2011) Redundancy and specificity of multiple trigger factor chaperones in *Desulfitobacteria*. Microbiology 157:2410–2421

Major DW, McMaster ML, Cox EE, Edwards EA, Dworatzek SM, Hendrickson ER, Starr MG, Payne JA, Buonamici LW (2002) Field demonstration of successful bioaugmentation to achieve dechlorination of tetrachloroethene to ethene. Environ Sci Technol 36:5106–5116

Marco-Urrea E, Nijenhuis I, Adrian L (2011) Transformation and carbon isotope fractionation of tetra- and trichloroethene to *trans*-dichloroethene by *Dehalococcoides* sp. strain CBDB1. Environ Sci Technol 45:1555–1562

Mayer F, Müller V (2014) Adaptations of anaerobic archaea to life under extreme energy limitation. FEMS Microbiol Rev 38:449–472

Maymó-Gatell X, Chien Y, Gossett JM, Zinder SH (1997) Isolation of a bacterium that reductively dechlorinates tetrachloroethene to ethene. Science 276:1568–1571

Miller E, Wohlfarth G, Diekert G (1996) Studies on tetrachloroethene respiration in *Dehalospirillum multivorans*. Arch Microbiol 166:379–387

Miller E, Wohlfarth G, Diekert G (1997) Comparative studies on tetrachloroethene reductive dechlorination mediated by *Desulfitobacterium* sp. strain PCE-S. Arch Microbiol 168:513–519

Miller E, Wohlfarth G, Diekert G (1998) Purification and characterization of the tetrachloroethene reductive dehalogenase of strain PCE-S. Arch Microbiol 169:497–502

Mohn WW, Tiedje JM (1990) Strain DCB-1 conserves energy for growth from reductive dechlorination coupled to formate oxidation. Arch Microbiol 153:267–271

Mohn WW, Tiedje JM (1991) Evidence for chemiosmotic coupling of reductive dechlorination and ATP synthesis in *Desulfomonile tiedjei*. Arch Microbiol 157:1–6

Moore SJ, Lawrence AD, Biedendieck R, Deery E, Frank S, Howard MJ, Rigby SE, Warren MJ (2013) Elucidation of the anaerobic pathway for the corrin component of cobalamin (vitamin B_{12}). Proc Natl Acad Sci U S A 110:14906–14911

Morita Y, Futagami T, Goto M, Furukawa K (2009) Functional characterization of the trigger factor protein PceT of tetrachloroethene-dechlorinating *Desulfitobacterium hafniense* Y51. Appl Microbiol Biotechnol 83:775–781

Morris RM, Fung JM, Rahm BG, Zhang S, Freedman DL, Zinder SH, Richardson RE (2007) Comparative proteomics of *Dehalococcoides* spp. reveals strain-specific peptides associated with activity. Appl Environ Microbiol 73:320–326

Müller JA, Rosner BM, Von Abendroth G, Meshulam-Simon G, McCarty PL, Spormann AM (2004) Molecular identification of the catabolic vinyl chloride reductase from *Dehalococcoides* sp. strain VS and its environmental distribution. Appl Environ Microbiol 70:4880–4888

Nachin L, Loiseau L, Expert D, Barras F (2003) SufC: an unorthodox cytoplasmic ABC/ATPase required for [Fe–S] biogenesis under oxidative stress. EMBO J 22:427–437

Neumann A, Scholz-Muramatsu H, Diekert G (1994) Tetrachloroethene metabolism of *Dehalospirillum multivorans*. Arch Microbiol 162:295–301

Neumann A, Wohlfarth G, Diekert G (1995) Properties of tetrachloroethene and trichloroethene dehalogenase of *Dehalospirillum multivorans*. Arch Microbiol 163:276–281

Neumann A, Wohlfarth G, Diekert G (1996) Purification and characterization of tetrachloroethene reductive dehalogenase from *Dehalospirillum multivorans*. J Biol Chem 271:16515–16519

Neumann A, Wohlfarth G, Diekert G (1998) Tetrachloroethene dehalogenase from *Dehalospirillum multivorans*: cloning, sequencing of the encoding genes, and expression of the *pceA* gene in *Escherichia coli*. J Bacteriol 180:4140–4145

Neumann A, Siebert A, Trescher T, Reinhardt S, Wohlfarth G, Diekert G (2002) Tetrachloroethene reductive dehalogenase of *Dehalospirillum multivorans*: substrate specificity of the native enzyme and its corrinoid cofactor. Arch Microbiol 177:420–426

Ni S, Fredrickson JK, Xun L (1995) Purification and characterization of a novel 3-chlorobenzoate-reductive dehalogenase from the cytoplasmic membrane of *Desulfomonile tiedjei* DCB-1. J Bacteriol 177:5135–5139

Nijenhuis I, Zinder SH (2005) Characterization of hydrogenase and reductive dehalogenase activities of *Dehalococcoides ethenogenes* strain 195. Appl Environ Microbiol 71:1664–1667

Nonaka H, Keresztes G, Shinoda Y, Ikenaga Y, Abe M, Naito K, Inatomi K, Furukawa K, Inui M, Yukawa H (2006) Complete genome sequence of the dehalorespiring bacterium *Desulfitobacterium hafniense* Y51 and comparison with *Dehalococcoides ethenogenes* 195. J Bacteriol 188:2262–2274

Nonnenberg C, van der Donk WA, Zipse H (2002) Reductive dechlorination of trichloroethylene: a computational study. J Phys Chem A 106:8708–8715

Outten FW, Djaman O, Storz G (2004) A *suf* operon requirement for Fe–S cluster assembly during iron starvation in *Escherichia coli*. Mol Microbiol 52:861–872

Padovani D, Labunska T, Palfey BA, Ballou DP, Banerjee R (2008) Adenosyltransferase tailors and delivers coenzyme B_{12}. Nat Chem Biol 4:194–196

Palmer T, Berks BC (2012) The twin-arginine translocation (Tat) protein export pathway. Nat Rev Microbiol 10:483–496

Parthasarathy A, Stich TA, Lohner ST, Lesnefsky A, Britt RD, Spormann AM (2015) Biochemical and EPR-spectroscopic Investigation into heterologously expressed vinyl chloride reductive dehalogenase (VcrA) from *Dehalococcoides mccartyi* strain VS. J Am Chem Soc 137:3525–32

Payne KAP, Quezada CP, Fisher K, Dunstan MS, Collins FA, Sjuts H, Levy C, Hay S, Rigby SEJ, Leys D (2015) Reductive dehalogenase structure suggests a mechanism for B_{12}-dependent dehalogenation. Nature 517:513–516

Reinhold A, Westermann M, Seifert J, von Bergen M, Schubert T, Diekert G (2012) Impact of vitamin B_{12} on formation of the tetrachloroethene reductive dehalogenase in *Desulfitobacterium hafniense* strain Y51. Appl Environ Microbiol 78:8025–8032

Renpenning J, Keller S, Cretnik S, Shouakar-Stash O, Elsner M, Schubert T, Nijenhuis I (2014) Combined C and Cl isotope effects indicate differences between corrinoids and enzyme (*Sulfurospirillum multivorans* PceA) in reductive dehalogenation of tetrachloroethene, but not trichloroethene. Environ Sci Technol 48:11837–11845

Rupakula A, Kruse T, Boeren S, Holliger C, Smidt H, Maillard J (2013) The restricted metabolism of the obligate organohalide respiring bacterium *Dehalobacter restrictus*: lessons from tiered functional genomics. Philos Trans R Soc Lond B Biol Sci 368:20120325

Sargent F (2007) Constructing the wonders of the bacterial world: biosynthesis of complex enzymes. Microbiology 153:633–651

Saunders NF, Houben EN, Koefoed S, de Weert S, Reijnders WN, Westerhoff HV, De Boer AP, Van Spanning RJ (1999) Transcription regulation of the *nir* gene cluster encoding nitrite reductase of *Paracoccus denitrificans* involves NNR and NirI, a novel type of membrane protein. Mol Microbiol 34:24–36

Schink B, Friedrich M (1994) Energetics of syntrophic fatty acid oxidation. FEMS Microbiol Rev 15:85–94

Schipp CJ, Marco-Urrea E, Kublik A, Seifert J, Adrian L (2013) Organic cofactors in the metabolism of *Dehalococcoides mccartyi* strains. Philos Trans R Soc Lond B Biol Sci 368:20120321

Schmitz RP, Wolf J, Habel A, Neumann A, Ploss K, Svatos A, Boland W, Diekert G (2007) Evidence for a radical mechanism of the dechlorination of chlorinated propenes mediated by the tetrachloroethene reductive dehalogenase of *Sulfurospirillum multivorans*. Environ Sci Technol 41:7370–7375

Scholz-Muramatsu H, Neumann A, Messmer M, Moore E, Diekert G (1995) Isolation and characterization of *Dehalospirillum multivorans* gen. nov., sp. nov., a tetrachloroethene-utilizing, strictly anaerobic bacterium. Arch Microbiol 163:48–56

Schrauzer GN, Holland RJ, Seck JA (1971) The mechanism of coenzyme B_{12} action in dioldehydrase. J Am Chem Soc 93:1503–1505

Schumacher W, Holliger C (1996) The proton/electron ratio of the menaquinone-dependent electron transport from dihydrogen to tetrachloroethene in "*Dehalobacter restrictus*". J Bacteriol 178:2328–2333

Schumacher W, Holliger C, Zehnder AJ, Hagen WR (1997) Redox chemistry of cobalamin and iron-sulfur cofactors in the tetrachloroethene reductase of *Dehalobacter restrictus*. FEBS Lett 409:421–425

Siebert A, Neumann A, Schubert T, Diekert G (2002) A non-dechlorinating strain of *Dehalospirillum multivorans*: evidence for a key role of the corrinoid cofactor in the synthesis of an active tetrachloroethene dehalogenase. Arch Microbiol 178:443–449

Silverstein TP (2014) An exploration of how the thermodynamic efficiency of bioenergetic membrane systems varies with *c*-subunit stoichiometry of F_1F_0 ATP synthases. J Bioenerg Biomembr 46:229–241
Sjuts H, Fisher K, Dunstan MS, Rigby SE, Leys D (2012) Heterologous expression, purification and cofactor reconstitution of the reductive dehalogenase PceA from *Dehalobacter restrictus*. Protein Expr Purif 85:224–229
Smidt H, de Vos WM (2004) Anaerobic microbial dehalogenation. Annu Rev Microbiol 58:43–73
Smidt H, van Leest M, van der Oost J, de Vos WM (2000) Transcriptional regulation of the *cpr* gene cluster in *ortho*-chlorophenol-respiring *Desulfitobacterium dehalogenans*. J Bacteriol 182:5683–5691
Stupperich E, Steiner I, Rühlemann M (1986) Isolation and analysis of bacterial cobamides by high-performance liquid chromatography. Anal Biochem 155:365–370
Suyama A, Yamashita M, Yoshino S, Furukawa K (2002) Molecular characterization of the PceA reductive dehalogenase of *Desulfitobacterium* sp. strain Y51. J Bacteriol 184:3419–3425
Takahashi Y, Tokumoto U (2002) A third bacterial system for the assembly of iron-sulfur clusters with homologs in archaea and plastids. J Biol Chem 277:28380–28383
Tang S, Chan WW, Fletcher KE, Seifert J, Liang X, Löffler FE, Edwards EA, Adrian L (2013) Functional characterization of reductive dehalogenases by using blue native polyacrylamide gel electrophoresis. Appl Environ Microbiol 79:974–981
Thauer RK, Jungermann K, Decker K (1977) Energy conservation in chemotrophic anaerobic bacteria. Bacteriol Rev 41:100–180
Thibodeau J, Gauthier A, Duguay M, Villemur R, Lépine F, Juteau P, Beaudet R (2004) Purification, cloning, and sequencing of a 3,5-dichlorophenol reductive dehalogenase from *Desulfitobacterium frappieri* PCP-1. Appl Environ Microbiol 70:4532–4537
Townsend GT, Suflita JM (1996) Characterization of chloroethylene dehalogenation by cell extracts of *Desulfomonile tiedjei* and its relationship to chlorobenzoate dehalogenation. Appl Environ Microbiol 62:2850–2853
Turner RJ, Papish AL, Sargent F (2004) Sequence analysis of bacterial redox enzyme maturation proteins (REMPs). Can J Microbiol 50:225–238
van de Pas BA, Smidt H, Hagen WR, van der Oost J, Schraa G, Stams AJ, de Vos WM (1999) Purification and molecular characterization of *ortho*-chlorophenol reductive dehalogenase, a key enzyme of halorespiration in *Desulfitobacterium dehalogenans*. J Biol Chem 274:20287–20292
van de Pas BA, Jansen S, Dijkema C, Schraa G, de Vos WM, Stams AJ (2001a) Energy yield of respiration on chloroaromatic compounds in *Desulfitobacterium dehalogenans*. Appl Environ Microbiol 67:3958–3963
van de Pas BA, Gerritse J, de Vos WM, Schraa G, Stams AJ (2001b) Two distinct enzyme systems are responsible for tetrachloroethene and chlorophenol reductive dehalogenation in *Desulfitobacterium* strain PCE1. Arch Microbiol 176:165–169
Vogel TM, Criddle CS, McCarty PL (1987) ES critical reviews: transformations of halogenated aliphatic compounds. Environ Sci Technol 21:722–736
Wagner DD, Hug LA, Hatt JK, Spitzmiller MR, Padilla-Crespo E, Ritalahti KM, Edwards EA, Konstantinidis KT, Löffler FE (2012) Genomic determinants of organohalide-respiration in *Geobacter lovleyi*, an unusual member of the *Geobacteraceae*. BMC Genom 13:200
Warren MJ, Raux E, Schubert HL, Escalante-Semerena JC (2002) The biosynthesis of adenosylcobalamin (vitamin B_{12}). Nat Prod Rep 19:390–412
Weber J, Senior AE (2003) ATP synthesis driven by proton transport in F_1F_0-ATP synthase. FEBS Lett 545:61–70
White DC, Geyer R, Peacock AD, Hedrick DB, Koenigsberg SS, Sung Y, He J, Löffler FE (2005) Phospholipid furan fatty acids and ubiquinone-8: lipid biomarkers that may protect *Dehalococcoides* strains from free radicals. Appl Environ Microbiol 71:8426–8433
Wohlfarth G, Diekert G (1997) Anaerobic dehalogenases. Curr Opin Biotechnol 8:290–295
Wood JM, Kennedy FS, Wolfe RS (1968) The reaction of multihalogenated hydrocarbons with free and bound reduced vitamin B_{12}. Biochemistry 7:1707–1713

Wunsch P, Zumft WG (2005) Functional domains of NosR, a novel transmembrane iron–sulfur flavoprotein necessary for nitrous oxide respiration. J Bacteriol 187:1992–2001

Yan J, Ritalahti KM, Wagner DD, Löffler FE (2012) Unexpected specificity of interspecies cobamide transfer from *Geobacter* spp. to organohalide-respiring *Dehalococcoides mccartyi* strains. Appl Environ Microbiol 78:6630–6636

Ye L, Schilhabel A, Bartram S, Boland W, Diekert G (2010) Reductive dehalogenation of brominated ethenes by *Sulfurospirillum multivorans* and *Desulfitobacterium hafniense* PCE-S. Environ Microbiol 12:501–509

Yi S, Seth EC, Men YJ, Stabler SP, Allen RH, Alvarez-Cohen L, Taga ME (2012) Versatility in corrinoid salvaging and remodeling pathways supports corrinoid-dependent metabolism in *Dehalococcoides mccartyi*. Appl Environ Microbiol 78:7745–7752

Zehnder AJ, Wuhrmann K (1976) Titanium (III) citrate as a nontoxic oxidation-reduction buffering system for the culture of obligate anaerobes. Science 194:1165–1166

Zheng L, Cash VL, Flint DH, Dean DR (1998) Assembly of iron–sulfur clusters. Identification of an *iscSUA-hscBA-fdx* gene cluster from *Azotobacter vinelandii*. J Biol Chem 273:13264–13272

Chapter 18
Evaluation of the Microbial Reductive Dehalogenation Reaction Using Compound-Specific Stable Isotope Analysis (CSIA)

Julian Renpenning and Ivonne Nijenhuis

Abstract In recent decades, concepts involving compound-specific stable isotope analysis have evolved allowing the assessment of organohalide biotransformation in situ as well as evaluating complex (bio)chemical reactions. The stable isotope composition can provide information on the source of a specific chemical and changes over time or space additionally allow to assess degradation pathways and reaction mechanisms involved during (bio)transformation. This chapter provides a basic introduction into stable isotope analysis, an overview of the application of compound-specific stable isotope analysis (CSIA) for investigation of the microbial reductive dehalogenation reaction of mainly chlorinated ethenes and summarizes recent advances and results.

18.1 Stable Isotopes

18.1.1 What Are Isotopes?

Isotopes of the same element have an identical number of protons (Z). However, they differ in their number of neutrons (N). All elements exist in different energetic state and can be divided into stable isotopes (nonradioactive)—as for instance or ^{12}C and ^{13}C, and unstable (radioactive) isotopes— ^{14}C. Isotopes of an element can be denoted by an atomic formula, i.e., carbon-12 $^{12}_{6}C$ with a total atomic mass of 12, including 6 protons. The remaining difference is the number of neutrons harbored by an isotope, in case of carbon-12 the number of neutrons is 6. Generally, isotopes with a similar number of protons and neutrons ($N/Z \leq 1.5$) tend to be stable. Around 300 stable isotopes are known thus far,

J. Renpenning · I. Nijenhuis (✉)
Department Isotope Biogeochemistry, Helmholtz Centre for Environmental Research—UFZ, Leipzig, Germany
e-mail: ivonne.nijenhuis@ufz.de

L. Adrian and F.E. Löffler (eds.), *Organohalide-Respiring Bacteria*,
DOI 10.1007/978-3-662-49875-0_18

however, the number of non-stable radioactive isotopes is about four times higher (Hoefs 2008). A unique isotopic composition can be found in a wide variety of materials, which provides an insightful technique to elucidate their origin, history, formation, or degradation based on the isotopic profile of their atoms and molecules. Isotope analysis can therefore be used in a wide range of applications, i.e., in the field of physical and organic chemistry, geochemistry, and hydrology (Thiemens 2006).

18.1.2 Definition and Delta Notation

The stable isotope composition of a substance is normally reported in delta notation (δ) in parts per thousand (per mil, ‰) or in the SI unit urey (mUr, which is equivalent to ‰) as recommended by Brand and Coplen (2012). The isotopic abundance of an element is usually determined relative to an international reference standard, which ensures the comparability of isotope analysis in different laboratories on an international scale (Coplen 2011). Variation in the natural abundance of isotopes can be calculated by Eqs. 18.1 and 18.2, with R_{sample} for the measured compound, $R_{\text{reference}}$ for reference standard, and E for the light and heavy elements.

$$\text{isotope ratio}(R) = \frac{\text{abundance}^{\text{heavy}}E}{\text{abundance}^{\text{ light}}E} \tag{18.1}$$

$$\delta(E)_{\text{sample}} = \left(\frac{R_{\text{sample}} - R_{\text{reference}}}{R_{\text{reference}}}\right) \times 1000 = \left(\frac{R_{\text{sample}}}{R_{\text{reference}}} - 1\right) \times 1000 \tag{18.2}$$

Corresponding reference material is provided by two organizations, the International Atomic Energy Agency (IAEA, www.iaea.org) and the National Institute of Standards and Technology (NIST, www.nist.gov). The standardization allows comparison of isotope signatures between different laboratories and experiments.

18.1.3 Isotopes of Interest for Chlorinated Hydrocarbons

Currently, compound-specific stable isotope analysis (CSIA) of hydrogen, carbon, and nitrogen are the most common routine applications (Elsner and Hofstetter 2011). Additional techniques are available for isotope analyses of a wide range of elements, as for instance, oxygen, sulfur, chlorine, bromine, but also noble gases and metals. This chapter will focus on three elements relevant for organohalide respiration of chlorinated hydrocarbons: hydrogen ($^{2}H/^{1}H$), carbon ($^{13}C/^{12}C$), and chlorine ($^{37}Cl/^{35}Cl$) isotopes.

Table 18.1 Relative abundance of hydrogen, carbon, and chlorine isotopes modified after Coplen (1994) and the corresponding (international) reference scales: VSMOW (Vienna Standard Mean Ocean Water), VPDB (Vienna Pee Dee Belemnite), and SMOC (standard mean ocean chlorine)

Element	Stable isotopes	Natural abundance (%)		Relative mass difference (%)	International reference standard
		$^{heavy}E$	$^{light}E$		
Hydrogen	$^{2}H/^{1}H$	0.0155	99.98	100	VSMOW
Carbon	$^{13}C/^{12}C$	1.1060	98.89	8.3	VPDB
Chlorine	$^{37}Cl/^{35}Cl$	24.220	75.78	5.7	SMOC

18.1.3.1 Hydrogen

The natural abundance of deuterium (^{2}H or D) is relatively low (0.0155 %) and $\delta^{2}H$ values are usually reported relative to IAEA reference water VSMOW (Vienna Standard Mean Ocean Water), which is assigned a $\delta^{2}H$ of 0 ‰ (Table 18.1) (Coplen et al. 2002c). Hydrogen undergoes the strongest kinetic isotope effects (KIEs), due to the highest relative mass difference between the two isotopes (Sect. 18.3.1), which results in isotope fractionation and is reflected in the largest range of isotope composition in comparison to other elements. Variation of hydrogen isotope composition is primarily caused by evaporation and condensation processes of meteoric water (water precipitation). Therefore, $\delta^{2}H$ values of water from different origin on earth can range from −495 ‰ in Atlantic ice shields to +129 ‰ in evaporated lakes (Jouzel et al. 1987; Coplen et al. 2002b). Isotope composition of plants usually reflects the isotope composition of regional water precipitates and is further incorporated in food and animals.

18.1.3.2 Carbon

The original reference material for carbon isotope analysis, the marine fossil Pee Dee Belemnite, is exhausted. Nevertheless, carbon stable isotope composition is still reported relative to the Vienna Pee Dee Belemnite (VPDB) international standard, which was assigned a $\delta^{13}C$ of 0 ‰ (Coplen et al. 2006) using alternative reference material. Variation of carbon isotope composition on earth was reported to range from the lowest $\delta^{13}C$ values of −130.3 ‰ for a material of natural terrestrial origin and the most positive $\delta^{13}C$ values of +37.5 ‰ reported for deep-sea pore waters (Elvert et al. 2000; Coplen et al. 2002c). The three main carbon reservoirs on earth are sedimentary organic material, the biosphere including atmospheric carbon, and sedimentary carbonates. The three reservoirs may differ from each other according to the isotope fractionation they undergo during formation. Subsequently, isotope composition incorporated in plants, food, and animal tissue is characteristic according to their carbon source and the metabolic pathway used for hydrocarbon formation (Coplen et al. 2002b).

18.1.3.3 Chlorine

Chlorine isotope composition (δ^{37}Cl) is usually expressed relative to SMOC (standard mean ocean chlorine), which is assumed to be isotopically homogeneous (Kaufmann et al. 1988). The use of chlorine stable isotopes analysis to evaluate hydrological processes and sources identification in hydrological environments is, to some extent, limited by the small range of observed isotopic ratios. Due to the small relative mass difference between ^{35}Cl and ^{37}Cl isotopes only minor isotope fractionation was observed for chlorine. In groundwater, surface and pore waters δ^{37}Cl values generally range from -8.00 to $+3.00$ ‰ (Coplen et al. 2002a). The lowest δ^{37}Cl values for naturally occurring pore water samples were measured to be -7.7 ‰ and the highest $+7.5$ ‰ for Cl in smectite from a Costa Rica Rift oceanic drill hole (Ransom et al. 1995; Magenheim et al. 1995). Chlorine isotope composition of organic solvents was reported to range between -6.0 and $+4.4$ ‰ (Tanaka and Rye 1991). However, due to the lack of commercially available techniques (discussed in Sect. 18.2), isotope fractionation of chlorine in chlorinated compounds is yet poorly investigated.

18.2 Isotope Analysis Techniques

For the analysis of the stable isotope composition the sample has to be quantitatively converted into a suitable purified gas that can be analyzed by a mass spectrometer (MS), for instance, in CO_2 for carbon and in H_2 for hydrogen (Fig. 18.1) (Elsner and Hofstetter 2011). The conversion of the organic sample into a gas can be done by an 'offline' method, where conversion of pure or bulk sample is usually done in a sealed tube, or an 'online' method, where a sample preparation line is connected directly to the MS. 'Offline' conversion techniques are however time-consuming and only a small number of samples can be analyzed per day. Moreover, 'offline' techniques are not suitable for compound mixtures found, for instance, in environmental samples. Today, continuous-flow isotope-ratio mass spectrometry (CF-IRMS) is frequently used as an 'online' method for CSIA. This type of system is appropriate for most organic tissue, sediment, and soil samples containing sufficient organic matter. To make this technique appropriate for sample mixtures, GC-IRMS analysis was introduced, where gas chromatographic (GC) separation is applied prior to the conversion of individual compounds and followed by subsequent isotope analysis via IRMS. Different conversion techniques are applied prior to MS analysis, as for instance, combustion for carbon or pyrolysis for hydrogen, which will be discussed in more detail below. Additionally, liquid chromatography in combination with isotope-ratio mass spectrometry (LC-IRMS) was introduced for compounds that are not GC compatible (Brenna et al. 1997b; Krummen et al. 2004; Godin and McCullagh 2011). Although polar substances may be made GC-amenable by derivatization, introduction of additional carbon or hydrogen into a molecule is not desirable as one needs to ascertain that the derivatization is free of isotope effect and later corrections are

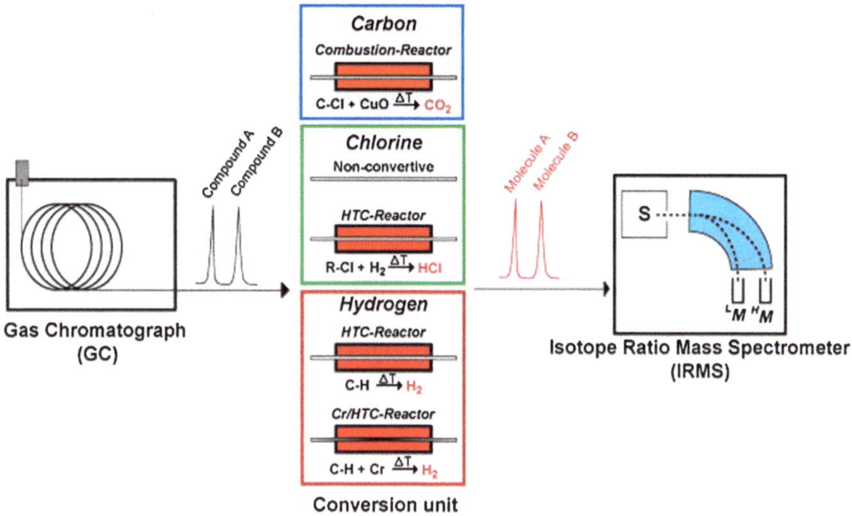

Fig. 18.1 Available continuous-flow methods in combination with gas chromatography and isotope-ratio mass spectrometry for carbon, chlorine, and hydrogen. The conversion technique is indicated with respect to the analyzed element and the conversion method: combustion, high-temperature conversion (HTC), chromium-based high-temperature conversion (Cr/HTC), and nonconvertive analysis

needed to correct for the dilution of the isotope signal. LC-IRMS approaches may provide the analysis of substance classes currently listed as 'emerging contaminants' such as pharmaceuticals and polar pesticides.

18.2.1 Carbon Stable Isotope Analysis (CSIA)

Carbon isotope analysis is nowadays the most frequent application in CSIA. Gas chromatography combustion isotope-ratio mass spectrometry (GC-C-IRMS) is commonly used for the compound-specific measurements of carbon isotopes. In GC-C-IRMS target compounds are first chromatographically separated and subsequently converted to CO_2 in the combustion unit at 980–1100 °C. The combustion unit consists of a ceramic tube with CuO/NiO/Pt as oxygen donor and catalyst for a quantitative conversion of carbon to CO_2 for isotope measurements (Elsner et al. 2012; Brenna et al. 1997a). After transfer to the IRMS the CO_2 becomes ionized and the different isotopes are separated in a magnetic field according to their mass-to-charge ratio (m/z). For detection, the IRMS is generally equipped with several Faraday cups (charged particle detectors). For carbon isotope analysis, the Faraday cups are adjusted to atomic mass units (amu) 44 [$^{12}C^{16}O_2$] and 45 [$^{13}C^{16}O_2$] for the measurement of the abundance of the different isotopes and the determination of the isotope composition. The isotope composition is typically given as delta notation (per mil or mUr) related to VPDB scale (Coplen 1995).

18.2.2 Chlorine Isotope Analysis

Although essential for numerous chlorinated compounds, like environmental pollutants (Cincinelli et al. 2012; Lebedev 2013; Sakaguchi-Soder et al. 2007; Zhang and Qi 2012), CSIA of chlorine was thus far limited by the difficulty of conversion of the sample into a simple chlorine-containing gas for isotope measurements by CF-IRMS (Elsner et al. 2012; Brenna et al. 1997a; Sessions 2006). Nowadays, chlorine isotope analysis is mainly restricted to traditional 'offline' methods, including time-consuming conversion to either chloromethane (Kaufmann et al. 1984; Holt et al. 1997) or cesium chloride (Holmstrand et al. 2004) in combination with dual inlet isotope-ratio mass spectrometry (DI-IRMS) or thermal ionization mass spectrometry (TIMS). Several techniques were applied during the last decade for 'online' analysis of chlorine isotopes, using a combination of GC-quadrupole-MS (GC-qMS) (Jin et al. 2011; Aeppli et al. 2010) or GC-IRMS (Shouakar-Stash et al. 2006). Those methods do not require the previous conversion of organochlorides (Bernstein et al. 2011; Palau et al. 2014; Renpenning et al. 2014; Cretnik et al. 2013, 2014; Audi-Miro et al. 2013; Wiegert et al. 2013, 2012; Hunkeler et al. 2011; Aeppli et al. 2010), but direct measurement of molecular or fragmented ions. However, due to specific cup configurations required for IRMS or moderate precision of qMS ($1\sigma \approx 0.5$ ‰) both methods can, thus far, only be applied for a limited range of compound classes (Elsner et al. 2012). An alternative application was introduced by Hitzfeld et al. (2011) involves a high-temperature conversion (HTC) of chlorinated hydrocarbons to hydrochloric acid (HCl) as target compound for chlorine isotope analysis. Recent application of GC-HTC-IRMS could extend the range of measurable compound classes (Renpenning et al. 2015a). Therefore, GC-HTC-IRMS is a potential method for chlorine isotope analysis in future.

18.2.3 Hydrogen Isotope Analysis

Though hydrogen isotope analysis is nowadays a routine application, analysis of hydrogen isotopes is still challenging for halogen-bearing compounds. Hydrogen isotopes are traditionally measured via HTC of the target compound to H_2 at 1400–1450 °C and the subsequent measurement of hydrogen isotopes in H_2 analyte gas (Gehre and Strauch 2003; Brenna et al. 1997a). Unfortunately, this technique is not applicable for chlorinated compounds, since HCl is produced from H and Cl in the molecule. Over the last decade only few techniques were targeting hydrogen isotope analysis of chlorinated compounds (Kuder and Philp 2013; Shouakar-Stash and Drimmie 2013; Armbruster et al. 2006). Only recently a chromium-based high-temperature conversion (Cr/HTC) was introduced in combination with an elemental analysis (EA) or gas chromatograph (GC). EA-Cr/HTC-IRMS as well as GC-Cr/HTC-IRMS were shown to be a promising technique for hydrogen isotope analysis of several heteroatom-bearing (N, Cl, S) compound classes (Gehre et al. 2015) (Renpenning et al. 2015c).

18.3 Quantification of Stable Isotope Fractionation

Stable isotope analysis can be used to obtain information on sources, transport routes, degradation pathways, and sinks of organic chemicals in the environment. (Bio)chemical degradation of contaminants leads to kinetic isotope fractionation, a result of the KIE, which may alter the isotope composition of organic chemicals. The isotope fractionation can be applied for characterization of (bio)chemical reaction mechanism by observing the transition state of a chemical bond cleavage (Northrop 1981; Paneth 2003; Dybala-Defratyka et al. 2007).

18.3.1 Kinetic Isotope Effects

Chemical behavior of two isotopes of the same element is qualitatively similar, since the number of electrons harbored in the outer shell of an atom is identical in both isotopes. However, both isotopes still differ in their reaction rate and bond strength. Their chemical kinetics is frequently rationalized using the transition state theory (TST). The difference in reaction kinetics (k) between the heavy (h) and light (l) isotopes of an element E in the transition state is called KIE (Eq. 18.3) (Northrop 1981).

$$\mathrm{KIE}_E = \frac{{}^{l}k}{{}^{h}k} \tag{18.3}$$

The KIE results from differences in the zero-point energy (ZPE) and the corresponding difference in the vibrational energy between heavy and light isotopes. Heavy atoms vibrate more slowly than lighter ones and therefore, have more stable and stronger bonds, since the ZPE of the molecule with the heavy isotope is lower. The corresponding reaction rate and bond strength differences among isotopes lead to isotopic differences between the source and product compounds of a chemical transformation and results in isotope fractionation (Northrop 1975). KIEs can be calculated when the mass of atoms in the bond and the vibrational frequency (i.e., its wave number) are given. The obtained values are known as Streitwieser semiclassical limits (Huskey 1991).

18.3.2 Isotope Enrichment Factor

The changes in isotope composition during a reaction can be described using the Rayleigh model, relating changes in isotope composition to changes in concentration under closed-system conditions (Hoefs 1987; Mariotti et al. 1981). The extent of isotope fractionation is usually expressed as the enrichment factor ε or the isotope fractionation factor α where $\varepsilon = (\alpha - 1) * 1000$. In the next sections, however, only the isotope enrichment factor ε is used. ε can be determined from the linearized logarithmic form of the Rayleigh equation, using measured isotope ratios (R) for element E and the corresponding concentration (C) of the analyte at time point 0 and t, as presented for carbon below (Eq. 18.4) (Meckenstock et al. 2004).

$$\ln\left(\frac{R_t}{R_0}\right) = \ln\left(\frac{\delta_t E_h + 1}{\delta_0 E_h + 1}\right) = \varepsilon \times \ln\left(\frac{C_t}{C_0}\right) \tag{18.4}$$

Since isotope enrichment factors are typically small, ε-values are reported in parts per thousand (‰). The determined enrichment factors can then be used to characterize reaction mechanisms and degradation pathways.

18.3.3 Dual-Element Isotope Analysis

Currently, stable isotope analysis of a single element is frequently applied for evaluation of a reaction mechanism. Examples from recent studies, however, showed that stable isotope fractionation for a single element must be interpreted very carefully in natural multistep processes. The additional analysis of two or more elements, however, enables a better characterization of a reaction mechanism (Zwank et al. 2005; Fischer et al. 2008). Usually, stable isotope analysis is done for elements involved in the reaction step, i.e., dual-element isotope analysis of carbon and chlorine during reductive dehalogenation. The results are usually presented as a dual-element slope m (or Λ), as shown below for carbon and chlorine (Elsner 2010) (Eq. 18.5) which represents the relative change in isotope enrichment in carbon (εC) and chlorine (εCl):

$$m = \frac{\Delta^{13}C}{\Delta^{37}Cl} \approx \frac{\varepsilon C}{\varepsilon Cl} \tag{18.5}$$

Dual-element isotope slopes can elucidate variations for the same mechanism that were previously masked in a single-element isotope analysis (see Sect. 18.4.3). Such variations may bear enormous potential to learn about the underlying processes (Elsner 2010). In addition to dual-element, multielement isotope analysis of chemically complex substances can be used to analyze transformation pathways, making use of isotope fractionation processes altering the reactive position, as well as to analyze the isotope composition of reactive and nonreactive entities of an organic molecule to track sources. Although analysis of hydrogen and chlorine isotopes of chlorinated hydrocarbons is limited by available analytical methods, first steps were taken to establish multielement analysis in the future (see Sect. 18.2).

18.4 Application of CSIA for the Evaluation of the Reductive Dehalogenation Reaction

18.4.1 Characterization of Bioremediation In Situ

Kinetic isotopic fractionations of (bio)chemical reaction processes may vary in magnitude depending on reaction type, reaction rates, concentrations of products and reactants, environmental conditions, and in the case of metabolic transformations species of the organism (Northrop 1981; Sherwood Lollar et al. 1999; Schmidt et al. 2004). However, evaluation of (bio)transformation in situ can be still achieved by monitoring changes in isotope composition of the degraded compound (Meckenstock et al. 2004). Over the last decades, stable isotope fractionation concepts have been developed which allow qualitative and quantitative assessment of the in situ biodegradation, source identification, and analysis of the reaction mechanism of organic contaminants. Concepts are available for the common groundwater contaminants such as BTEX, MTBE, and chlorinated ethenes. To develop this approach, laboratory and field studies were performed and evaluated (Meckenstock et al. 2004; Elsner et al. 2005; Bombach et al. 2010; Hunkeler et al. 1999; Sherwood Lollar et al. 2001, 1999; Bloom et al. 2000). These initial field investigations confirmed that a shift in carbon stable isotope composition could be used as an indicator for biodegradation. (Bloom et al. 2000; Slater et al. 2001). A more detailed application of bioremediation strategies in combination with CSIA is provided in Chap. 25.

18.4.2 Apparent Kinetic Isotope Effect (AKIE)

The enrichment factors calculated using Eq. 18.4 represents the observed isotope fractionation for the bulk of the molecule ($\varepsilon_{\text{bulk}}$) and do not consider the nonreacting positions and the intramolecular competition in the molecule. Calculation of the apparent kinetic isotope effect (AKIE) values, however, provides estimates of position-specific KIEs. Thereby n is the number of, for instance, carbon atoms, x is the number of atoms at a reacting positions, and z is the number of indistinguishable reactive sites (reactive positions with intramolecular isotopic competition) (Eq. 18.6). Depending on the postulated reaction mechanism(s) or possible reaction scenarios, different values for n, x, and z can be used in order to calculate the AKIE. The AKIE can be compared to the theoretical KIE values, i.e., Streitwieser semiclassical limits (Table 18.2) (Northrop 1981; Huskey 1991; Elsner et al. 2005).

$$\text{AKIE} = \frac{1}{1 + \left(z \times \left(\frac{n}{x}\right) \times \frac{\varepsilon_{\text{bulk}}}{1000}\right)} \qquad (18.6)$$

Table 18.2 Streitwieser semiclassical limits for KIE during bond cleavage at 25 °C for halogenated hydrocarbons (Cook 1991)

Bond	Frequency (cm^{-1})	Isotope	KIE
C–H	2900	$^{12}C/^{13}C$	1.021
C–H		$^{1}H/^{2}H$	6.44
C–C	1000	$^{12}C/^{13}C$	1.049
C–Cl	750	$^{12}C/^{13}C$	1.057
C–Cl		$^{35}Cl/^{37}Cl$	1.013

Table 18.3 Evaluation of $AKIE_C$ values for carbon in established cases of reduction by cleavage of one C–Cl bond according to Elsner et al. (2005) and Renpenning et al. (2014). In comparison, carbon kinetic isotope effect (KIE) expected from the Streitwieser Limit for cleavage of a C–Cl bond ~1.057 (see Table 18.2)

Type of reaction	Isotope	$AKIE_C$
Reduction of CCl_4 by Fe(II)	$^{12}C/^{13}C$	1.027–1.033
Reduction of CCl_4 by FeS	$^{12}C/^{13}C$	1.016
Reduction of PCE by vitamin B_{12}	$^{12}C/^{13}C$	1.033–1.053
Reduction of TCE by vitamin B_{12}	$^{12}C/^{13}C$	1.034–1.039

These corrections allow to compare compounds with, e.g., different numbers of carbon atoms which are subject to the same reaction and can be used to evaluate reaction mechanisms taking place by comparing the values to reactions with known mechanism (Table 18.2 [KIE] and Table 18.3 [AKIE]). Theoretical isotope effects can therefore be compared to experimental ones in order to elucidate reaction mechanisms. For example, theoretical KIE and AKIE for the reduction of tetrachloroethene (PCE) by vitamin B_{12} (Table 18.3) were similar suggesting that the KIE can be observed for abiotic dehalogenation of PCE in its full magnitude. For other reactions the AKIE was observed to be lower than the theoretical KIE suggesting that rate-limiting steps prior to the reaction step mask the real magnitude of the KIE. Additionally, calculation of AKIE will allow to evaluate similarity in reactions for different substrates, e.g., as done for dichloroethane and dichloropropane dichloroelimination (Fletcher et al. 2009; Schmidt et al. 2014).

18.4.3 Evaluation of Degradation Pathways

Isotope fractionation of compounds subject to degradation often reflects the reaction mechanism involved in its degradation. Therefore, enrichment factors can be used for identification of degradation pathways, as it was already used for MTBEs (Kuder et al. 2005; Zwank et al. 2005; Elsner et al. 2007). Halogenated hydrocarbons can be transformed in situ by different pathways, including abiotic

Table 18.4 Experimental carbon isotope enrichment factors (ε_C) determined for abiotic and enzymatic catalysis (Dayan et al. 1999; Slater et al. 2003; Nijenhuis et al. 2005; Cichocka et al. 2007, 2008; Elsner et al. 2008; Abe et al. 2009; Schmidt et al. 2010; Clingenpeel et al. 2012; Cretnik et al. 2013; Renpenning et al. 2014)

Compound	Abiotic		Biotic	
	Fe(0)	Corrinoids (reductive dehalogenation)	Anaerobic (reductive dehalogenation)	Aerobic (degradation)
PCE	−5.7 to −25.3	−13.0 to −25.3	−0.4 to −16.4	
TCE	−7.5 to −13.5	−15.0 to −21.3	−3.3 to −26.0	−11.6 to −14.7*
cis-DCE	−6.9 to −16.0		−14.9 to −29.7	−0.9 to −9.8
trans-DCE			−20.8 to −30.3	
1,1-DCE			−5.1 to −23.9	
VC	−6.9 to −19.3		−23.2 to −31.1	−3.2 to −8.2

*cometabolic reaction

and biotic (enzymatic) dehalogenation. Abiotic transformation can be mediated during reductive dehalogenation by zerovalent iron (ZVI) (Arnold and Roberts 2000; Elsner et al. 2008), mediated by corrinoids (Krone et al. 1989; Glod et al. 1997) or chemical oxidation using, for instance, permanganates or persulfates (Hrapovic et al. 2005; Tsitonaki et al. 2010). In contrast, biotic transformation occurs under oxic and anoxic conditions. Aerobic metabolic degradation, however, was observed only for lower chlorinated compounds, such as vinyl chloride (VC) and dichloroethene (DCE). Carbon isotope fractionation during aerobic degradation is usually small (Table 18.4) and can be explained by the catalytic reaction pathways, which do not involve a direct cleavage of the C–Cl bond (Chartrand et al. 2005; Abe et al. 2009; Mattes et al. 2010; Tiehm and Schmidt 2011; Tiehm et al. 2008; Clingenpeel et al. 2012). During organohalide respiration C–Cl bonds are sequentially cleaved leading to formation of lower chlorinated hydrocarbons (Scholz-Muramatsu et al. 1995; Maymo-Gatell et al. 1997) and carbon isotope fractionation is generally stronger. The corresponding isotope fractionation is usually larger for VC, *cis*-DCE (*cis*-DCE), and *trans*-DCE, while lower for 1,1-DCE, trichloroethene (TCE) and especially tetrachloroethene (PCE) and more variable isotope fractionation was measured (Table 18.4). The cause for the variability of microbial isotope fractionation will be discussed in detail in Sect. 18.5. The ranges for carbon isotope enrichment factors do not allow to distinguish abiotic from biotic, reductive dechlorination, reactions, however, allow distinguishing biodegradation pathways for VC and DCE in situ (Imfeld et al. 2010).

18.4.3.1 Assessment of the Reductive Dehalogenation Reaction Mechanism

To elucidate the reaction mechanisms of microbial strains capable of reductive dehalogenation, carbon stable isotope analysis was performed to investigate the involved (bio)catalytic step. Isotope fractionation patterns were investigated for several microorganisms, including members of δ-Proteobacteria, ε-Proteobacteria, Firmicutes, and Chloroflexi. Despite similarities of the reductive dehalogenase enzyme (RDase) in all microorganisms, isotope analysis of carbon revealed highly variable isotope fractionation for different strains during dehalogenation of PCE and TCE (Table 18.5), making characterization of the reaction mechanism difficult (Nijenhuis et al. 2005; Cichocka et al. 2008; Lee et al. 2007). Observed variability in carbon isotope fractionation was thought to be related to the specific corrinoids incorporated into the RDase enzymes as well as the microbial cell envelope properties or growth conditions (Nijenhuis et al. 2005; Cichocka et al. 2008; Mancini et al. 2006). Furthermore, effects of substrate properties, i.e., hydrophobicity, were suggested to be responsible for variability in isotope fractionation (Thullner et al. 2013; Cichocka et al. 2007).

Therefore, the observed variability of isotope fractionation during reductive dehalogenation was suspected to be a result of isotope masking. Isotope masking in microbial systems is a result of rate-limiting events prior to the actual catalytic reaction, as for instance, extracellular and intracellular mass transfers (Elsner 2010). The effect of extracellular mass transfer was demonstrated to affect observable isotope fractionation in substrate availability studies (Kampara et al. 2009; Thullner et al. 2013; Aeppli et al. 2009). Similarly, intracellular mass transfer was demonstrated to affect isotope fractionation, as discussed in Sect. 18.5 (Nijenhuis et al. 2005; Renpenning et al. 2015b).

18.4.3.2 Role of the Corrinoid Cofactor

The introduction of the dual-element approach offered a major step forward to overcome the bottlenecks of single-element isotope analysis. Simultaneous analysis of two (or more) elements involved in the bond cleavage potentially elucidates the real magnitude isotope fractionation, by excluding isotope-masking effects (Zwank et al. 2005).

Corrinoids are a key cofactor in almost all known RDases (Krone et al. 1989; Stupperich et al. 1990; Kräutler et al. 2003). Corrinoids, however, were found to differ in different microorganisms and to affect the reaction rates during reductive dehalogenation significantly (Neumann et al. 2002; Siebert et al. 2002; Kräutler et al. 2003; Keller et al. 2014). Therefore, corrinoids were suspected to affect the reaction mechanism. Recent application of dual-element (C/Cl) analysis however excluded corrinoid cofactors as the responsible reason for the observed variability in microbial isotope fractionation when carbon was used for single-element analysis only. Different corrinoid cofactors incorporated in RDase enzyme *Sulfurospirillum multivorans* (PceA-*norpseudo*-B_{12} and PceA-*nor*-B_{12}) were observed to result in similar dual-element isotope fractionation for TCE, as well as PCE. This observation confirmed the minor effect of the corrinoid structure on isotope fractionation (Renpenning et al. 2014).

Table 18.5 Enzymatic and abiotic carbon isotope enrichment factors for dehalogenation of PCE and TCE catalyzed by corresponding microbial strain or corrinoids

Enzymatic dehalogenation mediated by RDase				
Phyla	Organism	$\varepsilon_{C\text{-}PCE}$	$\varepsilon_{C\text{-}TCE}$	References
Firmicutes	*Desulfitobacterium PCE-S*	−5.2 to −8.9	−10.9 to −12.9	Nijenhuis et al. (2005), Cichocka et al. (2007)
	D. restrictus PER-K23	−4.0 to −6.3	−3.3 to −8.3	Lee et al. (2007), Renpenning et al. (2015b)
δ-Proteobacteria	*G. lovleyi SZ*	ns* to −2.3	−8.5 to −12.2	Cichocka et al. (2007), Cretnik et al. (2013), Renpenning et al. (2015b)
	D. michiganensis	−1.7 to −2.6	−3.5 to −7.1	Cichocka et al. (2007), Renpenning et al. (2015b)
ε-Proteobacteria	*S. halorespirans*	−0.5 to −3.2	−18.7 to −22.9	Cichocka et al. (2007)
	S. multivorans	−0.4 to −2.2	−16.2 to −26.0	Nijenhuis et al. (2005), Cichocka et al. (2007), Lee et al. (2007), Renpenning et al. (2014)
Chloroflexi	*Dhc strain 195*	−6.0	−9.6 to −13.7	Cichocka et al. (2008), Lee et al. (2007)
	Dhc strain CBDB1	−1.6	−11.2	Marco-Urrea et al. (2011)
Abiotic dehalogenation mediated by corrinoids				
Corrinoid type	Cyanocobalamin	−16.2 to −22.4	−15.0 to −16.5	Slater et al. (2003), Nijenhuis et al. (2005), Cichocka et al. (2007), Cretnik et al. (2013), Renpenning et al. (2014)
	Norpeudo-B12	−25.3	−18.5	Renpenning et al. (2014)
	Nor-B12	−23.7	−15.1	Renpenning et al. (2014)
	Dicyanocobinamide	−25.2	−16.5	Renpenning et al. (2014)
	Cobaloxime		−21.5	Cretnik et al. (2013)

*ns: not significant

Interestingly, variability in the dual-element slopes was still observed for structurally different pure corrinoids. Measured dual-element isotope fractionation could be distinguished into two groups: Corrinoids containing dimethylbenzamidazole (DMB) as ligand base with a dual-element slope of 4.6–5.0, and non-DMB containing corrinoids with a dual-element slope of 6.9–7.0. The significant differences were attributed to the difference in the lower ligand. Based on the only available crystal structure of a reductive dehalogenase of *S. multivorans*, the lower ligands in enzymes are thought to be bound by the enzyme structure and forced in a permanent base-off conformation (Bommer et al. 2014). Purified corrinoids, however, are able to change the conformation according to the redox state of the cobalt. Therefore, the dissociation of the ligand (base-on/off) during the reaction may be a rate-limiting step in abiotic dehalogenation reaction, affecting the overall rate of the reaction and masking the isotope fractionation (Renpenning et al. 2014). The effect of mass transfer on isotope fractionation will be discussed in Sect. 18.5.

18.4.3.3 Proton Transfer During Reductive Dehalogenation

Only few reports are available about proton transfer during dehalogenation and the corresponding isotope effects of hydrogen. The main reason for that is the difficulty of hydrogen stable isotope analysis within halogenated compounds (see Sect. 18.2.3). Thus far only one publication revealed strong isotope effects for hydrogen during dehalogenation of TCE, *cis*-DCE, and VC to ethene by a *Dehalococcoides* mixed culture (Kuder et al. 2013). The δ^2H isotope signature in the product was enriched in lighter isotopes by ~800 ‰ from +530 ‰ (TCE) to −270 ‰ (ethene). Each dehalogenation step resulted in an isotopic shift of several hundred per mil. Although hydrogen isotope fractionation effects during the protonation step are not yet well investigated, they promise a valuable improvement for characterization of reaction mechanisms by introducing a multidimensional stable isotope analysis (C, Cl and H).

18.5 Intracellular Mass Transfer and the Effect on Observed Isotope Fractionation

Variability in microbial isotope fractionation was considered be a result of rate-limiting steps during transport prior to that C–Cl bond cleavage (Nijenhuis et al. 2005; Cichocka et al. 2007; Thullner et al. 2013; Kampara et al. 2008). For example, mass transfer of the substrate to the enzyme may be affected by extracellular (solubility) and intracellular (membrane barriers, sorption at the membranes or enzymes) rate limitation, resulting in a dilution of the magnitude of isotope fractionation (Aeppli et al. 2009; Thullner et al. 2013). In microbial systems three main, potentially rate-limiting, barriers can be considered for the substrate: (1) the outer membrane or cell wall, (2) the cytoplasmic membrane in case of a cytoplasmic location of the enzyme, and (3) the structure and properties of reductive dehalogenase enzyme (Fig. 18.2). Furthermore, the properties of the substrate, such as solubility and hydrophobicity, may affect the extent of rate

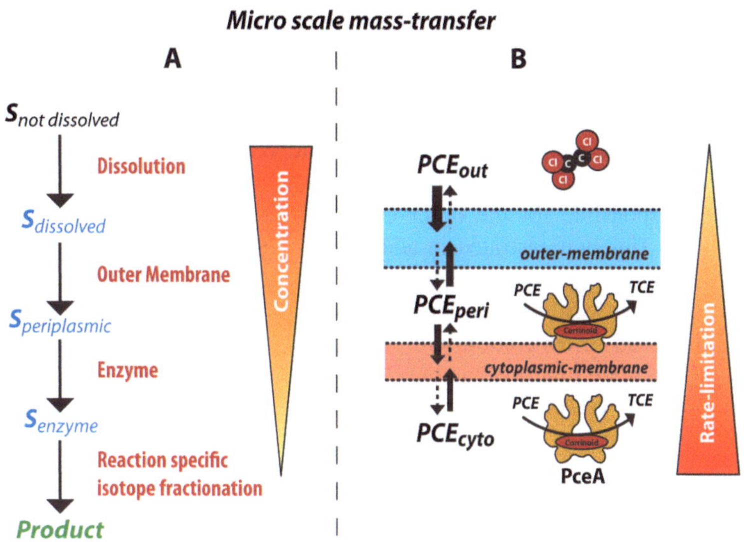

Fig. 18.2 **A** microscale mass transfer at microbial systems, as for instance, dissolution, transport trough the membranes, and enzyme-substrate association potentially affect isotope fractionation. Potentially rate-limiting barriers for the PCE were observed to be the *outer membrane* and the *cytoplasmic membrane*. **B** Furthermore, in case of a cytoplasmic location of the enzyme, the structure and the properties of RDase are suspected to have an additional effect on mass transfer limitation

limitation observed. Indeed, sorption tests to microbial biomass showed in general significantly higher sorption of PCE compared to TCE for the gram-negative *S. multivorans*, as well as a three times higher sorption capacity of PCE for the gram-negative *S. multivorans* in comparison to the gram-positive *Desulfitobacterium* (Renpenning et al. 2015b).

Although mass transfer limitation masks the real magnitude of the reaction-specific isotope enrichment, information can still be used for evaluation and interpretation intracellular microscale mass transfer processes. These masking effects, however, are only expected to be observed in cases where the catalytic rate at the enzyme higher is compared to the rate of mass transfer (Sherwood Lollar et al. 2010; Mancini et al. 2006).

18.5.1 Outer Membrane

Microscale mass transfer-induced isotope masking was reported in several bioavailability studies at low substrate concentration and concentration gradients (Thullner et al. 2008; Kampara et al. 2008). Effect of rate-limiting mass transfer on isotope fractionation was demonstrated in several studies using high biomass concentration (Staal et al. 2007; Templeton et al. 2006; Kampara et al. 2009). The first evidence for the membrane as a rate-limiting barrier during organohalide respiration was obtained from isotope fractionation studies with growing cells versus crude extracts (Cichocka et al. 2007; Nijenhuis et al. 2005). The destruction of the cell envelope increased the observed isotope effect significantly. Therefore, rate-limiting mass transfer at the outer membrane of *S. multivorans* was considered to be responsible for the observed isotope masking for PCE (Renpenning et al. 2015b). This demonstrated in addition a correlation between cell composition and variability of isotope fractionation in *S. multivorans*. The cultivation of *S. multivorans* with chlorinated ethenes resulted in a higher saturated fatty acid content compared to cultivation with fumarate. In addition, the cell surface was observed to be more hydrophobic during growth with fumarate compared to more hydrophilic with PCE or TCE. Subsequent dehalogenation experiments confirmed the contribution of the outer membrane to stronger isotope masking due to the higher hydrophobicity of the cell surface and higher sorption capacity at the cell membranes.

Firmicutes and Chloroflexi, however, do not possess an outer membrane, and therefore, isotope masking was supposed to be negligible for strains of these phyla. Still, isotope fractionation was determined to be significantly stronger for microbial crude extracts in comparison to growing cells for gram-positives, as *D. hafniense* and *D. restrictus* (Renpenning et al. 2015b). Therefore, variability of isotope fractionation could not be attributed to the outer membrane alone.

18.5.2 Cytoplasmic Membrane

Although *S. multivorans* RDases are all thought to face to the outside of the cytoplasmic membrane, initial studies on *S. multivorans* localized the RDase in the cytoplasm (Neumann et al. 1994). Later studies, however, showed that the cultivation conditions affected the location of the RDase (John et al. 2006). Partial location of the enzyme in the cytoplasm provided further evidence for membranes as rate-limiting barriers in the dehalogenation reaction, however, also provided evidence for the activity of the enzyme in the cytoplasm (Renpenning et al. 2015b). Though thus far only investigated for *S. multivorans* (John et al. 2006), active cytoplasmic dehalogenase may occur frequently in organohalide-respiring bacteria during the initial growth, affecting the observed isotope effect. Microscale mass transfer of chlorinated ethenes in this case will be limited by both, outer membrane and cytoplasmic membrane. The differences in relative distribution of dehalogenase cytoplasm and periplasm may explain the variability of isotope enrichment factors observed in different studies as result of differences in growth phase or conditions, as for instance, isotope enrichment factors for PCE dehalogenated by the mixed culture KB-1 (−2.6 to −5.5 ‰), *Desulfitobacterium* strain PCE-S (−5.2 to −8.9 ‰), and *Geobacter lovleyi* (not significant to −2.3 ‰) (Renpenning et al. 2014; Cichocka et al. 2008; Slater et al. 2001; Nijenhuis et al. 2005).

18.5.3 Rate Limitation at the RDase

The first crystal structure of PceA of *S. multivorans* was recently published by Bommer et al. (2014) and revealed an enzyme structure with an active site inside the core of the protein. To get access to the active site chlorinated hydrocarbons have to pass a 12 Å long and 3 × 5.5 Å wide hydrophobic channel. The channel forms a restriction filter and is thought to disfavor access for molecules larger than halogenated ethenes. Similarly to the isotope-masking effect of the outer and cytoplasmic membranes, Rdh enzymes may restrict the mass transfer for highly hydrophobic compounds to the active site and enhance isotope masking. Evidence for rate limitation at the active site of PceA RDase (*S. multivorans*) was provided by Renpenning et al. (2014). Using corrinoids, abiotic dehalogenation rates were observed to be about 10 times faster for PCE versus TCE, while enzyme-catalyzed dehalogenation rates were similar for both chlorinated ethenes. Therefore, the initial binding and transport of PCE toward the active center may be a rate-limiting step. Furthermore, dual-element analysis suggested a multistep reaction with different isotope effects of Cl versus C (Renpenning et al. 2014). This can only be explained by rate limitation if the association of PCE to the hydrophobic channel exhibits a pronounced isotope effect overlain by the isotope effects of the reaction or if the reaction involves two steps, e.g., binding of the substrate prior to the dehalogenation at the active center (Fig. 18.2). Experiments with pure corrinoids already indicated that rate-limiting events, such as the dissociation of the lower ligand (Sect. 18.4.3.2) may have a significant effect on the measured isotope fractionation. Moreover, the significant Cl isotope effect versus the insignificant carbon isotope effects during sorption of TCE (Shouakar-Stash et al. 2009) would suggest strong interaction of PCE with enzyme resulting in overlaying isotope effects. Therefore, rate limitation at the active site of the enzyme would explain the overall low isotope fractionation of hydrophobic PCE by several microbial strains capable of dehalogenation, whereas the less hydrophobic TCE was observed to be not or insignificantly affected.

18.6 Conclusion

Even with some limitations in microbial systems, CSIA is especially valuable to investigate a reaction without the need for a purified enzyme or crystal structure. Though CSIA is mainly applied for carbon and partly for chlorine, it helped to confirm similarity in reaction mechanisms for enzymatic and abiotic reductive dehalogenation mediated by pure corrinoids (Renpenning et al. 2014; Cretnik et al. 2013, 2014). Furthermore, CSIA could show that different corrinoids do not affect the reaction mechanism, as it was previously suggested (Nijenhuis et al. 2005; Yan et al. 2012). Different corrinoids types (DMB versus non-DMB ligand) were observed to change dual-element isotope fractionation, and differences were absent after incorporation of corresponding corrinoid types into the PceA RDase of *S. multivorans* (Renpenning et al. 2014). These results coincide with the first published crystal structure from PceA of *S. multivorans* (Bommer et al. 2014). For the reaction mechanism, however, preliminary results using dual-element analysis of carbon and chlorine do suggest a multistep reaction at the enzyme. Moreover, intracellular microscale mass transfer over membranes and at the enzyme can strongly affect the observed isotope fractionation as shown for *S. multivorans*. The extent of rate limitation is determined by growth conditions affecting cell composition but also by enzyme localization as well as by the substrate properties. Highest rate limitation can be expected for hydrophobic compounds such as PCE.

Overall, compound-specific isotope fractionation of organohalides remains a challenging task, though major steps were undertaken to overcome the limitations in isotope analysis of chlorine and hydrogen. Disregarding, CSIA of carbon and first investigations for chlorine already provided valuable information about the various steps of organohalide respiration. Therefore, extension of CSIA to a multielement stable isotope analysis, including carbon, chlorine, and hydrogen may reveal more and more detailed insight into the process of reductive dehalogenation.

Acknowledgments This study was funded by the Deutsche Forschungsgemeinschaft (Research Unit FOR 1530 NI 1329/1-1).

References

Abe Y, Aravena R, Zopfi J, Shouakar-Stash O, Cox E, Roberts JD, Hunkeler D (2009) Carbon and chlorine isotope fractionation during aerobic oxidation and reductive dechlorination of vinyl chloride and cis-1,2-dichloroethene. Environ Sci Technol 43(1):101–107. doi:10.1021/Es801759k

Aeppli C, Berg M, Cirpka OA, Holliger C, Schwarzenbach RP, Hofstetter TB (2009) Influence of mass-transfer limitations on carbon isotope fractionation during microbial dechlorination of trichloroethene. Environ Sci Technol 43(23):8813–8820. doi:10.1021/es901481b

Aeppli C, Holmstrand H, Andersson P, Gustafsson O (2010) Direct compound-specific stable chlorine isotope analysis of organic compounds with quadrupole GC/MS using standard isotope bracketing. Anal Chem 82(1):420–426. doi:10.1021/ac902445f

Armbruster W, Lehnert K, Vetter W (2006) Establishing a chromium-reactor design for measuring delta 2H values of solid polyhalogenated compounds using direct elemental analysis and stable isotope ratio mass spectrometry. Anal Bioanal Chem 384(1):237–243. doi:10.1007/s00216-005-0160-0

Arnold WA, Roberts AL (2000) Pathways and kinetics of chlorinated ethylene and chlorinated acetylene reaction with Fe(0) particles. Environ Sci Technol 34(9):1794–1805. doi:10.1021/es990884q

Audi-Miro C, Cretnik S, Otero N, Palau J, Shouakar-Stash O, Soler A, Elsner M (2013) Cl and C isotope analysis to assess the effectiveness of chlorinated ethene degradation by zero-valent iron: evidence from dual element and product isotope values. Appl Geochem 32:175–183. doi:10.1016/j.apgeochem.2012.08.025

Bernstein A, Shouakar-Stash O, Ebert K, Laskov C, Hunkeler D, Jeannottat S, Sakaguchi-Soder K, Laaks J, Jochmann MA, Cretnik S, Jager J, Haderlein SB, Schmidt TC, Aravena R, Elsner M (2011) Compound-specific chlorine isotope analysis: a comparison of gas chromatography/isotope ratio mass spectrometry and gas chromatography/quadrupole mass spectrometry methods in an interlaboratory study. Anal Chem 83(20):7624–7634. doi:10.1021/ac200516c

Bloom Y, Aravena R, Hunkeler D, Edwards E, Frape SK (2000) Carbon isotope fractionation during microbial dechlorination of trichloroethene, cis-1,2-dichloroethene, and vinyl chloride: implications for assessment of natural attenuation. Environ Sci Technol 34(13):2768–2772. doi:10.1021/Es991179k

Bombach P, Richnow HH, Kastner M, Fischer A (2010) Current approaches for the assessment of in situ biodegradation. Appl Microbiol Biotechnol 86(3):839–852. doi:10.1007/s00253-010-2461-2

Bommer M, Kunze C, Fesseler J, Schubert T, Diekert G, Dobbek H (2014) Structural basis for organohalide respiration. Science 346(6208):455–458. doi:10.1126/science.1258118

Brand WA, Coplen TB (2012) Stable isotope deltas: tiny, yet robust signatures in nature. Isot Environ Health Stud 48(3):393–409. doi:10.1080/10256016.2012.666977

Brenna JT, Corso TN, Tobias HJ, Caimi RJ (1997a) High-precision continuous-flow isotope ratio mass spectrometry. Mass Spectrom Rev 16(5):227–258. doi:10.1002/(SICI)1098-2787(1997)16:5<227:AID-MAS1>3.0.CO;2-J

Brenna JT, Corso TN, Tobias HJ, Caimi RJ (1997b) High-precision continuous-flow isotope ratio mass spectrometry. Mass Spectrom Rev 16(5):227–258. doi:10.1002/(sici)1098-2787(1997)16:5<227:aid-mas1>3.0.co;2-j

Chartrand MM, Waller A, Mattes TE, Elsner M, Lacrampe-Couloume G, Gossett JM, Edwards EA, Lollar BS (2005) Carbon isotopic fractionation during aerobic vinyl chloride degradation. Environ Sci Technol 39(4):1064–1070. doi:10.1021/es0492945

Cichocka D, Siegert M, Imfeld G, Andert J, Beck K, Diekert G, Richnow HH, Nijenhuis I (2007) Factors controlling the carbon isotope fractionation of tetra- and trichloroethene during reductive dechlorination by *Sulfurospirillum* ssp. and *Desulfitobacterium* sp. strain PCE-S. FEMS Microbiol Ecol 62(1):98–107. doi:10.1111/j.1574-6941.2007.00367.x

Cichocka D, Imfeld G, Richnow HH, Nijenhuis I (2008) Variability in microbial carbon isotope fractionation of tetra- and trichloroethene upon reductive dechlorination. Chemosphere 71(4):639–648. doi:10.1016/j.chemosphere.2007.11.013

Cincinelli A, Pieri F, Zhang Y, Seed M, Jones KC (2012) Compound specific isotope analysis (CSIA) for chlorine and bromine: a review of techniques and applications to elucidate environmental sources and processes. Environ Pollut 169:112–127. doi:10.1016/j.envpol.2012.05.006

Clingenpeel SR, Moan JL, McGrath DM, Hungate BA, Watwood ME (2012) Stable carbon isotope fractionation in chlorinated ethene degradation by bacteria expressing three toluene oxygenases. Front Microbiol 3:63. doi:10.3389/fmicb.2012.00063

Cook PF (1991) Enzyme mechanism from isotope effects. Crc Press

Coplen TB (1994) Reporting of stable hydrogen, carbon, and oxygen isotopic abundances. Pure Appl Chem 66(2):273–276. doi:10.1351/pac199466020273

Coplen TB (1995) New IUPAC guidelines for the reporting of stable hydrogen, carbon, and oxygen isotope-ratio data. J Res Natl Inst Stan 100(3):285. doi:10.6028/Jres.100.021

Coplen TB (2011) Guidelines and recommended terms for expression of stable-isotope-ratio and gas-ratio measurement results. Rapid Commun Mass Spectrom RCM 25(17):2538–2560. doi:10.1002/Rcm.5129

Coplen TB, Böhlke JK, De Bievre P, Ding T, Holden N, Hopple J, Krouse H, Lamberty A, Peiser H, Revesz K (2002a) Isotope-abundance variations of selected elements (IUPAC technical report). Pure Appl Chem 74(10):1987–2017

Coplen TB, Bohlke JK, De Bievre P, Ding T, Holden NE, Hopple JA, Krouse HR, Lamberty A, Peiser HS, Revesz K, Rieder SE, Rosman KJR, Roth E, Taylor PDP, Vocke RD, Xiao YK (2002b) Isotope-abundance variations of selected elements—(IUPAC technical report). Pure Appl Chem 74(10):1987–2017. doi:10.1351/pac200274101987

Coplen TB, Hopple J, Boehike J, Peiser H, Rieder S (2002c) Compilation of minimum and maximum isotope ratios of selected elements in naturally occurring terrestrial materials and reagents. US Geological Survey

Coplen TB, Brand WA, Gehre M, Groning M, Meijer HA, Toman B, Verkouteren RM, International Atomic Energy A (2006) After two decades a second anchor for the VPDB delta13C scale. Rapid communications in mass spectrometry. RCM 20 (21):3165–3166. doi:10.1002/rcm.2727

Cretnik S, Thoreson KA, Bernstein A, Ebert K, Buchner D, Laskov C, Haderlein S, Shouakar-Stash O, Kliegman S, McNeill K, Elsner M (2013) Reductive dechlorination of TCE by chemical model systems in comparison to dehalogenating bacteria: insights from dual element isotope analysis (13C/12C, 37Cl/35Cl). Environ Sci Technol 47(13):6855–6863. doi:10.1021/es400107n

Cretnik S, Bernstein A, Shouakar-Stash O, Loffler F, Elsner M (2014) Chlorine isotope effects from isotope ratio mass spectrometry suggest intramolecular C–Cl bond competition in trichloroethene (TCE) reductive dehalogenation. Molecules 19(5):6450–6473. doi:10.3390/molecules19056450

Dayan H, Abrajano T, Sturchio NC, Winsor L (1999) Carbon isotopic fractionation during reductive dehalogenation of chlorinated ethenes by metallic iron. Org Geochem 30(8A):755–763. doi:10.1016/S0146-6380(99)00058-3

Dybala-Defratyka A, Paneth P, Banerjee R, Truhlar DG (2007) Coupling of hydrogenic tunneling to active-site motion in the hydrogen radical transfer catalyzed by a coenzyme B12-dependent mutase. Proc Natl Acad Sci USA 104(26):10774–10779. doi:10.1073/pnas.0702188104

Elsner M (2010) Stable isotope fractionation to investigate natural transformation mechanisms of organic contaminants: principles, prospects and limitations. J Environ Monit JEM 12(11):2005–2031. doi:10.1039/c0em00277a

Elsner M, Hofstetter TB (2011) Current perspectives on the mechanisms of chlorohydrocarbon degradation in subsurface environments: insight from kinetics, product formation, probe molecules, and isotope fractionation. Aquat Redox Chem 1071:407–439. doi:10.1021/bk-2011-1071.ch019

Elsner M, Zwank L, Hunkeler D, Schwarzenbach RP (2005) A new concept linking observable stable isotope fractionation to transformation pathways of organic pollutants. Environ Sci Technol 39(18):6896–6916. doi:10.1021/es0504587

Elsner M, Mckelvie J, Couloume GL, Lollar BS (2007) Insight into methyl tert-butyl ether (MTBE) stable isotope fractionation from abiotic reference experiments. Environ Sci Technol 41(16):5693–5700. doi:10.1021/Es070531o

Elsner M, Chartrand M, Vanstone N, Couloume GL, Lollar BS (2008) Identifying abiotic chlorinated ethene degradation: characteristic isotope patterns in reaction products with nanoscale zero-valent iron. Environ Sci Technol 42(16):5963–5970. doi:10.1021/Es8001986

Elsner M, Jochmann MA, Hofstetter TB, Hunkeler D, Bernstein A, Schmidt TC, Schimmelmann A (2012) Current challenges in compound-specific stable isotope analysis of environmental organic contaminants. Anal Bioanal Chem 403(9):2471–2491. doi:10.1007/s00216-011-5683-y

Elvert M, Suess E, Greinert J, Whiticar MJ (2000) Archaea mediating anaerobic methane oxidation in deep-sea sediments at cold seeps of the eastern Aleutian subduction zone. Org Geochem 31(11):1175–1187. doi:10.1016/S0146-6380(00)00111-X

Fischer A, Herklotz I, Herrmann S, Thullner M, Weelink SA, Stams AJ, Schlomann M, Richnow HH, Vogt C (2008) Combined carbon and hydrogen isotope fractionation investigations for elucidating benzene biodegradation pathways. Environ Sci Technol 42(12):4356–4363. doi:10.1021/es702468f

Fletcher KE, Loffler FE, Richnow HH, Nijenhuis I (2009) Stable carbon isotope fractionation of 1,2-dichloropropane during dichloroelimination by Dehalococcoides populations. Environ Sci Technol 43(18):6915–6919. doi:10.1021/es900365x

Gehre M, Strauch G (2003) High-temperature elemental analysis and pyrolysis techniques for stable isotope analysis. Rapid Commun Mass Spectrom RCM 17(13):1497–1503. doi:10.1002/rcm.1076

Gehre M, Renpenning J, Gilevska T, Qi H, Coplen TB, Meijer HA, Brand WA, Schimmelmann A (2015) On-line hydrogen-isotope measurements of organic samples using elemental chromium: an extension for high temperature elemental-analyzer techniques. Anal Chem 87(10):5198–5205. doi:10.1021/acs.analchem.5b00085

Glod G, Angst W, Holliger C, Schwarzenbach RP (1997) Corrinoid-mediated reduction of tetrachloroethene, trichloroethene, and trichlorofluoroethene in homogeneous aqueous solution: reaction kinetics and reaction mechanisms. Environ Sci Technol 31(1):253–260. doi:10.1021/Es9603867

Godin JP, McCullagh JS (2011) Review: current applications and challenges for liquid chromatography coupled to isotope ratio mass spectrometry (LC/IRMS). Rapid Commun Mass Spectrom RCM 25(20):3019–3028. doi:10.1002/rcm.5167

Hitzfeld KL, Gehre M, Richnow HH (2011) A novel online approach to the determination of isotopic ratios for organically bound chlorine, bromine and sulphur. Rapid Commun Mass Spectrom RCM 25(20):3114–3122. doi:10.1002/rcm.5203

Hoefs J (1987) Stable isotope geochemistry, vol 116. Springer

Hoefs J (2008) Stable isotope geochemistry. Springer

Holmstrand H, Andersson P, Gustafsson O (2004) Chlorine isotope analysis of submicromole organochlorine samples by sealed tube combustion and thermal ionization mass spectrometry. Anal Chem 76(8):2336–2342. doi:10.1021/ac0354802

Holt BD, Sturchio NC, Abrajano TA, Heraty LJ (1997) Conversion of chlorinated volatile organic compounds to carbon dioxide and methyl chloride for isotopic analysis of carbon and chlorine. Anal Chem 69(14):2727–2733. doi:10.1021/Ac961096b

Hrapovic L, Sleep BE, Major DJ, Hood ED (2005) Laboratory study of treatment of trichloroethene by chemical oxidation followed by bioremediation. Environ Sci Technol 39(8):2888–2897. doi:10.1021/es049017y

Hunkeler D, Aravena R, Butler BJ (1999) Monitoring microbial dechlorination of tetrachloroethene (PCE) in groundwater using compound-specific stable carbon isotope ratios: microcosm and field studies. Environ Sci Technol 33(16):2733–2738. doi:10.1021/Es981282u

Hunkeler D, Abe Y, Broholm MM, Jeannottat S, Westergaard C, Jacobsen CS, Aravena R, Bjerg PL (2011) Assessing chlorinated ethene degradation in a large scale contaminant plume by dual carbon-chlorine isotope analysis and quantitative PCR. J Contam Hydrol 119(1–4):69–79. doi:10.1016/j.jconhyd.2010.09.009

Huskey W (1991) Origins and interpretations of heavy-atom isotope effects. Enzyme Mech Isot Eff 37–72

Imfeld G, Aragones CE, Fetzer I, Meszaros E, Zeiger S, Nijenhuis I, Nikolausz M, Delerce S, Richnow HH (2010) Characterization of microbial communities in the aqueous phase of a constructed model wetland treating 1,2-dichloroethene-contaminated groundwater. FEMS Microbiol Ecol 72(1):74–88. doi:10.1111/j.1574-6941.2009.00825.x

Jin B, Laskov C, Rolle M, Haderlein SB (2011) Chlorine isotope analysis of organic contaminants using GC-qMS: method optimization and comparison of different evaluation schemes. Environ Sci Technol 45(12):5279–5286. doi:10.1021/es200749d

John M, Schmitz RP, Westermann M, Richter W, Diekert G (2006) Growth substrate dependent localization of tetrachloroethene reductive dehalogenase in *Sulfurospirillum multivorans*. Arch Microbiol 186(2):99–106. doi:10.1007/s00203-006-0125-5

Jouzel J, Lorius C, Petit JR, Genthon C, Barkov NI, Kotlyakov VM, Petrov VM (1987) Vostok ice core—a continuous isotope temperature record over the last climatic cycle (160,000 Years). Nature 329(6138):403–408. doi:10.1038/329403a0

Kampara M, Thullner M, Richnow HH, Harms H, Wick LY (2008) Impact of bioavailability restrictions on microbially induced stable isotope fractionation. 2. Experimental evidence. Environ Sci Technol 42(17):6552–6558. doi:10.1021/es702781x

Kampara M, Thullner M, Harms H, Wick LY (2009) Impact of cell density on microbially induced stable isotope fractionation. Appl Microbiol Biotechnol 81(5):977–985. doi:10.1007/s00253-008-1755-0

Kaufmann R, Long A, Bentley H, Davis S (1984) Natural chlorine isotope variations. Nature 309(5966):338–340. doi:10.1038/309338a0

Kaufmann RS, Long A, Campbell DJ (1988) Chlorine isotope distribution in formation waters, texas and Louisiana: GEOLOGIC NOTE. AAPG Bull 72(7):839–844

Keller S, Ruetz M, Kunze C, Krautler B, Diekert G, Schubert T (2014) Exogenous 5,6-dimethylbenzimidazole caused production of a non-functional tetrachloroethene reductive dehalogenase in *Sulfurospirillum multivorans*. Environ Microbiol 16(11):3361–3369. doi:10.1111/1462-2920.12268

Kräutler B, Fieber W, Ostermann S, Fasching M, Ongania K-H, Gruber K, Kratky C, Mikl C, Siebert A, Diekert G (2003) The cofactor of tetrachloroethene reductive dehalogenase of *Dehalospirillum multivorans* is Norpseudo-B12, a new type of a natural corrinoid. Helv Chim Acta 86(11):3698–3716. doi:10.1002/hlca.200390313

Krone UE, Thauer RK, Hogenkamp HPC (1989) Reductive dehalogenation of chlorinated C1-hydrocarbons mediated by corrinoids. Biochemistry 28(11):4908–4914. doi:10.1021/Bi00437a057

Krummen M, Hilkert AW, Juchelka D, Duhr A, Schluter HJ, Pesch R (2004) A new concept for isotope ratio monitoring liquid chromatography/mass spectrometry. Rapid Commun Mass Spectrom RCM 18(19):2260–2266. doi:10.1002/rcm.1620

Kuder T, Philp P (2013) Demonstration of compound-specific isotope analysis of hydrogen isotope ratios in chlorinated ethenes. Environ Sci Technol 47(3):1461–1467. doi:10.1021/es303476v

Kuder T, Wilson JT, Kaiser P, Kolhatkar R, Philp P, Allen J (2005) Enrichment of stable carbon and hydrogen isotopes during anaerobic biodegradation of MTBE: microcosm and field evidence. Environ Sci Technol 39(1):213–220. doi:10.1021/es040420e

Kuder T, van Breukelen BM, Vanderford M, Philp P (2013) 3D-CSIA: carbon, chlorine, and hydrogen isotope fractionation in transformation of TCE to ethene by a *Dehalococcoides* culture. Environ Sci Technol 47(17):9668–9677. doi:10.1021/es400463p

Lebedev AT (2013) Environmental mass spectrometry. Ann Rev Anal Chem 6:163–189. doi:10.1146/annurev-anchem-062012-092604

Lee PK, Conrad ME, Alvarez-Cohen L (2007) Stable carbon isotope fractionation of chloroethenes by dehalorespiring isolates. Environ Sci Technol 41(12):4277–4285. doi:10.1021/es062763d

Magenheim AJ, Spivack AJ, Michael PJ, Gieskes JM (1995) Chlorine stable isotope composition of the oceanic crust: implications for Earth's distribution of chlorine. Earth Planet Sci Lett 131(3):427–432

Mancini SA, Hirschorn SK, Elsner M, Lacrampe-Couloume G, Sleep BE, Edwards EA, Lollar BS (2006) Effects of trace element concentration on enzyme controlled stable isotope fractionation during aerobic biodegradation of toluene. Environ Sci Technol 40(24):7675–7681. doi:10.1021/es061363n

Mariotti A, Germon JC, Hubert P, Kaiser P, Letolle R, Tardieux A, Tardieux P (1981) Experimental-determination of nitrogen kinetic isotope fractionation—some principles—illustration for the denitrification and nitrification processes. Plant Soil 62(3):413–430. doi:10.1007/Bf02374138

Marco-Urrea E, Nijenhuis I, Adrian L (2011) Transformation and carbon isotope fractionation of tetra- and trichloroethene to trans-dichloroethene by Dehalococcoides sp. strain CBDB1. Environ Sci Technol 45 (4):1555–1562. doi:10.1021/es1023459

Mattes TE, Alexander AK, Coleman NV (2010) Aerobic biodegradation of the chloroethenes: pathways, enzymes, ecology, and evolution. FEMS Microbiol Rev 34(4):445–475. doi:10.1111/j.1574-6976.2010.00210.x

Maymo-Gatell X, Chien Y, Gossett JM, Zinder SH (1997) Isolation of a bacterium that reductively dechlorinates tetrachloroethene to ethene. Science 276(5318):1568. doi:10.1126/science.276.5318.1568

Meckenstock RU, Morasch B, Griebler C, Richnow HH (2004) Stable isotope fractionation analysis as a tool to monitor biodegradation in contaminated acquifers. J Contam Hydrol 75(3–4):215–255. doi:10.1016/j.jconhyd.2004.06.003

Neumann A, Scholz-Muramatsu H, Diekert G (1994) Tetrachloroethene metabolism of *Dehalospirillum multivorans*. Arch Microbiol 162(4):295–301. doi:10.1007/BF00301854

Neumann A, Siebert A, Trescher T, Reinhardt S, Wohlfarth G, Diekert G (2002) Tetrachloroethene reductive dehalogenase of *Dehalospirillum multivorans*: substrate specificity of the native enzyme and its corrinoid cofactor. Arch Microbiol 177(5):420–426. doi:10.1007/s00203-002-0409-3

Nijenhuis I, Andert J, Beck K, Kastner M, Diekert G, Richnow HH (2005) Stable isotope fractionation of tetrachloroethene during reductive dechlorination by *Sulfurospirillum multivorans* and *Desulfitobacterium* sp. strain PCE-S and abiotic reactions with cyanocobalamin. Appl Environ Microbiol 71(7):3413–3419. doi:10.1128/AEM.71.7.3413-3419.2005

Northrop DB (1975) Steady-state analysis of kinetic isotope effects in enzymic reactions. Biochemistry 14(12):2644–2651. doi:10.1021/bi00683a013

Northrop DB (1981) The expression of isotope effects on enzyme-catalyzed reactions. Annu Rev Biochem 50:103–131. doi:10.1146/annurev.bi.50.070181.000535

Palau J, Cretnik S, Shouakar-Stash O, Hoche M, Elsner M, Hunkeler D (2014) C and Cl isotope fractionation of 1,2-dichloroethane displays unique delta(1)(3)C/delta(3)(7)Cl patterns for pathway identification and reveals surprising C–Cl bond involvement in microbial oxidation. Environ Sci Technol 48(16):9430–9437. doi:10.1021/es5031917

Paneth P (2003) Chlorine kinetic isotope effects on enzymatic dehalogenations. Acc Chem Res 36(2):120–126. doi:10.1021/ar010101h

Ransom B, Spivack AJ, Kastner M (1995) Stable Cl isotopes in subduction-zone pore waters—implications for fluid-rock reactions and the cycling of chlorine. Geology 23(8):715–718. doi:10.1130/0091-7613(1995)023<0715:Sciisz>2.3.Co;2

Renpenning J, Keller S, Cretnik S, Shouakar-Stash O, Elsner M, Schubert T, Nijenhuis I (2014) Combined C and Cl isotope effects indicate differences between corrinoids and enzyme (*Sulfurospirillum multivorans* PceA) in reductive dehalogenation of tetrachloroethene, but not trichloroethene. Environ Sci Technol 48(20):11837–11845. doi:10.1021/es503306g

Renpenning J, Hitzfeld KL, Gilevska T, Nijenhuis I, Gehre M, Richnow HH (2015a) Development and validation of an universal interface for compound-specific stable isotope analysis of chlorine (37Cl/35Cl) by GC-high-temperature conversion (HTC)-MS/IRMS. Anal Chem 87(5):2832–2839. doi:10.1021/ac504232u

Renpenning J, Rapp I, Nijenhuis I (2015b) Substrate hydrophobicity and cell composition influence the extent of rate limitation and masking of isotope fractionation during microbial reductive dehalogenation of chlorinated ethenes. Environ Sci Technol 49(7):4293–4301. doi:10.1021/es506108j

Renpenning J, Kummel S, Hitzfeld KL, Schimmelmann A, Gehre M (2015c) Compound-specific hydrogen isotope analysis of heteroatom-bearing compounds via gas chromatography-chromium-based high-temperature conversion (Cr/HTC)-isotope ratio mass spectrometry. Anal Chem 87 (18):9443–9450. doi:10.1021/acs.analchem.5b02475

Sakaguchi-Soder K, Jager J, Grund H, Matthaus F, Schuth C (2007) Monitoring and evaluation of dechlorination processes using compound-specific chlorine isotope analysis. Rapid Commun Mass Spectrom RCM 21(18):3077–3084. doi:10.1002/rcm.3170

Schmidt TC, Zwank L, Elsner M, Berg M, Meckenstock RU, Haderlein SB (2004) Compound-specific stable isotope analysis of organic contaminants in natural environments: a critical review of the state of the art, prospects, and future challenges. Anal Bioanal Chem 378(2):283–300. doi:10.1007/s00216-003-2350-y

Schmidt KR, Augenstein T, Heidinger M, Ertl S, Tiehm A (2010) Aerobic biodegradation of cis-1,2-dichloroethene as sole carbon source: stable carbon isotope fractionation and growth characteristics. Chemosphere 78(5):527–532. doi:10.1016/j.chemosphere.2009.11.033

Schmidt M, Lege S, Nijenhuis I (2014) Comparison of 1,2-dichloroethane, dichloroethene and vinyl chloride carbon stable isotope fractionation during dechlorination by two Dehalococcoides strains. Water Res 52:146–154. doi:10.1016/j.watres.2013.12.042

Scholz-Muramatsu H, Neumann A, Meßmer M, Moore E, Diekert G (1995) Isolation and characterization of *Dehalospirillum multivorans* gen. nov., sp. nov., a tetrachloroethene-utilizing, strictly anaerobic bacterium. Arch Microbiol 163(1):48–56. doi:10.1007/BF00262203

Sessions AL (2006) Isotope-ratio detection for gas chromatography. J Sep Sci 29(12):1946–1961. doi:10.1002/jssc.200600002

Sherwood Lollar B, Slater G, Ahad J, Sleep B, Spivack J, Brennan M, MacKenzie P (1999) Contrasting carbon isotope fractionation during biodegradation of trichloroethylene and toluene: implications for intrinsic bioremediation. Org Geochem 30(8):813–820. doi:10.1016/S0146-6380(99)00064-9

Sherwood Lollar B, Slater G, Sleep B, Witt M, Klecka G, Harkness M, Spivack J (2001) Stable carbon isotope evidence for intrinsic bioremediation of tetrachloroethene and trichloroethene at area 6, dover air force base. Environ Sci Technol 35(2):261–269. doi:10.1021/es001227x

Sherwood Lollar B, Hirschorn S, Mundle SO, Grostern A, Edwards EA, Lacrampe-Couloume G (2010) Insights into enzyme kinetics of chloroethane biodegradation using compound specific stable isotopes. Environ Sci Technol 44(19):7498–7503. doi:10.1021/es101330r

Shouakar-Stash O, Drimmie RJ (2013) Online methodology for determining compound-specific hydrogen stable isotope ratios of trichloroethene and 1,2-cis-dichloroethene by continuous-flow isotope ratio mass spectrometry. Rapid Commun Mass Spectrom RCM 27(12):1335–1344. doi:10.1002/rcm.6578

Shouakar-Stash O, Drimmie RJ, Zhang M, Frape SK (2006) Compound-specific chlorine isotope ratios of TCE, PCE and DCE isomers by direct injection using CF-IRMS. Appl Geochem 21(5):766–781. doi:10.1016/j.apgeochem.2006.02.006

Shouakar-Stash O, Stotler RL, Frape SK, Illman WA (2009) The effect of sorption on chlorine stable isotopes of TCE. Geochim Cosmochim Acta 73(13):A1217–A1217

Siebert A, Neumann A, Schubert T, Diekert G (2002) A non-dechlorinating strain of *Dehalospirillum multivorans*: evidence for a key role of the corrinoid cofactor in the synthesis of an active tetrachloroethene dehalogenase. Arch Microbiol 178(6):443–449. doi:10.1007/s00203-002-0473-8

Slater GF, Lollar BS, Sleep BE, Edwards EA (2001) Variability in carbon isotopic fractionation during biodegradation of chlorinated ethenes: implications for field applications. Environ Sci Technol 35(5):901–907. doi:10.1021/es001583f

Slater GF, Lollar BS, Lesage S, Brown S (2003) Carbon isotope fractionation of PCE and TCE during dechlorination by vitamin B12. Ground Water Monit Rem 23(4):59–67. doi:10.1111/j.1745-6592.2003.tb00695.x

Staal M, Thar R, Kuhl M, van Loosdrecht MCM, Wolf G, de Brouwer JFC, Rijstenbil JW (2007) Different carbon isotope fractionation patterns during the development of phototrophic freshwater and marine biofilms. Biogeosciences 4(4):613–626. doi:10.5194/bg-4-613-2007

Stupperich E, Eisinger HJ, Schurr S (1990) Corrinoids in anaerobic-bacteria. FEMS Microbiol Lett 87(3–4):355–359. doi:10.1111/j.1574-6968.1990.tb04936.x

Tanaka N, Rye DM (1991) Chlorine in the stratosphere. Nature 353(6346):707. doi:10.1038/353707a0

Templeton AS, Chu KH, Alvarez-Cohen L, Conrad ME (2006) Variable carbon isotope fractionation expressed by aerobic CH4-oxidizing bacteria. Geochim Cosmochim Acta 70(7):1739–1752. doi:10.1016/j.gca.2005.12.002

Thiemens MH (2006) History and applications of mass-independent isotope effects. Annu Rev Earth Planet Sci 34(1):217–262. doi:10.1146/annurev.earth.34.031405.125026

Thullner M, Kampara M, Richnow HH, Harms H, Wick LY (2008) Impact of bioavailability restrictions on microbially induced stable isotope fractionation. 1. Theoretical calculation. Environ Sci Technol 42(17):6544–6551. doi:10.1021/es702782c

Thullner M, Fischer A, Richnow HH, Wick LY (2013) Influence of mass transfer on stable isotope fractionation. Appl Microbiol Biotechnol 97(2):441–452. doi:10.1007/s00253-012-4537-7

Tiehm A, Schmidt KR (2011) Sequential anaerobic/aerobic biodegradation of chloroethenes—aspects of field application. Curr Opin Biotechnol 22(3):415–421. doi:10.1016/j.copbio.2011.02.003

Tiehm A, Schmidt KR, Pfeifer B, Heidinger M, Ertl S (2008) Growth kinetics and stable carbon isotope fractionation during aerobic degradation of cis-1,2-dichloroethene and vinyl chloride. Water Res 42(10–11):2431–2438. doi:10.1016/j.watres.2008.01.029

Tsitonaki A, Petri B, Crimi M, Mosbaek H, Siegrist RL, Bjerg PL (2010) In situ chemical oxidation of contaminated soil and groundwater using persulfate: a review. Crit Rev Environ Sci Technol 40(1):55–91. doi:10.1080/10643380802039303

Wiegert C, Aeppli C, Knowles T, Holmstrand H, Evershed R, Pancost RD, Machackova J, Gustafsson O (2012) Dual carbon-chlorine stable isotope investigation of sources and fate of chlorinated ethenes in contaminated groundwater. Environ Sci Technol 46(20):10918–10925. doi:10.1021/es3016843

Wiegert C, Mandalakis M, Knowles T, Polymenakou PN, Aeppli C, Machackova J, Holmstrand H, Evershed RP, Pancost RD, Gustafsson O (2013) Carbon and chlorine isotope fractionation during microbial degradation of tetra- and trichloroethene. Environ Sci Technol 47(12):6449–6456. doi:10.1021/es305236y

Yan J, Ritalahti KM, Wagner DD, Loffler FE (2012) Unexpected specificity of interspecies cobamide transfer from Geobacter spp. to organohalide-respiring *Dehalococcoides mccartyi* strains. Appl Environ Microbiol 78(18):6630–6636. doi:10.1128/AEM.01535-12

Zhang Y, Qi SH (2012) Techniques of stable chlorine isotope analysis and relevant applications in research of organochlorine pollutants. Prog Chem 24(12):2384–2390

Zwank L, Berg M, Elsner M, Schmidt TC, Schwarzenbach RP, Haderlein SB (2005) New evaluation scheme for two-dimensional isotope analysis to decipher biodegradation processes: application to groundwater contamination by MTBE. Environ Sci Technol 39(4):1018–1029. doi:10.1021/es049650j

Chapter 19
Corrinoid Metabolism in Dehalogenating Pure Cultures and Microbial Communities

Theodore C. Moore and Jorge C. Escalante-Semerena

Abstract Corrinoid cofactors are critical components of the electron transport chain for many organohalide-respiring bacteria (OHRB). This chapter examines the synthesis and metabolism of corrinoids, with a focus on studies in bacteria that express reductive dehalogenases (RDases). We discuss the physical characteristics of corrinoids that make them distinct from one another, and provide examples of the various corrinoids isolated from OHRB. We provide a brief review of the synthesis, salvaging, and transport of corrinoids as it is currently understood in non-dehalogenating model organisms. Wherever applicable, we draw parallels to the pathways present in OHRB. We present some recent examples of work studying the metabolism of corrinoids in mixed cultures of OHRB, and discuss how these bacteria may share and modify corrinoids at a community-based level.

19.1 Introduction

What are corrinoids? Corrinoids are complex organometallic compounds that belong to the family of cyclic tetrapyrroles, which are also known as "The Pigments of Life." Members of this family of biomolecules include hemes, chlorophylls, and factor F_{430} (Battersby 2000). The most obvious difference amongst cyclic tetrapyrroles is the transition metal ion chelated by the equatorial pyrrolic nitrogen atoms. Hemes contain iron, chlorophylls contain magnesium, factor F_{430} contains nickel, and corrinoids contain cobalt. Also important is the degree of reduction of the macrocycle, with cobamides containing six double bonds. Corrinoids are also set aside from the other cyclic tetrapyrroles by the lack of the methane bridge at position C20. The absence of such a bridge puckers the corrin

T.C. Moore · J.C. Escalante-Semerena (✉)
Department of Microbiology, University of Georgia, Athens, GA 30602, USA
e-mail: jcescala@uga.edu

L. Adrian and F.E. Löffler (eds.), *Organohalide-Respiring Bacteria*,
DOI 10.1007/978-3-662-49875-0_19

ring, in contrast to the planarity of heme. Last but not least, corrinoids are the only natural cyclic tetrapyrroles that contain axial ligands.
The tetrapyrrole ring of corrinoids is referred to as the corrin ring, reflecting the "core" of the structure. The chemical structure of the corrin ring is the same in all known corrinoids. However, the identity of the upper (beta, *Coβ*) and lower (alpha, *Coα*) axial ligands varies. We expand on the nature and proper nomenclature of ligands below.

Classes of corrinoids and distinguishing features There are two classes of corrinoids: incomplete corrinoids and complete corrinoids. The difference between incomplete and complete corrinoids is the absence or presence of the nucleotide loop. This loop structure tethers the nucleotide base to the corrin ring via an amide bond formed between the amino group of a 1-amino-2-propanol (AP) moiety with the carboxyl group of a propionyl substituent of ring D (Fig. 19.1a). The alcohol group of AP participates in the formation of a phosphodiester bond with the nucleotide. In fact, complete corrinoids are one of only three coenzymes containing a phosphodiester bond. The other two are coenzyme F_{420} (Eirich et al. 1978) and methanopterin (van Beelen et al. 1984).

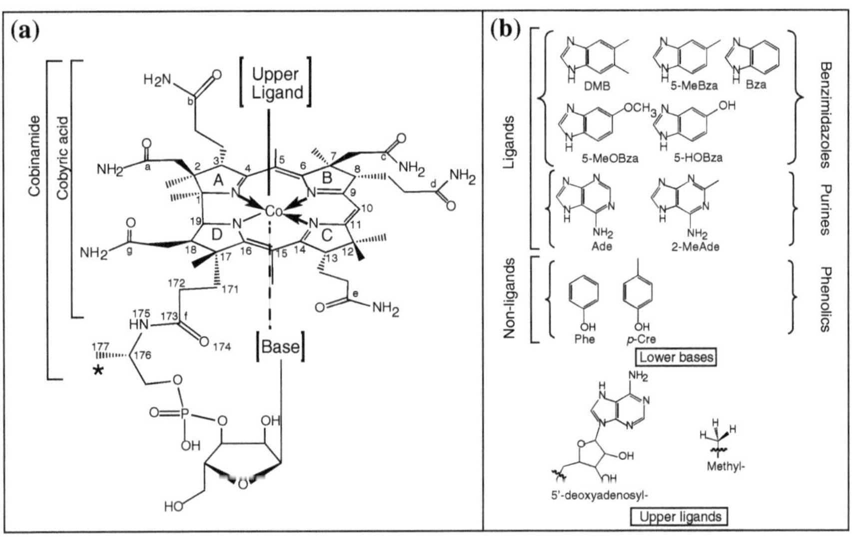

Fig. 19.1 Structural features of cobamides. **a** Structure of a cobamide. The *brackets* indicate the common precursors cobinamide and cobyric acid. Individual pyrroles are labeled *A–D*. Changing the starred methyl group 177 to a hydrogen results in a norcobamide. **b** A selection of biologically active axial ligands and bases of cobalamin that are mentioned in this chapter. The *brackets* around the *lower bases* indicate which are capable of ligating the Co ion. *Abbreviations* DMB, 5,6-dimethylbenzimidazole; 5-MeBza, 5-methylbenzimidazole; Bza, benzimidazole; 5-MeOBza, 5-methoxybenzimidazole; 5-HOBza, 5-hydroxybenzimidazole; Ade, adenine; 2-MeAde, 2-methyladenine; Phe, phenol; *p*-Cre, *para*-cresol. Wavy line denotes the Co–C covalent bond between upper ligands and cobalamin

Complete corrinoids can also vary amongst themselves. In this case, the structure that varies is the base of the nucleotide. We note that, although the majority of the bases found in corrinoids are nitrogenous benzimidazoles or purines that can directly interact with the Co ion of the ring via a coordination bond, there are bases, such as phenol and *p*-cresol, which cannot coordinate with Co because they lack unpaired electrons in their structures. Therefore, it is incorrect to refer to phenol or *p*-cresol as ligands. They should simply be referred to as "bases" since they are still part of the nucleotide loop.

Cobamides The term "cobamide" is used to refer to complete corrinoids, regardless of the base found in the nucleotide loop structure of the molecule. The best-studied cobamide is cobalamin (Cbl), which is the cobamide whose *Coα* (lower axial) ligand is 5,6-dimethylbenzimidazole (DMB) (Fig. 19.1b). Also shown in Fig. 19.1 are purine analogs that bacteria and archaea incorporate into cobamides, including adenine, which forms pseudocobalamin, another commonly studied cofactor. The list is not all-inclusive, but encompasses the lower ligands used by OHRB. To date, acetogenic bacteria are the only microorganisms known to incorporate phenol or *p*-cresol into their cobamides (Stupperich et al. 1989; Chan and Escalante-Semerena 2011; Newmister et al. 2012).

In cobamides containing purines or benzimidazoles, the base can be found in the "base-on" (coordinated to the Co ion) or "base-off" conformation. The base-on conformation is made possible by the presence of unpaired electrons in the nitrogen atom of the imidazole ring of the base that is not bonded to the ribosyl moiety of the nucleotide. As noted above, phenolyl cobamides cannot acquire the base-on conformation due to the absence of unpaired electrons in phenolic bases, thus restricting their use by some cobamide-dependent enzymes that require the base-on conformation of the cofactor to function (e.g., ethanolamine ammonia-lyase, EC 4.3.1.7; 1,2-propanediol dehydratase, EC 4.2.1.28) (Abend et al. 1999; Yamanishi et al. 2002).

Cobalamin The literature on cobalamin is extensive, covering many different aspects of this magnificent molecule. Some reviews cover the biosynthesis, metabolism, molecular biology, genetics, enzymology, and structure of cobalamin biosynthetic enzymes (Lawrence and Roth 1995; Roth et al. 1996; Warren et al. 2002; Roessner and Scott 2006; Escalante-Semerena 2007; Moore and Warren 2012), other reviews focus on the elucidation and chemical synthesis (Woodward 1973; Eschenmoser and Wintner 1977; Eschenmoser 1988), yet other reviews focus on chemical properties (Lexa and Saveant 1983; Banerjee and Ragsdale 2003; Brown 2005; Randaccio et al. 2010). This body of work provides a comprehensive understanding of the most complex coenzyme known and the largest molecule with biological activity that is not a polymer.

Cobalamin is defined by the coordination bond between a nitrogen atom of 5,6-dimethylbenzimidazole (DMB) and the Co ion on the α face of the corrin ring. The coenzymic form of cobalamin is adenosylcobalamin (AdoCbl). AdoCbl possesses a unique Co(III)-C covalent bond between the corrin ring and the *Coβ* (upper) ligand (Fig. 19.1A). This organometallic bond is pivotal to

AdoCbl-dependent chemistry, which proceeds via a homolytic, one-electron mechanism, generating a Co(II) cobalamin species and an adenosyl radical used by dehydratases, lyases, mutases, and reductases (Matthews 2001, 2009). Conversely, the Co–C bond of methylcobalamin (MeCbl, Fig. 19.1b) is exploited in a heterolytic, two-electron mechanism of methyl transfer involving a Co(I) cobalamin intermediate, used primarily in methanogenesis and methionine synthesis (Matthews 2001, 2009).

Cobamide biosynthesis To date, cobamide biosynthesis has only been observed in bacteria and archaea (Warren et al. 2002; Escalante-Semerena 2007), including genera of dehalogenating bacteria (e.g., *Desulfitobacterium, Dehalobacter, Sulfurospirillum*, and *Geobacter*), some of which use organohalides as electron acceptors in cellular respiration (Holliger et al. 1998). Since cobamide-dependent dehalogenases recognize narrow substrate spectra, bacteria encode several distinct dehalogenases to expand their range of usable electron acceptors. Case in point is *Dehalococcoides mccartyi* strain versus an obligate cobamide-salvaging strain, whose genome encodes the largest number (36) of cobamide-binding proteins known to occur in any prokaryote whose genome has been sequenced (McMurdie et al. 2009). Consequently, these microorganisms have a strong selective pressure exerted on them to maintain sufficient levels of cobamides to satisfy their physiological needs.

Some species of OHRB that do not synthesize cobamides de novo can salvage incomplete corrinoids that enter the late steps of the pathway (a.k.a. the nucleotide loop assembly pathway, NLA) yielding cobamides that can meet their physiological needs. In fact, many organisms that synthesize cobamides de novo also maintain salvaging pathways to save considerable energy on synthesis of the corrin ring.

Cobamides encountered in organohalide-respiring bacteria (OHRB) In OHRB, corrinoid-dependent dehalogenases serve an essential function in the electron transport chain as terminal reductases, and are known as reductive dehalogenases (RDases). The dependence of several RDases on cobamides has been studied biochemically via UV–Vis spectroscopy, electron paramagnetic resonance (EPR), inductively coupled plasma mass spectrometry (ICP-MS), and alkylating inhibitors which ligate the catalytically active β-position of cobamides (Neumann et al. 1998; Miller et al. 1998; Maillard et al. 2003; Adrian et al. 2007). However, the identity of the cobamide cofactor in most RDases is unknown. Prominent exceptions are the RDases of *Sulfurospirillum multivorans* (formerly *Dehalospirillum multivorans*). *S. multivorans* synthesizes the unusual cofactor norpseudocobalamin, that is, pseudocobalamin lacking a methyl group at the propionyl linkage of the nucleotide tail (C176, Fig. 19.1a) (Kräutler et al. 2003). The mechanistic preference of an enzyme for a particular type of cobamide is mostly unknown, both in regard to RDases as well as other cobamide-dependent enzymes.

In vitro analyses of RDases The cobamide content in representative RDases is not always known. Often the identity of the cobamide is confirmed by optical spectroscopy and alkyl halide inhibitors, which will selectively and reversibly inhibit cobamide activity by ligating to the upper ligand position (Neumann et al. 1995). Most cobamide-dependent RDases bind two 4Fe/4S iron–sulfur centers per molecule of corrinoid, which have been proposed to transfer electrons to the corrinoid cofactor to generate a reduced Co(I) state in vivo (Schumacher et al. 1997).

Many RDases are translocated through the inner membrane and are subsequently anchored to the membrane via a secondary protein (John et al. 2006; Neumann et al. 1998; Palmer et al. 2005). A recent study (Reinhold et al. 2012) characterized the effect of cobalamin on the maturation and localization of the PCE-degrading RDase PceA in *Desulfitobacterium hafniense* strain Y51 (EC 1.97.1.8). The authors found that when PceA was synthesized in the absence of cobalamin, the protein formed aggregates in the cytoplasm, whereas in the presence of cobalamin PceA was translocated to the exoplasmic face of the inner membrane. The study suggests that PceA is formed in a cofactor-free precursor state, referred to as prePceA, along with its chaperone PceT. PceT is proposed to keep prePceA in an open conformation, which allows cobalamin and iron–sulfur centers to be posttranslationally incorporated into prePceA (Morita et al. 2009; Maillard et al. 2011). Reconstitution with cofactors helps complete folding of PceA, which is translocated across the membrane via the Tat pathway.

Part of the barrier to obtaining clearer information on the nature of the cobamide cofactor has been the difficulty in purifying RDases to sufficient yield and homogeneity for structural studies. Recently, the norpseudocobalamin-binding PceA from *S. multivorans* was crystallized in complex with substrate and substrate analogs (Bommer et al. 2014). Key features include a unique base-off conformation in which the adenine lower ligand has been "curled" away from the Co(II) ion to generate an apparent four-coordinate corrinoid in complex with the substrate. Additionally, two cubane Fe-S centers are clearly visible in each monomer, and are positioned within a favorable range to transfer electrons to the corrin ring. Another recently identified and crystallized RDase, RdhA from *Nitratireductor pacificus*, revealed a similar conformation for the bound cobalamin cofactor, as well as two [4Fe-4S] clusters positioned favorably for reduction of cobalamin (Payne et al. 2014). For an in-depth description of the mechanism of RDases, please refer to Chaps. 17 and 18 in this book.

19.2 Corrin Ring Synthesis

Corrin ring biosynthesis has not been studied in OHRB. Most experiments regarding corrin ring assembly have been performed in *Salmonella enterica, Pseudomonas denitrificans*, and *Bacillus megaterium* (Cameron et al. 1989; Roth et al. 1993; Raux et al. 1998). Even though biochemical and genetic studies of de novo synthesis in OHRB are lacking, inferences can be made from genome sequencing efforts. Relevant parallels to studies in OHRB will be made throughout this section. Table 19.1 indicates the distribution of corrinoid synthesis

Table 19.1 Putative cobamide biosynthetic enzymes identified from available sequenced genomes of species containing RDase genes

	Transport				*Anaerobic pathway*																		
	BtuB	BtuF	BtuC	BtuD	CbiK	CbiX	CbiL	CbiH	CbiG	CbiF	CbiD	CbiJ	CbiE	CbiT	CbiC	CbiA	CobA[1]	CbiP	CbiB	CobU	CobS[2]	CobT[3]	CobC
R. pomeroyi DSS-3				x																			
R. sp. TM1040	x	x	x	s																			
P. profundum 3TCK	3x	x	x	x														x	x	2x	x	x	
Jannaschia sp. CCS1		x	x	x																			
Ahrensia sp. R2A130		x	x	x																			
S. sediminis	2x	3x			x		x	x	x	x	x	x	x	x	x	x	2x	2x	x	2x	x	x	x
Vibrio sp. RC586	x	x	x	4x												x	x	x	x	x	x	x	x
Anaeromyxobacter sp. K	x	2x	x	3x																			
A. dehalogenans 2CP-1	x	2x	x	4x																			
A. dehalogenans 2CP-C	x	2x	x	4x																			
Geobacter lovleyi SZ	2x	x	x	5x	x		x	x	x	x	x	x	x	x	x	x	2x	2x	x	x	x	2x	x
delta proteobacterium NaphS2	x	x	x	5x	x		2x	2x	x	x	x		x	x	x	2x	2x	x	x	x	x	2x	x
S. multivorans	x	x	x	x	x		x	x	x	x	x	x	x	x	x	x		x	x	x	x	x	x
D. mccartyi CBDB1		2x	x	x												x	2x	x	3x	x	x	x	x
D. mccartyi GT		x	x														x	x	2x	x	x	x	x
D. mccartyi BAV1		2x	x	x												x	2x	x	2x	x	x	x	x
D. mccartyi 195		3x	2x	2x												x	2x	x	4x	2x	2x	2x	2x
D. mccartyi VS		x	x	2x									x			x	x	x	x	x	x	x	x
D. lykanthroporepellens BL-DC-9		2x	2x	6x									x	x		x	x	x	x	x	x	x	x
D. tiedjei			2x	2x		x	x	x	x	x	x	x	x			x	3x	4x	x	x	x	x	
A. capsulatum	x			2x																			
D. alkaliphilus		2x	2x	6x		x	2x	x	x	x	x	x	x	x	x	x	x	2x	x	x	x	x	x
C. difficile 630		x	x	x	x		x	x	x	x	x	x	x	x	x	x		x	x	x	x	x	x
C. difficile R20291			2x	3x	x	x	x	x	x	x	x	x	x	x	x	x		x	x	x	x	x	x
D. hafniense Y51						x	x	x		x	x	x			x	x	4x	x		x	x	x	2x
D. hafniense DCB-2	x	5x	x				x	x		x	x	x			x	x	3x	x		x	x	x	x

(continued)

Table 19.1 (continued)

	Transport				*Anaerobic pathway*																		
	BtuB	BtuF	BtuC	BtuD	CbiK	CbiX	CbiL	CbiH	CbiG	CbiF	CbiD	CbiJ	CbiE	CbiT	CbiC	CbiA	CobA[1]	CbiP	CbiB	CobU	CobS[2]	CobT[3]	CobC
D. dehalogenans ATCC 51507		x	x	4x		x	x	x	x	x	x	x	x	x	x	2x	3x	x	x	x	x	x	x
D. dichloroeliminans LMG P-21439		x	x	2x		x	x	x	x	x	x	x	x	x	x	x	3x	x	x	x	x	x	x
D. restrictus DSM 9455									x	x	x	x	x	x	x	x	x	x					
Dehalobacter sp. CF		x	x	x		x	x		2x	x	2x	x	x	x	x	x	x	x	x		x	x	x
M. prima mesG1.Ag.4.2			x	x													x						
F. placidus DSM 10642		x	2x	6x			x	x	x	x	x			x	x	x		x	x		x	x	

	Aerobic pathway																	Miscellaneous							
	CobI	CobG	CobJ	CobM	CobF	CobK	CobL	CobH	CobB	CobN	CobS[4]	CobT[5]	CobR	CobO	CobQ	CobP	CobD	CobA[6]	CbiZ	CbiZ/BtuD	BluB	PduS	PduO	CblS	CblT
R. pomeroyi DSS-3	x	x	x	x	x	x	x	x		x	x	x	x								x				
R. sp. TM1040	2x	x	x	x		x	x	x		x	x	x	x		x	x					x				
P. profundum 3TCK											x	x	x				x	4x			x	x			
Jannaschia sp. CCS1	x	x	x	x		x	x	x		x	x	x	x	x	x	x									
Ahrensia sp. R2A130	x		x	x		x	x	x		x	x	x	x	x	x	x	x	x							
S. sediminis																		2x							
Vibrio sp. RC586																									
Anaeromyxobacter sp. K																		x					x		
A. dehalogenans 2CP-1																		x					x		
A. dehalogenans 2CP-C																		x					x		
Geobacter lovleyi SZ										x	x							x							
delta proteobacterium NaphS2																		2x							
S. multivorans																									
D. mccartyi CBDB1																	2x	x	2x						
D. mccartyi GT																	x		2x						

(continued)

Table 19.1 (continued)

	Aerobic pathway																Miscellaneous							
	CobI	CobG	CobJ	CobM	CobF	CobK	CobL	CobH	CobB	CobN	CobS[4]	CobT[5]	CobR	CobO	CobQ	CobP	CobD	CobA[6]	CbiZ	CbiZ/BtuD	BluB	PduS	PduO	CblS	CblT
D. mccartyi BAV1																			x						
D. mccartyi 195																	x		2x						
D. mccartyi VS																	2x	x							
D. lykanthroporepellens BL-DC-9																	2x	x	x						
D. tiedjei				x						x				x				x				x			
A. capsulatum														x									x		
D. alkaliphilus																									
C. difficile 630																		x							
C. difficile R20291																		x							
D. hafniense Y51																	x	x		x		2x		x	x
D. hafniense DCB-2																	x	2x		x				x	x
D. dehalogenans ATCC 51507																	x	x				x	x	x	x
D. dichloroeliminans LMG P-21439																		x							
D. restrictus DSM 9455																									
Dehalobacter sp. CF																	x	x							
M. prima mesG1.Ag.4.2																	x	x				x			
F. placidus DSM 10642													x					x	x				x		

The proteins are grouped by function, those involved in transport are in bold, those in the anaerobic, early Co insertion are in italic, and those involved in aerobic, late Co insertion are in bold italic. Proteins that are shared among the anaerobic and aerobic pathways, as well as those used as alternatives to parts of those pathways, are grouped under "Miscellaneous." The x indicates copy number. Superscript numbering indicates proteins with duplicate names: 1, ACAT; 2, cobalamin synthase; 3, α-R synthase; 4 and 5, chelators; 6, SUMT. Genes were identified using the SEED (Overbeek et al. 2005) and verified by homology to known cobamide synthesis proteins in *Salmonella enterica*, *Pseudomonas denitrificans*, *Listeria innocua*, *Bacillus megaterium*, or *Brucella melitensis*, as applicable, using BLAST (Altschul et al. 1990). The NCBI taxid numbers for the above strains are as follows: *Rugeria pomeroyi* DSS-3, 246200; *Ruegeria* sp. TM1040, 292414; *Photobacterium profundum* 3TCK, 314280; *Jannaschia* sp. CCS1, 290400; *Ahrensia* sp. R2A130, 744979; *Shewanella sediminis*, 271097; *Vibrio* sp. RC586, 675815; *Anaeromyxobacter* sp. K, 4472[illegible]7; *Anaeromyxobacter dehalogenans* 2CP-1, 455488; *Anaeromyxobacter dehalogenans* 2CP-C, 290397; *Geobacter lovleyi* SZ, 398767; delta proteobacterium NaphS2, 88274; *Sulfurospirillum multivorans*, 66821; *Dehalococcoides mccartyi* CBDB1, 255470; *Dehalococcoides mccartyi* GT, 633145; *D. mccartyi* BAV1, 216389; *Dehalococcoides mccartyi* 195, 243164; *Dehalococcoides mccartyi* VS, 311424; *Dehalogenimonas lykanthroporepellens* BL-DC-9, 552811; *Desulfuomonile tiedjei*, 2358; *Acidobacter capsulatum*, 33075; *Dethiobacter alkaliphilus*, 555088; *Clostridium difficile* 630; *Clostridium difficile* R20291; *Desulfitobacterium hafniense* Y51, 138119; *Desulfitobacterium hafniense* DCB-2, 272564; *Desulfitobacterium dehalogenans*, 36854; *Desulfitobacterium dichloroeliminans* LMG P-21439, 871963; *Dehalobacter restrictus* DSM 9455, 55583; *Dehalobacter* sp. CF, 11314[illegible]2; *Mesotoga prima* mesG1.Ag.4.2, 660470; *Ferroglobus placidus* DSM 10642, 589924. This information was reported by Hug et al. (2013)

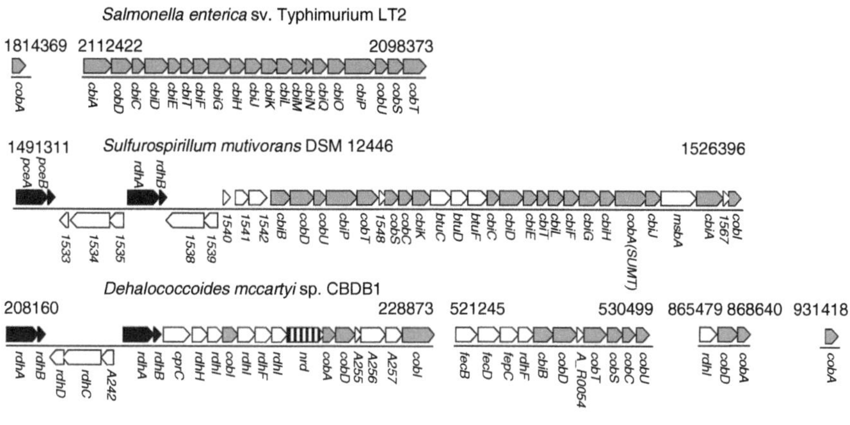

Fig. 19.2 Genetic organization of cobamide synthesis genes. Gene clusters relating to cobamide synthesis in non-organohalide-respiring *S. enterica*, organohalide-respiring de novo cobamide producer *S. multivorans*, and organohalide-respiring cobamide auxotroph *D. mccartyi* sp. CBDB1. Genes directly related to cobamide synthesis are colored gray. Genes encoding RDases are colored *black*. Other cobalamin-dependent genes are colored with *vertical stripes*. Gene name or locus number is shown above or below cartoon representation. Nucleotide position in the genome is shown above each gene cluster. Annotated gene clusters were downloaded from the Integrated Microbial Genomes database (Markowitz et al. 2012)

genes in the genomes of species with confirmed RDases (Hug et al. 2013). Figure 19.2 compares the organization of gene clusters related to cobamide synthesis in *D. mccartyi* sp. CBDB1 and *S. multivorans* to *S. enterica*. Notably, both the de novo cobamide synthesizer *S. multivorans* and the cobamide-salvaging bacterium *D. mccartyi* have cobamide synthetic genes located near genes encoding RDases.

The corrin ring of cobamides is similar to heme, siroheme, chlorophyll, and factor F_{430}. Unlike other corrinoids, the ring in cobamides is contracted between C1 and C19 (Fig. 19.1) to allow tighter metal binding, and is modified by additional methyl and amide groups. The corrin ring is more reduced than most tetrapyrroles, containing only six double bonds, compared to the 11 of heme, 10 of chlorophyll, and 9 of siroheme. Only factor F_{430} is more reduced, with five double bonds. Heme, siroheme, corrins, and other tetrapyrroles are all synthesized from the precursor uroporphyrinogen III (Warren et al. 2002). Uroporphyrinogen III is made up of four porphobilinogen moieties, which are each synthesized from the condensation of two molecules of δ-aminolevulinic acid (Bogorad and Granick 1953; Battersby et al. 1976; Leeper 1989). A *S*-adenosyl-L-methionine: Uroporphyrinogen III MethylTransferase (the SUMT family, CobA in *P. denitrificans*, EC 2.1.1.107) catalyzes methylations of carbons C2 and C7, resulting in precorrin-2 (Crouzet et al. 1991). Precorrin-2 refers to the number of methyl group decorations on the nascent corrin ring. At this point, the de novo synthesis diverges into one of two pathways, the "anaerobic" (early Co insertion route) or the "aerobic" (late Co insertion route) (Table 19.1) (Debussche et al. 1993; Roth et al. 1993). Certain enzymes in the two

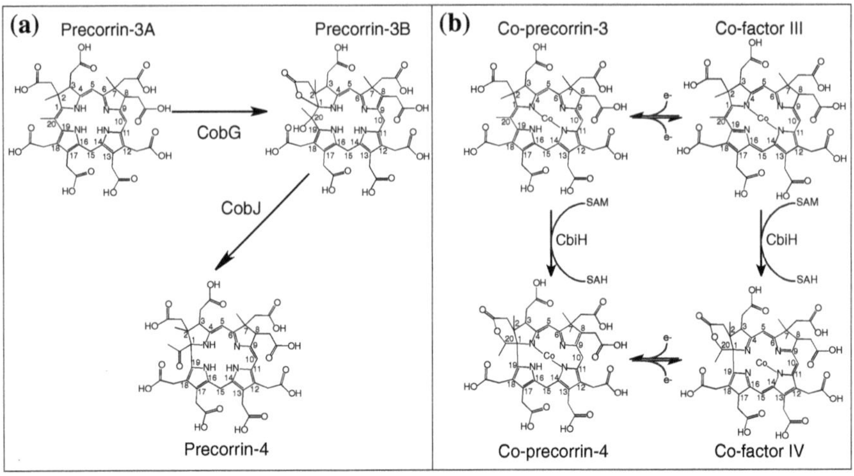

Fig. 19.3 Pathway for the ring contraction of precorrin-2. **a** Synthesis of the cobamide intermediate Precorrin-4 in the aerobic or "late Co insertion" pathway. Precorrin-2 undergoes methylation to precorrin-3A, lactone formation results in precorrin-3B. Subsequent lactone cleavage and extrusion of C20 result in precorrin-4. **b** Synthesis of the cobamide intermediate Co-precorrin-4 in the anaerobic or "early Co insertion" pathway. Co-precorrin-3 undergoes SAM-dependent methylation to Co-precorrin-4, resulting in the formation of a lactone moiety and the contraction of the corrin ring. The CbiH methylase can also catalyze the contraction of Co-factor III; the oxidized form of Co-precorrin-3

pathways have the same name, such as CobS and CobA. In Table 19.1 we attempt to clarify which pathway these enzymes are involved in.

The aerobic pathway is referred to as such because it utilizes a monooxygenase, CobG in the ring contraction step. After CobI (EC 2.1.1.130) methylates precorrin-2 at C20 to precorrin-3A (Thibaut et al. 1990), CobG (EC 1.14.13.83) reacts with precorrin-3A to form a lactone intermediate (precorrin-3B), which is acted upon by the *S*-adenosylmethionine-dependent CobJ (EC 2.1.1.131), catalyzing the ring contraction by releasing an acetic acid moiety when a methyl group attached to carbon C20 of the ring is excised (Fig. 19.3a), granting the corrin ring its signature "pucker." The resulting intermediate, precorrin-4, undergoes two subsequent methylations, on C2 by CobM (EC 2.1.1.271) yielding precorrin-5, and on C12 by CobF (EC 2.1.1.152). CobF also deacetylates C1 to produce precorrin-6A.

The NADPH-dependent reductase, CobK (EC 1.3.1.54), reduces a double bond between carbons C18 and C19 on precorrin-6A to produce precorrin-6B. CobL (EC 2.1.1.289/2.1.1.132, PDB 3EO5, 3NJR) methylates the C5 and C15 positions, as well as removing the carboxyl moiety at C12 to form precorrin-8. CobH (EC 5.4.99.61) isomerizes precorrin-8, moving the methyl at C11 to C12, resulting in hydrogenobyrinic acid (Thibaut et al. 1992). The CobB amidotransferase uses ATP and glutamine to amidate the acetic acid side chains *a* and *c*, producing hydrogenobyrinic acid *a,c*-diamide. The latter is the substrate for the CobNST cobalt ion chelatase (EC 6.6.1.2), which yields cobyrinic acid *a,c*-diamide upon insertion of the Co ion into the ring.

The anaerobic pathway was initially thought to be due to the inability of anaerobes to use molecular oxygen in the CobG reaction (Blanche et al. 1993). This included obligate anaerobic, cobamide-dependent organohalide-respiring species such as *D. hafniense*, *S. multivorans,* and others, which contain genes specific to the anaerobic pathway (Kruse et al. 2013). Some OHRB have incomplete anaerobic pathways that lack the ability to attach the nucleotide tail to cobamides, such as *Dehalobacter restrictus*. Other species with anaerobic pathways, such as several *D. mccartyi* strains, lack enzymes for insertion of cobalt into the corrin ring (Table 19.1). These species are characterized as being incapable of de novo corrin ring biosynthesis (Schipp et al. 2013).

In the anaerobic pathway, precorrin-2 is oxidized to factor II by SirC (EC 1.3.1.76, PDB 3dfz). At this point cobalt is inserted by either CbiX (EC 4.99.1.3, PDB 1TJN, 2DJ5, 2XWQ, 2XWS) as in *B. megaterium*, or by CbiK (EC 4.99.1.3, PDB 1QGO, 2XWP) as in *Salmonella enterica.* The multifunctional siroheme synthase/ferrochelatase enzyme CysG (EC 1.3.1.76/4.99.1.4, PDB 1PJQ, 1PJS) inserts cobalt in the absence of CbiK (Raux et al. 1997, 1998). This is the reason why the anaerobic pathway is also known as the "early" Co ion insertion pathway. The Co-factor II complex is methylated by the SAM-dependent CbiL (EC 2.1.1.136, PDB 2EOK, 2EON) to produce Co-factor III. This precursor is contracted by extrusion of the carbon atom C20, between the corrin ring C1 and C19. This carbon atom is incorporated into a lactone ring between C1 and C2 (Fig. 19.3b). The iron–sulfur protein CbiH (EC 2.1.1.272) catalyzes lactone formation and ring contraction via methyl donation from an *S*-adenosylmethionine (SAM) cofactor to form Co-factor IV. This was determined recently in a *Bacillus* system, where it was noted that CbiH could also act on the Co-precorrin-3, the reduced form of Co-factor III (Moore et al. 2013a). Difficulties in isolating the intermediates delayed our understanding of the steps leading from Co-precorrin-4 to cobyrinic acid. These difficulties were recently overcome (Moore et al. 2013b). CbiF (EC 2.1.1.271, PDB 1CBF, 2CBF) methylates the C11 position of Co-precorrin-4, and the lactone is opened and removed by CbiG (EC 3.7.1.12) to form Co-precorrin-5B. CbiD (EC 2.1.1.195) catalyzes the methylation of carbon C1 using *S*-adenosylmethionine as the methyl donor yielding Co-precorrin-6A. The NADH reductase CbiJ (EC 1.3.1.106) is expected to reduce the double bond between C18 and C19 in a manner similar to CobK in the aerobic pathway, but to date, the resulting intermediate Co-precorrin-6B has not been detected when the reaction is performed in vitro. CbiE (EC 2.1.1.196) and CbiT (EC 2.1.1.289) catalyze additionally two methylation reactions at carbons C5 and C15, along with a deacetylation at C12, resulting in Co-precorrin-8. CbiC (EC 5.4.99.60) converts precorrin-8 to cobyrinic acid by transferring the C11 methyl to C12.

19.3 Final Corrin Ring Amidations and Corrinoid Adenosylation

The aerobic and anaerobic pathways converge with the production of cobyric acid. Either CobQ (oxically, EC 6.3.5.10) or CbiA and CbiP (anoxically, EC 6.3.5.11, EC 6.3.5.10, respectively) catalyze the final corrin ring amidations to form cobyric acid. All three enzymes belong to the same family of glutamine amidotransferases (Fresquet et al. 2004). CobQ acts on cobyrinic acid *a,c*-diamide, and amidates the acetic acid side chains *b, d, e,* and *g* in an ATP and glutamine-dependent process (Blanche et al. 1991a). The substrate of CbiA is cobyrinic acid, which is amidated at side chains *c* and *a*, in that order (Fresquet et al. 2004). CbiP is responsible for side chains *b, d, g,* and *e*.

Cobyrinic acid and cobyrinic acid *a,c*-diamide likely have a Co(II) charge due to nonenzymatic reduction from cytosolic flavoproteins (Fonseca and Escalante-Semerena 2000). In *S. enterica,* CobA [the housekeeping ATP:Co(I)rrinoid adenosyltransferases (ACATs)] adds the *Coβ*-adenosyl moiety to Co(I) corrinoids and cobamides (Suh and Escalante-Semerena 1993; Fonseca and Escalante-Semerena 2000; Fonseca et al. 2002). In the aerobic pathway, cobyrinic acid *a,c*-diamide is adenosylated by CobO (EC 2.1.5.17) (Debussche et al. 1991). In the anaerobic pathway present in *S. enterica*, CobA (EC 2.1.5.17; PDB 1G5R, 1G5T, 1G64, 4HUT) can adenosylate cobyric acid, cobinamide, and cobalamin (Escalante-Semerena et al. 1990). In *Escherichia coli*, the CobA homologue is known as BtuR (Lundrigan and Kadner 1989). CobA has been annotated as being capable of adenosylating cobyrinic acid *a,c*-diamide due to its homology to CobO, but data in support of that have not been published. The arrest of de novo corrin ring biosynthesis in *S. enterica cobA* strains indicates that the final product of this branch of the pathway needs to be adenosylated prior to entering the late steps of the pathway, the branch of the pathway also known as the NLA (Escalante-Semerena et al. 1990; Jeter et al. 1984).

In *S. enterica,* CobA facilitates the thermodynamically unfavorable reduction of the Co ion from its 2+ to its 1+ oxidation state, which is "supernucleophilic" and capable of reacting with bound ATP. *S. enterica* CobA (and its homologue *P. denitrificans* CobO) are also necessary for salvaging cobamide precursors, and they can catalyze the reduction of wide variety of corrinoids (Debussche et al. 1991; Suh and Escalante-Semerena 1995; Anderson et al. 2008). The genomes of several organohalide-respiring *D. mccartyi* strains, which cannot synthesize cobalamin de novo, contain multiple *cobA* copies, presumably to facilitate efficient corrinoid salvaging (Table 19.1). Experimental data demonstrating that *D. mccartyi cobA* genes encode proteins with ATP:Co(I)rrinoid adenosyltransferase activity have not been reported.

The *S. enterica* genome (and those of other prokaryotes) also encodes non-homologous enzymes with CobA-like activity (i.e., PduO and EutT, ECs: 2.1.5.17) (Johnson et al. 2001; Buan et al. 2004; Buan and Escalante-Semerena 2006). Three-dimensional crystal structures of several PduO-type enzymes from different prokaryotes have been reported (PDB: 3GAH, 3GAI, 3GAJ, 2ZHY, 2R6X, 2NT8, 2R6T, 3CI3, 3CI4, 3CI1, 1WVT, 2G2D, 2ZHZ, 3KE4, 3KE5, 1WOZ) (Johnson et al. 2001; Buan et al. 2004; Buan and Escalante-Semerena 2006). The structure of EutT-type enzymes has not been solved, but it is known that the *S. enterica* enzyme is a ferroprotein whose activity is oxygen labile (Buan and Escalante-Semerena 2006; Moore et al. 2014). PduO and EutT are not involved in the de novo biosynthesis of cobamides, but they are necessary to maintain adequate levels of adenosylcobalamin for the degradation of specific diols (e.g., 1,2-propanediol) and alcoholamines (e.g., ethanolamine) (Mera and Escalante-Semerena 2010b; Moore et al. 2012). Collectively, these enzymes are known as ACATs.

PduO and EutT are used in *S. enterica* exclusively in 1,2-propanediol and ethanolamine catabolism (Johnson et al. 2004; Buan et al. 2004; Buan and Escalante-Semerena 2006; Mera et al. 2007), respectively. However, certain organisms lacking CobA will use PduO to adenosylate de novo or salvaged corrinoids. A bioinformatics study of cobalamin-utilizing prokaryotes identified ACATs in all but two *Clostridium* species and three archaeal species (Rodionov et al. 2003), although it was not experimentally determined whether or not these species could bypass the requirement for adenosylated corrinoid precursors. The organohalide-respiring bacterium *S. multivorans* also lacks an ACAT, although it contains an otherwise complete set of cobamide synthesis genes and is capable of synthesizing norpseudocobalamin de novo (Goris et al. 2014). The role of adenosylated corrinoids in *S. multivorans* is unclear. A study in *S. enterica* determined that mutations in *cobU* allowed the strain to utilize nonadenosylated cobamide precursors (O'Toole and Escalante-Semerena 1993), however these mutations are absent in *S. multivorans cobU*.

Both cob(I) and cob(II)alamin catalyze reductive dehalogenation reactions in the absence of an enzyme and in the absence of an upper ligand (Krone et al. 1989, 1991; Gantzer and Wackett 1991; Assaf-Anid et al. 1994; Chiu and Reinhard 1996; Lesage et al. 1998). However, genomics studies indicate that most OHRB contain CobA homologues. Assuming these genes are functional, this information suggests that there is a role for these enzymes in the attachment of the upper ligand. The mechanistic role of the upper ligand has yet to be defined in RDases. Recent RDase crystal structures contained cobamides rather than adenosylcobamides (Payne et al. 2014; Bommer et al. 2014), suggesting that the adenosyl group is removed prior to binding.

19.4 Corrinoid Transport System

Alternative to de novo synthesis, many organisms, including OHRB, can import and assimilate corrinoids. This is often advantageous to de novo synthesizers as well, due to the resource-intensive process of assembling the corrin ring, an estimated 2 % of the *S. enterica* genome is dedicated to synthesis and metabolism of cobalamin (Price-Carter et al. 2001). In Gram-negative bacteria, corrinoid uptake is performed by the well-characterized BtuBCDF system (DeVeaux et al. 1986; Cadieux et al. 2002; Rees et al. 2009) (PDB 1L7V, 2QI9, 4DBL, 4FL3). BtuB is a TonB-dependent, outer-membrane corrinoid transporter (Chimento et al. 2003; Noinaj et al. 2010). BtuF is a periplasmic binding protein (Borths et al. 2002), which shuttles the corrinoid from BtuB to the inner membrane ABC transporter, BtuCD (EC 3.6.3.33) (Lewinson et al. 2010). Gram-positive bacteria lack the outer-membrane BtuB protein, and it is thought that their homologous BtuF protein is membrane anchored (Köster 2001; Braun and Hantke 2007). Although the BtuBFCD system is the only known family of corrinoid transporter, the existence of paralogous copies has been described in *Bacteroides thetaiotamicron* (Degnan et al. 2014), and the proposed function of these multiple copies has been to recognize different types of corrinoids and cobamides. A recent study in *D. mccartyi* showed that the *btuFCD* genes of the standard cobamide transporter were upregulated in cells grown in medium containing phenolyl cobamides plus DMB (Men et al. 2014a). In addition, a nonhomologous, putative iron/cobalamin ABC transporter, DET1174-1176, was upregulated in the presence of exogenous cobalamin and downregulated in its absence. Although it was not shown that this transporter specifically bound cobalamin, these findings raised the intriguing possibility that *D. mccartyi* uses differential transporters to selectively take up cobamides from its environment. It is possible that other OHRB may also use such a strategy. A variety of corrinoids are known to exist in environments such as the human gut (Allen and Stabler 2008; Girard et al. 2009), but surveys of cobamides at sites contaminated by halogenated organics are lacking. However, Men et al. (2014b) recently detailed an LC/MS/MS method for differentiating and quantifying cobamides from dehalogenating mixed cultures extracted from two contaminated sites and enriched in the lab. These authors found the majority of the cobamides to be cobalamin and *p*-cresol cobamide, with trace amounts of pseudocobalamin (adenylyl-cobamide), 2-methyladenylyl-cobamide, 5-hydroxybenzimidazoyl-cobamide, and 5-methylbenzimidazoyl-cobamide.

Upon import, corrinoids must be adenosylated by an ACAT before further modification can take place (Escalante-Semerena et al. 1990). This necessitates the removal of inactive upper ligands, such as cyanide. The existence of corrinoid-specific reductases is controversial. One such enzyme, CblC, has been characterized in humans and other animals (Koutmos et al. 2011) (EC 1.16.1.5, PDB 3SBY, 3SBZ, 3SC0), and recently Warren and coworkers reported detailed studies in support of the bacterial corrinoid reductase CobR as a Co(II) reductase in *Brucella melitensis* (EC 1.14.14.9, PDB 4IRA, 3CBO), which utilizes the aerobic or late Co insertion pathway (Lawrence et al. 2014). In bacteria, reduction of salvaged corrinoids is more likely accomplished by nonenzymatic electron transfer by cytosolic reduced flavoproteins (Fonseca and Escalante-Semerena 2000) or free dihydroflavins (Mera and Escalante-Semerena 2010a). Such nonspecific reductions result in the removal of the native *Coβ* ligand of the corrinoid, generating a five-coordinate species that can be acted upon by CobA or PduO (Mera and Escalante-Semerena 2010a).

19.5 Assembly of the Nucleotide Loop

After acquisition of an adenosylated corrinoid, either through de novo synthesis or import through the BtuBFCD system, the next step in the pathway is the generation of the nucleotide loop that will ultimately coordinate the cobalt center on the α face of the ring. There are several pathways by which this can occur, for simplicity we will start by discussing the de novo pathway for cobalamin synthesis.

In *S. enterica*, CbiB synthesizes adenosylcobinamide-phosphate (AdoCbi-P, EC 6.3.1.10), the first intermediate in the assembly of the nucleotide loop. CbiB catalyzes the addition of aminopropanol phosphate to the *f* propionic acid side chain of adenosylcobyric acid (Crouzet et al. 1991; Zayas et al. 2007). In *S. enterica*, it was shown that CbiB could use the alternative substrate, ethanolamine-phosphate, to catalyze the eventual synthesis of norcobalamin, i.e., cobalamin lacking the methyl group at C176 (Zayas et al. 2007). The similar molecule norpseudocobalamin, which has an adenine in place of DMB, was isolated from *S. multivorans* (Kräutler et al. 2003). It would not be surprising that, in this organism, the CbiB homologue preferentially uses ethanolamine-phosphate instead of aminopropanol phosphate.

The multifunctional AdoCbi kinase/AdoCbi-P guanylyltransferase enzyme CobU (characterized in *S. enterica*, (Thompson et al. 1999; Thomas et al. 2000), EC 2.7.1.156/2.7.7.62 PDB 1C9K, 1CBU) or the homologous CobP in *P. denitrificans* (Blanche et al. 1991b) plays two separate roles. It can salvage and phosphorylate AdoCbi, in effect replacing the role CbiB plays in the de novo pathway, and it can attach a guanosyl moiety to the aminopropanol phosphate tail of AdoCbi-P, forming AdoCbi-GDP. The latter is necessary for the addition of the base, which with the exception of phenolic bases, will form a coordination bond with the cobalt ion of the ring. In the case of cobalamin, the base is 5,6-dimethylbenzimidazole, which is activated by CobT to its ribotide form, α-ribazole phosphate (α-RP) (EC 2.4.2.21, PDB 1L4E, 1L4F, 1L4B), before it can be condensed with AdoCbi-GDP by CobS to yield AdoCbl-5′P (O'Toole et al. 1993; Maggio-Hall and Escalante-Semerena 1999; Maggio-Hall et al. 2004) (EC 2.7.8.26). The phosphate moiety of AdoCbl-5′P is released by CobC to form the final product, AdoCbl (EC 3.1.3.73, PDB 3HJG).

19.6 Synthesis of the Lower Ligand

The diversity of cobamides found in nature is due to their lower bases. These bases fall into three general categories: benzimidazoles, purines, and phenolics (Renz 1999). The rationale behind the requirement of different enzymes for different lower bases has yet to be resolved. Several OHRB incapable of de novo synthesis require exogeneous cobalamin, at least one species, *S. multivorans*, has been demonstrated to synthesize norpseudocobalamin (Kräutler et al. 2003).

Some enzymes can use alternative corrinoids with minimal loss to activity, yet others are greatly affected by nonoptimal corrinoids. For instance, pseudocobalamin, which utilizes adenine instead of DMB as a lower ligand, can support growth of *S. enterica* under all conditions that require cobalamin (Trzebiatowski and Escalante-Semerena 1997). 5′-Methybenzimidazolylcobamide, which is synthesized by some sulfate-reducing bacteria (Kräutler et al. 1988; Aaron et al. 2007) can support PCE dechlorination of *D. mccartyi* even though that bacterium preferentially salvages cobalamin (Yi et al. 2012). However, the RDase in crude extracts of *Sulfospirillum multivorans* has 50-fold lower activity when using cobalamin instead of its native norpseudocobalamin (Keller et al. 2013).

5,6-Dimethylbenzimidazole is the most studied of the cobamide lower ligands, yet many questions remain about how DMB is synthesized. DMB can be synthesized by one of two pathways. An oxygen-dependent mechanism has been reported in *Propionibacterium shermanii*, in which isotopic labeling experiments revealed that DMB was synthesized from the flavin group of flavin mononucleotide (Horig and Renz 1980; Kolonko et al. 1992). The enzyme that catalyzes this $FMNH_2$-dependent synthesis, BluB, was characterized in *Rhodospirillum rubrum* and *Sinorhizobium meliloti* (Gray and Escalante-Semerena 2007; Taga et al. 2007). The enzymes catalyzing the anaerobic route of DMB synthesis have not been identified, although the process had been described *Eubacterium limosum* using isotopic labeling (Lamm et al. 1982). It was determined that formate, erythrose, glutamine, glycine, and methionine make up the component carbon and nitrogen sources for DMB synthesis.

There is little known about the synthesis of benzimidazole bases other than DMB. Glycine was shown to be incorporated into the same positions of 5-hydroxybenzimidazole (Scherer et al. 1984) and 5-methoxybenzimidazole (Hollriegl et al. 1982) as in DMB, suggesting a partially shared pathway as DMB synthesis. However, differences have been identified as well, such as the inability to use erythrose in 5-hydroxybenzimidazole (Eisenreich and Bacher 1991) or methionine in 5-methylbenzimidazole synthesis (Endres 1997).

Crofts et al. recently described a high-throughput bioassay for detection of free benzimidazoles and derivatives from environmental samples (Crofts et al. 2014). The authors detected picomolar amounts of benzimidazole, but not DMB, from two geographically different soil samples. The availability of free benzimidazole in the environment is an intriguing potential source of lower ligands for organisms that synthesize or salvage cobamides.

Purines, particularly adenine, are another type of lower ligand often found in cobamides. The source of purine ligands in wild-type cells is unknown, although it has been proposed that adenine, hypoxanthine, and guanine are redirected to corrinoid synthesis directly from the purine biosynthetic pathway, which assumes the existence of an appropriate glycosidase to convert the 5′-nucleotides from that pathway to free bases (Renz 1999). Anderson and coworkers isolated several mutant strains in *S. enterica* that increased the intercellular concentrations of free adenine in a DMB-free environment, which resulted in pseudocobalamin synthesis. The synthesis of pseudocobalamin was reverted to cobalamin by adding exogenous DMB (Anderson et al. 2008).

Phenolic lower bases also exist in the environment, produced most notably by *Sporomusa ovata* (Stupperich et al. 1989; Stupperich and Eisinger 1989a, b). Tyrosine degradation has been identified as the source of the *p*-cresol base in *S. ovata*, and has been suggested to be the source of the phenol base as well (Stupperich and Eisinger 1989a, b). The phenolic lower bases lack the nitrogen moiety that coordinates the central cobalt ion, and it is unable to support Cbl-dependent growth of *S. enterica* under conditions that require one-electron, AdoCbl chemistry.

The lower ligand must be activated to an α-nucleotide form before it can be attached to AdoCbi-GDP. In the case of DMB, this process involves the catabolism of an existing nucleotide, namely nicotinate mononucleotide, to attach the ribosyl-P moiety of NaMN to DMB resulting in an unusual α-*N*-glycosidic linkage yielding α-ribazole phosphate (α-RP). In *S. enterica*, this reaction is performed by CobT (Trzebiatowski et al. 1994). The CobT enzyme also phosphoribosylates DMB using nicotinamide adenine dinucleotide (NAD^+), forming the α-DMB-adenine dinucleotide intermediate called α-DAD (Maggio-Hall and Escalante-Semerena 2003). It is thought that α-DAD is converted to α-RP through an unidentified hydrolase, yet the identity of this putative hydrolase remains unknown. In the case of pseudocobalamin, it was proposed that adenine is activated to an α-nucleoside by the same enzymes, CobU, S, T, and C, that are responsible for activating DMB in *S. enterica* (Anderson et al. 2008), although biochemical evidence to support this idea has not been reported to date.

Sporomusa ovata uses the CobT homologues ArsAB (EC 2.4.2.21/2.4.2.55), which, unlike CobT, form a heterodimer capable of activating phenolic lower bases, such as *p*-cresol and phenol (Chan and Escalante-Semerena 2011) (PDB 4HDM, 4HDN, 4HDR, 4HDS). However, ArsAB is not specific for phenolics. The enzyme can also activate DMB and purines, despite a clear preference for phenolic substrates (Newmister et al. 2012). Notably, when *S. ovata* is grown with excess benzimidazoles in the medium, it incorporates them into cobamides at the expense of phenolic groups, resulting in growth inhibition (Mok and Taga 2013), suggesting that the enzyme may have high in vivo affinity for benzimidazole and its derivatives. ArsAB can support growth of a *S. enterica cobT* strain in the presence of cobinamide and DMB at nearly the same doubling time as a *cobT*$^+$ strain (Chan et al. 2014).

Like the *S. ovata* ArsAB enzyme, *S. multivorans* CobT uses exogenous DMB at the expense of its native lower ligand base. Micromolar amounts of DMB added to *S. multivorans* growth medium on PCE inhibit growth and produces norcobalamin instead of the native cofactor, norpseudocobalamin. Addition of *p*-cresol to the growth medium had no effect on the growth rate (Keller et al. 2013).

Listeria innocua lacks CobT, and instead makes α-RP through the nonhomologous CblT and CblS enzymes (Gray and Escalante-Semerena 2010). CblT transports α-R from the environment, and CblS phosphorylates it to α-RP (EC: 2.7.1.-), suggesting that α-R may be environmentally available to some organisms. Recently, CblT and CblS were identified in *D. hafniense*. However, unlike *L. innocua*, *D. hafniense* contains CobT (Choudhary et al. 2013).

The precursor of the RDase PceA in *D. hafniense* strain Y51 was shown to aggregate in the absence of cobalamin, which is necessary to complete folding of PceA and subsequent membrane translocation (Reinhold et al. 2012). PrePceA aggregates contain CobT among other enzymes. It is an intriguing possibility that the final steps of cobamide synthesis are taking place near the membrane to efficiently reconstitute and translocate PceA. It has been shown that CobS is a membrane-bound protein in *S. enterica* and *Methanobacterium thermoautotrophicum* (Maggio-Hall et al. 2004) and it is possible that other Cob enzymes are components of a macromolecular complex.

19.7 Lower Ligand Remodeling

Cobamide-salvaging organisms may not always encounter the exact cobamide they require, or the cobamide may be incomplete. Such organisms must be able to remove the lower ligand and replace it. In *S. enterica*, the kinase activity associated with CobU is needed to salvage cobinamide from the environment. The first step after translocation of Cbi into the cell is its conversion to AdoCbi by the housekeeping adenosyltransferase CobA (Escalante-Semerena et al. 1990). Once AdoCbi is available, CobU uses it to generate AdoCbi-P. The final step in the activation of Cbi is the conversion of AdoCbi-P to AdoCbi-GDP also by CobU (Blanche et al. 1991b; O'Toole and Escalante-Semerena 1995; Thompson et al. 1999; Thomas et al. 2000). A non-orthologous replacement for CobU was found in archaea (Thomas and Escalante-Semerena 2000). The archaeal non-orthologous CobU replacement was named CobY, and it should be noted that, while it retains the CobU-like guanylyltransferase activity, it is devoid of AdoCbi kinase activity. To make up for the lack of kinase activity, archaea evolved a different strategy for the assimilation of Cbi from the environment. Archaea possess an amidohydrolase, CbiZ (EC 3.5.1.90) that hydrolyzes the amide bond in the nucleotide tail of AdoCbi, generating Ado-cobyric acid (AdoCby) (Fig. 19.4; Woodson et al. 2003; Woodson and Escalante-Semerena 2004). Surprisingly, *cbiZ* can be found in some bacteria. For example, it is known that the genome of *Rhodobacter sphaeroides* encodes a *cbiZ* homologue that encodes a functional enzyme (Gray et al. 2008; Gray and Escalante-Semerena 2009a, b). In *R. sphaeroides*, cobinamide and cobamides such as pseudocobalamin are substrates of CbiZ. In sharp contrast, CbiZ in this bacterium had no hydrolytic activity when either cyanocobalamin or adenosylcobalamin was the substrate (Gray and Escalante-Semerena 2009a). The CbiZ reaction results in AdoCby, the substrate for CbiB. The resulting AdoCbi-P product then enters the late steps of the pathway that assemble the nucleotide loop.

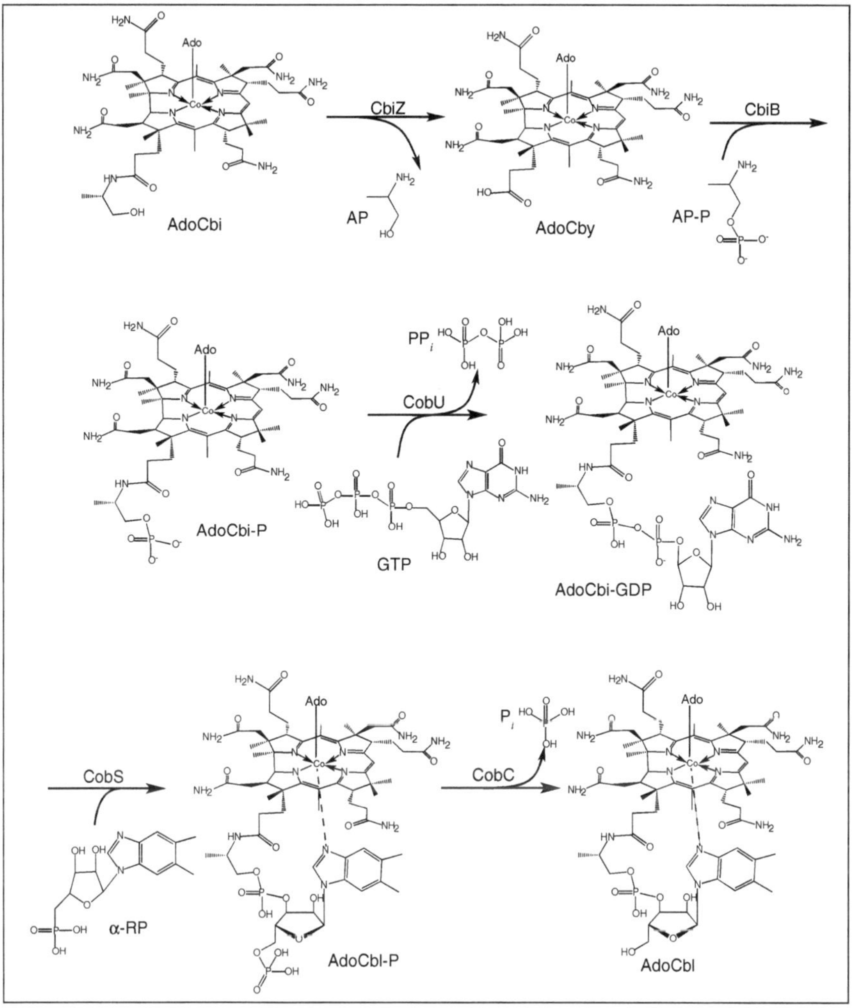

Fig. 19.4 Pathway for the synthesis of adenosylcobalamin via CbiZ-mediated salvaging of adenosylcobinamide. *Abbreviations* Ado, adenosyl group; AdoCbi, adenosylcobinamide; AP, aminopropanol; AdoCby, adenosylcobyric acid; AP-P, aminopropanol phosphate; GTP, guanosine triphosphate; PP, pyrophosphate; AdoCbi-GTP, adenosylcobinamide-guanosine diphosphate; α-RP, alpha-ribazole phosphate; AdoCbl-P, adenosylcobalamin-phosphate; P, phosphate; AdoCbl, adenosylcobalamin

When *D. mccartyi* strain 195 is grown under conditions requiring it to salvage phenolic cobamides from a coculture strain, a microarray analysis indicated *cbiZ* transcription is upregulated (Men et al. 2014a). Also upregulated are the *cbiB* and *cobD* genes, which synthesize AdoCbi-P from Cby generated by CbiZ, which subsequently is channeled into the nucleotide loop assembly pathway. Together, these data indicate a dependence on CbiZ activity under simulated environmental conditions.

19.8 Cobalamin Riboswitches

AdoCbl-binding riboswitches are important regulators of corrinoid metabolism in many organisms. For instance, expression of cobalamin-synthesizing genes in *S. enterica* is repressed by a riboswitch in the 5′ leader of the *cob* operon mRNA, and expression of cobalamin transporters in *S. enterica* and *E. coli* is also repressed by a riboswitch in the 5′ untranslated region of the *btuB* gene (Richter-Dahlfors and Andersson 1992; Richter-Dahlfors et al. 1994; Ravnum and Andersson 1997; Nahvi et al. 2004). Recently, several riboswitches have been characterized in *D. hafniense* strains Y51, DCB-2, and TCE1 (Choudhary et al. 2013). Eighteen riboswitches were described in total, with predicted functions primarily to regulate de novo corrinoid synthesis and corrinoid transporter genes. Additionally, corrinoid-dependent transcriptional repression was observed for the transporter *btuF*, the phosphoribosyltransferase *cobT*, and the precorrin methyltransferase *cbiET*.

Reinhold and coworkers observed that adding cobalamin to the growth medium of *D. hafniense* stabilized the dechlorinating *pceA* gene (Reinhold et al. 2012). When cobalamin was absent from the PCE-free medium, *pceA* was lost after eight passages. When cobalamin was present in PCE-free medium, *pceA* levels were detectable for 60 passages. Whether this effect is due to direct or indirect interactions between cobalamin and *pceA* is unknown.

19.9 Environmental Supply of Cobalamin

A study by Guerrero-Barajas and coworkers indicated that supplying the corrin ring precursor porphobilinogen to carbon tetrachloride-dehalogenating sludge enhanced the dehalogenation rate by severalfold over controls in a concentration-dependent manner (Guerrero-Barajas and Field 2006). Similarly, although smaller effects were also observed with the addition of other precursors, including methionine, adenine, threonine, and δ-aminolevulinic acid, these effects may have been unrelated to cobamide synthesis. Overall the results indicated that corrin ring synthesis is a limiting factor for reductive dechlorination among certain consortia.

Although almost all OHRB require corrinoids to utilize halogenated compounds as electron acceptors, not all are capable of de novo corrin ring biosynthesis. For instance, a comparative genomics study of cobalamin utilization pathways identified 15 species out of 539 analyzed that contain genes for corrinoid-dependent RDases (Zhang et al. 2009). Of these, six species lacked genes for de novo corrin ring biosynthesis. Given that many are dependent on organohalides for respiration, it is of interest to learn how corrinoid auxotrophs acquire their cobamides.

As of this writing, no protein capable of exporting corrinoids has been defined, and thus the method by which corrinoids enter the environment is unclear. Some corrinoids are presumably released via cell lysis, but coculture experiments suggest "cross-feeding" of cobalamin between mutualistic strains (Seth and Taga 2014).

A number of studies have defined interactions among the cobalamin-dependent, organohalide-respiring *D. mccartyi* and strains in its biome (Yan et al. 2012, 2013; Men et al. 2014a). *D. mccartyi* lacks the genes for de novo cobalamin synthesis. *Geobacter lovleyi,* a cobalamin-synthesizing species capable of organohalide respiration with tetrachloroethene (PCE) and trichloroethene (TCE) (Wagner et al. 2012), could support growth of *D. mccartyi* in cobalamin-free media, indicating *D. mccartyi* feeds off cobalamin produced by *G. lovleyi* (Yan et al. 2012). *Geobacter sulfurreducens*, a nondehalogenating species producing an unspecified cobamide, could only support growth of *D. mccartyi* when exogenous DMB was provided. This indicates that either *G. sulfurreducens* is incorporating DMB into its cobamide, or that *D. mccartyi* is utilizing CbiZ to remodel an unusable cobamide into cobalamin. Another study had determined that pure culture *D. mccartyi* could also incorporate benzimidazole, 5-methylbenzimidazole and 5-methoxybenzimidazole into a cobamide to support growth on halogenated compounds (Yi et al. 2012). *G. lovleyi* respires PCE to *cis*-dichloroethene (*cis*-DCE) (Sung et al. 2006) whereas some *D. mccartyi* strains can dechlorinate *cis*-DCE and vinyl chloride (VC). It was proposed that the co-occurrence of these organisms at sites contaminated by PCE would be beneficial to *D. mccartyi*, as *G. lovleyi* could supply both *cis*-DCE and cobalamin.

Additional coculture experiments between *D. mccartyi* strains and *Methanosarcina barkeri*, *S. ovata* and *Sporomusa* sp. strain KB-1 produced similar results (Yan et al. 2013). Production of factor III (*Co*α-5-hydroxybenzimidazolyl-cobamide) from *M. barkeri* or phenol and *p*-cresol cobamide from the *Sporomusa* strains did not support cobalamin-dependent growth of *D. mccartyi* on PCE. Exogenous DMB supported growth under these coculture conditions, as was the case with *G. sulfurreducens*, providing further evidence that *D. mccartyi* is remodeling cobamides into cobalamin using exogenous DMB.

Recently, a defined consortium of three species was shown to be capable of supporting *D. mccartyi* growth when supplied lactate, PCE, and DMB (Men et al. 2014a). *Desulfovibrio vulgaris* Hildenborough and *Pelosinus fermentans* R7 fermented lactate and supplied acetate to *D. mccartyi*. Additionally, *D. vulgaris* supplied H_2, the electron donor for the reductive dechlorination of PCE, while *P. fermentans* supplied corrinoids, either phenyl cobamide or *p*-cresol cobamide. It was assumed that *D. mccartyi* was supplying fermentable products to the coculture species to complete the cycle. *D. mccartyi* significantly upregulated *cbiZ* in response to cobamide exposure, indicating that it was importing and remodeling the inactive cobamide into cobalamin through the CbiZ pathway.

19.10 Outlook

Much remains unknown about the metabolism of corrinoids in the microbial reductive dechlorination process. Genome sequencing efforts have expanded our knowledge of the potential for cobamide synthesis in OHRB, however the lack of tractable genetic systems in these organisms has made it difficult to tease apart the pathways in the de novo synthesizers. Thus the possibility that OHRB might be making novel corrinoids, such as the norpseudocobalamin of *S. multivorans*, remains largely unexplored.

The diversity of lower bases figures foremost in many of the questions regarding cobamides in OHRB as it does in the wider study of cobalamin. We have a limited understanding of what drives the preference of one base over the other. Although we know that bacteria can remodel cobamides using the CbiZ system to use a different base, we do not know how diverse or available cobamides are in the environment of OHRB. Likewise, we are only beginning to understand the source of the cobamides in the environment, and whether or not dehalogenating communities share corrinoids to metabolize available resources.

Acknowledgments This work was supported by USPHS grant R37 GM040313 from the National Institutes of General Medical Sciences to J.C.E.-S.

References

Aaron M, Charbon G, Lam H, Schwarz H, Vollmer W, Jacobs-Wagner C (2007) The tubulin homologue FtsZ contributes to cell elongation by guiding cell wall precursor synthesis in *Caulobacter crescentus*. Mol Microbiol 64:938–952

Abend A, Bandarian V, Nitsche R, Stupperich E, Retey J, Reed GH (1999) Ethanolamine ammonia-lyase has a "base-on" binding mode for coenzyme B12. Arch Biochem Biophys 370:138–141

Adrian L, Hansen SK, Fung JM, Gorisch H, Zinder SH (2007) Growth of Dehalococcoides strains with chlorophenols as electron acceptors. Environ Sci Technol 41:2318–2323

Allen RH, Stabler SP (2008) Identification and quantitation of cobalamin and cobalamin analogues in human feces. Am J Clin Nutr 87:1324–1335

Altschul SF, Gish W, Miller W, Myers EW (1990) Basic local alignment search tool. J Mol Biol 215:403–410

Anderson PJ, Lango J, Carkeet C, Britten A, Kräutler B, Hammock BD, Roth JR (2008) One pathway can incorporate either adenine or dimethylbenzimidazole as an alpha-axial ligand of B_{12} cofactors in *Salmonella enterica*. J Bacteriol 190:1160–1171

Assaf-Anid N, Hayes KF, Vogel TM (1994) Reductive dechlorination of carbon tetrachloride by cobalamin (II) in the presence of dithiotreitol: Mechanistic study, effect of redox potential and pH. Environ Sc Technol 28:246–252

Banerjee R, Ragsdale SW (2003) The many faces of vitamin B12: catalysis by cobalamin-dependent enzymes. Annu Rev Biochem 72:209–247. doi:10.1146/annurev.biochem.72.121801.161828

Battersby AR (2000) Tetrapyrroles: the pigments of life. Natural product reports 17:507–526

Battersby AR, McDonald E, Cornforth JW, Frydman B (1976) Biosynthesis of porphyrins and corrins [and discussion]. Phil Transac Royal Soc London Series B, Biol Sci 273(924):161–180. doi:10.2307/2417458

Blanche F, Couder M, Debussche L, Thibaut D, Cameron B, Crouzet J (1991a) Biosynthesis of vitamin B12: stepwise amidation of carboxyl groups b, d, e, and g of cobyrinic acid a, c-diamide is catalyzed by one enzyme in Pseudomonas denitrificans. J Bacteriol 173:6046–6051

Blanche F, Debussche L, Famechon A, Thibaut D, Cameron B, Crouzet J (1991b) A bifunctional protein from *Pseudomonas denitrificans* carries cobinamide kinase and cobinamide phosphate guanylyltransferase activities. J Bacteriol 173:6052–6057

Blanche F, Thibaut D, Debussche L, Hertle R, Zipfel F, Muller G (1993) Paralleles and decisive differences in vitamin B12 biosynthesis. Angew Chem Int Ed Engl 32:1651–1653

Bogorad L, Granick S (1953) The enzymatic synthesis of porphyrins from porphobilinogen. Proc Natl Acad Sci USA 39:1176

Bommer M, Kunze C, Fesseler J, Schubert T, Diekert G, Dobbek H (2014) Structural basis for organohalide respiration. Science 346:455–458. doi:10.1126/science.1258118

Borths EL, Locher KP, Lee AT, Rees DC (2002) The structure of Escherichia coli BtuF and binding to its cognate ATP binding cassette transporter. Proc Natl Acad Sci USA 99:16642–16647. doi:10.1073/pnas.262659699

Braun V, Hantke K (2007) Acquisition of iron by bacteria. Microbiology Monographs. In: Molecular microbiology of heavy metals, vol 6. Springer, Berlin

Brown KL (2005) Chemistry and enzymology of vitamin B_{12}. Chem Rev 105:2075–2149. doi:10.1021/cr030720z

Buan NR, Escalante-Semerena JC (2006) Purification and initial biochemical characterization of ATP:cob(I)alamin adenosyltransferase (EutT) enzyme of *Salmonella enterica*. J Biol Chem 281:16971–16977

Buan NR, Suh SJ, Escalante-Semerena JC (2004) The *eutT* gene of *Salmonella enterica* encodes an oxygen-labile, metal-containing ATP: corrinoid adenosyltransferase enzyme. J Bacteriol 186:5708–5714. doi:10.1128/JB.186.17.5708-5714.2004

Cadieux N, Bradbeer C, Reeger-Schneider E, Koster W, Mohanty AK, Wiener MC, Kadner RJ (2002) Identification of the periplasmic cobalamin-binding protein BtuF of *Escherichia coli*. J Bacteriol 184:706–717

Cameron B, Briggs K, Pridmore S, Brefort G, Crouzet J (1989) Cloning and analysis of genes involved in coenzyme B12 biosynthesis in *Pseudomonas denitrificans*. J Bacteriol 171:547–557

Chan CH, Escalante-Semerena JC (2011) ArsAB, a novel enzyme from Sporomusa ovata activates phenolic bases for adenosylcobamide biosynthesis. Mol Microbiol 81:952–967. doi:10.1111/j.1365-2958.2011.07741.x

Chan CH, Newmister SA, Talyor K, Claas KR, Rayment I, Escalante-Semerena JC (2014) Dissecting cobamide diversity through structural and functional analyses of the base-activating CobT enzyme of *Salmonella enterica*. Biochim Biophys Acta 1840:464–475. doi:10.1016/j.bbagen.2013.09.038

Chimento DP, Mohanty AK, Kadner RJ, Wiener MC (2003) Crystallization and initial X-ray diffraction of BtuB, the integral membrane cobalamin transporter of *Escherichia coli*. Acta Crystallogr D Biol Crystallogr 59:509–511

Chiu P-C, Reinhard M (1996) Transformation of carbon tetrachloride by reduced vitamin B12 in aqueous cysteine solution. Environ Sci Technol 30:1882–1889. doi:10.1021/es950477o

Choudhary PK, Duret A, Rohrbach-Brandt E, Holliger C, Sigel RK, Maillard J (2013) Diversity of cobalamin riboswitches in the corrinoid-producing organohalide respirer *Desulfitobacterium hafniense*. J Bacteriol 195:5186–5195. doi:10.1128/JB.00730-13

Crofts TS, Men Y, Alvarez-Cohen L, Taga ME (2014) A bioassay for the detection of benzimidazoles reveals their presence in a range of environmental samples. Front Microbiol 5:592

Crouzet J, Levy-Schil S, Cameron B, Cauchois L, Rigault S, Rouyez MC, Blanche F, Debussche L, Thibaut D (1991) Nucleotide sequence and genetic analysis of a 13.1-kilobase-pair *Pseudomonas denitrificans* DNA fragment containing five *cob* genes and identification of structural genes encoding Cob(I)alamin adenosyltransferase, cobyric acid synthase, and bifunctional cobinamide kinase-cobinamide phosphate guanylyltransferase. J Bacteriol 173:6074–6087

Debussche L, Couder M, Thibaut D, Cameron B, Crouzet J, Blanche F (1991) Purification and partial characterization of cob(I)alamin adenosyltransferase from *Pseudomonas denitrificans*. J Bacteriol 173:6300–6302

Debussche L, Thibaut D, Cameron B, Crouzet J, Blanche F (1993) Biosynthesis of the corrin macrocycle of coenzyme B_{12} in *Pseudomonas denitrificans*. J Bacteriol 175:7430–7440

Degnan PH, Barry NA, Mok KC, Taga ME, Goodman AL (2014) Human gut microbes use multiple transporters to distinguish vitamin B12 analogs and compete in the gut. Cell Host Microbe 15:47–57. doi:10.1016/j.chom.2013.12.007

DeVeaux LC, Clevenson DS, Bradbeer C, Kadner RJ (1986) Identification of the BtuCED polypeptides and evidence for their role in vitamin B12 transport in *Escherichia coli*. J Bacteriol 167:920–927

Eirich LD, Vogels GD, Wolfe RS (1978) Proposed structure for coenzyme F420 from Methanobacterium. Biochemistry 17:4583–4593

Eisenreich W, Bacher A (1991) Biosynthesis of 5-hydroxybenzimidazolylcobamid (factor III) in Methanobacterium thermoautotrophicum. J Biol Chem 266:23840–23849

Endres B (1997) Untersuchungen zur biosynthese des Basenteils von Vitamin B12 bei *Eubacterium limosum* und von 5-methylbenzimidazolylcobamid bei *Desulfobulbus propionicus*

Escalante-Semerena JC (2007) Conversion of cobinamide into adenosylcobamide in bacteria and archaea. J Bacteriol 189:4555–4560

Escalante-Semerena JC, Suh SJ, Roth JR (1990) *cobA* function is required for both de novo cobalamin biosynthesis and assimilation of exogenous corrinoids in *Salmonella typhimurium*. J Bacteriol 172:273–280

Eschenmoser A (1988) Vitamin B12: experiments concerning the origin of its molecular structure. Angew Chem Int Ed Engl 27:5–40

Eschenmoser A, Wintner CE (1977) Natural product synthesis and vitamin B_{12}. Science 196(4297):1410–1420

Fonseca MV, Escalante-Semerena JC (2000) Reduction of cob(III)alamin to cob(II)alamin in *Salmonella enterica* Serovar Typhimurium LT2. J Bacteriol 182:4304–4309. doi:10.1128/JB.182.15.4304-4309.2000

Fonseca MV, Buan NR, Horswill AR, Rayment I, Escalante-Semerena JC (2002) The ATP:co(I)rrinoid adenosyltransferase (CobA) enzyme of *Salmonella enterica* requires the 2'-OH Group of ATP for function and yields inorganic triphosphate as its reaction byproduct. J Biol Chem 277:33127–33131. doi:10.1074/jbc.M203893200

Fresquet V, Williams L, Raushel FM (2004) Mechanism of cobyrinic acid a, c-diamide synthetase from Salmonella typhimurium LT2. Biochemistry 43:10619–10627

Gantzer CJ, Wackett LP (1991) Reductive dechlorination catalyzed by bacterial transition-metal coenzymes. Environ Sci Technol 25:715–722. doi:10.1021/es00016a017

Girard CL, Santschi DE, Stabler SP, Allen RH (2009) Apparent ruminal synthesis and intestinal disappearance of vitamin B12 and its analogs in dairy cows. J Dairy Sci 92:4524–4529. doi:10.3168/jds.2009-2049

Goris T, Schubert T, Gadkari J, Wubet T, Tarkka M, Buscot F, Adrian L, Diekert G (2014) Insights into organohalide respiration and the versatile catabolism of *Sulfurospirillum multivorans* gained from comparative genomics and physiological studies. Environ Microbiol 16:3562–3580. doi:10.1111/1462-2920.12589

Gray MJ, Escalante-Semerena JC (2007) Single-enzyme conversion of $FMNH_2$ to 5,6-dimethylbenzimidazole, the lower ligand of B_{12}. Proc Natl Acad Sci USA 104:2921–2926. doi:10.1073/pnas.0609270104

Gray MJ, Escalante-Semerena JC (2009a) The cobinamide amidohydrolase (cobyric acid-forming) CbiZ enzyme: a critical activity of the cobamide remodelling system of *Rhodobacter sphaeroides*. Mol Microbiol 74:1198–1210. doi:10.1111/j.1365-2958.2009.06928.x MMI6928 [pii]

Gray MJ, Escalante-Semerena JC (2009b) In vivo analysis of cobinamide salvaging in *Rhodobacter sphaeroides* strain 2.4.1. J Bacteriol 191:3842–3851. doi:10.1128/JB.00230-09 JB.00230-09 [pii]

Gray MJ, Escalante-Semerena JC (2010) A new pathway for the synthesis of alpha-ribazole-phosphate in *Listeria innocua*. Mol Microbiol 77:1429–1438. doi:10.1111/j.1365-2958.2010.07294.x MMI7294 [pii]

Gray MJ, Tavares NK, Escalante-Semerena JC (2008) The genome of *Rhodobacter sphaeroides* strain 2.4.1 encodes functional cobinamide salvaging systems of archaeal and bacterial origins. Mol Microbiol 70:824–836. doi:10.1111/j.1365-2958.2008.06437.x MMI6437 [pii]

Guerrero-Barajas C, Field J (2006) Enhanced anaerobic biotransformation of carbon tetrachloride with precursors of vitamin B12 biosynthesis. Biodegradation 17:317–329. doi:10.1007/s10532-005-9001-2

Holliger C, Wohlfarth G, Diekert G (1998) Reductive dechlorination in the energy metabolism of anaerobic bacteria. FEMS Microbiol Rev 22:383–398

Hollriegl V, Lamm L, Rowold J, Horig J, Renz P (1982) Biosynthesis of vitamin B12. Different pathways in some aerobic and anaerobic microorganisms. Arch Microbiol 132:155–158

Horig JA, Renz P (1980) Biosynthesis of vitamin B12. Some properties of the 5,6-dimethylbenzimidazole-forming system of *Propionibacterium freudenreichii* and *Propionibacterium shermanii*. Eur J Biochem 105:587–592

Hug LA, Maphosa F, Leys D, Loffler FE, Smidt H, Edwards EA, Adrian L (2013) Overview of organohalide-respiring bacteria and a proposal for a classification system for reductive dehalogenases. Philos Trans R Soc Lond B Biol Sci 368:20120322. doi:10.1098/rstb.2012.0322

Jeter RM, Olivera BM, Roth JR (1984) *Salmonella typhimurium* synthesizes cobalamin (vitamin B_{12}) de novo under anaerobic growth conditions. J Bacteriol 159:206–213

John M, Schmitz RP, Westermann M, Richter W, Diekert G (2006) Growth substrate dependent localization of tetrachloroethene reductive dehalogenase in *Sulfurospirillum multivorans*. Arch Microbiol 186:99–106

Johnson CL, Pechonick E, Park SD, Havemann GD, Leal NA, Bobik TA (2001) Functional genomic, biochemical, and genetic characterization of the *Salmonella pduO* gene, an ATP:cob(I)alamin adenosyltransferase gene. J Bacteriol 183:1577–1584

Johnson CL, Buszko ML, Bobik TA (2004) Purification and initial characterization of the *Salmonella enterica* PduO ATP:Cob(I)alamin adenosyltransferase. J Bacteriol 186:7881–7887

Keller S, Ruetz M, Kunze C, Krautler B, Diekert G, Schubert T (2013) Exogenous 5,6-dimethylbenzimidazole caused production of a non-functional tetrachloroethene reductive dehalogenase in *Sulfurospirillum multivorans*. Environ Microbiol 16:3361–3369. doi:10.1111/1462-2920.12268

Kolonko B, Hörig JA, Renz P (1992) Transformation of tritiated (5′) riboflavin into 5,6-dimethylbenzimidazole. Zeits fuer Naturf Sect C Biosc 47:171–176

Köster W (2001) ABC transporter-mediated uptake of iron, siderophores, heme and vitamin B12. Res Microbiol 152:291–301. doi:10.1016/S0923-2508(01)01200-1

Koutmos M, Gherasim C, Smith JL, Banerjee R (2011) The structural basis of multifunctionality in a vitamin B12-processing enzyme. J Biol Chem 286:29780–29787. doi:10.1074/jbc.M111.261370

Kräutler B, Kohler HP, Stupperich E (1988) 5′-Methylbenzimidazolyl-cobamides are the corrinoids from some sulfate-reducing and sulfur-metabolizing bacteria. Eur J Biochem 176:461–469

Kräutler B, Fieber W, Osterman S, Fasching M, Ongania K-H, Gruber K, Kratky C, Mikl C, Siebert A, Diekert G (2003) The cofactor of tetrachloroethene reductive dehalogenase of *Dehalospirillum multivorans* is Norpseudo-B12, a new type of natural corrinoid. Helv Chim Acta 86:3698–3716

Krone UE, Thauer RK, Hogenkamp HPC (1989) Reductive dehalogenation of chlorinated C1-hydrocarbons mediated by corrinoids. Biochemistry 28:4908–4914. doi:10.1021/bi00437a057

Krone UE, Thauer RK, Hogenkamp HPC, Steinbach K (1991) Reductive formation of carbon monoxide from carbon tetrachloride and FREONS 11, 12, and 13 catalyzed by corrinoids. Biochemistry 30:2713–2719. doi:10.1021/bi00224a020

Kruse T, Maillard J, Goodwin L, Woyke T, Teshima H, Bruce D, Detter C, Tapia R, Han C, Huntemann M, Wei CL, Han J, Chen A, Kyrpides N, Szeto E, Markowitz V, Ivanova N, Pagani I, Pati A, Pitluck S, Nolan M, Holliger C, Smidt H (2013) Complete genome sequence of Dehalobacter restrictus PER-K23(T.). Stand Genom Sci 8:375–388

Lamm L, Heckmann G, Renz P (1982) Biosynthesis of vitamin B12 in anaerobic bacteria. Mode of incorporation of glycine into the 5,6-dimethylbenzimidazole moiety in *Eubacterium limosum*. Eur J Biochem 122:569–571

Lawrence JG, Roth JR (1995) The cobalamin (coenzyme B12) biosynthetic genes of *Escherichia coli*. J Bacteriol 177:6371–6380

Lawrence AD, Taylor SL, Scott A, Rowe ML, Johnson CM, Rigby SE, Geeves MA, Pickersgill RW, Howard MJ, Warren MJ (2014) FAD binding, cobinamide binding and active site communication in the corrin reductase CobR. Biosci Rep 34:e00120. doi:10.1042/BSR20140060

Leeper FJ (1989) The biosynthesis of porphyrins, chlorophylls and vitamin B12. Natural product reports 6:171–203

Lesage S, Brown S, Millar K (1998) A different mechanism for the reductive dechlorination of chlorinated ethenes: kinetic and spectroscopic evidence. Environ Sci Technol 32:2264–2272. doi:10.1021/es970959v

Lewinson O, Lee AT, Locher KP, Rees DC (2010) A distinct mechanism for the ABC transporter BtuCD-BtuF revealed by the dynamics of complex formation. Nat Struct Mol Biol 17:332–338. doi:10.1038/nsmb.1770

Lexa D, Saveant J-M (1983) The electrochemistry of vitamin B12. Acc Chem Res 16:235–243

Lundrigan MD, Kadner RJ (1989) Altered cobalamin metabolism in *Escherichia coli btuR* mutants affects *btuB* gene regulation. J Bacteriol 171:154–161

Maggio-Hall LA, Escalante-Semerena JC (1999) In vitro synthesis of the nucleotide loop of cobalamin by *Salmonella typhimurium* enzymes. Proc Natl Acad Sci USA 96:11798–11803

Maggio-Hall LA, Escalante-Semerena JC (2003) Alpha-5,6-dimethylbenzimidazole adenine dinucleotide (alpha-DAD), a putative new intermediate of coenzyme B12 biosynthesis in *Salmonella typhimurium*. Microbiology 149:983–990

Maggio-Hall LA, Claas KR, Escalante-Semerena JC (2004) The last step in coenzyme B(12) synthesis is localized to the cell membrane in bacteria and archaea. Microbiology 150:1385–1395

Maillard J, Schumacher W, Vazquez F, Regeard C, Hagen WR, Holliger C (2003) Characterization of the corrinoid iron-sulfur protein tetrachloroethene reductive dehalogenase of *Dehalobacter restrictus*. Appl Environ Microbiol 69:4628–4638. doi:10.1128/aem.69.8.4628-4638.2003

Maillard J, Genevaux P, Holliger C (2011) Redundancy and specificity of multiple trigger factor chaperones in Desulfitobacteria. Microbiology 157:2410–2421. doi:10.1099/mic.0.050880-0

Markowitz VM, Chen IM, Palaniappan K, Chu K, Szeto E, Grechkin Y, Ratner A, Jacob B, Huang J, Williams P, Huntemann M, Anderson I, Mavromatis K, Ivanova NN, Kyrpides NC (2012) IMG: the integrated microbial genomes database and comparative analysis system. Nucleic Acids Res 40(Database issue):D115–D122. doi:10.1093/nar/gkr1044

Matthews RG (2001) Cobalamin-dependent methyltransferases. Acc Chem Res 34:681–689

Matthews RG (2009) Cobalamin- and corrinoid-dependent enzymes. Metal Ions Life Sci 6:53–114

McMurdie PJ, Behrens SF, Muller JA, Goke J, Ritalahti KM, Wagner R, Goltsman E, Lapidus A, Holmes S, Loffler FE, Spormann AM (2009) Localized plasticity in the streamlined genomes of vinyl chloride respiring Dehalococcoides. PLoS Genet 5:e1000714. doi:10.1371/journal.pgen.1000714

Men Y, Seth EC, Yi S, Allen RH, Taga ME, Alvarez-Cohen L (2014a) Sustainable growth of *Dehalococcoides mccartyi* 195 by corrinoid salvaging and remodeling in defined lactate-fermenting consortia. Appl Environ Microbiol. doi:10.1128/aem.03477-13 (in press)

Men Y, Seth EC, Yi S, Crofts TS, Allen RH, Taga ME, Alvarez-Cohen L (2014b) Identification of specific corrinoids reveals corrinoid modification in dechlorinating microbial communities. Environ Microbiol. doi:10.1111/1462-2920.12500

Mera PE, Escalante-Semerena JC (2010a) Dihydroflavin-driven adenosylation of 4-coordinate Co(II) corrinoids: are cobalamin reductases enzymes or electron transfer proteins? J Biol Chem 285:2911–2917. doi:10.1074/jbc.M109.059485 M109.059485 [pii]

Mera PE, Escalante-Semerena JC (2010b) Multiple roles of ATP:cob(I)alamin adenosyltransferases in the conversion of B(12) to coenzyme B (12). Appl Microbiol Biotechnol 88:41–48. doi:10.1007/s00253-010-2773-2

Mera PE, Maurice MS, Rayment I, Escalante-Semerena JC (2007) Structural and functional analyses of the human-type corrinoid adenosyltransferase (PduO) from *Lactobacillus reuteri*. Biochemistry 46:13829–13836. doi:10.1021/bi701622j

Miller E, Wohlfarth G, Diekert G (1998) Purification and characterization of the tetrachloroethene reductive dehalogenase of strain PCE-S. Arch Microbiol 169:497–502. doi:10.1007/s002030050602

Mok KC, Taga ME (2013) Growth inhibition of *Sporomusa ovata* by incorporation of benzimidazole bases into cobamides. J Bacteriol 195:1902–1911. doi:10.1128/JB.01282-12

Moore SJ, Warren MJ (2012) The anaerobic biosynthesis of vitamin B12. Biochem Soc Trans 40:581–586. doi:10.1042/BST20120066

Moore TC, Newmister SA, Rayment I, Escalante-Semerena JC (2012) Structural insights into the mechanism of four-coordinate cob(II)alamin formation in the active site of the *Salmonella enterica* ATP:co(I)rrinoid adenosyltransferase (CobA) enzyme: critical role of residues Phe91 and Trp93. Biochemistry 51:9647–9657. doi:10.1021/bi301378d

Moore SJ, Biedendieck R, Lawrence AD, Deery E, Howard MJ, Rigby SE, Warren MJ (2013a) Characterization of the enzyme CbiH60 involved in anaerobic ring contraction of the cobalamin (vitamin B12) biosynthetic pathway. J Biol Chem 288:297–305. doi:10.1074/jbc.M112.422535

Moore SJ, Lawrence AD, Biedendieck R, Deery E, Frank S, Howard MJ, Rigby SE, Warren MJ (2013b) Elucidation of the anaerobic pathway for the corrin component of cobalamin (vitamin B12). Proc Natl Acad Sci USA 110:14906–14911. doi:10.1073/pnas.1308098110

Moore TC, Mera PE, Escalante-Semerena JC (2014) The Eutt enzyme of Salmonella enterica is a unique ATP:Cob(I)alamin adenosyltransferase metalloprotein that requires ferrous ions for maximal activity. J Bacteriol 196(4):903–910. doi:10.1128/JB.01304-13

Morita Y, Futagami T, Goto M, Furukawa K (2009) Functional characterization of the trigger factor protein PceT of tetrachloroethene-dechlorinating *Desulfitobacterium hafniense* Y51. Appl Microbiol Biotechnol 83:775–781

Nahvi A, Barrick JE, Breaker RR (2004) Coenzyme B12 riboswitches are widespread genetic control elements in prokaryotes. Nucl Acids Res 32:143–150. doi:10.1093/nar/gkh167

Neumann A, Wohlfarth G, Diekert G (1995) Properties of tetrachloroethene and trichloroethene dehalogenase of *Dehalospirillum multivorans*. Arch Microbiol 163:276–281

Neumann A, Wohlfarth G, Diekert G (1998) Tetrachloroethene dehalogenase from *Dehalospirillum multivorans*: cloning, sequencing of the encoding genes, and expression of the *pceA* gene in *Escherichia coli*. J Bacteriol 180:4140–4145

Newmister SA, Chan CH, Escalante-Semerena JC, Rayment I (2012) Structural insights into the function of the nicotinate mononucleotide: phenol/p-cresol phosphoribosyltransferase (ArsAB) enzyme from Sporomusa ovata. Biochemistry 51:8571–8582. doi:10.1021/bi301142h

Noinaj N, Guillier M, Barnard TJ, Buchanan SK (2010) TonB-dependent transporters: regulation, structure, and function. Annu Rev Microbiol 64:43–60. doi:10.1146/Annurev.Micro.112408.134247

O'Toole GA, Rondon MR, Escalante-Semerena JC (1993) Analysis of mutants of defective in the synthesis of the nucleotide loop of cobalamin. J Bacteriol 175:3317–3326

O'Toole GA, Escalante-Semerena JC (1993) *cobU*-dependent assimilation of nonadenosylated cobinamide in cobA mutants of *Salmonella typhimurium*. J Bacteriol 175:6328–6336

O'Toole GA, Escalante-Semerena JC (1995) Purification and characterization of the bifunctional CobU enzyme of *Salmonella typhimurium* LT2. Evidence for a CobU-GMP intermediate. J Biol Chem 270:23560–23569

Overbeek R, Begley T, Butler RM, Choudhuri JV, Chuang HY, Cohoon M, de Crecy-Lagard V, Diaz N, Disz T, Edwards R, Fonstein M, Frank ED, Gerdes S, Glass EM, Goesmann A, Hanson A, Iwata-Reuyl D, Jensen R, Jamshidi N, Krause L, Kubal M, Larsen N, Linke B, McHardy AC, Meyer F, Neuweger H, Olsen G, Olson R, Osterman A, Portnoy V, Pusch GD, Rodionov DA, Ruckert C, Steiner J, Stevens R, Thiele I, Vassieva O, Ye Y, Zagnitko O, Vonstein V (2005) The subsystems approach to genome annotation and its use in the project to annotate 1000 genomes. Nucleic Acids Res 33 (17):5691–5702. doi:10.1093/nar/gki866 33/17/5691 [pii]

Palmer T, Sargent F, Berks BC (2005) Export of complex cofactor-containing proteins by the bacterial Tat pathway. Trends Microbiol 13:175–180. doi:10.1016/j.tim.2005.02.002

Payne KAP, Quezada CP, Fisher K, Dunstan MS, Collins FA, Sjuts H, Levy C, Hay S, Rigby SEJ, Leys D (2014) Reductive dehalogenase structure suggests a mechanism for B12-dependent dehalogenation. Nature advance online publication. doi:10.1038/nature13901. http://www.nature.com/nature/journal/vaop/ncurrent/abs/nature13901.html#/supplementary-information

Price-Carter M, Tingey J, Bobik TA, Roth JR (2001) The alternative electron acceptor tetrathionate supports B12-dependent anaerobic growth of *Salmonella enterica* serovar Typhimurium on ethanolamine or 1,2-propanediol. J Bacteriol 183:2463–2475

Randaccio L, Geremia S, Demitri N, Wuerges J (2010) Vitamin B12: unique metalorganic compounds and the most complex vitamins. Molecules 15:3228–3259. doi:10.3390/molecules15053228 15053228 [pii]

Raux E, Thermes C, Heathcote P, Rambach A, Warren MJ (1997) A role for *Salmonella typhimurium cbiK* in cobalamin (vitamin B12) and siroheme biosynthesis. J Bacteriol 179:3202–3212

Raux E, Lanois A, Warren MJ, Rambach A, Thermes C (1998) Cobalamin (vitamin B_{12}) biosynthesis: identification and characterization of a *Bacillus megaterium cobI* operon. Biochem J 335:159–166

Ravnum S, Andersson DI (1997) Vitamin B12 repression of the *btuB* gene in *Salmonella typhimurium* is mediated via a translational control which requires leader and coding sequences. Mol Microbiol 23:35–42

Rees DC, Johnson E, Lewinson O (2009) ABC transporters: the power to change. Nat Rev Mol Cell Biol 10:218–227. doi:10.1038/nrm2646

Reinhold A, Westermann M, Seifert J, von Bergen M, Schubert T, Diekert G (2012) Impact of vitamin B12 on formation of the tetrachloroethene reductive dehalogenase in *Desulfitobacterium hafniense* strain Y51. Appl Environ Microbiol 78:8025–8032. doi:10.1128/aem.02173-12

Renz P (1999) Biosynthesis of the 5,6-dimethylbenzimidazole moiety of cobalamin and of other bases found in natural corrinoids. In: Banerjee R (ed) Chemistry and biochemistry of B12. Wiley, New York, pp 557–575

Richter-Dahlfors AA, Andersson DI (1992) Cobalamin (vitamin B12) repression of the Cob operon in *Salmonella typhimurium* requires sequences within the leader and the first translated open reading frame. Mol Microbiol 6:743–749

Richter-Dahlfors AA, Ravnum S, Andersson DI (1994) Vitamin B12 repression of the *cob* operon in *Salmonella typhimurium*: translational control of the *cbiA* gene. Mol Microbiol 13:541–553

Rodionov DA, Vitreschak AG, Mironov AA, Gelfand MS (2003) Comparative genomics of the vitamin B_{12} metabolism and regulation in prokaryotes. J Biol Chem 278:41148–41159. doi:10.1074/jbc.M305837200 M305837200 [pii]

Roessner CA, Scott AI (2006) Fine-tuning our knowledge of the anaerobic route to cobalamin (vitamin B_{12}). J Bacteriol 188:7331–7334

Roth JR, Lawrence JG, Rubenfield M, Kieffer-Higgins S, Church GM (1993) Characterization of the cobalamin (vitamin B_{12}) biosynthetic genes of *Salmonella typhimurium*. J Bacteriol 175:3303–3316

Roth JR, Lawrence JG, Bobik TA (1996) Cobalamin (coenzyme B12): synthesis and biological significance. Annu Rev Microbiol 50:137–181

Scherer P, Höllriegl V, Krug C, Bokel M, Renz P (1984) On the biosynthesis of 5-hydroxybenzimidazolylcobamide (vitamin B12-factor III) in *Methanosarcina barkeri*. Arch Microbiol 138:354–359

Schipp CJ, Marco-Urrea E, Kublik A, Seifert J, Adrian L (2013) Organic cofactors in the metabolism of Dehalococcoides mccartyi strains. Philos Trans R Soc Lond B Biol Sci 368:20120321. doi:10.1098/rstb.2012.0321

Schumacher W, Holliger C, Zehnder AJ, Hagen WR (1997) Redox chemistry of cobalamin and iron-sulfur cofactors in the tetrachloroethene reductase of *Dehalobacter restrictus*. FEBS Lett 409:421–425

Seth EC, Taga ME (2014) Nutrient cross-feeding in the microbial world. Front Microbiol 5:350. doi:10.3389/fmicb.2014.00350

Stupperich E, Eisinger HJ (1989a) Biosynthesis of *para*-cresolyl cobamide in *Sporomusa ovata*. Arch Microbiol 151:372–377

Stupperich E, Eisinger HJ (1989b) Function and the biosynthesis of unusual corrinoids by a novel activation mechanism of aromatic compounds in anaerobic bacteria. Adv Space Res 9:117–125

Stupperich E, Eisinger HJ, Kräutler B (1989) Identification of phenolyl cobamide from the homoacetogenic bacterium *Sporomusa ovata*. Eur J Biochem 186:657–661

Suh SJ, Escalante-Semerena JC (1993) Cloning, sequencing and overexpression of *cobA* which encodes ATP: corrinoid adenosyltransferase in *Salmonella typhimurium*. Gene 129:93–97

Suh S, Escalante-Semerena JC (1995) Purification and initial characterization of the ATP: corrinoid adenosyltransferase encoded by the *cobA* gene of *Salmonella typhimurium*. J Bacteriol 177:921–925

Sung Y, Fletcher KE, Ritalahti KM, Apkarian RP, Ramos-Hernández N, Sanford RA, Mesbah NM, Löffler FE (2006) Geobacter lovleyi sp. nov. strain SZ, a novel metal-reducing and tetrachloroethene-dechlorinating bacterium. Appl Environ Microbiol 72:2775–2782. doi: 10.1128/aem.72.4.2775-2782.2006

Taga ME, Larsen NA, Howard-Jones AR, Walsh CT, Walker GC (2007) BluB cannibalizes flavin to form the lower ligand of vitamin B12. Nature 446:449–453

Thibaut D, Debussche L, Blanche F (1990) Biosynthesis of vitamin B12: isolation of precorrin-6x, a metal-free precursor of the corrin macrocycle retaining five S-adenosylmethionine-derived peripheral methyl groups. Proc Natl Acad Sci USA 87:8795–8799

Thibaut D, Couder M, Famechon A, Debussche L, Cameron B, Crouzet J, Blanche F (1992) The final step in the biosynthesis of hydrogenobyrinic acid is catalyzed by the *cobH* gene product with precorrin-8x as the substrate. J Bacteriol 174:1043–1049

Thomas MG, Escalante-Semerena JC (2000) Identification of an alternative nucleoside triphosphate: 5′-deoxyadenosylcobinamide phosphate nucleotidyltransferase in *Methanobacterium thermoautotrophicum* ΔH. J Bacteriol 182:4227–4233

Thomas MG, Thompson TB, Rayment I, Escalante-Semerena JC (2000) Analysis of the adenosylcobinamide kinase/adenosylcobinamide-phosphate guanylyltransferase (CobU) enzyme of Salmonella typhimurium LT2. Identification of residue His-46 as the site of guanylylation. J Biol Chem 275(36):27576–27586. doi:10.1074/jbc.M000977200

Thompson TB, Thomas MG, Escalante-Semerena JC, Rayment I (1999) Three-dimensional structure of adenosylcobinamide kinase/adenosylcobinamide phosphate guanylyltransferase (CobU) complexed with GMP: evidence for a substrate-induced transferase active site. Biochemistry 38(40):12995–13005

Trzebiatowski JR, Escalante-Semerena JC (1997) Purification and characterization of CobT, the nicotinate-mononucleotide: 5,6-dimethylbenzimidazole phosphoribosyltransferase enzyme from *Salmonella typhimurium* LT2. J Biol Chem 272:17662–17667

Trzebiatowski JR, O'Toole GA, Escalante-Semerena JC (1994) The *cobT* gene of *Salmonella typhimurium* encodes the NaMN: 5,6-dimethylbenzimidazole phosphoribosyltransferase responsible for the synthesis of N^1-(5-phospho-alpha-D-ribosyl)-5,6-dimethylbenzimidazole, an intermediate in the synthesis of the nucleotide loop of cobalamin. J Bacteriol 176:3568–3575

van Beelen P, Stassen AP, Bosch JW, Vogels GD, Guijt W, Haasnoot CA (1984) Elucidation of the structure of methanopterin, a coenzyme from *Methanobacterium thermoautotrophicum*, using two-dimensional nuclear-magnetic-resonance techniques. Eur J Biochem 138:563–571

Wagner DD, Hug LA, Hatt JK, Spitzmiller MR, Padilla-Crespo E, Ritalahti KM, Edwards EA, Konstantinidis KT, Löffler FE (2012) Genomic determinants of organohalide-respiration in *Geobacter lovleyi*, an unusual member of the Geobacteraceae. BMC Genom 13:200

Warren MJ, Raux E, Schubert HL, Escalante-Semerena JC (2002) The biosynthesis of adenosylcobalamin (vitamin B12). Natural Prod Rep 19:390–412

Woodson JD, Escalante-Semerena JC (2004) CbiZ, an amidohydrolase enzyme required for salvaging the coenzyme B_{12} precursor cobinamide in archaea. Proc Natl Acad Sci USA 101:3591–3596

Woodson JD, Zayas CL, Escalante-Semerena JC (2003) A new pathway for salvaging the coenzyme B_{12} precursor cobinamide in archaea requires cobinamide-phosphate synthase (CbiB) enzyme activity. J Bacteriol 185:7193–7201. doi:10.1128/JB

Woodward RB (1973) The total synthesis of vitamin B_{12}. Pure Appl Chem 33:145–177

Yamanishi M, Yunoki M, Tobimatsu T, Sato H, Matsui J, Dokiya A, Iuchi Y, Oe K, Suto K, Shibata N, Morimoto Y, Yasuoka N, Toraya T (2002) The crystal structure of coenzyme B12-dependent glycerol dehydratase in complex with cobalamin and propane-1,2-diol. Eur J Biochem 269:4484–4494

Yan J, Ritalahti KM, Wagner DD, Loffler FE (2012) Unexpected specificity of interspecies cobamide transfer from Geobacter spp. to organohalide-respiring *Dehalococcoides mccartyi* strains. Appl Environ Microbiol 78:6630–6636. doi:10.1128/AEM.01535-12

Yan J, Im J, Yang Y, Loffler FE (2013) Guided cobalamin biosynthesis supports Dehalococcoides mccartyi reductive dechlorination activity. Philos Trans R Soc Lond B Biol Sci 368:20120320. doi:10.1098/rstb.2012.0320

Yi S, Seth EC, Men YJ, Stabler SP, Allen RH, Alvarez-Cohen L, Taga ME (2012) Versatility in corrinoid salvaging and remodeling pathways supports corrinoid-dependent metabolism in Dehalococcoides mccartyi. Appl Environ Microbiol 78:7745–7752. doi:10.1128/AEM.02150-12

Zayas CL, Claas K, Escalante-Semerena JC (2007) The CbiB protein of *Salmonella enterica* is an integral membrane protein involved in the last step of the de novo corrin ring biosynthetic pathway. J Bacteriol 189:7697–7708. doi:10.1128/JB.01090-07

Zhang Y, Rodionov DA, Gelfand MS, Gladyshev VN (2009) Comparative genomic analyses of nickel, cobalt and vitamin B12 utilization. BMC Genomics 10:78. doi:10.1186/1471-2164-10-78 1471-2164-10-78 [pii]

Chapter 20
Insights into Reductive Dehalogenase Function Obtained from Crystal Structures

Holger Dobbek and David Leys

Abstract Reductive dehalogenases are a class of corrinoid and [4Fe–4S] cluster-dependent enzymes that are arguably key to organohalide respiration. Recently, the first crystal structures for these enzymes were reported, including one representative of both a respiratory as well as a nonrespiratory catabolic reductive dehalogenase. The comparison made between both structures establishes two highly conserved elements: the configuration of the redox chain within the protein and the Tyr–Lys/Arg active site dyad involved in proton transfer to the substrate. In contrast, the substrate binding elements are highly distinct. These insights serve to guide further study of RdhA structure–function relationships.

20.1 Introduction

The chemical industry makes extensive use of organohalide chemistry, with a wide range of organohalides serving as intermediate and/or end-products. Organohalides are, however, not exclusively of anthropogenic origin, with natural biotic and abiotic processes contributing to a steady production and breakdown of organohalides, thereby driving the biogeochemical halogen cycles (Oberg 2002). The versatility of nature as an organic chemist is attested by more than 3800 naturally produced halogenated compounds (Gribble 1998).

H. Dobbek
Institut für Biologie, Strukturbiologie/Biochemie,
Humboldt-Universität zu Berlin, Unter den Linden 6, 10099 Berlin, Germany
e-mail: holger.dobbek@biologie.hu-berlin.de

D. Leys (✉)
MIB, University of Manchester, Princess Street 131, M1 7DN Manchester, UK
e-mail: David.Leys@manchester.ac.uk

L. Adrian and F.E. Löffler (eds.), *Organohalide-Respiring Bacteria*,
DOI 10.1007/978-3-662-49875-0_20

It has become apparent that the remarkable utility of organohalides is somewhat offset by their negative properties, as their production and use often result in the release of xenobiotic compounds in the environment. The recalcitrant nature of most organohalides leads to rapid accumulation, with the potential to create a major problem given their toxicity. Compounds such as tetrachloroethene (PCE), trichloroethene (TCE), and polychlorinated biphenyls persist in the environment as they are especially recalcitrant to aerobic degradation (Smidt and de Vos 2004).

Anaerobic microorganisms employ a very wide range of different terminal electron acceptors such as nitrate, sulfate, iron, and also organohalides (Smidt and de Vos 2004). Respiration using organohalides is found in distantly related bacteria (Hug and Edwards 2013) and requires reductive dehalogenases (RDases or RdhAses) as catalysts. These enzymes break the otherwise stable C–halide bond through reduction, leading to the corresponding halide anion and the nonhalogenated product (Holliger et al. 1998; Smidt and de Vos 2004). Therefore, RDases are able to convert some of the most noxious environmental pollutants including halogenated phenols, dioxins, biphenyls, and aliphatic hydrocarbons (Bunge et al. 2003; Wohlfarth and Diekert 1997; Adrian et al. 2000). As of now, RDases are the only terminal respiratory enzymes known to depend on a corrinoid cofactor. The cobalamin (B12)-dependent enzymes can be roughly divided into the isomerases, B12-dependent methyltransferases, and the RDases (Banerjee and Ragsdale 2003). RDases are typically composed of two subunits: RdhA and RdhB (Futagami et al. 2008). The catalytically active RdhA-subunit contains a corrinoid and two [4Fe–4S] clusters and is typically associated to the periplasmic side of the cytoplasmic membrane (John et al. 2006; Futagami et al. 2008), presumably via the second subunit, RdhB, which is a small integral membrane protein. In contrast to respiratory RdhAs, which are inactivated by molecular oxygen, certain catabolic RdhAs are found in facultative aerobic microorganisms and have, despite being only distantly related to the respiratory chlorophenol reductases, the same cofactor content and principal architecture as the respiratory RdhAs (Chen et al. 2013). In the latter case, these enzymes are involved in the degradation of organohalides and enable catabolism through reductive dehalogenation.

B12-dependent enzymes are very versatile catalysts, due to different chemical properties and ligation states of the corrinoid cofactor (Banerjee and Ragsdale 2003). The AdoCbl-dependent radical enzymes and the B12-dependent methyltransferases typically rely on Co–C chemistry, which led to the suggestion that Co–C bond formation could also play a role in reductive dehalogenation (Banerjee and Ragsdale 2003).

Structural and in-depth biochemical studies were long hampered by the distinct lack of sufficient RdhA material, related to the low cell yield associated with organohalide reducing bacteria (Hug et al. 2013). The recombinant heterologous and homologous production of functional, holo-RdhAs were only recently reported, and allow now mg-scale production of homogenous dehalogenase preparations (Sjuts et al. 2012; Payne et al. 2015; Mac Nelly et al. 2014; Parthasarathy et al. 2015). These advances finally allowed structural studies using X-ray crystallography (Bommer et al. 2014; Payne et al. 2015). More details on the biochemistry and heterologous production of RdhAs can be found in other chapters of this book.

20.2 Structure of Reductive Dehalogenases

In this chapter we summarize insights gained from recent structural studies of two very distinct RdhAs (Bommer et al. 2014; Payne et al. 2015). These RdhAs, the respiratory RDase PceA from *Sulfurospirillum multivorans* (Neumann et al. 1996) and a catabolic NpRdhA from *Nitratireductor pacificus* pht-3B (Payne et al. 2015), allow a first glimpse on how RdhAs bind halogenated alkenes and aromatics selectively and catalyze the reductive conversion of these rather unreactive class of compounds. Given the fact that each enzyme represents a distinct class of RdhAs, a detailed comparison is likely to illustrate those key features common to all RdhA enzymes. For clarity, we will use the term RdhA when referring to those structural features conserved or similar in both PceA and NpRdhA structures, while we will refer explicitly to either enzyme by name when highlighting features particular to each.

Previous sequence alignments of various *rdhA* genes or their translated amino acid sequences revealed a putative B12-binding N-terminal domain fused to a bacterial 2[4Fe–4S] ferredoxin module (Smidt and de Vos 2004; Hug et al. 2013). Unlike some other B12-dependent enzyme classes, RdhA sequences contain, with some exceptions (Holscher et al. 2004), no obvious B12-binding signature motif such as the DXHXXG motif. Furthermore, the catabolic *RdhA* genes all contained a duplication of the N-terminal domain, but only one ferredoxin module (Chen et al. 2013). The PceA crystal structures clearly revealed the enzyme to be a dimer, with a large interface made predominantly by the N-terminal B12-binding module (Fig. 20.1). In contrast, the NpRdhA structure is clearly monomeric. In the latter, the duplicated N-terminal domains are positioned such that they mimic the PceA dimer. In NpRdhA, only one domain actually binds the corrinoid cofactor, the other is a likely consequence of gene duplication and divergence. It might be that dimerization of the corrinoid-binding domains is an integral feature of RdhAs, although this will require further study.

Most surprisingly, and despite the lack of any obvious sequence similarity, the RdhA corrinoid-binding module is structurally similar to a previously reported B12-binding enzyme structure: the human CblC (Koutmos et al. 2011) (Fig. 20.1). The similarity in fold, cofactor position, and sequence similarity as detected from a structural superimposition all strongly argue for a common ancestry between reductive dehalogenases and CblC. CblC is a B12-trafficking chaperone that catalyses reductive removal of the upper axial ligands from cyano- or alkylcobalamins. In both RdhA and the CblC, the corrinoid is bound by a nitroreductase fold, in the so-called base-off conformation, likely reflecting a common evolutionary origin. In case of the RDases, the lower ligand (dimethylbenzimidazole in the cobalamin-binding NpRdhA, adenosine in the norpseudovitamin-B12-binding PceA) is firmly held by the protein in a position away from the cobalt coordination sphere. As with many other metal ions, a change in Co-ligation dramatically alters the redox potential (Brown 2005), and the pentacoordinate cobalamin redox potential is significantly higher than for the corresponding base-on, hexacoordinate form.

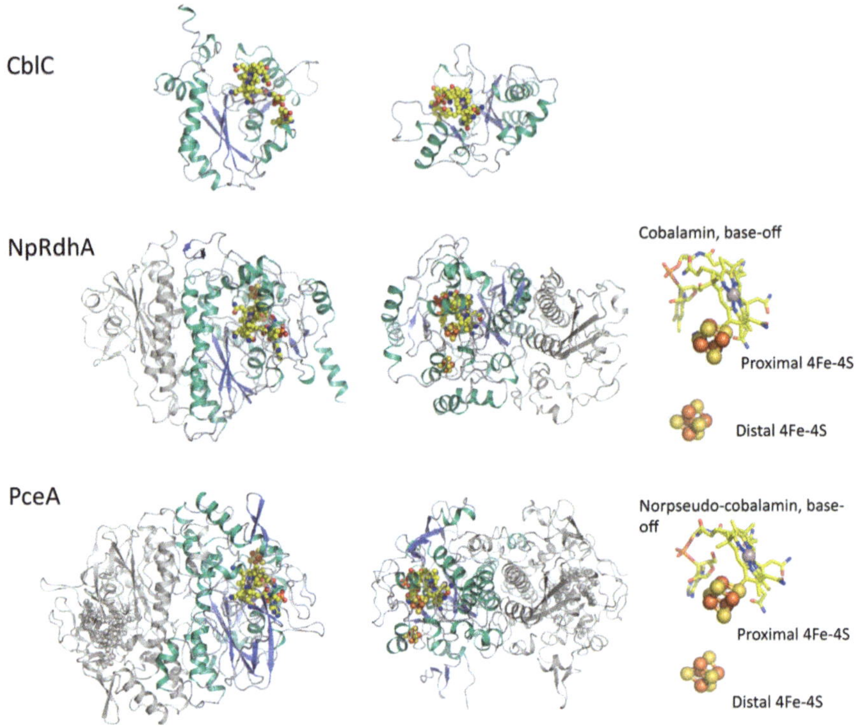

Fig. 20.1 Comparison between CblC and the reductive dehalogenases PceA and NpRdhA. The *left* and *middle* columns show a cartoon representation of the three proteins in similar orientation, in two distinct views. Only one monomer of the PceA dimer is shown in *color*, while the vestigial N-terminal corrinoid-binding domain of NpRdhA is shown in *gray*. The *right* column shows a detailed view of the reductive dehalogenase redox chains, in the same orientation as for the *middle* column

In both RdhA structures, the cobalamin is found to be bound in the same base-off conformation, which provides a rationale for the relatively high redox potentials measured for the RdhA Co(II)/Co(I) couple (Maillard et al. 2003; Krasotkina et al. 2001; Neumann et al. 2002; van de Pas et al. 1999) While the overall features of corrinoid binding in CblC, PceA, and NpRdhA are indeed similar (Fig. 20.1), the detailed binding interactions made between corrin ring and enzyme are in fact distinct for each pair. This provides further rationale for the lack of a dedicated corrinoid-binding motif in the RdhA family.

While CblC lacks an associated redox module, the RdhA enzymes contain a C-terminal bacterial ferredoxin domain. The latter domain is tightly packed against the corrinoid-binding module, and even contributes some residues to the binding pocket of the lower base. As with other ferredoxin structures, both RdhA [4Fe–4S] clusters are in close proximity, approximately 10 Å (closest Fe–Fe distance) apart (Figs. 20.1 and 20.2). This is likely to ensure rapid

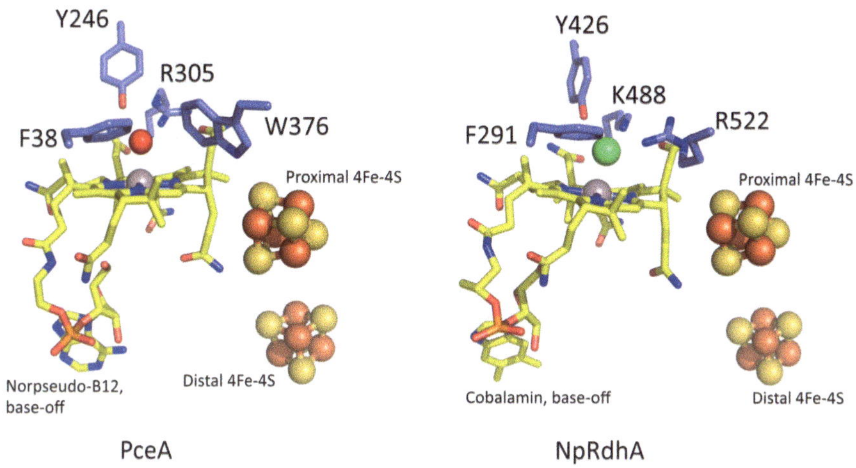

Fig. 20.2 Side-by-side comparison of the reductive dehalogenase active sites. Key residues in close proximity of the Co fifth ligand (respectively water in PceA, shown as a *red sphere*; and a chloride atom in NpRdhA, shown as a *green sphere*) are shown as atom colored sticks (*C yellow, N blue, O red* and *S* in *green*)

intraprotein electron transfer between both clusters. Interprotein electron transfer to the RdhAs can, in principle, occur via either iron–sulfur cluster, as both are positioned relatively close to the protein surface (Fig. 20.3). While certain catabolic *rdhA* genes consist of an RdhA module fused to a putative reductase (Chen et al. 2013), the physiological redox donor to the membrane associated-PceA and other respiratory RdhAs is as yet unknown. In case of NpRdhA, genes encoding a putative ferredoxin-reductase and an associated ferredoxin are found in close proximity of the *NpRdhA* gene (Lai et al. 2012). It remains to be established whether these serve as physiological electron donors. It is therefore unclear at this stage whether one or both of the iron–sulfur clusters serves as an entry point for electrons. However, it is attractive to speculate that the distal (i.e., further away from the cobalamin) iron–sulfur cluster would serve this purpose.

In both structures, the proximal iron–sulfur cluster is positioned in direct van der Waals contact with the RdhA cobalamin, an unusually close interaction. Furthermore, in both RdhA structures, a direct hydrogen bond interaction between the cobalamin and one of the bridging S atoms is observed. Taken together with the fact that an extended loop region from the ferredoxin module contributes to the lower ligand binding pocket, this suggests a close link between cobalamin and [4Fe–4S] cluster binding during RdhA maturation. This might explain in part the difficulties encountered when attempting heterologous production of holo-RdhAs (Sjuts et al. 2012; Mac Nelly et al. 2014; Parthasarathy et al. 2015; Payne et al. 2015), and explains why several RdhA genes are linked with a variety of chaperones. The reason for the unusually close corrin-proximal [4Fe–4S] cluster

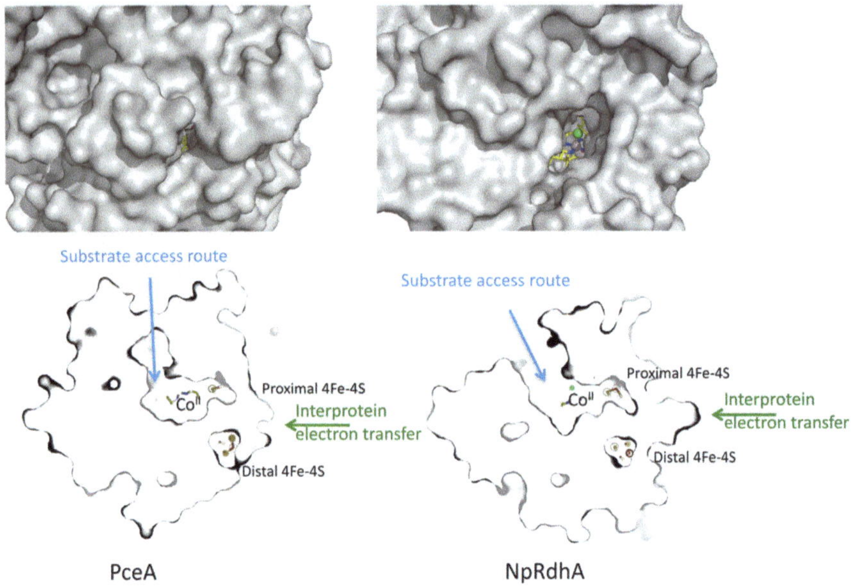

Fig. 20.3 Side-by-side comparison of reductive dehalogenase substrate access channels. The *top* panels show the solvent accessible surface, revealing the corrinoid cofactor of NpRdhA is readily accessible in comparison to the PceA cofactor. The *bottom* panels show a cross section through the solvent accessible surface centered on the corrinoid cobalt atom, which is roughly perpendicular to the corrin ring. This clearly reveals distinct substrate binding pockets in both enzymes, but also shows clearly that both [4Fe–4S] clusters are positioned close to the surface

interaction is not immediately clear. It might merely serve to ensure sufficiently rapid intraprotein electron transfer between both redox cofactors during turnover. On the other hand, its purpose might be to allow a direct electron transfer between the proximal [4Fe–4S] cluster and the substrate.

The RdhA active site can be easily identified as the solvent filled cavity directly above the cobalamin. The shape, access channel and majority of residues that line the active site, are distinct for the two RdhA structures (Figs. 20.2 and 20.3). This likely reflects both the evolutionary distance between these enzymes as well as the contrast in their substrate specificity. PceA reduces small, hydrophobic chlorinated ethenes while NpRdhA serves to reduce aromatic dibromophenols. Despite these differences, a few residues are conserved: a tyrosine is found to point directly towards the cobalt ion in both RdhA structures. The phenol group is within hydrogen bonding distance of a second conserved and positively charged residue (Lys in Pce; Arg in NpRdhA). This RdhA Tyr-Lys/Arg motif is likely involved in proton transfer coupled to substrate reduction and can be readily identified in the vast majority of RdhA proteins. For NpRdhA, mutations to either residue abolish enzyme activity (Payne et al. 2015).

The distinct nature of the substrate binding pockets for both enzymes reflects the distinct substrate specificity. While the structure of the ligand–bound complex

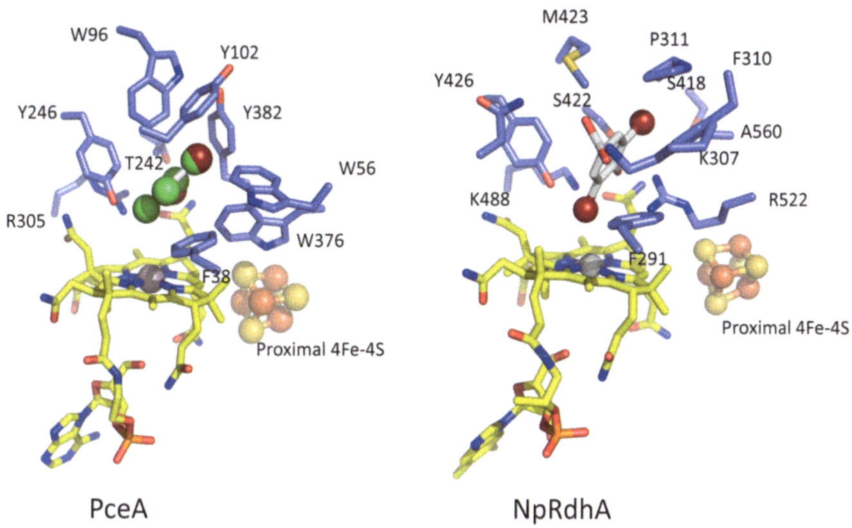

Fig. 20.4 Side-by-side comparison of reductive dehalogenase ligand complexes. Key residues in close proximity of the ligand are shown in atom colored (see Fig. 20.1 for color code) sticks. The PceA active site contains an overlay of the substrate trichloroethene and cis-dibromoethene (a product like molecule) complex structures while the NpRdhA active site shows the modeled dibromohydroxybenzoic acid substrate

could be obtained for PceA, the NpRdhA–substrate complex could only be modeled (Fig. 20.4). This model predicted a direct interaction between the halogenated substrate and the cobalt ion, and was corroborated by EPR spectroscopy. In contrast, PceA ligand structures reveal the substrate or product molecules to bind in close proximity to the cobalamin, but not in direct contact. While halogen ligation of the cobalt could in fact be detected in both RdhA enzymes, it remains to be established whether a substrate–Co(II) interaction is a common feature for Rdhs.

Prior to structural information on RdhA, mechanistic proposals drew heavily on those established for other, unrelated B12-dependent enzymes (Banerjee and Ragsdale 2003). These either invoked the formation of a direct Co(III)–C bond with the substrate, or the formation of transient substrate radical species. The formation of a Co(III)–C bond appears incompatible with both RdhA structures. The active site is relatively narrow in the direct vicinity of the cobalt ion and would need to undergo drastic reorganization to allow for such a species to form. Mechanism(s) that invoke transient formation of substrate radical species appear compatible, with direct electron transfer to the substrate either exclusively occurring via the cobalamin or the proximal [4Fe–4S] cluster in addition (Fig. 20.5). A third type of mechanism was proposed on the basis of the halogen–cobalamin interactions observed for NpRdhA. In the latter case, C–halogen bond cleavage is proposed to be accompanied by Co–halogen bond formation.

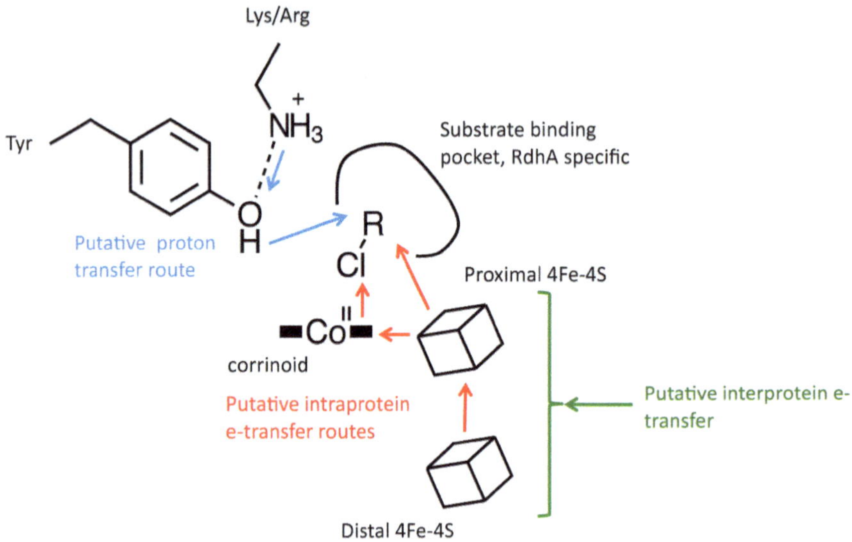

Fig. 20.5 A schematic representation of electron and proton transfer networks in reductive dehalogenases. The comparison of the available RdhA structures reveals both the redox chain as well as the putative proton transferring Tyr-Lys/Arg dyad are conserved in RdhA. In contrast, the substrate binding pocket is clearly tailored to select for distinct substrates in each individual RdhA enzyme. While certain mechanisms (i.e. those based on direct Co–C bond formation) appear incompatible with the available structural data, a variety of electron transfer and concomitant proton transfer steps can now be proposed. Putative proton (*blue arrows*) and electron transfer (*red arrows*) routes are shown

20.3 Outlook

The first structures of two RdhA proteins reviewed in this chapter should help to inform and guide future exploitations of reductive dehalogenases (Bommer et al. 2014; Payne et al. 2015). The rapidly increasing number of sequenced bacterial genomes is leading to a significant increase in RdhA sequences. Sequence similarity and substrate specificity of RdhAs are not strictly correlated, making the prediction of substrate specificity based on sequence similarity alone very difficult (Hug et al. 2013). The two structures show how RdhAs combine conserved elements involved in electron and proton transport with a variable substrate binding pocket (Fig. 20.5). In that regard, the RdhA enzyme family might come to mimic the wide variation observed for cytochrome P450s, oxidative enzymes that appear to have evolved a common scaffold to bind almost any type of substrate (Munro et al. 2007). Hence, now that the RdhA variable substrate binding regions are identified, these may be exploited to deduce existing and possibly engineer new substrate specificities for biotechnological applications (Richardson 2013).

Despite the recent progresses, several open questions concerning the structure and function of RdhAs remain. The suggested formation of a cobalamin–halide complex via oxidative addition to the Co–ion indicates that the catalytic power of RdhAs could rely on a new paradigm in organohalide and cobalamin biochemistry (Payne et al. 2015). However, this may not be a general feature of RdhAs. The missing close proximity between Co and the nearest chlorine atom in the trichloroethene complexes of PceA does not support a direct Co–substrate interaction, unless the substrate rearranges in the active site to bind to Co (Bommer et al. 2014).

Free cob(I)alamin is known to react rapidly with alkyl halides like methyl iodide forming alkyl cobalamins, but surprisingly alkyl cobalamins appear to be excluded as intermediates due to steric hindrance. It will be interesting to see, if this is a general feature of RdhAs. CmuA and CmuB, for example, which are more closely related to tetrahydrofolate-dependent methyltransferases than to other RdhAs, reductively dehalogenate chloromethane in *Methylobacterium* sp. Studer et al. (1999, 2001). Similar to the two RdhAs discussed here, a base-off mode for binding cobalamin is present, but iron/sulfur clusters appear to be absent from CmuA. However, sequence similarity and type of substrate would suggest that CmuA forms a Co–CH_3 bond after reductive dechlorination, as the generated methyl group is transferred to tetrahydrofolate, for which it needs to be retained in the active site and activated for the transfer (Studer et al. 2001). This would indicate that different RdhAs, acting on different substrates, employ the corrinoid cofactor for fundamentally different mechanisms.

The hallmark of a respiratory enzyme is the coupling of electron transfer to the generation of a proton-motive force at the membrane. How this is achieved by respiratory RdhAs is unknown, thus the primary bioenergetic role of RdhA function is not yet understood. Similarly it is still puzzling why some of the RdhAs, like PceA of *S. multivorans*, catalyze the conversion of tetra- and trichlorinated ethenes to *cis*–dichloroethene, while others also catalyze the further conversions down to ethene (Smidt and de Vos 2004).

RdhAs are infamous for their use of different corrinoids, some of which, like norpseudovitamin-B12 (Krautler et al. 2003), have only been found in RdhAs. Why some RdhAs have a preference for the more uncommon corrinoids is unclear, and we are only beginning to see how the enzymes achieve cofactor selection by their architecture.

Future structural data on these and other RdhAs, especially from distinct bacterial branches, will tell us more about how these enzymes evolved, how they function, and how each specifically selects for a subset of the different organohalides, ranging from small aliphatic to large polycyclic aromatic compounds. These open questions attest that structure–function studies are still at their infancy, and further studies into this fascinating class of enzymes will hopefully reveal how the versatility and variability in the specific reductive dehalogenations are achieved.

References

Adrian L, Szewzyk U, Wecke J, Gorisch H (2000) Bacterial dehalorespiration with chlorinated benzenes. Nature 408(6812):580–583. doi:10.1038/35046063

Banerjee R, Ragsdale SW (2003) The many faces of vitamin B-12: catalysis by cobalamin-dependent enzymes. Annu Rev Biochem 72:209–247. doi:10.1146/annurev.biochem.72.121801.161828

Bommer M, Kunze C, Fesseler J, Schubert T, Diekert G, Dobbek H (2014) Structural basis for organohalide respiration. Science 346(6208):455–458. doi:10.1126/science.1258118

Brown KL (2005) Chemistry and enzymology of vitamin B-12. Chem Rev 105(6):2075–2149

Bunge M, Adrian L, Kraus A, Opel M, Lorenz WG, Andreesen JR, Gorisch H, Lechner U (2003) Reductive dehalogenation of chlorinated dioxins by an anaerobic bacterium. Nature 421(6921):357–360. doi:10.1038/nature01237

Chen K, Huang LL, Xu CF, Liu XM, He J, Zinder SH, Li SP, Jiang JD (2013) Molecular characterization of the enzymes involved in the degradation of a brominated aromatic herbicide. Mol Microbiol 89(6):1121–1139

Futagami T, Goto M, Furukawa K (2008) Biochemical and genetic bases of dehalorespiration. Chem Rec 8(1):1–12. doi:10.1002/tcr.20134

Gribble GW (1998) Naturally occurring organohalogen compounds. Acc Chem Res 31(3):141–152. doi:10.1021/ar9701777

Holliger C, Wohlfarth G, Diekert G (1998) Reductive dechlorination in the energy metabolism of anaerobic bacteria. FEMS Microbiol Rev 22(5):383–398. doi:10.1111/j.1574-6976.1998.tb00377.x

Holscher T, Krajmalnik-Brown R, Ritalahti KM, Von Wintzingerode F, Gorisch H, Loffler FE, Adrian L (2004) Multiple nonidentical reductive-dehalogenase-homologous genes are common in dehalococcoides. Appl Environ Microbiol 70(9):5290–5297. doi:10.1128/AEM.70.9.5290-5297.2004

Hug LA, Edwards EA (2013) Diversity of reductive dehalogenase genes from environmental samples and enrichment cultures identified with degenerate primer PCR screens. Front Microbiol 4:341. doi:10.3389/fmicb.2013.00341

Hug LA, Maphosa F, Leys D, Loffler FE, Smidt H, Edwards EA, Adrian L (2013) Overview of organohalide-respiring bacteria and a proposal for a classification system for reductive dehalogenases. Philos Trans R Soc B 368(1616):2012. doi:10.1098/rstb.2012.0322 (ARTN0322)

John M, Schmitz RP, Westermann M, Richter W, Diekert G (2006) Growth substrate dependent localization of tetrachloroethene reductive dehalogenase in *Sulfurospirillum multivorans*. Arch Microbiol 186(2):99–106. doi:10.1007/s00203-006-0125-5

Koutmos M, Gherasim C, Smith JL, Banerjee R (2011) Structural basis of multifunctionality in a vitamin B-12-processing enzyme. J Biol Chem 286(34):29780–29787

Krasotkina J, Walters T, Maruya KA, Ragsdale SW (2001) Characterization of the B12- and iron–sulfur-containing reductive dehalogenase from *desulfitobacterium chlororespirans*. J Biol Chem 276(44):40991–40997. doi:10.1074/jbc.M106217200

Krautler B, Fieber W, Ostermann S, Fasching M, Ongania KH, Gruber K, Kratky C, Mikl C (2003) The cofactor of tetrachloroethene reductive dehalogenase of *Dehalospirillum multi vorans* is norpseudo-B-12, a new type of a natural corrinoid. Helv Chim Acta 86(11):3698–3716. doi:10.1002/hlca.200390313

Lai QL, Li GZ, Shao ZZ (2012) Genome Sequence of *Nitratireductor pacificus* type strain pht-3B. J Bacteriol 194(24):6958

Mac Nelly A, Kai M, Svatos A, Diekert G, Schubert T (2014) Functional heterologous production of reductive dehalogenases from *Desulfitobacterium hafniense* strains. Appl Environ Microbiol 80(14):4313–4322. doi:10.1128/AEM.00881-14

Maillard J, Schumacher W, Vazquez F, Regeard C, Hagen WR, Holliger C (2003) Characterization of the corrinoid iron–sulfur protein tetrachloroethene reductive dehalogenase of *Dehalobacter restrictus*. Appl Environ Microbiol 69(8):4628–4638

Munro AW, Girvan HM, McLean KJ (2007) Variations on a (t)heme—novel mechanisms, redox partners and catalytic functions in the cytochrome P450 superfamily. Nat Prod Rep 24(3):585–609

Neumann A, Wohlfarth G, Diekert G (1996) Purification and characterization of tetrachloroethene reductive dehalogenase from *Dehalospirillum multivorans*. J Biol Chem 271(28):16515–16519

Neumann A, Siebert A, Trescher T, Reinhardt S, Wohlfarth G, Diekert G (2002) Tetrachloroethene reductive dehalogenase of *Dehalospirillum multivorans*: substrate specificity of the native enzyme and its corrinoid cofactor. Arch Microbiol 177(5):420–426. doi:10.1007/s00203-002-0409-3

Oberg G (2002) The natural chlorine cycle–fitting the scattered pieces. Appl Microbiol Biotechnol 58(5):565–581. doi:10.1007/s00253-001-0895-2

Parthasarathy A, Stich TA, Lohner ST, Lesnefsky A, Britt RD, Spormann AM (2015) Biochemical and EPR-spectroscopic investigation into heterologously expressed vinyl chloride reductive dehalogenase (VcrA) from *Dehalococcoides mccartyi* strain VS. J Am Chem Soc 137(10):3525–3532. doi:10.1021/ja511653d

Payne KA, Quezada CP, Fisher K, Dunstan MS, Collins FA, Sjuts H, Levy C, Hay S, Rigby SE, Leys D (2015) Reductive dehalogenase structure suggests a mechanism for B12-dependent dehalogenation. Nature 517(7535):513–516. doi:10.1038/nature13901

Richardson RE (2013) Genomic insights into organohalide respiration. Curr Opin Biotechnol 24(3):498–505. doi:10.1016/j.copbio.2013.02.014

Sjuts H, Fisher K, Dunstan MS, Rigby SE, Leys D (2012) Heterologous expression, purification and cofactor reconstitution of the reductive dehalogenase PceA from *Dehalobacter restrictus*. Protein Expr Purif 85(2):224–229. doi:10.1016/j.pep.2012.08.007

Smidt H, de Vos WM (2004) Anaerobic microbial dehalogenation. Annu Rev Microbiol 58:43–73

Studer A, Vuilleumier S, Leisinger T (1999) Properties of the methylcobalamin:H4folate methyltransferase involved in chloromethane utilization by *Methylobacterium* sp. strain CM4. Eur J Biochem 264(1):242–249

Studer A, Stupperich E, Vuilleumier S, Leisinger T (2001) Chloromethane: tetrahydrofolate methyl transfer by two proteins from *Methylobacterium chloromethanicum* strain CM4. Eur J Biochem 268(10):2931–2938. doi:10.1046/j.1432-1327.2001.02182.x

van de Pas BA, Smidt H, Hagen WR, van der Oost J, Schraa G, Stams AJ, de Vos WM (1999) Purification and molecular characterization of ortho-chlorophenol reductive dehalogenase, a key enzyme of halorespiration in *Desulfitobacterium dehalogenans*. J Biol Chem 274(29):20287–20292

Wohlfarth G, Diekert G (1997) Anaerobic dehalogenases. Curr Opin Biotechnol 8(3):290–295

Part VI
Applications

Chapter 21
Redox Interactions of Organohalide-Respiring Bacteria (OHRB) with Solid-State Electrodes: Principles and Perspectives of Microbial Electrochemical Remediation

Federico Aulenta, Simona Rossetti, Bruna Matturro, Valter Tandoi, Roberta Verdini and Mauro Majone

Abstract Recent studies have revealed that a number of organohalide-respiring bacteria (OHRB) are capable to establish redox interactions with solid-state electrodes by using them as direct or indirect electron donors in their energy metabolism. Although the biochemical, ecological, and evolutionary significance of electron transfer capabilities in OHRB remain largely unknown, they are increasingly being considered for bioremediation applications. In principle, bioelectrochemical remediation systems which use insoluble electrodes to drive the microbial reduction of chlorinated compounds offer numerous advantages compared to conventional approaches, such as the possibility to fine-tune the rate of electron delivery and consumption, avoid injection of chemicals to the subsurface environment and ultimately gain a more direct control over the biodegradation reactions taking place at the electrodes. In spite of that, however, the technology is still in its infancy and further research and extensive field testing is needed to prove its actual potential for site remediation.

List of Abbreviations

AQDS	Anthraquinone-2,6-disulfonate
BES	Bioelectrochemical systems
cis-DCE	*cis*-Dichloroethene

F. Aulenta (✉) · S. Rossetti · B. Matturro · V. Tandoi
Water Research Institute (IRSA), National Research Council (CNR), Rome, Italy
e-mail: aulenta@irsa.cnr.it

R. Verdini · M. Majone
Department of Chemistry, Sapienza University of Rome, Rome, Italy

L. Adrian and F.E. Löffler (eds.), *Organohalide-Respiring Bacteria*,
DOI 10.1007/978-3-662-49875-0_21

DET	Direct electron transfer
DIET	Direct interspecies electron transfer
EET	Extracellular electron transfer
ETH	Ethene
MV	Methyl viologen
OHRB	Organohalide-respiring bacteria
PCE	Tetrachloroethene
SHE	Standard hydrogen electrode
TCE	Trichloroethene
VC	Vinyl chloride

21.1 Background

In recent years, microbial bioelectrochemical systems (BES) have attracted considerable attention as an effective methodology for the treatment of a variety of organic and inorganic contaminants, such as (chlorinated) hydrocarbons, heavy metals, and nitrate (Harnisch et al. 2011). However, the idea of using solid-state electrodes to stimulate reductive or oxidative metabolisms is not new. As an example, over 15 years ago, researchers were able to demonstrate the electric current-dependent reductive dechlorination of 2,6-dichlorophenol to phenol by a mixed microbial culture in a bioelectrochemical reactor (Skadberg et al. 1999). According to the authors, formation of H_2 on the cathode (i.e., a stainless steel mesh) generated highly reducing conditions that induced the microbial reductive dechlorination of 2,6-dichlorophenol.

In essence, a bioelectrochemical remediation method utilizes electrochemical measures (e.g., application of a current or a voltage) to enhance biodegradative processes in the subsurface (Lohner et al. 2011). The basic idea behind the use of electrochemical means to stimulate biological reductions (or oxidations) is that the delivery (or withdrawal) of electrons can, in principle, be continuously monitored/controlled in terms of current and potential; no chemicals need to be injected, thereby eliminating the need for transport, storage, dosing, and post-treatment. Moreover, since the microorganisms responsible for the biodegradation can be grown directly at the surface of the electrodes or at their proximity, an effective usage of the electric current can be expected.

With specific reference to the application of BES to stimulate reductive dechlorination processes, the increased awareness of the central role of H_2 in the metabolism of the majority of organohalide-respiring bacteria (OHRB), as well the increased understanding of the importance of fine-tuning H_2 levels in subsurface environments to selectively stimulate reductive dechlorination processes over competing metabolisms (Kassenga and Pardue 2006; Löffler et al. 1999; Luijten et al. 2004a; Aulenta et al. 2005; Fennell et al. 1995; Smatlak et al. 1996), have enforced the scientific interest for electrolysis-driven bioremediation systems.

A number of recent laboratory investigations (Aulenta et al. 2008b; Zhang et al. 2013; Lohner et al. 2011) have, indeed, clearly demonstrated that in BES the rate of electrolytic H_2 generation can be fine-tuned by properly controlling the cathode potential or the electric current. This feature could be, at least in principle, exploited to channel most of the produced H_2 to the desired dechlorination process, ultimately resulting in a highly efficient bioremediation process. Overall, in situ generation of H_2, driven by low electric currents possibly supplied by solar panels, would represent a sustainable and environment-friendly alternative to H_2 sparging or other more sophisticated methods for H_2 supply to groundwater (e.g., bubble-less supply through gas permeable membranes).

A major scientific breakthrough in the application of microbial BES for groundwater remediation was the discovery that, under certain conditions, electrodes can also serve as direct electron donors for the microbially catalyzed reductive dechlorination of chlorinated hydrocarbons without the mediation of electrolytic H_2 (Aulenta et al. 2007a, 2009; Strycharz et al. 2008). Figure 21.1 shows representative microscopy images of a trichloroethene (TCE)-dechlorinating biocathode capable of using a polarized graphite electrode as direct electron donor (Fig. 21.1).

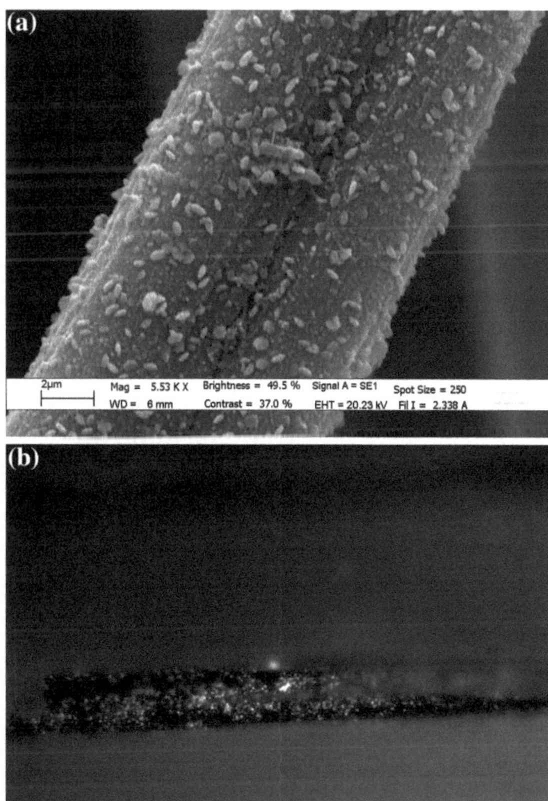

Fig. 21.1 **a** Scanning electron microscopy micrographs of a dechlorinating biocathode; **b** Appearance of bacteria attached onto the graphite electrode surface by fluorescence in situ hybridization (FISH). Reprinted with permission from Aulenta et al. (2010). Biosensors and Bioelectronics 25(7): 1796–1802. Copyright (2010) Elsevier

Bypassing H_2 productions eliminates a number of problems such as groundwater alkalinization (due to protons consumption at the electrode) which may disrupt biochemical soil functions, and growth of H_2-consuming non-dechlorinating microorganisms such as methanogens or sulfate-reducers which accumulate near the point of H_2 generation, often resulting in electrode fouling problems, aquifer clogging, low efficiencies of H_2 utilization by dechlorinators.

Furthermore, with OHRBs adhering on the surface of electrodes, it is potentially feasible to specifically co-localize the electron donor and the dechlorinating microorganisms in specific locations in the subsurface and control the flux of electrons into the site in order to fine-tune the rate of biostimulation. Taken as a whole, this approach could eliminate the substantial growth of undesired, competing organisms and the productions of deleterious end products that are often associated with the addition of organic electron donors to the subsurface. Clearly, this exciting opportunity will have to be verified under field conditions, whereby a fierce competition between OHRB and other microorganisms may establish.

More generally, the enforcing scientific interest for bioelectrochemical remediation systems, notwithstanding the underlying electron transfer mechanisms (direct vs. mediated by, for example, H_2), is primarily due to the following advantages they hold compared to conventional (bio)remediation approaches, including: (i) no need for chemical introduction into the surface, making them environmentally benign; (ii) no in situ mixing of reagents or nutrients is required; (iii) electrical power cost associated with driving transformations are low; (iv) electrode materials appear to be resilient under standard treatment conditions (Ho et al. 1995), hence comparing favorably against other in situ treatment technologies such as permeable reactive barriers, in which reactive materials are consumed over time, and finally (v) low operational and maintenance costs. However, taking into consideration that BES technology is in its early stage of development (especially in the field of remediation) and has been so far assessed almost exclusively at the laboratory scale, a more detailed evaluation of its pros and cons will require further investigations and extensive field testing.

21.2 Electron Transfer Mechanisms

Although an increasing number of biocathode studies were published in recent years, little fundamental information is still available on the mechanisms of microbial electron uptake from a (polarized) cathode (Rosenbaum et al. 2011; Huang et al. 2011). This is particularly true for OHRBs, whose application in BES has been, so far, only marginally explored and exploited. In general, in microbial BES, extracellular electron transfer (EET) mechanisms are commonly classified in two broad groups: indirect (or mediated) and direct EET, depending on the way microorganisms access electrons available from a cathode (Huang et al. 2011).

Fig. 21.2 Schematic representation of direct and mediated extracellular electron transfer mechanisms in dechlorinating biocathodes

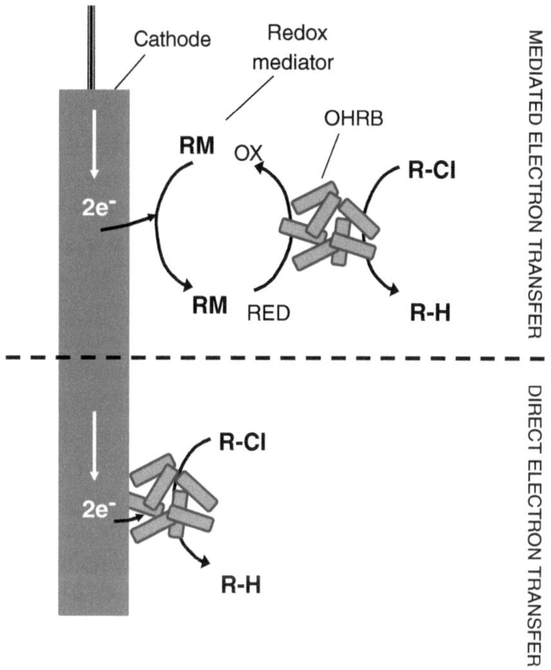

21.2.1 Indirect or Mediated Extracellular Electron Transfer

Indirect, or mediated EET, typically involves the redox shuttling of a molecule, often referred to as redox mediator or redox shuttle, between the electrode (cathode) and the microorganisms, according to the mechanism schematically depicted in Fig. 21.2. While the electrochemical reduction of the mediator necessarily takes place at the surface of the electrode, its bacterial reoxidation can possibly take place at a distance from the electrode, upon diffusive or convective transport of the reduced mediator. Accordingly, the process involves a number of reactive and transport steps, each one possibly controlling the overall performance of the dechlorination process, depending on its intrinsic kinetics and efficiency.

In aqueous environments, the most abundant molecule which could function as a redox mediator is H_2. H_2 can be abiotically generated from proton reduction at the surface of cathodes maintained at sufficiently low redox potentials. Indeed, although the standard biochemical redox potential (i.e., reactants and products at 1 M or 1 atm and pH 7) of the redox couple H_2/H^+ is −414 mV versus the standard hydrogen electrode (SHE), substantially more reducing potentials are typically required for the reaction to proceed at appreciable rates over non noble metal electrodes. As an example, previous studies have shown that cathode potentials more negative than −550 mV are required to drive H_2 formation over carbon-based electrodes, such as granular graphite (Aulenta et al. 2008b). Whenever H_2 is involved (either when electrochemically generated or produced through conventional fermentation processes), hydrogenases are the most likely "entrance points" of electrons into the OHRB cells (Morris et al. 2006; Nijenhuis and Zinder 2005). Although H_2 can be regarded as a "ubiquitous" mediator, its bioelectrochemical exploitation presents some specific drawbacks. Indeed, the very low redox potentials needed to generate H_2 may also cause the nondiscriminatory reduction of other redox-active compounds in the subsurface, ultimately resulting in a waste of electric energy. Furthermore, the reduction of protons to H_2 increases the groundwater pH, which can disrupt biological, physical, and chemical soil functions. Finally, if not properly tuned, H_2 generation may cause the growth of non-dechlorinating biomass and the accumulation of organic byproducts (such as acetate and other organic acids) which deteriorate groundwater quality (Aulenta et al. 2007b).

Besides H_2, other "natural" or synthetic compounds may also serve as redox mediators in the bioelectrochemical reductive dechlorination process. As an example in BES experiments, methyl viologen (MV) an artificial redox mediator with a low standard redox potential ($E^{0'} = -446$ mV vs. SHE), at concentrations as low as 25 μmol/L, was found to stimulate the reductive dechlorination of TCE (to VC and ethene) by a mixed culture containing *Dehalococcoides mccartyi* and *Desulfitobacterium* spp. as the putative OHRB (Aulenta et al. 2007a, 2008a). Since MV is considered to be unable to cross the cytoplasmatic membrane of bacteria (Nijenhuis and Zinder 2005), it was suggested that it could donate electrons directly to periplasmic hydrogenases of the OHRB. The electrochemical interaction of reduced MV with hydrogenases of OHRB was further suggested by the fact that in the absence of chlorinated solvents, the dechlorinating culture produced substantial amounts of H_2, possibly as a bypass strategy to dispose of the excess of electrons (deriving from electrochemically reduced MV) which could not be rapidly diverted to the physiological electron acceptor (Aulenta et al. 2008a; Villano et al. 2011). The poor (electro)chemical stability of MV coupled to its environmental toxicity represents a major obstacle to the field scale applicability of this mediator for remediation applications.

Naturally occurring humic substances are ubiquitous redox-active compounds in the environment. In recent years, evidence has been accumulated that humic acids, and particularly their quinoid moieties, can play an important role as redox mediators in the reductive dechlorination of chlorinated hydrocarbons, also in BES (Cervantes et al. 2013). As an example, electrically reduced anthraquinone-2,6-disulfonate (AQDS), a humic acid analog, was found to serve as a redox mediator (with a cathode set at −250 mV) in the reductive dechlorination of TCE-to *cis*-DCE by a *Desulfitobacterium*-enriched culture (Aulenta et al. 2010). This finding is consistent with a previous study which pointed out that the ability of (chemically reduced) AQDS to serve as an electron donor for TCE-to-*cis*-DCE dechlorination is widespread among *Desulfitobacterium* species (Luijten et al. 2004b). In the case of this microorganism, electrons possibly entered the electron transport chain at a cytochrome level and then were transferred to the reductive dechlorination enzymes. Indeed, due to their slightly higher redox potential, cytochromes are suitable in vivo redox partners of AQDS (Lovley 2008).

Preliminary studies have also suggested that, upon extended acclimation in a bioelectrochemical system, a mixed dechlorinating culture is even capable of producing and releasing into the solution its own redox mediator (Aulenta et al. 2009). This self-produced (yet unidentified) redox mediator, displaying a redox potential of around −400 mV versus SHE, most likely contributed to the shuttling of electrons from the polarized electrode to OHRB (either present in solution or attached at the electrode surface) driving the reductive dechlorination of TCE to VC and ethene. Since this redox mediator was not found in the supernatant of the same culture when it was grown on H_2 and TCE, it was suggested that it could have been produced and excreted in response to the electrode polarization.

21.2.2 Direct EET

The occurrence of reductive dechlorination activity at cathode potentials where there is no abiotic hydrogen production, together with the undetectable presence of soluble redox mediators in the medium, suggested the possibility that OHRB could directly accept electrons from a polarized cathode. Direct electron transfer (DET) between a microorganism and a cathode requires a physical contact of the membrane or, more precisely, of a redox-active molecule or appendage (e.g., cytochromes, nanowires, membrane-bound hydrogenases) with the BES electrodes (Fig. 21.2). Without any redox-active diffusive species being involved, these membrane-bound cellular redox centers allow the electron transfer between the microbial cells and the BES electrode.

So far, the only OHRBs available in pure culture, known to perform the reductive dechlorination process with a polarized cathode serving as direct electron donor are *Geobacter lovleyi* and *Anaeromyxobacter dehalogenans* being capable of reductively dechlorinating PCE (to *cis*-DCE) and chlorophenol (to phenol), respectively, with a graphite cathode polarized at −300 mV versus SHE serving as direct electron donor (Strycharz et al. 2008, 2010). However, in spite of the extensive work currently being made to elucidate DET mechanisms especially in *Geobacter* species, the cell redox components of these two microorganisms possibly involved in the electron transfer mechanism still remain unidentified.

Although, preliminary bioelectrochemical tests (Aulenta et al., unpublished data) on a pure culture of *Dehalococcoides mccartyi* strain 195 indicated the inability of this key bacterium to use a polarized (i.e., −300 mV vs. SHE) graphite electrode as direct electron donor for the reductive dechlorination of TCE, numerous mixed culture investigations have shown bioelectrochemical dechlorination of chloroethenes beyond *cis*-DCE, also at potentials whereby H_2 generation is unfeasible. Furthermore, in most of these studies no biogenic redox mediators could be detected in the culture supernatant, thereby pointing to a DET-driven bioelectrochemical dechlorination.

As an example, a recent study analyzed by CARD-FISH the microbial composition of the cathode compartment of a bioelectrochemical reactor, continuously

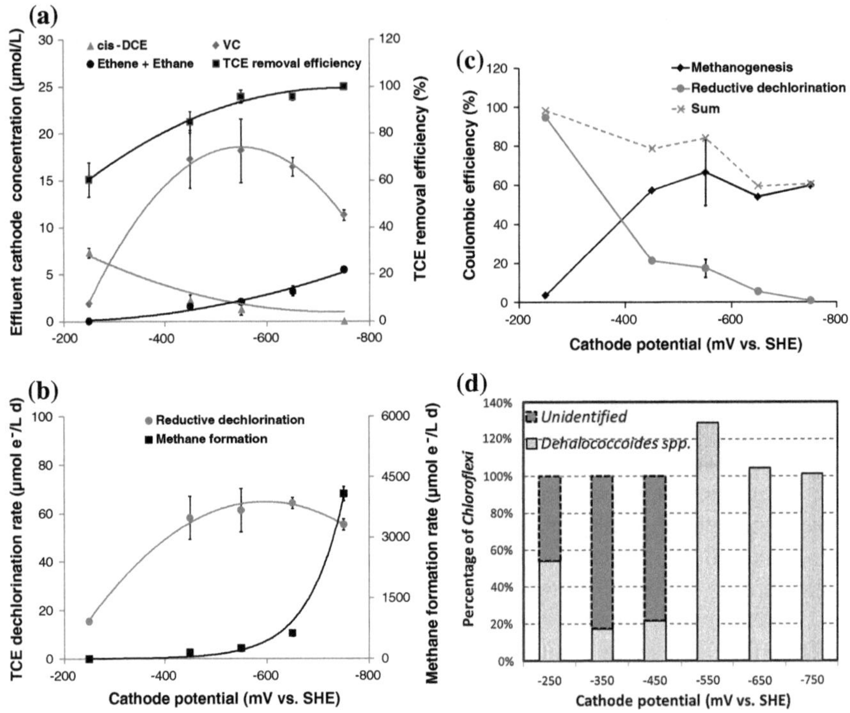

Fig. 21.3 Performance of a bioelectrochemical reactor as a function of the set cathode potential: **a** Average concentration (±1 standard deviation) of dechlorination products in the effluent of the cathodic chamber, and TCE removal efficiency; **b** Average rates (±1 standard deviation) of reductive dechlorination and methane formation; **c** Average Coulombic efficiency (±1 standard deviation) of the reductive dechlorination and methanogenesis (i.e., percentage of electric current recovered as dechlorination products or methane) as a function of the set cathode potential; **d** Percentage of total *Chloroflexi* being hybridized with the CARD-FISH probe specific for *Dehalococcoides* species. Panels **a**–**c** are reprinted with permission from Aulenta et al. (2011b). Copyright (2011) American Chemical Society. Panel **d** is reprinted with permission from Di Battista et al. (2012). Copyright (2012) Elsevier

fed with TCE, and operated at different (potentiostatically-controlled) cathode potentials, spanning from −250 to −750 mV versus SHE (Di Battista et al. 2012; Aulenta et al. 2011b).

In the system, the rate and extent of TCE dechlorination, as well as the competition for the available electrons were highly dependent on the set cathode potential (Fig. 21.3) (Aulenta et al. 2011b). When the cathode was controlled at −250 mV, no abiotic hydrogen production occurred and TCE dechlorination (predominantly to *cis*-DCE and VC), most probably sustained via DET, proceeded at an average rate of 15.5 ± 1.2 µmol e^-/L d (Fig. 21.3a, b). At this cathode, potential methanogenesis was almost completely suppressed and dechlorination accounted for 94.7 ± 0.1 % of the electric current (15.0 ± 0.8 µA) flowing in the system (Fig. 21.3c). A higher rate of TCE dechlorination (up to 64 ± 2 μmol e$^-$/L d) was achieved at cathode potentials lower than −450 mV, though in the presence of a very active methanogenesis which accounted for over 60 % of the electric current (Fig. 21.3c).

Interestingly, *Dehalococcoides* spp. was the dominant dechlorinating bacterial genus (from 65 % up to the totality of Bacteria) in the range from −550 to −750 mV, whereas it was abruptly outcompeted by other (yet unidentified) members of the *Chloroflexi* phylum, when the cathode was controlled in the range from −250 to −450 mV (Fig. 21.3d). It was argued that, the observed changes in the microbial composition of the biocathode were driven by changes in the dominant mechanisms of electron transfer to TCE: mediated by the electrolytic production of H_2 gas (in the range from −550 to −750 mV), or direct (in the range of cathode potentials from −250 to −450 mV).

21.2.3 Ecological Significance of Extracellular Electron Transfer in OHRB

Despite the capability of OHRB to function in BES is increasingly being recognized and exploited, the ecological and evolutionary significance of DET in OHRB remains poorly elucidated. One hypothesis is that, the capacity of accepting electrons directly from electrodes may derive from the capacity to use reduced soil geochemical components as "insoluble" electron donors, such as reduced iron-bearing minerals. So far, however, this hypothesis has not found confirmations by laboratory investigations. Another possibility is that in their natural habitats, OHRB have developed the ability to accept electrons from other microorganisms, serving as "extracellular" electron donors. In this context, the capacity of microorganisms to "wire up" each other and exchange metabolic electrons, a process often referred to as direct interspecies electron transfer (DIET), is increasingly being reported (Lovley 2011; Morita et al. 2011; Summers et al. 2010). So far, a number of syntrophic cultures relying on DIET have been described, with most of them involving microorganisms belonging to the *Geobacter* group (Lovley 2011). Remarkably, recent studies have demonstrated that DIET also plays a role in dechlorinating communities and that the process is greatly enhanced by the presence of electrically conductive magnetite particles (Aulenta et al. 2013). Multiple lines of evidence, including the direct visualization of microbial cells and magnetite particles via confocal scanning laser microscopy (CLSM), suggest that electrically conductive particles promoted a cooperative metabolism, based on DIET, between dechlorinating and non-dechlorinating microorganisms (Aulenta et al. 2014). It is interesting to note that, while *Desulfuromonas* spp. was found to be substantially enriched in microcosms supplemented with magnetite particles, *Dehalococcoides mccartyi* spp. was found to be markedly inhibited or outcompeted. These findings provide additional lines of evidence of lack of capability of *Dehalococcoides* to engage in DET or DIET (Aulenta et al. 2014).

21.3 Bioelectrochemical Dechlorination of Aromatics and Involved OHRBs

Besides chlorinated ethenes (whose bioelectrochemical dechlorination potential has been extensively discussed in the previous paragraphs), very few other classes of chlorinated contaminants have been studied in (lab-scale) microbial BES, with the majority of the studies conducted so far focusing on chlorinated phenols (Table 21.1).

A pure culture study was carried out by Strycharz et al. (2010) who have conducted experiments with *A. dehalogenans*, a microorganism which had been previously reported to reductively dechlorinate 2-chlorophenol to phenol using acetate as the electron donor. Interestingly, *A. dehalogenans*, could also oxidize acetate with a graphite electrode (set to +500 mV vs. SHE) as direct electron acceptor, demonstrating its ability to directly donate electrons to an electrode. When the electrode potential was switched to negative values (i.e., −300 mV vs. SHE), and acetate was removed from the medium, electrons started to flow from the electrode to microorganisms and chlorophenol was concomitantly dechlorinated to phenol, presumably via DET.

As far as mixed culture biocathodes are concerned, a recent work analyzed the performance of different biocathode materials for 4-chlorophenol dechlorination (Kong et al. 2014). A composite cathode consisting of a stainless steel basket filled with graphite granules and carbon brush yielded the best performance.

In an early study, researchers (Skadberg et al. 1999) investigated the stimulation of biological reduction of 2,6-chlorophenol using a cathodic current of 1–10 mA, in the presence of an anaerobic sludge. Electrolytic formation of H_2 on the cathode stimulated the biological dechlorination process. Higher chlorinated phenols, like pentachlorophenol, were also dechlorinated with microbial biocathodes in BES either via direct or mediated electron transfer, depending on the applied cathode potential and involved microbial composition (Zhang et al. 2014; Huang et al. 2012; Liu et al. 2013). Typically, reductive dechlorination via DET has been reported to occur at cathode potentials in the range from +200 to −300 mV versus SHE, whereas H_2-mediated dechlorination has been reported for lower (more reducing) cathode potentials (Table 21.1). Interestingly, one of these studies reported the reductive dechlorination of pentachlorophenol by a microbial biocathode utilizing immobilized solid-phase humin as a stable redox mediator. A microbial community analysis based on 16S rRNA genes showed that *Dehalobacter* and *Desulfovibrio* grew on the immobilized humin as potential dechlorinators.

In addition to chlorophenols, other researchers have attempted to use BES to enhance the reductive dechlorination of polychlorobiphenyls (PCBs) (Chun et al. 2013) or tetrachlorobenzene (Sun et al. 2010). In both cases, due to the hydrophobic nature of these compounds, the treatment primarily addressed contaminated sediments and involved the electrolytic generation of H_2 (and O_2) at the surface of carbon electrodes to drive the reductive (and oxidative) dechlorination process. In particular, in the PCB study, the bioelectrochemical stimulation resulted in the enrichment of anaerobic dechlorinating *Chloroflexi*.

Table 21.1 Overview of BES utilized for the treatment of chlorinated contaminants (other than chlorinated aliphatic hydrocarbons)

Chlorinated contaminant	End-product(s)	Cathode potential (mV vs. SHE)/material	Microorganisms/inoculum	Putative electron transfer mechanism	Dechlorination rate	Refs.
2-Chlorophenol	Phenol	−300/graphite	*Anaeromyxobacter dehalogenans*	Direct	40 μmol/Ld	Strycharz et al. (2010)
4-Chlorophenol	Phenol	−900/stainless steel-graphite	Anaerobic sludge	N.R.	0.04 1/h[a]	Kong et al. (2014)
2,6-dichlorophenol	2-Chlorophenol	N.R./stainless steel	Anaerobic sludge	Mediated (via H_2)	N.R.	Skadberg et al. (1999)
Pentachlorophenol	N.R.	−400 mV/graphite	*Proteobacteria*; *Bacteroidetes*, *Firmicutes*	Direct	19.2 μmol/L d	Liu et al. (2013)
Pentachlorophenol	Lower chlorinated phenols	+100 to +200/carbon felt	Domestic wastewater	Direct	23.7 μmol/L d	Huang et al. (2012)
Pentachlorophenol	Phenol	−500 mV/graphite	*Dehalobacter/Desulfovibrio*	Mediated (via immobilized humin)	11.6 μmol/L d (or 0.36 μmol Cl^-/cm^2 d)	Zhang et al. (2014)
Aroclor 1242	Lower chlorinated congeners	1.5–3.0 V (applied voltage)/titanium	*Chloroflexi*	Mediated (via H_2)	N.R.	Chun et al. (2013)
1,2,3,5-tetrachlorobenzene	N.R.	1.5–5.0 V (applied voltage)/graphite	N.R.	Mediated (via H_2)	N.R.	Sun et al. (2010)

N.R. Not reported
[a]First order kinetic constant

21.4 Perspectives for Field Scale Application of BES

Although microbial biocathodes are increasingly being investigated worldwide, there is, so far, virtually no information regarding full-scale applications and serious concerns have actually been raised regarding their scalability from the laboratory to the field.

Although the scalability of any (bio)electrochemical system is undoubtedly challenging, it should be considered that a number of well-established electrochemical remediation technologies are already available in the market, and are widely applied for in situ treatment of a variety of inorganic and organic pollutants (Trombly 1994). These technologies typically apply relatively low-voltage direct current, by means of electrode arrays, to favor the electrokinetic movement of contaminants from soils into "treatment zones" where they are removed from the water by adsorption, immobilization, or (bio)degradation. Electric current flows between pairs of anodes and cathodes placed in the ground, either in the vadose zone or in the aquifer. Current density is generally on the order of few Amperes per square meter (A/m^2) or an electric potential difference on the order of a few Volts per centimeter across electrodes. One example is the "Lasagna Process" (the name derives from the layered appearance of electrodes and treatment zones) (Ho et al. 1995), originally proposed by Monsanto and further developed by a consortium of industry (DuPont, General Electrics, and Monsanto) in collaboration with US Federal Agencies (EPA; DOE), which provided large-scale demonstrations of the feasibility of the technology for in situ treatment of a variety of contaminants, including TCE (Ho et al. 1999a, b).

In theory, similar configurations could also be adopted for bioelectrochemical remediation systems. Along this line, one of the first configurations being proposed in the literature, for in situ treatment of chloroethenes-contaminated groundwater, is shown in Fig. 21.4.

The electrodes (cathode and anode), consisting of conductive and non-corrosive carbon materials (e.g., graphite granules), are placed within the aquifer as to form permeable reactive barriers which intercepts (and treats) the contamination plume. OHRB reductively dechlorinate chloroethenes using the cathode as electron donor. Products of incomplete reductive dechlorination, such as *cis*-DCE and VC, migrates downgradient where, in principle can be oxidized by microorganisms which use the anode as direct or indirect (i.e., via O_2 generation) electron acceptor. In principle, besides lower chlorinated hydrocarbons, other pollutants (e.g., petroleum hydrocarbons) could also be oxidized at the anode. If needed, the potential generated by the reaction can be augmented using an external power supply.

This sequential (cathodic–anodic) treatment has, however, some specific limitations. Indeed, the distance between cathode and anode should be kept as small as possible to minimize voltage and energy losses due the Ohmic resistance of groundwater. As an example, assuming a typical groundwater conductivity of 1000 μS/cm (Pous et al. 2013) and a current density of 1 A/m^2 of cathode surface area, the Ohmic loss would be 10 V for each meter of distance between cathode

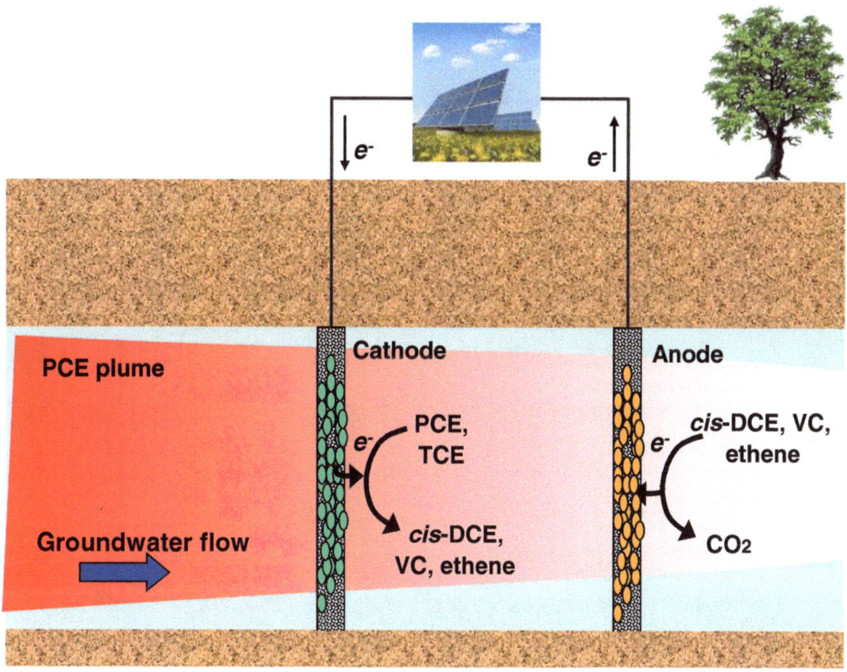

Fig. 21.4 Conceptual illustration of a permeable reactive barrier-like bioelectrochemical system for in situ treatment of chlorinated solvents. Redrafted from Aulenta et al. (2009). Journal of Chemical Technology and Biotechnology 84 (6):864–870. Copyright (2009) Society of Chemical Industry

and anode (Rozendal et al. 2008). Based on this calculation, it is apparent that electrode spacing higher than a few meters would ultimately result in unacceptable energy losses. On the other hand, if the bioremediation process is driven by water electrolysis (H_2 generation at the cathode and oxygen generation at the anode), maintaining a certain spacing between electrodes is essential to prevent the back diffusion of oxygen to the cathode which could lead to the inhibition of anaerobic reductive dechlorination. In principle, a better scenario would probably result from the stimulation of reductive and oxidative dechlorination via DET since lower electric currents would be involved, as well as, no (or minimal) H_2 and O_2 would be produced. To overcome these limitations, a number of alternative configurations (to the permeable reactive barrier-like one) are presently being considered, whereby the distance between electrodes is kept as small as possible without adversely affecting the reductive dechlorination performance.

Besides the treatment of contaminated groundwater, BESs have also been proposed for the treatment of sediments contaminated by hydrophobic chlorinated compounds, such as chlorobenzenes or PCBs. In this context, in a previous work, researchers (Sun et al. 2010) proposed to utilize geotextiles with polarized carbon

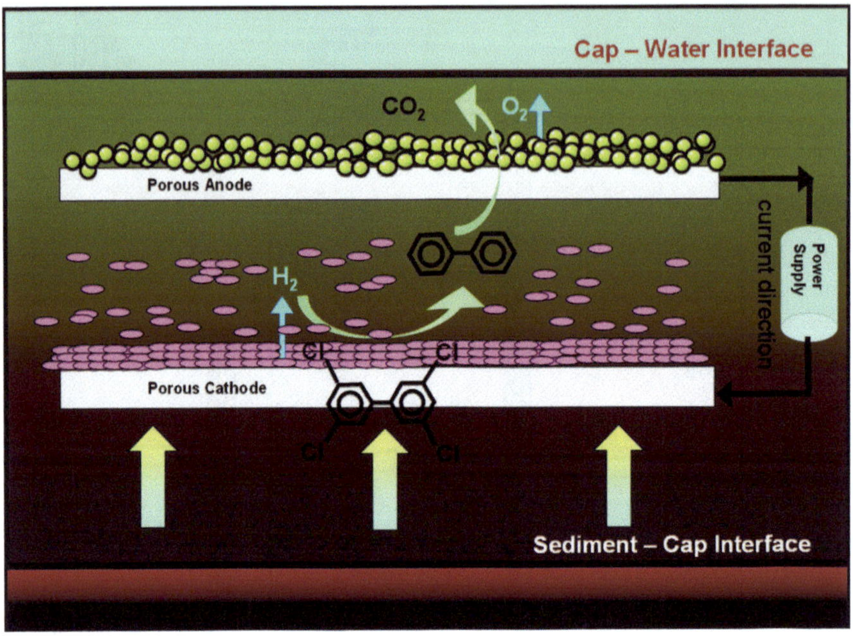

Fig. 21.5 Conceptual model for an electrode-based reactive sediment cap. Reprinted with permission from Sun et al. (2010). Copyright (2010) American Chemical Society

electrodes in a sediment cap to impose a desired electrical potential gradient to stimulate biodegradation of contaminants within the cap (Fig. 21.5).

In brief, the electrodes are placed perpendicular to contaminant transport through the cap, and polarized at low potentials to accelerate contaminant degradation by, (1) rapidly establishing and maintaining a redox gradient within the cap, which may be varied in real time, and (2) supplying the cap with electron acceptor and electron donor to stimulate microbial growth. As contaminants migrate from the deeper sediments, they will be exposed to the electrode biobarrier maintained by microbial growth and respiration, which is scalable in both magnitude and thickness of the respective reducing and oxidizing zones created by the electrodes.

This configuration enables the application of low voltage (<5 V) and low current (in the range of mA) to large area electrodes to enhance and control appropriate microbial activity in a thin horizontal layer within a cap. In principle, an electrode-based biobarrier for reactive sediment capping has the potential to address the unique challenges imposed by the sediment environment. First, biodegradation processes in sediment caps are commonly limited by the availability of either electron donor or electron acceptor, depending on the contaminant. The electrode cap will supply both spatially and temporally controlled electron donor and acceptor. Second, complete mineralization of some contaminants (e.g., PCBs and some chlorinated solvents) only occurs through sequential reduction and oxidation, which may not develop under natural conditions. Through the electrode imposed and microbially maintained redox gradient, residence times through reductive and oxidative conditions for specific contaminant transformation may be engineered. Third, the common observance of contamination in mixtures which degrade or detoxify under disparate redox conditions confounds single-approach remedial or sequestration designs. As before, an electrode stimulated biobarrier, redox gradient may be engineered to address the specific contaminants of concern on a site by site basis.

21.5 Outlook

The exciting discovery of the ability of OHRB to engage in EET with solid-state electrodes has recently prompted investigations on the application of BES for soil and groundwater remediation. BES offer unique opportunities to steer and control microbial dechlorination reactions taking place at the electrodes by fine-tuning the electrode(s) potential or electric current. Due to these specific features, they hold a great promise to increase the reliability and sustainability of in situ bioremediation systems and in turn to promote their wider application and public acceptance. For BES to become a marketable remediation technology, major research efforts and extensive field testing are, however, still necessary. These will primarily have to address (1) fundamental aspects of cathodic EET mechanisms in OHRB which remain largely unknown; (2) identification of scalable BES configurations appropriate for field scale application. Finally, besides subsurface remediation, direct or mediated EET processes in OHRB could find noticeable applications also in the field of subsurface sensing. Indeed, in BES current generation is directly linked to substrate concentration/availability and microbial respiratory activity (Aulenta et al. 2011a); therefore, upon miniaturization these devices could be used to in situ monitoring of contaminants concentration as well as to evaluating bioremediation potential (Sun et al. 2015; Friedman et al. 2012).

References

Aulenta F, Gossett JM, Papini MP, Rossetti S, Majone M (2005) Comparative study of methanol, butyrate, and hydrogen as electron donors for long-term dechlorination of tetrachloroethene in mixed anerobic cultures. Biotechnol Bioeng 91(6):743–753

Aulenta F, Catervi A, Majone M, Panero S, Reale P, Rossetti S (2007a) Electron transfer from a solid-state electrode assisted by methyl viologen sustains efficient microbial reductive dechlorination of TCE. Environ Sci Technol 41(7):2554–2559. doi:10.1021/es0624321

Aulenta F, Pera A, Rossetti S, Papini MP, Majone M (2007b) Relevance of side reactions in anaerobic reductive dechlorination microcosms amended with different electron donors. Water Res 41(1):27–38

Aulenta F, Canosa A, Majone M, Panero S, Reale P, Rossetti S (2008a) Trichloroethene dechlorination and H_2 evolution are alternative biological pathways of electric charge utilization by a dechlorinating culture in a bioelectrochemical system. Environ Sci Technol 42(16):6185–6190

Aulenta F, Reale P, Catervi A, Panero S, Majone M (2008b) Kinetics of trichloroethene dechlorination and methane formation by a mixed anaerobic culture in a bio-electrochemical system. Electrochim Acta 53(16):5300–5305

Aulenta F, Canosa A, Reale P, Rossetti S, Panero S, Majone M (2009) Microbial reductive dechlorination of trichloroethene to ethene with electrodes serving as electron donors without the external addition of redox mediators. Biotechnol Bioeng 103(1):85–91

Aulenta F, Di Maio V, Ferri T, Majone M (2010) The humic acid analogue antraquinone-2,6-disulfonate (AQDS) serves as an electron shuttle in the electricity-driven microbial dechlorination of trichloroethene to *cis*-dichloroethene. Bioresour Technol 101(24):9728–9733

Aulenta F, Ferri T, Nicastro D, Majone M, Papini MP (2011a) Improved electrical wiring of microbes: anthraquinone-modified electrodes for biosensing of chlorinated hydrocarbons. New Biotechnol 29(1):126–131

Aulenta F, Tocca L, Verdini R, Reale P, Majone M (2011b) Dechlorination of trichloroethene in a continuous-flow bioelectrochemical reactor: effect of cathode potential on rate, selectivity, and electron transfer mechanisms. Environ Sci Technol 45(19):8444–8451

Aulenta F, Rossetti S, Amalfitano S, Majone M, Tandoi V (2013) Conductive magnetite nanoparticles accelerate the microbial reductive dechlorination of trichloroethene by promoting interspecies electron transfer processes. Chemsuschem 6(3):433–436

Aulenta F, Fazi S, Majone M, Rossetti S (2014) Electrically conductive magnetite particles enhance the kinetics and steer the composition of anaerobic TCE-dechlorinating cultures. Process Biochem. doi:10.1016/j.procbio.2014.09.015

Cervantes FJ, Martinez CM, Gonzalez-Estrella J, Marquez A, Arriaga S (2013) Kinetics during the redox biotransformation of pollutants mediated by immobilized and soluble humic acids. Appl Microbiol Biotechnol 97(6):2671–2679

Chun CL, Payne RB, Sowers KR, May HD (2013) Electrical stimulation of microbial PCB degradation in sediment. Water Res 47(1):141–152

Di Battista A, Verdini R, Rossetti S, Pietrangeli B, Majone M, Aulenta F (2012) CARD-FISH analysis of a TCE-dechlorinating biocathode operated at different set potentials. New Biotechnol 30(1):33–38

Fennell DE, Stover MA, Zinder SH, Gossett JM (1995) Comparison of alternative electron donors to sustain PCE anaerobic reductive dechlorination. Bioremediat Chlorinated Solvents 3(4):9–16

Friedman ES, Rosenbaum MA, Lee AW, Lipson DA, Land BR, Angenent LT (2012) A cost-effective and field-ready potentiostat that poises subsurface electrodes to monitor bacterial respiration. Biosens Bioelectron 32(1):309–313

Harnisch F, Aulenta F, Schroeder U (2011) Microbial fuel cells and bioelectrochemical systems: industrial and environmental biotechnologies based on extracellular electron transfer. In: Moo-Young M (ed) Comprehensive biotechnology, 2nd edn. Academic Press, Burlington, pp 643–659

Ho SV, Sheridan PW, Athmer CJ, Heitkamp MA, Brackin JM, Weber D, Brodsky PH (1995) Integrated in-situ soil remediation technology—the lasagna process. Environ Sci Technol 29(10):2528–2534

Ho SV, Athmer C, Sheridan PW, Hughes BM, Orth R, McKenzie D, Brodsky PH, Shapiro A, Thornton R, Salvo J, Schultz D, Landis R, Griffith R, Shoemaker S (1999a) The lasagna technology for in situ soil remediation. 1. Small field test. Environ Sci Technol 33(7):1086–1091

Ho SV, Athmer C, Sheridan PW, Hughes BM, Orth R, McKenzie D, Brodsky PH, Shapiro AM, Sivavec TM, Salvo J, Schultz D, Landis R, Griffith R, Shoemaker S (1999b) The lasagna technology for in situ soil remediation. 2. Large field test. Environ Sci Technol 33(7):1092–1099

Huang LP, Regan JM, Quan X (2011) Electron transfer mechanisms, new applications, and performance of biocathode microbial fuel cells. Bioresour Technol 102(1):316–323

Huang LP, Chai XL, Quan X, Logan BE, Chen GH (2012) Reductive dechlorination and mineralization of pentachlorophenol in biocathode microbial fuel cells. Bioresour Technol 111:167–174

Kassenga GR, Pardue JH (2006) Effect of competitive terminal electron acceptor processes on dechlorination of *cis*-1,2-dichloroethene and 1,2-dichloroethane in constructed wetland soils. FEMS Microbiol Ecol 57(2):311–323

Kong FY, Wang AJ, Ren HY (2014) Improved 4-chlorophenol dechlorination at biocathode in bioelectrochemical system using optimized modular cathode design with composite stainless steel and carbon-based materials. Bioresour Technol 166:252–258

Liu D, Lei LC, Yang B, Yu QN, Li ZJ (2013) Direct electron transfer from electrode to electrochemically active bacteria in a bioelectrochemical dechlorination system. Bioresour Technol 148:9–14

Löffler FE, Tiedje JM, Sanford RA (1999) Fraction of electrons consumed in electron acceptor reduction and hydrogen thresholds as indicators of halorespiratory physiology. Appl Environ Microbiol 65(9):4049–4056

Lohner ST, Becker D, Mangold KM, Tiehm A (2011) Sequential reductive and oxidative biodegradation of chloroethenes stimulated in a coupled bioelectro-process. Environ Sci Technol 45(15):6491–6497

Lovley DR (2008) Extracellular electron transfer: wires, capacitors, iron lungs, and more. Geobiology 6(3):225–231

Lovley DR (2011) Reach out and touch someone: potential impact of DIET (direct interspecies energy transfer) on anaerobic biogeochemistry, bioremediation, and bioenergy. Rev Environ Sci Bio-Technol 10(2):101–105

Luijten MLGC, Roelofsen W, Langenhoff AAM, Schraa G, Stams AJM (2004a) Hydrogen threshold concentrations in pure cultures of halorespiring bacteria and at a site polluted with chlorinated ethenes. Environ Microbiol 6(6):646–650

Luijten MLGC, Weelink SAB, Godschalk B, Langenhoff AAM, van Eekert MHA, Schraa G, Stams AJM (2004b) Anaerobic reduction and oxidation of quinone moieties and the reduction of oxidized metals by halorespiring and related organisms. FEMS Microbiol Ecol 49(1):145–150

Morita M, Malvankar NS, Franks AE, Summers ZM, Giloteaux L, Rotaru AE, Rotaru C, Lovley DR (2011) Potential for direct interspecies electron transfer in methanogenic wastewater digester aggregates. Mbio 2(4):e00159-11

Morris RM, Sowell S, Barofsky D, Zinder S, Richardson R (2006) Transcription and mass-spectroscopic proteomic studies of electron transport oxidoreductases in *Dehalococcoides ethenogenes*. Environ Microbiol 8(9):1499–1509. doi:10.1111/j.1462-2920.2006.01090.x

Nijenhuis I, Zinder SH (2005) Characterization of hydrogenase and reductive dehalogenase activities of Dehalococcoides ethenogenes strain 195. Appl Environ Microbiol 71(3):1664–1667

Pous N, Puig S, Coma M, Balaguer MD, Colprim J (2013) Bioremediation of nitrate-polluted groundwater in a microbial fuel cell. J Chem Technol Biotechnol 88(9):1690–1696

Rosenbaum M, Aulenta F, Villano M, Angenent LT (2011) Cathodes as electron donors for microbial metabolism: which extracellular electron transfer mechanisms are involved? Bioresour Technol 102(1):324–333

Rozendal RA, Hamelers HVM, Rabaey K, Keller J, Buisman CJN (2008) Towards practical implementation of bioelectrochemical wastewater treatment. Trends Biotechnol 26(8):450–459

Skadberg B, Geoly-Horn SL, Sangamalli V, Flora JRV (1999) Influence of pH, current and copper on the biological dechlorination of 2,6-dichlorophenol in an electrochemical cell. Water Res 33(9):1997–2010

Smatlak CR, Gossett JM, Zinder SH (1996) Comparative kinetics of hydrogen utilization for reductive dechlorination of tetrachloroethene and methanogenesis in an anaerobic enrichment culture. Environ Sci Technol 30(9):2850–2858

Strycharz SM, Woodard TL, Johnson JP, Nevin KP, Sanford RA, Loffler FE, Lovley DR (2008) Graphite electrode as a sole electron donor for reductive dechlorination of tetrachlorethene by *Geobacter lovleyi*. Appl Environ Microbiol 74(19):5943–5947

Strycharz SM, Gannon SM, Boles AR, Franks AE, Nevin KP, Lovley DR (2010) Reductive dechlorination of 2-chlorophenol by *Anaeromyxobacter dehalogenans* with an electrode serving as the electron donor. Environ Microbiol Rep 2(2):289–294

Summers ZM, Fogarty HE, Leang C, Franks AE, Malvankar NS, Lovley DR (2010) Direct exchange of electrons within aggregates of an evolved syntrophic coculture of anaerobic bacteria. Science 330(6009):1413–1415

Sun M, Yan F, Zhang RL, Reible DD, Lowry GV, Gregory KB (2010) Redox control and hydrogen production in sediment caps using carbon cloth electrodes. Environ Sci Technol 44(21):8209–8215

Sun JZ, Kingori GP, Si RW, Zhai DD, Liao ZH, Sun DZ, Zheng T, Yong YC (2015) Microbial fuel cell-based biosensors for environmental monitoring: a review. Water Sci Technol 71(6):801–809

Trombly J (1994) Electrochemical remediation takes to the field. Environ Sci Technol 28(6):A289–A291

Villano M, De Bonis L, Rossetti S, Aulenta F, Majone M (2011) Bioelectrochemical hydrogen production with hydrogenophilic dechlorinating bacteria as electrocatalytic agents. Bioresour Technol 102(3):3193–3199

Zhang RL, Lu XX, Reible DD, Jiao GZ, Qin SY (2013) Cathodic hydrogen as electron donor in enhanced reductive dechlorination. Chin J Chem Eng 21(12):1386–1390

Zhang DD, Zhang CF, Li ZL, Suzuki D, Komatsu DD, Tsunogai U, Katayama A (2014) Electrochemical stimulation of microbial reductive dechlorination of pentachlorophenol using solid-state redox mediator (humin) immobilization. Bioresour Technol 164:232–240

Chapter 22
Current and Future Bioremediation Applications: Bioremediation from a Practical and Regulatory Perspective

Robert J. Steffan and Charles E. Schaefer

Abstract Chlorinated solvents have been a primary focus of the remediation industry since the 1980s, and many remedial technologies have been developed, tested, and applied to remove these constituents from contaminated aquifers. The relative ease of stimulating organohalide-respiring bacteria in situ and the availability of low cost electron donor substrates and effective bioaugmentation cultures have allowed in situ bioremediation technologies to be applied successfully at thousands of sites around the world. Typically, the success of the remediation is dependent more on the site characteristics (e.g., geochemistry, geology, hydrology, contaminant concentration, etc.) than the fidelity of the microbes. As we begin to address the most challenging contaminated sites that remain to be remediated, including those with free product contamination, complicated geologies (e.g., low permeability soils or fractured rock), or complex contaminant mixtures, in situ bioremediation may not be the sole technology applied at these sites but it will likely be an important component of many remedies. Therefore, fundamental understandings of microbiology and the development of novel application approaches remain essential to ensure continued success in remediation of the most difficult chlorinated solvent-contaminated sites.

22.1 Introduction

Chlorinated ethenes and ethanes, including tetrachloroethylene (PCE), trichloroethylene (TCE), 1,1,1-trichloroethane (1,1,1-TCA), 1,1,2-trichloroethane (1,1,2-TCA), and dichloroethane (DCA), are collectively referred to as chlorinated solvents. They have been used extensively as industrial solvents and cleaning agents for many decades, and they have been used commercially at all levels of industry; from small

R.J. Steffan (✉) · C.E. Schaefer
CB&I Federal Services, LLC, Lawrenceville, NJ 08648, USA
e-mail: Rob.Steffan@CBIFederalServices.com

L. Adrian and F.E. Löffler (eds.), *Organohalide-Respiring Bacteria*,
DOI 10.1007/978-3-662-49875-0_22

local drycleaners and machine shops, to large-scale manufacturing and at large government installations. Their chemical properties, widespread use, and improper disposal practices have led to them becoming the most common groundwater contaminants throughout the developed world. Because of this, they also became a primary focus of the remediation industry, from individual local consultants to large international remediation firms, and many remedial technologies have been developed and applied to remove chlorinated solvents from contaminated aquifers.

Early developments in remediation technologies for chlorinated solvents included the use of pump-and-treat approaches where groundwater was pumped to the surface and the chlorinated solvent pollutants were removed by adsorption onto activated carbon or by air stripping. The treated water was then disposed of or discharged into an acceptable receptor (e.g., a local sewage treatment system). Although effective, experience showed that because of the low solubility and adsorption of chlorinated solvents to aquifer solids, large volumes of the contaminated groundwater had to be recovered and treated, and decades of operation were often required to achieve treatment goals (Rivett et al. 2006). Other approaches such as air sparging (Bass et al. 2000), which involved injecting large volumes of air into the groundwater to strip the contaminants, also were widely applied and often effective, but high energy consumption and the inability to apply the technology in many aquifers because of site geology limited the application of this approach.

In the early 1990s bioremediation emerged as a potentially viable alternative for treating chlorinated solvent-contaminated aquifers. Early work involved the development of aerobic cometabolic treatment approaches that employed monooxygenase-producing microbes that fortuitously degraded the contaminants by using enzymes involved in metabolizing a primary substrate. Common approaches used microorganisms that grew on methane (Wilson and Wilson 1985), toluene (McCarty et al. 1998; Steffan et al. 1999), or propane (Kim et al. 2004), and the necessary biocatalyst could be enriched from the indigenous microbiota (i.e., biostimulation), or appropriate microorganisms could be added to the aquifer (i.e., bioaugmentation). Issues of competitive inhibition, product toxicity, insufficient transformation capacity and the need to maintain oxic conditions in aquifers created challenges for these technologies, but in some cases effective results were achieved (for review see, Frascari et al. 2015; Semprini 1997).

The discovery of microbes that could completely dechlorinate chlorinated ethenes under anaerobic conditions (Freedman and Gossett 1989), and the subsequent identification of the responsible biocatalysts, especially *Dehalococcoides* mccartyi (*Dhc*) (Maymó-Gatell et al. 1997, 2001; Löffler et al. 2013), led to a revolutionary change in the way bioremediation was applied to treat chlorinated solvent contamination. Anaerobic processes are relatively easy, and inexpensive, to apply (Lee et al. 1998), and in many cases the necessary biocatalysts already existed in aquifers (Hendrickson et al. 2002). These organisms could be stimulated in situ simply by the addition of an organic compound (termed electron donor) that could be fermented to produce the H_2 and acetate needed to support their growth and activity (Fig. 22.1). Advances in anaerobic treatment also led to the development, and ultimately the commercial availability, of bioaugmentation

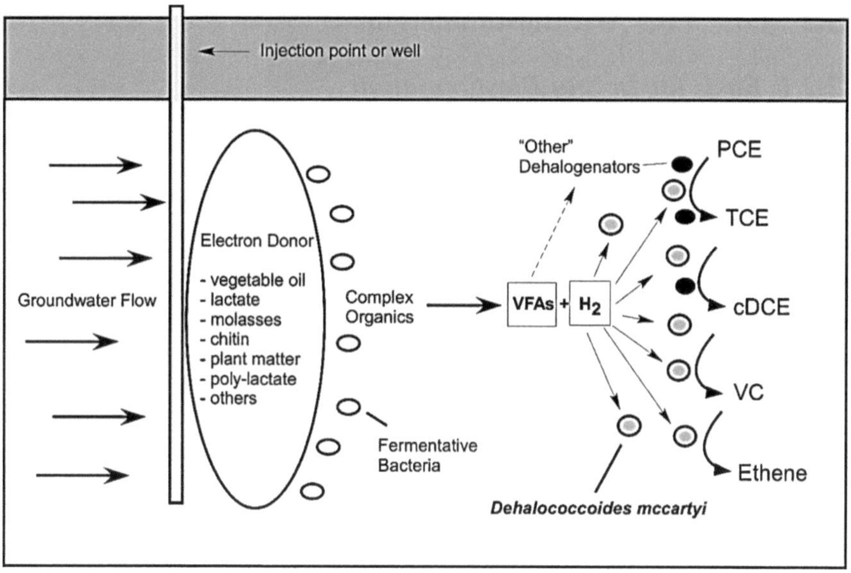

Fig. 22.1 Schematic of the most common approach for in situ biotreatment of chlorinated solvent contamination. Complex organic substrates (i.e., "electron donors") are injected into the aquifer where they are fermented to volatile fatty acids and H_2 by fermentative bacteria. The H_2 is then used as an electron donor by organohalide-respiring bacteria, including *Dhc* that reductively dechlorinate the contaminants to innocuous products like ethene

cultures that contained *Dhc* strains (Major et al. 2002; Vainberg et al. 2009) that could be applied to sites where the native microbial population did not promote rapid or complete reductive dechlorination of the contaminants.

In situ bioremediation, supported by two decades of fundamental research on organohalide-respiring bacteria, has now been applied to treat thousands of chlorinated solvent-contaminated sites. The relative ease of stimulating organohalide-respiring bacteria in situ, and the availability of low cost electron donor substrates and effective bioaugmentation cultures, has allowed the technology to be applied successfully even by remediation practitioners with little or no microbiology training. Typically, the success of the remediation is dependent more on the site characteristics (e.g., geochemistry, geology, hydrology, contaminant concentration, etc.) than the fidelity of the microbes. As we begin to address the most challenging contaminated sites that remain to be remediated, including those with free product contamination, complicated geologies (e.g., low permeability soils or fractured rock), or complex contaminant mixtures (NRC 2013), fundamental understandings of microbiology, and the development of novel application approaches remain essential to ensure success. In this chapter we discuss some of the current challenges of applying anaerobic bioremediation approaches for remediating aquifers contaminated with chlorinated ethenes, and we provide some insights into future remediation challenges.

22.2 State of the Art: Biostimulation

22.2.1 Basis for In Situ Biostimulation

Although statistics are not available, in situ biostimulation is likely the most widely used approach for remediating chlorinated solvent-contaminated aquifers. As discussed above and in much detail in earlier chapters in this volume, the predominant biodegradation pathway for chlorinated ethenes under anaerobic conditions is via bacterial-mediated reductive dechlorination (Fig. 22.1). During this process, PCE is typically dehalogenated to TCE, *cis*-1,2-dichloroethene (*c*DCE), vinyl chloride (VC), and finally the desired end product, ethene. In the practice of site remediation, therefore, the presence of any of the identified daughter products is used as an indicator of ongoing processes at the site. In some cases, site data indicated that the ongoing rate of contaminant degradation is sufficient to prevent risks to human health, and aggressive action is not needed to enhance the rate of contaminant degradation. In these cases, the contaminants may be allowed to attenuate naturally, whereby the remedy is referred to as "monitored natural attenuation" (McGuire et al. 2004; Weidemeier et al. 1998). In other cases, there are indicators of complete degradation (e.g., production of ethene), or incomplete degradation (e.g., accumulation of *c*DCE or VC), and steps are taken to stimulate the indigenous microbial population to achieve effective site remediation.

Notably, *Dhc* require chlorinated compounds as obligate electron acceptors (He et al. 2003; Maymó-Gatell et al. 1997, 2001; Löffler et al. 2013), and their obligate electron donor for chlorinated solvent reduction is H_2. In the *Dhc* strains studied to date, acetate serves as a carbon source for growth of the organisms (He et al. 2003; Maymó-Gatell et al. 1997). Consequently, remedial technologies developed to stimulate the growth and activity of these organisms in situ (i.e., biostimulation) rely on the addition of chemicals or materials that support these nutritional needs. In most cases, the materials added are selected to provide a source of H_2, and thus they are termed "electron donors". In some cases other additives including inorganic nutrients (i.e., nitrogen and phosphorous) or organic nutrients like yeast extracts also are added to enhance in situ activity (Schaefer et al. 2010a).

22.2.2 Electron Donors for In Situ Biostimulation

The most common approach for supplying metabolic H_2 for reductive dechlorination is through the addition of a carbon source that can be fermented to H_2. Early studies utilized sodium benzoate (Beeman and Bleckmann 2002), molasses (Wu et al. 1998), whey (Semkiw and Barcelona 2011) or sodium lactate (Lendvay et al. 2003; Schaefer et al. 2010a), and even toluene was shown to support reductive dechlorination (Sewell and Gibson 1991). Other applications have used direct injection of H_2 (Ma et al. 2006), and more recent activities have evaluated electrolytic production of H_2 in situ (Lohner and Tiehm 2009; Lohner et al. 2011; Steffan and Sewell 2011; Zhang et al. 2001). Currently, a variety of commercial electron donor products are marketed to support in situ bioremediation of chlorinated solvents. These include water-soluble substrates like sodium lactate, molasses, alcohols and glycerol, and a host of insoluble substrates including emulsified vegetable oils, polylactates, and plant-based materials. Many of the commercially available substrate formulations are routinely modified by their developers to improve their performance in the field, or to address particular common issues like pH buffering, the need to better distribute the substrate, or the need to rapidly initiate fermentation, and H_2 formation.

22.2.3 Selection of an Appropriate Electron Donor

The selection of an appropriate electron donor is based primarily on site conditions, remedial goals, and cost. At sites where rapid remediation is a goal, soluble substrates often are used because they can be quickly distributed in situ and they are rapidly fermented to support dehalogenation. The cost of using a soluble fast-fermenting substrate can be high, however, if repeated substrate applications are required. At other sites where prolonged treatment is needed because of the large mass of contaminants present, or because of the low permeability of the aquifer, insoluble electron donors can be used because they slowly release H_2 as the large organic polymers are slowly fermented (Borden and Rodriguez 2007; Borden et al. 2007). The higher cost of the substrates themselves may be offset by the need to add substrate infrequently, thereby reducing overall deployment costs. Because of all of these considerations, calculating the long term cost of the selected electron donor is complicated, and it requires a detailed site conceptual model and clear remedial goals. Many of the vendors who provide these substrates to remediation practitioners sell multiple electron donor products, and many of the vendors are highly experienced and able to provide specific insights and examples to help determine which substrate type is best for a particular application.

22.2.4 Electron Donor Application and Distribution

The success of in situ bioremediation for chlorinated solvents is often dependent upon the ability to effectively distribute electron donor in the subsurface. Distribution issues consist of both the ability to deliver electron donor radially away from the injection point(s), as well as the ability to deliver electron donor into low (or, lower) permeability zones. For soluble electron donors (e.g., lactate, ethanol), distribution along the groundwater flow path is typically proportional to the injection volume. Borden (2007) showed that this also was true for emulsified vegetable oil, and that there was little impact of injection flow rate or water dilution on the oil distribution. Delivery of short-lived electron donors (lactate, ethanol) is typically performed in multiple batches, or via continuous/intermittent injections. Injection of long-lived electron donors (e.g., vegetable oils) may occur only once, or at periodic (1–5 year) intervals. Enhancement of the hydraulic gradient and flow via groundwater pumping has been used to facilitate soluble electron donor distribution following injection, providing a means to enhance migration of the electron donor through the aquifer before the substrate is consumed (Schaefer et al. 2010a; Major et al. 2002; Ellis et al. 2000a, b).

When designing and implementing in situ bioremediation for chlorinated solvents, it is important that electron donor is sufficiently distributed along the groundwater flow path to ensure a sufficient treatment residence time (i.e., the organohalide-respiring *Dhc* strains are not H_2-limited) so that complete dechlorination occurs. Thus, whether by increasing the injection volume, using groundwater recirculation, or by adding additional injection points, the effective residence time of the target chlorinated ethenes within the treatment zone (defined by the presence of sufficient substrate) needs to be sufficiently large to ensure complete biodegradation of the target pollutants. Specific design issues related to injection well spacing, electron donor distribution along the groundwater flowpath (which determines the treatment residence time), electron donor dosage, and the required frequency of injection are best determined (or, at least estimated) prior to full-scale implementation. Groundwater fate and transport modeling (c.f., Clement 1997; Clement et al. 1997) can be used to determine electron donor distribution, as well as the persistence of the electron donor, in the subsurface. Laboratory bench scale or small-scale field pilot testing also can be used to provide site-specific assessments of electron donor longevity and distribution. This information can then be used to design the required electron donor distribution approach.

The lifetime of a rapidly fermentable substrate such as lactate is typically very short. In these cases, lactate is often not distributed far beyond the injection point. However, lactate fermentation by-products (i.e., slowly fermentable organic acids and hydrogen) can persist and be observed significant distances from the injection well (Schaefer et al. 2010a; Scheutz et al. 2008). Using groundwater recirculation to distribute amendments, propionate, acetate, and hydrogen were effectively distributed up to 25 meters from the injection well (Scheutz et al. 2008).

Because most aquifers are heterogeneous, care also must be taken to ensure electron donor is properly distributed vertically across the target treatment interval. It is noted that even in aquifers that appear (based on visual inspection of intact soil cores) to be homogenous, hydraulic conductivities can vary by 1–2 orders of magnitude along a vertical transect. To ensure lower permeability zones are not by-passed, injection intervals are typically limited to 5 or 10 ft (1.5–3 m). If direct push injection techniques are applicable for the site, amendments can be directly injected in 5 or 10 ft discrete intervals at a single borehole location. Otherwise, nested wells with well screens located at different depth intervals, can be used to target the delivery intervals. If a groundwater recirculation approach is employed, there is less concern about minimizing the injection interval because enhanced hydraulic flow will facilitate mixing and dispersion of electron donor into the lower permeability intervals, but prolonged operation of recirculation systems often result in well screen fouling and other maintenance issues.

For insoluble electron donors, fracturing techniques, both pneumatic and hydraulic, have been employed to distribute amendments. The advantage of hydraulic fractures is that proppant materials can be added to sustain the fractures for subsequent injections (Strong et al. 2004). The ability to reinject electron donor can be important if the electron donor is rapidly consumed (e.g., lactate), or if the chlorinated solvent source is persistent (e.g., DNAPL sources, back-diffusion from low permeability materials).

One of the greatest challenges with electron donor distribution is the presence of low permeability zones, such as clays and silts, in the subsurface. Pneumatic and hydraulic fracturing have been employed to deliver electron donor in low permeability zones such as silt, clay, and weathered bedrock (Lee et al. 2013; Swift et al. 2012; Christiansen et al. 2010). However, challenges with these approaches have been noted, due to limited or unpredictable propagation and distribution of the fractures (Christiansen et al. 2010). Analyses of clay cores following hydraulic fracturing and subsequent electron donor delivery showed that electron donor migrated only 5–6 cm into the adjacent clay matrix (Scheutz et al. 2010), thereby emphasizing the need for a high fracture density when using fracturing approaches. Electrokinetics also have been employed to deliver electron donor into silty and clayey materials with some success (Mao et al. 2012; Rabbi et al. 2000), but more field scale demonstrations are needed to confirm the efficacy of this approach. Use of sheer-thinning fluids also recently has received attention as a means to facilitate amendment delivery into lower permeability zones (Silva et al. 2012).

Because distribution of electron donor is often a key factor determining the overall success of in situ bioremediation, the appropriate monitoring techniques and tools to verify amendment distribution should be employed. High resolution vertical multilevel sampling wells have proven to be very useful for understanding amendment and mass distribution in heterogeneous systems (Smith et al. 1991; Thomson et al. 2007; Wang et al. 2014). Postinjection collection of soil cores also can be employed to verify distribution of amendments (Scheutz et al. 2010).

22.3 State of the Art: Bioaugmentation

The use of bioaugmentation to enhance or ensure in situ bioremediation of chlorinated solvents has been a topic of some debate (c.f., Nyer et al. 2003). Some of the uncertainty about bioaugmentation likely arose from a period in the late 1980s where bioaugmentation cultures, particularly for remediating petroleum hydrocarbon contamination, were greatly oversold even though there were no conclusive studies that could confirm that bioaugmentation cultures could persist in the environment and compete with native organisms (DeFlaun and Steffan 2002). Early studies with *Dhc*-containing cultures, however, demonstrated that these organisms do persist in situ, and they do support complete dehalogenation of chlorinated solvents in situ (Ellis et al. 2000a, b; Harkness et al. 1999; Lendvay et al. 2003; Major et al. 2002; Schaefer et al. 2010a). Their survival and effectiveness may be linked to the unique niche that they occupy which is determined by their ability to use halogenated hydrocarbons as terminal electron acceptors. Because of the success of these early demonstrations, several companies began growing and marketing bioaugmentation cultures containing *Dhc* strains, and to date more than 1000 applications of bioaugmentation cultures have been performed for chlorinated solvent remediation in the United States (Simon Vainberg, CB&I Federal Services LLC, personal communication). As the production cost and purchase price of these cultures have been reduced significantly over the last several years, many remediation practitioners now simply add a bioaugmentation culture whenever they perform in situ chlorinated solvent treatment because the cost of the culture is such a small part of the overall remediation cost (Stroo et al. 2012). Adding the culture at the beginning of the treatment may help ensure remedial performance, and it may eliminate the need for an additional site mobilization to bioaugment if the initial biostimulation-only effort is unsuccessful. Bioaugmentation at the beginning of treatment also may reduce the loss of electron donor to methanogenesis, and bioaugmentation may help to repopulate an aquifer that has been thermally treated or treated with chemical oxidants or reductants that can reduce the indigenous *Dhc* population.

22.3.1 Large-Scale Production of Dhc Cultures

One of the challenges of performing bioaugmentation at a commercial scale is the large size of some contaminant plumes and the large volume of culture needed to facilitate timely and successful remediation. Recent studies have suggested that about 10^7 *Dhc*/L of groundwater are needed to support acceptable organohalide degradation rates (Lu et al. 2006; Schaefer et al. 2010a). Steffan and Vainberg (2013) calculated that a 0.4 ha (one-acre) aquifer with a 3 m (10 ft) thick saturated zone and a 25 % porosity would contain ~3×10^6 L of groundwater and, therefore, would require 3×10^{13} *Dhc* to achieve a final average *Dhc* concentration of 10^7/L. At the reported *Dhc* concentrations of early bioaugmentation cultures (10^9 *Dhc*/L; Major et al. 2002), as much as 10^4 L of culture would be required to treat a one-acre site. Most commercially available *Dhc*-containing bioaugmentation cultures sold to date, however, contain at least 1×10^{11} *Dhc*/L (Vainberg et al. 2009; Steffan and Vainberg 2013), so the same site could now be treated with only 100 L of a commercially available culture. A recent report demonstrated that a high cell density *Dhc* culture could be reliably produced in a continuous flow fermentation to allow even larger scale culture production to meet culture volume needs at very large sites (Delgado et al. 2014).

As one of the largest producers and distributors of *Dhc* cultures for site remediation, we (CB&I Federal Services, LLC [formerly Shaw Environmental Inc.])

Table 22.1 Deliveries of the SDC-9™ bioaugmentation culture to aquifer remediation sites between 2008 and 2014

Year	Volume delivered (L)	Mean delivered volume (L)	Standard deviation (L)	Number of deliveries	Deliveries ≥100 L (%)	Deliveries ≤20 L (%)
2008	4165	71	79	59	19	29
2009	8051	134	178	60	42	27
2010	8949	110	138	81	31	22
2011	11,344	117	274	97	26	43
2012	14,259	100	127	145	36	27
2013	8809	64	94	137	20	43
2014	11,030	46	53	237	14	49

have kept records of our *Dhc* culture deliveries since about 2005. Some of that data is summarized in Table 22.1. The data presented represent culture deliveries where the purchased cultures were packaged and shipped to a site from our facility for injection into a contaminated aquifer. Because many of the culture purchases were made through licensed distributors, we have no information about the size or complexity of the treated site, but we still can observe some interesting trends in the data. Notably, the number of orders of culture increased steadily from 2008 to 2014, with only 2013 having fewer deliveries than the previous year. This suggests that the acceptance of bioaugmentation for chlorinated solvent remediation has increased with time. We also can observe that there is a general decrease in volume of culture ordered for each application over time. The mean culture volume data are somewhat skewed because of periodic large culture applications, as indicated by the large standard deviations, but this decreasing volume per order trend becomes increasingly apparent in 2013 and 2014.

Several factors could account for the trend towards smaller culture orders. For example, because of the current relatively low cost of bioaugmentation cultures, bioaugmentation may now be more commonly applied at smaller sites. This has been substantiated through feedback from our licensed culture distributors. In addition, although the commercial SDC-9™ (Vainberg et al. 2009) culture always contained at least 10^{11} *Dhc*/L, for business reasons it was initially market as containing from 1 to 5×10^{10} *Dhc*/L. As competitors acquired the ability to produce high density *Dhc* cultures, we began to guarantee that the culture contained at least 1×10^{11} *Dhc*/L. Thus, those applying SDC-9™ could buy less culture to achieve the same final amount of *Dhc*. This certainly accounts for some of the decrease in cell culture volume in 2013 and 2014.

Another factor that contributed to this trend toward smaller culture orders was the marketing strategies of some *Dhc* culture vendors. Each of our six licensed distributors of SDC-9™ also sold their own electron donor product(s). It became apparent in 2013 and 2014 that the frequency of culture orders from some of these distributors was increasing, and the reason for this was primarily attributable to internal sales strategies that promoted the use of bioaugmentation whenever possible. It is important to note that as the *Dhc* concentration in the commercial cultures increased as much as tenfold, the retail prices of the cultures remained the same or decreased, effectively reducing the per cell cost of *Dhc* by a factor of at least 10. If, as in 2014, a site required an average volume of only 46 L (Table 22.1), this usually could be applied for less than $7000 (assuming an average retail price of $150/L of *Dhc* culture) during substrate injection, which was often less expensive than mobilizing to a site at a later date if biostimulation alone was unsuccessful.

We have not yet encountered applications where the amount of culture needed was so great that continuous culturing was warranted or necessary and we have found that the ability to produce 3000 L batches of *Dhc* culture that can be concentrated and stored (Vainberg et al. 2009) has allowed us to meet current demands. The need to produce larger volumes of culture may arise, however, as efforts are made to remediate some very large and complex sites (NRC 2013).

22.3.2 Bioaugmentation Culture Dosage in the Field

There have been very few studies to examine the effect of bioaugmentation dosage on treatment effectiveness, and on how the bioaugmentation dosage impacts the overall remedial timeframe. As discussed in the previous section, by measuring *Dhc* present in non-bioaugmented groundwater, Lu et al. (2006) noted that *Dhc* cell numbers of at least 10^7 cell/L were needed for "useful" rates of dechlorination. This approach assumed that aqueous phase (i.e., planktonic) *Dhc* is an appropriate surrogate for dechlorination activity. Bench scale studies performed by Schaefer et al. (2009) and field data obtained by Fennell et al. (2001) are consistent with this observation, suggesting that planktonic *Dhc* are an appropriate metric of activity (with the exception of localized biofilm in the immediate vicinity of culture injection). However, studies performed using silica sand (Amos et al. 2009) and fractured bedrock (Schaefer et al. 2010b) suggest that *Dhc* attached to surfaces or associated with biofilms likely were responsible for the observed dechlorination activity following bioaugmentation. Results from a recent study by Cápiro et al. (2014) suggest that the relative importance of planktonic and attached *Dhc* is related to soil type, and that measurements of *Dhc* in groundwater may underestimate the total cell abundance in the aquifer by several orders of magnitude. These uncertainties make determination of a theoretical required *Dhc* dosage difficult, because while measuring *Dhc* abundance in groundwater is relatively easy and inexpensive, measuring the amount of *Dhc* attached to soils is more challenging and much more expensive because soil samples must be collected.

In a controlled field experiment evaluating multiple *Dhc* bioaugmentation dosages, Schaefer et al. (2010a) showed that bioaugmenting with *Dhc* at levels greater than an initial dosage of 10^7 *Dhc* cells/L of groundwater within a treatment zone did not have any significant remedial benefit over targeting a dosage of 10^7 *Dhc* cell/L; both planktonic and solid phase *Dhc* were shown to contribute to the observed TCE dechlorination. Although bioaugmentation effectiveness was well predicted by a dosing model (Schaefer et al. 2009, 2010a), this controlled field experiment represents only one test case where *Dhc* dosage was carefully evaluated. A *Dhc* bioaugmentation dosage of 10^7 cells/L was also used for the effective treatment *c*DCE and VC by Scheutz et al. (2008).

There are several factors that can impact the required dosage at actual remediation sites. These factors include groundwater velocity, contaminant concentration (and the potential presence of co-contaminants), soil type, electron donor, and *Dhc* delivery methods. Increased groundwater velocity and more aggressive injection techniques (i.e., closer injection well spacing, higher injection pressures, and/or fracturing) may facilitate *Dhc* distribution in the subsurface, thus enhancing overall treatment rates and reducing dosage requirements. As noted by Cupples et al. (2004), very low chlorinated solvent levels may limit the ability of *Dhc* to grow and migrate, thus requiring a larger bioaugmentation dosage. Similarly, elevated chlorinated solvent levels may be inhibitory to microbial growth and dechlorination (Yu et al. 2005; Adamson et al. 2003). To overcome the limited rate of *Dhc* growth, kinetic models (Schaefer et al. 2010a; Torlapati et al. 2012) indicate that a greater *Dhc* dosage can be applied to compensate and provide increased *Dhc* levels within the aquifer; this has yet to be verified at the field scale. While all of these parameters will likely impact dosage requirements and remedial success, fully developed and validated conceptual and mathematical models are currently lacking. Thus, bioaugmentation dosage typically is estimated based on targeting a particular *Dhc* dosage per volume of groundwater (e.g., 10^7 *Dhc* cells/L) or based on field experience with the particular culture and site.

While the target bioaugmentation dosages discussed in the previous paragraphs provide a basis for initial dosage, post bioaugmentation monitoring is needed to confirm if bioaugmentation was successful (with respect to remedial goals or treatment milestones). Unfortunately, due to the multiple factors that control *Dhc* distribution and dechlorination rates, it often is difficult to determine if increased bioaugmentation dosage will remedy any observed poor remedial performance at a site. Thus, potential issues associated with *Dhc* distribution (discussed in the following section) need to be considered equally with Dhc dosage when determining if an increase or supplemental *Dhc* injections are warranted.

22.3.3 Culture Distribution

As with electron donor substrates, the success of bioaugmentation is largely dependent on successful distribution of the added biocatalysts. In early bioaugmentation studies with *Dhc*, the organisms, remarkably, appeared to be distributed over long distances (Major et al. 2002). Consequently, simple conceptual models showing the bacteria being pushed farther radially with each subsequent amendment (e.g., electron donor, anoxic "push water", etc.) were used to describe how to distribute the cultures at sites (Aziz et al. 2013). Laboratory studies, however, clearly show that most of the *Dhc* added to soil columns are efficiently filtered near the injection point, and that the ultimate distribution of the organisms relies heavily on soil texture and *Dhc* growth and detachment (Schaefer et al. 2009). Several models have now been developed that describe these growth-attachment-detachment mechanisms for microbial migration through sand and soil (Clement et al. 1997; Phanikumar et al. 2005; Schaefer et al. 2009; Torlapati et al. 2012). Consistent with this conceptual model, several field studies have now shown that *Dhc* cells are transported with groundwater under growth conditions, although in some cases at a rate slower than the groundwater velocity (Schaefer et al. 2010a; Santharam et al. 2011; Haest et al. 2010). For this reason, injection of *Dhc* into groundwater wells typically still is followed by several well-volumes of anoxic water to facilitate detachment and provide some nominal initial distribution of bacteria into the aquifer, but, to the best of our knowledge, detailed assessment of such flushing has not been performed to determine its effectiveness on *Dhc* distribution and overall remedial performance.

The distribution of bioaugmented *Dhc* also may be impacted by other site conditions. For example, increases in groundwater velocity, for example by pumping or recirculating groundwater, may facilitate *Dhc* distribution by increasing the shear and enhancing detachment from biofilms. Laboratory studies have shown that high contaminant concentrations can limit *Dhc* distribution and growth (Amos et al. 2009), even though there have been several successful field applications of bioaugmentation or biostimulation to treat PCE or TCE dense nonaqueous phase liquids (DNAPLs) in unconsolidated materials (USEPA 2004; ITRC 2007; Hood et al. 2008). Another recent study suggested that the presence of other microbial populations and the specific electron donor used may play a role in the growth and subsequent distribution of *Dhc* within an aquifer following bioaugmentation (Schneidewind et al. 2014).

22.4 Future Challenges

Despite the excellent track record of in situ bioremediation for chlorinated solvents, several future challenges exist. It could be argued, for example, that many of the "easy" chlorinated solvent sites have been remediated, or can be remediated by many practitioners, while the most difficult to remediate sites remain to be treated. The National Research Council (NRC 2013) recently evaluated and discussed many of the remaining challenges for addressing the most difficult sites in the United States. Many of these sites are large, and they often contain mixtures of contaminants, including organic compounds and heavy metals, and high contaminant concentrations that can be inhibitory to *Dhc*. Although the scope of this chapter does not allow a full discussion of all challenges to successful bioremediation of chlorinated solvent-contaminated groundwater, we have discussed below some of the more common and challenging issues frequently encountered by remediation practitioners.

22.4.1 Fractured Bedrock

One of the challenges for bioremediation of chlorinated ethenes is treatment of fractured bedrock aquifers. Fractured rock matrices create many intrinsic challenges that are problematic for most remedial technologies; these challenges include the following:

- *Complex fracture connectivity, resulting in a complex flow field.* This makes distribution of bioremediation amendments difficult, and it complicates performance monitoring.
- *Difficulty identifying contaminant sources.* Many of the tools that have been demonstrated to be useful for locating and quantifying DNAPL sources in unconsolidated materials are not appropriate, or have not been demonstrated, for fractured bedrock. Uncertainty regarding sources in bedrock is a challenge for implementing in situ bioremediation.
- *The rock matrix.* In many cases, the majority of the contaminant mass in a given fractured rock aquifer may reside in the low permeability rock matrix itself, where it is essentially inaccessible to bioremediation amendments.
- *Amendment migration and retention through the fracture network.* Migration and retention of remedial amendments within fractured bedrock may differ from that in unconsolidated material. For example, the migration of emulsified vegetable oil (a commonly used electron donor for bioremediation of chlorinated ethenes) is not nearly as well understood in fractured bedrock as it is in sandy aquifers. Migration of *Dhc* also may differ substantially in fractured bedrock compared to sands (Schaefer et al. 2010b).

There have been few reports of successful bioremediation of chlorinated ethenes in fractured rock. Field studies have shown that complete reductive dechlorination of dissolved chlorinated ethenes can be enhanced when electron donor and *Dhc* can be effectively delivered (Mora et al. 2008; Pérez de Mora et al. 2014; Révész et al. 2014). Because of the low fracture porosity in these studies, amendment distribution up to 30 m was observed. Révész et al. (2014) showed molar conversion of TCE to ethene, with no observed rebound, in wells that showed good hydraulic connection.

While the studies noted above show promise for bioremediation of chlorinated ethenes in fractured bedrock, more detailed studies examining the long-term potential challenges of addressing contaminant storage (and subsequent back-diffusion) in rock matrices and identification and treatment of DNAPL sources within fractured bedrock currently are not available. Thus, optimizing and demonstrating biostimulation/bioaugmentation treatment for addressing such challenges remain a goal of the industry.

Contaminants present in the rock matrix represent a challenge similar to the uptake of contaminants in low permeability clays. That is, as the contaminants diffuse into the low permeability matrix, they become inaccessible to treatment amendments, and treatment efficiency is controlled by the rate at which the contaminants diffuse back out of the matrix. Fracturing technologies may provide a means to reduce diffusional mass transfer lengths (Swift et al. 2012), but the effectiveness of such an approach has yet to be demonstrated for bioremediation in fractured bedrock; appropriate control of fracture density and direction, as well as costs, would likely be challenging. Electrokinetic approaches for distributing electron donor into the rock matrix may hold some promise, but the feasibility of doing this in rock also is unknown. To date, the only bioremediation approach for addressing contaminant mass in the rock matrix is to implement a long-term strategy where bioactive conditions can be maintained in the conductive fractures for extended periods of time (Schaefer et al. 2010b; Torlapati et al. 2012).

While applying biostimulation/bioaugmentation to address DNAPL sources in unconsolidated materials has been successful (USEPA 2004; ITRC 2007; Hood et al. 2008; McGuire et al. 2006), effective treatment of DNAPL sources in fractured bedrock at the field scale has not yet been demonstrated. Bench scale studies (Schaefer et al. 2010b), however, have shown that bioaugmentation has potential to enhance the dissolution of DNAPL present in conductive fractures. Demonstrating and implementing such treatment at the field scale are much more challenging because of the difficulty of identifying and quantifying DNAPL sources in a complex fracture network. Thus, DNAPL in fractured bedrock remains a significant challenge for the industry.

22.4.2 Low pH Aquifers

Most successful bioremediation applications have been performed in aquifers with circum-neutral pH. Bioremediation in low pH aquifers, however, is usually ineffective, presumably because dehalogenating organisms do not dechlorinate well below pH ~ 6 (Fig. 22.2; McCarty et al. 2007; Sung et al. 2003; Vainberg et al. 2009; Zhuang and Pavlostathis 1995). In many cases low pH is a function of the natural site geochemistry, but low pH conditions also can occur as a result of substrate fermentation and reductive dechlorination of target cVOCs (Adamson et al. 2004; Amos et al. 2008; Chu et al. 2004; Lee et al. 1998).

Aquifer buffering, while effective at small scale, often is impractical for full-scale remedial efforts due to the quantity of buffer required to overcome the aquifer acidity and the necessity to distribute this buffer throughout the aquifer. For example, during an in situ demonstration conducted in Indian Head, MD, nearly 4500 L of a 6.7 % solution of sodium carbonate buffer was required to increase the pH of groundwater in a small pilot site (3.7 m sq. area × 3 m of groundwater depth deep) from an average of pH ~ 4.7 to pH ~ 6.3 (Hatzinger et al. 2006). This equates to 1.4 million L of a similar buffer for a 0.4 ha (1 acre) remediation site with only 3 m of saturated zone. Similarly, during a bioaugmentation demonstration at Fort Dix, NJ (Schaefer et al. 2010a), 3048 kg of sodium bicarbonate and 4354 kg of sodium carbonate were needed to sustain a suitable pH in a 12 m × 12 m × 3 m thick pilot test plot in a naturally low pH aquifer. Clearly, standard buffering approaches are both expensive and challenging for a large-scale remediation.

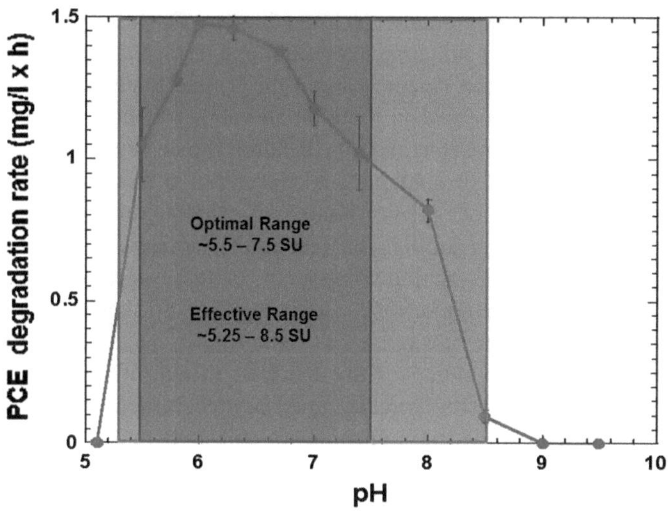

Fig. 22.2 Effect of pH on PCE dehalogenation by the *Dhc*-containing culture SDC-9 (derived from Vainberg et al. 2009, with permission). *SU* standard units, *h* hour

Current approaches for treating low pH aquifers still rely on adjusting the pH of the aquifer, and some commercial products have been developed and tested for this application. In some recent cases, pH has been adjusted with strong base (e.g., KOH) followed by a buffer (e.g., potassium or sodium bicarbonate) to maintain a desired pH level (David Lippincott, CB&I Federal Services, LLC, personal communication). Others, in a novel approach, have investigated the use of silicate minerals for long term pH control, with forsterite providing the best buffering capacity of the mineral tested (Lacroix et al. 2014a, b). Although such solid silicate mineral particles may be difficult to distribute from injection wells, they may be appropriate for distribution during hydraulic fracturing, or in permeable reactive barriers, where they could provide long-term buffering. Future approaches for treating acidic aquifers also might include the use of low pH adapted organohalide-respiring cultures. This would require that the entire microbial community that supports reductive dehalogenation, including fermenters, organohalide-respiring bacteria and possibly others, be able to tolerate the low pH. Our efforts to create such an adapted culture have met with only limited success, with dechlorination rates more than an order of magnitude lower at pH 5 than pH 6 (unpublished data). Clearly more work is needed to improve treatment of low pH aquifers.

22.5 Emerging Regulatory Issues

The widespread use of in situ bioremediation for chlorinated solvents, by either biostimulation or bioaugmentation, suggests that the regulatory community that is responsible for approving remedial measures generally views this approach as an acceptable method for groundwater remediation. The use of these products and techniques, however, is not completely immune to regulatory challenges. Two issues that have arisen in the recent years in the United States are (1) the impact of reductive treatment processes on the ultimate quality of the treated water (i.e., secondary water quality impacts); and, (2) the ability to achieve regulatory cleanup standards without the risk of contaminant concentration rebound after treatment is complete. Secondary water quality issues are centered around the creation of highly reduced conditions in aquifers that lead to the formation of hydrogen sulfide and other odor- and taste-impacting compounds, or that can lead to the mobilization of heavy metals including arsenic (Boopathy 2000). Likewise, the formation of large amounts of methane in situ creates some safety concerns, especially when treated plumes are under buildings. Although we will not fully explore secondary water quality impacts of in situ bioremediation in this chapter, we will address the issue of achieving regulatory contaminant levels and contaminant rebound.

The ultimate goal of most site remediation activities is to achieve some remedial action objective (RAO) that results in a minimum risk to health and the environment. The RAOs can be either containment or restoration, or a combination of both, and in the United States they are usually regulated by either federal or state authorities (i.e., "regulators"). Once these RAOs are achieved, the site can be "closed", and no additional action will be needed to further remediate the site. The USEPA and several states are now increasingly faced with the decision of whether or not RAOs have been met, and whether it is safe to close a site. The EPA has recently provided guidance on how to better plan for site closure early in the remediation process (USEPA 2014). The document recommends that upon completion of treatment two important questions must be answered: (1) Has the contaminant cleanup level for each contaminant of concern (COC) been met?; and, (2) Will the groundwater continue to meet the contaminant cleanup level for each COC in the future?. Measuring COC levels to determine whether they are below some established regulatory threshold is straight forward and usually involves simply analyzing the groundwater for the target COCs. Ensuring that the COC levels will not exceed the regulatory levels in the future is more complicated, however, because it requires certainty that contaminant concentrations will not "rebound" sometime in the future. Contaminant rebound can occur at some sites, especially if some contaminants remain sequestered in low permeability soils where that can slowly diffuse back into the groundwater over time (Damgaard et al. 2013; Manoli et al. 2012; Mundle et al. 2007; Chambon et al. 2010a, b).

As an example of these challenges, the authors recently worked on a project where the state regulations dictated that site closure could be achieved if the contaminant concentrations were reduced to below the state water quality standard levels, and remained below these levels for 12 consecutive quarters (i.e. 3 years) following cessation of treatment. If no rebound occurred above the state groundwater standards during this monitoring period, site closure could be obtained. Problems arose, however, around the definition of "cessation of treatment". Cessation of treatment is rather certain when using traditional remedial approaches like pump-and-treat or air sparging where a switch can be flipped to stop treatment, but it is less certain during in situ biotreatment which can remain active for long periods following active treatment. Although the long-term persistence of enhanced reductive dechlorination is one of the attractive features of this remedial approach, the regulators needed assurance that once this beneficial dechlorination reaction ceased contaminant concentrations would not increase above target remedial levels any time in the future. That is, the regulators were concerned that ongoing biodegradation would prevent the reliable evaluation of contaminant rebound potential because it is uncertain when treatment ceases, thereby triggering the 3-year monitoring period. The approach we selected to address this issue is described below.

One of the greatest challenges to attaining remedial goals is aquifer heterogeneity. It is well recognized that in situ contaminant treatment, whether by chemical, biological, or physical (e.g., air sparging or pump-and-treat) methods, may occur at rates that are ultimately controlled by mass transfer processes associated with heterogeneity in the subsurface. In general terms, these contaminant mass transfer limitations can be caused by several factors, including the presence of slowly dissolving nonaqueous phases, slow desorption from aquifer solids, diffusion out of dead-end fractures, and/or heterogeneity in subsurface permeability (i.e., "high" and "low" flow zones such as clays, silts, or rock matrices).

To assess the ultimate performance of an in situ remedial approach, contaminant concentrations in groundwater typically are measured after active treatment has ceased. For example, delivery of air to air sparging injection points would cease, or oxidant levels in groundwater would dissipate. This cessation of treatment would allow for steady-state conditions to reestablish within the aquifer, as any impacts from contaminant mass within nonaqueous phases, low permeability zones, etc., would be observed as concentration rebound within the hydraulically conductive zones (which are what is typically measured at groundwater monitoring wells) during the three-year monitoring period. However, for in situ bioremediation, reactions facilitating the dechlorination of compounds such as PCE and TCE may persist long after amendment delivery has ceased, due, for example, to the exogenous decay of biomass (Adamson and Newell 2009; Sleep et al. 2006), which can sustain reductive dechlorination for years after active treatment has ceased. Other studies have shown that reduction of oxidized iron and sulfur species that occurs during active bioremediation can facilitate the dechlorination of chlorinated ethenes after carbon additions have ceased and substrate has been consumed (He et al. 2008, 2010; Kennedy et al. 2006a, b). The persistence of these long-term dechlorination reactions can delay the observed rebound of the parent compound within the hydraulically conductive zones due to back-diffusion from low permeability zones. Regulators are concerned that once biological dechlorination does eventually cease, which could be long after site closure has been awarded, contaminant concentrations could rebound to above target levels. Thus, methods are needed to reliably predict rebound potential even as residual biological activity is occurring.

Several recent studies (Damgaard et al. 2013; Manoli et al. 2012; Mundle et al. 2007; Chambon et al. 2010a, b) show that the extent of this rebound can be predicted even during ongoing microbial reductive dechlorination. To quantitatively describe mass transfer-limited contaminant removal during application of in situ remedial technologies (i.e., technologies that enhance the chemical or biological degradation of the parent compound), conceptual and mathematical models have been developed. The overall process can be described using a 1-dimensional modeling approach (Mundle et al. 2007; Chambon et al. 2010a, b). Chambon et al. (2010a, b) and Manoli et al. (2012) showed that the model could be applied to describe the sequential reductive dechlorination of TCE (with *c*DCE, VC, ethene and ethane as daughter products) in a hydraulically conductive zone surrounded by low permeability materials by substituting the reaction term of the model with Monod kinetic expressions to describe the sequential reductive dechlorination of TCE and its chlorinated daughter products. This reaction model shows that the TCE diffuses out of the low permeability zones which is rapidly dechlorinated and converted to an equimolar mass of daughter products (sum of *c*DCE, VC, ethene, and sometimes ethane); only these daughter products are observed in the hydraulically conductive zone. Furthermore, Manoli et al. (2012) demonstrated through their simulations that TCE concentrations rebound after the cessation of microbial dechlorination activity in the hydraulically conductive zones due to the back-diffusion of TCE from low permeability zones (assuming sufficient contaminant mass was stored in the low permeability material). It is noted that in the context of this discussion on emerging regulatory issues, that the cessation of microbial dechlorination activity might not occur until long after substrate addition had ceased and become depleted within the aquifer. The extent of TCE rebound that occurred after cessation of microbial reductive dechlorination was approximately equal to the molar sum of the chlorinated ethene and ethene/ethane daughter products that were observed while microbial reductive dechlorination was ongoing. Therefore, based on this model, the molar concentrations of the daughter products are directly related to the molar concentration of TCE that would be present following rebound in the absence of biological degradation, and thus can be used to predict what the TCE rebound will be in absence of microbial dechlorination. These model results are corroborated by field scale data, where microbially enhanced reductive dechlorination for treatment of TCE was performed, that were consistent with the model simulations (Damgaard et al. 2013).

Thus, based on the mass balance and modeling efforts described above, TCE or PCE daughter products observed while microbial reductive dechlorination is ongoing can be used as a surrogate to estimate the equilibrium rebound (due to contaminant mass remaining in low permeability materials) in parent compound concentrations that would occur in hydraulically conductive zones once all dechlorination reactions have ceased. This should give regulatory authorities greater certainty about the potential for contaminant rebound, and allow them to award closure at sites even though the cessation of biological activity is uncertain. The acceptance of this rationale and conceptual model by regulatory authorities, however, is likely to occur on a case-by-case basis, and it will be bolstered by examples of successful remedies where the approach has been applied to correctly predict the absence of contaminant rebound.

22.6 Conclusions

Bioremediation of chlorinated solvents by employing organohalide-respiring bacteria, whether by stimulating indigenous organisms or by adding exogenous cultures, is the most common and successful remedial approach for restoring contaminated aquifers. The rapid and widespread acceptance and application of this technology could not have been predicted at the time of the discovery of these processes and the organohalide-respiring microorganisms. Fortunately, the in situ robustness of organohalide-respiring bacteria and the ease at which they can often be applied has led to the apparently successful treatment of thousands of contaminated aquifers around the world. The legacy of poor chemical disposal practices, however, has ensured that many site remediation challenges still exist, and many scientific and engineering advances are still needed to complete our mission of restoring the World's contaminated waters.

References

Adamson DT, Newell CJ (2009) Support of source zone bioremediation through endogenous biomass decay and electron donor recycling. Bioremediat J 13:29–40

Adamson DT, McDade JM, Hughes JB (2003) Inoculation of a DNAPL source zone to initiate reductive dechlorination of PCE. Environ Sci Technol 37:2525–2533

Adamson DT, Lyon DY, Hughes JB (2004) Flux and product distribution during biological treatment of tetrachloroethene dense non-agueous-phase liquid. Environ Sci Technol 38:2021–2028

Amos BK, Suchomel EJ, Pennell KD, Löffler FE (2008) Microbial activity and distribution during enhanced contaminant dissolution from a NAPL source zone. Water Res 42:2963–2974

Amos BK, Suchomel EJ, Pennell KD, Löffler FE (2009) Spatial and temporal distributions of *Geobacter lovleyi* and *Dehalococcoides* spp. during bioenhanced PCE-DNAPL dissolution. Environ Sci Technol 43:1977–1985

Aziz CE, Wymore RA, Steffan RJ (2013) Bioaugmentation considerations. In: Stroo HF, Leeson A, Ward CH (eds) Bioaugmentation for groundwater remediation. Springer Science+Business Media, New York, pp 141–169

Bass DH, Hastings NA, Brown RA (2000) Performance of air sparging systems: a review or case studies. J Hazard Mater 72:101–119

Beeman RE, Bleckmann CA (2002) Sequential anaerobic–aerobic treatment of an aquifer contaminated by halogenated organics: field results. J Contam Hydrol 57:147–159

Boopathy R (2000) Factors limiting bioremediation technologies. Bioresour Technol 74:63–67

Borden RC (2007) Effective distribution of emulsified edible oil for enhanced anaerobic bioremediation. J Contam Hydrol 94:1–12

Borden RC, Rodriguez BX (2007) Evaluation of slow release substrates for anaerobic bioremediation. Bioremediat J 10:59–69

Borden RC, Beckwith WJ, Lieberman MT, Akladiss N, Hill SR (2007) Enhanced anaerobic bioremediation of a TCE source at the Tarheel Army Missile Plant using EOS. Remediation 17:5–19

Cápiro NL, Wang Y, Hatt JK, Lebrón CA, Pennell KD, Löffler FE (2014) Distribution of organohalide-respiring bacteria between solid and aqueous phases. Environ Sci Technol 48:10878–10887

Chambon JC, Broholm MM, Binning PJ, Bjerg PL (2010a) Modeling multi-component transport and enhanced anaerobic dechlorination processes in a single fracture–clay matrix system. J Contam Hydrol 112:77–90

Chambon JC, Broholm MM, Binning PJ, Bjerg PL (2010b) Modeling multi-component transport and enhanced anaerobic dechlorination processes in a single fracture–clay matrix system. J Contam Hydrol 112:77–90

Christiansen CM, Damgaard I, Broholm M, Kessler T, Klint KE, Niulsson B, Bjerg PL (2010) Comparison of delivery methods for enhanced in situ remediation in clay till. Ground Water Monitor Remediat 30:107–122

Chu M, Kitanidis PK, McCarty PL (2004) Possible factors controlling the effectiveness of bioenhanced dissolution of non-aqueous phase tetrachloroethene. Adv Water Resour 27:601–615

Clement TP (1997) RT3D—a computer code for simulating reactive multi-species transport in 3-dimensional groundwater aquifers. Pacific Northwest National Laboratory Report, PNNL-11720. http://www.pnl.gov/main/publications/external/technical_reports/pnnl-11720.pdf

Clement TP, Peyton BM, Skeen RS, Jennings DA, Petersen JN (1997) Microbial growth and transport in porous media under denitrification conditions: Experiments and simulations. J Contam Hydrol 24:269–285

Cupples AM, Spormann AM, McCarty PL (2004) Vinyl chloride and *cis*-dichloroethene dechlorination kinetics and microorganism growth under substrate limiting conditions. Environ Sci Technol 38:1102–1107

Damgaard I, Bjerg PL, Bælum J, Scheutz C, Hunkeler D, Jacobsen CS, Tuxen N, Broholm MM (2013) Identification of chlorinated solvents degradation zones in clay till by high resolution chemical, microbial and compound specific isotope analysis. J Contam Hydrol 146:37–50

DeFlaun MF, Steffan RJ (2002) Bioaugmentation. In: Bitton G (ed) Encyclopedia of environmental microbiology. Wiley, New York, NY, pp 434–442

Delgado AG, Fajardo-Williams D, Popat SC, Torres CI, Krajmalnik-Brown R (2014) Successful operation of continuos reactors at short retention times results in high-density, fast-rate *Dehalococcoides* dechlorinating cultures. Appl Microbiol Biotechnol 98:2729–2737

Ellis DE, Lutz EJ, Odom JM, Buchanan RJ Jr, Bartlett CL, Lee MD, Harkness MR, Deweerd KA (2000a) Bioaugmentation for accelerated in situ anaerobic bioremediation. Environ Sci Technol 34:2254–2260

Ellis DE, Lutz EJ, Odom JM, Buchanan RJ, Bartlett CL, Lee MD, Harkness MR, Deweerd KA (2000b) Bioaugmentation for accelerated in situ anaerobic bioremediation. Environ Sci Technol 34:2254–2260

Fennell DE, Carroll AB, Gossett JM, Zinder SH (2001) Assessment of indigenous reductive dechlorinating potential at a TCE contaminated site using microcosms, polymerase chain reaction analysis, and site data. Environ Sci Technol 35:1830–1839

Frascari D, Zanaroli G, Danko AS (2015) In situ cometabolism of chlorinated solvents: a review. J Hazard Mater 283:382–399

Freedman DL, Gossett JM (1989) Biological reductive dechlorination of tetrachloroethylene and trichloroethylene to ethylene under methanogenic conditions. Appl Environ Microbiol 55:2144–2151

Haest PJ, Springael D, Smolders E (2010) Dechlorination kinetics of TCE at toxic TCE concentrations: assessment of different models. Water Res 44:331–339

Harkness MR, Bracco AA, Brennan MJ Jr, DeWeerd KA, Spivack JL (1999) Use of bioaugmentation to stimulate complete reductive dechlorination of trichloroethene in Dover soil columns. Environ Sci Technol 33:1100–1109

Hatzinger PB, Diebold J, Yates CA, Cramer RJ (2006) Field demonstration of in situ perchlorate bioremediation in groundwater (Chap. 14). In: Gu B, Coates JC (eds) perchlorate: environment occurrence, interactions, and treatment. Springer, New York, pp 311–341

He J, Ritalhati KM, Yang KL, Koenigsberg SS, Löffler FE (2003) Detoxification of vinyl chloride to ethene coupled to growth of an anaerobic bacterium. Nature 424:62–65

He YT, Wilson JT, Wilkin RT (2008) Transformation of reactive iron minerals in a permeable reactive barrier (biowall) used to treat TCE in groundwater. Environ Sci Technol 42:6690–6696
He YT, Wilson JT, Wilkin RT (2010) Impact of iron sulfide transformation on trichloroethylene degradation. Geochim Cosmochim Acta 74:2025–2039
Hendrickson ER, Payne JA, Young RM, Starr MG, Perry MP, Fahnestock S, Ellis DE, Ebersole RC (2002) Molecular analysis of Dehalococcoides 16S ribosomal DNA from chloroethene-contaminated sites throughout North America and Europe. Appl Environ Microbiol 68(2):485–495
Hood ED, Major DW, Quinn JW, Yoon W-S, Gavaskar A, Edwards EA (2008) Demonstration of enhanced bioremediation in a TCE source area at Launch Complex 34, Cape Canaveral Air Force Station. Ground Water Monitor Remediat 28:98–107
Interstate Technology and Regulatory Council (ITRC) (2007) In situ bioremediation of chlorinated ethene DNAPL source zones: case studies. Washington, D.C
Kennedy LG, Everett JW, Gonzales J (2006a) Assessment of biogeochemical natural attenuation and treatment of chlorinated solvents, Altus Air Force Base, Altus, Oklahoma. J Contam Hydrol 83:221–236
Kennedy LG, Everett JW, Gonzales J (2006b) Assessment of biogeochemical natural attenuation and treatment of chlorinated solvents, Altus Air Force Base, Altus, Oklahoma. J Contam Hydrol 83:221–236
Kim Y, Istok JD, Semprini L (2004) Push-pull tests for assessing in situ aerobic cometabolism. Groundwater 42:329–337
Lacroix E, Brovelli A, Barry DA, Holliger C (2014a) Use of silicate minerals for pH control during reductive dechlorination of chloroethenes in batch cultures of different microbial consortia. Appl Environ Microbiol 80:3858–3867
Lacroix E, Brovelli A, Maillard J, Rhorbach-Brandt E, Barry DA, Holliger C (2014b) Use of silicate minerals for long-term pH control during reductive dechlorination of high tetrachloroethene concentrations in continuous flow-through columns. Sci Total Environ 482–483:23–35
Lee MD, Odom JM, Buchman RJ (1998) New perspectives on microbial dehalogenation of chlorinated solvents: insights from the field. Ann Rev Microbiol 52:423–452
Lee A, Abrams SH, Moskal E, Ciambruschini S, Moss D (2013) Chemical and biological reduction of PCE by pneumatic fracturing and injection of ZVI in saprolite. Remediat J 23:7–21
Lendvay JM, Löffler FE, Dollhopf M, Aiello MR, Daniels G, Fathepure BZ, Gebhard M, Heine R, Helton R, Shi J, Krajmalnik-Brown R, Major CL Jr, Barcelona MJ, Petrovskis E, Hickey R, Tiedje JM, Adriaens P (2003) Bioreactive barriers: a comparison of bioaugmentation and biostimulation for chlorinated solvent remediation. Environ Sci Technol 37:1422–1431
Lohner ST, Tiehm A (2009) Application of electrolysis to stimulate microbial reductive PCE dechlorination and oxidative VC biodegradation. Environ Sci Technol 43:7098–7104
Lohner ST, Becker D, Mangold K-M, Tiehm A (2011) Sequential reductive and oxidative biodegradation of chloroethenes stimulated in a coupled bioelectro-process. Environ Sci Technol 45:6191–6197
Löffler FE, Ritalahti KM, Zinder SH (2013) Dehalococcoides and reductive dechlorination of chlorinated solvents. In: Stroo HF, Leeson A, Ward CH (eds) Bioaugmentation for groundwater remediation, vol 5. SERDP ESTCP Environmental Remediation Technology. Springer New York, pp 39–88. doi:10.1007/978-1-4614-4115-1_2
Lu X, Wilson JT, Kampbell DH (2006) Relationship between *Dehalococcoides* DNA in ground water and rates of reductive dechlorination at field scale. Water Res 40:3131–3140
Ma X, Novak PJ, Semmens MJ, Clapp LW, Hozalaski RM (2006) Comparison of pulsed and continuous addition of H_2 gas via membranes for stimulating PCE biodegradation in soil columns. Water Res 40:1155–1166
Major DW, McMaster ML, Cox EE, Edwards EA, Dworatzek SM, Hendrickson ER, Starr MG, Payne JA, Buonamici LW (2002) Field demonstration of successful bioaugmentation to achieve dechlorination of tetrachloroethene to ethene. Environ Sci Technol 36:5106–5116

Manoli G, Chambon JC, Bjerg PL, Scheutz C, Binning PJ, Broholm MM (2012) A remediation performance model for enhanced metabolic reductive dechlorination of chloroethenes in fractured clay till. J Contam Hydrol 131:64–78

Mao X, Wang J, Ciblak A, Cox EE, Riis C, Terkelsen M, Gent DB, Alshawabkeh AN (2012) Electrokinetic-enhanced bioaugmentation for remediation of chlorinated solvents contaminated clay. J Hazard Mater 213–14:311–317

Maymó-Gatell X, Chien Y-T, Gossett JM, Zinder SH (1997) Isolation of a bacterium that reductively dechlorinates tetrachloroethene to ethene. Science 276:1568–1571

Maymó-Gatell X, Nijenhuis I, Zinder SH (2001) Reductive dechlorination of *cis*-dichloroethene and vinyl chloride by *Dehalococcoides ethenogenes* strain 195. Environ Sci Technol 35:516–521

McCarty PL, Goltz MN, Hopkins GD, Dolan ME, Allan JP, Kawakami BT, Carrothers TJ (1998) Full-scale evaluation of in situ cometabolic degradation of trichloroethylene in groundwater through toluene injection. Environ Sci Technol 32:88–100

McCarty PL, Chu MY, Kitanidis PK (2007) Electron donor and pH relationships for biologically enhanced dissolution of chlorinated solvent DNAPL in groundwater. Eur J Soil Biol 43:276–282

McGuire TM, Newell CJ, Looney BB, Vangelas KM, Sink CH (2004) Historical analysis of monitored natural attenuation: a survey of 191 chlorinated solvent sites and 45 solvent plumes. Remediat J 15:99–112

McGuire TM, McDade JM, Newell CJ (2006) Performance of DNAPL source depletion technologies at 59 chlorinated solvent-impacted sites. Groundwater Monitor Remediat 26:73–84

Mora RH, Macbeth TW, MacHarg T, Gundarlahalli J, Holbrook H, Schiff P (2008) Enhanced bioremediation using whey powder for a trichloroethene plume in a high-sulfate, fractured granitic aquifer. Remediat J 18:7–30

Mundle K, Renyolds DA, West MR, Kueper BH (2007) Concentration rebound following in situ chemical oxidation in fractured clay. Ground Water 35:1077–1088

National Research Council (NRC) (2013) Alternatives for managing the nation's complex contaminated groundwater sites. National Academies Press, Washington, DC, 422 pp

Nyer EK, Payne FC, Sutherson S (2003) Discussion of environment vs. bacteria or let's play, 'name that bacteria'. Ground Water Monitor Remediat 23:36–45

Pérez-de-Mora A, Zila A, McMaster ML, Edwards EA (2014) Bioremediation of chlorinated ethenes in fractured bedrock and associated changes in dechlorinating and nondechlorinating microbial populations. Environ Sci Technol 48:5770–5779

Phanikumar MS, Hyndman DW, Zhao X, Dybas MJ (2005) A three-dimensional model of microbial transport and biodegradation at the Schoolcraft, Michigan, site. Water Resour Res 41:W05011. doi:10.1029/2004WR003376

Rabbi MF, Clark B, Gale RJ, Ozsu-Acar E, Pardue J, Jackson A (2000) In situ TCE bioremediation study using electrokinetic cometabolite injection. Waste Manage 20:279–286

Révész KM, Lollar BS, Kirshtein JD, Tiedeman CR, Imbrigiotta TE, Goode DJ, Shapiro AM, Voytek MA, Lacombe PJ, Busenberg E (2014) Integration of stable carbon isotope, microbial community, dissolved hydrogen gas, and $^{2}H_{H2O}$ tracer data to assess bioaugmentation for chlorinated ethene degradation in fractured rocks. J Contam Hydrol 156:62–77

Rivett MO, Chapman SW, Allen-King RM, Feenstra S, Cherry JA (2006) Pump-and-treat remediation of chlorinated solvent contamination at a controlled field-experiment site. Environ Sci Technol 40:6770–6781

Santharam S, Ibbini J, Davis LC, Erickson LE (2011) Field study of biostimulation and bioaugmentation for remediation of tetrachloroethene in groundwater. Remediat J 21:51–68

Schaefer CE, Vainberg S, Condee C, Steffan RJ (2009) Bioaugmentation for chlorinated ethenes using *Dehalococcoides* sp.: comparison between batch and column experiments. Chemosphere 75:141–148

Schaefer CE, Lippincott D, Steffan RJ (2010a) Field scale evaluation of bioaugmentation dosage for treating chlorinated ethenes. Ground Water Monitor Remediat 30:113–124

Schaefer CE, Towne RM, Vainberg S, McCray JE, Steffan RJ (2010b) Bioaugmentation for treatment of dense non-aqueous phase liquid in fractured sandstone blocks. Environ Sci Technol 44:4958–4964

Scheutz C, Durant ND, Dennis P, Hansen MH, Jorgensen T, Jakobsen R, Cox EE, Bjerg PL (2008) Concurrent ethene generation and growth of *Dehalococcoides* containing vinyl chloride reductive dehalogenase genes during an enhanced reductive dechlorination field demonstration. Environ Sci Technol 42:9302–9309

Scheutz C, Broholm MM, Durant ND, Weeth EB, Jorgensen TH, Dennis P, Jacobsen CS, Cox EE, Chambon JC, Bjerg PL (2010) Field evaluation of biological enhanced reductive dechlorination of chloroethenes in clayey till. Environ Sci Technol 44:5134–5141

Schneidewind U, Haest PJ, Atashgahi S, Maphosa F, Hamonts K, Maesen M, Calderer M, Seuntjens P, Smidt H, Springael D, Dejonghe W (2014) Kinetics of dechlorination by *Dehalococcoides mccartyi* using different carbon sources. J Contam Hydrol 157:25–36

Semprini L (1997) Strategies for the aerobic co-metabolism of chlorinated solvents. Curr Opin Biotechnol 8:296–308

Semkiw ES, Barcelona MJ (2011) Field study of enhanced tce reductive dechlorination by a full-scale whey prb. Ground Water Monitoring & Remediation 31(1):68-78. doi:10.1111/j.1745-6592.2010.01321.x

Sewell GW, Gibson SA (1991) Stimulation of the reductive dechlorination of tetrachloroethene in anaerobic aquifer microcosms by the addition of toluene. Environ Sci Technol 25:982–984

Silva JAK, Smith MM, Munakata-Marr J, McCray JE (2012) The effect of system variables on in situ sweep-efficiency improvements via viscosity modification. J Contam Hydrol 136–137:117–130

Sleep BE, Seepersad DJ, Mo K, Heidorn CM, Hrapovic L, Morrill PL, McMaster ML, Hood ED, LeBron C, Sherwood Lollar B, Major DW, Edwards EA (2006) Biological enhancement of tetrachloroethene dissolution and associated microbial community changes. Environ Sci Technol 40:3623–3633

Smith RL, Harvey RW, LeBlanc DR (1991) Importance of closely spaced vertical sampling in delineating chemical and microbiological gradients in groundwater studies. J Contam Hydrol 7:285–300

Steffan RJ, Sewell GW (2011) Advances in bioremediation of aquifers (Chap. 11). In: Quercia FF, Vidojevic D (eds) Clean soil and safe water, NATO science for peace and security series C: environmental security. Springer Science+Business Media B.V., pp 143–151. ISBN 978-94-007-2239-2

Steffan RJ, Vainberg S (2013) Production and handling of *Dehalococcoides* bioaugmentation cultures. In: Stroo HF, Leeson A, Ward CH (eds) Bioaugmentation for groundwater remediation. Springer Science+Business Media, New York, pp 89–113

Steffan RJ, Sperry KL, Walsh MT, Vainberg S, Condee CW (1999) Field-scale evaluation of in situ bioaugmentation for remediation of chlorinated solvents in groundwater. Environ Sci Technol 33:2771–2781

Strong M, Owens D, Ventura, B, Smith L, Liskowitz JJ (2004) Comparison of pneumatic and hydraulic fracturing for emplacement of treatment materials in low-permeability formations. In: Proceedings of the fourth international conference on remediation of chlorinated and recalcitrant compounds. Battelle Press, Monterey, CA

Stroo HF, Leeson A, Marqusee JA, Johnson PC, Ward CH, Kavanaugh MC, Sale TC, Newell CJ, Pennell KD, Lebrón CA, Unger M (2012) Chlorinated ethene source remediation: Lessons learned. Environ Sci Technol 46(12):6438-6447. doi:10.1021/es204714w

Sung Y, Ritalahti KM, Sanford RA, Urbance JW, Flynn SJ, Tiedje JM, Löffler FE (2003) Characterization of two tetrachloroethene-reducing, acetate-oxidizing anaerobic bacteria and their description as *Desulfuromonas michiganensis* sp. nov. Appl Environ Microbiol 69:2964–2974

Swift D, Rothermel J, Peterson L, Orr B, Bures GH, Weidhaas J (2012) Remediating TCE-contaminated groundwater in low-permeability media using hydraulic fracturing to emplace zero-valent iron/organic carbon amendment. Remediat J 22:49–67

Thomson NR, Hood ED, Farquhar GJ (2007) Permanganate treatment of an emplaced DNAPL source. Ground Water Monitor Remediat 27:74–85

Torlapati J, Clement TP, Schaefer CE, Lee KK (2012) Modeling *Dehalococcoides* sp. augmented bioremediation in a single fracture system. Ground Water Monitor Remediat 32:75–83

United States Environmental Protection Agency (USEPA) (2004) Demonstration of bioaugmentation of DNAPL through biostimulation and bioaugmentation at Launch Complex 34 Cape Canaveral Air Force station, Florida. EPA/540/R-07/007. USEPA, Washington DC

United States Environmental Protection Agency (USEPA) (2014) Groundwater remedy completion strategy. OSWER9200.2-144. USEPA, Washington DC

Vainberg S, Condee CW, Steffan RJ (2009) Large scale production of *Dehalococcoides* sp.—containing cultures for bioaugmentation. J Indust Microbiol Biotechnol 36:1189–1197

Wang F, Annable MD, Schaefer CE, Ault TD, Cho J, Jawitz JW (2014) Enhanced queous dissolution of a DNAPL source to characterize the source strength function. J Contam Hydrol 169:75–89

Weidemeier TH, Swanson MA, Moutoux DE, Gordon EK, Wilson JT, Wilson BH, Kampbell DH, Haas PE, Miller RN, Hansen JE, Chapelle FW (1998) Technical protocol for evaluating natural attenuation of chlorinated solvents in groundwater. EPA/600/R-98/128. USEPA, Washington DC

Wilson JT, Wilson BH (1985) Biotransformation of trichloroethylene in soil. Appl Environ Microbiol 49:242–243

Wu W-M, Nye J, Jain MK, Hickey RF (1998) Anaerobic dechlorination of trichloroethylene (TCE) to ethylene using complex organic materials. Water Res 32:1445–1454

Yu S, Dolan ME, Semprini L (2005) Kinetics and inhibition of reductive dechlorination of chlorinated ethylenes by two different mixed cultures. Environ Sci Technol 39:195–205

Zhang X-H, Sewell GW, Cui S-Y (2001) An improved method of hydrogen production as electron donor for anaerobic bioremediation. J Environ Sci Health A36:1661–1670

Zhuang P, Pavlostathis SG (1995) Effect of temperature, pH, and electron donor on the microbial reductive dechlorination of chloroalkenes. Chemosphere 31:3537–3548

Chapter 23
The Microbiology of Anaerobic PCB Dechlorination

Jianzhong He and Donna L. Bedard

Abstract The last few years have seen a great deal of progress in our understanding of microbial dechlorination of polychlorinated biphenyls (PCBs). Four new strains of *Dehalococcoides mccartyi*, representing all three phylogenetic subgroups, and a strain of "Dehalobium chlorocoercia" have been isolated and, together with two previously isolated strains of *D. mccartyi*, demonstrated to dechlorinate the commercial PCB mixture Aroclor 1260. Complete genomes for five of these isolates have been published. In addition, members of the genera *Dehalogenimonas* and *Dehalobacter* have been implicated in the reductive dechlorination and respiration of PCBs. It is clear that *D. mccartyi* strains capable of dechlorinating Aroclor 1260 are widespread in freshwater environments, having been found in PCB-impacted sites in China, Germany, Singapore, and the USA. Pure strains of *D. mccartyi* that dechlorinate Aroclor 1260 by following different sets of dechlorination routes, i.e., PCB Dechlorination Processes H, N,Z, and variations of these are now available. A member of the *Chloroflexi* belonging to the m1/SF1 clade appears to be responsible for the dechlorination of Aroclor 1254 in a marine site. The discovery and characterization of the first three PCB reductive dehalogenases constitute a new milestone in the field. PcbA1, PcbA4, and PcbA5 dechlorinate Aroclor 1260 with distinct regiospecificities and prove that individual RDases can carry out the complex dechlorination of dozens of PCB congeners described by the PCB dechlorination processes. Each of these three PCB dechlorinases is bifunctional and can also dechlorinate tetrachloroethene (PCE). PCB dechlorinators with such

J. He (✉)
Department of Civil and Environmental Engineering,
National University of Singapore, Singapore 117576, Singapore
e-mail: jianzhong.he@nus.edu.sg

D.L. Bedard (✉)
Department of Biological Sciences, Rensselaer Polytechnic Institute,
Troy, NY 12180, USA
e-mail: Donna.L.Bedard@gmail.com

L. Adrian and F.E. Löffler (eds.), *Organohalide-Respiring Bacteria*,
DOI 10.1007/978-3-662-49875-0_23

bifunctional PCB/PCE RDases can be selectively enriched, transferred repeatedly, and grown to high cell densities with PCE as the sole electron acceptor with no possibility of losing their ability to dechlorinate PCBs. This property makes them ideal candidates for use in bioremediation of PCBs.

23.1 Introduction

Reductive dechlorination of the commercial polychlorinated biphenyl (PCB) mixture Aroclor 1242 was first reported 30 years ago (Brown et al. 1984) and subsequently PCB dechlorination was found to be occurring in several aquatic sediments (Brown et al. 1987; Brown and Wagner 1990). Even before the phenomenon had been confirmed in the laboratory, Brown et al. (1987) proposed that the dechlorination was carried out by anaerobic bacteria and that the PCBs might be used as electron acceptors. Quensen et al. (1988, 1990) in the Tiedje lab quickly confirmed that anaerobic bacteria were responsible for the observed PCB dechlorination. However, despite studies in multiple laboratories [reviewed in (Bedard and Quensen 1995; Wiegel and Wu 2000)], the bacterial agents responsible for PCB dechlorination remained unidentified for many years. Nevertheless, it was determined that mixtures, known as Aroclors in the USA and Great Britain, are dechlorinated by one or more dechlorination processes. Nine different

Table 23.1 Brief summary of PCB dechlorination processes (Brown et al. 1987; Bedard and Quensen 1995; Bedard et al. 2005; Zhen et al. 2014, Sect. 23.2.4)

Dechlorination process	Targeted chlorines	Reactive chlorophenyl group (underlined chlorine is removed by dechlorination)
P	Flanked *para*	$3\underline{4}$, $23\underline{4}$, $2\underline{4}5$, $23\underline{4}5$, $23\underline{4}56$
H	Flanked *para* and doubly flanked *meta* of 234	$3\underline{4}$, $2\underline{3}4$, $2\underline{4}5$, $23\underline{4}5$
H′	Flanked *para,* doubly flanked *meta* of 234 and flanked *meta* of 23	$2\underline{3}$, $3\underline{4}$, $2\underline{3}4$, $2\underline{4}5$, $23\underline{4}5$
N	Flanked *meta*	$2\underline{3}4$, $23\underline{5}$, $2\underline{3}6$, $24\underline{5}$, $2\underline{3}4$, $2\underline{3}46$, $23\underline{5}6$, $234\underline{5}6$
M	Flanked and unflanked *meta*	$\underline{3}$, $2\underline{3}$, $2\underline{5}$, $\underline{3}4$, $2\underline{3}4$, $2\underline{3}6$
Q	Flanked and unflanked *para*, doubly flanked *meta* of 234, and flanked *meta* of 23	$\underline{4}$, $2\underline{3}$, $2\underline{4}$, $3\underline{4}$, $2\underline{3}4$, $2\underline{4}5$, $2\underline{4}6$
LP	Flanked and unflanked *para*, *ortho*-flanked *meta* of 23, 234, and 235	$\underline{4}$, $2\underline{3}$, $2\underline{4}$, $3\underline{4}$, $2\underline{3}4$, $2\underline{3}5$, $2\underline{4}5$, $2\underline{4}6$
T	Doubly flanked *meta* of 2345	$2\underline{3}45$
Z	Doubly flanked *meta* of $2\underline{3}4$, $2\underline{3}46$ and doubly flanked *para* of $23\underline{4}5$ and $23\underline{4}56$	$2\underline{3}4$, $2\underline{3}46$, $23\underline{4}5$, and $23\underline{4}56$

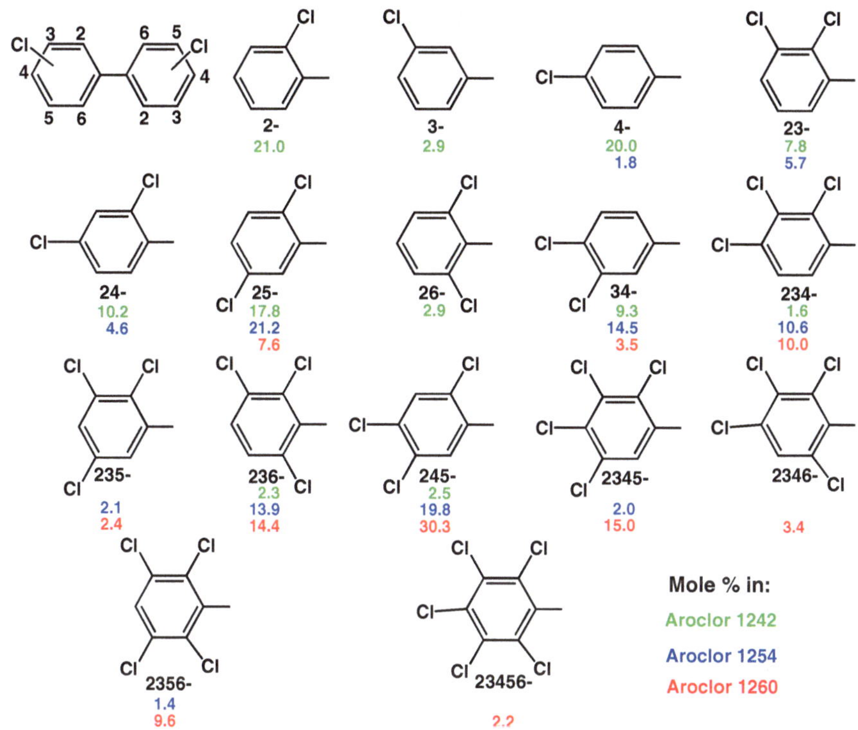

Fig. 23.1 PCB structure, numbering scheme, and chlorophenyl structure and distribution in Aroclors 1242, 1254, and 1260. Mole percent amounts of chlorophenyl rings ≥1 % are shown. Data are from Frame et al. (1996)

PCB dechlorination processes observed in mixed cultures or in situ have been described: M, Q, H′, H, P, N, LP, T, and Z (Table 23.1).

Aroclors 1242, 1254, and 1260 contain 42, 54, and 60 % chlorine by weight, respectively. The average numbers of chlorines per biphenyl molecule in these three PCB mixtures are 3.5, 5.1, and 6.2. Commercial PCBs are highly complex mixtures composed of as many as 90 different molecular forms of PCB (congeners) with different numbers and configurations of chlorine atoms. Some chlorine configurations are much more common than others, but their relative proportions differ in the various Aroclors as shown in Fig. 23.1. Because of this complexity, commercial PCBs present a substantial challenge for bioremediation. Reductive dechlorination offers the best hope for remediation because it reduces the toxicity and bioaccumulation potential of PCBs, and the products, which have fewer chlorines, are more susceptible to degradation by other organisms. In addition, PCB dechlorination is an attractive remedy because it is a natural process.

23.2 Bacteria Involved in the Dechlorination of Commercial PCB Mixtures in Mixed Cultures

Previous reviews of PCB dechlorination have extensively described works published prior to 2008 and we refer the reader to those reviews (Bedard 2008; Field and Sierra-Alvarez 2008; Hiraishi 2008), and for details, primary literature reports (Cutter et al. 2001; Wu et al. 2002a; Fennell et al. 2004; Fagervold et al. 2007; Yan et al. 2006b; Bedard et al. 2007). Here we will focus on more recent discoveries. Earlier studies regarding reductive dechlorination of PCBs, most using single congeners, revealed the involvement of *Dehalococcoides*, now *Dehalococcoides mccartyi* (Löffler et al. 2013), two related but distinct *Chloroflexi* clades designated the *o*-17/DF-1 group (Fagervold et al. 2005) and the m1/SF1 group (Fagervold et al. 2005), and *Dehalobacter* (also see Sect. 23.6 on *Dehalobacter*) (Fig. 23.2).

More recent studies, which we describe in the following sections, have demonstrated the importance of *D. mccartyi* and other *Chloroflexi* in the dechlorination of Aroclors.

23.2.1 Chloroflexus Phylotype VL-CHL1

Zanaroli et al. (2012) have identified a non-*Dehalococcoides Chloroflexi* phylotype as the agent responsible for dechlorinating Aroclor 1254 in enrichments developed under marine conditions. They used cultures developed from PCB-contaminated sediments of the Brenta Canal of the Venice Lagoon (Italy), and previously enriched with co-planar PCBs, to inoculate autoclaved sediment from the same site suspended in filter-sterilized site water (Zanaroli et al. 2010, 2012). The cultures were spiked with Aroclor 1254 (1000 mg kg^{-1}) and incubated at 28 °C. Strong sulfate reduction was observed for seven weeks, but PCB dechlorination also occurred and continued until week 30.

Two more transfers into slurries of sterile site water and sediment were carried out with and without PCBs. DGGE analysis of the community DNA using bacteria-specific primers revealed many sulfate reducers, but no candidates for PCB dechlorinators. The addition of vancomycin, ampicillin, or both enhanced PCB dechlorination, but greatly diminished sulfate reduction. This finding suggested that *Chloroflexi* were responsible for the PCB dechlorination because *Chloroflexi* lack cell walls and are not inhibited by these antibiotics. Therefore, Zanaroli et al. chose to amplify the community genomic DNA with Chl 348F/Dehal 884R, primers targeting *D. mccartyi* and the *o*-17/DF1 and m1/SF1 groups of *Chloroflexi* (Fagervold et al. 2005). With these primers the authors identified four *Chloroflexi* phylotypes in the enrichments. One of these, VL-CHL1, was specifically associated with PCB dechlorination. DNA sequencing revealed that a 466 bp-long 16S rRNA gene fragment of VL-CHL1 was 100 % identical to that of the putative PCB dechlorinator represented by *Chloroflexi* clone m1. PCR amplification of

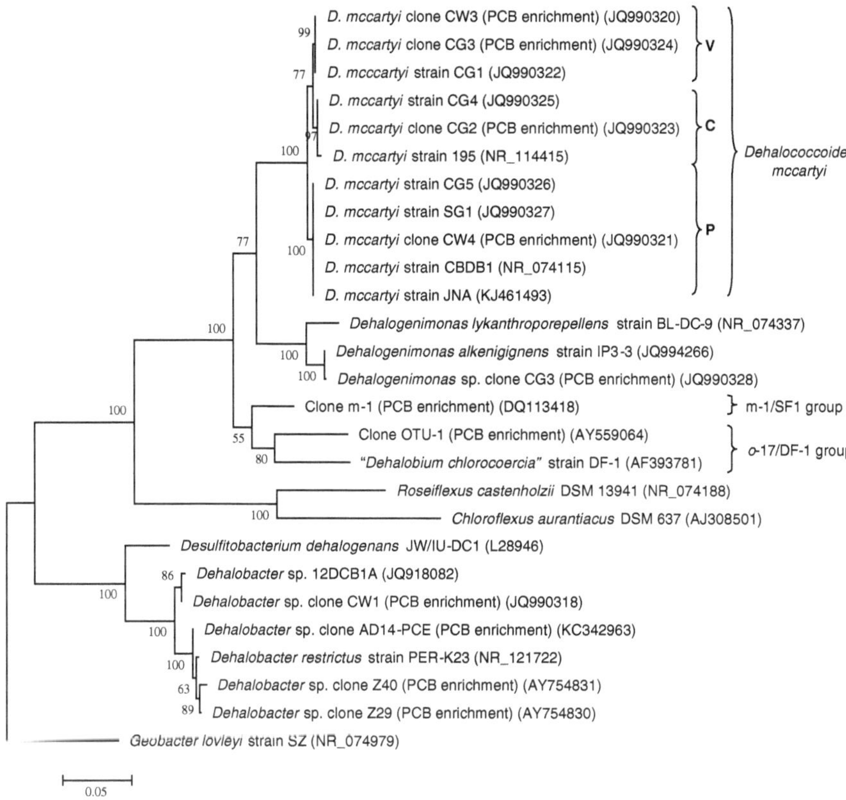

Fig. 23.2 Phylogenetic analysis of the 16S rRNA genes of PCB dechlorinators and putative PCB dechlorinators from PCB enrichments. V, C, and P refer to the phylogenetic subgroups of *D. mccartyi*, Victoria, Cornell, and Pinellas, respectively. The evolutionary history of the 16S rRNA genes of PCB dechlorinating strains was inferred by using the maximum likelihood method based on the General Time Reversible model (Nei and Kumar 2000). The 16S rRNA gene sequences of *Dehalogenimonas* strains BL-DC-9 and P3-3, *Dehalobacter* strains PER-K23 and 12DCB1A, and the *Roseiflexus* and *Chloroflexus* strains were included for comparison, and the 16S rRNA sequence of *Geobacter lovleyi* strain SZ was used to root the tree. All other strains and clones shown have been associated with PCB dechlorination. All clones were obtained from PCB enrichment cultures. Phylotypes VL-CHL1, , and *o*-17 are not shown because their published sequences are too short. However, their positions on the tree are represented by clone m-1 (identical to clones VL-CHL1 and SF1 over 466 and 470 bp, respectively) and clone OTU-1 (only 4 bp differences from *o*-17 over 714 bp)

the community DNA with the *D. mccartyi* specific primers DHC1F/DHC1377R (Hendrickson et al. 2002) did not detect any *D. mccartyi* 16 S rRNA genes in these enrichments. Zanaroli et al. concluded that phylotype VL-CHL1 represents the bacterial agent responsible for the dechlorination of Aroclor 1254 in these enrichments. This is the first time that the dechlorination of an Aroclor has been exclusively attributed to a member of the *Chloroflexi* other than *D. mccartyi*.

The tree with the highest log likelihood (−6038.3051) is shown. The percentage of trees in which the associated taxa clustered together (out of 100 replicates) is shown next to the branches. Initial tree(s) for the heuristic search were obtained automatically by applying Neighbor-Joining and BioNJ algorithms to a matrix of pairwise distances estimated using the maximum composite likelihood (MCL) approach, and then selecting the topology with superior log likelihood value. The tree is drawn to scale, with branch lengths measured in the number of substitutions per site. The analysis involved 27 nearly full-length nucleotide sequences. All positions containing gaps and missing data were eliminated. There were a total of 1036 positions in the final dataset. Evolutionary analyses were conducted in MEGA6 (Tamura et al. 2013).

VL-CHL1 removed ~75 % of the penta-, hexa-, and heptachlorobiphenyls in Aroclor 1254, converting these to tri- and tetrachlorobiphenyls. The dechlorination removed about 20 % of the chlorine substituents in 30 weeks and was characterized by the removal of flanked *meta* chlorines from $2\underline{3}$- and $2\underline{3}4$-chlorophenyl groups and the removal of flanked *para* chlorines from $3\underline{4}$- and $2\underline{4}5$-chlorophenyl groups (underscores here and throughout the chapter indicate the chlorines removed). The most prominent products were 2,2′,4,5′-chlorobiphenyl (24-25-CB) and 25-25-CB, and to a lesser extent, 25-3-CB. These characteristics match PCB Dechlorination Process H′ which was previously reported in the Acushnet Estuary of New Bedford Harbor, Massachusetts, USA (Brown and Wagner 1990).

23.2.2 *D. mccartyi* Mixed Cultures

The low bioavailability of PCBs results in a correspondingly low abundance of PCB dechlorinating *Dehalococcoides* in mixed communities obtained from contaminated sites. In addition to hampering PCB bioremediation efforts, the low abundance of PCB respiring bacteria in available samples poses a challenge in subsequent bacterial enrichment, isolation, and characterization.

Wang and He (2013a) screened the commercial PCB dechlorination activities in sediment and soil samples originating from four Asian countries (China, Indonesia, Malaysia, and Singapore), ultimately establishing nine PCB dechlorinating microcosms. To further elucidate PCB dechlorination processes, isolation of PCB dechlorinators is necessary, which requires the development of sediment-free cultures first. The nine microcosms were set up with soils, sediments or sludge, but without addition of any sediment substitutes. In each microcosm, only a single 16S rRNA *Dehalococcoides* genotype was identified. The *D. mccartyi* bacteria in these cultures are affiliated to all three phylogenetic subgroups: Cornell (cultures CG-2 and CG-4), Victoria (cultures CW-3, CG-1, and CG-3), and Pinellas (cultures CW-2, CW-4, CG-5 and SG-1) (Wang and He 2013b).

Unlike previous Aroclor 1260 dechlorinating microcosms, serial transfers of these nine microcosms were carried out in defined medium amended with

Aroclor 1260 and lactate, but without any substitutes for soil, sediment or sludge, and yielded six sediment-free PCB dechlorinating cultures: CW-4, CG-1, CG-3, CG-4, CG-5 and SG-1 (Wang and He 2013b). (Throughout this chapter we will use the term sediment-free to refer to cultures containing no soil, sludge, or sediment). The *D. mccartyi* organisms in each of these enrichment cultures coupled their growth with dechlorination of Aroclor 1260. The cell yield of *D. mccartyi* supported by PCB respiration reached ~3.3×10^{14} cells per mole of chlorine removed in these sediment-free cultures, which is comparable to the cell yield of *D. mccartyi* supported by respiration of chlorinated ethenes which ranges from ~7.8×10^{13} to 5.6×10^{14} cells per mole of chlorine removed (Löffler et al. 2013).

Previously reported mixed cultures exhibiting PCB dechlorination activity were all ultimately shown to contain PCB dechlorinating *Chloroflexi*, either *D. mccartyi* or the phylogenetically related, but distinct bacteria (e.g., *o*-17, DF-1, m1, SF1) (Wu et al. 2002a; Fagervold et al. 2007; Bedard et al. 2007). PCR amplification with *o*-17/DF-1 specific primers did not detect the presence of *o*-17/DF-1-type or m1/SF1-type *Chloroflexi* in any of the six sediment-free enrichment established by Wang and He (2013b).

In cultures CW-4, CG-1, CG-3, CG-4, CG-5, and SG-1, several distinct PCB dechlorination patterns were observed, as determined by PCB congener profile changes from the dechlorination of Aroclor 1260 and by the dechlorination products of two individual PCB congeners, 2345-245-CB and 234-245-CB. Process H is the dominant PCB dechlorination pattern observed in cultures CW-4 and SG-1. Dechlorination Process H removes flanked *para* and doubly flanked *meta* chlorines (Table 23.1). In culture SG-1 the dechlorination of the 245-chlorophenyl group was diminished compared to culture CW-4, resulting in more accumulation of 245-25-CB and less accumulation of 25-25-CB.

The dechlorination observed in culture CG-4 shared some elements of Process H, but was either a different dechlorination process or a combination of a less robust Process H and Process T. Either of the doubly flanked chlorines of the 2345-chlorophenyl group could be removed to yield both 235- and 245-chlorophenyl groups. The doubly flanked *meta* chlorine of $2\underline{3}4$- groups and the flanked *para* chlorine of $2\underline{4}5$-groups were also removed, but these latter two activities were much less prominent than in Process H. The dechlorination primarily converted heptachlorobiphenyls to pentachlorobiphenyls. The CG-5 culture showed the most extensive dechlorination of Aroclor 1260 via Dechlorination Process N.

Both cultures CG-1 and CG-3 exhibited novel PCB dechlorination patterns attacking primarily doubly flanked chlorines on heptachlorobiphenyls bearing 2345- and 234-chlorophenyl groups (Wang and He 2013b). Culture CG-3 primarily removed the *meta* chlorine of $2\underline{3}4$-chlorophenyl groups, and either the *meta* or *para* chlorine of 2345-chlorophenyl groups (both are doubly flanked). Culture CG-1 primarily attacked *meta* chlorines of $2\underline{3}4$- and $2\underline{3}45$-chlorophenyl groups (where the underlined chlorines are removed).

23.2.3 Mixed Culture AD14 (*D. mccartyi* and *Dehalobacter* sp.)

Members of *D. mccartyi* have also been found to dechlorinate PCBs in mixed cultures containing other obligate dechlorinating bacteria. A possible synergistic relationship between *D. mccartyi* and *Dehalobacter* was observed in culture AD14, a sediment-free Aroclor 1260 dechlorinating culture amended with lactate (Wang and He 2013a). This culture was established with sludge from an anaerobic digester in a wastewater treatment plant in which concentrations of PCBs, polybrominated diphenyl ethers, chlorophenols, chlorinated ethenes, and chlorinated ethanes were below the detection limit ($<0.1\ \mu M$). The PCB dechlorination pattern of culture AD14 most closely resembles Process H (Table 23.1) (Wang and He 2013a).

High throughput pair-end Illumina sequencing of 16S rRNA genes was performed in order to obtain a snapshot of the microbial community structure of culture AD14. *D. mccartyi* and *Dehalobacter* sp. were present in low abundance, 2.1 and 2.2 %, respectively, of the total sequences (Wang and He 2013a). The growth of both organisms was correlated with chlorine removal from PCBs, as determined by quantitative polymerase chain reaction (qPCR) analysis of 16S rRNA genes during dechlorination of Aroclor 1260. The qPCR data also showed that the *Dehalobacter* sp. had a longer lag phase than the *D. mccartyi* genotype, suggesting a possible requirement for intermediate PCB dechlorination products generated by *D. mccartyi*. The Illumina sequencing data (34,724 pair-end reads) showed the absence of other known reductive dechlorinating bacteria such as *o*-17/DF-1 type or m1/SF1 type Chloroflexi, *Desulfitobacterium*, *Geobacter*, *Sulfurospirillum*, or *Anaeromyxobacter*.

The low proportions of potential PCB dechlorinators in culture AD14 suggested that further enrichment of the dechlorinating bacteria was necessary prior to any attempt to characterize the RDase genes and gene products responsible for dechlorination of Aroclor 1260 in this culture. The PCB dechlorinating bacteria were enriched via addition of more bioavailable organohalides such as tetrachloroethene (PCE), 1,2-dichloroethane, and 2,4,6-trichlorophenol. The low relative abundance of both the *Dehalobacter* and *D. mccartyi* (each ~2 %) in the original culture AD14 increased to more than 50 % when grown with PCE. Along with the relative increase in abundance of certain populations, this highly enriched PCE fed subculture AD14-PCE retained PCB dechlorination activity. This provides confirmation that *D. mccartyi* and *Dehalobacter* were responsible for the Aroclor 1260 dechlorination in the original microcosm, consistent with the original microcosm Illumina sequencing result (Wang and He 2013a).

A significant finding is that PCB dechlorination was not inhibited by the presence of other organohalides that are found as co-contaminants with Aroclor 1260: octabromodiphenyl ether mixture, PCE, 1,2-dichloroethane, and 2,4,6-trichlorophenol. This may be important for the development of effective in situ bioremediation technologies.

D. mccartyi strains AD14-1 and AD14-2 were isolated from the sediment-free enrichment culture AD14-PCE. However, neither of these isolates was capable of dechlorinating PCB congeners in Aroclor 1260. This loss of metabolic ability may be attributed to loss of the PCB dechlorinators, loss of functional reductive dehalogenase genes for PCB dechlorination during the isolation process, or to PCB dechlorination requiring the cooperation of both *Dehalobacter* and *D. mccartyi.*

23.2.4 *D. mccartyi* Strain in Mixed Culture Dechlorinates Aroclor 1260 Exclusively by Removal of Doubly Flanked Chlorines

D. mccartyi strain 195 grows to much higher cell densities when grown in mixed culture with butyrate as the electron donor and carbon source and with fermented yeast extract as a supplement; therefore those conditions were used in the following experiments. Zhen et al. (2014) tested the ability of strain 195, the only known dechlorinator in the culture, to dechlorinate 1 μg/ml of Aroclor 1260, Aroclor 1254, or Aroclor 1242 in the presence, or absence, of periodic supplements of 25 μM 1,2,3,4-tetrachlorobenzene. This chlorobenzene is dechlorinated to 1,2,3- and 1,2,4-trichlorobenzene and appears to support growth of strain 195 by organohalide respiration (Fennell et al. 2004).

In 250 days, strain 195 dechlorinated 13 of the 24 major PCB congeners in Aroclor 1260. These congeners constitute 44 % of the total PCBs in Aroclor 1260. In the absence of 1,2,3,4-tetrachlorobenzene these congeners were decreased by 42 % in 250 days, but when 1,2,3,4-tetrachlorobenzene was added on days 0, 65, 108, and 156, the same congeners were decreased by 84 %. The congeners dechlorinated were primarily hepta-, octa-, and nonachlorobiphenyls which showed decreases of 21.5, 6.5, and 0.6 mol%, respectively; corresponding increases occurred in penta- and hexachlorobiphenyls (Zhen et al. 2014).

The congeners that were dechlorinated were composed mainly of 234-, 2345-, 2346-, and 23456-chlorophenyl rings (the targeted chlorines are underlined). The primary products were 235-245-CB, 245-24-CB, 235-236-CB/2356-25-CB, and 235-24-CB which increased by 8.2, 6.1, 5.6, and 4.9 mol%, respectively (Zhen et al. 2014). Three additional congeners, 235-235-CB, 245-246-CB, and 235-25-CB increased by 2.4 to 3.0 mol%. The authors showed stoichiometric mass balances for dechlorination substrates and products. On the basis of these, they concluded that the 23456-chlorophenyl group, which has three doubly flanked chlorines, and the 2345-chlorophenyl group which has two doubly flanked chlorines, are both primarily attacked at the *para* chlorine to yield 2356- and 235-chlorophenyl groups, respectively. The latter conclusion was confirmed by an experiment using 2345-4-CB as a substrate. Both 235-4-CB and 245-4-CB were products, but they were produced in a ratio of 49:1 (Zhen et al. 2014). This well characterized dechlorination pattern is novel and we assign it the name Dechlorination Process Z (Table 23.1).

Dechlorination experiments of strain 195 with the less-chlorinated Aroclor 1254, which has an average of about 5.1 chlorines per biphenyl, showed dechlorination of hexa- and heptachlorobiphenyls with doubly flanked chlorines to tetra- and pentachlorobiphenyls (Zhen et al. 2014). However, the impact of the dechlorination was far less than that for Aroclor 1260 because the Aroclor 1254 has a much smaller proportion of congeners with doubly flanked chlorines than Aroclor 1260 (Fig. 23.1). Dechlorination experiments with Aroclor 1242, which has an average of about 3.5 chlorines per biphenyl, showed very little dechlorination.

Several attempts to determine if strain 195 can use PCB congeners in Aroclor 1260 for respiration failed, as did an attempt using 2345-4-CB as an electron acceptor. The authors concluded that strain 195 most likely does not use any of the PCBs in Aroclor 1260 for organohalide respiration (Zhen et al. 2014).

23.3 Pure Culture of "*Dehalobium chlorocoercia*" Strain DF-1 Exclusively Removes Doubly Flanked Chlorines

Strain DF-1, informally named "*Dehalobium chlorocoercia*", was the first PCB respiring bacterium to be isolated (May et al. 2008b). It was isolated from sediments of Charleston Harbor (South Carolina, USA). Strain DF-1 is a member of the *Chloroflexi* related to the *Dehalococcoides*, and like them appears to be an obligate organohalide respirer, but it is significantly smaller, with a mean size of 137 ± 51 nm, and it can only be grown as a co-culture with cells of, or cell extract from, a *Desulfovibrio* sp. (May et al. 2008b). Similar to *D. mccartyi* strain 195, its PCB dechlorinating specificity, as determined by incubation with single congeners substituted on only one ring, is limited to removal of doubly flanked *meta* chlorines from 234- and 2346-chlorophenyl rings and doubly flanked *para* chlorines from 345- and 2345-chlorophenyl rings (where the underlined chlorines are removed) (Wu et al. 2002b).

DF-1 was grown with 2345-CB and added to nonsterile sediment contaminated with 4.62 μg/g of weathered Aroclor 1260 in order to determine if bioaugmentation with strain DF-1 would dechlorinate weathered PCBs in the presence of indigenous bacteria. The addition of DF-1 resulted in significant losses of hepta- and octachlorobiphenyls with doubly flanked chlorines (May et al. 2008a). Specifically, 2345-245-CB (PCB 180), 2345-234-CB (PCB 170), and 2346-234-CB (PCB 171) plus 2345-34-CB (PCB 156), where underscores indicate the chlorines targeted, decreased by 4.90, 2.55, and 2.12 mol%, respectively. However, there were no corresponding increases in the expected products: 235-245-CB (PCB 146), 235-24-CB (PCB 90), 234-246-CB (PCB 140), 2346-24-CB (PCB 139), 246-24-CB (PCB 100), and 235-34-CB (PCB 109) (May et al. 2008a). Instead, there were large increases in 235-4-CB (PCB 63), and in the peak containing 235-25-CB (PCB 92). The authors proposed that dechlorination products from DF-1 may have stimulated indigenous bacteria to carry out additional dechlorination (May et al. 2008a). Indeed, further dechlorination of the putative DF-1 products 235-34-CB and 235-245-CB, two congeners that DF-1 should not be able to dechlorinate, would form 235-4-CB and 235-25-CB, respectively.

The chlorophenyl ring specificity of a PCB dechlorinator is not always exactly the same for highly chlorinated PCBs as it is for congeners substituted on only one ring, because the chlorine configuration on the opposite ring can affect the position of dechlorination (Adrian et al. 2009; LaRoe et al. 2014) (see Sect. 23.4.1). Strain DF-1 is a unique and interesting PCB dechlorinator. It would be of interest to the field to know its precise specificity for Aroclor 1260 as has been determined for the six PCB dechlorinating *D. mccartyi* strains.

23.4 Pure Cultures of *D. mccartyi* Exhibit Diverse Complex Patterns of Dechlorination of Commercial PCB Mixtures

23.4.1 D. mccartyi Strain CBDB1

Strains of *D. mccartyi* that harbor different suites of RDase genes frequently have identical or nearly identical 16S rRNA genes (Löffler et al. 2013). Therefore, none of the previously described studies could determine whether the complex PCB dechlorination patterns observed in mixed cultures resulted from the action of a single strain or several strains of *D. mccartyi*. Definitive proof that a single *D. mccartyi* strain in pure culture can exhibit a complex pattern of dechlorination was first demonstrated with strain CBDB1, a strain originally isolated by growth with trichlorobenzenes as sole electron acceptor (Adrian et al. 2000, 2009).

Adrian et al. (2009) identified 43 PCB congeners with 3–8 chlorines that were dechlorinated by CBDB1. Most of these congeners are components of Aroclor 1260, although a few are components of Aroclor 1248 and some are not present in either Aroclor and were tested as single congeners (Adrian et al. 2009). Seven chlorophenyl rings were dechlorination substrates: singly flanked and doubly flanked *para* chlorines were removed from 34-, 245-, 2345-, and 345-chlorophenyl rings; primarily doubly flanked *meta* chlorines were removed from 234 and 2346-chlorophenyl rings; and either the doubly flanked *meta* (23456-) or *para* chlorines (23456-) from 23456-chlorophenyl rings (Adrian et al. 2009). The primary dechlorination products from Aroclor 1260 were 235-25-CB, 25-25-CB, and 24-25-CB. The observed dechlorination corresponds to PCB Dechlorination Process H which was observed in situ in sediments of the Acushnet Estuary of New Bedford Harbor (Massachusetts, USA) and the Hudson River (New York, USA) (Brown and Wagner 1990). This conclusively proved that a single *D. mccartyi* strain is capable of carrying out a complex PCB dechlorination pattern that occurs in the environment.

Further experiments with single congeners proved that the dechlorination of 234-chlorophenyl rings was strongly influenced by the chlorine configuration on the opposite ring; usually the doubly flanked *meta* chlorine was removed, but the singly flanked *para* chlorine could also be removed. Hence 234-234-CB and 234-245-CB were each dechlorinated by four distinct pathways comprising, respectively, six and seven different dechlorination reactions (Adrian et al. 2009). This was a level of complexity not previously appreciated.

23.4.2 *D. mccartyi* Strain JNA

D. mccartyi strain JNA, which, like CBDB1, belongs to the Pinellas subgroup, was recently isolated from the JN enrichment that metabolically dehalogenates Aroclor 1260 (LaRoe et al. 2014; Bedard et al. 2007). It had not previously been determined which of the congeners in Aroclor 1260 serve as respiratory electron acceptors. LaRoe et al. (2014) tested multiple PCB congeners and found that 236-236-CB was the best substrate for respiration. This was an unexpected finding because 236-236-CB has four *ortho* chlorines and exists as two separate enantiomers. Despite this, strain JNA stoichiometrically dechlorinated 236-236-CB to 236-26-CB and then 26-26-CB. Using 236-236-CB as sole electron acceptor, LaRoe et al. (2014) carried out multiple serial transfers to extinction to isolate strain JNA.

Strain JNA almost exclusively removes flanked *meta* chlorines from PCB congeners with 3–8 chlorines. Eight chlorophenyl groups are confirmed substrates for JNA: $\underline{3}4$-, $2\underline{3}4$-, $2\underline{3}5$-, $2\underline{3}6$-, $24\underline{5}$-, $2\underline{3}45$-, $2\underline{3}46$-, and $23\underline{5}6$-CB (underscores indicate the chlorines removed). These chlorophenyl groups constitute 88.7 mol% of the chlorophenyl groups in Aroclor 1260 (Frame et al. 1996), explaining why this dechlorination is so extensive. The major dechlorination products are 24-24-CB, 24-26-CB, 24-25-CB, and 25-26-CB. This dechlorination matches PCB Dechlorination Process N which occurs in the Housatonic River (Lenox, Massachusetts, USA). LaRoe et al. (2014) determined that JNA carries out 85 distinct PCB dechlorination reactions and utilizes 56 different dechlorination pathways, demonstrating the complexity of dechlorination that this strain is capable of. Genome sequencing is in progress, but it has already been determined that this strain harbors at least 19 putative reductive dehalogenase genes (Fricker et al. 2014).

23.4.3 Optimizing Enrichment and Isolation of PCB Dechlorinating *D. mccartyi* Strains

Like bacteria that reductively dechlorinate other halogenated compounds, PCB dechlorinating *D. mccartyi* strains employ reductive dehalogenases to catalyze chlorine removal from biphenyl rings. Wang et al. set about isolating PCB

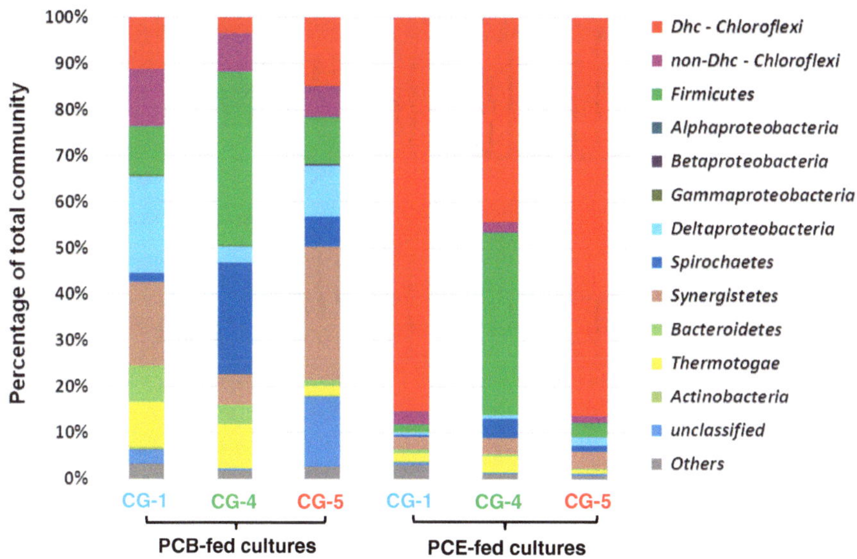

Fig. 23.3 Microbial community profiling of the enrichment process in mineral salts medium with lactate as the sole carbon source shows enrichment of *Dehalococcoides mccartyi* (*Dhc*, in *red*) in Aroclor 1260 dechlorinating cultures via a single transfer with PCE as an alternative electron acceptor. The stacked bar graph depicts microbial community compositions (obtained by Illumina sequencing of 16S rRNA genes) at the phylum level. Reprinted with permission from PNAS (Wang et al. 2014)

dechlorinating strains from three of their sediment-free cultures with the specific objective of identifying the functional genes responsible for PCB dechlorination. One difficulty in obtaining a pure culture of PCB dechlorinators is that even after enrichment with PCBs, the putative dechlorinators, in this case strains of *D. mccartyi*, represent a very small fraction of the total population. This is likely because the solubility of PCBs is extremely low. Therefore, Wang et al. (2014) used PCE, which is much more soluble than PCBs, as an alternative electron acceptor in order to increase the proportion and total biomass of PCB dechlorinating bacteria. As seen in Fig. 23.3, this strategy substantially increased the proportion of *D. mccartyi* cells and facilitated the isolation of *D. mccartyi* strains CG1, CG4, and CG5.

23.4.4 *D. mccartyi* Strains CG1, CG4, and CG5

Phylogenetic analysis showed that strains CG1, CG4, and CG5 (see Sect. 23.4.3) cluster into distinct *D. mccartyi* subgroups, Victoria, Cornell, and Pinellas (Fig. 23.2), represented by previously sequenced strains VS, 195, and CBDB1, respectively (Wang et al. 2014). Overall, each strain retained the PCB

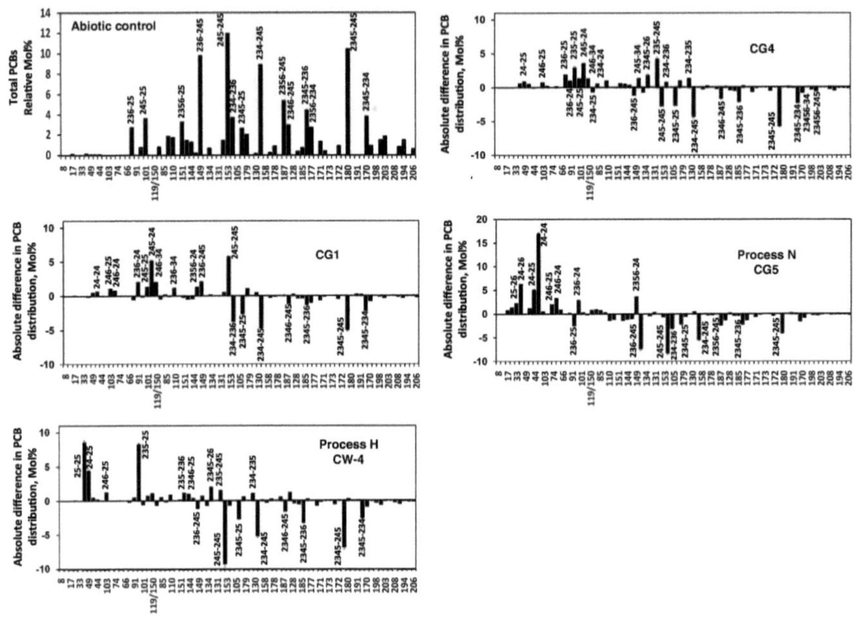

Fig. 23.4 Reductive dechlorination of Aroclor 1260 in three pure cultures and one sediment-free enrichment after 6 months of incubation. Absolute difference in the PCB congener distribution of Aroclor 1260 between the abiotic control and pure cultures CG1, CG4, and CG5, and mixed culture CW-4. Negative mol% indicates the amount of PCBs dechlorinated; positive numbers represent PCB congeners produced by dechlorination. Dechlorination Processes H and N are illustrated. The X-axes indicate the predominant PCB congeners in each peak; PCB congeners are designated by IUPAC number. This figure is adapted from figures previously published in PNAS and Plos One and is published here with permission (Wang et al. 2014; Wang and He 2013b)

dechlorinating specificity exhibited by its parent mixed culture (see Sect. 23.2.2) (Fig. 23.4) (Wang et al. 2014; Wang and He 2013b). Strain CG1 removed the doubly flanked *meta* chlorines of 234-, 2345-, and 2346-chlorophenyl groups. We are now able to describe the activity of strain CG4 in more detail and it appears to be a novel dechlorination process. Strain CG4 preferentially removed doubly flanked chlorines such as the *meta* chlorine of 234- and 2346-chlorophenyl groups, the *para* chlorine of 2345-, and to a lesser extent, the *meta* chlorine of 2345-chlorophenyl groups. Strain CG4 also removed the flanked *para* chlorines of 2346-, and to a much lesser extent, of 245-chlorophenyl groups. Strain CG4 removed nearly equal numbers of *meta* and *para* chlorines from Aroclor 1260, 0.13 and 0.19 chlorines per biphenyl, respectively (Wang et al. 2014). Strain CG5 removed primarily flanked and doubly flanked *meta* chlorines from 245-, 234-, 236-, 2345-, and 2346-chlorophenyl groups.

When grown with Aroclor 1260, strain CG1 removed 9 % of the *meta* chlorines, strain CG4 removed 5 % of the *meta* chlorines and 14.4 % of the *para* chlorines, and strain CG5 removed 35.6 % of the *meta* chlorines and 6 % of the *para* chlorines (Wang et al. 2014).

Continued growth with PCE increased the biomass of strains CG1, CG4, and CG5 and enabled Wang et al. to sequence the genomes revealing that the three strains have, respectively, 35, 15, and 32 putative RDase genes (Wang et al. 2014).

23.5 Identification and Characterization of Three PCB Reductive Dehalogenases

23.5.1 Identification of PCB Reductive Dechlorinases

RDase genes in *D. mccartyi* strains are usually identified through a combination of transcriptional analysis and enzymatic activity assays with purified RDases. However, the transcriptional and proteomic descriptions required for positive functional characterization require a high biomass of cells that cannot typically be reached by PCB respiring isolates (Wang et al. 2014). A single identical RDase gene dominated in both PCB- and PCE-fed cultures making the use of PCE to increase the biomass of the cells possible, which was critical (Fig. 23.3) for harvesting enough biomass for subsequent enzymatic activity tests. This enabled Wang and colleagues to use transcriptomic and in vitro enzyme activity assays to identify three distinct PCB dechlorinating enzymes PcbA1, PcbA4 and PcbA5 from each of *D. mccartyi* strains CG1, CG4, and CG5 (Wang et al. 2014). Cell-free enzyme assays with the crude cell lysates confirmed that the dechlorination specificity was the same as observed in the pure cultures. The three enzymes encoded by these genes each exhibit different regiospecificities in catalyzing dechlorination of a broad range of PCB congeners in Aroclor 1260.

23.5.2 Regiospecificity of the PCB RDases

Each PCB RDase gene was highly transcribed in pure culture amended with either PCE or Aroclor 1260. Dechlorination activities on Aroclor 1260 and highly chlorinated congeners were measured in crude cell lysates. The cell lysate containing PcbA1 from strain CG1 primarily removed doubly flanked *meta* chlorines from 2<u>3</u>45-, 2<u>3</u>46-, and 2<u>3</u>4-chlorophenyl rings. The cell lysate containing PcbA4 had the same specificity as the pure culture, preferential removal of doubly flanked chlorines and, to a much lesser extent, of the *para* chlorine of 245-chlorophenyl groups. The cell lysate containing PcbA5 had the most extensive PCB dechlorination capability which is very similar to Process N, removing flanked and doubly flanked *meta* chlorines from 2<u>3</u>45-, 2<u>3</u>4-, 2<u>3</u>5-, 2<u>3</u>6-, and 24<u>5</u>-chlorophenyl rings.

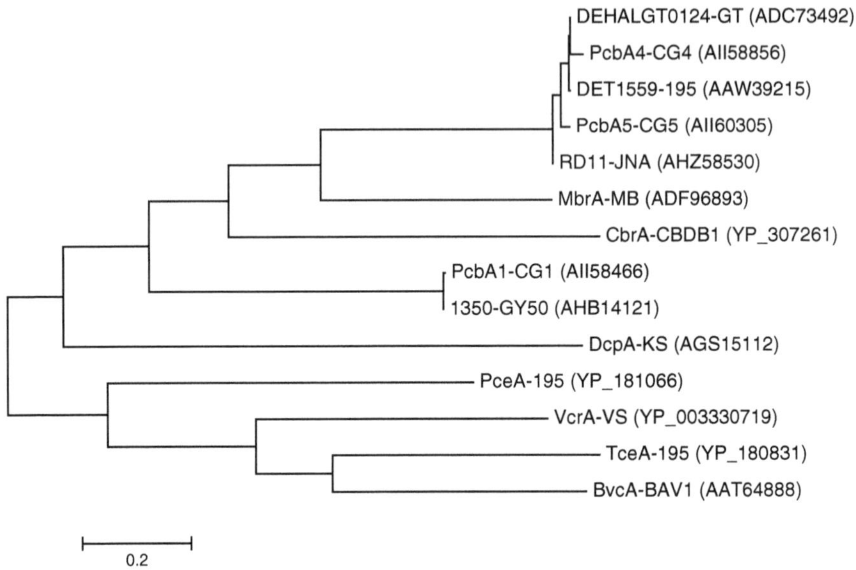

Fig. 23.5 Phylogenetic analysis of functionally characterized RDases in *D. mccartyi* including orthologs of the PCB reductive dehalogenases. DEHALGT0124, Det1559, and RD11 are orthologs of PcbA4 and PcbA5 in *D. mccartyi* strains GT, 195, and JNA; they have not yet been functionally characterized. Likewise, 1350-GY50 is an ortholog of Pcb1 in *D. mccartyi* strain GY50 but has not been functionally characterized. The tree was constructed with MEGA 6 (Tamura et al. 2013) using the maximum likelihood method in the Jones-Taylor-Thornton (JTT) model. Branch lengths indicate the number of substitutions per site

23.5.3 Phylogenetic Lineage of PCB RDases

Two of the PCB RDase enzymes, PcbA4 and PcbA5, which attack *para*- and *meta* chlorines, respectively, are phylogenetically similar, sharing 97 % amino acid sequence identity. The other, PcbA1, clusters in a distant clade and shares only 38 % amino acid sequence identity with PcbA4 and PcbA5. No PCB RDase has been identified in strains 195 or JNA yet, but both of these strains have orthologs of PcbA4 and PcbA5 that are potential candidates (Fig. 23.5) (Seshadri et al. 2005; Fricker et al. 2014). CBDB1 has no orthologs of either PcbA1 or the other two PCB RDases, so it must use an enzyme of yet another lineage. These findings suggest the existence of at least three different lineages of the PCB RDases in *D. mccartyi*. Moreover, the results suggest considerable diversity of regiospecificity within lineages.

23.5.4 PCB Reductive Dechlorinases also Dechlorinate PCE

Transcriptome and enzyme activity assays prove that in addition to dechlorinating PCBs, PcbA1, PcbA4, and PcbA5 dechlorinate PCE to TCE and then to *cis*- and *trans*-DCE (Wang et al. 2014). This substrate flexibility may provide an insight into the ability of *D. mccartyi* strains to evolve and metabolize anthropogenic compounds that did not exist prior to their industrial production. PCE and TCE are produced in volcanic gases (Jordan et al. 2000), hence these substrates have been present on earth for billions of years, whereas PCBs were first manufactured in 1929.

The discovery of the PCB RDases (Wang et al. 2014) increases the functional diversity of RDases identified in *D. mccartyi* which include the previously characterized RDases PceA, MbrA, TceA, VcrA, BvcA, DcpA, and CbrA (Fig. 23.5).

23.5.5 Bifunctional PCB/PCE RDases May Facilitate Bioaugmentation for PCB Remediation

D. mccartyi strains CG1, CG4, and CG5 utilize PcbA1, PcbA4, and PcbA5, respectively, for PCE respiration when grown with PCE as the sole electron acceptor, yet at the same time they can respire and dechlorinate PCBs (Wang et al. 2014). When grown with PCE as the electron acceptor, the cell density is 12.5- to 22-fold higher than when grown with PCBs. In addition, the cells grow much faster with PCE than with PCBs (several weeks versus several months) (Wang et al. 2014). PCB dechlorinators with bifunctional PCB/PCE RDases can be selectively enriched, transferred repeatedly, and grown to high cell densities with PCE as the sole electron acceptor with no possibility of losing their ability to dechlorinate PCBs, i.e., with no risk of losing their PCB RDase gene(s) (Wang et al. 2014). This property makes PCB dechlorinators with such bifunctional PCB/PCE RDases ideal candidates for use in bioremediation of PCBs because they can be easily grown in large amounts for bioaugmentation using PCE as the electron acceptor and the PCB RDases will be expressed.

23.6 Other Genera Implicated in PCB Dechlorination

23.6.1 Dehalobacter spp. Involved in PCB Dechlorination

The first evidence that *Dehalobacter* spp. were involved in PCB dechlorination was reported by Yan et al. (2006a) who implicated three putative PCB dechlorinating populations in enrichment established from Hudson River sediment and amended with 2345-CB. These investigators studied the impact of various amounts of aqueous CO_2 (supplied as sodium bicarbonate at final concentrations of 0, 100, 500, or 1000 mg/L) on PCB dechlorination and on the microbial populations. The primary route of dechlorination was 2345-CB → 235-CB → 25-CB → 2-CB, but small amounts of 245-CB, 23-CB, and 24-CB were also observed (Yan et al. 2006a), all of which are consistent with results typically reported for enrichments with Hudson River sediment. A single Pinellas-type *D. mccartyi* population was present in all enrichment cultures and was not affected by the sodium bicarbonate concentration. In contrast, two distinct *Dehalobacter* populations were much more prominent in the enrichment cultures amended with 100 mg/L sodium bicarbonate. These same cultures also exhibited the most rapid and extensive dechlorination, i.e., more production of 2-CB (Yan et al. 2006a). Because increased removal of the unflanked *meta* chlorine from 25-CB to produce 2-CB (a characteristic unique to Dechlorination Process M) was observed in the cultures that had the most prominent *Dehalobacter* populations, the authors proposed that this particular dechlorination reaction might be catalyzed by the *Dehalobacter* populations. The 1445 bp 16S rRNA gene fragments of these *Dehalobacter* phylotypes were 99.1 and 99.6 % identical to the 16S rRNA gene of *Dehalobacter restrictus* strain PER-K23.

In support of the idea that *Dehalobacter* may play an active role in dechlorination of PCBs with unflanked chlorines, Nelson et al. (2014) have recently described three *Dehalobacter* strains that can dehalogenate dichlorobenzenes with flanked and unflanked chlorines.

Further evidence implicating *Dehalobacter* in PCB dechlorination was reported by Yoshida et al. (2009b) who demonstrated reductive dechlorination of PCBs and 1,2,3-trichlorodibenzo-*p*-dioxin in a sediment-free environmental enrichment culture containing two *Dehalobacter* phylotypes; a novel dechlorination pathway: 2,3,4-trichlorobiphenyl to 3,4-dichlorobiphenyl and then to 4-chlorobiphenyl was attributed to the *Dehalobacter*. In addition, *Dehalobacter* was the only dechlorinator detected in an Aroclor 1260 dechlorinating microcosm CW-1, which was developed by Wang et al. from silt and clay near the Yang Tze River in China (Wang and He 2013b). The pattern of dechlorination in the CW-1 microcosm was typical of Process H. The 16S rRNA gene sequence of *Dehalobacter* sp. clone CW1 (JQ990318) is 99.3 % identical (over 1524 bp) to that of *Dehalobacter* sp. strain 12DCB1A which dechlorinates 1,2-dichlorobenzene (Nelson et al. 2014), and *Dehalobacter* sp. clone FTH2 which was implicated in the dechlorination of 4,5,6,7-tetrachlorophthalide (Yoshida et al. 2009a).

Finally, a *Dehalobacter* sp. was found in a subculture of PCB dechlorinating enrichment culture AD14 (see Sect. 23.2.3). The 1423 bp 16S rRNA gene fragment of *Dehalobacter* sp. clone AD14-PCE shares 99.2 % sequence identity with that of *Dehalobacter restrictus* sp. strain PER-K23 (DSM 9455). *Dehalobacter* sp. AD14-PCE couples growth with dechlorination of Aroclor 1260 (Wang and He, 2013b). Interestingly, *Dehalobacter* sp. clone AD14-PCE had a longer lag phase than the PCB dechlorinating D. mccartyi in the same culture, suggesting the possible synergistic involvement of the *Dehalobacter* in removal of chlorines from the less-chlorinated PCB congeners produced by the *D. mccartyi* in the same culture.

Dehalobacter spp. have recently been found to harbor multiple RDase genes, with up to 25 RDase genes present on a single genome (Kruse et al. 2013). The *Dehalobacter* spp. are currently the only non-Chloroflexi bacteria known to carry out PCB dechlorination activities.

23.6.2 *Dehalogenimonas alkenigignens*

The *Dehalogenimonas* genus of the *Chloroflexi* phylum, which is closely related to *Dehalococcoides,* also appears capable of dechlorinating and respiring PCBs. For example, Illumina sequencing analysis and qPCR suggested the involvement of members of the *Dehalogenimonas* genus as well as *D. mccartyi* in the reductive dechlorination of Aroclor 1260 in enrichment culture CG-3 (see Sect. 23.2.2) (Wang and He 2013b). Dechlorination in this sediment-free enrichment most closely approximates a mixture of Dechlorination Processes H and T, mainly attacking the doubly flanked *meta* and *para* chlorines (Wang and He 2013a). Subsequent two-step denaturing gradient gel electrophoresis (Wang and He 2012) and sequencing showed that the full-length 16S rRNA gene sequence of the *Dehalogenimonas* in culture CG-3 (Accession no. JQ990328) differs from that of *Dehalogenimonas alkenigignens* strain IP3-3(T), by a single nucleotide over 1493 bp. qPCR analysis showed that growth of *Dehalogenimonas alkenigignens* strain CG-3 was coupled to the dechlorination of Aroclor 1260 in this sediment-free culture, and that *D. alkenigignens* strain CG-3 was the dominant dechlorinator in the culture, representing 2.16 % of the total microbial population versus 0.37 % for *D. mccartyi* (Wang and He 2013a).

23.7 Conclusions and Outlook

The last seven years have been exciting and enlightening for those of us interested in PCB dechlorination. There are now six pure strains of *D. mccartyi*, representing all three phylogenetic subgroups, and a pure strain of "*Dehalobium chlorocoercia*", all of which have been demonstrated to dechlorinate the commercial PCB mixture Aroclor 1260, and at least four of which can use Aroclor 1260 for respiration. Complete genomes have been published for five of the *D. mccartyi* strains. In addition, there is strong evidence that both *Dehalogenimonas* spp. and *Dehalobacter* spp. can dechlorinate and respire PCBs.

It is now clear from many examples that *D. mccartyi* organisms capable of dechlorinating Aroclor 1260 are widespread at freshwater sites including locations in China, Germany, Singapore, and the USA. In contrast, non-*Dehalococcoides Chloroflexi* of the *o*-17/DF-1 and m1/SF1 groups have not been found in any freshwater sites to date even though they have been targeted with specific primers. On the other hand, a member of the *Chloroflexi* with a 446 bp 16S rRNA gene fragment sequence identical to that of clone m1 is apparently responsible for the dechlorination of Aroclor 1254 in a marine site (Zanaroli et al. 2012). Now that we know that the *Dehalobacter* and *Dehalogenimonas* genera also play a role in Aroclor dechlorination, it will be interesting to see the precise PCB dechlorination processes that they catalyze.

Pure strains of *D. mccartyi* that dechlorinate Aroclor 1260 by Processes H, N, Z, and variations of these are now available. However, no organisms capable of removing unflanked chlorines as in PCB Dechlorination Processes M, LP, and Q, have yet been identified. It will be important to identify and isolate such organisms because they hold the key to further dechlorinating the less chlorinated PCBs generated by Processes H, H′, N, P, T, and Z.

The discovery and characterization of the first three PCB reductive dehalogenases constitute a new milestone in the field. PcbA1, PcbA4, and PcbA5 dechlorinate Aroclor 1260 with distinct regiospecificities and prove that individual RDases can carry out the complex dechlorination of dozens of PCB congeners described by PCB dechlorination processes (Wang et al. 2014). PcbA4 and PcbA5 differ by only 14 amino acids out of 482, yet their specificity is entirely different (Wang et al. 2014). Studying and perhaps altering the amino acid sequences of these RDases should yield new insights into PCB dechlorination and perhaps increase their substrate range (Bedard 2014).

The discovery that PCB dechlorination and PCE dechlorination are both catalyzed by all three PCB dechlorinases has important implications for PCB remediation because it means that PCE can be used to grow large amounts of PCB dechlorinators actively expressing PCB RDases for use in bioaugmentation. This makes PCB remediation much more feasible.

References

Adrian L, Szewzyk U, Wecke J, Görisch H (2000) Bacterial dehalorespiration with chlorinated benzenes. Nature 408(6812):580–583

Adrian L, Dudková V, Demnerová K, Bedard DL (2009) "*Dehalococcoides*" sp. strain CBDB1 extensively dechlorinates the commercial polychlorinated biphenyl mixture Aroclor 1260. Appl Environ Microbiol 75(13):4516–4524

Bedard DL (2008) A case study for microbial biodegradation: anaerobic bacterial reductive dechlorination of polychlorinated biphenyls—from sediment to defined medium. Ann Rev Microbiol 62:253–270

Bedard DL (2014) PCB dechlorinases revealed at last. Proc Natl Acad Sci USA 111(33):11919–11920

Bedard DL, Quensen JF III (1995) Microbial reductive dechlorination of polychlorinated biphenyls. In: Young LY, Cerniglia CE (eds) Microbial transformation and degradation of toxic organic chemicals. Wiley, New York, pp 127–216

Bedard DL, Pohl EA, Bailey JJ, Murphy A (2005) Characterization of the PCB substrate range of microbial Dechlorination Process LP. Environ Sci Technol 39:6831–6839

Bedard DL, Ritalahti KM, Löffler FE (2007) The *Dehalococcoides* population in sediment-free mixed cultures metabolically dechlorinates the commercial polychlorinated biphenyl mixture Aroclor 1260. Appl Environ Microbiol 73(8):2513–2521

Brown JF, Wagner RE (1990) PCB movement, dechlorination, and detoxication in the Acushnet Estuary. Environ Toxicol Chem 9(10):1215–1233

Brown JF Jr, Wagner RE, Bedard DL, Brennan MJ, Carnahan JC, May RJ, Tofflemire TJ (1984) PCB transformations in upper Hudson river, USA, sediments. Northeast Environ Sci 3(3–4):167–179

Brown JF Jr, Bedard DL, Brennan MJ, Carnahan JC, Feng H, Wagner RE (1987) Polychlorinated biphenyl dechlorination in aquatic sediments. Science 236(4802):709–711

Cutter LA, Watts JEM, Sowers KR, May HD (2001) Identification of a microorganism that links its growth to the reductive dechlorination of 2,3,5,6-chlorobiphenyl. Environ Microbiol 3(11):699–709

Fagervold SK, Watts JEM, May HD, Sowers KR (2005) Sequential reductive dechlorination of *meta*-chlorinated polychlorinated biphenyl congeners in sediment microcosms by two different *Chloroflexi* phylotypes. Appl Environ Microbiol 71(12):8085–8090

Fagervold SK, May HD, Sowers KR (2007) Microbial reductive dechlorination of Aroclor 1260 in Baltimore Harbor sediment microcosms is catalyzed by three phylotypes within the phylum *Chloroflexi*. Appl Environ Microbiol 73(9):3009–3018

Fennell DE, Nijenhuis I, Wilson SF, Zinder SH, Häggblom MM (2004) *Dehalococcoides ethenogenes* strain 195 reductively dechlorinates diverse chlorinated aromatic pollutants. Environ Sci Technol 38(7):2075–2081

Field JA, Sierra-Alvarez R (2008) Microbial transformation and degradation of polychlorinated biphenyls. Environ Pollut 155(1):1–12

Frame GM, Cochran JW, Bøwadt SS (1996) Complete PCB congener distributions for 17 Aroclor mixtures determined by 3 HRGC systems optimized for comprehensive, quantitative, congener-specific analysis. J High Resol Chromatogr 19(12):657–668

Fricker AD, LaRoe SL, Shea ME, Bedard DL (2014) *Dehalococcoides mccartyi* strain JNA dechlorinates multiple chlorinated phenols including pentachlorophenol and harbors at least 19 reductive dehalogenase homologous genes. Env Sci Technol 48(24):14300–14308. doi:10.1021/es503553f

Hendrickson ER, Payne JA, Young RM, Starr MG, Perry MP, Fahnestock S, Ellis DE, Ebersole RC (2002) Molecular analysis of *Dehalococcoides* 16S ribosomal DNA from chloroethene-contaminated sites throughout North America and Europe. Appl Environ Microbiol 68(2):485–495

Hiraishi A (2008) Biodiversity of dehalorespiring bacteria with special emphasis on polychlorinated biphenyl/dioxin dechlorinators. Microbes Environ 23(1):1–12

Jordan A, Harnisch J, Borchers R, Le Guern F, Shinohara H (2000) Volcanogenic halocarbons. Environ Sci Technol 34(6):1122–1124

Kruse T, Maillard J, Goodwin L, Woyke T, Teshima H, Bruce D, Detter C, Tapia R, Han C, Huntemann M, Wei CL, Han J, Chen A, Kyrpides N, Szeto E, Markowitz V, Ivanova N, Pagani I, Pati A, Pitluck S, Nolan M, Holliger C, Smidt H (2013) Complete genome sequence of *Dehalobacter restrictus* PER-K23(T.). Stand Genomic Sci 8(3):375–388

LaRoe SL, Fricker AD, Bedard DL (2014) *Dehalococcoides mccartyi* strain JNA in pure culture extensively dechlorinates Aroclor 1260 according to polychlorinated biphenyl (PCB) Dechlorination Process N. Environ Sci Technol 48(16):9187–9196. doi:10.1021/es500872t

Löffler FE, Yan J, Ritalahti KM, Adrian L, Edwards EA, Konstantinidis KT, Muller JA, Fullerton H, Zinder SH, Spormann AM (2013) *Dehalococcoides mccartyi gen. nov., sp nov.*, obligately organohalide-respiring anaerobic bacteria relevant to halogen cycling and bioremediation, belong to a novel bacterial class, *Dehalococcoidia classis nov.*, order *Dehalococcoidales ord. nov* and family *Dehalococcoidaceae fam. nov.*, within the phylum Chloroflexi. Int J Syst Evol Microbiol 63(Pt 2):625–635

May HD, Miller GS, Kjellerup BV, Sowers KR (2008a) Author's correction. Dehalorespiration with polychlorinated biphenyls by an anaerobic ultramicrobacterium. Appl Environ Microbiol 74:6169–6170

May HD, Miller GS, Kjellerup BV, Sowers KR (2008b) Dehalorespiration with polychlorinated biphenyls by an anaerobic ultramicrobacterium. Appl Environ Microbiol 74(7):2089–2094

Nei M, Kumar S (2000) Molecular evolution and phylogenetics. Oxford University Press, New York

Nelson JL, Jiang J, Zinder SH (2014) Dehalogenation of chlorobenzenes, dichlorotoluenes, and tetrachloroethene by three *Dehalobacter* spp. Environ Sci Technol 48(7):3776–3782

Quensen JF III, Tiedje JM, Boyd SA (1988) Reductive dechlorination of polychlorinated biphenyls by anaerobic microorganisms from sediments. Science 242(4879):752–754
Quensen JF III, Boyd SA, Tiedje JM (1990) Dechlorination of four commercial polychlorinated biphenyl mixtures (Aroclors) by anaerobic microorganisms from sediments. Appl Environ Microbiol 56(8):2360–2369
Seshadri R, Adrian L, Fouts DE, Eisen JA, Phillippy AM, Methe BA, Ward NL, Nelson WC, Deboy RT, Khouri HM, Kolonay JF, Dodson RJ, Daugherty SC, Brinkac LM, Sullivan SA, Madupu R, Nelson KT, Kang KH, Impraim M, Tran K, Robinson JM, Forberger HA, Fraser CM, Zinder SH, Heidelberg JF (2005) Genome sequence of the PCE-dechlorinating bacterium *Dehalococcoides ethenogenes*. Science 307(5706):105–108
Tamura K, Stecher G, Peterson D, Filipski A, Kumar S (2013) MEGA6: molecular evolutionary genetics analysis version 6.0. Mol Biol Evol 30(12):2725–2729. doi:10.1093/molbev/mst197
Wang S, He J (2012) Two-step denaturing gradient gel electrophoresis (2S-DGGE), a gel-based strategy to capture full-length 16S rRNA gene sequences. Appl Microbiol Biotechnol 95(5):1305–1312
Wang S, He J (2013a) Dechlorination of commercial PCBs and other multiple halogenated compounds by a sediment-free culture containing *Dehalococcoides* and *Dehalobacter*. Environ Sci Technol 47(18):10526–10534
Wang S, He J (2013b) Phylogenetically distinct bacteria involve extensive dechlorination of Aroclor 1260 in sediment-free cultures. Plos One 8(3):e59178
Wang S, Chng KR, Wilm A, Zhao S, Yang K-L, Nagarajan N, He J (2014) Genomic characterization of three unique *Dehalococcoides* that respire on persistent polychlorinated biphenyls. PNAS 111(33):12103–12108. doi:10.1073/pnas.1404845111
Wiegel J, Wu Q (2000) Microbial reductive dehalogenation of polychlorinated biphenyls. FEMS Microbiol Ecol 32(1):1–15
Wu Q, Watts JEM, Sowers KR, May HD (2002a) Identification of a bacterium that specifically catalyzes the reductive dechlorination of polychlorinated biphenyls with doubly flanked chlorines. Appl Environ Microbiol 68(2):807–812
Wu Q, Milliken CE, Meier GP, Watts JEM, Sowers KR, May HD (2002b) Dechlorination of chlorobenzenes by a culture containing bacterium DF-1, a PCB dechlorinating microorganism. Environ Sci Technol 36(15):3290–3294
Yan T, LaPara TM, Novak PJ (2006a) The effect of varying levels of sodium bicarbonate on polychlorinated biphenyl dechlorination in Hudson River sediment cultures. Environ Microbiol 8(7):1288–1298
Yan T, LaPara TM, Novak PJ (2006b) The reductive dechlorination of 2,3,4,5-tetrachlorobiphenyl in three different sediment cultures: evidence for the involvement of phylogenetically similar *Dehalococcoides*-like bacterial populations. FEMS Microbiol Ecol 55(2):248–261
Yoshida N, Ye L, Baba D, Katayama A (2009a) A novel *Dehalobacter* species is involved in extensive 4,5,6,7-tetrachlorophthalide dechlorination. Appl Environ Microbiol 75(8):2400–2405
Yoshida N, Ye L, Baba D, Katayama A (2009b) Reductive dechlorination of polychlorinated biphenyls and dibenzo-*p*-dioxins in an enrichment culture containing *Dehalobacter* species. Microbes Environ 24(4):343–346
Zanaroli G, Balloi A, Negroni A, Daffonchio D, Young LY, Fava F (2010) Characterization of the microbial community from the marine sediment of the Venice lagoon capable of reductive dechlorination of coplanar polychlorinated biphenyls (PCBs). J Hazard Mater 178(1–3):417–426
Zanaroli G, Balloi A, Negroni A, Borruso L, Daffonchio D, Fava F (2012) A *Chloroflexi* bacterium dechlorinates polychlorinated biphenyls in marine sediments under in situ-like biogeochemical conditions. J Hazard Mater 209:449–457
Zhen H, Du S, Rodenburg LA, Mainelis G, Fennell DE (2014) Reductive dechlorination of 1,2,3,7,8-pentachlorodibenzo-*p*-dioxin and Aroclor 1260, 1254 and 1242 by a mixed culture containing *Dehalococcoides mccartyi* strain 195. Water Res 52C:51–62

Chapter 24
"*Dehalobium chlorocoercia*" DF-1—from Discovery to Application

Harold D. May and Kevin R. Sowers

Abstract "*Dehalobium chlorocoercia*" strain DF-1 is an organohalide respiring ultramicrobacterium isolated from a tidal estuary of Charleston Harbor using a polychlorinated biphenyl (PCB) congener as the sole electron acceptor. Organohalide respiration occurs by dechlorination of PCB congeners with doubly flanked chlorines, but this strain is also capable of dechlorinating chlorobenzenes with doubly flanked chlorines and tetra- and tri-chloroethene to a mixture of *cis*- and *trans*-1,2-dichloroethene. The range of PCB congeners dechlorinated from an Aroclor is limited in comparison with other PCB respiring strains; however, "*D. chlorocoercia*" strain DF-1 is capable of dechlorinating PCBs at environmentally relevant concentrations that are typically below saturation in water. In sediment-free medium an unidentified water-soluble factor from a *Desulfovibrio* sp. is required for growth. "*D. chlorocoercia*" strain DF-1 is osmotolerant, enabling it to grow and dechlorinate PCBs in sediments ranging from freshwater to marine. What follows is a description of "*D. chlorocoercia*" strain DF-1 and some of its related PCB respiring species from the perspective of environmental detection, dechlorination pathways and kinetics, biostimulation, electrostimulation, and finally bioaugmentation to enhance PCB degradation in sediments.

H.D. May
Marine Medicine and Environmental Science Center, Department of Microbiology and Immunology, Medical University of South Carolina, Charleston, SC, USA
e-mail: mayh@musc.com

K.R. Sowers (✉)
Institute of Marine and Environmental Technology, Department of Marine Biotechnology, University of Maryland Baltimore County, Baltimore, MD, USA
e-mail: sowers@umbc.edu

L. Adrian and F.E. Löffler (eds.), *Organohalide-Respiring Bacteria*,
DOI 10.1007/978-3-662-49875-0_24

24.1 The Discovery of PCB Dechlorinating Microorganisms

Polychlorinated biphenyls (PCBs) remain among the more difficult to treat environmental pollutants due to their chemical stability and low aqueous solubility, particularly as more chlorines are added to the biphenyl ring. The complexity of commercial PCB mixtures, which contain combinations of 209 possible congeners, further contributes to their recalcitrance in the environment. Identifying microorganisms that are capable of attacking these chemical structures proved to be a daunting task and began with the discovery of aerobic bacteria that could cleave the biphenyl ring of lesser-chlorinated congeners (Ahmed and Focht 1973; Furukawa 1976). Transformation of higher chlorinated PCB congeners was also observed in anaerobic sediments as early as 1984 (Brown et al. 1984). Identifying the microorganisms that dechlorinated PCBs proved to be difficult, but as described below anaerobic bacteria that could link their growth to PCB respiration were eventually identified as *Dehalococcoides* and closely related microorganisms within the Chloroflexi.

Growing PCB dechlorinating microorganisms remains difficult for several reasons, including the low solubility of PCBs in aqueous growth medium, the slow growth rates of organohalide respiring bacteria on PCBs and the requirement for anoxic, low redox growth conditions. This impeded research on anaerobic dechlorination of PCBs for many years, but the advent of molecular tools to detect microbes without cultivation in pure culture provided the first indication that the anaerobic microorganisms responsible for PCB dechlorination belonged to the Chloroflexi, which included *Dehalococcoides mccartyi* (formerly *Dehalococcoides ethenogenes*), known at that time to dechlorinate chlorinated ethenes. This initial discovery was accomplished by sequential transfer of estuarine sediment in minimal medium with short chain fatty acids as electron donors and a single PCB congener as an electron acceptor (Pulliam Holoman et al. 1998). Cloning and sequencing of 16S rRNA genes were then used to identify the predominant microorganisms within highly enriched sediment microcosms. Phylotype *o*-17 was identified as the microbe responsible for the *ortho* dechlorination of 2,3,5,6-tetrachlorobiphenyl to 3,5-dichlorobiphenyl through a combination of restriction and comparative sequence analyses of PCR-amplified 16S rRNA genes (Cutter et al. 1998, 2001). Despite identification of a PCB dechlorinator within the enriched microbial community, isolation continued to elude investigators due to the inability to grow the PCB dechlorinator in the absence of sediment. At the time, reductive dechlorination of PCBs was only observed in enrichment cultures that contained sediment, which also provided nutrients that supported the growth of non-PCB dechlorinating microorganisms. Cutter et al. (1998) finally developed a sediment-free culture of strain *o*-17 by incrementally diluting out sediment with sequential transfer in sediment-free minimal medium that contained acetate and 2,3,5,6-tetrachlorobiphenyl. An initial lag of 100 days in sediment-free medium was decreased to less than 50 days and dechlorinating activity was maintained throughout subsequent transfers. Ultimately, a coculture was obtained containing the 2,3,5,6-tetrachlrobiphenyl-respiring strain *o*-17 and a non-dechlorinating *Desulfovibrio* sp. Similarly, sequential dilutions in sediment-free minimal medium with formate and 2,3,4,5-tetrachlorobiphenyl were used to identify and isolate “*D. chlorocoercia*” strain DF-1 as a dechlorinator of PCBs with doubly flanked chlorines (Wu et al. 2002b; May et al. 2008a, b). A suite of molecular approaches was subsequently used to identify other PCB dechlorinating bacteria (Fagervold et al. 2005, 2007) with broader dechlorinating capabilities. The subsequent discovery that pure cultures of *D. mccartyi* dechlorinated PCBs expanded the diversity of PCB organohalide respiring bacteria even further (Fennell et al. 2004; Zhen et al. 2014; LaRoe et al. 2014; Wang et al. 2014; Adrian et al. 2009). Combined, these investigations indicated that a broad range of extensively chlorinated PCB congeners could be anaerobically dechlorinated, including Aroclor mixtures, by organohalide respiring microorganisms within the Chloroflexi. Many of the lesser-chlorinated congeners that remain following the transformative action of organohalide respiring anaerobes have the potential to be subsequently degraded via ring cleavage and mineralization by aerobic PCB degrading bacteria (Bedard 2003).

24.2 Isolation and Characterization of “*Dehalobium chlorocoercia*” Strain DF-1

“*D. chlorocoercia*” strain DF-1 was initially enriched in microcosms containing sediment from the Ashley River, a tributary of Charleston Harbor, SC, and minimal medium with fumarate as electron donor and 2,3,4,5-tetrachlorobiphenyl as the sole electron acceptor (Wu et al. 2000). Initial attempts to isolate strain DF-1 on solidified medium or in agar shake tubes were unsuccessful. Serial dilutions in sediment-free minimal medium with formate and 2,3,4,5-tetrachlorobiphenyl as the sole electron donor and acceptor, respectively, yielded a coculture with a *Desulfovibrio* sp. (Wu et al. 2002b). Strain DF-1 could only be maintained in coculture with the *Desulfovibrio* sp. throughout subsequent serial transfers. “*D. chlorocoercia*” strain DF-1 was finally isolated after serial dilution in minimal medium with titanium (III) nitrilotriacetate (TiNTA) substituted for cysteine as a chemical reductant, thereby minimizing partially oxidized sulfur product that could be used by the *Desulfovibrio* sp. as an electron acceptor (May et al. 2008a, b). However, growth and activity by strain DF-1 was no longer detected after three sequential serial dilution series and could only be restored and maintained by reinoculating cells or adding cell-free extracts from the *Desulfovibrio* sp. to the medium. Growth of strain DF-1 could also be maintained with cells or cell-free extracts from *Desulfovibrio vulgaris* but nutrients such as yeast extract, peptone, B-vitamins including cyanocobalamin, which is reported to stimulate growth of *D. mccartyi* strains (Löffler et al. 2013), could not be substituted. The factor in *Desulfovibrio* spp. required for maintaining growth of “*D. chlorocoercia*” strain DF-1 is water-soluble and heat-stable, but the structure is unknown at this time.

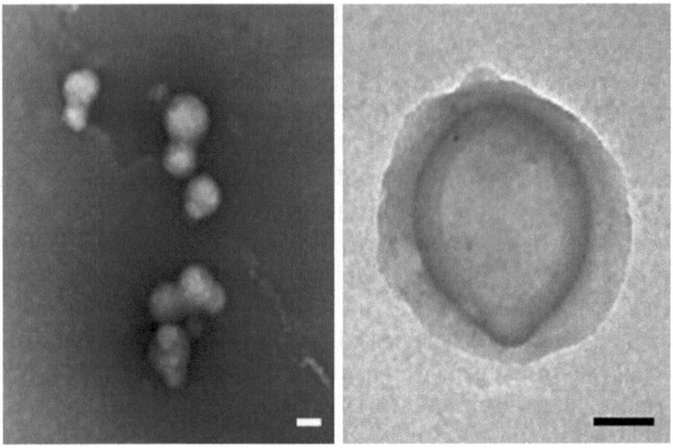

Fig. 24.1 TEM micrographs of negatively stained "*D. chlorocoercia*" strain DF-1. Bar = 0.1 μm. *Right panel* reprinted with permission from (May et al. 2008). Copyright 2008 American Society for Microbiology

"*D. chlorocoercia*" strain DF-1 is closely related to *D. mccartyi* within the organohalide respiring Chloroflexi. Strain DF-1 is an ultramicrobacterium with cells 75–339 nm in diameter when grown with 2,3,4,5-tetrachlorobiphenyl and often observed in clusters (Fig. 24.1). The small cell size may reflect the slow growth rates, but it would also maximize the membrane surface area-to-volume ratio, which would be an advantage for uptake of hydrophobic compounds such as a PCBs. Electron micrographs of the organism also reveal a possible matrix structure surrounding the cells, which if hydrophobic, would be consistent with the tendency of the organisms to clump or cluster. "*D. chlorocoercia*" strain DF-1 was the first organohalide respiring bacterium shown to link its growth to reductive dechlorination of PCBs in pure culture (May et al. 2008a, b). This strain requires hydrogen or formate as an electron donor and select organohalides as electron acceptors (May et al. 2008a, b). Growth and organohalide respiration by strain DF-1 is restricted to the dechlorination of PCB congeners with doubly flanked *meta* and *para* chlorines, but the microorganism is also capable of dechlorinating chlorobenzenes with doubly flanked chlorines (Wu et al. 2002a) and PCE and TCE to a mixture of *cis*- and *trans*-1,2-dichloroethene in a ratio of 1.2-1.7 (Miller et al. 2005). More recently, *D. mccartyi* strain CBDB1 was also reported to dechlorinate TCE to a mixture of *cis*- and *trans*-1,2-dichloroethene in a ratio of 0.3 (Marco-Urrea et al. 2011). Prior to these reports high amounts of *trans*-dichloroethene had only been observed in sediment enrichments and environmental samples. This observation suggests that organohalide respiring microorganisms similar to strain DF-1 and *D. mccartyi* strain CBDB1 are a potential source of *trans*-DCE, which is often detected in the environment. The optimal temperature range for growth is 30–33 °C with no growth or dechlorination observed at 10 or 35 °C. The pH optimum is 6.8 with growth observed in the range of 6.5–8.0. "*D. chlorocoercia*" strain DF-1 is osmotolerant and will grow and dechlorinate in the presence of NaCl concentrations ranging from 0.05 to 0.75 M and optimally in NaCl concentrations of 0.1–0.5 M (May et al. 2008a, b). The latter may be expected since the microorganism originated from a tidal estuary of Charleston Harbor that daily receives fresh water from the Ashley River and marine water from the Atlantic Ocean. The genome of "*D. chlorocoercia*" strain DF-1 (JGI project Gp0001256; NCBI Sequence Read Archive SRX018006) has been sequenced, but is not yet closed. Current efforts to annotate the genome have revealed the presence of multiple putative reductive dehalogenase genes, one of which is described below as a target for species specific monitoring of "*D. chlorocoercia*" strain DF-1 in environmental samples.

24.3 Stable Carbon Isotopic Fractionation of PCBs

Stable carbon isotopic fractionation conducted with "*D. chlorocoercia*" strain DF-1 shows that the $\delta^{13}C$ values during the microbial reductive dechlorination of 2,3,4,5-tetrachlorobiphenyl to 2,3,5-trichlorobiphenyl in enrichment cultures were relatively constant indicating no measurable isotopic effect (Drenzek et al. 2001). In the same study, it was reported that compound-specific $\delta^{13}C$ analysis performed for every congener in Aroclor 1268 showed an isotopic trend of decreasing ^{13}C abundance chlorine content increased, which is similar to observations for other commercial PCB mixtures. Overall, the study indicated that microbial dechlorination of PCBs in contaminated sediments will generate congener products with more depleted $\delta^{13}C$ values than those in the original Aroclor mixtures of similar chlorination. Similarly, no isotopic fractionation was observed for $\delta^{37}Cl$ during the dechlorination of 2,3,4,5-tetrachlorobiphenyl by enrichment cultures with "*D. chlorocoercia*" (Drenzek et al. 2004). The bulk $\delta^{37}Cl$ composition should be insensitive to the preferential loss of less chlorinated (more soluble) congeners before sedimentary deposition. In this case, other factors that might produce an isotope effect, such as sequential phase partitioning or chemical breakdown are implicated. This provides a means to discriminate between alteration of PCB congener profiles produced from abiotic weathering (depleted bulk $\delta^{37}Cl$) and reductive dechlorination (unaffected bulk $\delta^{37}Cl$). The combined results indicate that transformation of PCBs by reductive dechlorination can be discriminated from abiotic transformations by the absence of a measurable isotope effect during microbial PCB dechlorination. This absence of isotopic fractionation may prove equally valuable for discriminating these processes because of systematic internal variations in the $\delta^{13}C$ and $\delta^{37}Cl$ values of congeners in Aroclors and other PCB mixtures. The $\delta^{13}C$ and values of congeners in Aroclor 1242 and 1254 as well as other commercial PCB mixtures generally decrease with increasing chlorine content (Jarman et al. 1998). If laboratory results with "*D. chlorocoercia*" strain DF-1 accurately reflect activities in the field, then microbial reductive dechlorination will result in congeners with more negative $\delta^{13}C$ values than indigenous PCBs exhibiting the same degree of chlorination (Fig. 24.2).

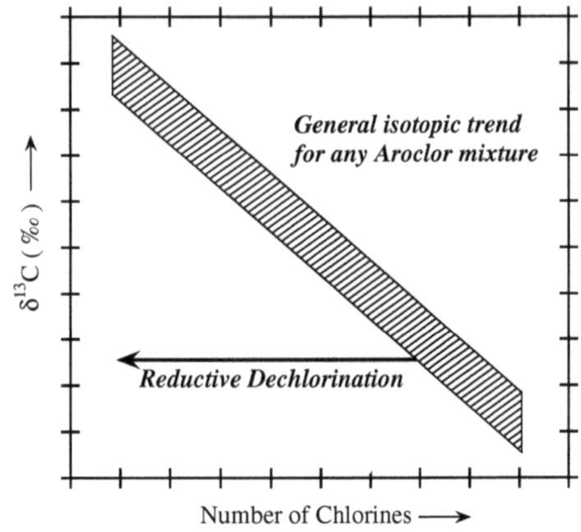

Fig. 24.2 Conceptual plot showing how PCBs formed from reductive dechlorination will be isotopically depleted relative to native PCBs of similar chlorination. Reprinted with permission from (Drenzek et al. 2001). Copyright 2001 American Chemical Society

24.4 Detection and Monitoring of "*D. chlorocoercia*" Strain DF-1

"*D. chlorocoercia*" strain DF-1 can be monitored in cultures and environmental samples using selective primers and polymerase chain reaction (PCR) amplified gene targets from total DNA. DNA from organohalide respiring bacteria can be extracted from 0.1 ml of sediment-free culture with Instagene matrix (Bio-Rad Laboratories) (Lombard et al. 2014) or from 0.25 g of sediment (wet wt) in a PowerBead microfuge tube with a Power Soil DNA Isolation Kit (MOBIO Laboratories, Inc.) (Payne et al. 2011). PCR primers Chl348F and Dehal884R amplify 16S rRNA gene fragments from strain *o*-17, "*D. chlorocoercia*" strain DF-1, phylotype DEH10, and *D. mccartyi* strain 195, but not *Chloroflexus aurantiacus,* which is outside of the known clade of organohalide respiring Chloroflexi (Fagervold et al. 2005). PCR products also are not detected for species outside of the Chloroflexi, including those from several bacterial and archaeal phyla. This primer set is effective for qualitative detection and monitoring of "*D. chlorocoercia*" strain DF-1 in mixed communities using denaturing gradient gel electrophoresis (DGGE) (Fagervold et al. 2005) and denaturing HPLC (DHPLC) (Kjellerup et al. 2008). Since this primer set is not strain DF-1 selective both assays require a PCR amplified product of the 16S rRNA gene from strain DF-1 as a standard to determine the correct migration distance for DGGE or elution time for DHPLC. This primer set has also been used for enumeration of "*D. chlorocoercia*" strain DF-1 in pure or mixed cultures without other organohalide respiring bacteria present using quantitative real-time polymerase chain reaction (qPCR) (Payne et al. 2011; Lombard et al. 2014). A more specific primer, Dehal1265R, was developed with high similarity values to strains *o*-17 and DF-1 16S rRNA gene sequences and was tested in silico (Watts et al. 2005). This specific primer paired with a universal forward primer (Edwards et al. 1989) yields a PCR product of 1215 base pairs with "*D. chlorocoercia*" strain DF-1. PCR amplicons are detected only with

Table 24.1 PCR primers for detecting organohalide respiring bacteria

Primer	Sequence 5′-3′	Target gene	References
Univ14F	AGAGTTTGATCCTGGCTCAG	Universal 16S rRNA	Edwards et al. (1989)
Dehal1265R	GCTATTCCTACCTGCTGTACC	Non-*Dehalococcoides* organohalide respiring Chloroflexi 16S rRNA	Watts et al. (2005)
Chl348F	GAGGCAGCAGCAAGGAA	Organohalide respiring Chloroflexi 16S rRNA	Fagervold et al. (2005)
Dehal884R	GGCCGGACACTTAAAGCG	Organohalide respiring Chloroflexi 16S rRNA	Fagervold et al. (2005)
SKFPat9F	GACAATGAGGACCCGGAATT	DF1 reductive dehalogenase	Payne et al. (2013)
SKFPat9R	TCCGCCAAAATAACGAACTG	DF1 reductive dehalogenase	Payne et al. (2013)

DNA from isolates and phylotypes within the non-*Dehalococcoides* organohalide respiring Chloroflexi, including strains *o*-17 or "*D. chlorocoercia*" strain DF-1. No PCR products are detected with other bacteria, including closely related *D. mccartyi* strains, indicating that the primers are suitable for detecting the *o*-17/DF-1 group in a microbial community. However, neither of these primer sets is effective for enumerating "*D. chlorocoercia*" strain DF-1 in an environmental sample if other non-*Dehalococcoides* organohalide respiring bacteria are present. An assay that selectively enumerates "*D. chlorocoercia*" strain DF-1 within a mixed community was developed with primers SKFPat9F and SKFPat9R, which target a gene encoding a putative reductive dehalogenase in "*D. chlorocoercia*" strain DF-1 (Payne et al. 2013). This assay was shown to selectively enumerate "*D. chlorocoercia*" strain DF-1 after bioaugmentation in PCB-impacted sediment mesocosms within a population of indigenous organohalide respiring bacteria. A summary of relevant primer sets is shown in Table 24.1. The lower detection limit of the PCR-based assays is approximately 100 target gene copies per ml culture or gram sediment (wet wt) based on 0.1 ml and 0.25 g samples, respectively.

24.5 Kinetics and Threshold Levels of PCB Organohalide Respiration

The low rates of natural attenuation for PCBs observed in the environment are often attributed to low bioavailability caused by several factors (Schwarzenbach et al. 2002). First, PCBs tend to adsorb strongly to particles, especially organic, due to high partitioning coefficients that are 1–5 orders of magnitude greater than that of chlorinated ethenes, which are frequent groundwater contaminants. This results in a low rate of mobilization into the aqueous phase. Second, depending on the congener, PCB solubility is up to 550 times lower than the solubility of chlorinated ethenes, depending on the specific congeners. Third, PCBs accumulate and remain relatively stable in soils and aqueous sediments because of their low vapor pressures. As a result of these characteristics the time required for a PCB-contaminated site to recover cannot yet be predicted due in part to a lack of quantitative information on rates of PCB dechlorination in the pore water phase. Although rates of reductive dechlorination in sediments depend upon the specific activities and abundance of organohalide respiring microbes, in situ activity will also be influenced by the aqueous concentrations of the PCB congeners. Several published reports suggest that substrates in nonaqueous phase solids or liquids are unavailable for microbial uptake (Zhang et al. 1998). In early studies, attempts to estimate dechlorination rates and the minimal threshold concentrations for organohalide respiration of PCBs involved adding Aroclors above the aqueous saturation range to sediment microcosms and assaying the rates of reductive dechlorination (Fish 1996; Rhee et al. 2001; Cho et al. 2002, 2003; Abramowicz et al. 1993). The minimal threshold Aroclor concentration for reductive dechlorination in these studies ranged from 10 to 40 mg kg^{-1}. The range of threshold values observed in the reports are a reflection of the specific indigenous dechlorinating populations, the different Aroclors added (Aroclor 1242 or 1248) and the sediment characteristics from different sources, which would affect the bioavailability. In contrast, Payne et al. (2011, 2013) observed dechlorination with as low as 1.3 mg kg^{-1} weathered PCBs in sediments after bioaugmentation with "*D. chlorocoercia*" strain DF-1, which indicated that low concentrations of PCBs typically observed in the environment were indeed available for direct microbial uptake. As these studies indicate, a major challenge with relating dechlorination rate to PCB concentration in sediment has been accounting for bioavailability differences caused by the association of PCBs to different types of organic matter (Ghosh et al. 2003). Perhaps a more appropriate metric that accounts for bioavailability to organisms in different sediment matrixes is to measure the dissolved concentrations of PCBs in the pore water (Peijnenburg and Jager 2003; Friedman et al. 2009).

Lombard et al. (2014) took advantage of recent advances in the use of polymer phase passive samplers for measurement, and for passive dosing of compounds, to measure PCB dechlorination rates at low, environmentally relevant aqueous concentrations. Dechlorination rates of 2,3,4,5-tetrachlorobiphenyl to 2,3,5-trichlorobiphenyl by "*D. chlorocoercia*" strain DF-1 were measured over a range of 1–500 ng L^{-1} in sediment-free medium using a steady-state concentration of cells (10^6 cells mL^{-1}). The dechlorination rates of 2,3,4,5-tetrachlorobiphenyl over a range of initial concentrations were linear indicating first order rate kinetics (Fig. 24.3). In addition, a minimum concentration threshold for 2,3,4,5-tetrachlorobiphenyl dechlorination was not detected with the size of inoculum used. Previous studies (Fish 1996; Rhee et al. 2001; Cho et al. 2002, 2003) also reported first order rate kinetics, but the apparent minimal threshold was several orders of magnitude greater since measurment included both PCBs in the pore water and those adsorbed to sediment. Furthermore, Lombard et al. (2014) observed higher rates up to 1000 fold more than reported previously. These rate

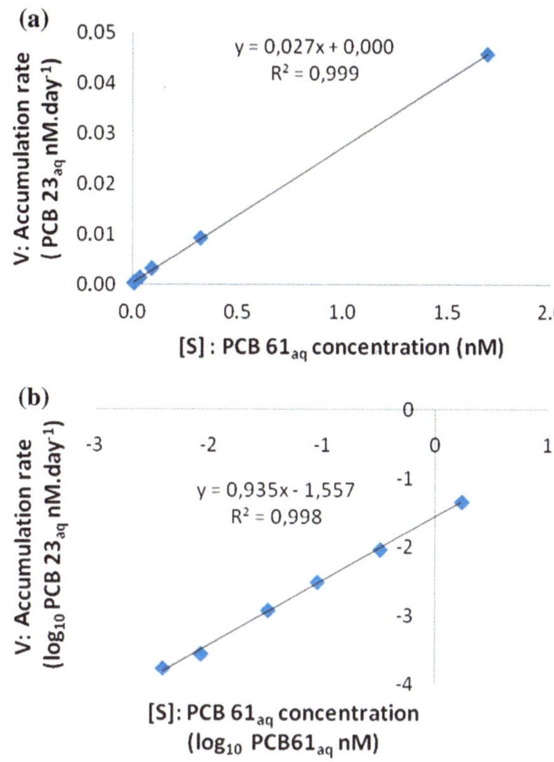

Fig. 24.3 Accumulation rate of product PCB 23_{aq} (2,3,5-trichlorobiphenyl) plotted against concentration of the substrate PCB 61_{aq} (2,3,4,5-tetrachlorobiphenyl), **a** normal scale **b** logarithmic scale at aqueous concentrations of: 2.9, 2.9×10^{-1}, 7.2×10^{-2}, 2.9×10^{-2}, 7.2×10^{-3} and 2.9×10^{-3} nM. Reprinted with permission from (Lombard et al. 2014). Copyright 2014 American Chemical Society

variations can be explained in part by differences in number and types of dechlorinating microorganisms present. Cho et al. (2003) reported that a fivefold difference in rates observed between two independent studies was negligible; when rates were normalized to the number of microorganisms, slope variations could be attributed to the cell (or more specifically the enzyme) affinity for a specific substrate. However, rate differences might also be explained by differences in buffering capacity of the associated sediments. Since only total PCBs were measured and the pore water PCB concentrations were unknown in these earlier studies, the kinetics of dechlorination for the bioavailable fraction of PCBs could not be determined.

Lombard et al. (2014) observed no net population growth of "*D. chlorococercia*" strain DF-1 at a cell density of approximately 1×10^6 cells mL^{-1} and 2,3,4,5-tetrachlorobiphenyl concentrations ranging from 1 to 500 ng L^{-1}. The thermodynamic cell yield of "*D. chlorocoercia*" strain DF-1 based on the estimated cell yield from oxidation of formate (Heijnen and van Dijken 1992) would require 2.4×10^{-8} mol of 2,3,4,5-tetrachlorobiphenyl reduction to support one doubling of 6×10^7 "*D. chlorocoercia*" strain DF-1 cells in a 50 mL microcosm. At the highest PCB concentration tested in the study only 3.3×10^{-11} mol of 2,3,4,5-tetrachlorobiphenyl was dechlorinated, which is consistent with the lack

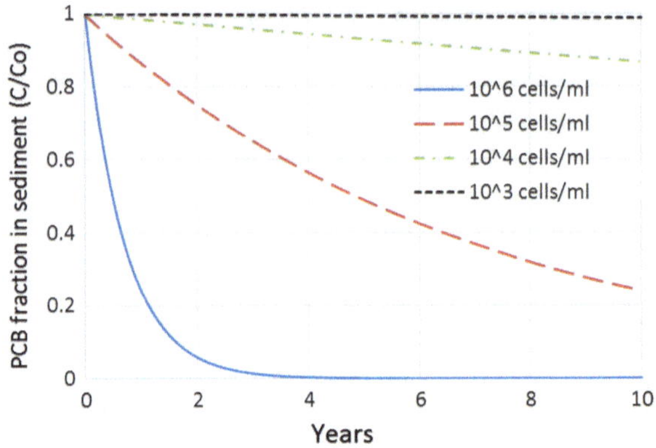

Fig. 24.4 Simulation of dechlorination profiles for bioavailable PCBs in sediment for different cell densities based on aqueous phase dechlorination rates for PCB 61 (2,3,4,5-tetrachlorobiphenyl). Reprinted with permission from (Lombard et al. 2014). Copyright 2014 American Chemical Society

of detectable growth. The combined results of these studies suggest that bioavailability was not a factor in the apparent inhibition of dechlorinating activity at relatively high PCB concentrations observed in earlier studies, but rather was due to low numbers of indigenous organohalide respiring microorganisms. Based on the dechlorination rate for 2,3,4,5-tetrachlorobiphenyl by 10^6 cells mL^{-1}, the estimated dechlorination rates were extrapolated for a range of organohalide respiring population densities in a typical organic sediment matrix (Fig. 24.4). The results show that although dechlorination likely occurs with low cell numbers, the rates would be too low for short-term detection in many environments. Higher concentrations of PCBs or alternative organohalide substrates are required for sustained growth of the organisms to reach population levels where substantial dechlorination can be observed within days or months. These kinetics and threshold data indicate that "*D. chlorocoercia*" strain DF-1, and presumably other PCB organohalide respiring bacteria, are capable of dechlorinating PCBs at environmentally relevant concentrations that are typically below saturation in water. Furthermore, sufficient cell numbers of "*D. chlorocoercia*" strain DF-1 reductively dechlorinate substantial levels of PCBs in days or months rather than years or decades. Using passive sampling to measure the dissolved aqueous concentrations of different PCB congeners and rates of PCB desorption from the sediment matrix, combined with knowledge of the congener specificity of the organohalide respiring bacteria used for bioaugmentation, it may be possible to project the rate and threshold levels of PCB dechlorination for a specific sediment site.

24.6 Dechlorination and Degradation of PCBs in Contaminated Sediments and Soils

Aerobic bioaugmentation studies with bacteria, fungi, and plants to degrade PCBs in soils and sediments have been done on a laboratory scale, but these processes generally attack PCBs without substitutions in the 2,3 and 3,4 positions and have a limited capacity to attack highly chlorinated congeners often found in the environment (Abraham et al. 2002). In contrast, microbial reductive dechlorination attacks more extensively chlorinated PCB congeners with flanked chlorines ultimately complementing the aerobic process by removing 2,3 and 3,4 substitutions. Prior to identification and isolation of PCB respiring bacteria, attempts to treat PCB contaminated sediments by bioaugmentation with PCB-enriched sediment yielded mixed results. Wu and Wiegel (1997) observed no significant stimulation and Bedard et al. (1997) observed only slight stimulation of weathered Aroclor 1260 dechlorination in Housatonic River sediment bioaugmented with PCB-enriched sediment slurries, although this activity could be increased by addition of 2,3,4,5,6-pentachlorobiphenyl as a biostimulant. Natarajan et al. (1997) reported dechlorination of weathered Aroclor 1242 and 1248 in Raisin River sediment microcosms inoculated with microbial granules from an upflow anaerobic sludge blanket digestor. However, the granules were not grown or maintained with PCBs or other organochlorides. The study did not rule out the possibility that the observed activity resulted from hydrogen generated from fermentative bacteria in the granules, which would stimulate indigenous PCB dechlorinating bacteria, rather than bioaugmentation with organohalide respiring bacteria. At the time these reports were published PCB respiring bacteria had not yet been identified as *Dehalococcoides* and related bacteria within the Chloroflexi. More recently, the commercial PCB mixture Aroclor 1260 was reported to be significantly dechlorinated by a consortium consisting of one or more phylotypes within the Chloroflexi enriched from sediment microcosms (Fagervold et al. 2007), sediment-free microcosms (Bedard et al. 2007) and by an individual strains of *D. mccartyi* (Fennell et al. 2004; LaRoe et al. 2014; Adrian et al. 2009; Wang et al. 2014). The different dechlorination patterns observed in the environment have been attributed to the diversity of indigenous PCB dechlorinators that attack different chlorine substituted positions within a congener. This conclusion is further supported in a report by Fagervold et al. demonstrating that the overall pathway for dechlorination of spiked Aroclor 1260 could be altered by adding different combinations of PCB respiring bacteria to sediment microcosms (Fagervold et al. 2011). Since PCBs generally accumulate in soils and sediments the challenges for their remediation differ from those developed for volatile halogenated contaminants such as chlorinated ethenes in groundwater. Efforts to enhance the dechlorination of PCBs in contaminated sediments with strain DF-1 and other organohalide respiring bacteria are discussed below.

24.6.1 Biostimulation

There have been efforts to identify factors affecting dechlorination and degradation activities in laboratory microcosms (Abramowicz et al. 1993; Tiedje et al. 1993; Berkaw et al. 1996; Wu et al. 1996; Cho et al. 2004) with the goal of accelerating the natural processes in the environment. Although PCB respiring bacteria can be enriched in laboratory microcosms, they require PCBs at concentrations ranging from 10 to 100 mg mL^{-1} to observe growth and activity, which is significantly greater than concentrations often detected in most contaminated sites. As the addition of high concentrations of PCBs as a biostimulant to a contaminated site is not a practical solution for treating PCB contamination, investigators have attempted to enhance dechlorination of PCB contaminated sediments by addition of an alternative halogenated electron acceptor, often referred to as a haloprimer, to stimulate the activity of indigenous organohalide respiring bacteria. The first in situ stimulation of PCB dechlorination was reported over 20 years ago (Bedard et al. 1995). In that study, the application of a high concentration of 2,6-dibromobiphenyl (109 mg kg^{-1}) as a haloprimer successfully stimulated a 74 % decrease in hexa- to octachlorobiphenyls in 1 year. Other strategies have been effective for halopriming PCB contaminated sediments in laboratory microcosms including the addition of 2,3,4,5,6-pentachlorobiphenyl (Van Dort et al. 1997), halobenzoates (Deweerd and Bedard 1999) and pentachloronitrobenzene (Park et al. 2011). Chlorobenzenes and chlorophenols stimulated reductive dechlorination in sediments spiked with a high concentration of Aroclor 1248 (ca. 100 mg kg^{-1}), but were not tested with sediments contaminated with low levels of PCBs (Cho et al. 2002). Wu et al. (1999) used a most probable number method to show that halopriming with 2,6-dibromobiphenyl increased the number of PCB- and 2,6-dibromobiphenyl dehalogenators, suggesting that increased dechlorination rates observed resulted from increasing the population size of the indigenous organohalide respiring population. Although biostimulation shows potential for treating PCB-impacted sediments, all of the effective electron acceptors known at this time are halogenated aromatic compounds and their release into the environment is subject to regulatory restrictions. Polybrominated biphenyls, often used in flame retardants, are restricted under the Restriction of Hazardous Substances Directive in the EU and in several other countries (http://eur-lex.europa.eu). Although dibromobiphenyls are not found in commercial polybrominated biphenyl mixtures they could be subject to the same regulatory restrictions. Chlorobenzenes, chlorophenols and halogenated benzoates, used in a wide range of industrial applications are a large group of potentially toxic environmental pollutants (Bhatt et al. 2007). Pentachloronitrobenzene, which is widely used as a fungicide in several countries, is perhaps the most tractable option for biostimulation. However, application of pentachloronitrobenzene as a fungicide has been subject to periodic restrictions by the U.S. Environmental Protection Agency. Several non-organohalide biostimulants such as lactate and iron have a stimulatory effect on the anaerobic reductive dechlorination of organohalides (Bruce and Henry 2010). However, the addition of sodium lactate and zerovalent iron as potential electron donors did not have a significant effect on dechlorination of PCB-impacted sediment mesocosms bioaugmented with "*D. chlorocoercia*" strain DF-1 (Payne et al. 2013) or a highly enriched organohalide respiring consortium from the Raisin River (Winchell and Novak 2008). $FeSO_4$, was reported to stimulate PCB dechlorination in sediment microcosms spiked with a high concentration of Aroclor 1242 (Zwiernik et al. 1998). The authors suggested that the stimulation resulted from PCB dechlorinating sulfate reducing bacteria (SRB), however, they did not rule out the possibility that stimulation of SRB indirectly increased the size of the organohalide respiring population by increasing hydrogen production by fermentation and interspecies hydrogen exchange. Since organohalide respiring bacteria have a lower K_s for hydrogen uptake, an increased hydrogen pool would be available to them without competition with SRB (Fennell and Gossett 1998). That study was conducted with highly enriched cultures spiked with Aroclor 1242 and the effectiveness of $FeSO_4$ as a biostimulant for PCB dechlorination in situ with low PCB concentrations remains untested. At this time, a non-organohalide biostimulant for use in situ that specifically targets "*D. chlorocoercia*" strain DF-1 or other PCB respiring bacteria has not been identified.

24.6.2 Electrostimulation

An intriguing approach to bioremediation is the electrostimulation of biodegradative and dechlorinating microorganisms (Jin and Fallgren 2014; Lu et al. 2014). The application of electrochemistry to the reductive dechlorination of organohalides is reviewed by Aulenta elsewhere in this book. By applying electrodes to contaminated groundwater, sediment, or soil, one may supply electron donor and/or acceptor possibly through direct interactions with the electrodes or by the generation of other chemical reductants or oxidants. In addition, oxygen and hydrogen may be supplied to the degradative/dechlorinating microorganisms through water electrolysis. In the case of PCBs, both reductive and oxidative processes may be required for full and effective biodegradation. A two-electrode application, perhaps with alternating polarity, has the potential to stimulate microbial degradation of PCBs in contaminated sediment.

"*D. chlorocoercia*" strain DF-1 grown in a defined minimum medium provides the opportunity to examine the donation of electrons for PCB reductive dechlorination with a cathode to an organohalide respiring microbe under controlled conditions. However, a successful experiment as such has not been reported. Incubating strain DF-1 at a graphite cathode of a 3-electrode bioelectrochemical cell that was poised at 600 mV versus the standard hydrogen potential did not result in detectable dechlorination of 2,3,4,5-tetrachlorobiphenyl (unpublished data, C. Chun, K. Sowers, H. May). At this potential a graphite cathode will produce H_2, but the amount is not appreciable unless the system is sealed and the H_2 is allowed to accumulate. These tests were conducted with N_2: CO_2 (4:1) sparged through the cell. Therefore, these results indicate that PCB dechlorination by "*D. chlorocoercia*" strain DF-1 is not likely to be facilitated through direct electron transfer of electrons from an electrode. However, one may expect that PCB dechlorination by the microbe would occur if H_2 were allowed to accumulate in the electrochemical cell or at depth in sediment. Furthermore, the addition of a soluble electron mediator or sediment, which may supply a mediator, could potentially support PCB dechlorination by "*D. chlorocoercia*" strain DF-1 or a consortium containing such organohalide respiring microorganisms.

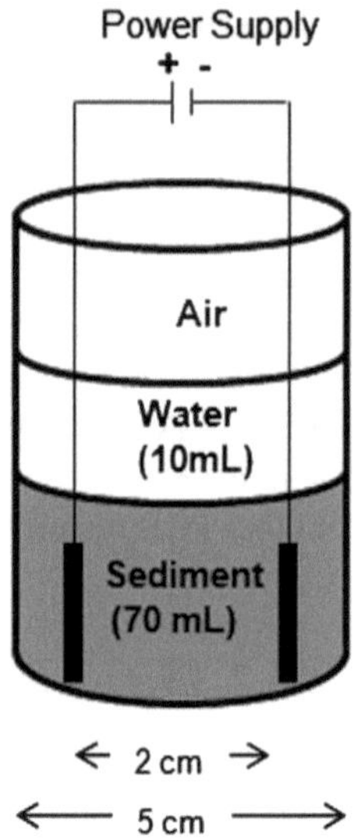

Fig. 24.5 Schematic representation of a bioelectrochemical reactor for testing the effect of current on PCB transformation in sediment microcosms. Reprinted with permission from (Chun et al. 2013). Copyright 2013 Elsevier

Chun et al. (2013) examined the application of voltage to live sediment with weathered PCBs (Aroclor 1242 in sediment from Fox River, Neenah, WI) with and without the addition of 2-monochlorobiphenyl and 2,3,4,5-tetrachlorobiphenyl. The sediment was maintained in sediment bioelectrochemical reactors (SBRs) that were maintained open to the atmosphere in order to mimic field conditions (Fig. 24.5). Two graphite electrodes were applied vertically into the sediment and voltage was applied at 1.5, 2.2, and 3.0 V. After 88 days the entire contents of the SBR were analyzed for PCBs and compared with sediment in control reactors that were identically prepared but did not receive power (Fig. 24.6). Even with only 1.5 V applied to the system, which was below the voltage required to electrolytically generate O_2 at the cathode, greater than ~50–60 % of the total PCBs were eliminated. Small amounts of chlorobenzoates, products of PCB oxidation and most likely transient, were detected with (2-, 2,6-, and 2,3,5-chlorobenzoate) and without (2-, 2,3-, and 3,4-chlorobenzoate) the addition of 2-monochlorobiphenyl and 2,3,4,5-tetrachlorobiphenyl. Aerobic PCB degrading microbes and PCB dechlorinators were enriched from the sediments in experiments independent of the electrochemical experiments within this study.

Reductive dechlorination of PCBs may have been supported by H_2 that would be formed with this amount of voltage, but the reduction in total PCBs was so extensive that PCB dechlorination was not discernible. Since no PCBs were lost from the controls, it was concluded that oxidative PCB biodegradation must have occurred due to either the introduction of oxygen, perhaps by benthic worms whose mobility was observed to be increased when voltage was applied, or to the anaerobic oxidation of the PCBs coupled to electrode reduction. The oxidation of aromatic compounds coupled to electrode reduction by microbes has been reported (Bond et al. 2002; Zhang et al. 2010). In addition, the cycling of

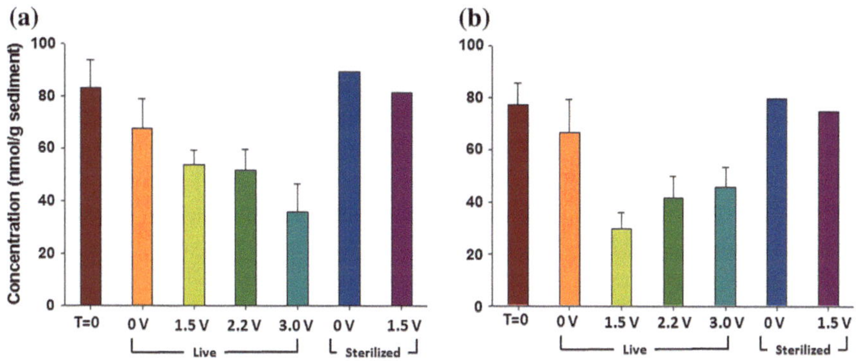

Fig. 24.6 Total concentration of weathered Aroclors in sterilized and live Sediment Bioelectrochemical Reactors (SBR) with (**a**) and without (**b**) the addition of 2-monochlorobiphenyl and 2,3,4,5-tetrachlorobiphenyl. The sample size of T = 0 and 0 V at 88 days is 18 (triplicate samples from 6 reactors) and 5 (triplicate samples from one reactor and one sample from two reactors), respectively. For SBRs with the applied voltage at 88 days, triplicate samples were analyzed from the homogenized contents of each. Each datum point represents the mean and standard deviation of three replicates samples

inorganic and organic molecules in the SBRs due to abiotic and biotic reactions driven by the application of the voltage may support the observed elimination of the PCBs. These results are promising but a better understanding of the mechanism at work will require further research, particularly to know if and how microbial dechlorination may participate under the conditions tested. Finally, it remains to be seen if electrostimulation may further stimulate the treatment of weathered PCBs following bioaugmentation.

24.6.3 Bioaugmentation

Bedard et al. (2007) observed that a critical mass of *Dehalococcoides* cells was required before reductive dechlorination was detected in a sediment-free enrichment culture spiked with Aroclor 1260 and proposed that low indigenous numbers of organohalide respiring bacteria might explain why substantial attenuation of PCBs is rarely observed in the environment. Growth of organohalide respiring bacteria is linked to reductive dechlorination of PCBs in laboratory cultures containing high concentrations of PCBs (Fagervold et al. 2007; Adrian et al. 2009). However, Lombard et al. (2014) calculated that the aqueous concentrations of PCBs typically found in environmental samples is insufficient to support the critical mass of cells necessary to cause high rates of PCB dechlorination. An alternative strategy for stimulating in situ PCB dechlorination in sediment impacted with low levels of PCBs is to bioaugment sediment with high numbers of organohalide respiring microorganisms. Bedard (1997) demonstrated that bioaugmentation with sediment slurries enriched for PCB dechlorinating bacteria by sequential transfer with 2,3,4,5,6-pentachlorobiphenyl as a haloprimer reduced hexa- through nona-chlorobiphenyls of relatively high levels (50 ppm) of weathered Aroclor 1260 in Housatonic River sediment by 19.7 % in 312 days. Krumins et al. (2009) found that the addition of *D. mccartyi* strain 195 stimulated the dechlorination of PCBs in sediment contaminated with a low concentration (2.1 ppm) of Aroclors 1248, 1254 and 1260 in laboratory microcosms, although pentachloronitrobenzene was also added as a haloprimer. May et al. (2008a, b) demonstrated that bioaugmentation with a pure culture of "*D. chlorocoercia*" strain DF-1 stimulated the reductive dechlorination of low concentrations of Aroclor 1260 (4.6 ppm) by 8.9 mol% in contaminated soil microcosms containing indigenous organohalide respiring bacteria and in the absence of a haloprimer. Based on the latter results (Payne et al. 2011) tested the efficacy of scaling up bioaugmentation with "*D. chlorocoercia*" strain DF-1 in 2-L laboratory mesocosms containing sediment contaminated with weathered Aroclor 1260 (1.3 ppm) and water from Baltimore Harbor, MD (Fig. 24.7). In this study total penta- and higher chlorinated PCBs decreased by approximately 56 % (by mass) in bioaugmented mesocosms after 120 days in contrast to un-amended controls that showed no measureable activity. Bioaugmentation with "*D. chlorocoercia*" strain DF-1 enhanced the dechlorination of doubly flanked chlorines as expected, but also stimulated the dechlorination of singly flanked chlorines as a result of an apparent synergistic effect on the indigenous population. Furthermore, although "*D. chlorocoercia*" strain DF-1" is not indigenous to Baltimore Harbor, the inoculum was sustained throughout the dechlorination process. This study affirmed the feasibility of using bioaugmentation with a PCB respiring microorganism to stimulate in situ reductive dechlorination of PCBs in contaminated sediments.

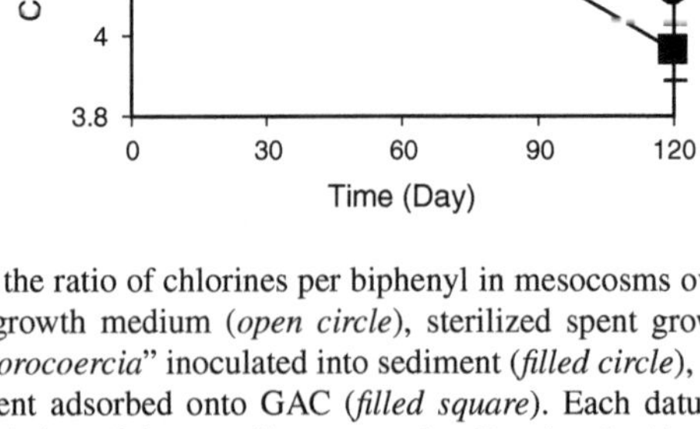

Fig. 24.7 Changes in the ratio of chlorines per biphenyl in mesocosms over time after treatment with sterilized spent growth medium (*open circle*), sterilized spent growth medium and GAC (*open square*), "*D. chlorocoercia*" inoculated into sediment (*filled circle*), and "*D. chlorocoercia*" inoculated into sediment adsorbed onto GAC (*filled square*). Each datum point represents the mean and standard deviation of three replicates samples. Reprinted with permission from (Payne et al. 2011). Copyright 2011 American Chemical Society

24.6.4 Coupling Reductive Dechlorination by "D. chlorocoercia" Strain DF-1 with Aerobic Degradation

Natural microbial attenuation of PCBs in the environment is presumed to occur by the complementary processes of anaerobic dechlorination of more highly chlorinated congeners and subsequent aerobic degradation of those dechlorination products (Abramowicz 1995), and this has been suggested to be a potential treatment strategy for PCB-impacted sediment. Anaerobic incubation of PCB-impacted sediment in a microcosm with an Aroclor followed by transfer into an aerobic culture containing *Burkholderia xenovorans* strain LB400 has been reported to effectively degrade Aroclors by as much a 70 % (Evans et al. 1996; Master et al. 2002). Although creating sequential anaerobic–aerobic conditions are possible in closed laboratory microcosms, it is difficult to reproduce these conditions in situ. To address this problem Payne et al. (2013) tested the efficacy of bioaugmentation with anaerobic PCB respiring "*D. chlorocoercia*" strain DF-1 and aerobic PCB degrading *B. xenovorans* strain LB400 added concurrently in 2 L laboratory mesocosms containing sediments historically contaminated with Aroclor 1260, but also contaminated with relatively high concentrations of di-, tri- and tetra-chlorobiphenyls. In contrast to prior studies that employed sequential anaerobic and aerobic treatments, Payne et al. (2013) employed static sediment mesocosms open to the air to simultaneously create both aerobic and anaerobic zones, employed only indigenous water rather than culture medium, and did not employ aromatic biostimulants, such as biphenyl, brominated biphenyls, or chlorobenzoates. The premise of the study was that under in situ conditions anaerobic and aerobic regions coexist throughout the sediments column as a result of microniches within particles and dynamic cycling of redox conditions by bioturbation. Bioaugmentation with "*D. chlorocoercia*" and *B. xenovorans* together resulted in an 80 % decrease by mass of PCBs, from 8 to less than 2 mg/kg after 120 days (Fig. 24.8). The mesocosm inoculated with both the aerobe and anaerobe did not show a significant increase in the extent of degradation compared with the aerobe alone. However, in the mesocosm bioaugmented with both microorganisms, there was a 20 % greater decrease in mass of hexa- through nona-chlorobiphenyl homologs and there was a shift in the overall congener pattern, compared with bioaugmentation with the aerobe alone. This observation indicates that both anaerobic dechlorination of highly chlorinated congeners and aerobic degradation of less chlorinated congeners occurred concurrently. In contrast, non-bioaugmented controls containing filtered culture supernatant showed only a 25 % decrease in total levels of PCBs after 365 days. Most aerobic degradation was detected in the first 120 days and then continued to day 365 at a reduced rate. However, both the PCB transforming anaerobe and aerobe were viable at the end of the study. This result indicates that "*D. chlorocoercia*" and aerobic *B. xenovorans* could successfully compete with the indigenous populations and the long-term viability suggests that enhanced dechlorination has the potential to continue beyond 365 days.

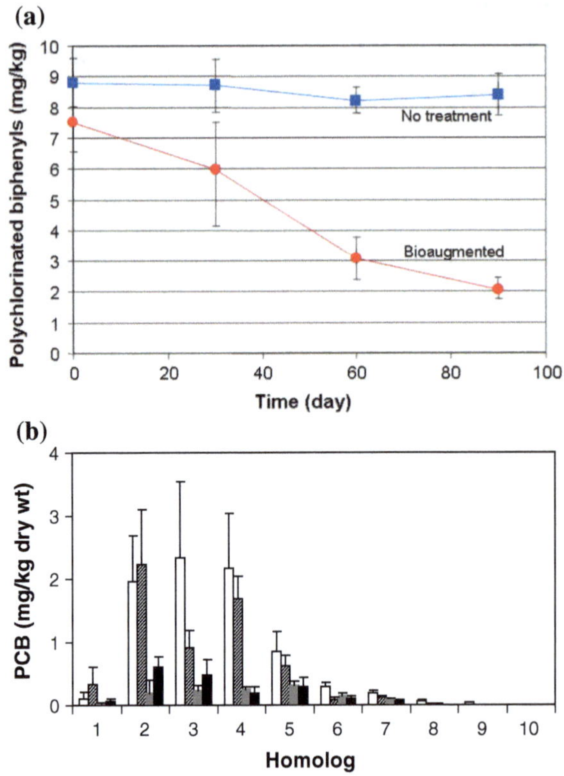

Fig. 24.8 Effect of bioaugmentation with "*D. chlorocoercia*" strain DF-1 with the aerobic PCB degrader *B. xenovorans* strain LB400. Panel A shows total reduction in levels of PCBs in bioaugmented and non-bioaugmented Baltimore Harbor sediment mesocosms after 90 days. Panel B PCB homolog distribution in Baltimore Harbor sediment mesocosms at day 0 (white bars, $n = 9$) and day 365 in mesocosms augmented with filtered growth medium (slashed bars, $n = 3$), LB400 (*gray bars*, $n = 3$), and LB400 plus DF1 (*black bars*, $n = 3$). Each datum point represents the mean and standard deviation of three replicates samples Reprinted with permission from (Payne et al. 2013). Copyright 2013 American Chemical Society

Even at a reduced rate this enhanced dechlorination could result in long-term dechlorination/degradation after the initial transformation of the most bioavailable congeners. Overall, the results indicate that in situ treatment employing the simultaneous application of anaerobic and aerobic microorganisms could be an effective and environmentally sustainable strategy to reduce PCB levels in contaminated sediments.

24.6.5 In Situ Bioremediation with "*D. chlorocoercia*" Strain DF-1—Future Challenges

One current challenge with in situ bioremediation is that Aroclors contain varying percentages of *ortho*-substituted PCB congeners. Payne et al. (2011, 2013) reported an accumulation of *ortho*-substituted tri- and tetra-chlorinated biphenyls after bioaugmentation with "*D. chlorocoercia*" strain DF-1 and *B. xenovorans* strain LB400. Reductive dechlorination PCB congeners in the *ortho* position has been reported infrequently in the environment and tri- and tetra-*ortho* chlorinated congeners are often recalcitrant to aerobic degradation (Bedard et al. 2003). One potential approach for further reducing the residual concentrations of PCBs after in situ treatment of a PCB-impacted site is by bioaugmentation with an *ortho*-dechlorinating organohalide respiring bacterium in order to prevent a buildup of more recalcitrant *ortho*-PCBs. Fagervold et al. (2011) reported that a coculture of "*D. chlorocoercia*" strain DF-1 and organohalide respiring strain *o*-17, which is capable of reducing PCBs in *ortho*-substituted positions, significantly reduced the accumulation of *ortho*-substituted PCBs in sediment microcosms. Although this approach has the potential to lead to more complete degradation of Aroclors it has not been tested with low concentrations of Aroclors either in sediment mesocosms or in situ.

Another issue is production of adequate biomass of PCB-dechlorinating bacteria for full-scale treatment of PCB-impacted sediment. Payne et al. (2011; Payne et al.) showed that approximately 10^5 cells g^{-1} (wet wt) sediment was required to effectively stimulate dechlorination of weathered Aroclor 1260. "*D. chlorococercia*" strain DF-1 can thus far be grown to a maximum cell density of only 10^8 cells ml^{-1}, therefore large-scale culturing in facilities with capacities of thousands of liters would be required for full-scale treatment of PCB-impacted sites of an acre or more (Sowers and May 2013). Postproduction removal of residual organohalides after cell production before use of the biomass in bioaugmentation must also be addressed. The only electron acceptors known to support growth of PCB dechlorinating bacteria are halogenated aliphatic or aromatic compounds that are also considered to be environmental contaminants. Unless a nontoxic electron acceptor is identified, methods need to be developed to readily remove halogenated compounds from the bioaugmentation inocula. Miller et al. (2005) reported that "*D. chlorocoercia*" strain DF-1 pre-grown with tetrachloroethene showed no significant lag in growth when transferred to 2,3,4,5-tetrachlorobiphenyl, which suggests that residual volatile substrates such as chlorinated ethenes could be removed from cultures by gas sparging before harvesting. Alternatively, substituting limited amounts of more readily biodegraded electron acceptors such as brominated biphenyls (discussed above) as a growth substrate and depleting most of the brominated biphenyls during harvesting might be a viable approach for preparation of bioaugmentation inocula (Bedard et al. 1998).

Finally, a means of deploying bioamendments into PCB-impacted sediments is required for effective bioremediation. More soluble organohalides such as chlorinated ethenes, which are common contaminants of groundwater, have been successfully bioaugmented by pumping microorganisms and nutrients into groundwater plumes. In contrast, PCBs are hydrophobic and tend to become immobilized by adsorption to sediment particles that settle in open water bodies such as lakes and oceans. Effective bioaugmentation of PCB-impacted sediments will require a method for inoculating sediment either by direct injection or deployment on solid particles that will pass through a water column. Payne et al. (2011, 2013) demonstrated that bioaugmentation with "*D. chlorocoercia*" strain DF-1 adsorbed to granulated activated carbon was effective for reducing PCB concentrations in mesocosms containing sediments contaminated with Aroclor 1260. Although organic particles such as clay or granulated activated carbon strongly sorb PCBs in an aqueous environment, they also provide a substrate for biofilm formation in close proximity to the hydrophobic PCBs. The ability to use a solid substrate such as clay or GAC particles for inoculation of cells offers a possible solution for dispersing cells into sediments.

24.7 Summary

"*D. chlorocoercia*" strain DF-1 grows by organohalide respiration of PCB congeners and chlorinated benzenes with doubly flanked chlorines, and both tetra- and tri-chloroethene. This was the first organohalide respiring isolate demonstrated to dechlorinate weathered commercial mixtures of PCBs as a bioamendment. Although its range of dechlorination for an Aroclor is limited in comparison with other organohalide respiring species, it is capable of facilitating extensive overall degradation of weathered PCBs in sediments co-augmented with an aerobic PCB degrader. Kinetic studies demonstrate that "*D. chlorocoercia*" strain DF-1 dechlorinates PCBs at subsaturating aqueous concentrations typically found in sediments impacted with low levels of PCBs and it is osmotolerant enabling it to be active in both freshwater and estuarine sediments. The fortuitous ability to apply this strict anaerobe with an aerobe to treat PCBs comprehensively in weathered sediment is now being tested in the field.

References

Abraham W-R, Nogales B, Golyshin PN, Pieper DH, Timmis KN (2002) Polychlorinated biphenyl-degrading microbial communities in soils and sediments. Curr Opin Microbiol 5:246–253

Abramowicz DA (1995) Aerobic and anaerobic PCB biodegradation in the environment. Environ Health Perspect 103(Suppl 5):97–99

Abramowicz DA, Brennan MJ, Van Dort HM, Gallagher EL (1993) Factors influencing the rate of polychlorinated biphenyl dechlorination in Hudson River sediments. Environ Sci Technol 27:1125–1131

Adrian L, Dudkova V, Demnerova K, Bedard DL (2009) "*Dehalococcoides*" sp. strain CBDB1 extensively dechlorinates the commercial polychlorinated biphenyl mixture Aroclor 1260. Appl Environ Microbiol 75(13):4516–4524

Ahmed M, Focht DD (1973) Degradation of polychlorinated biphenyls by two species of *Achromobacter*. Can J Microbiol 19:47–52

Bedard DL (2003) Polychlorinated biphenyls in aquatic sediments: environmental fate and outlook for biological treatment. In: Häggblom M, Bossert I (eds) In Dehalogenation: microbial processes and environmental applications. Kluwer Press, The Netherlands, pp 443–465

Bedard DL, Smullen LA, DeWeerd KA, Dietrich DK, Frame GM, May RJ, Principe JM, Rouse TO, Fessler WA, Nicholson JS (1995) Chemical activation of microbially-mediated PCB dechlorination: A field study. Organohalog Compd 24:23–28

Bedard DL, VanDort HM, May RJ, Smullen LA (1997) Enrichment of microorganisms that sequentially *meta*, *para*-dechlorinate the residue of Aroclor 1260 in Housatonic River sediment. Environ Sci Technol 31(11):3308–3313

Bedard DL, VanDort H, Deweerd KA (1998) Brominated biphenyls prime extensive microbial reductive dehalogenation of Aroclor 1260 in Housatonic River sediment. Appl Environ Microbiol 64(5):1786–1795

Bedard DL, Van Slyke Jerzak G, Bailey JJ (2003) Strategies for the selective enrichment of microorganisms carrying out reductive dechlorination of polychlorinated biphenyls in freshwater sediments. Fresenius Environ Bull 12(3):276–285

Bedard DL, Ritalahti KM, Löffler FE (2007) The *Dehalococcoides* population in sediment-free mixed cultures metabolically dechlorinates the commercial polychlorinated biphenyl mixture Aroclor 1260. Appl Environ Microbiol 73(8):2513–2521

Berkaw M, Sowers KR, May HD (1996) Anaerobic *ortho* dechlorination of polychlorinated biphenyls by estuarine sediments from Baltimore Harbor. Appl Environ Microbiol 62(7):2534–2539

Bhatt P, Kumar MS, Mudliar S, Chakrabarti T (2007) Biodegradation of chlorinated compounds—a review. Crit Rev Environ Sci Technol 37(2):165–198

Bond DR, Holmes DE, Tender LM, Lovley DR (2002) Electrode-reducing microorganisms that harvest energy from marine sediments. Science 295(5554):483–485

Brown JJF, Wagner RE, Bedard DL, Brennan MJ, Carnahan JC, May RJ (1984) PCB transformations in upper Hudson sediments. Northeast Environ Sci 3:167–179

Bruce M, Henry PG (2010) Biostimulation for anaerobic bioremediation of chlorinated solvents. In: Stroo HF, Ward CH (eds) In Situ remediation of chlorinated solvent plumes. Springer Science + Business Media, LLC, pp 357–423. doi:10.1007/978-1-4419-1401-9 12

Cho Y-C, Sokol RC, Rhee G-Y (2002) Kinetics of polychlorinated biphenyl dechlorination by Hudson River, New York, USA, sediment microorganisms. Environ Toxicol Chem 21(4):715–719

Cho YC, Sokol RC, Frohnhoefer RC, Rhee GY (2003) Reductive dechlorination of polychlorinated biphenyls: threshold concentration and dechlorination kinetics of individual congeners in Aroclor 1248. Environ Sci Technol 37:5651–5656

Cho Y-C, Oostrofsky EB, Rhee G-Y (2004) Effects of a rhamnolipid biosurfactant on the reductive dechlorination of polychlorinated biphenyls by St. Lawrence River (North America) microorganisms. Environ Toxicol Chem 23(6):1425–1430

Chun CL, Payne RB, Sowers KR, May HD (2013) Electrical stimulation of microbial PCB degradation in sediment. Wat Res 47(1):141–152

Cutter L, Sowers KR, May HD (1998) Microbial dechlorination of 2,3,5,6-tetrachlorobiphenyl under anaerobic conditions in the absence of soil or sediment. Appl Environ Microbiol 64(8):2966–2969

Cutter LA, Watts JEM, Sowers KR, May HD (2001) Identification of a microorganism that links its growth to the reductive dechlorination of 2,3,5,6-chlorobiphenyl. Environ Microbiol 3(11):699–709

Deweerd KA, Bedard DL (1999) Use of halogenated benzoates and other halogenated aromatic compounds to stimulate the microbial dechlorination of PCBs. Environ Sci Technol 33(12):2057–2063

Drenzek NJ, Eglinton TI, Wirsen CO, May HD, Wu Q, Sowers KR, Reddy CM (2001) The absence and application of stable carbon isotopic fractionation during the reductive dechlorination of polychlorinated biphenyls. Environ Sci Tech 35:3310–3313

Drenzek NJ, Eglinton TI, Wirsen CO, Sturchio NC, Heraty LJ, Sowers KR, Wu Q, May HD, Reddy CM (2004) Invariant chlorine isotopic signatures during microbial PCB reductive dechlorination. Environ Pollut 128(3):445–448

Edwards U, Rogall T, Blocker H, Emde M, Bottger EC (1989) Isolation and direct complete nucleotide determination of entire genes characterization of a gene coding for 16S ribosomal RNA. Nuc Acids Res 17:7843–7853

Evans BS, Dudley CA, Klasson KT (1996) Sequential anaerobic-aerobic biodegradation of PCBs in soil slurry microcosms. Appl Biochem Biotechnol 57–58:885–894

Fagervold SK, Watts JEM, May HD, Sowers KR (2005) Sequential reductive dechlorination of *meta*-chlorinated polychlorinated biphenyl congeners in sediment microcosms by two different Chloroflexi phylotypes. Appl Environ Microbiol 71(12):8085–8090

Fagervold SK, May HD, Sowers KR (2007) Microbial reductive dechlorination of Aroclor 1260 in Baltimore Harbor sediment microcosms is catalyzed by three phylotypes within the phylum Chloroflexi. Appl Environ Microbiol 73(9):3009–3018

Fagervold SK, Watts JEM, May HD, Sowers KR (2011) Effects of bioaugmentation on indigenous PCB dechlorinating activity in sediment microcosms. Wat Res 45:3899–3907

Fennell DE, Gossett JM (1998) Modeling the production of and competition for hydrogen in a dechlorinating culture. Environ Sci Technol 32(16):2450–2460

Fennell DE, Nijenhuis I, Wilson SF, Zinder SH, Häggblom MM (2004) *Dehalococcoides ethenogenes* strain 195 reductively dechlorinates diverse chlorinated aromatic pollutants. Environ Sci Technol 38(7):2075–2081

Fish KM (1996) Influence of Aroclor 1242 concentration on polychlorinated biphenyl biotransformations in Hudson River test tube microcosms. Appl Environ Microbiol 62(8):3014–3016

Friedman C, Burgess R, Cantwell M, Ho K, Lohmann R (2009) Comparing polychaete and polyethylene uptake to assess sediment resuspension effects on PCB bioavailability. Environ Sci Technol 43(8):2865–2870

Furukawa K (1976) Microbial metabolism of polychlorinated biphenyls: studies on relative degradability of polychlorinated biphenyl components by *Alcaligenes* sp. J Agric Food Chem 24:251–256

Ghosh U, Zimmerman J, Luthy RG (2003) PCB and PAH speciation among particle types in contaminated sediments and effects on PAH bioavailability. Environ Sci Technol 37:2209–2217

Heijnen JJ, van Dijken JP (1992) In search of a thermodynamic description of biomass yields for the chemotrophic growth of microorganisms. Biotechnol Bioeng 39:833–858

Jarman WM, Hilkert A, Bacon CE, Collister JW, Ballschmiter K, Risebrough RW (1998) Compound-specific carbon isotopic analysis of Aroclors, Clophens, Kaneclors, and Phenoclors. Environ Sci Technol 32(6):833–836

Jin S, Fallgren PH (2014) Feasibility of using bioelectrochemical systems for bioremediation. In: Das S (ed) Microbial biodegradation and bioremediation. Elsevier, Amsterdam, pp 389–406

Kjellerup BV, Sun X, Ghosh U, May HD, Sowers KR (2008) Site-specific microbial communities in three PCB-impacted sediments are associated with different in situ dechlorinating activities. Environ Microbiol 10:1296–1309

Krumins V, Park JW, Son EK, Rodenburg LA, Kerkhof LJ, Häggblom MM, Fennell DE (2009) PCB dechlorination enhancement in Anacostia River sediment microcosms. Wat Res 43(18):4549–4558. doi:10.1016/j.watres.2009.08.003

LaRoe SL, Fricker AD, Bedard DL (2014) *Dehalococcoides mccartyi* strain JNA in pure culture extensively dechlorinates Aroclor 1260 according to polychlorinated biphenyl (PCB) dechlorination Process N. Environ Sci Technol 48:9187–9196

Löffler FE, Yan J, Ritalahti KM, Adrian L, Edwards EA, Konstantinidis KT, Müller JA, Fullerton H, Zinder SH, Spormann AM (2013) *Dehalococcoides mccartyi* gen. nov., sp. nov., obligate organohalide-respiring anaerobic bacteria, relevant to halogen cycling and bioremediation, belong to a novel bacterial class, Dehalococcoidia classis nov., within the phylum Chloroflexi. Int J Syst Evol Microbiol 63:625–635

Lombard NJ, Ghosh U, Kjellerup BV, Sowers KR (2014) Kinetics and threshold level of 2,3,4,5-tetrachlorobiphenyl dechlorination by an organohalide respiring bacterium. Environ Sci Technol 48(8):4353–4360

Lu L, Yazdi H, Jin S, Zuo Y, Fallgren PH, Ren ZJ (2014) Enhanced bioremediation of hydrocarbon-contaminated soil using pilot-scale bioelectrochemical systems. J Haz Mat 274:8–15

Marco-Urrea E, Nijenhuis I, Adrian L (2011) Transformation and carbon isotope fractionation of tetra- and trichloroethene to trans-dichloroethene by *Dehalococcoides* sp. strain CBDB1. Environ Sci Technol 45:1555–1562

Master ER, Lai VW, Kuipers B, Cullen WR, Mohn WW (2002) Sequential anaerobic-aerobic treatment of soil contaminated with weathered Aroclor 1260. Environ Sci Technol 36:100–103

May HD, Miller GS, Kjellerup BV, Sowers KR (2008a) Dehalorespiration with polychlorinated biphenyls by an anaerobic ultramicrobacterium. Appl Environ Microbiol 74(7):2089–2094

May HD, Miller GS, Kjellerup BV, Sowers KR (2008b) Author's Correction—Dehalorespiration with polychlorinated biphenyls by an anaerobic ultramicrobacterium. Appl Environ Microbiol 74(19):6169–6170

Miller GS, Milliken CE, Sowers KR, May HD (2005) Reductive dechlorination of tetrachloroethene to *trans*-dichloroethene and *cis*-dichloroethene by PCB-dechlorinating bacterium DF-1. Environ Sci Technol 39(8):2631–2635

Natarajan MR, Nye J, Wu W-M, Wang H, Jain MK (1997) Reductive dechlorination of PCB contaminated Raisin River sediments by anaerobic microbial granules. Biotechnol Bioeng 55:181–190

Park J-W, Krumins V, Kjellerup BV, Fennell DE, Rodenburg LA, Sowers KR, Kerkhof LJ, Häggblom MM (2011) The effect of co-substrate activation on indigenous and bioaugmented PCB dechlorinating bacterial communities in sediment microcosms. Appl Microbiol Biotechnol 89:2005–2017

Payne RB, Chun C, May HD, Sowers KR (2011) Enhanced reductive dechlorination of polychlorinated biphenyl impacted sediment by bioaugmentation with a dehalorespiring bacterium. Environ Sci Technol 45:8772–8779

Payne RB, Fagervold SK, May HD, Sowers KR (2013) Remediation of polychlorinated biphenyl impacted sediment by concurrent bioaugmentation with anaerobic halorespiring and aerobic degrading bacteria. Environ Sci Technol 47:3807–3815

Peijnenburg WJGM, Jager T (2003) Monitoring approaches to assess bioaccessibility and bioavailability of metals: matrix issues. Ecotoxicol Environ Saf 56(1):63–77

Pulliam Holoman TR, Elberson MA, Cutter LA, May HD, Sowers KR (1998) Characterization of a defined 2,3,5,6-tetrachlorobiphenyl-*ortho*-dechlorinating microbial community by comparative sequence analysis of genes coding for 16S rRNA. Appl Environ Microbiol 64:3359–3367

Rhee GY, Sokol RC, Bethoney CM, Cho YC, Frohnhoefer RC, Erkkila T (2001) Kinetics of polychlorinated biphenyl dechlorination and growth of dechlorinating microorganisms. Environ Toxicol Chem 20:721–726

Schwarzenbach RP, Gschwend PM, Imboden DM (2002) Environmental organic chemistry. Wiley, Hoboken

Sowers KR, May HD (2013) In situ treatment of PCBs by anaerobic microbial dechlorination in aquatic sediment: are we there yet? Curr Opin Biotech 24:482–488

Tiedje JM, Quensen JF 3rd, Chee-Sanford J, Schimel JP, Boyd SA (1993) Microbial reductive dechlorination of PCBs. Biodegradation 4(4):231–240

Van Dort HM, Smullen LA, May RJ, Bedard DL (1997) Priming microbial *meta*-dechlorination of polychlorinated biphenyls that have persisted in Housatonic River sediments for decades. Environ Sci Technol 31(11):3300–3307

Wang S, Chng KR, Wilm A, Zhao S, Yang K-L, Nagarajan N, He J (2014) Genomic characterization of three unique *Dehalococcoides* that respire on persistent polychlorinated biphenyls. Proc Natl Acad Sci USA 111(33):12103–12108

Watts JEM, Fagervold SK, Sowers KR, May HD (2005) A PCR based specific assay reveals a population of bacteria within the *Chloroflexi* associated with the reductive dehalogenation of polychlorinated biphenyls. Microbiology 151:2039–2046

Winchell LJ, Novak PJ (2008) Enhancing polychlorinated biphenyl dechlorination in fresh water sediment with biostimulation and bioaugmentation. Chemosphere 71:176–182

Wu Q, Wiegel J (1997) Two anaerobic polychlorinated biphenyl-dehalogenating enrichments that exhibit different *para*-dechlorination specificities. Appl Environ Microbiol 63(12):4826–4832

Wu QZ, Bedard DL, Wiegel J (1996) Influence of incubation temperature on the microbial reductive dechlorination of 2,3,4,6-tetrachlorobiphenyl in two freshwater sediments. Appl Environ Microbiol 62(11):4174–4179

Wu QZ, Bedard DL, Wiegel J (1999) 2,6-dibromobiphenyl primes extensive dechlorination of Aroclor 1260 in contaminated sediment at 8–30 °C by stimulating growth of PCB-dehalogenating microorganisms. Environ Sci Technol 33(4):595–602

Wu Q, Sowers KR, May HD (2000) Establishment of a polychlorinated biphenyl-dechlorinating microbial consortium, specific for doubly flanked chlorines in a defined, sediment-free medium. Appl Environ Microbiol 66(1):49–53

Wu Q, Milliken CE, Meier GP, Watts GEM, Sowers KR, May HD (2002a) Dechlorination of chlorobenzenes by a culture containing bacterium DF-1, a PCB dechlorinating microorganism. Environ Sci Technol 36(15):3290–3294

Wu Q, Watts JEM, Sowers KR, May HD (2002b) Identification of a bacterium that specifically catalyzes the reductive dechlorination of polychlorinated biphenyls with doubly flanked chlorines. Appl Environ Microbiol 68:807–812

Zhang W, Bouwer EJ, Ball WP (1998) Bioavailability of hydrophobic organic contaminants: effects and implications of sorption-related mass transfer on bioremediation. Groundwater Monit Rem 18(1):126–138

Zhang T, Gannon SM, Nevin KP, Franks AE, Lovley DR (2010) Stimulating the anaerobic degradation of aromatic hydrocarbons in contaminated sediments by providing an electrode as the electron acceptor. Environ Microbiol 12(4):1011–1020

Zhen H, Du S, Rodenburg LA, Mainelis G, Fennell DE (2014) Reductive dechlorination of 1,2,3,7,8-pentachlorodibenzo-p-dioxin and Aroclor 1260, 1254 and 1242 by a mixed culture containing *Dehalococcoides mccartyi* strain 195. Wat Res 52:51–62

Zwiernik M, Quensen JF III, Boyd SA (1998) $FeSO_4$ amendments stimulate extensive anaerobic PCB dechlorination. Environ Sci Technol 32(21):3360–3365

Chapter 25
Use of Compound-Specific Isotope Analysis (CSIA) to Assess the Origin and Fate of Chlorinated Hydrocarbons

Daniel Hunkeler

Abstract This chapter provides a comprehensive presentation of how compound-specific-isotope analysis (CSIA) can be used to evaluate the origin and fate of chlorinated hydrocarbons at the laboratory or field scale. It lays the foundation by introducing concepts to quantify isotope fractionation associated with reactive processes, explaining what causes such changes and presenting current methods to determine C, Cl and H isotope ratios. It then discusses how the transformation of a compound by various mechanisms can be differentiated using isotope ratios of multiple elements. It also summarizes the current knowledge about isotope fractionation during the transformation of common classes of chlorinated hydrocarbons. Finally, strategies to apply isotope methods at the field scale to track different sources of contamination or the type and progress of reactive processes are outlined.

25.1 Introduction

Compound-specific isotope analysis (CSIA) can provide unique insight into the fate of chlorinated hydrocarbons. A particular strength of the method is that information about specific transformation processes is directly imprinted in the molecule. Stable isotope methods can potentially be used to distinguish between different transformation mechanisms of a given compound, to quantify the extent of transformation, to establish links between parent and daughter compounds, and to distinguish between different sources of the same contaminant. So far, stable isotope methods have mainly been applied to chlorinated ethenes but there is a growing body of literature about isotope fractionation patterns associated with

D. Hunkeler (✉)
Centre for Hydrogeology and Geothermics (CHYN), Emile-Argand Rue 11,
CH-2000 Neuchâtel, Switzerland
e-mail: Daniel.Hunkeler@unine.ch

L. Adrian and F.E. Löffler (eds.), *Organohalide-Respiring Bacteria*,
DOI 10.1007/978-3-662-49875-0_25

Table 25.1 Natural abundance of carbon, chlorine and hydrogen isotopes

Element	Natural abundance of stable isotopes (%)	Isotope ratio (R)	International standard	International standard ratio ($R_{standard}$)
Carbon	^{12}C (98.93) ^{13}C (1.07)	$^{13}C/^{12}C$	Carbonate from Vienna Pee Dee Belemnite (VPDB)	0.011237
Chlorine	^{35}Cl (75.76) ^{37}Cl (24.24)	$^{37}Cl/^{35}Cl$	Chloride ion in ocean water standard mean ocean chlorine (SMOC)	0.324
Hydrogen	^{1}H (99.9885) ^{2}H (0.0115)	$^{2}H/^{1}H$	Vienna standard mean ocean water (VSMOW)	1.5575×10^{-4}

other chlorinated hydrocarbons. This chapter starts with an introduction to principles of isotope fractionation and stable isotope analysis. It then provides an overview of the current knowledge about isotope fractionation of different chlorinated hydrocarbons during reactive processes and finally discusses strategies to apply isotope methods and evaluate field data. More detailed treatment of specific topics can be found in several textbooks. Textbooks are available on isotope methods in biology and chemistry (Kohen and Limbach 2006; Wolfsberg et al. 2010), in geosciences (Hoefs 2009), in environmental sciences (Aelion et al. 2010) and in hydrogeology (Clark 2015; Clark and Fritz 1997).

25.2 Expressing and Quantifying Isotope Fractionation

All the three elements that are commonly present in chlorinated hydrocarbons, carbon, chlorine and hydrogen, occur naturally in form of different stable isotopes. The abundance of the heavy isotope, however, strongly varies from 0.012 % for hydrogen and 1.07 % for carbon to 24.2 % for chlorine (Table 25.1). Furthermore, for carbon and hydrogen, the mass difference between light and heavy isotope is one, while for chlorine it is two. As will be discussed in the subsequent section, this has some bearing for analytical methods and the precision of isotope measurements. In its most simple form, the isotope composition of an element can be expressed as an absolute ratio

$$R_{\mathrm{E}} = \frac{^{\mathrm{H}}\mathrm{E}}{^{\mathrm{L}}\mathrm{E}} \tag{25.1}$$

where

R_{E} is the isotope ratio of the element E
$^{\mathrm{H}}\mathrm{E}$ is the abundance of the heavy isotope of element E
$^{\mathrm{L}}\mathrm{H}$ is the abundance of the light isotope of element E

However, most commonly, the isotope composition is reported relative to a reference compound using the delta notation

$$\delta^{m_H}E = \frac{R_E - R_{E,ref}}{R_{E,ref}} = \left(\frac{R}{R_{E,ref}} - 1\right) \tag{25.2}$$

where

$R_{E,ref}$ is the isotope ratio of a reference compound; usually an international standard

m_H is the atomic mass of the heavy isotope

The calculated values are usually reported in per mille units by multiplying the expression in Eq. 25.2 by 1000. While absolute isotope ratios typically vary in the third or fourth decimal position, delta values expressed in per mille often differ by tenths or whole numbers, which facilitates the discussion of isotope patterns.

The reaction rate or equilibrium distribution of a compound can be influenced by its isotopic composition, which is denoted as a kinetic isotope effect (KIE) or equilibrium isotope effect (EIE). Isotope effects lead to uneven distribution of light and heavy isotopes in a forming product relative to the reactant, or among a compound present in different phases (e.g. air and water), which is denoted as kinetic or equilibrium isotope fractionation. The extent of isotope fractionation is commonly expressed by the isotopic fractionation factor α. For an equilibrium process, the equilibrium isotopic fractionation factors reflect that the isotope ratios of a compound present in two phases or of a reactant and product in reversible reaction can differ by a fixed ratio

$$\alpha_{eq} = \frac{R_{E,i}}{R_{E,j}} \tag{25.3}$$

where

$R_{E,i}$ isotope ratio of element in compartment or compound i

$R_{E,j}$ isotope ratio of the element in compartment or compound j

α_{eq} equilibrium isotopic fractionation factor

The kinetic isotopic fractionation factor α_{kin} relates the isotope ratio of the product that is formed at specific moment in time (instantaneous product), to the isotope ratio of the reactant at the same moment as follows:

$$\alpha_{kin} = \frac{R_{E,IP}}{R_{E,R}} \tag{25.4}$$

$R_{E,IP}$ isotope ratio of element E in the instantaneous product

$R_{E,R}$ isotope ratio of element E in the reactant

α_{kin} kinetic isotopic fractionation factor

As isotopic fractionation factors tend to be close to one, the extent of isotope fractionation is often reported as isotopic enrichment factor ε

$$\varepsilon = (\alpha - 1) \tag{25.5}$$

In analogy to the delta notation, the calculated values are usually reported on a per mille scale.

Starting from the definition of the isotope fractionation factor, an equation can be developed that describes the isotope evolution during a kinetic process having a constant α_{kin}.

$$\alpha_{\mathrm{kin}} = \frac{R_{\mathrm{E,IP}}}{R_{\mathrm{E,R}}} = \frac{\mathrm{d}^{\mathrm{H}}\mathrm{E}_{\mathrm{P}}/\mathrm{d}^{\mathrm{L}}\mathrm{E}_{\mathrm{P}}}{^{\mathrm{H}}\mathrm{E}_{\mathrm{R}}/^{\mathrm{L}}\mathrm{E}_{\mathrm{R}}} \tag{25.6}$$

where

$\mathrm{d}^{\mathrm{H}}\mathrm{E}_{\mathrm{P}}$, $\mathrm{d}^{\mathrm{L}}\mathrm{E}_{\mathrm{P}}$ are increments of product with heavy and light isotopes of element E formed at time t

The amount of product that is formed is equal to the amount of rectant consumed; hence

$$\alpha_{\mathrm{kin}} = \frac{\mathrm{d}^{\mathrm{H}}\mathrm{E}_{\mathrm{R}}/\mathrm{d}^{\mathrm{L}}\mathrm{E}_{\mathrm{R}}}{^{\mathrm{H}}\mathrm{E}_{\mathrm{R}}/^{\mathrm{L}}\mathrm{E}_{\mathrm{R}}} \tag{25.7}$$

Rearrangement leads to

$$\frac{\mathrm{d}^{\mathrm{H}}\mathrm{E}_{\mathrm{R}}}{^{\mathrm{H}}\mathrm{E}_{\mathrm{R}}} = \alpha_{\mathrm{kin}} \cdot \frac{\mathrm{d}^{\mathrm{L}}\mathrm{E}_{\mathrm{R}}}{^{\mathrm{L}}\mathrm{E}_{\mathrm{R}}} \tag{25.8}$$

Integration from time 0 to t leads to

$$\ln \frac{^{\mathrm{H}}\mathrm{E}_{\mathrm{R},t}}{^{\mathrm{H}}\mathrm{E}_{\mathrm{R},0}} = \alpha_{\mathrm{kin}} \cdot \ln \frac{^{\mathrm{L}}\mathrm{E}_{\mathrm{R},t}}{^{\mathrm{L}}\mathrm{E}_{\mathrm{R},0}} \tag{25.9}$$

Or:

$$\frac{^{\mathrm{H}}\mathrm{E}_{\mathrm{R},t}}{^{\mathrm{H}}\mathrm{E}_{\mathrm{R},0}} = \left[\frac{^{\mathrm{L}}\mathrm{E}_{\mathrm{R},t}}{^{L}\mathrm{E}_{\mathrm{R},0}}\right]^{\alpha_{\mathrm{kin}}} \tag{25.10}$$

Dividing both sides by $^{\mathrm{L}}\mathrm{E}_{\mathrm{R},t}/^{\mathrm{L}}\mathrm{E}_{\mathrm{R},0}$ leads to

$$\frac{^{\mathrm{H}}\mathrm{E}_{\mathrm{R},t}/^{\mathrm{L}}\mathrm{E}_{\mathrm{R},t}}{^{\mathrm{H}}\mathrm{E}_{\mathrm{R},0}/^{\mathrm{L}}\mathrm{E}_{\mathrm{R},0}} = \left[\frac{^{\mathrm{L}}\mathrm{E}_{\mathrm{R},t}}{^{\mathrm{L}}\mathrm{E}_{\mathrm{R},0}}\right]^{\alpha_{\mathrm{kin}}-1} \tag{25.11}$$

Or, if the left hand-side is expressed in terms of isotope ratios

$$\frac{R_{\mathrm{R},t}}{R_{\mathrm{R},0}} = \left[\frac{^{\mathrm{L}}\mathrm{E}_{\mathrm{R},t}}{^{\mathrm{L}}\mathrm{E}_{\mathrm{R},0}}\right]^{\alpha_{\mathrm{kin}}-1} \tag{25.12}$$

The term in brackets corresponds in approximation to the reaction progress f

$$f = \frac{{}^{L}E_{R,t} + {}^{H}E_{R,t}}{{}^{L}E_{R,0} + {}^{H}E_{R,0}} \approx \frac{{}^{L}E_{R,t}}{{}^{L}E_{R,0}} \tag{25.13}$$

The approximation is valid if either the abundance of the heavy isotope is very small, as for C or H, or the extent of isotope fractionation is limited.

Inserting Eq. 25.13 into Eq. 25.12 leads to

$$\frac{R_{R,t}}{R_{R,0}} = f^{\alpha_{kin}-1} \tag{25.14}$$

The obtained equation is commonly known as the Rayleigh equation as it is analogous to an equation published in 1894 by Lord Rayleigh for distillation of two liquids with vapour pressures that have a fixed offset.

Inserting a rearranged version of Eq. 25.2 into Eq. 25.14 and taking natural logarithm leads to

$$\ln\left(\delta E_{R,t} + 1\right) - \ln\left(\delta E_{R,0} + 1\right) = (\alpha_{kin} - 1) \cdot \ln f \tag{25.15}$$

As the arguments of the logarithms are close to 1, the following approximation holds:

$$\ln(\alpha_{kin} + 1) \approx \alpha_{kin}$$

Hence, Eq. 25.15 can be simplified to

$$\delta E_{R,t} - \delta E_{R,0} = (\alpha_{kin} - 1) \cdot \ln f = \varepsilon_{kin} \cdot \ln f$$

or

$$\delta E_{R,t} = \delta E_{R,0} + \varepsilon_{kin} \cdot \ln f \tag{25.16}$$

which is the widely utilized Rayleigh equation in the delta notation.

For a one-step reaction, the instantaneous product is always offset by ε_{kin}, as the Rayleigh equation is based on the assumption that isotope fractionation is constant. The average isotope ratio of the accumulated product can be obtained using an isotope mass balance equation and inserting Eq. 25.16

$$\delta E_{R,0} = f \cdot \delta E_{R,t} + (1-f) \cdot \delta E_{\overline{P},t} = f \cdot \left(\delta E_{R,0} + \varepsilon_{kin} \cdot \ln f\right) + (1-f) \cdot \delta E_{\overline{P},t} \tag{25.17}$$

Rearrangement of Eq. 25.17 leads to

$$\delta E_{\overline{P},t} = \delta E_{R,0} - \frac{f \cdot \varepsilon_{kin}}{1-f} \cdot \ln f \tag{25.18}$$

Equation 25.18 is valid if all atoms of the element of interest are transferred from the reactant to the product, as for example for carbon during reductive dechlorination of chlorinated ethenes. If the atoms are distributed among several compounds, Eq. 25.18 applies to the average of all compounds.

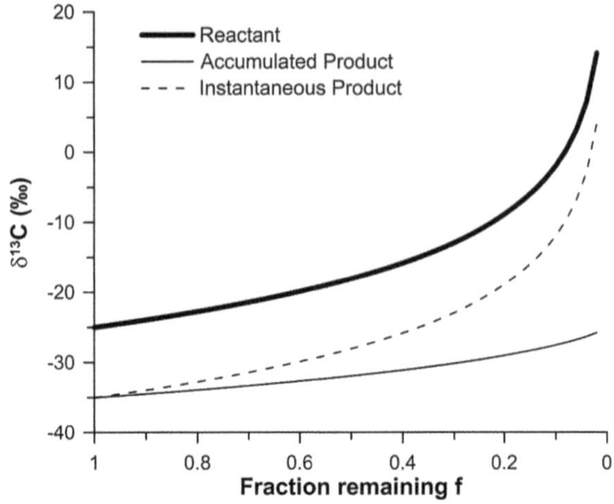

Fig. 25.1 Rayleigh plot illustrating the carbon isotope evolution of a reactant and the corresponding product assuming that all carbon atoms are preserved in the product. The isotope ratio of the product that is formed in certain moment of time (instantaneous product) and the average isotope ratio of all formed product (accumulated product) is shown

The evolution of the isotope ratio of reactant and instantaneous/accumulated product, according to Eqs. 25.16 and 25.18 as a function of reaction progress f, is illustrated in Fig. 25.1. The reactant becomes increasingly enriched in the heavy isotope and the instantaneous product follows this trend with an offset corresponding to ε_{kin}. In contrast, the accumulated product evolves towards the initial isotope ratio of the reactant, which is expected for mass balance reasons.

25.3 Stable Isotope Analysis of Chlorinated Hydrocarbons

This section describes the general principles of stable isotope analysis of chlorination hydrocarbons. More in depth treatment of current trends and challenges in isotope analysis can be found in a recent review (Elsner et al. 2012). The most commonly applied methods for C, H and Cl isotope analysis in chlorinated hydrocarbons and their typical detection limits and precision are summarized in Table 25.2. Stable isotope ratios of light isotopes are generally determined with isotope ratio mass-spectrometers (IRMS). In order to reach a high measurement precision, the isotope ratios are usually measured in form of simple standard molecules, such as CO_2 for C and H_2 for H, for which the instrument is optimized. Traditionally, these molecules have been generated by laborious offline sample preparation followed by IRMS measurements. Using this approach, it is difficult to obtain compound-specific information. The direct coupling of GC (Meier-Augenstein 1999), and recently HPLC (Abramson et al. 2001), to IRMS systems has greatly simplified analytical methods, lowered detection limits and thus, lead to a much wider adoption of isotope methods. Initially, most studies have focussed on carbon isotopes in chlorinated hydrocarbons, due to the relative ease with which these compounds can be transformed to CO_2. The measurement of hydrogen

Table 25.2 Analytical methods for carbon, chlorine and hydrogen isotope analysis in chlorinated hydrocarbons and typical detection limits and precisions

Element	Measurement technique	Sample extraction/ injection	Typical detection limits	Typical uncertainty	Reference
C	GC-C-IRMS	P&T	2–10 μg/L	0.3–0.5 ‰	Zwank et al. (2003)
Cl	GC-IRMS (direct ionization)	SPME	5–20 μg/L	0.1–0.3 ‰	Shouakar-Stash et al. (2006)
	GC-qMS	Headspace injection	50–100 μg/L	0.3–0.8 ‰	Bernstein et al. (2011), Sakaguchi-Soder et al. (2007)
	GC-HTC-IRMS	Headspace/ liquid injection	n.d.	<0.3 ‰	(Renpenning et al. 2015a)
H	GC-IRMS	SPME	200–400 μg/L	2–7 ‰	Shouakar-Stash and Drimmie (2013)

With Purge&Trap lower detection limits can be reached for Cl isotopes with GC-qMS and H isotopes with GC-IRMS

HTC High temperature conversion, *n.d.* not determined

isotopes is more challenging. During pyrolysis, HCl gas is formed which introduces variability in the isotope signature of H_2 due to the large fractionation associated with the process (Shouakar-Stash and Drimmie 2013). Furthermore, HCl might damage the instruments. It was proposed to trap the HCl but the feasibility of the method was only demonstrated for high concentrations of the analytes (Chartrand et al. 2007). Thanks to the development of new Cr reactors (Shouakar-Stash and Drimmie 2013; Renpenning et al. 2015b), H-isotope analysis in chlorinated hydrocarbons is now more widely applied. However, detection limits are higher than for C due to the low abundance of the heavy isotope and the precision is lower. The large mass difference between the heavy and light isotope in case of hydrocarbon not only leads to larger isotopic fractionation during physico-chemical and reactive processes but also introduces a higher variability in the measurements.

It might be surprising that studies of Cl-isotope ratios have only emerged in the past years although chlorine is as per default present in chlorinated hydrocarbons. Traditionally, methyl-chloride has been used as a measurement gas in IRMS instruments. There is, however, no method to directly transform chlorinated hydrocarbons to this compound. It was finally discovered that no conversion interface at all is necessary and that chlorine isotope ratios can be measured reliably by introducing them directly into the ion source of IRMS instruments (Shouakar-Stash et al. 2006). While this greatly simplifies the measurement, the mass of the measured ions will vary depending on the compound unlike, for example, for carbon isotopes, where always the same ions of CO_2 will be measured. For this reason, usually specialized IRMS instruments are used with a modified layout of ion detectors, which makes the method more costly. Recently, it was demonstrated that even classical quadrupole MS are suitable for Cl isotope ratio analysis (Sakaguchi-Soder et al. 2007; Aeppli et al. 2010). The high abundance heavy isotope makes it possible to record the molecules and fragment ions of the heavy isotope at sufficient precision. Thanks to the mass difference of two, protonation of compounds during ionization, which adds a mass of one, does not interfere. However, to reach a sufficiently high precision, a higher number of repeats are necessary than for IRMS, typically 5–10 (Bernstein et al. 2011). Another approach is to use HCl as a measurement gas in an IRMS instrument after high temperature conversion of chlorinated hydrocarbons to this compound (Hitzfeld et al. 2011; Renpenning et al. 2015a).

In classical IRMS, there is a relatively simple link between the abundance of the isotopologues of the measurement gas and the isotope ratio, as the measurement gas only contains one or two atoms of the element of interest. Furthermore, for compounds with more than one atom of the element of interest, isotopologues (=molecules that differ only in their isotopic composition) with more than one heavy isotopes occur at such a low abundance that they can be neglected. For chlorine isotopes measured by direct ionization, the situation is different. Several chlorine atoms can be present in the ion that is recorded, for example, four in the tetrachloroethene (PCE) molecular ion. Furthermore, due to high abundance of the heavy chlorine isotope (around 24 %), isotopologues with several heavy isotopes occur at relatively high frequency. During the measurement, often only some of the isotopologues are tracked, which raises the question if isotope ratios can be calculated reliably from such partial information. It was demonstrated, that all pairs of isotopologues differing in one heavy isotopes fractionate proportionally (Elsner and Hunkeler 2008). Hence, even if only the abundance of one isotopologue pair is recorded, the abundances of the others are implicitly known and thus the isotope ratio can be calculated.

In order to obtain reliable data, a robust calibration and validation method is required. For classical GC/HPLC–IRMS systems, calibration procedures are well established. The measurement is usually carried out relative to a reference gas that is chemically identical to the measurement gas and has a known isotope signature. The reference gas is included at the beginning and end of each run. In addition, it has to be verified that during transfer of the compound to the GC and its conversion to the reference gas, no bias or excessive uncertainty is introduced. This is usually accomplished by including it in the sequence standards containing the target compounds with a known isotope signature. For direction ionization methods for chlorine, the approach is different as usually reference gases are not available. Furthermore, quadrupole GCs are not designed for reference gas injection and might show more drift over time. In addition, IRMS systems are designed to reach a high linearity for isotope ratio analysis (independence of isotope ratio from signal intensity), which might not be the case for GC-quadrupole systems. For such systems, samples are usually bracketed with molecularly identical reference compounds included in the measurement sequence (Aeppli et al. 2010; Bernstein et al. 2011). As measured difference in isotope ratios are proportional to effective differences but not necessary with a 1:1 ratio, standards with two different isotope ratios have to be included that cover the measurement range (Bernstein et al. 2011). The availability of suitable standards still limits the application of the methods as only few laboratories are equipped to characterize the isotope ratios of such standards and special procedures are required to generate reference material with a suitable isotope value.

25.4 Mechanisms of Isotope Fractionation of Chlorinated Hydrocarbons

Isotope fractionation originates from differences in the energy level of a compound as a function of which isotopes are present in the compound. Energy occurs in different forms in molecules, including translational, rotational and vibrational energy. Among these forms, the vibrational energy is most sensitive to isotope substitution. While the bond force constant is independent of the isotope substitution as the electron configuration remains the same, the vibrational frequency and thus the vibration energy varies. Isotope fractionation occurs if the bond stiffness is modified in a process or the bond is completely broken. Heavy isotopes tend to accumulate in stiffer bonds with a higher bond force constant whereas light isotopes tend to accumulate in looser bond. The stronger the change in bond stiffness, the larger is the isotope fractionation. When molecules partition among different phase (gas, water, solid), changes in bond stiffness are usually fairly small. Small changes in bond stiffness can, for example, occur during evaporation from an organic liquid. In the organic phase, bonds can be slightly loosened due to van der Waals interactions, which are absent in the gas phase. During reactive processes, changes in bond stiffness are much larger and as a result, much larger isotope fractionation is expected. Light isotopes are present at a higher abundance in the looser transition state compared to the stiffer ground state. Or in other words, the energy difference for going from the ground to the transition state is smaller for the light isotope compared to the heavy isotope as illustrated in Fig. 25.2, leading to a higher reaction rate for bonds with light isotopes that is denoted as normal isotope effect. In some cases, some of the bonds in the molecule get stiffer in the transition state. Hence, heavy isotopes are preferred in the transition state and thus react faster—denoted as inverse isotope effect.

As can be deduced from above, during a reactive process, strong isotope fractionation is mainly expected if a heavy isotope is present in the bond that is cleaved (primary isotope effect). A difference in reaction rate due to the presence of a heavy isotope in the cleaved bond is denoted as primary isotope effect. However, as the molecules move from the ground to the transition state, the stiffness of bonds adjacent to the reactive site might be modified as well but to a smaller degree. Hence, the presence of a heavy isotope can influence the reaction rate even if it is not present in the cleaved bond, which is denoted as secondary isotope effect. Depending on the position of heavy isotope relative to the cleaved bond, alpha- (heavy isotope at reaction center) and beta- (heavy isotope adjacent to it) secondary isotope effects can be further distinguished. The difference between primary and

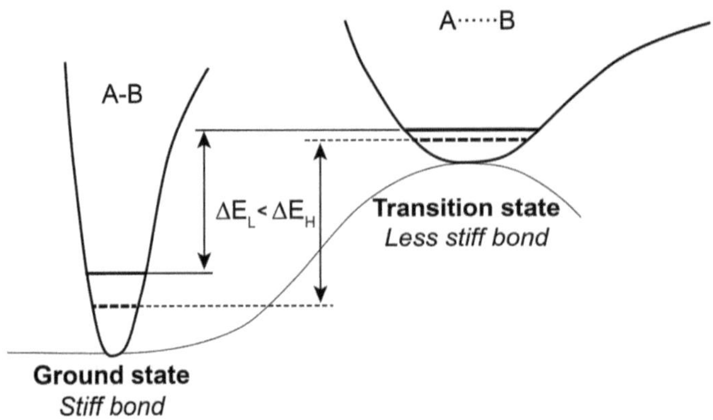

Fig. 25.2 Schematic illustration of the energetic of a reaction during which a bond A–B becomes looser when going from the ground to the transition state. In the ground state the vibrational energy level of bonds with the light (*solid line*) and heavy (*dashed line*) isotope are further apart than in the transition state. As a result, the energy difference for going from the ground to transition state is smaller for the bond with the light than the bond with the heavy isotope. Thus, the bond containing the light isotope reacts faster compared to the bond with the heavy isotope

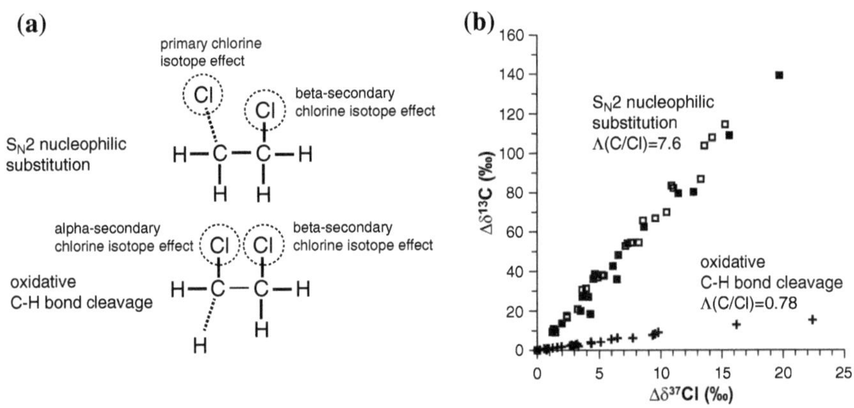

Fig. 25.3 Chlorine isotope effects during aerobic biodegradation of 1,2-dichloroethane via nucleophilic substitution and oxidative C–H bond cleavage as initial transformation step (**a**) and dual-element isotope plot for the two mechanisms (modified from Palau et al. 2014a) with dual isotope slopes Λ(C/Cl) (**b**)

secondary isotope effect can be illustrated for chlorine isotope fractionation during the transformation of 1,2-dichloroethane (1,2-DCA) by two different pathways, S_N2 nucleophilic substitution and oxidative C–H bond cleavage (Fig. 25.3a). In the S_N2 mechanism, a C–Cl bond is cleaved and thus a primary chlorine isotope effect is expected, and possibly a beta-secondary isotope effect. In contrast, for oxidative C–H bond cleavage, only an alpha- and/or beta-secondary chlorine isotope effects might occur. Note that with CSIA methods, it is generally not possible to differentiate between isotope effects at different positions. Only the average isotope effect over all positons is obtained.

In the past, large secondary isotope effects were mainly reported for hydrogen, due to the large mass differences between the two isotopes, while for heavier isotopes they were assumed to be small. However, for chlorine isotopes, some surprisingly large secondary isotope effects were reported as illustrated by the example of 1,2-DCA. Chlorine isotope fractionation for C–H bond oxidation (−3.8 ‰) was almost as large as for an S_N2 process although in the former case only a secondary isotope effect is expected while in the latter a primary isotope effect occurs (Palau et al. 2014a).

25.5 Relating Isotope Fractionation to Isotope Effects and Reaction Mechanisms

For the interpretation of isotope data, it is important to know how magnitudes of KIE and isotope fractionation factors are related. Two factors have to be considered (Elsner et al. 2005). Transport and binding steps preceding the reactive step can impact the observed isotope fractionation even if they are not associated by isotope fractionation themselves, which sometimes denotes as "masking" of isotope effects. Furthermore, the presence of heavy isotopes at non-reactive position can dampen isotope fractionation, sometimes denoted as "dilution" of the isotope effect. In the following, these two effects are discussed in more detailed.

"Masking" of KIEs can especially occur during enzyme-catalysed reactions due to the difference between the locations where isotope fractionation occurs and where it is recorded. In a microbial process, the KIE is caused by a specific reaction step catalysed by an enzyme located within the cell, while isotope analysis reflects the isotope composition in the solution outside the cell. Transport and binding steps preceding the reactive step can impact the observed isotope fractionation even if they are not associated by isotope fractionation themselves. In the following, this effect is illustrated for the case of a single equilibration process that precedes the reactive step, e.g. the binding of a substrate to the enzyme (Fig. 25.4).

The KIE corresponds to the difference in the rate with which the enzyme-bound substrate is transformed

$$\mathrm{KIE} = \frac{{}^{L}k_2}{{}^{H}k_2} \tag{25.19}$$

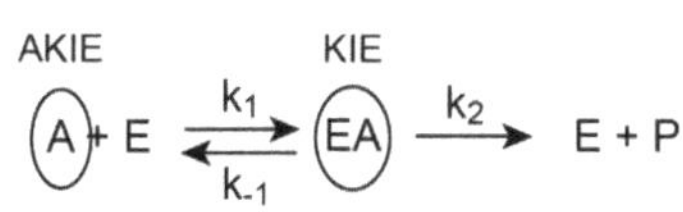

Fig. 25.4 Simple enzymatic process consisting of substrate-enzyme binding followed by a reactive step catalysed by the enzyme

where

$^{L}k_2$ is the rate constant for the reactive step when a light isotope is present at the reacting site

$^{H}k_2$ is the rate constant for the reactive step when a heavy isotope is present at the reacting site

This difference in the rate constant of the reactive step is the fundamental cause of isotope fractionation and therefore often denoted as the intrinsic KIE. The apparent kinetic isotope effect (AKIE) reflects differences in rates that are detected when the substrate A outside the cell is considered, and thus might be altered by rate-limiting steps preceding the reaction. The AKIE can be expressed as follows (Cleland 2006; Hunkeler and Elsner 2010):

$$\mathrm{AKIE} = \frac{\frac{^{L}k_2}{^{H}k_2} + C}{1 + C} = \frac{\mathrm{KIE} + C}{1 + C} \tag{25.20}$$

where C is the commitment to catalysis, which reflects the tendency of the enzyme-substrate complex to undergo reaction versus dissociation

$$C = \frac{^{L}k_2}{k_{-1}} \tag{25.21}$$

A high C corresponds to the situation where each molecule that binds reacts immediately independent of its isotope substitution. In this case, AKIE converges towards 1 and no isotope fractionation is detectable because the molecules do not get back into solution where measurement takes place. In reverse, if C is small, the full KIE is detectable because the molecules that have undergone isotope fractionation partition back into the bulk solution. This effect of steps preceding the reaction is one of the reasons why isotope fractionation factors can vary somewhat even if the compound is degraded by the same mechanisms having the same KIE.

The second effect ("dilution") occurs because KIEs reflect the presence of a heavy isotope at a specific position in the molecule while CSIA and isotope fractionation factors represent the average behaviour of a heavy isotope located at any position in the molecule. Therefore, if a heavy isotope is present at a non-reacting position it would not influence the reaction rate but it would damp the observed isotope effect. Two situations can be distinguished. A molecule can have several identical reactive positions such as the two C–Cl bonds in 1,2-DCA that compete for reaction (intramolecular competition). For example, during nucleophilic substitution of 1,2-DCA, a heavy chlorine isotope might be present in the reacting bond causing primary isotope effect, or in the adjacent position having little effect. Furthermore, the heavy isotopes might be present at positions that never react. For example, in vinyl chloride some of the heavy carbon atoms will be at the position without chlorine and hence will never participate in reactive dechlorination. The AKIE can be related to the isotope fractionation factor using the following equation (Elsner et al. 2005):

$$\text{AKIE} = \frac{1}{1 + z \cdot \frac{n}{x} \cdot (\alpha - 1)} \tag{25.22}$$

where

n is the number of atoms of the considered element that are present in the molecule
x is the number of atoms located at reactive sites
z is the number of atoms located at reactive sites that are in intramolecular competition

Using carbon isotope fractionation during degradation of 1,2-DCA by nucleophilic substitution as an example ($\alpha = 0.968$), n is two, both atoms are at reactive position ($x = 2$) and both of them are in competition for reaction ($z = 2$). Using Eq. 25.22, an AKIE of 1.068 is obtained, which is close to the expected KIE for cleavage of a C–Cl bond indicating that commitment to catalysis is small. While the calculation of AKIE values using Eq. 25.22 provides considerable insight into the causes of variations in isotope fractionation; it has some limitations. It is based on the assumption that the molecule can be subdivided into reactive and non-reactive positions, which is not always the case. In some reactions, the stiffness of several bonds might vary causing isotope effects at multiple positions. Which bonds change by how much is often not known unless methods of computational chemistry are applied to characterize the transition statc.

In the context of research projects or practical applications, it can be of considerable interest to understand by which mechanism a compound is transformed. For this purpose, observed isotope fractionation factors can be compared to those from well-constrained laboratory studies or with calculated values. However, as discussed above, the magnitude of isotope fractionation can vary somewhat for a given mechanism, for example due to masking effects. Furthermore, under field conditions, it is usually difficult to quantify isotope fractionation factor as concentrations are influenced by other processes than the transformation reaction. Therefore, multi-element isotope approaches are generally preferred to identify reaction mechanisms. As different mechanisms usually act on different bonds in the molecules, the relative shift of the isotope composition among different elements varies, as illustrated in Fig. 25.3 for degradation of 1,2-DCA by two different mechanisms. Most often, such differences are illustrated in the form of dual-element isotope plots and referred as a dual-element isotope approach. Sometimes also the terms 2D plots and 2D isotope approach are used, which can however create confusion, as in environmental studies 1D/2D/3D is generally used to characterize the spatial dimension of a study or a modelling approach. A key advantage of the method is that dual-element isotope slopes are generally not sensitive to masking because commitment to catalysis affects the isotopes of all elements to the same degree.

25.6 Isotope Fractionation of Chlorinated Hydrocarbons

In this section, the effect of reactive processes on the isotope ratio of chlorinated hydrocarbons is illustrated for different compound classes that commonly occur as environmental contaminants.

25.6.1 Chlorinated Ethenes

Among chlorinated hydrocarbons, isotope fractionation has been studied most extensively for chlorinated ethenes due to the prevalence of PCE and trichloroethene (TCE) as environmental contaminants. Isotope fractionation factors are known for both biotic and abiotic transformation for parent compounds as well as the most common intermediates. Biotic hydrogenolysis is particularly well investigated as it commonly occurs in natural systems and is increasingly used as a remediation method. The magnitude of carbon isotope fractionation (Table 25.3) tends to increase from PCE to vinyl chloride (VC). In addition, for PCE and TCE carbon isotope fractionation is more variable than for *cis*-1,2-dichloroethene (*cis*-1,2-DCE) and VC. This likely reflects that PCE and TCE can be transformed by a broader range of bacteria that might use different reaction mechanisms and/or show various degree of "masking". The enrichment factor for hydrogenolysis of cis-1,2-DCE and VC correspond to AKIE that are close to the expected value for cleavage of a C–Cl bond without masking. Less data are available for chlorine isotopes as analytical methods have only become available more recently. The reported chlorine enrichment factors for hydrogenolysis (Table 25.3) are smaller than for carbon, which is expected, given the smaller relative mass difference for chlorine compared to carbon. Aerobic oxidation of VC and *cis*-1,2-DCE is associated with fairly consistent carbon isotope fractionation. The smaller isotope enrichment factors compared to hydrogenolysis can be explained by more limited bond-modifications during formation of an epoxide intermediate compared to bond cleavage during hydrogenolysis. Carbon and chlorine isotope fractionation also occur during abiotic reductive dechlorination. The variable carbon isotope enrichment factors probably originate from masking by rate limiting steps preceding reaction, whose rates may vary depending on the characteristics of metal surfaces. Finally, also oxidation of chlorinated ethenes by permanganate is associated by substantial carbon isotope fractionation. Theoretical simulations have indicated that the large carbon isotope fractionation for TCE originates from a concerted reaction at both carbons by a 3 + 2 electrocyclic addition (Adamczyk et al. 2011).

For identifying transformation processes, dual-element isotope slopes are of particular interest (Fig. 25.5). Dual-element isotope slopes are fairly variable for PCE degradation, likely again reflecting the diversity of bacteria and reaction mechanisms involved in their transformation. In contrast, the slopes seem to be more consistent for TCE, *cis*-1,2-DCE and VC although the data set is still small. For hydrogenolysis and abiotic reductive dechlorination of TCE, the slopes are fairly

Table 25.3 Carbon and chlorine isotope fractionation during biotic and abiotic transformation of chlorinated ethenes by different mechanisms

Compound		Biotic		Abiotic		
		Reductive dechlorination	Oxidation (metabolic/cometabolic)	Reductive dechlorination (Fe_0, FeS, FeS_2, green rust)	Oxidation (permanganate)	Oxidation (persulfate)
Tetrachloroethene	C	−3.6 (−0.4 to −19, $n = 31$)		−15.5 (−5.7 to −30.2, $n = 9$)	−17.0 ($n = 1$)	−4.9 ($n = 1$)
	Cl	−2.0 (−0.9 to −10.0, $n = 5$)				
Trichloroethene	C	−12.2 (−2.2 to −18.9, $n = 31$)	−1.1 and −18.2, $n = 2$	−16.7 (−7.5 to −33.4, $n = 17$)	−25.1 (−21.4 to −26.8, $n = 3$)	−3.6 ($n = 1$)
	Cl	−4.3 (−2.7 to −5.7, $n = 7$)		−2.6 ($n = 1$)		
cis-1,2-Dichloroethene	C	−18.8 (−12 to −29.7, $n = 17$)	−8.4 (−0.4 to −9.8, $n = 6$)	−15.9 (−6.9 to −21.7, $n = 7$)	−21.1 ($n = 1$)	−7.6 ($n = 1$)
	Cl	−1.5 ($n = 1$)		−6.2 ($n = 1$)		
Vinyl chloride	C	−23.5 (−19.9 to −31.1, $n = 14$)	−6.3 (−3.2 to −8.2, $n = 17$)	−16.6 (−6.9 to −19.4, $n = 5$)		
	Cl	−1.7 ($n = 1$)	−0.3 ($n = 1$)			

The median and range of the reported isotope enrichment factors are provided (Nijenhuis et al. 2005; Badin et al. 2014; Abe et al. 2009b; Audí-Miró et al. 2013; Barth et al. 2002; Chartrand et al. 2005; Chu et al. 2004; Cichocka et al. 2007, Cretnik et al. 2013, 2014; Elsner et al. 2008; Hunkeler et al. 1999, 2002, 2003; Lee et al. 2007; Numata et al. 2002; Poulson and Naraoka 2002; VanStone et al. 2004; Dong et al. 2009; Fletcher et al. 2011; Liang et al. 2007; Liu et al. 2014; Marchesi et al. 2012; Tiehm et al. 2008)

similar while they are more distinct for *cis*-1,2-DCE. Clearly, different slopes occur for hydrogenolysis versus aerobic oxidation of *cis*-1,2-DCE and VC (Fig. 25.5).

The classical Rayleigh equation describes stable isotope patterns for the reactant and product of a one-step reaction. However, hydrogenolysis of chlorinated ethenes proceeds step-wise with accumulation of less-chlorinated intermediates. Isotope patterns of intermediates are often influenced by formation and degradation of the compound. Isotope patterns for sequential hydrogenolysis of

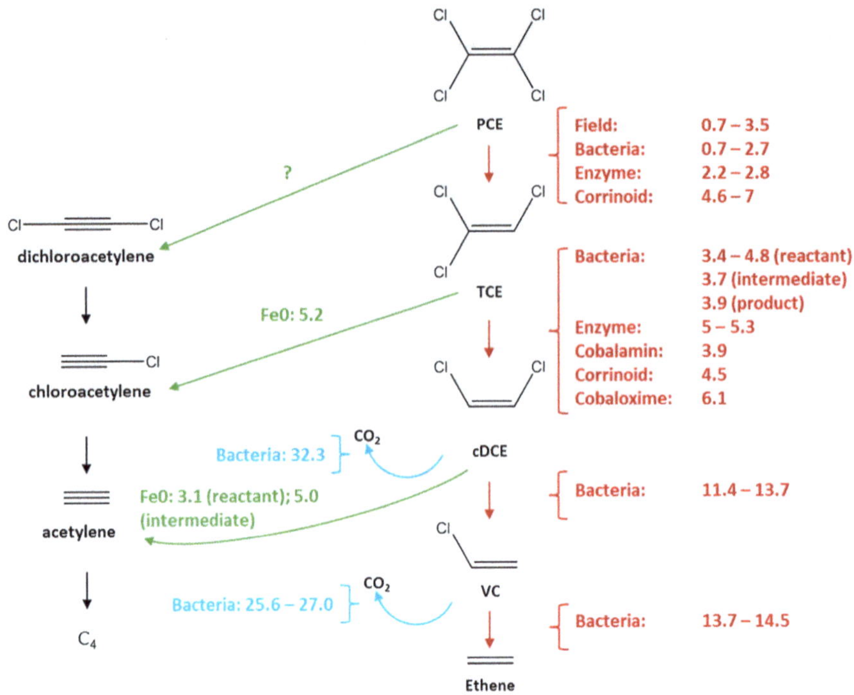

Fig. 25.5 Dual-element isotope slopes Λ (C/Cl) for various transformation processes of chlorinated ethenes (Abe et al. 2009a; Audí-Miró et al. 2013; Cretnik et al. 2014, 2013; Kuder and Philp 2013; Renpenning et al. 2014)

chlorinated ethenes are distinctly different depending on the element as for carbon, all atoms remain conserved, for chlorine, atoms are cleaved, while for hydrogen, atoms are added. For carbon, each intermediate is depleted in light isotopes initially compared to the precursor (Hunkeler et al. 1999), but depending on the relative rate of degradation and the enrichment factors can get heavier than the precursor as degradation advances (Van Breukelen et al. 2005). If an intermediate is not degraded further, its carbon isotope ratio approaches that of the initial compound for mass balance reasons. Only if the intermediate gets degraded further can it become "heavier" than the initial compound. For chlorine isotopes, the isotope pattern is more complex as chlorine atoms are released (Hunkeler et al. 2009). As the released chlorine atoms are depleted in heavy isotopes, the intermediate becomes enriched in heavy isotopes relative to the starting point, even if the intermediate does not degrade further. In addition, a secondary isotope effect during hydrogenolysis can further shift the isotope ratio of the remaining chlorine atoms in the intermediate (Cretnik et al. 2014). As for hydrogen, the incorporated atoms are fairly depleted in heavy isotope. As a result, each intermediate has more depleted isotope signature compared to the precursor (Kuder et al. 2013).

25.6.2 Chlorinated Ethanes

Among the chlorinated ethanes, 1,2-DCA, 1,1,1-trichloroethane (1,1,1-TCA) and 1,1,2,2-tetrachloroethane (1,1,2,2-PCA) occur frequently as environmental contaminants. Under oxic conditions, 1,2-DCA can be biodegraded by two different mechanisms, via an S_N2 nucleophilic substation or C–H bond oxidation. As already discussed above, these mechanisms are associated with distinctly different isotope fractionation factors and dual-element isotope slopes (Fig. 25.3, Table 25.4). Under anoxic conditions, biodegradation proceeds via dichloroelimination to ethene, a process that is associated with strong carbon isotope fractionation originating from cleavage of the two carbon–chlorine bonds (Table 25.4). Alternatively, 1,2-DCA might biodegrade by hydrogenolysis to chloroethane or undergo abiotic dehydrohalogenation to vinyl chloride. However, for these pathways, isotope fractionation has not been characterized yet.

1,1,1-TCA can be degraded abiotically and biotically. For abiotic degradation, distinctly different dual-element isotope slopes were observed for hydrolysis/dehydrohalogenation, reductive dechlorination with Fe(0) and oxidation by heat-activated persulfate (Table 25.4). It is interesting to note that the former two mechanisms are associated with significantly different slopes although both of them involve cleavage of a C–Cl bond in the initial transformation step. Secondary isotope effects might contribute to isotope fractionation to a different degree.

25.6.3 Hexachlorohexane

Hexachlorohexane (HCH) degradation is a good example to illustrate how the molecule size ("dilution") and rate-limiting steps ("masking") influence the magnitude of isotope fractionation. Generally, carbon isotope enrichment factors (−0.7 to −7.6 ‰) tend to be smaller than for chlorinated ethanes (Table 25.4). This can be explained by the larger number of carbon atoms in the molecules which increases the probability that the heavy isotope is located in a non-reacting position. Two reaction mechanisms, dehydrochlorination and dichloroelimination occur during abiotic as well as biotic degradation of the compound. In both cases, the enrichment factors are larger during abiotic than biotic degradation. Rate-limiting transport to the enzyme or enzyme-substrate binding might partly mask the KIE.

25.6.4 Chlorinated Benzenes

Chlorinated benzenes have been used as degreasers, solvents, pesticides and for chemical synthesis. Under aerobic conditions, chlorinated benzenes are generally degraded via dihydroxylation by a ring dioxygenase (Van der Meer et al. 1991; Field and Sierra-Alvarez 2008). During this process, carbon isotope fractionation is

Table 25.4 Isotope enrichment factors and dual-element isotope slopes for environmentally relevant degradation processes of common chlorinated ethanes and hexachlorohexane

	Mechanism	ε_C (‰)	ε_{Cl} (‰)	Λ (C/Cl)	Refs.
1,2-Dichloroethane	Aerobic oxidation S_N2	−28.7 to −33.4	−4.2 to −4.4	7.6–7.7	Hirschorn et al. (2004), Hunkeler and Aravena (2000)
	Aerobic oxidation C–H cleavage	−3.0 to −3.8	−3.8	0.78	Hirschorn et al. (2004), Palau et al. (2014a)
	Dichloroelimination	−32.1	–	–	Hunkeler et al. (2002)
1,1,1-Trichloroethane	Hydrolysis and dehydrochlorination	−1.6	−4.7	0.33	Palau et al. (2014b)
	Abiotic reduction by Fe(0)	−7.8 to −13.6	−5.2	1.5	Elsner et al. (2007), Palau et al. (2014b)
	Oxidation by heat-activated persulfate	−4.0	No significant fractionation	∞	Palau et al. (2014b)
	Oxidation by base-activated persulfate	−7.0	–	–	Marchesi et al. (2013)
	Biotic hydrogenolysis	1.8	–	–	Sherwood Lollar et al. (2010)
1,1,2,2-Tetrachloroethane	Dehydrochlorination	−20.8 to −27.1	–	–	Elsner et al. (2007), Hofstetter et al. (2007), Neumann et al. (2009)
	Abiotic reduction by Fe(0)	−19.3	–	–	Elsner et al. (2007)
α-Hexachlorohexane	Alkaline hydrolysis via dehydrochlorination	−7.6 bulk −1.7 (+) enantiomer −2.1 (−) enantiomer	–	–	Zhang et al. (2014)
	Abiotic reduction by Fe(0) via dichloroelimination	−4.9 bulk −5.1 (+) enantiomer −4.8 (−) enantiomer	–	–	Zhang et al. (2014)
	Aerobic biodegradation via dehydrochlorination	−1.0 to −1.6 bulk −2.4 to −2.5 (+) enantiomer −0.7 to −1.0 (−) enantiomer	–	–	Bashir et al. (2013)
	Anaerobic biodegradation via dichloroelimination	−3.7 bulk	–	–	Badea et al. (2011)
γ-Hexachlorohexane	Anaerobic biodegradation	−3.4 to −3.9	–	–	Badea et al. (2009)

Table 25.5 Isotope enrichment factors for biodegradation of chlorinated benzenes

Compound	Mechanism/conditions	ε_C (‰)	Refs.
Monochlorobenzene	Aerobic—pure culture	−0.1 to −0.4	Kaschl et al. (2005)
Monochlorobenzene	Anaerobic	−5.0	Liang et al. (2011)
1,2-Dichlorobenzene	Anaerobic	−0.8	Liang et al. (2014)
1,3-Dichlorobenzene	Anaerobic	−5.4	Liang et al. (2014)
1,4-Dichlorobenzene	Anaerobic	−6.3	Liang et al. (2014)
1,2,3-Trichlorobenzene	Anaerobic	−3.5	Griebler et al. (2004), Liang et al. (2011)
1,2,4-Trichlorobenzene	Anaerobic	−3.0 to −3.2	Griebler et al. (2004), Liang et al. (2011)

absent or very small, as shown for monochlorobenzene and 1,2,3-trichlorobenzene (Table 25.5). Similar observations have also been made for the degradation of other aromatic compounds, such as toluene, by ring dioxygenases (Vogt et al. 2008). The small or absent carbon isotope fractionation can be explained by the absence of complete bond cleavage during the dihydroxylation process. Under anaerobic conditions, mono- and polychlorinated benzenes can be degraded by hydrogenolysis similarly as chlorinated ethenes (Adrian et al. 2000; Field and Sierra-Alvarez 2008). Significant carbon isotope fractionation occurs with enrichment factors reaching from −3.0 to −6.3 ‰ (Table 25.5), except for 1,2-dichlorobenzene, which shows less isotope fractionation (−0.8 %). The fairly large isotope fractionation can be explained by C–Cl bond cleavage in the rate limiting step. The magnitude of carbon isotope fractionation is, however, smaller than for most chlorinated ethenes. This can again be explained by the larger number of carbon atoms in non-reactive positions in chlorinated benzenes compared to chlorinated ethenes.

25.6.5 Chlorinated Methanes

Chlorinated methanes have been widely used as solvents and reagents in organic synthesis. In addition, some chlorinated methanes can also be naturally produced or occur as a by-product during chlorination of drinking water. Chloromethane is the most abundant volatile chlorinated hydrocarbon in the atmosphere, most of which originates from terrestrial vegetation (Keppler et al. 2005). Most of the natural chloroform originates from biogenic sources such as plants, insects and probably fungi, with comparable contributions by terrestrial and oceanic environments (Laturnus et al. 2002). In groundwater below coniferous forests, chloroform can occur naturally at concentrations up to 10 μg/L (Hunkeler et al. 2012). Biodegradation and abiotic transformation of chlorinated methanes tends to be associated with a large carbon isotope fractionation (Table 25.6). As only one carbon is present, no reduction of the KIE by non-reactive positions occurs. For chloromethane, a C–Cl bond is broken by a methyl transfer reaction in the initial step (Vannelli et al. 1998). This

Table 25.6 Isotope enrichment factors for degradation of chlorinated methanes

Compound	Mechanism/ conditions	ε_C (‰)	ε_{Cl} (‰)	ε_H (‰)	Refs.
Chloromethane	Biotic Aerobic by methylotrophic bacteria	−38 to −50	–	−27 to −29	Nadalig et al. (2013), Miller et al. (2001)
Dichloromethane	Biotic Aerobic by methylotrophic bacteria	−42.4 to −66.3	−3.8	–	Heraty et al. (1999), Nikolausz et al. (2006)
Dichloromethane	Biotic Denitrifying by methylotrophic bacteria	−45.8 to −61.0	–	–	Nikolausz et al. (2006)
Chloroform	Biotic Dehalobacter-containing enrichment culture	−27.5	–	–	Chan et al. (2012)
Carbon tetrachloride	Abiotic reductive dechlorination by Fe(II) in smectites	−10.9 to −13.3			Neumann et al. (2009)
	Abiotic reductive dehalogenation by iron (hydr)oxides or siderite	−25.6 to −32			Zwank et al. (2005)
	Abiotic reductive dehalogenation by mackinawite	−15.9			Zwank et al. (2005)
	Alkaline hydrolysis	−49 to −56			Torrento et al. (2014)

also explains why only small hydrogen isotope fractionation occurs likely due to a secondary hydrogen isotope effect. For dichloromethane degradation by methylotrophic bacteria, the initial transformation likely occurs via a biologically mediated nucleophilic substitution which also involves cleavage of a C–Cl bond. This proposed mechanism is consistent with the relatively large chlorine isotope enrichment factor of −3.8 ‰ (Heraty et al. 1999). For carbon tetrachloride, large but variable carbon isotope fractionation was observed for reductive dechlorination by solid phases containing reduced iron or sulfur. The variation in isotope fractionation might be due to "masking" by rate-limiting transport to reactive solid phase.

25.7 Application of Isotope Methods in Field Studies

Stable isotope methods can be applied to address different questions related to the origin and fate of chlorinated hydrocarbons in the environment (Table 25.7). Applications can broadly be divided in those that aim at identifying different

Table 25.7 Common applications of isotope methods in field studies

	Objective	Approach	Example
Tracking sources of contaminants	Evaluating if different sources of the same compound are present at a site and linking downgradient contamination to their source	Comparison of multi-element isotope signatures in source area and downgradient plumes	Dual carbon–chloride isotope analysis indicated presence of multiple trichloroethene sources in bedrock aquifer (Lojkasek-Lima et al. 2012)
	Distinguishing natural and anthropogenic sources of a compound	Comparing isotope signatures to known ranges for natural and anthropogenic compounds	Chloroform from natural production in forest soils versus anthropogenic chloroform (Hunkeler et al. 2012)
Tracking reactive processes	Identification by which process a contaminant is degraded at a site	Determination of dual-element isotope slopes for field samples and comparison with laboratory value	Field dual carbon-chlorine isotope slopes suggested that cis-1,2-dichloroethene was subject to abiotic biodegradation (Hunkeler et al. 2011a)
	Tracking progress of a process qualitatively	Evaluation of shifts in isotope ratios	Demonstrating natural attenuation of chlorinated ethenes based on enrichment of heavy carbon isotopes (Hunkeler et al. 1999).
	Linking reactants to products of degradation for contaminant mixtures where multiple pathways can lead to the same product	Evaluation of isotope patterns of reactants and products	Carbon isotope analysis indicated that TCE originated partly from dehydrochlorination of 1,1,2,2- tetrachloroethane. Vinyl chloride and ethene were identified as products of dehydrochlorination of 1,1,2-trichloroethane and dichloroelimination of 1,2-dichloroethane rather than reductive dechlorination of chlorinated ethenes (Hunkeler et al. 2005)
	Simplified estimation of degree of contaminant transformation	Application of the Rayleigh equation based on known isotope enrichment factors from laboratory studies	Estimation of the amount of chlorinated hydrocarbon transformation by natural attenuation (Vieth et al. 2003)
	Estimation of first order transformation rate	Combining Rayleigh equation with first order rate expression to estimate rate constants relying on laboratory enrichment factors	Estimation of rate of cis-1,2-dichloroethene biodegradation at site with bioaugmentation (Morrill et al. 2005)
	Quantification of reaction progress and prediction of future evolution	Reactive transport model that incorporates isotope data	Isotope data of tetrachloroethene and trichloroethene made it possible to distinguish between two reactive transport models that reproduced product patterns well but gave different predictions for future trends (Atteia et al. 2008)

sources of chlorinated hydrocarbons, and those that target transformation of contaminants. In the following, these two application types are discussed separately.

25.7.1 Tracking Different Sources of Contamination

At industrial sites, often multiple sources of the same compounds are present due to the widespread use of chlorinated hydrocarbons for various industrial applications. For a successful remediation and for cost-attribution, it is of considerable interest which source contributes to how much of the contamination. Chlorinated hydrocarbons from different manufacturers and production batches can have a different isotope signature (Hunkeler et al. 2004; Van Wanderdam et al. 1995). In addition, compounds that result from environmental transformation processes can have a distinctly different isotope signature compared to industrial substance. For example, TCE that results from reductive dechlorination has a distinctly different hydrogen isotope signature compared to industrial TCE (Shouakar-Stash et al. 2003). These variations open the possibility to distinguish different sources of the same compound. However, potential changes of isotope signatures during the migration of the compounds through environmental systems have to be considered. In water-saturated zone of the subsurface, transport processes generally do not alter isotope signatures significantly. In contrast, in the vadose zone, gas-phase diffusion can shift isotope signatures considerably, especially under transient conditions (Hunkeler et al. 2011b; Jeannottat and Hunkeler 2012). Sorption seems to have only a small effect on carbon isotope ratios. It could potentially have a larger effect on Cl and H that interact with surfaces, but very little information is available on this topic so far. Differentiating sources becomes more challenging when reactive processes occur, which have a larger effect on isotope signature. However, if a multi-element isotope approach is used, it can still be possible to differentiate among sources. Isotope shifts due to reactive processes rather different sources can be identified based on their characteristic dual-element isotope trajectories (Lojkasek-Lima et al. 2012; Lutz and Van Breukelen 2014).

In the case of chloroform, it is not only of interest to differentiate between anthropogenic sources but also between anthropogenic versus natural sources. While naturally produced chloroform has a carbon isotope composition similarly to natural organic matter (−22 to −28 ‰), anthropogenic chloroform is strongly depleted in ^{13}C (−40 to −65 ‰) as methane is used for its synthesis. The large difference in carbon isotope ratio makes it possible to reliably distinguish the two sources (Hunkeler et al. 2012).

25.7.2 Tracking Reactive Processes

Isotope methods can be used to differentiate between degradation pathways, to relate reactants to products, to track processes qualitatively and to quantify the progress of degradation.

25.7.2.1 Identification of Degradation Pathways

Especially when implementing remediation methods, it is important to know if a compound is transformed by the intended process as often considerable means are invested to promote a specific reactive process. For example, the question arises if aerobic metabolism is initiated after implementation of air sparging, or if the emplacement of Fe(0) really promotes abiotic transformation. For some compounds, isotope enrichment factors strongly differ among degradation pathways. For example for VC, carbon isotope enrichment factors are between −3.2 and −8.2 ‰ for aerobic oxidation and between −19.9 and −31.1 ‰ for reductive dechlorination (Table 25.3). A potential approach to differentiate between reaction mechanisms would be to apply the Rayleigh equation to the field data and compare field-based enrichment factors to laboratory determined factors. However, such an approach is associated with considerable uncertainty as concentrations are also influenced by physical processes (dilution) in addition to reactive process. A more robust approach is to compare field dual-element isotope slopes, which are not affected by dilution, to laboratory-derived slopes. The method is fairly widely applicable as slopes for different reactive processes affecting a compound vary as discussed above. However, in some cases (e.g. biotic hydrogenolysis versus abiotic reduction of TCE), the slopes might be too close for a distinction.

25.7.2.2 Tracking Progress of Transformation

Once a process is identified, the question often arises to what extent it occurs at a site or whether a remediation method successfully increased the amount of transformation. Concentration data alone are often not sufficient as concentrations are often also influenced by physical processes such as dilution or sorption. The first approach is usually to simply track spatial or temporal trends of enrichment of the heavy isotope. The amount of enrichment of the heavy isotope thereby depends on both the isotope enrichment factor as well as the transformation progress. According to Eq. 25.16, the shift of isotope ratio is given by

$$\Delta\delta E_R = \delta E_{R,t} - \delta E_{R,0} - \varepsilon \cdot \ln f = 2.3 \cdot \varepsilon \cdot \log f \qquad (25.23)$$

According to this equation, for each order of magnitude of mass reduction of a compound, its isotope ratio shifts by 2.3 times the isotope enrichment factor. Compared to other compound classes (e.g. petroleum hydrocarbons), the isotope enrichment factors for chlorinated hydrocarbons tend to be large, which makes isotope analysis a very sensitive approach for tracking the progress of degradation. For example, for reductive dechlorination of VC, the median of reported carbon isotope enrichment factors is −23.5 ‰. If the mass of the compound is reduced by an order of magnitude its carbon isotope composition shifts by 54 ‰, which is very large relative to the uncertainty of the measurement (0.3–0.5 ‰). In addition to qualitative data interpretation, the degree of transformation can be quantified using the Rayleigh equation. However, it has to be taken into account that the calculation is only approximate. The Rayleigh equation strictly holds only for well-mixed reservoirs of compounds which all degrade at the same rate. In contrast, a groundwater sample usually represents different flow paths that are associated with varying travel times and degradation kinetics. As result, the Rayleigh equation applied to field sample tends to underestimate the actual degree of degradation (Abe and Hunkeler 2006).

25.7.2.3 Quantification of Reactive Processes Using Reactive Transport Models

An alternative approach to make use of isotope data for process quantification is the use of reactive transport models (Van Breukelen et al. 2005). Such models take into account differences in flow paths and spatial variations of reaction rates to some degree. In its most rigorous form, all isotopologue are modelled as a separate species in the reactive transport model attributing specific degradation rates to each of them (Jin et al. 2013). However, for polychlorinated hydrocarbons, this leads to a large number of species if isotopes of several elements are considered, which complicates or even prevents the incorporation of isotope data into reactive transport models. Therefore, approximate approaches are usually warranted for practical application. Carbon and chlorine isotopes can be treated in good approximation separately (Hunkeler et al. 2009). For carbon, often only two species are considered which correspond to light and heavy carbon. Given that the heavy carbon isotope has a low abundance, these species can also be considered to represent molecules with light isotopes only and those with one heavy isotope, as species with several heavy isotopes are rare. For chlorine, isopologues may be simulated separately given the high abundance of isotopologues with several heavy isotopes. However, even for chlorine, isotopes rather than isotopologues can be considered with greatly simplifies the calculations (Hunkeler et al. 2009).

25.7.2.4 Relating Reactants to Products

At sites with complex mixtures of chlorinated hydrocarbons, often multiple pathways lead to the same product or a compound may be present as a primary substance or a degradation product. For example at sites with chlorinated ethenes and 1,2-dichloroethane, vinyl chloride and ethene can originate from both of these compounds. The measurement of isotope ratios of all potential precursors and products often makes it possible to link precursors to products based on several principles. If a precursor and product are linked, their isotope ratios evolve according a well-defined pattern. Furthermore, compounds that are produced in situ often have a different isotope signature than industrial chemicals. For example, at a landfill site, carbon isotope analysis suggested that TCE likely originated from dehydrochlorination of 1,1,2,2-PCA rather than being a primary contaminant.

Furthermore, vinyl chloride and ethene were likely produced by dichloroelimination of 1,1,2-trichloroethane and 1,2-DCA rather than reductive dechlorination of chlorinated ethenes (Hunkeler et al. 2005).

25.8 Conclusion and Research Needs

In the past decade, significant advances have been made in understanding isotope effects associated with the transformation of chlorinated hydrocarbons. On the one hand, isotope data have provided additional insight into degradation mechanisms of these compounds. On the other hand, isotope methods are increasingly used as a tool to track the origin and fate of chlorinated hydrocarbons in the environment. Isotope methods work particularly well for chlorinated hydrocarbons. Isotope enrichment factors are usually fairly large as the number of atoms of each element that is present in the molecules is limited and underlying isotope effects tend to be fairly large. Even for chlorine, isotope effects can be substantial despite the larger mass as often both primary and secondary isotope effects occur. Many chlorinated hydrocarbons contain three elements, which opens additional possibilities to differentiate between processes and sources of compounds. Despite these significant advances, there are still a number of research needs. Isotope fractionation is best documented for carbon. In comparison, there is much less information on chlorine and hydrogen isotope fractionation as efficient analytical methods only became available recently. While such methods are now available, there is a need for more reference material to ensure that comparable results are obtained in different laboratories. For many reaction mechanisms relevant for field studies, dual-element isotopes slopes have not been characterized yet. There is a need to better understand how variable dual-element isotope slopes are for specific processes. There are also still some gaps in knowledge regarding the effect of physical processes on isotope fractionation, especially the effect of sorption on Cl and H isotope ratios. In addition, advances are required when it comes to quantitative data evaluation at different levels. A wider use of methods of computational chemistry could provide additional insight into how isotope effects are related to specific reaction mechanisms. When it comes to application of models to field isotope data, a better understanding is required which level of model complexity is required under what conditions. Thus, while isotope methods are increasing used as a tool to support the management of contaminated sites and often provide unique insight into the origin and fate of contaminants, there are also still ample research opportunities on various aspects of the methodology reaching from process understanding to field applications.

References

Abe Y, Hunkeler D (2006) Does the Rayleigh equation apply to evaluate field isotope data in contaminant hydrogeology? Environ Sci Technol 40(5):1588–1596

Abe Y, Aravena R, Parker BL, Hunkeler D (2009a) Evaluating the fate of chlorinated ethenes in streambed sediments by combining stable isotope, geochemical and microbial methods. J Contam Hydrol 107(1–2):10–21

Abe Y, Aravena R, Zopfi J, Shouakar-Stash O, Cox E, Roberts JD, Hunkeler D (2009b) Carbon and chlorine isotope fractionation during aerobic oxidation and reductive dechlorination of vinyl chloride and cis-1,2-dichloroethene. Environ Sci Technol 43(1):101–107

Abramson FP, Black GE, Lecchi P (2001) Application of high-performance liquid chromatography with isotope-ratio mass spectrometry for measuring low levels of enrichment of underivatized materials. J Chromatogr A 913(1–2):269–273. doi:10.1016/s0021-9673(00)01032-3

Adamczyk P, Dybala-Defratyka A, Paneth P (2011) DFT study of trichloroethene reaction with permanganate in aqueous solution. Environ Sci Technol 45(7):3006–3011. doi:10.1021/es103251u

Adrian L, Szewzyk U, Wecke J, Görisch H (2000) Bacterial dehalorespiration with chlorinated benzenes. Nature 408(6812):580–583. doi:10.1038/35046063

Aelion MC, Höhener P, Hunkeler D, Aravena R (eds) (2010) Environmental isotopes in biodegradation and bioremediation. CRC Press, Boca Raton

Aeppli C, Holmstrand H, Andersson P, Gustafsson O (2010) Direct compound-specific stable chlorine isotope analysis of organic compounds with quadrupole GC/MS using standard isotope bracketing. Anal Chem 82(1):420–426. doi:10.1021/ac902445f

Atteia O, Franceschi M, Dupuy A (2008) Validation of reactive model assumptions with isotope data: application to the Dover case. Environ Sci Technol 42(9):3289–3295. doi:10.1021/es071269m

Audí-Miró C, Cretnik S, Otero N, Palau J, Shouakar-Stash O, Soler A, Elsner M (2013) Cl and C isotope analysis to assess the effectiveness of chlorinated ethene degradation by zero-valent iron: evidence from dual element and product isotope values. Appl Geochem 32:175–183. doi:10.1016/j.apgeochem.2012.08.025

Badea S-L, Vogt C, Weber S, Danet A-F, Richnow H-H (2009) Stable isotope fractionation of gamma-hexachlorocyclohexane (lindane) during reductive dechlorination by two strains of sulfate-reducing bacteria. Environ Sci Technol 43(9):3155–3161. doi:10.1021/es801284m

Badea S-L, Vogt C, Gehre M, Fischer A, Danet A-F, Richnow H-H (2011) Development of an enantiomer-specific stable carbon isotope analysis (ESIA) method for assessing the fate of alpha-hexachlorocyclohexane in the environment. Rapid Commun Mass Spectrom 25(10):1363–1372. doi:10.1002/rcm.4987

Badin A, Buttet G, Maillard J, Holliger C, Hunkeler D (2014) Multiple dual C–Cl isotope patterns associated with reductive dechlorination of tetrachloroethene. Environ Sci Technol 48(16):9179–9186. doi:10.1021/es500822d

Barth JAC, Slater G, Schueth C, Bill M, Downey A, Larkin M, Kalin RM (2002) Carbon isotope fractionation during aerobic biodegradation of trichloroethene by Burkholderia cepacia G4: a tool to map degradation mechanisms. Appl Environ Microbiol 68(4):1728–1734

Bashir S, Fischer A, Nijenhuis I, Richnow H-H (2013) Enantioselective carbon stable isotope fractionation of hexachlorocyclohexane during aerobic biodegradation by Sphingobium spp. Environ Sci Technol 47(20):11432–11439. doi:10.1021/es402197s

Bernstein A, Shouakar-Stash O, Ebert K, Laskov C, Hunkeler D, Jennottat S, Sakaguchi-Soder K, Cretnik S, Jager J, Haderlein SB, Aravena R, Elsner M (2011) Compound-specific chlorine isotope analysis: An interlaboratory comparison of GC-IRMS and GC-qMS methods. Anal Chem 83(20):7624–7634

Chan CCH, Mundle SOC, Eckert T, Liang X, Tang S, Lacrampe-Couloume G, Edwards EA, Lollar BS (2012) Large carbon isotope fractionation during biodegradation of chloroform by Dehalobacter cultures. Environ Sci Technol 46(18):10154–10160. doi:10.1021/es3010317

Chartrand MMG, Waller A, Mattes TE, Elsner M, Lacrampe-Couloume G, Gossett JM, Edwards EA, Sherwood Lollar B (2005) Carbon isotopic fractionation during aerobic vinyl chloride degradation. Environ Sci Technol 39:1064–1070

Chartrand MMG, Hirschorn SK, Lacrampe-Couloume G, Lollar BS (2007) Compound specific hydrogen isotope analysis of 1,2-dichloroethane: potential for delineating source and fate of chlorinated hydrocarbon contaminants in groundwater. Rapid Commun Mass Spectrom 21(12):1841–1847. doi:10.1002/rcm.3026

Chu K-H, Mahendra S, Song DL, Conrad ME, Alverez-Cohen L (2004) Stable carbon isotope fractionation during aerobic biodegradation of chlorinated ethenes. Environ Sci Technol 38:3126–3130

Cichocka D, Siegert M, Imfeld G, Andert J, Beck K, Diekert G, Richnow HH, Nijenhuis I (2007) Factors controlling the carbon isotope fractionation of tetra- and trichloroethene during reductive dechlorination by Sulfurospirillum ssp and Desulfitobacterium sp strain PCE-S. FEMS Microbiol Ecol 62(1):98–107

Clark I (2015) Groundwater geochemistry and isotopes. CRC Press, Boca Raton

Clark ID, Fritz P (1997) Environmental isotopes in hydrogeology. Lewis Publishers, Boca Raton

Cleland WW (2006) Enzyme mechanisms from isotope effects. In: Kohen A, Limbach HH (eds). CRC Press, Boca Raton, pp 915–930

Cretnik S, Thoreson KA, Bernstein A, Ebert K, Buchner D, Laskov C, Haderlein S, Shouakar-Stash O, Kliegman S, McNeill K, Elsner M (2013) Reductive dechlorination of TCE by chemical model systems in comparison to dehalogenating bacteria: Insights from dual element isotope analysis (13C/12C, 37Cl/35Cl). Environ Sci Technol 47(13):6855–6863. doi:10.1021/es400107n

Cretnik S, Bernstein A, Shouakar-Stash O, Löffler F, Elsner M (2014) Chlorine isotope effects from isotope ratio mass spectrometry suggest intramolecular C–Cl bond competition in trichloroethene (TCE) reductive dehalogenation. Molecules 19(5):6450–6473. doi:10.3390/molecules19056450

Dong YR, Liang XM, Krumholz LR, Philp RP, Butler EC (2009) The Relative contributions of abiotic and microbial processes to the transformation of tetrachloroethylene and trichloroethylene in anaerobic microcosms. Environ Sci Technol 43(3):690–697. doi:10.1021/es801917p

Elsner M, Hunkeler D (2008) Evaluating chlorine isotope effects from isotope ratios and mass spectra of polychlorinated molecules. Anal Chem 80(12):4731–4740

Elsner M, Zwank L, Hunkeler D, Schwarzenbach RP (2005) A new concept linking observable stable isotope fractionation to transformation pathways of organic pollutants. Environ Sci Technol 39:6896–6916

Elsner M, Cwiertny DM, Roberts AL, Lollar BS (2007) 1,1,2,2-tetrachloroethane reactions with OH-, Cr(II), granular iron, and a copper-iron bimetal: insights from product formation and associated carbon isotope fractionation. Environ Sci Technol 41(11):4111–4117

Elsner M, Chartrand M, Vanstone N, Couloume GL, Lollar BS (2008) Identifying abiotic chlorinated ethene degradation: characteristic isotope patterns in reaction products with nanoscale zero-valent iron. Environ Sci Technol 42(16):5963–5970. doi:10.1021/es8001986

Elsner M, Jochmann A, Hofstetter TB, Hunkeler D, Bernstein A, Schmidt TC, Schimmelmann A (2012) Current challenges in compound-specific stable isotope analysis of environmental organic contaminants. Anal Bioanal Chem 403(9):2471–2491. doi:10.1007/s00216-011-5683-y

Field JA, Sierra-Alvarez R (2008) Microbial degradation of chlorinated benzenes. Biodegradation 19(4):463–480. doi:10.1007/s10532-007-9155-1

Fletcher KE, Nijenhuis I, Richnow HH, Löffler FE (2011) Stable carbon isotope enrichment factors for cis-1,2-Dichloroethene and vinyl chloride reductive dechlorination by Dehalococcoides. Environ Sci Technol 45(7):2951–2957. doi:10.1021/es103728q

Griebler C, Adrian L, Meckenstock RU, Richnow HH (2004) Stable carbon isotope fractionation during aerobic and anaerobic transformation of trichlorobenzene. FEMS Microbiol Ecol 48(3):313–321. doi:10.1016/j.femsec.2004.02.005

Heraty LJ, Fuller ME, Huang L, Abrajano T, Sturchio NC (1999) Isotope fractionation of carbon and chlorine by microbial degradation of dichloromethane. Org Geochem 30:793–799

Hirschorn SK, Dinglasan MJ, Elsner M, Mancini SA, Lacrampe-Couloume G, Edwards EA, Sherwood Lollar B (2004) Pathway dependent isotopic fractionation during aerobic biodegradation of 1,2-Dichloroethane. Environ Sci Technol 38:4775–4781

Hitzfeld KL, Gehre M, Richnow H-H (2011) A novel online approach to the determination of isotopic ratios for organically bound chlorine, bromine and sulphur. Rapid Commun Mass Spectrom 25(20):3114–3122. doi:10.1002/rcm.5203

Hoefs J (2009) Stable isotope geochemistry. Springer, Berlin

Hofstetter TB, Reddy CM, Heraty LJ, Berg M, Sturchio NC (2007) Carbon and chlorine isotope effects during abiotic reductive dechlorination of polychlorinated ethanes. Environ Sci Technol 41(13):4662–4668

Hunkeler D, Aravena R (2000) Evidence of substantial carbon isotope fractionation between substrate, inorganic carbon, and biomass during aerobic mineralization of 1,2-dichloroethene by *Xanthobacter autotrophicus*. Appl Environ Microbiol 66(11):4870–4876

Hunkeler D, Elsner M (2010) Principles and mechanisms of isotope fractionation. In: Aelion MC, Höhener P, Hunkeler D, Aravena R (eds) Environmental isotopes in biodegradation and bioremediation. CRC Press, Boca Raton, pp 43–77

Hunkeler D, Aravena R, Butler BJ (1999) Monitoring microbial dechlorination of tetrachloroethene (PCE) using compound-specific carbon isotope ratios: microcosms and field experiments. Environ Sci Technol 33(16):2733–2738

Hunkeler D, Aravena R, Cox E (2002) Carbon isotopes as a tool to evaluate the origin and fate of vinyl chloride: laboratory experiments and modeling of isotope evolution. Environ Sci Technol 36:3378–3384

Hunkeler D, Aravena R, Cherry JA, Parker B (2003) Carbon isotope fractionation during chemical oxidation of chlorinated ethenes by permanganate: laboratory and field studies. Environ Sci Technol 37:798–804

Hunkeler D, Chollet N, Pittet X, Aravena R, Cherry JA, Parker BL (2004) Effect of source variability and transport processes on carbon isotope ratios of TCE and PCE in two sandy aquifers. J Contam Hydrol 74:265–282

Hunkeler D, Aravena R, Berry-Spark K, Cox E (2005) Assessment of degradation pathways at a site with mixed chlorinated hydrocarbon contamination using stable isotope analysis. Environ Sci Technol, vol 39

Hunkeler D, Van Breukelen BM, Elsner M (2009) Modeling chlorine isotope trends during reductive sequential transformation of chlorinated ethenes. Environ Sci Technol 43(17):6750–6756

Hunkeler D, Abe Y, Broholm MM, Jeannotat S, Aravena R, Westergaard C, Jacobsen CS, Bjerg PL (2011a) Assessing chlorinated ethene degradation in a large scale contaminant plume by dual carbon-chlorine isotope analysis and quantitative PCR. J Contam Hydrol 119:69–79

Hunkeler D, Aravena R, Shouakar-Stash O, Weisbrod N, Nasser A, Netzer L, Ronen D (2011b) Carbon and chlorine isotope ratios of chlorinated ethenes migrating through a thick unsaturated zone of a sandy aquifer. Environ Sci Technol 45(19):8247–8253. doi:10.1021/es201415k

Hunkeler D, Laier T, Breider F, Jacobsen OS (2012) Demonstrating a natural origin of chloroform in groundwater using carbon isotopes. Environ Sci Technol 46(11):6096–6101

Jeannottat S, Hunkeler D (2012) Chlorine and carbon isotopes fractionation during volatilization and diffusive transport of trichloroethene in the unsaturated zone. Environ Sci Technol 46(6):3169–3176. doi:10.1021/es203547p

Jin B, Haderlein SB, Rolle M (2013) Integrated carbon and chlorine isotope modeling: applications to chlorinated aliphatic hydrocarbons dechlorination. Environ Sci Technol 47(3):1443–1451. doi:10.1021/es304053h

Kaschl A, Vogt C, Uhlig S, Nijenhuis I, Weiss H, Kastner M, Richnow HH (2005) Isotopic fractionation indicates anaerobic monochlorobenzene biodegradation. Environ Toxicol Chem 24(6):1315–1324. doi:10.1897/04-321r.1

Keppler F, Harper DB, Rockmann T, Moore RM, Hamilton JTG (2005) New insight into the atmospheric chloromethane budget gained using stable carbon isotope ratios. Atmos Chem Phys 5:2403–2411

Kohen A, Limbach HH (eds) (2006) Isotope effects in chemistry and biology. CRC Press, Boca Raton

Kuder T, Philp P (2013) Demonstration of compound-specific isotope analysis of hydrogen isotope ratios in chlorinated ethenes. Environ Sci Technol 47(3):1461–1467. doi:10.1021/es303476v

Kuder T, van Breukelen BM, Vanderford M, Philp P (2013) 3D-CSIA: Carbon, chlorine, and hydrogen isotope fractionation in transformation of ICE to ethene by a *Dehalococcoides* culture. Environ Sci Technol 47(17):9668–9677. doi:10.1021/es400463p

Laturnus F, Haselmann KF, Borch T, Gron C (2002) Terrestrial natural sources of trichloromethane (chloroform, $CHCl_3$)—an overview. Biogeochemistry 60(2):121–139. doi:10.1023/a:1019887505651

Lee PKH, Conrad ME, Alvarez-Cohen L (2007) Stable carbon isotope fractionation of chloroethenes by dehalorespiring isolates. Environ Sci Technol 41(12):4277–4285

Liang X, Dong Y, Kuder T, Krumholz LR, Philp RP, Butler EC (2007) Distinguishing abiotic and biotic transformation of tetrachloroethylene and trichloroethylene by stable carbon isotope fractionation. Environ Sci Technol 41(20):7094–7100. doi:10.1021/es070970n

Liang X, Howlett MR, Nelson JL, Grant G, Dworatzek S, Lacrampe-Couloume G, Zinder SH, Edwards EA, Lollar BS (2011) Pathway-dependent isotope fractionation during aerobic and anaerobic degradation of monochlorobenzene and 1,2,4-trichlorobenzene. Environ Sci Technol 45(19):8321–8327. doi:10.1021/es201224x

Liang X, Mundle SOC, Nelson JL, Passeport E, Chan CCH, Lacrampe-Couloume G, Zinder SH, Lollar BS (2014) Distinct carbon isotope fractionation during anaerobic degradation of dichlorobenzene isomers. Environ Sci Technol 48(9):4844–4851. doi:10.1021/es4054384

Liu YD, Gan YQ, Zhou AG, Liu CF, Li XQ, Yu TT (2014) Carbon and chlorine isotope fractionation during Fenton-like degradation of trichloroethene. Chemosphere 107:94–100. doi:10.1016/j.chemosphere.2014.03.011

Lojkasek-Lima P, Aravena R, Parker BL, Cherry JA (2012) Fingerprinting TCE in a bedrock aquifer using compound-specific isotope analysis. Ground Water 50(5):754–764. doi:10.1111/j.1745-6584.2011.00897.x

Lutz SR, Van Breukelen BM (2014) Combined source apportionment and degradation quantification of organic pollutants with CSIA: 1. Model derivation. Environ Sci Technol 48(11):6220–6228. doi:10.1021/es405400w

Marchesi M, Aravena R, Sra KS, Thomson NR, Otero N, Soler A, Mancini S (2012) Carbon isotope fractionation of chlorinated ethenes during oxidation by Fe^{2+} activated persulfate. Sci Total Environ 433:318–322. doi:10.1016/j.scitotenv.2012.06.051

Marchesi M, Thomson NR, Aravena R, Sra KS, Otero N, Soler A (2013) Carbon isotope fractionation of 1,1,1-trichloroethane during base-catalyzed persulfate treatment. J Hazard Mater 260:61–66. doi:10.1016/j.jhazmat.2013.05.011

Meier-Augenstein W (1999) Applied gas chromatography coupled to isotope ratio mass spectrometry. J Chromatogr A 842:351–371

Miller LG, Kalin RM, McCauley SE, Hamilton JTG, Harper DB, Millet DB, Oremland RS, Goldstein AH (2001) Large carbon isotope fractionation associated with oxidation of methyl halide by methylotrophic bacteria. Proc Natl Acad Sci 98(10):5833–5837

Morrill PL, Lacrampe-Couloume G, Slater GF, Sleep BE, Edwards EA, McMaster ML, Major DW, Lollar BS (2005) Quantifying chlorinated ethene degradation during reductive dechlorination at Kelly AFB using stable carbon isotopes. J Contam Hydrol 76(3–4):279–293

Nadalig T, Greule M, Bringel F, Vuilleumier S, Keppler F (2013) Hydrogen and carbon isotope fractionation during degradation of chloromethane by methylotrophic bacteria. Microbiology Open 2(6):893–900. doi:10.1002/mbo3.124

Neumann A, Hofstetter TB, Skarpeli-Liati M, Schwarzenbach RP (2009) Reduction of polychlorinated ethanes and carbon tetrachloride by structural Fe(II) in smectites. Environ Sci Technol 43(11):4082–4089. doi:10.1021/es9001967

Nijenhuis I, Andert J, Beck K, Kästner M, Diekert G, Richnow HH (2005) Stable isotope fractionation of tetrachloroethene during reductive dechlorination by *Sulfurospirillum multivorans* and *Desulfitobacterium* sp. strain PCE-S and abiotic reactions with cyanocobalamin. Appl Environ Microbiol 71(7):3413–3419

Nikolausz M, Nijenhuis I, Ziller K, Richnow HH, Kastner M (2006) Stable carbon isotope fractionation during degradation of dichloromethane by methylotrophic bacteria. Environ Microbiol 8(1):156–164. doi:10.1111/j.1462-2920.2005.00878.x

Numata M, Nakamura N, Koshikawa H, Terashima Y (2002) Chlorine isotope fractionation during reductive dechlorination of chlorinated ethenes by anaerobic bacteria. Environ Sci Technol 36:4389–4394

Palau J, Cretnik S, Shouakar-Stash O, Hoeche M, Elsner M, Hunkeler D (2014a) C and Cl isotope fractionation of 1,2-dichloroethane displays unique $\delta^{13}C/\delta^{37}Cl$ patterns for pathway identification and reveals surprising C-Cl bond involvement in microbial oxidation. Environ Sci Technol 48(16):9430–9437. doi:10.1021/es5031917

Palau J, Shouakar-Stash O, Hunkeler D (2014b) Carbon and chlorine isotope analysis to identify abiotic degradation pathways of 1,1,1-trichloroethane. Environ Sci Technol 48(24):14400–14408. doi:10.1021/es504252z

Poulson SR, Naraoka H (2002) Carbon isotope fractionation during permanganate oxidation of chlorinated ethylenes (cDCE, TCE, PCE). Environ Sci Technol 36:3270–3274

Renpenning J, Keller S, Cretnik S, Shouakar-Stash O, Elsner M, Schubert T, Nijenhuis I (2014) Combined C and Cl isotope effects indicate differences between corrinoids and enzyme (*Sulfurospirillum multivorans* PceA) in reductive dehalogenation of tetrachloroethene, but not trichloroethene. Environ Sci Technol 48(20):11837–11845. doi:10.1021/es503306g

Renpenning J, Hitzfeld KL, Gilevska T, Nijenhuis I, Gehre M, Richnow H-H (2015a) Development and validation of an universal interface for compound-specific stable isotope analysis of chlorine ($^{37}Cl/^{35}Cl$) by GC-high-temperature conversion (HTC)-MS/IRMS. Anal Chem 87(5):2832–2839. doi:10.1021/ac504232u

Renpenning J, Kuemmel S, Hitzfeld KL, Schimmelmann A, Gehre M (2015b) Compound-specific hydrogen isotope analysis of heteroatom-bearing compounds via gas chromatography-chromium-based high-temperature conversion (Cr/HTC)-isotope ratio mass spectrometry. Anal Chem 87(18):9443–9450. doi:10.1021/acs.analchem.5b02475

Sakaguchi-Soder K, Jager J, Grund H, Matthaus F, Schuth C (2007) Monitoring and evaluation of dechlorination processes using compound-specific chlorine isotope analysis. Rapid Commun Mass Spectrom 21(18):3077–3084

Sherwood Lollar B, Hirschorn S, Mundle SOC, Grostern A, Edwards EA, Lacrampe-Couloume G (2010) Insights into enzyme kinetics of chloroethane biodegradation using compound specific stable isotopes. Environ Sci Technol 44(19):7498–7503. doi:10.1021/es101330r

Shouakar-Stash O, Drimmie RJ (2013) Online methodology for determining compound-specific hydrogen stable isotope ratios of trichloroethene and 1,2-cis-dichloroethene by continuous-flow isotope ratio mass spectrometry. Rapid Commun Mass Spectrom 27(12):1335–1344. doi:10.1002/rcm.6578

Shouakar-Stash O, Frape SK, Drimmie RJ (2003) Stable hydrogen, carbon and chlorine isotope measurements of selected chlorinated solvents. J Contam Hydrol 60(3–4):211–228

Shouakar-Stash O, Drimmie RJ, Zhang M, Frape SK (2006) Compound-specific chlorine isotope ratios of TCE, PCE and DCE isomers by direct injection using CF-IRMS. Appl Geochem 21:766–781

Tiehm A, Schmidt KR, Pfeifer B, Heidinger M, Ertl S (2008) Growth kinetics and stable carbon isotope fractionation during aerobic degradation of cis-1,2-dichloroethene and vinyl chloride. Water Res 42(10–11):2431–2438. doi:10.1016/j.watres.2008.01.029

Torrento C, Audi-Miro C, Bordeleau G, Marchesi M, Rosell M, Otero N, Soler A (2014) The use of alkaline hydrolysis as a novel strategy for chloroform remediation: the feasibility of using construction wastes and evaluation of carbon isotopic fractionation. Environ Sci Technol 48(3):1869–1877. doi:10.1021/es403838t

Van Breukelen BM, Hunkeler D, Volkering F (2005) Quantification of sequential chlorinated ethene degradation using a reactive transport model incorporating isotope fractionation. Environ Sci Technol 39:4189–4197

Van der Meer JR, van Neerven ARW, Devries EJ, Devos WM, Zehnder AJB (1991) Cloning and characterization of plasmid-encoded genes for the degradation of 1,2-dichlorobenzene, 1,4-dichlorobenzene, and 1,2,4-trichlorobenzene of *Pseudomonas sp.* strain P51. J Bacteriol 173(1):6–15

Van Wanderdam EM, Frape SK, Aravena R, Drimmie RJ, Flatt H, Cherry JA (1995) Stable chlorine and carbon isotope measurements of selected organic solvents. Appl Geochem 10:547–552

Vannelli T, Studer A, Kertesz M, Leisinger T (1998) Chloromethane metabolism by *Methylobacterium sp.* strain CM4. Appl Environ Microbiol 64(5):1933–1936

VanStone NA, Focht RM, Mabury SA, Sherwood Lollar B (2004) Effect of iron type on kinetics and carbon isotopic enrichment of chlorinated ethylenes during abiotic reduction on Fe(0). Ground Water 42(2):268–276

Vieth A, Müller J, Strauch G, Kästner M, Gehre M, Meckenstock RU, Richnow HH (2003) In-situ biodegradation of tetrachloroethene and trichloroethene in contaminated aquifers monitored by stable isotope fractionation. Isot Environ Health Stud 39(2):113–124

Vogt C, Cyrus E, Herklotz I, Schlosser D, Bahr A, Herrmann S, Richnow HH, Fischer A (2008) Evaluation of toluene degradation pathways by two-dimensional stable isotope fractionation. Environ Sci Technol 42(21):7793–7800

Wolfsberg M, Van Hook WA, Paneth P (2010) Isotope effects in chemical, geological and bio sciences. Springer, Dortrecht

Zhang N, Bashir S, Qin J, Schindelka J, Fischer A, Nijenhuis I, Herrmann H, Wick LY, Richnow HH (2014) Compound specific stable isotope analysis (CSIA) to characterize transformation mechanisms of alpha-hexachlorocyclohexane. J Hazard Mater 280:750–757. doi:10.1016/j.jhazmat.2014.08.046

Zwank L, Berg M, Schmidt TC, Haderlein SB (2003) Compound-specific carbon isotope analysis of volatile organic compounds in the low-microgram per liter range. Anal Chem 20:5575–5583

Zwank L, Elsner M, Aeberhard A, Schwarzenbach RP, Haderlein SB (2005) Carbon isotope fractionation in the reductive dehalogenation of carbon tetrachloride at iron (hydr)oxide and iron sulfide minerals. Environ Sci Technol 39(15):5634–5641

Part VII
Outlook

Chapter 26
Outlook—The Next Frontiers for Research on Organohalide-Respiring Bacteria

Lorenz Adrian and Frank E. Löffler

Abstract Research efforts over the last two decades have substantially advanced the understanding of organohalide-respiring bacteria (OHRB), and this progress has enabled successful bioremediation applications at chlorinated solvent-contaminated sites. Yet, major knowledge gaps remain, and detailed biochemical, genetic, regulatory, evolutionary, taxonomic, and ecological questions should be explored to reveal the underlying principles of organohalide respiration, to better define the roles of OHRB in natural microbial communities, and to fully exploit their activities for contaminated site cleanup. The chapters in this book summarize the various advances that have been achieved following the discovery, physiological description, and practical application of OHRB. But where will the field go next? Which major topics will be targeted in the coming decade? What are the major unresolved questions? What new discoveries will be made overcoming insufficient concepts and leading to new questions and hypotheses? What new techniques will drive research in the near- and midterm future? Will scientists be able to convince funding agencies to invest in this field to enable further transformative discoveries? Will environmental scientists and

L. Adrian (✉)
Department Isotope Biogeochemistry, Helmholtz Centre for Environmental Research—UFZ, Leipzig, Germany
e-mail: lorenz.adrian@ufz.de

F.E. Löffler
Biosciences Division, Oak Ridge National Laboratory, Oak Ridge, TN 37831, USA

F.E. Löffler
Joint Institute for Biological Sciences (JIBS), University of Tennessee and Oak Ridge National Laboratory (UT-ORNL), Oak Ridge, TN 37831, USA

F.E. Löffler (✉)
Center for Environmental Biotechnology, Department of Microbiology, Department of Civil and Environmental Engineering, University of Tennessee, Knoxville, TN 37996, USA
e-mail: frank.loeffler@utk.edu

L. Adrian and F.E. Löffler (eds.), *Organohalide-Respiring Bacteria*,
DOI 10.1007/978-3-662-49875-0_26

engineers be successful in demonstrating that the current achievements are just the beginning, and that support for developing precision bioremediation treatment can substantially improve the current practice and realize considerable benefits to society, including safe drinking water, a cleaner environment and financial benefits to the taxpayer? We expect major progress in the biological understanding as well as more sophisticated engineering applications of organohalide respiration in the coming years.

26.1 Microbiology

As outlined in several book chapters, bacteria from diverse taxa conserve energy associated with reductive dechlorination reactions. There is little doubt that a larger diversity of microbes sharing this metabolic capability awaits discovery. As an example, enrichment and isolation efforts for *Dehalococcoidia* strains always included the addition of cell wall antibiotics to the growth medium. While this approach is productive for members of the *Dehalococcoidia*, this approach selects against OHRB with a peptidoglycan cell wall. Not surprisingly, diverse reductive dehalogenation reactions observed in initial microcosms by many researchers were lost during further enrichment. Isolation work is tedious and short-term rewards are unlikely; however, isolates are very important and will remain the cornerstone of microbiology. Different and innovative enrichment and isolation techniques should be pursued to increase the chances to find yet-unknown organisms. The value of these efforts cannot be overemphasized, especially in the current situation where more and more research is done exclusively on a computer. As observed in many other biological fields, a huge amount of data is now available describing dehalogenating microbial populations, including (meta) genomes, (meta) transcriptomes, and (meta) proteomes, allowing the construction of detailed metabolic models. It is important to keep in mind that these models must be used with caution and that all hypotheses need to be experimentally verified, ideally with pure or defined cultures.

As described in various chapters in this book, we can distinguish two groups of OHRB. Opportunistic OHRB (i.e., generalists) that can grow by other means but take advantage of chlorinated compounds as electron acceptors when they are available, and obligate OHRB (i.e., specialists) that strictly depend on chlorinated compounds and cannot derive energy for growth by any other means. The environmental constraints and evolution of OHRB specialists and OHRB generalists are poorly understood. Is the occurrence of OHRB recent and a consequence of the introduction of organohalogens from anthropogenic sources? Is organohalide respiration an evolutionary old process driven by naturally occurring organohalogens? And what are the reasons for the specialization of *Dehalobacter* compared to the related opportunistic *Desulfitobacterium* strains? Recently, single-cell genome analyses have shown that organohalide respiration is not ubiquitous in the *Dehalococcoidia* and that other modes of energy conservation exist in this bacterial class. What is the reason for the specialization of the obligate organohalide-respiring *Dehalococcoides mccartyi* strains? Reductive dehalogenase genes appear to be enriched among the *Deltaproteobacteria* but are these organisms growing by organohalide respiration? Only detailed physiological characterization of existing and new isolates can answer these questions.

The systematic identification and categorization of naturally occurring halogenated electron acceptors would enable approaches to match the full complement of reductive dehalogenase genes with different classes of electron acceptors. Such an effort could also provide significant information for the fate prediction of halogenated contaminants, which so far have not been sufficiently investigated as electron acceptors for OHRB, including many organohalogen pesticides, pharmaceuticals, and the large group of (per)fluorinated compounds. Also, alkanes with single halogen substituents seem to resist transformation by OHRB, whereas vicinally halogenated alkanes can be transformed by dihaloelimination reactions. Finally, organohalogens with complicated or those with very simple structures (e.g., chlorinated methanes) are insufficiently investigated. In essence, many needs and opportunities exist for extending research that would greatly expand our knowledge of OHRB biology and the roles these organisms play to maintain ecosystem services, foremost the recycling of organohalogens from anthropogenic and natural sources.

In ecological terms, what is the overall contribution of OHRB to the halogen cycle on Earth? Can the masses of halogenated materials turned over by OHRB be estimated? Although several cross-feeding partners in complex communities have been identified, general guidelines on the specifications and importance of accompanying bacteria have been unsatisfactorily described. How do other bacteria contribute to make suitable carbon source available, which growth factors do they provide, how do they influence the redox potential and pH in the microenvironment? It has often been described that mixed cultures more efficiently dehalogenate organohalogens than pure cultures. What are the biomolecular underpinnings for this observation? Do syntrophic interactions play relevant roles and what type of syntrophy is occurring? Do we observe higher resistance against stress or higher resilience after stress on the single cell or on the population level? Finally, the basis for toxicity of single organohalogens or mixtures of organohalogens on OHRB as well as essential members of the community, should be investigated to establish causal effects for recalcitrance.

26.2 Biochemistry and Genes Encoding Reductive Dehalogenases

Large steps forward have recently been achieved in the description of the respiratory machinery involved in organohalide respiration. Such progress is important as in-depth information about the reductive dehalogenase enzyme systems may lead to biotechnological applications, both in controlled systems and in the environment.

The possibility to heterologously express active reductive dehalogenases opens up a wide field for new research. With more than 2000 putative reductive dehalogenase sequences identified, the majority without functional assignment, heterologous expression is an important tool to characterize substrate specificity, substrate affinity, cofactor requirement, and to provide structural information. A high-throughput expression pipeline is desirable to analyze the hundreds, or possibly thousands of distinct reductive dehalogenases. Likely, the traditional approach to induce, enrich, and purify reductive dehalogenases by chromatographic or electrophoretic techniques will still have value. Conventional biochemical approaches will also be needed for the investigation of a recently identified larger respiratory complex in *Dehalococcoides mccartyi* (Kublik et al. 2016), for which heterologous expression and reconstitution may not be feasible.

With more structural information of reductive dehalogenases, comparative analyses can reveal structure-function relationships, relevant binding motifs, and contribute to predictive understanding of substrate range and reaction specificity. Among the reductive dehalogenases, a system of orthologous clusters has been established (Hug et al. 2013), and it will be a crucial task to refine this system as new information becomes available. Also, the role of accessory proteins involved in dehalogenase maturation should be studied (e.g., cofactor incorporation, folding, transport across the cytoplasmic membrane, assembly of larger complexes). Most promising are integrated approaches, and cross-disciplinary team research is most likely to produce transformative discoveries that advance and broaden the field.

Although some progress has been made elucidating the controls of reductive dehalogenase gene expression (Wagner et al. 2013; Kemp et al. 2013), the regulatory cascade is poorly understood. Regulatory genes are located in the vicinity of reductive dehalogenase genes but virtually nothing is known about the inducing molecules, how they interact with the transcription regulator(s), and how they trigger physiological responses. Also, it is unclear why different types of regulators are involved in the transcriptional regulation of different reductive dehalogenase genes. Understanding the regulation of reductive dehalogenation expression will likely reveal a new conceptual understanding of regulatory circuits in prokaryotes.

Recently, it has been shown that reductive dehalogenases can display strong substrate promiscuity and that relative reductive dechlorination rates follow electron density properties in the electron acceptor (Cooper et al. 2015). Apparently, the traditional lock-and-key or induced-fit models for enzymes might not be appropriate for reductive dehalogenases, and a more dynamic substrate binding concept is needed. Further investigations are required using structural, kinetic, and quantum chemistry information to analyze enzyme–substrate interplay and electron transfer. This will also provide more detailed mechanistic insights into reductive dehalogenation reactions, but more widely, will also contribute to better understanding of coenzyme B_{12} dependent reactions.

Another research area of interest will be the elucidation of the components and/or complexes involved in proton translocation across the membrane, and if mechanistic differences distinguish phylogenetically distinct OHRB taxa. Especially the electron transfer between the primary oxidizing protein complex (e.g., a hydrogenase) and the reductive dehalogenase will need further attention to identify electron carriers and their biochemical characteristics. These studies will reveal if obligate OHRB (e.g., members of the *Dehalococcoidia*) and facultative OHRB (e.g., members of the *Deltaproteobacteria*) share electron transport components or have distinct machineries to capture energy released in reductive dechlorination reactions.

Reductive dehalogenases of OHRB share characteristic features including a Tat leader peptide and require the so-called B protein with a putative membrane-anchoring function. Genome analyses revealed that diverse taxa possess reductive dehalogenase genes not encoding these characteristic features. Experiments with *Comamonas* sp. 7D-2 (Chen et al. 2013) and *Nitratireductor pacificus* strain pht-3B (Payne et al. 2015) showed that such enzymes have reductive dehalogenase function albeit they are not directly involved in respiration. Similar reductive dehalogenase genes were found in marine sediments. Possibly, these nonrespiratory reductive dehalogenases are part of degradation pathways that enable the host to oxidize the chloroorganic compound and utilize an alternate electron acceptor such as oxygen. Have organisms with nonrespiratory reductive dehalogenase lost the ability to grow via organohalide respiration, or are they possibly the ancestors that gave rise to the evolution of (obligate) OHRB?

26.3 Bioremediation Applications

Without extensive contamination of the environment with chlorinated chemicals and ensuing public awareness and pressure, OHRB may not have been studied. The reductive dechlorination process is a good example how practical needs enable fundamental scientific discoveries while at the same time delivering solutions for pressing environmental problems. It is hoped that funding resources for multidisciplinary team efforts will be available in the future to advance the science, generate economic opportunities, and elevate environmental cleanup from an empirical practice to a science with predictable outcomes.

Bioaugmentation, the delivery of OHRB consortia into aquifers impacted with chlorinated contaminants, can initiate or accelerate degradation and detoxification, as documented at many chlorinated solvent-contaminated sites (Ellis et al. 2000; Lendvay et al. 2003; Major et al. 2002; Löffler et al. 2013). *D. mccartyi* appears to be crucial and only strains of this bacterial species have been demonstrated to detoxify chlorinated ethenes and produce environmentally benign ethene. Interestingly, *D. mccartyi* strains are often present in contaminated aquifers but efficient ethene formation does not occur, presumably because the resident *Dehalococcoides* populations lack the *bvcA* and/or *vcrA* reductive dehalogenase genes required for vinyl chloride reductive dechlorination (Krajmalnik-Brown et al. 2004; Müller et al. 2004). Thus, the complement of reductive dehalogenase genes determines if the resident *Dehalococcoides* population efficiently degrades the target contaminant(s). An interesting question is if bioaugmentation successes really rely on the proliferation of the *D. mccartyi* strains introduced with the inoculum, or if the introduction of the genetic material encoding the vinyl chloride reductive dehalogenase(s) is sufficient. Mounting evidence suggests that members of the *Dehalococcoidia* acquire reductive dehalogenase genes via horizontal gene transfer (McMurdie et al. 2011; Padilla-Crespo et al. 2014), which may offer alternative bioremediation strategies.

Detailed laboratory studies unravelled the complicated nutritional requirements of *D. mccartyi* strains. In addition to hydrogen, *Dehalococcoides* requires other growth factors, foremost corrinoid, which is needed to assemble functional reductive dehalogenases. The recent observation that the type or corrinoid (i.e., cobamides with different lower bases) has distinct effects on the reductive dechlorination performance of *D. mccartyi* strains expressing different vinyl chloride reductive dehalogenases emphasizes the need to understand the roles of the community to support *Dehalococcoides* activity (Yan et al. 2015).

Metagenomics and metaproteomics enable a census of the genetic and actual catalytic potential, respectively, of entire microbial communities. Such approaches have not been effectively brought to bear at bioremediation sites but may be ideal tools to develop systems understanding, which is needed to assess the complicated interactions that govern activity of OHRB. Such detailed knowledge can inform about reductive dechlorination potential, measure actual activity, and reveal interspecies dependencies, nutritional limitations, and possible synergistic effects, and thus offer opportunities to refine bioremediation to efficiently achieve the desired outcomes.

References

Chen K, Huang L, Xu C, Liu X, He J, Zinder SH, Li S, Jiang J (2013) Molecular characterization of the enzymes involved in the degradation of a brominated aromatic herbicide. Mol Microbiol 89(6):1121–1139. doi:10.1111/mmi.12332

Cooper M, Wagner A, Wondrousch D, Sonntag F, Sonnabend A, Brehm M, Schüürmann G, Adrian L (2015) Anaerobic microbial transformation of halogenated aromatics and fate prediction using electron density modelling. Environ Sci Technol 49(10):6018–6028. doi:10.1021/acs.est.5b00303

Ellis D, Lutz E, Odom J, Buchanan R, Bartlett C, Lee M, Harkness M, Deweerd K (2000) Bioaugmentation for accelerated in situ anaerobic bioremediation. Environ Sci Technol 34(11):2254–2260. doi:10.1021/es990638e

Hug LA, Maphosa F, Leys D, Löffler FE, Smidt H, Edwards EA, Adrian L (2013) Overview of organohalide-respiring bacteria and a proposal for a classification system for reductive dehalogenases. Philos Trans R Soc Lond B Biol Sci 368(1616):20120322. doi:10.1098/rstb.2012.0322

Kemp LR, Dunstan MS, Fisher K, Warwicker J, Leys D (2013) The transcriptional regulator CprK detects chlorination by combining direct and indirect readout mechanisms. Philos Trans R Soc B 368(1616):20120323. doi:10.1098/rstb.2012.0323

Krajmalnik-Brown R, Hölscher T, Thomson IN, Saunders FM, Ritalahti KM, Löffler FE (2004) Genetic identification of a putative vinyl chloride reductase in *Dehalococcoides* sp. strain BAV1. Appl Environ Microbiol 70(10):6347–6351. doi:10.1128/AEM.70.10.6347-6351.2004

Kublik A, Deobald D, Hartwig S, Schiffmann C, Andrades A, von Bergen M, Sawers RG, Adrian L (2016) Identification of a multiprotein reductive dehalogenase complex in *Dehalococcoides mccartyi* strain CBDB1 suggests a protein-dependent respiratory electron transport chain obviating quinone involvement. Environ Microbiol. doi:10.1111/1462-2920.13200

Lendvay JM, Löffler FE, Dollhopf M, Aiello MR, Daniels G, Fathepure BZ, Gebhard M, Heine R, Helton R, Shi J, Krajmalnik-Brown R, Major CL, Barcelona MJ, Petrovskis E, Hickey R, Tiedje JM, Adriaens P (2003) Bioreactive barriers: a comparison of bioaugmentation and biostimulation for chlorinated solvent remediation. Environ Sci Technol 37(7):1422–1431. doi:10.1021/es025985u

Löffler FE, Ritalahti KM, Zinder SH (2013) *Dehalococcoides* and reductive dechlorination of chlorinated solvents. In: Stroo HF, Leeson A, Ward CH (eds) Bioaugmentation for groundwater remediation, vol 5. SERDP ESTCP environmental remediation technology. Springer, New York, pp 39–88. doi:10.1007/978-1-4614-4115-1_2

Major DW, McMaster ML, Cox EE, Edwards EA, Dworatzek SM, Hendrickson ER, Starr MG, Payne JA, Buonamici LW (2002) Field demonstration of successful bioaugmentation to achieve dechlorination of tetrachloroethene to ethene. Environ Sci Technol 36:5106–5116. doi:10.1021/es0255711

McMurdie P, Hug L, Edwards E, Holmes S, Spormann A (2011) Site-specific mobilization of vinyl chloride respiration islands by a mechanism common in *Dehalococcoides*. BMC Genom 12(1):287. doi:10.1186/1471-2164-12-287

Müller JA, Rosner BM, von Abendroth G, Meshulam-Simon G, McCarty PL, Spormann AM (2004) Molecular identification of the catabolic vinyl chloride reductase from *Dehalococcoides* sp. strain VS and its environmental distribution. Appl Environ Microbiol 70(8):4880–4888. doi:10.1128/AEM.70.8.4880-4888.2004

Padilla-Crespo E, Yan J, Swift C, Wagner DD, Chourey K, Hettich RL, Ritalahti KM, Löffler FE (2014) Identification and environmental distribution of *dcpA*, which encodes the reductive dehalogenase catalyzing the dichloroelimination of 1,2-dichloropropane to propene in organohalide-respiring *Chloroflexi*. Appl Environ Microbiol 80(3):808–818. doi:10.1128/AEM.02927-13

Payne KAP, Quezada CP, Fisher K, Dunstan MS, Collins FA, Sjuts H, Levy C, Hay S, Rigby SEJ, Leys D (2015) Reductive dehalogenase structure suggests a mechanism for B12-dependent dehalogenation. Nature 517(7535):513–516. doi:10.1038/nature13901

Wagner A, Segler L, Kleinsteuber S, Sawers G, Smidt H, Lechner U (2013) Regulation of reductive dehalogenase gene transcription in *Dehalococcoides mccartyi*. Philos Trans R Soc B 368:20120317. doi:10.1098/rstb.2012.0317

Yan J, Simsir B, Farmer AT, Bi M, Yang Y, Campagna SR, Löffler FE (2015) The corrinoid cofactor of reductive dehalogenases affects dechlorination rates and extents in organohalide-respiring *Dehalococcoides mccartyi*. ISME J. doi:10.1038/ismej.2015.197

Index

L. Adrian and F.E. Löffler (eds.), *Organohalide-Respiring Bacteria*,
DOI 10.1007/978-3-662-49875-0

MIX
Papier aus verantwortungsvollen Quellen
Paper from responsible sources
FSC® C105338

If you have any concerns about our products,
you can contact us on
ProductSafety@springernature.com

In case Publisher is established outside the EU,
the EU authorized representative is:
Springer Nature Customer Service Center GmbH
Europaplatz 3, 69115 Heidelberg, Germany

Printed by Libri Plureos GmbH
in Hamburg, Germany